Student Solutions Manual
to Accompany

Technical Mathematics with Calculus

Second Edition

Dale Ewen
Parkland Community College

Joan S. Gary
Parkland Community College

James E. Trefzger
Parkland Community College

Upper Saddle River, New Jersey
Columbus, Ohio

Editor in Chief: Stephen Helba
Senior Acquisitions Editor: Gary Bauer
Editorial Assistant: Natasha Holden
Media Development Editor: Michelle Churma
Production Editor: Louise N. Sette
Design Coordinator: Diane Ernsberger
Cover Designer: Eric Davis
Production Manager: Pat Tonneman
Marketing Manager: Leigh Ann Sims

Pearson Education Ltd.
Pearson Education Singapore Pte. Ltd.
Pearson Education Canada, Ltd.
Pearson Education—Japan
Pearson Education Australia Pty. Limited
Pearson Education North Asia Ltd.
Pearson Educación de Mexico, S.A. de C.V.
Pearson Education Malaysia Pte. Ltd.

ISBN 0-13-118742-2

Contents

SOLUTIONS TO ODD NUMBERED EXERCISES

CHAPTER 1

Exercises 1.1

1. $|-15| = -(-15) = 15$ **3.** $|+11| = 11$ **5.** $|-8| = -(-8) = 8$ **7.** $|3-5| = |-2| = -(-2) = 2$

9. $(-3)+(-6) = -9$ **11.** $(+7)+(-9) = -2$ **13.** $(+6)+(+4) = 10$ **15.** $(-5)+(+8) = 3$

17. $\left(-\frac{1}{3}\right)+\left(-\frac{3}{4}\right)=\left(-\frac{4}{12}\right)+\left(-\frac{9}{12}\right)=\frac{-13}{12}=-1\frac{1}{12}$

19. $\left(-3\frac{4}{9}\right)+\left(+4\frac{5}{12}\right)=\left(-3\frac{16}{36}\right)+\left(4\frac{15}{36}\right)=\left(-3\frac{16}{36}\right)+\left(3\frac{51}{36}\right)=\frac{35}{36}$

21. $(-13)+(+3)+(-7)+(+6)+(-2)$

Step 1: $(+3)+(+6) \quad = +9$

Step 2: $(-13)+(-7)+(-2) = -22$

Step 3: $\quad -13$

23. $(-2)+(-3)+(+6)+(-4)+(+1)+(-5)$

Step 1: $(+6)+(+1) \quad = +7$

Step 2: $(-2)+(-3)+(-4)+(-5) = -14$

Step 3: $\quad -7$

25. $(+4)-(+9)=(+4)+(-9)=-5$ **27.** $(-2)-(-8)=(-2)+(+8)=6$

29. $(+7)-(-2)=(+7)+(+2)=9$ **31.** $(-6)-(+8)=(-6)+(-8)=-14$

33. $4-7=4+(-7)=-3$ **35.** $-12.5-3.2=-12.5+(-3.2)=-15.7$

37. $\left(-\frac{2}{3}\right)-\left(-\frac{1}{4}\right)=\left(-\frac{2}{3}\right)+\left(+\frac{1}{4}\right)=\left(-\frac{8}{12}\right)+\left(+\frac{3}{12}\right)=\frac{-5}{12}$

39. $\left(+\frac{1}{12}\right)-\left(+1\frac{2}{3}\right)=\left(+\frac{1}{12}\right)+\left(-1\frac{2}{3}\right)=\left(+\frac{1}{12}\right)+\left(-1\frac{8}{12}\right)=-1\frac{7}{12}$

41. $(+3)-(-6)-(+9)-(-6)=(+3)+(+6)+(-9)+(+6)=+6$

43. $(+3)+(-4)-(-7)-(+2)+(+7)=(+3)+(-4)+(7)+(-2)+(+7)=11$

45. $8+9-16+4-5-6+1=8+9+(-16)+4+(-5)+(-6)+1=-5$

47. $9-7+4+3-8-6-6+2=9+(-7)+4+3+(-8)+(-6)+(-6)+2=-9$

49. $(-5)(-6)=+30$ **51.** $(+2)(-12)=-24$ **53.** $(-8)\div(-2)=+4$ **55.** $\frac{+54}{-30}=\frac{9}{-5}=-1\frac{4}{5}$

57. $\left(-\frac{2}{3}\right)\left(-\frac{9}{16}\right)=\frac{3}{8}$ **59.** $\left(+1\frac{3}{8}\right)\div\left(-1\frac{5}{6}\right)=\left(+\frac{11}{8}\right)\div\left(-\frac{11}{6}\right)=\left(+\frac{11}{8}\right)\left(-\frac{6}{11}\right)=-\frac{3}{4}$

61. $\dfrac{-\frac{7}{16}}{\frac{21}{32}} = \left(-\dfrac{7}{16}\right)\left(\dfrac{32}{21}\right) = -\dfrac{2}{3}$

63. $(+2)(-7)(-6)(-1) = -84$

65. $(-2)(+3)(-1)(+5)(+2) = +60$

67. $\dfrac{(-6)(+4)(-7)}{(-12)(-2)} = +7$

69. $\dfrac{(+36)(-18)(+5)}{(-4)(-8)(+30)} = -\dfrac{27}{8}$

71. $-15^\circ + 35^\circ = 20^\circ$

73. $18^\circ - (-12^\circ) = 18^\circ + (+12)^\circ = 30^\circ$

75. $-65 + (-25) + (+15) = -75$
The third diver is 75 feet below the surface.

Exercises 1.2

1. $0 - 3 = 0 + (-3) = -3$

3. $6 \cdot 0 = 0$

5. $\dfrac{9 \cdot 0}{3} = \dfrac{0}{3} = 0$

7. $\dfrac{13}{6-6} = \dfrac{13}{0}$ is meaningless

9. $\dfrac{7+(-7)}{6-6(3+2)} = \dfrac{0}{6-6(5)} = \dfrac{0}{6+(-30)} = \dfrac{0}{-24} = 0$

11. $\dfrac{5 \cdot 0}{-6-(-6)} = \dfrac{0}{-6+6} = \dfrac{0}{0}$ is indeterminate

13. $x - 2 = 0$ implies $x = 2$

15. $x(3x+4) = 0$ implies $x = 0$ or $3x + 4 = 0$ implies $x = 0$ or $3x = -4$ implies $x = 0$ or $x = -\dfrac{4}{3}$

17. $(x+1)(2x-1) = 0$ implies $x + 1 = 0$ or $2x - 1 = 0$ implies $x = -1$ or $2x = 1$ implies $x = -1$ or $x = \dfrac{1}{2}$

19. Numerator and denominator must equal zero. $x^2 = 0$ and $x = 0$ for $x = 0$.

21. $\dfrac{12x-10}{(3+x)(5-6x)}$
The numerator, $12x - 10$, is zero when $12x = 10$ or when $x = \dfrac{5}{6}$. The denominator, $(3+x)(5-6x)$, is zero when $x = -3$ or $x = \dfrac{5}{6}$. The numerator and denominator are both zero when $x = \dfrac{5}{6}$.

23. $\dfrac{(2x+1)(x-3)}{(6-2x)(2x+1)}$
The numerator, $(2x+1)(x-3)$, is zero when $x = -\dfrac{1}{2}$ or $x = 3$. The denominator, $(6-2x)(2x+1)$, is zero when $x = 3$ or $-\dfrac{1}{2}$. Both are zero for $x = 3$ or $x = -\dfrac{1}{2}$.

25. $15 + 2 \cdot 4 = 15 + 8 = 23$

27. $3 - 8(5-2) = 3 - 8(3) = 3 - 24 = -21$

29. $12 + 4 \div 2 = 12 + 2 = 14$

31. $6 - 7(4+1) = 6 - 7(5) = 6 - 35 = -29$

33. $5-[4+6(2-7)-3]=5-[4+6(-5)-3]=5-[4+(-30)-3]=5-[-29]=5+29=34$

35. $26\cdot 2\div 13+4-7\cdot 2=52\div 13+4-7\cdot 2=4+4-7\cdot 2=4+4-14=8-14=-6$

37. $3(3-7)\div 6-2=3(-4)\div 6-2=-12\div 6-2=-2-2=-4$

39.
$$\begin{gathered}6\cdot 8\div 2\cdot 72\div 24+4=\\48\div 2\cdot 72\div 24+4=\\24\cdot 72\div 24+4=\\1728\div 24+4=\\72+4=\\76\end{gathered}$$

41.
$$\begin{gathered}18\div 6\cdot 24\div 4\div 6=\\3\cdot 24\div 4\div 6=\\72\div 4\div 6=\\18\div 6=\\3\end{gathered}$$

43.
$$\begin{gathered}7+6(3+2)-6-5(4-2)=\\7+6(5)-6-5(2)=\\7+30-6-10=\\37-6-10=\\31-10=\\21\end{gathered}$$

45.
$$\begin{gathered}\frac{4-7(6-2)}{9\div 3+7}=\\\frac{4-7(4)}{3+7}=\\\frac{4-28}{10}=\\\frac{-24}{10}=\\-\frac{12}{5}\end{gathered}$$

47.
$$\begin{gathered}\frac{5\cdot 12\div 6\cdot 2-(-4)}{6-2(5+4)}=\\\frac{60\div 6\cdot 2-(-4)}{6-2(9)}=\\\frac{10\cdot 2-(-4)}{6-18}=\\\frac{20-(-4)}{-12}=\\\frac{20+4}{-12}=\\\frac{24}{-12}=\\-2\end{gathered}$$

Exercises 1.3

1. $10^3\cdot 10^8=10^{3+8}=10^{11}$

3. $(10^3)^3=10^{3(3)}=10^9$

5. $(10^{-4})^2=10^{-4(2)}=10^{-8}$

7. $10^{-2}\div 10^{-5}=10^{-2-(-5)}=10^3$

9. $\frac{1}{10^{-4}}=10^4$

11. $\frac{10^{-2}\cdot 10^{-7}\cdot 10^{-3}\cdot 10^0}{10^3\cdot 10^4\cdot 10^{-2}}=\frac{10^{-2-7-3+0}}{10^{3+4-2}}=\frac{10^{-12}}{10^5}=10^{-12-5}=10^{-17}$

13. $(10^{-2}\cdot 10^3\cdot 10^{-5})^{-4}=(10^{-2+3-5})^{-4}=(10^{-4})^{-4}=10^{-4(-4)}=10^{16}$

15. $\left(\frac{10^2\cdot 10^{-4}}{10^{-3}\cdot 10^6}\right)^3=\left(\frac{10^{2-4}}{10^{-3+6}}\right)^3=\left(\frac{10^{-2}}{10^3}\right)^3=(10^{-2-3})^3=(10^{-5})^3=10^{-15}$

17.

$$\left(\frac{10^{-3}\cdot 10^{5}\cdot 10^{-7}}{10^{2}\cdot 10^{-1}\cdot 10^{-5}}\right)^{-4}=$$

$$\left(\frac{10^{-3+5-7}}{10^{2-1-5}}\right)^{-4}=$$

$$\left(\frac{10^{-5}}{10^{-4}}\right)^{-4}=$$

$$\left(10^{-5-(-4)}\right)^{-4}=$$

$$\left(10^{-1}\right)^{-4}=10^{-1(-4)}=10^{4}$$

19. $2_{\wedge}070=2.07\times 10^{3}$

21. $0.09_{\wedge}1=9.1\times 10^{-2}$

23. $5.61=5.61\times 10^{0}$

25. $8_{\wedge}500\ 000=8.5\times 10^{6}$

27. $0.000006_{\wedge}=6\times 10^{-6}$

29. $1_{\wedge}0,060=1.006\times 10^{4}$

31. $1.27\times 10^{2}=127_{\wedge}=127$

33. $6.14\times 10^{-5}=0_{\wedge}0000614=0.0000614$

35. $9.24\times 10^{6}=9240000_{\wedge}=9,240,000$

37. $6.96\times 10^{-9}=0_{\wedge}00000000696=0.00000000696$

39. $9.66\times 10^{0}=9.66$

41. $5.03\times 10^{4}=50300_{\wedge}=50,300$

43.

$$\left(6.43\times 10^{8}\right)\left(5.16\times 10^{10}\right)=$$

$$(6.43)(5.16)\left(10^{8}\times 10^{10}\right)=$$

$$33.1788\times 10^{18}=$$

$$3.32\times 10^{1}\times 10^{18}=$$

$$3.32\times 10^{19}$$

45.

$$\left(1.456\times 10^{12}\right)\left(-4.69\times 10^{-18}\right)=$$

$$(1.456)(-4.69)\left(10^{12}\times 10^{-18}\right)=$$

$$-6.83\times 10^{-6}$$

47.

$$\left(7.46\times 10^{8}\right)\div\left(8.92\times 10^{18}\right)=$$

$$\frac{7.46\times 10^{8}}{8.92\times 10^{18}}=$$

$$\frac{7.46}{8.92}\times 10^{8-18}=$$

$$0.836\times 10^{-10}=$$

$$8.36\times 10^{-1}\times 10^{-10}=$$

$$8.36\times 10^{-11}$$

49.

$$\frac{-6.19\times 10^{12}}{7.755\times 10^{-8}}=$$

$$\frac{-6.19}{7.755}\times 10^{12-(-8)}=$$

$$-0.798\times 10^{20}=$$

$$-7.98\times 10^{-1}\times 10^{20}=$$

$$-7.98\times 10^{19}$$

51. $\dfrac{(5.26\times10^{-8})(8.45\times10^{6})}{(-6.142\times10^{9})(1.056\times10^{-12})}=$

$\dfrac{(5.26\times8.45)(10^{-8+6})}{(-6.142\times1.056)(10^{9-12})}=$

$\dfrac{44.4\times10^{-2}}{-6.49\times10^{-3}}=$

$\dfrac{44.44}{-6.486}\times10^{-2-(-3)}=$

-6.85×10^{1}

53. $\dfrac{(4.68\times10^{-15})(5.19\times10^{-7})}{(-7.27\times10^{-16})(4.045\times10^{-8})(1.68\times10^{24})}=$

$\dfrac{(4.68\times5.19)(10^{-15-7})}{(-7.27\times4.045\times1.68)(10^{-16-8+24})}=$

$\dfrac{24.3\times10^{-22}}{-49.4\times10^{0}}=$

$\dfrac{24.3}{-49.4}\times10^{-22}=$

$-0.492\times10^{-22}=$

$-4.92\times10^{-1}\times10^{-22}=$

-4.92×10^{-23}

Exercises 1.4

1. 3, use rules 1 and 2

3. 3, use rules 4, 2 and 1

5. 4, use rules 4, 2 and 1

7. 3, use rules 1 and 3

9. 2, use rules 1 and 4

11. 3, use rules 1 and 3

13. Use the position of the last significant digit. 0.1 cm

15. 0.01 cm

17. 1 mm

19. 0.01 m

21. 10Ω

23. 0.0001 A

25. For (a) name the measurement with the most number of significant digits.
For (b) name the measurement with the smallest position for its last significant digit.
(a) 15.2 m (b) 0.023 m

27. (a) 14.02 cm (b) 0.642 m

29. (a) 0.0270 A (b) 0.00060 A

31. (a) 305,000Ω (b) 305,000Ω, 38,000Ω

33. For (a) name the measurement with least number of significant digits.
For (b) name the measurement with the largest position for its last significant digit.
(a) 0.08 m (b) 13.2 m

35. (a) 0.52 km (b) 16.8 km

37. (a) 0.00009A (b) 0.41A

39. (a) 500,000Ω (b) 500,000Ω

Exercises 1.5

1. 15.7 in. + 6.4 in. = 22.1 in.

3. 45.6 cm + 13.41 cm +1.407 cm + 24.4 cm = 84.817 cm, round to 84.8 cm

5. 1.0443 g + 0.00134 g + 0.08986 g + 0.001359 g = 1.136859 g, round to 1.1369 g

7. 14 V + 1.005 V + 0.018 V + 3.5 V = 18.523 V, round to 19 V

9. 10.505 cm + 9.35 mm + 13.65 cm = 10.505 cm + 0.935 cm + 13.65 cm = 25.09 cm, round to 25.09 cm

11. 16.3 cm – 12.4 cm = 3.9 cm

13. 15.02 mm – 12.6 mm = 2.42 mm, round to 2.4 mm

15. 16.61 oz. – 11.372 oz. = 5.238 oz., round to 5.24 oz.

17. 6.000 in. – 2.004 in. = 3.996 in.

19. 0.149 in. + 0.407 in. + 1.028 in. + 0.77 in. = 2.354 in., round to 2.35 in.

21.
$$
\begin{aligned}
0.48\text{A} + 0.209\text{A} + 1.005\text{A} &= 0.84\text{A} + R_5 \\
1.694\text{A} &= 0.84\text{A} = R_5 \\
1.694\text{A} - 0.84\text{A} &= R_5 \\
0.854\text{A} &= R_5 \\
\text{round to } 0.85\text{A} &= R_5
\end{aligned}
$$

23. (17.7 m)(48.2 m) = 853.14 m^2, round to 853 m^2

25. (4.6 in.)(0.0285 in.) = 0.1311 in.^2, round to 0.13 in.^2

27. (34.2 cm) (26.1 cm)(28.9 cm) = $25{,}796.718 \text{ cm}^3$, round to $25{,}800 \text{ cm}^3$

29. $19.4\text{m}^3 \div 9.3\text{m}^2 = 2.086$ m, round to 2.1 m

31. $\dfrac{490 \text{ cm}}{6.73 \text{ s}^2} = 72.808 \text{ cm}/s^2$, round to $73 \text{ cm}/s^2$

33. $\dfrac{0.447\ N}{(1.43 \text{ m})(4.0 \text{ m})} = 0.0781\ N/\text{m}^2$, round to $0.078\ N/\text{m}^2$

35. $\dfrac{(4\bar{0} \text{ kg})(3.0 \text{ m}/s)^2}{5.50 \text{ m}} = 65 \text{ kgm}/s^2$

37. A = (6.2 cm)(17.5 cm) = 110 cm^2

39. $V = (8.50 \text{ cm})^3 = 614 \text{ cm}^3$

41. $K.E. = \dfrac{1}{2}(2.37 \times 10^6 \text{ kg})(10.4 \text{ m}/s)^2 = 1.28 \times 10^8 \text{ kg m}^2/s^2$

43. 6(2.08 mm) = 12.48 mm, or add six 2.08 mm and find precision to hundredths.

45. Yield in 1970: $\dfrac{4.20 \times 10^9 \text{ bu}}{6.68 \times 10^7 \text{ acres}} = 62.9$ bu/acre

Yield in 2000: $\dfrac{1.02 \times 10^{10} \text{ bu}}{7.31 \times 10^7 \text{ acres}} = 14\bar{0}$ bu/acre

Increase is: 2000 yield – 1970 yield = $14\bar{0}$ bu/acre – 62.9 bu/acre = 77 bu/acre

Exercises 1.6

1. two terms, binomial

3. three terms, trinomial

5. one term, monomial

7. two terms, binomial

9. three terms, trinomial

11. two terms, binomial

13. exponent is 2, degree 2

15. exponent is 4, degree 4

17. exponent is 10, degree 10

19. sum of the exponents is 5, degree 5

21. sum of the exponents is 6, degree 6

23. sum of the exponents is 4, degree 4

25. $3x^2+5x+2$; degree 2

27. $9x^8-5x^4+6x^3+5x^2$; degree 8

29. $4y^5+3y^3-3y+5$; degree 5

31. $4x+2x^3-3x^4$; degree 4

33. $-7+c+5c^3+3c^4-8c^5$; degree 5

35. $2+2y+5y^3-8y^4-6y^6$; degree 6

37. The term degrees are 3, 3, and 2 respectively; polynomial degree is 3.

39. The term degrees are 4, 4, 5, and 0 respectively; polynomial degree is 5.

41. The term degrees are 7, 8 and 8 respectively; polynomial degree is 8.

43. The term degrees are 8, 6, 9, and 2 respectively; polynomial degree is 9.

45. $\left(3x^2-4x+8\right)+\left(6x^2-x-3\right)=3x^2+6x^2-4x-x+8-3=9x^2-5x+5$

47. $\left(5x^2+7x-9\right)+\left(-x^2-6x+4\right)=5x^2-x^2+7x-6x-9+4=4x^2+x-5$

49. $\left(-3x^2-5x-3\right)-\left(5x^2-2x-7\right)=3x^2-5x-3-5x^2+2x+7=-8x^2-3x+4$

51. $\left(-6x^2+7\right)-\left(4x^2+x-3\right)=-6x^2+7-4x^2-x+3=-10x^2-x+10$

53. $\left(3x^2+4x-4\right)+\left(-x^2-x+2\right)-\left(-2x^2+2x+8\right)=3x^2+4x-4-x^2-x+2+2x^2-2x-8=4x^2+x-10$

55. $\left(-4x^2+6x-2\right)-\left(5x^2+7x-4\right)+\left(5x^2-6x+1\right)=-4x^2+6x-2-5x^2-7x+4+5x^2-6x+1=-4x^2-7x+$

57. $\left(3x^2-1+2x\right)-\left(9x^2+3-9x\right)-\left(3x-4-2x^2\right)=3x^2+2x-1-9x^2+9x-3+2x^2-3x+4=-4x^2+8x$

59. $\left(5x^2-12x-1\right)-\left(11x^2+4\right)+\left(4x+7\right)-\left(3x-2\right)=5x^2-12x-1-11x^2-4+4x+7-3x+2=-6x^2-11x+4$

61. $\left(3x^3+5x-2\right)+\left(6x^2-10x+1\right)-\left(4x^2-1\right)-\left(-5x^3-3\right)=3x^3+5x-2+6x^2-10x+1-4x^2+1+5x^3+3=$
$8x^3+2x^2-5x+3$

63. $\left(3x^2+2x-1\right)-\left(1-5x\right)-\left(3x^2+x\right)+\left(3x-4x^2\right)-\left(6x^2+x^3\right)=3x^2+2x-1-1+5x-3x^2-x+3x-4x^2$
$-6x^2-x^3=-x^3-10x^2+9x-2$

65. $-(x-3y)-[(x+2y)+(3y-2x)]=$
$-(x-3y)-[-x+5y]=$
$-x+3y+x-5y=$
$-2y$

67. $-\{(5x+3y)-(2x+5y)-[3x+(4y+x)]\}=$
$-\{(5x+3y)-(2x+5y)-[4x+4y]\}=$
$-\{5x+3y-2x-5y-4x-4y\}=$
$-\{-x-6y\}=x+6y$

69. $a-b=$
$(-2)-(3)=$
-5

71. $3a-2c=$
$3(-2)-2(-1)=$
$-6+2=-4$

73. $3a^2b^3c^2=$
$3(-2)^2(3)^3-(-1)^2=$
$3(4)(27)(1)=$
324

75. $(-b+3cd)^3=$
$(-3+3(-1)(1))^3=$
$(-3-3)^3=$
$(-6)^3=-216$

77. $b-a(c+d)=$
$3-(-2)(-1+1)=$
$3-(-2)(0)=$
$3-0=3$

79. $\frac{12a+6b^2}{9c+10a}=$
$\frac{12(-2)+6(3)^2}{9(-1)+10(-2)}=$
$\frac{12(-2)+6(9)}{-9-20}=$
$\frac{-24+54}{-29}=\frac{30}{-29}$

81. $(4a^2bc)^3=$
$(4(-2)^2(3)(-1))^3=$
$(4(4)(3)(-1))^3=$
$(-48)^3=$
$-110{,}592$

83. $\frac{5}{ab}-\frac{6a}{5b}-\frac{cd}{b}=$
$\frac{5}{(-2)(3)}-\frac{6(-2)}{5(3)}-\frac{(-1)(1)}{3}=$
$\frac{5}{-6}+\frac{12}{15}+\frac{1}{3}=$
$\frac{-5}{6}+\frac{4}{5}+\frac{1}{3}=$
$\frac{-5(5)}{6(5)}+\frac{4(6)}{5(6)}+\frac{1(10)}{3(10)}=$
$\frac{-25}{30}+\frac{24}{30}+\frac{10}{30}=$
$\frac{9}{30}=\frac{3}{10}$

Exercises 1.7

1. $x^5\cdot x^7=x^{5+7}=x^{12}$

3. $(3a^2)(4a^3)=3(4)a^{2+3}=12a^5$

5. $\frac{m^9}{m^3}=m^{9-3}=m^6$

7. $\frac{x^2}{x^6}=\frac{1}{x^{6-2}}=\frac{1}{x^4}$

9. $\frac{12x^8}{4x^4}=3x^{8-4}=3x^4$

11. $\frac{15x^2}{3x^5}=\frac{5}{x^{5-2}}=\frac{5}{x^3}$

13. $(a^2)^3 = a^{2(3)} = a^6$ **15.** $(c^4)^4 = c^{4(4)} = c^{16}$ **17.** $(9a)^2 = 9^2a^2 = 81a^2$

19. $(2x^2)^5 = 2^5x^{2(5)} = 32x^{10}$ **21.** $\left(\frac{3}{4}\right)^2 = \frac{3^2}{4^2} = \frac{9}{16}$ **23.** $\left(\frac{2}{a^3}\right)^4 = \frac{2^4}{a^{3(4)}} = \frac{16}{a^{12}}$

25. $4^o = 1$ (by definition) **27.** $3x^o = 3(1) = 3$ **29.** $(-3x)^2 = (-3)^2x^2 = 9x^2$

31. $(-t^3)^4 = (-1)^4(t^3)^4 = t^{3(4)} = t^{12}$ **33.** $(-a^2)^3 = (-1)^3(a^2)^3 = -1a^{2(3)} = -a^6$

35. $(-2a^2b)^3 = (-2)^3(a^2)^3(b)^3 = -8a^{2(3)}b^3 = -8a^6b^3$ **37.** $(3x^2y^3)^2 = 3^2x^{2(2)}y^{3(2)} = 9x^4y^6$

39. $(-3x^3y^4z)^3 = (-3)^3x^{3(3)}y^{4(3)}z^{1(3)} = -27x^9y^{12}z^3$ **41.** $\left(\frac{2x^2}{3y^3}\right)^2 = \frac{2^2x^{2(2)}}{3^2y^{3(2)}} = \frac{4x^4}{9y^6}$

43. $\left(\frac{-4x}{3y^2}\right)^2 = \frac{(-4)^2x^2}{3^2y^{2(2)}} = \frac{16x^2}{9y^4}$ **45.** $\left(\frac{-1}{6y^3}\right)^2 = \frac{(-1)^2}{6^2y^{3(2)}} = \frac{1}{36y^6}$

47. $\sqrt{4} = \sqrt{2^2} = 2$ **49.** $\sqrt{64} = \sqrt{8^2} = 8$ **51.** $\sqrt{121} = \sqrt{11^2} = 11$

53. $\sqrt{5^{16}} = \sqrt{(5^8)^2} = 5^8$ **55.** $\sqrt[3]{125} = \sqrt[3]{5^3} = 5$ **57.** $\sqrt[3]{-216} = \sqrt[3]{-6^3} = -6$

59. $\sqrt[3]{512} = \sqrt[3]{8^3} = 8$ **61.** $\sqrt{45} = \sqrt{9 \cdot 5} = \sqrt{9}\sqrt{5} = 3\sqrt{5}$ **63.** $\sqrt{50} = \sqrt{25 \cdot 2} = \sqrt{25}\sqrt{2} = 5\sqrt{2}$

65. $\sqrt{72} = \sqrt{36 \cdot 2} = \sqrt{36}\sqrt{2} = 6\sqrt{2}$ **67.** $\sqrt{4^2 + 32} = \sqrt{16 + 32} = \sqrt{48} = \sqrt{16}\sqrt{3} = 4\sqrt{3}$

69. $\sqrt{3 \cdot 4^2 - 4 \cdot 2^2} = \sqrt{3(16) - 4(4)} = \sqrt{48 - 16} = \sqrt{32} = \sqrt{16 \cdot 2} = \sqrt{16}\sqrt{2} = 4\sqrt{2}$

71. $\sqrt{329} \approx 18.1$ **73.** $\sqrt{2596} \approx 51.0$ **75.** $\sqrt{0.00472} \approx 0.0687$

77. $\sqrt{16 + 36} = \sqrt{52} \approx 7.21$ **79.** $\sqrt{5^2 + 8^2} = \sqrt{25 + 64} = \sqrt{89} \approx 9.43$

81. $\sqrt{(2.73 \times 10^4)^2 + (1.00 \times 10^5)^2} = \sqrt{2.73^2 \times 10^8 + 10^{10}} \approx \sqrt{1.0745 \times 10^{10}} \approx 1.04 \times 10^5$

83. $\sqrt{(115)^2 + (15.5 - 84.6)^2} = \sqrt{115^2 + (-69.1)^2} = \sqrt{17999.81} \approx 134$

Exercises 1.8

1. $(4x^2)(8x^3) = 4(8)x^{2+3} = 32x^5$ 3. $(-4a^2b)(6a^3b^2) = -4(6)a^{2+3}b^{1+2} = -24a^5b^3$

5. $(12a^2bc^3)(-4ac^2) = 12(-4)a^{2+1}bc^{3+2} = -48a^3bc^5$

7. $\left(-3a^2b^4c\right)\left(2ab^2c^5\right)\left(-4ab^3\right)=(-3)(2)(-4)a^{2+1+1}b^{4+2+3}c^{1+5}=24a^4b^9c^6$

9. $3a(4a-7b)=3a(4a)+3a(-7b)=12a^2-21ab$

11. $3x\left(2x^2+4x-5\right)=3x\left(2x^2\right)+3x(4x)+3x(-5)=6x^3+12x^2-15x$

13. $-5x^2\left(3x^2-5x+8\right)=-5x^2\left(3x^2\right)-5x^2(-5x)-5x^2(8)=-15x^4+25x^3-40x^2$

15. $6ab^3\left(4a^2b-8a^3b^4\right)=6ab^3\left(4a^2b\right)+6ab^3\left(-8a^3b^4\right)=24a^3b^4-48a^4b^7$

17. $-3a^2b^4\left(-a^4b^3+3ab-b^2\right)=-3a^2b^4\left(-a^4b^3\right)-3a^2b^4(3ab)-3a^2b^4\left(-b^2\right)=3a^6b^7-9a^3b^5+3a^2b^6$

19. $(3x-7)(2x+5)$

F O I L

$6x^2+15x-14x-35=$

$6x^2+x-35$

21. $(6x+3)(8x+5)$

F O I L

$48x^2+30x+24x+15=$

$48^2+54x+15$

23. $(3x+4)(3x-4)$

F O I L

$9x^2-12x+12x-16=$

$9x^2-16$

25. $(4x+1)(6x-1)$

F O I L

$24x^2-4x+6x-1=$

$24x^2+2x-1$

27. $(3x-7)(2x-3)$

F O I L

$6x^2-9x-14x+21=$

$6x^2-23x+21$

29. $(3x+8y)(6x+4y)$

F O I L

$18x^2+12xy+48xy+32y^2=$

$18x^2+60xy+32y^2$

31. $(5s-9t)(8s+2t)$

F O I L

$40s^2+10st-72st-18t^2=$

$40s^2-62st-18t^2$

33. $(-3x+4)(5x+6)$

F O I L

$-15x^2-18x+20x+24=$

$-15x^2+2x+24$

35. $\left(3x^2-1\right)\left(2x^2+7\right)$

F O I L

$6x^4+21x^2-2x^2-7=$

$6x^4+19x^2-7$

37. $\left(5x^2-6\right)\left(6x^2-5\right)$

F O I L

$30x^4-25x^2-36x^2+30=$

$30x^4-61x^2+30$

39.

$$\begin{array}{r} 3x^2+4x-3 \\ 2x+5 \\ \hline 6x^3+8x^2-6x \\ 15x^2+20x-15 \\ \hline 6x^3+23x^2+14x-15 \end{array}$$

41.

$$\begin{array}{r} x^2+x+2 \\ x^2-x+3 \\ \hline 3x^2+3x+6 \\ -x^3-x^2-2x \\ x^4+x^3+2x^2 \\ \hline x^4+4x^2+x+6 \end{array}$$

43.

$$\begin{array}{r} x+y-7 \\ x-y+4 \\ \hline 4x+4y-28 \\ -xy-y^2+7y \\ x^2+xy-7x \\ \hline x^2-3x-y^2+11y-28 \end{array}$$

45.

$$\begin{array}{r} 3x^2+2x-6 \\ 5x^2-4x-1 \\ \hline -3x^2-2x+6 \\ -12x^3-8x^2+24x \\ 15x^4+10x^3-30x^2 \\ \hline 15x^4-2x^3-41x^2+22x+6 \end{array}$$

47.

$$\begin{array}{r} 3x^2+5x+2 \\ 4x^2-3 \\ \hline -9x^2-15x-6 \\ 12x^4+20x^3+8x^2 \\ \hline 12x^4+20x^3-x^2-15x-6 \end{array}$$

49. $(2x-5)^2=(2x-5)(2x-5)=4x^2-20x+25$

51. $(3x+8)^2=(3x+8)(3x+8)=9x^2+48x+64$

Section 1.8

53. $(-5x+2)^2=(-5x+2)(-5x+2)=25x^2-20x+4$

55.
$$(3x-4)^2(x^2-2x+1)=$$
$$(3x-4)(3x-4)(x^2-2x+1)=$$
$$(9x^2-24x+16)(x^2-2x+1)=$$
$$\begin{array}{r} 9x^2-24x+16 \\ x^2-2x+1 \\ \hline 9x^2-24x+16 \\ -18x^3+48x^2-32x \\ 9x^4-24x^3+16x^2 \\ \hline 9x^4-42x^3+73x^2-56x+16 \end{array}$$

57.
$$(2x-1)^3=(2x-1)(2x-1)(2x-1)=$$
$$(4x^2-4x+1)(2x-1)=$$
$$\begin{array}{r} 4x^2-4x+1 \\ 2x-1 \\ \hline -4x^2+4x-1 \\ 8x^3-8x^2+2x \\ \hline 8x^3-12x^2+6x-1 \end{array}$$

59.
$$(2a+5b)^3=(2a+5b)(2a+5b)(2a+5b)=$$
$$(4a^2+20ab+25b^2)(2a+5b)=$$
$$\begin{array}{r} 4a^2+20ab+25b^2 \\ 2a+5b \\ \hline 20a^2b+100ab^2+125b^3 \\ 8a^3+40a^2b+50ab^2 \\ \hline 8a^3+60a^2b+150ab^2+125b^3 \end{array}$$

Exercises 1.9

1. $\dfrac{24x^2}{4x}=6x^{2-1}=6x$

3. $\dfrac{36a^3b^5}{-9a^2b^2}=-4a^{3-2}b^{5-2}=-4ab^3$

5. $\dfrac{45x^6y}{72x^3y^2}=\dfrac{5x^{6-3}}{8y^{2-1}}=\dfrac{5x^3}{8y}$

7. $\dfrac{-25x^3}{20x^5y^2}=\dfrac{-5}{4x^{5-3}y^2}=\dfrac{-5}{4x^2y^2}$

9. $\dfrac{3a(2b^2)^2}{(12ab)^2}=\dfrac{3a(4b^4)}{144a^2b^2}=\dfrac{12ab^4}{144a^2b^2}=\dfrac{b^2}{12a}$

11. $\dfrac{(4st^2)^2(3t^2)^3}{t^2(9st)^2}=\dfrac{16s^2t^4(27t^6)}{t^2(81s^2t^2)}=\dfrac{16(27)s^2t^{10}}{81s^2t^4}=\dfrac{16t^6}{3}$

13. $\dfrac{24x^2-16x+8}{8}=\dfrac{24x^2}{8}-\dfrac{16x}{8}+\dfrac{8}{8}=3x^2-2x+1$

15. $\dfrac{15x^5-20x^4+10x^2}{5x}=\dfrac{15x^5}{5x}-\dfrac{20x^4}{5x}+\dfrac{10x^2}{5x}=3x^4-4x^3+2x$

17. $\dfrac{-28x^5+35x^4-49x^3}{7x^3}=\dfrac{-28x^5}{7x^3}+\dfrac{35x^4}{7x^3}-\dfrac{49x^3}{7x^3}=-4x^2+5x-7$

19. $\dfrac{-64x^7+48x^5+36x^3+24x}{8x^4}=\dfrac{-64x^7}{8x^4}+\dfrac{48x^5}{8x^4}+\dfrac{36x^3}{8x^4}+\dfrac{24x}{8x^4}=-8x^3+6x+\dfrac{9}{2x}+\dfrac{3}{x^3}$

21. $\dfrac{4a^2b-6a^2b^2+8ab}{2ab}=\dfrac{4a^2b}{2ab}-\dfrac{6a^2b^2}{2ab}+\dfrac{8ab}{2ab}=2a-3ab+4$

23. $\dfrac{9m^2n^2+12mn^3-15m^3}{-3mn}=\dfrac{9m^2n^2}{-3mn}+\dfrac{12mn^3}{-3mn}-\dfrac{15m^3}{-3mn}=-3mn-4n^2+\dfrac{5m^2}{n}$

25. $\dfrac{224x^4y^2z^5-168x^3y^3z^4-112xy^4z^2}{28xy^2z^2}=\dfrac{224x^4y^2z^5}{28xy^2z^2}+\dfrac{-168x^3y^3z^4}{28xy^2z^2}-\dfrac{112xy^4z^2}{28xy^2z^2}=8x^3z^3-6x^2yz^2-4y^2$

27.

$$\begin{array}{r}
2x-5 \\
x+3\overline{)2x^2+x-15} \\
\underline{2x^2+6x} \\
-5x-15 \\
\underline{-5x-15} \\
0
\end{array}$$

Thus: $\dfrac{2x^2+x-15}{x+3}=2x-5$

29.

$$\begin{array}{r}
x^2-x+3 \\
x-2\overline{)x^3-3x^2+5x-6} \\
\underline{x^3-2x^2} \\
-x^2+5x \\
\underline{-x^2+2x} \\
3x-6 \\
\underline{3x-6} \\
0
\end{array}$$

Thus: $\dfrac{x^3-3x^2+5x-6}{x-2}=x^2-x+3$

31.

$$\begin{array}{r}
x^2-2x+3 \\
2x-1\overline{)2x^3-5x^2+8x+1} \\
\underline{2x^3-x^2} \\
-4x^2+8x \\
\underline{-4x^2+2x} \\
6x+1 \\
\underline{6x-3} \\
4
\end{array}$$

Thus: $\dfrac{2x^3-5x^2+8x+1}{2x-1}=x^2-2x+3+\dfrac{4}{2x-1}$

33.

$$\begin{array}{r}
2x^2+7x-3 \\
3x+5\overline{)6x^3+31x^2+26x-15} \\
\underline{6x^3+10x^2} \\
21x^2+26x \\
\underline{21x^2+35x} \\
-9x-15 \\
\underline{-9x-15} \\
0
\end{array}$$

Thus: $\dfrac{2x^2+7x-3}{3x+5}=2x^2+7x-3$

35. Use a zero place holder for the missing x^2 term.

$$\begin{array}{r} -4x^2-3x+7 \\ 4x-3\overline{)-16x^3+0x^2+37x-24} \\ \underline{-16x^3+12x^2} \\ -12x^2+37x \\ \underline{-12x^2+9x} \\ 28x-24 \\ \underline{28x-21} \\ -3 \end{array}$$

Thus: $\dfrac{-16x^3+37x-24}{4x-3}=-4x^2-3x+7+\dfrac{-3}{4x-3}$

37.

$$\begin{array}{r} 3x^3-2x^2+0x+4 \\ 3x+1\overline{)9x^4-3x^3-2x^2+12x+4} \\ \underline{9x^4+3x^3} \\ -6x^3-2x^2 \\ \underline{-6x^3-2x^2} \\ 0x^2+12x \\ \underline{0x^2+0x} \\ 12x+4 \\ \underline{12x+4} \\ 0 \end{array}$$

Thus: $\dfrac{9x^4-3x^3-2x^2+12x+4}{3x+1}=3x^3-2x^2+4$

39. Put terms in descending powers of the variables.

$$\begin{array}{r} x^2+3x-5 \\ x^2-x+2\overline{)x^4+2x^3-6x^2+11x-10} \\ \underline{x^4-x^3+2x^2} \\ 3x^3-8x^2+11x \\ \underline{3x^3-3x^2+6x} \\ -5x^2+5x-10 \\ \underline{-5x^2+5x-10} \\ 0 \end{array}$$

Thus: $\dfrac{x^4+2x^3-6x^2+11x-10}{x^2-x+2}=x^2+3x-5$

41. Use zero place holders for the missing terms.

$$\begin{array}{r} x^2-4x+16 \\ x+4\overline{)x^3+0x^2+0x-64} \\ \underline{x^3+4x^2} \\ -4x^2+0x \\ \underline{-4x^2-16x} \\ 16x-64 \\ \underline{16x+64} \\ -128 \end{array}$$

Thus: $\dfrac{x^3-64}{x+4}=x^2-4x+16+\dfrac{-128}{x+4}$

43. Use zero place holders for the missing terms.

$$\begin{array}{r} 4x^2-2x+1 \\ 2x+1\overline{)8x^3+0x^2+0x+1} \\ \underline{8x^3+4x^2} \\ -4x^2+0x \\ \underline{-4x^2-2x} \\ 2x+1 \\ \underline{2x+1} \\ 0 \end{array}$$

Thus: $\dfrac{8x^3+1}{2x+1}=4x^2-2x+1$

Exercises 1.10

1.
$$x+9=7$$
$$x+9-9=7-9$$
$$x=-2$$

3.
$$x-6=10$$
$$x-6+6=10+6$$
$$x=16$$

5.
$$5x=35$$
$$5x\div 5=35\div 5$$
$$x=7$$

7.
$$\frac{x}{8}=-9$$
$$\frac{x}{8}(8)=-9(8)$$
$$x=-72$$

9.
$$4x-6=14$$
$$4x-6+6=14+6$$
$$4x=20$$
$$4x\div 4=20\div 4$$
$$x=5$$

11.
$$5y+6=16$$
$$5y+6-6=16-6$$
$$5y=10$$
$$5y\div 5=10\div 5$$
$$y=2$$

13.
$$8-3x=14$$
$$8-3x-8=14-8$$
$$-3x=6$$
$$-3x\div(-3)=6\div(-3)$$
$$x=-2$$

15.
$$3x-2=5x+8$$
$$3x-2-5x=5x+8-5x$$
$$-2x-2=8$$
$$-2x-2+2=8+2$$
$$-2x=10$$
$$-2x\div(-2)=10\div(-2)$$
$$x=-5$$

17.
$$8-6x=5x+25$$
$$8-6x-5x=5x+25-5x$$
$$8-11x=25$$
$$8-11x-8=25-8$$
$$-11x=17$$
$$-11x\div(-11)=17\div(-11)$$
$$x=\frac{-17}{11}$$

19.
$$3(x-5)=27$$
$$3x-15=27$$
$$3x-15+15=27+15$$
$$3x=42$$
$$3x\div 3=42\div 3$$
$$x=14$$

21.
$$4(y+2)=30-(y-3)$$
$$4y+8=30-y+3$$
$$4y+8=33-y$$
$$4y+8+y=33-y+y$$
$$5y+8=33$$
$$5y+8-8=33-8$$
$$5y=25$$
$$5y\div 5=25\div 5$$
$$y=5$$

23.
$$6(3x+1)+(2x+2)=5x$$
$$18x+6+2x+2=5x$$
$$20x+8=5x$$
$$20x+8-20x=5x-20x$$
$$8=-15x$$
$$\frac{-8}{15}=x$$

25.
$$16(x+3)=7(x-5)-9(x+4)-7$$
$$16x+48=7x-35-9x-36-7$$
$$16x+48=-2x-78$$
$$16x+48-48=-2x-78-48$$
$$16x=-2x-126$$
$$16x+2x=-2x-126+2x$$
$$18x=-126$$
$$x=-7$$

27.
$$4(2x-1)-2x=3-5(2x-5)$$
$$8x-4-2x=3-10x+25$$
$$6x-4=-10x+28$$
$$6x-4+10x=-10x+28+10x$$
$$16x-4=28$$
$$16x-4+4=28+4$$
$$16x=32$$
$$16x\div 16=32\div 16$$
$$x=2$$

29.
$$1.5x-3.8+0.4(2x-5)=8$$
$$1.5x-3.8+0.8x-2=8$$
$$2.3x-5.8=8$$
$$2.3x-5.8+5.8=8+5.8$$
$$2.3x=13.8$$
$$x=6$$

31.
$$-[x-(4x+5)]=2+3(2x+3)$$
$$-[x-4x-5]=2+6x+9$$
$$-[-3x-5]=6x+11$$
$$3x+5=6x+11$$
$$3x+5-6x=6x+11-6x$$
$$-3x+5=11$$
$$-3x+5-5=11-5$$
$$-3x=6$$
$$-3x\div(-3)=6\div(-3)$$
$$x=-2$$

33.
$$\frac{2}{3}y-4=\frac{3}{4}y-5$$
L.C.D=12
$$\frac{2}{3}y(12)-4(12)=\frac{3}{4}y(12)-5(12)$$
$$8y-48=9y-60$$
$$8y-48-8y=9y-60-8y$$
$$-48=y-60$$
$$-48+60=y-60+60$$
$$12=y$$

35.
$$2-\frac{3}{7}x=\frac{x}{4}-\frac{15}{2}$$
L.C.D. = 28
$$2(28)-\frac{3}{7}x(28)=\frac{x}{4}(28)-\frac{15}{2}(28)$$
$$56-12x=7x-210$$
$$56-12x+12x=7x-210+12x$$
$$56=19x-210$$
$$56+210=19x-210+210$$
$$266=19x$$
$$266\div19=19x\div19$$
$$14=x$$

37.
$$3\left(\frac{2}{5}x-7\right)=4\left(\frac{1}{3}x-5\right)$$
$$\frac{6}{5}x-21=\frac{4}{3}x-20$$
L.C.D.=15
$$\frac{6}{5}x(15)-21(15)=\frac{4}{3}x(15)-20(15)$$
$$18x-315=20x-300$$
$$18x-315-18x=20x-300-18x$$
$$-315=2x-300$$
$$-315+300=2x-300+300$$
$$-15=2x$$
$$-15\div2=2x\div2$$
$$\frac{-15}{2}=x$$

39.
$$\frac{3(y+2)}{4}-\frac{y-3}{2}=\frac{3}{4}y$$
L.C.D.=4
$$\frac{3(y+2)}{4}(4)-\frac{(y-3)}{2}(4)=\frac{3}{4}y(4)$$
$$3(y+2)-2(y-3)=3y$$
$$3y+6-2y+6=3y$$
$$y+12=3y$$
$$y+12-y=3y-y$$
$$12=2y$$
$$12\div2=2y\div2$$
$$6=y$$

41.

$$\frac{4(x-3)}{9}-\frac{2x}{3}=1$$
$$\text{L.C.D.}=9$$
$$\frac{4(x-3)}{9}(9)-\frac{2x}{3}(9)=1(9)$$
$$4(x-3)-6x=9$$
$$4x-12-6x=9$$
$$-2x-12=9$$
$$-2x-12+12=9+12$$
$$-2x=21$$
$$-2x\div(-2)=21\div(-2)$$
$$x=-\frac{21}{2}$$

43.

$$\frac{3}{5}\left(\frac{2}{3}y+\frac{1}{5}\right)=\frac{2}{3}\left(\frac{3}{4}y-\frac{1}{3}\right)$$
$$\frac{2}{5}y+\frac{3}{25}=\frac{1}{2}y-\frac{2}{9}$$
$$\text{L.C.D.}=450$$
$$\frac{2}{5}y(450)+\frac{3}{25}(450)=\frac{1}{2}y(450)-\frac{2}{9}(450)$$
$$180y+54=225y-100$$
$$180y+54-180y=225y-100-180y$$
$$54=45y-100$$
$$54+100=45y-100+100$$
$$154=45y$$
$$154\div45=45y\div45$$
$$\frac{154}{45}=y$$

45.

$$18.7x+253=28.6x$$
$$18.7x+253-18.7x=28.6x-18.7x$$
$$253=9.9x$$
$$253\div9.9=9.9x\div9.9$$
$$25.6=x$$

47.

$$4.12x+6.18=12.6x-3.6$$
$$4.12x+6.18-12.6x=12.6x-3.6-12.6x$$
$$-8.48x+6.18=-3.6$$
$$-8.48x+6.18=6.18=-3.6-6.18$$
$$-8.48x=-9.78$$
$$-8.48x\div(-8.48)=-9.78\div(-8.48)$$
$$x=1.15$$

49.

$$6.3-0.4(9.36-2x)=1.2x$$
$$6.3-3.744+0.8x=1.2x$$
$$2.556+0.8x=1.2x$$
$$2.556+0.8x-0.8x=1.2x-0.8x$$
$$2.556=0.4x$$
$$2.556\div(0.4)=0.4x\div0.4$$
$$6.39=x$$

51.

$$2.76(1.81x+59.2)+1.67=763$$
$$4.9956x+163.392+16.7=763$$
$$4.9956x+180.092=763$$
$$4.9956x+180.092-180.092=763-180.092$$
$$4.9956x=582.908$$
$$4.9956x\div4.9956=582.908\div4.9956$$
$$x=117$$

53.

$$2\pi\left(2.50x10^{6}\right)L=2590$$
$$\frac{2\pi\left(2.50x10^{6}\right)L}{2\pi\left(2.50x10^{6}\right)}=\frac{2590}{2\pi\left(2.50x10^{6}\right)}$$
$$L=1.65x10^{-4}$$

55.

$$2.50\times10^{-3}=\frac{\left(9.00\times10^{9}\right)\left(8.50\times10^{-7}\right)q}{(0.15)^{2}}$$
$$\frac{2.50\times10^{-3}\times(0.15)^{2}}{\left(9.00\times10^{9}\right)\left(8.50\times10^{-7}\right)}=q$$
$$q=7.35\times10^{-9}$$

Exercises 1.11

1.

$$W = JQ \text{ for } J$$
$$\frac{W}{Q} = \frac{JQ}{Q}$$
$$\frac{W}{Q} = J$$

3.

$$R_T = R_1 + R_2 + R_3 \text{ for } R_2$$
$$R_T - R_1 - R_3 = R_1 + R_2 + R_3 - R_1 - R_3$$
$$R_T - R_1 - R_3 = R_2$$

5.

$$E = IR \text{ for } R$$
$$\frac{E}{I} = \frac{IR}{I}$$
$$\frac{E}{I} = R$$

7.

$$C = \frac{Q}{V} \textit{ for } Q$$
$$C(V) = \frac{Q}{V}(V)$$
$$CV = Q$$

9.

$$V = \frac{W}{Q} \textit{ for } Q$$
$$V(Q) = \frac{W}{Q}(Q)$$
$$VQ = W$$
$$\frac{VQ}{V} = \frac{W}{V}$$
$$Q = \frac{W}{V}$$

11.

$$Q = \frac{I^2Rt}{J} \text{ for } R$$
$$Q(J) = \frac{I^2Rt}{J}(J)$$
$$QJ = I^2Rt$$
$$\frac{QJ}{I^2t} = \frac{I^2Rt}{I^2t}$$
$$\frac{QJ}{I^2t} = R$$

13.

$$P = \frac{(O.D.)}{N+2} \text{ for } N$$
$$P(N+2) = \frac{(O.D.)}{N+2}(N+2)$$
$$PN + 2P = (O.D.)$$
$$PN + 2P - 2P = (O.D.) - 2P$$
$$PN = (O.D.) - 2P$$
$$\frac{PN}{P} = \frac{(O.D.) - 2P}{P}$$
$$N = \frac{(O.D.) - 2P}{P}$$

15.

$$R = \frac{kL}{D^2} \text{ for } L$$
$$R(D^2) = \frac{kL}{D^2}(D^2)$$
$$RD^2 = kL$$
$$\frac{RD^2}{k} = \frac{kL}{k}$$
$$\frac{RD^2}{k} = L$$

17.

$$R = \frac{\pi}{2P} \text{ for } P$$
$$R(2P) = \frac{\pi}{2P}(2P)$$
$$2RP = \pi$$
$$\frac{2RP}{2R} = \frac{\pi}{2R}$$
$$P = \frac{\pi}{2R}$$

19.

$$\frac{V}{V'} = \frac{T}{T'} \text{ for } T$$
$$\frac{V}{V'}(T') = \frac{T}{T'}(T')$$
$$\frac{VT'}{V'} = T$$

21. $C = \frac{5}{9}(F - 32)$ for F

$$9C = 9\left(\frac{5}{9}(F-32)\right)$$
$$9C = 5(F-32)$$
$$9C = 5F - 160$$
$$9C + 160 = 5F - 160 + 160$$
$$9C + 160 = 5F$$
$$\frac{9C+160}{5} = \frac{5F}{5}$$
$$\frac{9C}{5} + \frac{160}{5} = F$$
$$\frac{9}{5}C + 32 = F$$

23. $\frac{I_s}{I_p} = \frac{N_p}{N_s}$ for N_s

$$\frac{I_s}{I_p}(I_p N_s) = \frac{N_p}{N_s}(I_p N_s)$$
$$I_s N_s = N_p I_p$$
$$\frac{I_s N_s}{I_s} = \frac{N_p I_p}{I_s}$$
$$N_s = \frac{N_p I_p}{I_s}$$

25. $\frac{\Delta d}{d} = 1.22\frac{\lambda}{a}$ for a

$$\frac{\Delta d}{d}(da) = 1.22\frac{\lambda}{a}(da)$$
$$a\Delta d = 1.22\lambda d$$
$$\frac{a\Delta d}{\Delta d} = \frac{1.22\lambda d}{\Delta d}$$
$$a = \frac{1.22\lambda d}{\Delta d}$$

27. $l = n\frac{\lambda}{2}$ for λ

$$2l = n\frac{\lambda}{2}(2)$$
$$2l = n\lambda$$
$$\frac{2l}{n} = \frac{n\lambda}{n}$$
$$\frac{2l}{n} = \lambda$$

29. $\Delta L = \alpha L(T - T_o)$ for T

$$\Delta L = \alpha LT - \alpha LT_o$$
$$\Delta L + \alpha LT_o = \alpha LT - \alpha LT_o + \alpha LT_o$$
$$\Delta L + \alpha LT_o = \alpha LT$$
$$\frac{\Delta L + \alpha LT_o}{\alpha L} = \frac{\alpha LT}{\alpha L}$$
$$\frac{\Delta L + \alpha LT_o}{\alpha L} = T$$

31. $f' = f\left(\frac{v+v_o}{v-v_s}\right)$ for v

$$f'(v-v_s) = f\frac{(v+v_o)}{(v-v_s)}\cdot(v-v_s)$$
$$f'(v-v_s) = f(v+v_o)$$
$$f'v - f'v_s = fv + fv_o$$
$$f'v - f'v_s - fv = fv + fv_o - fv$$
$$f'v - f'v_s - fv = fv_o$$
$$f'v - f'v_s - fv + f'v_s = fv_o + f'v_s$$
$$f'v - fv = fv_o + f'v_s$$
$$(f'-f)v = fv_o + f'v_s$$
$$\frac{(f'-f)v}{(f'-f)} = \frac{fv_o + f'v_s}{f'-f}$$
$$v = \frac{fv_o + f'v_s}{f'-f}$$

33. $$F = \frac{q_1 q_2}{4\pi\varepsilon_o r^2} \text{ for } q_1$$

$$F\left(4\pi\varepsilon_o r^2\right) = \frac{q_1 q_2}{4\pi\varepsilon_o r^2} \cdot 4\pi\varepsilon_o r^2$$

$$4\pi F\varepsilon_o r^2 = q_1 q_2$$

$$\frac{4\pi F\varepsilon_o r^2}{q_2} = \frac{q_1 q_2}{q_2}$$

$$\frac{4\pi F\varepsilon_o r^2}{q_2} = q_1$$

35. $$E - IR - \frac{q}{C} = 0 \text{ for } R$$

$$E - IR - \frac{q}{C} + IR = 0 + IR$$

$$E - \frac{q}{C} = IR$$

$$\left(E - \frac{q}{C}\right)\left(\frac{1}{I}\right) = IR\left(\frac{1}{I}\right)$$

$$\frac{E}{I} - \frac{q}{CI} = R$$

37. $$\frac{1}{R_T} = \frac{1}{R_1} + \frac{1}{R_2} \text{ for } R_2$$

$$\frac{1}{R_T}\left(R_T R_1 R_2\right) = \left(\frac{1}{R_1} + \frac{1}{R_2}\right)\left(R_T R_1 R_2\right)$$

$$R_1 R_2 = R_T R_2 + R_T R_1$$

$$R_1 R_2 - R_T R_2 = R_T R_2 + R_T R_1 - R_T R_2$$

$$R_1 R_2 - R_T R_2 = R_T R_1$$

$$\left(R_1 - R_T\right) R_2 = R_T R_1$$

$$\frac{\left(R_1 - R_T\right) R_2}{\left(R_1 - R_T\right)} = \frac{R_T R_1}{\left(R_1 - R_T\right)}$$

$$R_2 = \frac{R_T R_1}{R_1 - R_T}$$

39. $$\frac{1}{R_T} = \frac{1}{R_1} + \frac{1}{R_2} + \frac{1}{R_3} \text{ for } R_T$$

For right side, $L.C.D. = R_1 R_2 R_3$

$$\frac{1}{R_T} = \frac{R_2 R_3}{R_1 R_2 R_3} + \frac{R_1 R_3}{R_1 R_2 R_3} + \frac{R_1 R_2}{R_1 R_2 R_3}$$

$$\frac{1}{R_T} = \frac{R_2 R_3 + R_1 R_3 + R_1 R_2}{R_1 R_2 R_3}$$

Since we have a single fraction on each side, the reciprocals are equal.

$$R_T = \frac{R_1 R_2 R_3}{R_2 R_3 + R_1 R_3 + R_1 R_2}$$

41. $$\frac{1}{f} = \frac{1}{s_o} + \frac{1}{s_i} \text{ for } s$$

$$\frac{1}{f} - \frac{1}{s_i} = \frac{1}{s_o} + \frac{1}{s_i} - \frac{1}{s_i}$$

$$\frac{1}{f} - \frac{1}{s_i} = \frac{1}{s_o}$$

$$\frac{s_i}{fs_i} - \frac{f}{fs_i} = \frac{1}{s_o}$$

$$\frac{s_i - f}{fs_i} = \frac{1}{s_o}$$

$$\frac{fs_i}{s_i - f} = s_o$$

43. $$\frac{1}{f} = (n-1)\left(\frac{1}{R'} - \frac{1}{R''}\right) \text{ for } n$$

$$\frac{1}{f} = (n-1)\left(\frac{R''}{R'R''} - \frac{R'}{R'R''}\right)$$

$$\frac{1}{f} = (n-1)\left(\frac{R'' - R'}{R'R''}\right)$$

$$\frac{1}{f}\left(\frac{R'R''}{R'' - R'}\right) = (n-1)\left(\frac{R'' - R'}{R'R''}\right)\left(\frac{R'R''}{R'' - R'}\right)$$

$$\frac{R'R''}{f(R'' - R')} = n - 1$$

$$\frac{R'R''}{f(R'' - R')} + 1 = n$$

$$\text{or } \frac{R'R'' + fR'' - fR'}{fR'' - fR'} = n$$

45.

$$eV = hf - \phi \text{ for } f$$
$$eV + \phi = hf - \phi + \phi$$
$$eV + \phi = hf$$
$$\frac{eV + \phi}{h} = \frac{hf}{h}$$
$$\frac{eV + \phi}{h} = f$$

47.

$$I_2 = \frac{Z_3}{Z_2 + Z_3} I_T \text{ for } Z_2$$
$$I_2(Z_2 + Z_3) = \frac{Z_3 I_T}{(Z_2 + Z_3)} \cdot (Z_2 + Z_3)$$
$$I_2 Z_2 + I_2 Z_3 = Z_3 I_T$$
$$I_2 Z_2 + I_2 Z_3 - I_2 Z_3 = Z_3 I_T - I_2 Z_3$$
$$I_2 Z_2 = Z_3 I_T - I_2 Z_3$$
$$\frac{I_2 Z_2}{I_2} = \frac{Z_3 I_T - I_2 Z_3}{I_2}$$
$$Z_2 = \frac{Z_3 I_T - I_2 Z_3}{I_2}$$
$$\text{or } Z_2 = \frac{Z_3(I_T - I_2)}{I_2}$$

49.

$$R_A = \frac{R_1 R_3}{R_1 + R_2 + R_3} \text{ for } R_1$$
$$R_A(R_1 + R_2 + R_3) = \frac{R_1 R_3}{(R_1 + R_2 + R_3)} \cdot (R_1 + R_2 + R_3)$$
$$R_A R_1 + R_A R_2 + R_A R_3 = R_1 R_3$$
$$R_A R_1 + R_A R_2 + R_A R_3 - R_A R_1 = R_1 R_3 - R_A R_1$$
$$R_A R_2 + R_A R_3 = R_1(R_3 - R_A)$$
$$\frac{R_A R_2 + R_A R_3}{R_3 - R_A} = \frac{R_1(R_3 - R_A)}{(R_3 - R_A)}$$
$$\frac{R_A R_2 + R_A R_3}{R_3 - R_A} = R_1$$
$$\frac{R_A(R_2 + R_3)}{R_3 - R_A} = R_1$$

Exercises 1.12

1.

$$A = bh$$
$$\frac{A}{h} = b$$
$$\frac{24.0 \text{ in}^2}{6.00 \text{ in}} = b$$
$$b = 4.00 \text{ in}$$

3.

$$A = \frac{1}{2}bh$$
$$2A = bh$$
$$\frac{2A}{b} = h$$
$$\frac{2(144)m^2}{8.00 \text{ m}}$$
$$h = 36.0 \text{ m}$$

5.

$$V = \pi r^2 h$$
$$\frac{V}{\pi r^2} = h$$
$$\frac{1950 \text{ m}^3}{\pi(12.6 \text{ m})^2} = h$$
$$h = 3.91 \text{ m}$$

7.

$$V = lwh$$
$$\frac{V}{lh} = w$$
$$\frac{2.50 \times 10^4 \text{ ft}^3}{44.5 \text{ ft}(19.7) \text{ ft}}$$

9.

$$C = \frac{5}{9}(F - 32)$$
$$9C = 5F - 160$$
$$9C + 160 = 5F$$
$$\frac{9C + 160}{5} = F$$
$$\frac{9(55) + 160}{5} = F$$
$$F = 131°$$

11.

$$Q = mc\Delta T$$
$$Q = 50\bar{0}g(0.214)\text{cal}/g°C\ (40.0°C)$$
$$Q = 4280 \text{ cal}$$

13.

$$X_L = 2\pi f L$$
$$\frac{X_L}{2\pi f} = L$$
$$\frac{75.0\Omega}{2\pi 60.0\text{ Hz}} = L$$
$$L = 0.199\Omega/\text{Hz}$$

15.

$$P = \frac{Fd}{550\frac{\text{ft-lb}}{s}(t)}$$
$$P = \frac{8\bar{0}(2000)\text{lb}(2\bar{0}0\text{ ft})}{550\frac{\text{ft-lb}}{s}(60(60))s}$$
$$P = 16hp$$

17.

$$\frac{P_1}{P_2} = \frac{V_2}{V_1}$$
$$\frac{P_1V_1}{P_2} = V_2$$
$$V_2 = \frac{270kPa(750\text{ cm}^3)}{210kPa}$$
$$V_2 = 960\text{ cm}^3$$

19.

$$\frac{\Delta d}{d} = 1.22\frac{\lambda}{a}$$
$$\Delta d = \frac{1.22\lambda d}{a}$$
$$\frac{a\Delta d}{1.22d} = d$$
$$d = \frac{2.00\times10^{-3}\text{m}(1.75)\text{mm}}{1.22(5.50\times10^{-7}\text{ m})}$$
$$d = 5220\text{ mm}$$
$$d = 5220\ (.001\text{m})$$
$$d = 5.22\text{ m}$$

21.

$$\frac{V_1P_1}{T_1} = \frac{V_2P_2}{T_2}$$
$$\frac{T_1P_1T_2}{T_1P_2} = V_2$$
$$V_2 = \frac{125m^3(13.5)MPa(305)K}{295K(18.4)MPa}$$
$$V_2 = 94.8m^3$$

23.

$$f = \frac{nv}{2l}$$
$$2fl = nv$$
$$l = \frac{nv}{2f}$$
$$l = \frac{1(348)m/s}{2(384/s)}$$
$$l = 0.453\text{ m}$$
$$l = 0.453\ (100\text{ cm})$$
$$l = 45.3\text{ cm}$$

25.

$$f' = f\left(\frac{v - v_o}{v + v_s}\right)$$
$$f' = 425/s\left(\frac{343m/s - 0}{343m/s + 25.0m/s}\right)$$
$$f' = 396/s = 396\text{ Hz}$$

27.

$$\frac{1}{R} = \frac{1}{R_1} + \frac{1}{R_2}$$
$$\frac{1}{R} - \frac{1}{R_2} = \frac{1}{R_1}$$
$$\frac{R_2 - R}{RR_2} = \frac{1}{R_1}$$
$$\frac{RR_2}{R_2 - R} = R_1$$
$$R_1 = \frac{60.0\Omega(24\bar{0}\Omega)}{24\bar{0}\Omega - 60.0\Omega}$$
$$R_1 = 80.0\Omega$$

29.

$$\frac{1}{R} = \frac{1}{R_1} + \frac{1}{R_2} + \frac{1}{R_3}$$
$$\frac{1}{R} = \frac{R_2R_3}{R_1R_2R_3} + \frac{R_1R_3}{R_1R_2R_3} + \frac{R_1R_2}{R_1R_2R_3}$$
$$\frac{1}{R} = \frac{R_2R_3 + R_1R_3 + R_1R_2}{R_1R_2R_3}$$
$$R = \frac{R_1R_2R_3}{R_2R_3 + R_1R_3 + R_1R_2}$$
$$R = \frac{3\bar{0}\Omega(4\bar{0})\Omega(5\bar{0})\Omega}{4\bar{0}\Omega(5\bar{0})\Omega + 3\bar{0}\Omega(5\bar{0})\Omega + 3\bar{0}\Omega(4\bar{0})\Omega}$$
$$R = 13\Omega$$

Section 1.12

31.

$$\frac{1}{f}=\frac{1}{s_o}+\frac{1}{s_i}$$
$$\frac{1}{f}-\frac{1}{s_o}=\frac{1}{s_i}$$
$$\frac{s_o-f}{fs_o}=\frac{1}{s_i}$$
$$\frac{fs_o}{s_o-f}=s_i$$

33.

$$s=v_it+\frac{1}{2}at^2$$
$$s-\frac{1}{2}at^2=v_it$$
$$\frac{s-\frac{1}{2}at^2}{t}=v_i$$
$$v_i=\frac{40.0m-\frac{1}{2}\left(-6.00m/s^2\right)\left(3.70s\right)^2}{3.70s}$$
$$v_i=21.9m/s$$

35.

$$I_1=\frac{R_2}{R_1+R_2}I_2$$
$$I_1\left(\frac{R_1+R_2}{R_2}\right)=I_2$$
$$I_2=0.125mA\left(\frac{5.0\Omega+25.0\Omega}{25.0\Omega}\right)$$
$$I_2=0.150mA$$

Exercises 1.13

1.

Let x = one part

$3x$ = other part

$x+3x=\$216$

$4x=\$216$

$x=\$54$

and $3x=\$162$

3.

There are 8 shelves, so there are 9 spaces,

$$\frac{\text{Floor to ceiling space}-\text{shelf space}}{\text{number of spaces}}=$$

space between each shelf

$$\frac{9.5\text{ ft}\left(12\text{ in/ft}\right)-8\left(0.75\text{ in}\right)}{9\text{ spaces}}=$$

12 inches per space

5.

Let x = width

$x+4$ = length

p = 2w+2 1

$56\text{m}=\left(2x+2\left(x+4\right)\right)\text{m}$

$56=2x+2x+8$

$48=4x$

$12=x$

width is 12 m

length is 16 m

7.

Let x = no. acres @ \$650

$40-x$ = no. acres @ \$450

$650x+450\left(40-x\right)=20{,}400$

$650x+18{,}000-450x=20{,}400$

$200x=2400$

$x=12$

$40-x=28$

12 acres @ \$650 and

28 acres @ \$450

9.

The salary increase

is $\$1660 - \$1440 = \$220$

Let x = no. weeks to cover

moving costs

$220x = 4840$

$x = 22$

11.

Find time for distances

to be equal. Use $d = rt$.

Let x = time of passenger train

$x + 3$ = time of freight train

$$75x = 45(x+3)$$

$$75x = 45x + 135$$

$$30x = 135$$

$$x = 4.5$$

(a) $4.5h$

(b) $d = 75 \text{ mph}(4.5)h$

$d = 337.5$ mi

13.

Distance plane 1 + distance plane 2 = 760 mi

Let x = speed of plane 1

$x+20$=speed of plane 2

$$4x+4(x+20)=760$$

$$4x+4x+80=760$$

$$8x=680$$

$$x=85$$

plane 1 travels at 85 mph

plane 2 travels at 105 mph

15.

Pure tin from 1st +

pure tin from 2nd =

pure tin in mixture

Let x = volume of 30%

$20 - x$ = volume of 70%

$$.30(x) + .70(20 - x) = .45(20)$$

$$.30x + 14 - .70x = 9$$

$$-.40x = -5$$

$$x = 12.5$$

12.5 lb. of 30%

$20 - 12.5 = 7.5$ lb of 70%

17.

Let x = the number of qt

to be drained and added

$$.25(12) - .25x + 1.00x = .50(12)$$

$$3 + .75x = 6$$

$$.75x = 3$$

$x = 4$ qt

19.

Let x = least amount current

$4x$ = greatest amount current

$x + 0.75$ = third amount current

$$x+4x+x+0.75A = 4.65A$$

$$6x+0.75A = 4.65A$$

$$6x = 3.9A$$

$$x = 0.65A$$

$$4x = 2.60A$$

$$x+0.75A = 1.40A$$

21.

$$\text{Let } x = \text{width}$$
$$2x = \text{length}$$
$$P = 2\,1 + 2w$$
$$\text{Fencing needed} =$$
$$2\,1 + 2w + \text{extra}$$
$$\text{width to divide.}$$
$$280 = 2(2x) + 2x + x$$
$$280 = 4x + 3x$$
$$280 = 7x$$
$$x = 40m$$
$$40m \times 80m$$

23.

$$\text{Let } x = \text{first } R$$
$$x + 15 = \text{second } R$$
$$x + 15 + 60 = \text{third } R$$
$$x + x + 15 + x + 75 = 540$$
$$3x + 90 = 540$$
$$3x = 450$$
$$x = 150$$
$$\text{first } R = 150\Omega$$
$$\text{second } R = 165\Omega$$
$$\text{third } R = 225\Omega$$

Exercises 1.14

1. $\dfrac{36 \text{ mi}}{8 \text{ mi}} = \dfrac{9}{2}$

3. $\dfrac{2500 \text{ m}}{25 \text{ km}} = \dfrac{2500(.001) \text{ km}}{25 \text{ km}} = \dfrac{1}{10}$

5. $\dfrac{12 \text{ ft}^2}{24 \text{ in}^2} = \dfrac{12(144 \text{ in}^2)}{24 \text{ in}^2} = \dfrac{72}{1}$

7.

$$\frac{1.75 \text{ m}^2}{35 \text{ cm}^2} = \frac{1.75(100 \text{ cm})^2}{35 \text{ cm}^2} = \frac{1.75(100)^2 \text{ cm}^2}{35 \text{ cm}^2} = \frac{500}{1}$$

9. $\dfrac{(28 \text{ in})^2}{(7 \text{ in})^2} = \dfrac{28^2 \text{ in}^2}{7^2 \text{ in}^2} = \dfrac{16}{1}$

11. $\text{MA} = \dfrac{\text{resistance force}}{\text{effort force}} = \dfrac{7500 \text{ lb}}{250 \text{ lb}} = \dfrac{30}{1}$

13. $\dfrac{3/4 \text{ in}}{2500 \text{ ft}} = \dfrac{3/4 \text{ in}}{2500(12) \text{ in}} = \dfrac{3}{4(2500)(12)} = \dfrac{1}{40{,}000}$

15. $\dfrac{72 \text{ teeth}}{18 \text{ teeth}} = \dfrac{4}{1}$

17. $\dfrac{\$80{,}475}{1850 \text{ ft}^2} = \dfrac{\$43.50}{\text{ft}^2}$

19. $\dfrac{350 \text{ gal}}{14 \text{ acres}} = 25 \text{ gal/acre}$

21. $\dfrac{1125 \text{ rpm}}{500 \text{ rpm}} = \dfrac{9}{4}$

23. $\dfrac{72 \text{ teeth}}{15 \text{ teeth}} = \dfrac{24}{5}$

25. $\dfrac{6075 \text{ bu}}{45 \text{ acre}} = 135 \text{ bu/acre}$

27. $\dfrac{25 \text{ ft/min}}{15 \text{ ft/min}} = \dfrac{5}{3}$

29.
$$\frac{x}{9}=\frac{81}{27}$$
$$\frac{x}{9}(9)=\frac{81(9)}{27}$$
$$x=27$$

31.
$$\frac{64}{2x}=\frac{16}{84}$$
$$2x(16)=64(84)$$
$$\frac{2(16)x}{32}=\frac{64(84)}{32}$$
$$x=168$$

33.
$$\frac{x}{5-x}=\frac{2}{3}$$
$$3x=2(5-x)$$
$$3x=10-2x$$
$$3x+2x=10-2x+2x$$
$$5x=10$$
$$\frac{5x}{5}=\frac{10}{5}$$
$$x=2$$

35.
$$\frac{x}{28-x}=\frac{3}{4}$$
$$4x=3(28-x)$$
$$4x=84-3x$$
$$4x+3x=84-3x+3x$$
$$7x=84$$
$$\frac{7x}{7}=\frac{84}{7}$$
$$x=12$$

37.
$$\frac{m}{x}=\frac{n}{p}$$
$$nx=mp$$
$$\frac{nx}{n}=\frac{mp}{n}$$
$$x=\frac{mp}{n}$$

39.
$$\frac{x}{2a}=\frac{4}{a}$$
$$ax=4(2a)$$
$$ax=8a$$
$$\frac{ax}{a}=\frac{8a}{a}$$
$$x=8$$

41.
$$\frac{30.2}{276}=\frac{85.6}{x}$$
$$30.2x=276(85.6)$$
$$\frac{30.2x}{30.2}=\frac{276(85.6)}{30.2}$$
$$x=782$$

43.
$$\frac{32.5}{115}=\frac{2450}{x}$$
$$32.5x=115(2450)$$
$$\frac{32.5x}{32.5}=\frac{115(2450)}{32.5}$$
$$x=8670$$

45.
$$\frac{x}{0.477}=\frac{2.75}{16.1}$$
$$\frac{x}{0.477}(0.477)=\frac{2.75}{16.1}(0.477)$$
$$x=0.0815$$

47.(a) *Let* x = cost *of* 750 *bolts*
$$\frac{\$240}{2000\ bolts}=\frac{x}{750}$$
$$2000x=240(750)$$
$$x=\frac{240(750)}{2000}$$
$$x=\$90$$

47.(b) *Let* y = cost *of* 2800 *bolts*
$$\frac{\$240}{2000}=\frac{y}{2800}$$
$$2000y=240(2800)$$
$$y=\frac{240(2800)}{2000}$$
$$y=\$336$$

49. Let x = pay for 36 h
$$\frac{\$182.40}{30\ h}=\frac{x}{36}$$
$$30x=36(182.40)$$
$$x=\frac{36(182.40)}{30}$$
$$x=\$218.88$$

51. Let x = no. strokes *for* 75 mL
$$\frac{540 \text{ strokes}}{45 \text{ mL}}=\frac{x}{75}$$
$$45x=540(75)$$
$$x=\frac{540(75)}{45}$$
$$x=900 \text{ strokes}$$

Section 1.14

53. Let x = price of 2100 ft²

$$\frac{\$94,500}{1500\text{ ft}^2} = \frac{x}{2100}$$
$$1500x = 2100(94500)$$
$$x = \frac{2100(94500)}{1500}$$
$$x = \$132,300$$

55. Let x = length

1260 - x = width

$$\frac{x}{1260 - x} = \frac{5}{2}$$
$$2x = 5(1260 - x)$$
$$2x = 6300 - 5x$$
$$2x + 5x = 6300 - 5x + 5x$$
$$7x = 6300$$
$$x = 900$$
$$90\bar{0}\text{ ft} \times 36\bar{0}\text{ ft}$$

57. Let x = no. parts in group 1

$68340 - x$ = no. parts in group 2

$$\frac{x}{68340 - x} = \frac{5}{12}$$
$$12x = 5(68340 - x)$$
$$12x = 341700 - 5x$$
$$12x + 5x = 341700 - 5x + 5x$$
$$17x = 341700$$
$$x = 20,1\bar{0}0$$
$$68340 - x = 48,240$$

59. Let x = amt. copper

$2500 - x$ = amt. zinc

$$\frac{x}{2500 - x} = \frac{3}{2}$$
$$2x = 3(2500 - x)$$
$$2x = 7500 - 3x$$
$$2x + 3x = 7500 - 3x + 3x$$
$$5x = 7500$$
$$x = 1500$$

1500 pounds copper

$1\bar{0}00$ pounds zinc

61. Let x = amt. taxes

$$\frac{\$638.40}{\$45000} = \frac{x}{78000}$$
$$45000x = 638.40(78000)$$
$$x = \frac{638.40(78000)}{45000}$$
$$x = \$1106.56$$

63. Let x = amt. needed

$$\frac{150\text{ lb}}{40\text{ acre}} = \frac{x}{180}$$
$$40x = 180(150)$$
$$x = \frac{180(150)}{40}$$
$$x = 675\text{ pounds}$$

65. Let x = expected yield

$$\frac{38400 \text{ bu}}{320 \text{ acre}} = \frac{x}{520}$$
$$320x = 38400(520)$$
$$x = \frac{38400(520)}{320}$$
$$x = 62,400 \text{ bu}$$

67. Let x = voltage drop sought

$$\frac{52V}{28\Omega} = \frac{x}{63}$$
$$28x = 52(63)$$
$$x = \frac{52(63)}{28}$$
$$x = 117V$$

Rounded to 2 significant digits is 120V.

69. Let x = no. secondary turns

$$\frac{\text{sec. voltage}}{\text{prim voltage}} = \frac{\text{sec. turns}}{\text{prim turns}}$$
$$\frac{15,000}{8V} = \frac{x}{250}$$
$$8x = 15000(250)$$
$$x = \frac{15000(250)}{8}$$
$$x = 468,750$$

Rounded to 2 significant digits is 470,000.

71.

$$\frac{\text{linkage force}}{\text{driver force}} = \frac{30}{1}$$

Let x = force to pressure plate

$$\frac{30}{1} = \frac{x}{15N}$$
$$1x = 30(15)N$$
$$x = 450N$$

73. Triangles whose sides are in the ratio of 3:4:5 are proportional to a 3, 4, 5 triangle. This means there is a constant, k, so that the hypotenuse is $5k$, the smaller leg is $3k$, and the other leg is $4k$. Let k be the constant of proportion.

Then $\frac{5}{1} = \frac{85}{k}$ or $5k = 85$, thus $k = 17$. Let x = length of the smallest leg.

Then $\frac{3}{1} = \frac{x}{17}$ or $x = 3(17) = 51$. Let y be the length of the other leg. Then $\frac{4}{1} = \frac{y}{17}$ or $y = 4(17)$. That is $y = 68$. The legs are 51 m and 68 m.

75. See solution to 73. Let k = constant of proportion. Then the sides of the lot are $3k$, $4k$, and $5k$.

sum sides = fence length

$$3k + 4k + 5k = 360$$
$$12k = 360$$
$$k = 30$$

So the sides are

$$3k = 3(30) = 90 \text{ ft}$$
$$4k = 4(30) = 120 \text{ ft}$$
$$5k = 5(30) = 150 \text{ ft}$$

Exercises 1.15

1. Direct. Each ratio of y to x is $\frac{3}{4}$.

3. Inverse. Each product of x and y is 1.5.

5. Neither. Ratios of y to x are not the same, and products of x and y are not the same.

7. $y = kz$

9. $a = kbc$

11. $r = \frac{ks}{\sqrt{t}}$

13. $f = \frac{kgh}{j^2}$

15.

$$y = kx \qquad y = \frac{1}{3}x$$
$$8 = k(24) \qquad y = \frac{1}{3}(36)$$
$$\frac{8}{24} = k$$
$$k = \frac{1}{3} \qquad y = 12$$

17.

$$y = \frac{k}{x} \qquad y = \frac{54}{x}$$
$$9 = \frac{k}{6} \qquad y = \frac{54}{18}$$
$$9(6) = k$$
$$k = 54 \qquad y = 3$$

19.

$$y = k\sqrt{x} \qquad y = 6\sqrt{x}$$
$$24 = k\sqrt{16} \qquad y = 6\sqrt{36}$$
$$24 = 4k \qquad y = 6(6)$$
$$\frac{24}{4} = k$$
$$k = 6 \qquad y = 36$$

21.

$$y = kst \qquad y = 0.25st$$
$$7.2 = k(36)(0.8) \qquad y = 0.25(52)(1.5)$$
$$k = \frac{7.2}{36(0.8)}$$
$$k = 0.25 \qquad y = 19.5$$

23.

$$p = \frac{kq}{r^2} \qquad p = \frac{32q}{r^2}$$
$$40 = \frac{k(20)}{4^2} \qquad p = \frac{32(24)}{6^2}$$
$$40(16) = 20k \qquad p = 21\frac{1}{3} \text{ or}$$
$$\frac{640}{20} = k \qquad p = \frac{64}{3}$$
$$32 = k$$

25.

$$F = \frac{km_1m_2}{r^2}$$
$$1.98 \times 10^{20}\,N = \frac{k(5.97 \times 10^{24}\,kg)(7.35 \times 10^{22}\,kg)}{(3.84 \times 10^{8}\,m)^2}$$
$$k = \frac{(1.98 \times 10^{20}\,N)(3.84 \times 10^{8}\,m)^2}{(5.97 \times 10^{24}\,kg)(7.35 \times 10^{22}\,kg)}$$
$$k = 6.65 \times 10^{-11}\,\frac{Nm^2}{kg^2}$$
$$F = \frac{6.65 \times 10^{-11}\,m_1m_2}{r^2} \cdot \frac{Nm^2}{kg^2}$$
$$F = \frac{6.65 \times 10^{-11}(5.97 \times 10^{24}\,kg)(1.90 \times 10^{27}\,kg)}{(7.95 \times 10^{11}\,m)^2} \cdot \frac{Nm^2}{kg^2}$$
$$F = 1.19 \times 10^{18}\,N$$

27.

$$V = kT \qquad V = 6.00\text{ cm}^3/KT$$
$$15\bar{0}0\text{ cm}^3 = k(25\bar{0}k) \qquad V = 6.00\,\frac{\text{cm}^3}{K}(30\bar{0})K$$
$$\frac{15\bar{0}0\text{ cm}^3}{25\bar{0}K} = k$$
$$6.00\text{ cm}^3/K = k \qquad V = 18\bar{0}0\text{ cm}^3$$

29.

$$V = \frac{kT}{p}$$

$$15,\bar{0}00 \text{ ft}^3 = \frac{k(40\bar{0})°R}{20.0 \text{ lb/in.}^2}$$

$$15,\bar{0}00 \text{ ft}^3\left(20.0\frac{\text{lb}}{\text{in.}^2}\right) = k(40\bar{0}°)R$$

$$\frac{30\bar{0},000 \text{ ft}^3\text{lb}}{40\bar{0}°R \text{ in.}^2} = k$$

$$k = 75\bar{0}\frac{\text{ft}^3 \text{ lb}}{°R \text{ in.}^2}$$

$$V = \frac{75\bar{0} \text{ ft}^3\text{lb}}{°R \text{ in.}^2}\cdot\frac{T}{p}$$

$$V = \frac{75\bar{0} \text{ ft}^3\text{lb}}{°R \text{ in.}^2}\cdot\frac{575°R}{50.0 \text{ lb/in.}^2}$$

$$V = 8630 \text{ ft}^3$$

31.

$$p = \frac{kV^2}{R}$$

$$180w = \frac{k(9\bar{0}v)^2}{45\Omega}$$

$$\frac{180w(45\Omega)}{(9\bar{0})^2 V^2} = k$$

$$k = \frac{1.0w\Omega}{V^2}$$

$$p = \frac{1.0w\Omega}{V^2}\cdot\frac{(120V)^2}{3\bar{0}\Omega}$$

$$p = \frac{120^2}{3\bar{0}}w$$

$$p = 480w$$

33.

$$\text{number teeth} = \frac{k}{\text{no. rev/time}}$$

Let x = teeth in second gear.

$$50 = \frac{\text{k}}{40\bar{0} \text{ rpm}}$$

$$50(40\bar{0}) \text{ rpm} = \text{k}$$

$$20\bar{0}00 \text{ rpm} = \text{k}$$

$$\text{x} = \frac{20\bar{0}00 \text{ rpm}}{\text{no. rpm}}$$

$$x = \frac{20\bar{0}00 \text{ rpm}}{125 \text{ rpm}}$$

$$x = 160 \text{ teeth}$$

35.

$$d = \frac{k}{n \text{ rp time}}$$

$$25 \text{ cm} = \frac{\text{k}}{72 \text{ rpm}}$$

$$k = 1800 \text{ cm rpm}$$

$$\text{d} = \frac{1800 \text{ cm rpm}}{n \text{ rpm}}$$

$$75 \text{ cm} = \frac{1800 \text{ cm rpm}}{n \text{ rpm}}$$

$$n = \frac{1800 \text{ cm rpm}}{75 \text{ cm rpm}}$$

$$n = 24 \text{ rpm}$$

37.(a) $d = \dfrac{k}{n\ rpm}$

$15 \text{ in.} = \dfrac{\text{k}}{150 \text{ rpm}}$

$k = 15 \text{ in.}(150)\text{rpm}$

$\text{k} = 2250 \text{ in. rpm}$

So $d = \dfrac{2250 \text{ in. rpm}}{n}$

$45 \text{ in.} = \dfrac{2250 \text{ in. rpm}}{n}$

$n = \dfrac{2250 \text{ in. rpm}}{45 \text{ in.}}$

$n = 5\bar{0} \text{ rpm}$

37.(b) $d = \dfrac{k}{n \text{ rpm}}$

$15 \text{ in.} = \dfrac{\text{k}}{300 \text{ rpm}}$

$k = 3\bar{0}0(15)\text{in. rpm}$

$\text{k} = 4500 \text{ in. rpm}$

$45 \text{ in.} = \dfrac{4500 \text{ in. rpm}}{n}$

$n = \dfrac{4500 \text{ in. rpm}}{45 \text{ in.}}$

$n = 1\bar{0}0 \text{ rpm}$

Chapter 1 Review

1. $(-3)+(+6)+(-8)=-5$

2. $(+4)-(+7)+(-3)-(-6)=(+4)+(-7)+(-3)+(+6)=0$

3. $(-3)(+5)(-7)(-2)=-210$

4. $\dfrac{(+6)(-12)(+4)}{(+9)(-8)(-2)}=-2$

5. $8-3\cdot 4=8-12=8+(-12)=-4$

6. $4+10\div 2=4+5=9$

7. $6-(8+4)\div 6=6-(12)\div 6=6-2=4$

8. $5+2(6-9)=5+2(-3)=5+(-6)=-1$

9. $48\div 8\cdot 3+2\cdot 5-17=6\cdot 3+2\cdot 5-17=18+10-17=11$

10. $3\cdot 4\div 6-8+6(-3)=12\div 6-8+6(-3)=2-8+6(-3)=2-8+(-18)=-24$

11.
$$\frac{3\cdot 5+6(4\div 2)}{8\div 4\cdot 8-3}=$$
$$\frac{3\cdot 5+6(2)}{2\cdot 8-3}=$$
$$\frac{15+12}{16-3}=$$
$$\frac{27}{13}$$

12.
$$\frac{18\div 9\cdot 2-3}{5-(2-3)}=$$
$$\frac{2\cdot 2-3}{5-(-1)}=$$
$$\frac{4-3}{5+1}=$$
$$\frac{1}{6}$$

13. $\dfrac{4\cdot 0}{8}=0$

14. $\dfrac{-5-(15-20)}{9\cdot 0+(-4-(-4))}=\dfrac{-5-(-5)}{0+(-4+4)}=\dfrac{-5+5}{-4+4}=\dfrac{0}{0}$ is indeterminate.

15. $\dfrac{5-6(2+4)}{-16-(-16)}=\dfrac{5-6(6)}{-16+16}=\dfrac{5-36}{0}=\dfrac{-31}{0}$ is meaningless.

16. $(10^2)^4=10^{2(4)}=10^8$

17. $\dfrac{10^{-2}}{10^{-6}}=10^{-2(-6)}=10^4$

18. $10^3\cdot 10^4=10^{3+4}=10^7$

19. 3.42×10^6

20. 0.000561

21. $(8.54\times 10^7)(4.97\times 10^{-14})=(8.54\times 4.97)(10^7\times 10^{-14})=42.4\times 10^{-7}=4.24\times 10^1\times 10^{-7}=4.24\times 10^{-6}$

22. $\dfrac{1.85\times 10^{12}}{6.17\times 10^{-18}}=\dfrac{1.85}{6.17}\times 10^{12-(-18)}=0.300\times 10^{30}=3.00\times 10^{-1}\times 10^{30}=3.00\times 10^{29}$

23. 307 m **(a)** 3 significant digits **(b)** 1 m

24. 0.050A **(a)** 2 significant digits **(b)** .001A

25. **(a)** 3 significant digits **(b)** 1000V **26.** 57.63L rounded to 57.6L

27. $12.600 \text{ cm}+10.40(0.1 \text{ cm})+16.75 \text{ cm}+7.005(100 \text{ cm})=730.89 \text{ cm}$ rounded to 730.9 cm

28. $(18.5 \text{ m})(21.6 \text{ m})=399.6 \text{ m}^2$ rounded to $40\bar{0} \text{ m}^2$

29. $\dfrac{49.7 \text{ m}^3}{16.0 \text{ m}^2}=3.106 \text{ m}$ rounded to 3.11 m

30. $\dfrac{680 \text{ lb}}{(14.5 \text{ in.})(18.6 \text{ in.})}=2.5213\ 16/\text{in.}^2$ rounded to 2.5 lb/in.^2

31. trinomial, degree 3 **32.** $(3x^2+5x-7)+(13x^2-x-14)=16x^2+4x-21$

33. $(-5x^2-2x-3)+(12x^2+x-8)=7x^2-x-11$

34. $(-4x^2-3x+1)-(-5x^2+2x-1)=-4x^2-3x+1+5x^2-2x+1=x^2-5x+2$

35. $(7x^2+x-5)-(2x^2-3x-5)=7x^2+x-5-2x^2+3x+5=5x^2+4x$

36. $-[(2a-3b)-(-4a+b)+(-a-4b)]=-[2a-3b+4a-b-a-4b]=-[5a-8b]=-5a+8b$

37. $-3[-(2a-b)-(a-b)+(6a+5b)]=-3[-2a+b-a+b+6a+5b]=-3[3a+7b]=-9a-21b$

38. $\dfrac{3a^2b-6b}{2b+1}=\dfrac{3(-2)^2(-3)-6(-3)}{2(-3)+1}=\dfrac{3(4)(-3)-6(-3)}{2(-3)+1}=\dfrac{-36+18}{-6+1}=\dfrac{-18}{-5}=\dfrac{18}{5}$

39. $y^5\cdot y^4=y^{5+4}=y^9$ **40.** $\dfrac{b^{12}}{b^3}=b^{12-3}=b^9$

41. $(3x^2)(-5x^4)=(3)(-5)x^{2+4}=-15x^6$ **42.** $\dfrac{18m^6}{3m^2}=\dfrac{18}{3}m^{6-2}=6m^4$

43. $(a^3)^4=a^{3(4)}=a^{12}$ **44.** $(5a^3)^2=5^2a^{3(2)}=25a^6$

45. $\left(\dfrac{y}{x^2}\right)^3=\dfrac{y^3}{x^{2(3)}}=\dfrac{y^3}{x^6}$ **46.** $(5x^2)^0=1$

47. $(-5^3)^3=(-1)^3(5^{3(3)})=-5^9$ **48.** $(x^5\cdot x^4)^2=(x^{5+4})^2=(x^9)^2=x^{9(2)}=x^{18}$

49. $\left(2a^3b^2\right)^3 = 2^3a^{3(3)}b^{2(3)} = 8a^9b^6$

50. $\left(\dfrac{-3x^6}{6x^2}\right)^2 = \left(\dfrac{-x^{6-2}}{2}\right)^2 = \left(\dfrac{-x^4}{2}\right)^2 = \dfrac{(-1)^2x^{4(2)}}{2^2} = \dfrac{x^8}{4}$

51. $\left(\dfrac{-4x^8}{2x^2}\right)^3 = \left(-2x^{8-2}\right)^3 = \left(-2x^6\right)^3 = (-2)^3\left(x^{6(3)}\right) = -8x^{18}$

52. $\sqrt{49} = 7$

53. $\sqrt[3]{27} = 3$

54. $\sqrt{63} = \sqrt{9}\sqrt{7} = 3\sqrt{7}$

55. $\sqrt{108} = \sqrt{36}\sqrt{3} = 6\sqrt{3}$

56. $\sqrt{3147} \approx 56.1$

57. $\sqrt{0.0205} \approx 0.143$

58. $\left(-3a^2bc^3\right)\left(8a^3bc\right) = -3(8)a^2a^3bbc^3c = -24a^5b^2c^4$

59. $5x(2x-6y) = 10x^2 - 30xy$

60. $-3x^2\left(4x^3 - 3x^2 - x + 4\right) = -12x^5 + 9x^4 + 3x^3 - 12x^2$

61. $3a^2b\left(-5a^2b^3 + 3a^3 - 5b^2\right) = -15a^4b^4 + 9a^5b - 15a^2b^3$

62. $(2x+7)(3x-4) = 6x^2 - 8x + 21x - 28 = 6x^2 + 13x - 28$

63. $(5x-3)(6x-9) = 30x^2 - 45x - 18x + 27 = 30x^2 - 63x + 27$

64. $(4x+6)(5x+2) = 20x^2 + 8x + 30x + 12 = 20x^2 + 38x + 12$

65. $(2x-3)(3x+2) = 6x^2 + 4x - 9x - 6 = 6x^2 - 5x - 6$

66. $(8x-5)^2 = (8x-5)(8x-5) = 64x^2 - 40x - 40x + 25 = 64x^2 - 80x + 25$

67.

$$\begin{array}{r} x+y-7 \\ \underline{2x-y+3} \\ 3x+3y-21 \\ -xy-y^2+7y \\ \underline{2x^2+2xy-14x} \\ 2x^2+xy-11x+10y-y^2-21 \end{array}$$

68. $\dfrac{81x^2y^3z}{27x^2yz^2} =$

$\dfrac{3y^2}{z}$

69. $\dfrac{(4a)^2\left(5a^2b\right)^2}{50a^3b^2} =$

$\dfrac{4^2a^25^2a^4b^2}{50a^3b^2} =$

$\dfrac{16(25)a^6b^2}{50a^3b^2} =$

$8a^3$

70. $\dfrac{48x^2 - 24x + 15}{3x} =$

$\dfrac{48x^2}{3x} - \dfrac{24x}{3x} + \dfrac{15}{3x} =$

$16x - 8 + \dfrac{5}{x}$

71. $\dfrac{15x^3 - 25x + 35}{5x^2}$

$\dfrac{15x^3}{5x^2} - \dfrac{25x}{5x^2} + \dfrac{35}{5x^2}$

$3x - \dfrac{5}{x} + \dfrac{7}{x^2}$

72.

$$\frac{18m^2n^5 + 24mn - 8m^6n^2}{6m^2n^3}$$

$$\frac{18m^2n^5}{6m^2n^3} + \frac{24mn}{6m^2n^3} - \frac{8m^6n^2}{6m^2n^3}$$

$$3n^2 + \frac{4}{mn^2} - \frac{4m^4}{3n}$$

73.

$$\begin{array}{r} 3x+4 \\ 2x-3\overline{)6x^2 - x - 12} \\ \underline{6x^2 - 9x} \\ 8x - 12 \\ \underline{8x - 12} \\ 0 \end{array}$$

Thus $\dfrac{6x^2 - x - 12}{2x - 3} = 3x + 4$

74.

$$\begin{array}{r} 3x^2 - 3x + 8 \\ x+1\overline{)3x^3 + 0x^2 + 5x - 3} \\ \underline{3x^3 + 3x^2} \\ -3x^2 + 5x \\ \underline{-3x^2 - 3x} \\ 8x - 3 \\ \underline{8x + 8} \\ -11 \end{array}$$

Thus $\dfrac{3x^3 + 5x - 3}{x + 1} = 3x^2 - 3x + 8 + \dfrac{-11}{x + 1}$

75.

$$\begin{array}{r} x^2 - 2x + 3 \\ 2x-1\overline{)2x^3 - 5x^2 + 8x + 7} \\ \underline{2x^3 - x^2} \\ -4x^2 + 8x \\ \underline{-4x^2 + 2x} \\ 6x + 7 \\ \underline{6x - 3} \\ 10 \end{array}$$

Thus $\dfrac{2x^3 - 5x^2 + 8x + 7}{2x - 1} = x^2 - 2x + 3 + \dfrac{10}{2x - 1}$

77.

$$\begin{aligned} 8 - 3x &= 12 - 2x \\ 8 - 3x + 2x &= 12 - 2x + 2x \\ 8 - x &= 12 \\ 8 - x - 8 &= 12 - 8 \\ -x &= 4 \\ -x(-1) &= 4(-1) \\ x &= -4 \end{aligned}$$

78.

$$\begin{aligned} 3(x+8) &= 2(x-3) \\ 3x + 24 &= 2x - 6 \\ 3x + 24 - 2x &= 2x - 6 - 2x \\ x + 24 &= -6 \\ x + 24 - 24 &= -6 - 24 \\ x &= -30 \end{aligned}$$

79.

$$\begin{aligned} 3(2x-1) - 2(3x-1) &= 3x + 1 \\ 6x - 3 - 6x + 2 &= 3x + 1 \\ -1 &= 3x + 1 \\ -1 - 1 &= 3x + 1 - 1 \\ -2 &= 3x \\ \frac{-2}{3} &= \frac{3x}{3} \\ \frac{-2}{3} &= x \end{aligned}$$

80.

$$\frac{1}{4}x + 8 = \frac{3}{4}x - \frac{7}{2} \quad \text{L.C.D.} = 4$$

$$\begin{aligned} \frac{1}{4}x(4) + 8(4) &= \frac{3}{4}x(4) - \frac{7}{2}(4) \\ x + 32 &= 3x - 14 \\ x + 32 - 3x &= 3x - 14 - 3x \\ -2x + 32 &= -14 \\ -2x + 32 - 32 &= -14 - 32 \\ -2x &= -46 \\ \frac{-2x}{-2} &= \frac{-46}{-2} \\ x &= 23 \end{aligned}$$

81.

$$\frac{2}{3}x-\frac{4}{5}=\frac{1}{5}x+\frac{4}{3}$$
$$\text{L.C.D.}=15$$
$$\frac{2}{3}x(15)-\frac{4}{5}(15)=\frac{1}{5}x(15)+\frac{4}{3}(15)$$
$$10x-12=3x+20$$
$$10x-12-3x=3x+20-3x$$
$$7x-12=20$$
$$7x-12+12=20+12$$
$$7x=32$$
$$\frac{7x}{7}=\frac{32}{7}$$
$$x=\frac{32}{7}$$

82.

$$\frac{5(x-2)}{6}-\frac{3x}{2}=\frac{5}{3}x$$
$$\text{L.C.D.}=6$$
$$\frac{5(x-2)}{6}(6)-\frac{3x}{2}(6)=\frac{5}{3}x(6)$$
$$5(x-2)-9x=10x$$
$$5x-10-9x=10x$$
$$-4x-10=10x$$
$$-4x-10+4x=10x+4x$$
$$-10=14x$$
$$\frac{-10}{14}=\frac{14x}{14}$$
$$\frac{-5}{7}=x$$

83.

$$\frac{3}{4}(2x-3)-\frac{2}{3}(6x-2)=x$$
$$\text{L.C.D.}=12$$
$$12\left(\frac{3}{4}\right)(2x-3)-12\left(\frac{2}{3}\right)(6x-2)=12x$$
$$9(2x-3)-8(6x-2)=12x$$
$$18x-27-48x+16=12x$$
$$-30x-11=12x$$
$$-30x-11+30x=12x+30x$$
$$-11=42x$$
$$-\frac{11}{42}=\frac{42x}{42}$$
$$\frac{-11}{42}=x$$

84.

$$24.6x-45.2=0.4(39.2x+82.5)$$
$$24.6x-45.2=15.68x+33$$
$$24.6x-45.2-15.68x=15.68x+33-15.68x$$
$$8.92x-45.2=33$$
$$8.92x-45.2+45.2=33+45.2$$
$$8.92x=78.2$$
$$\frac{8.92x}{8.92}=\frac{78.2}{8.92}$$
$$x=8.77$$

85.

$$S=\frac{\pi D}{12}\text{ for }D$$
$$12S=\pi D$$
$$\frac{12S}{\pi}=D$$

86.

$$C=\frac{Q}{V}\ \textit{for}\ V$$
$$CV=Q$$
$$V=\frac{Q}{C}$$

87.

$$v=v_o-gt\text{ for }t$$
$$v-v_o=-gt$$
$$\frac{v-v_o}{-g}=t$$
$$\frac{v_o-v}{g}=t$$

88.

$$(\Delta V) = \beta V(T - T_o) \text{ for } T_o$$

$$\frac{(\Delta V)}{\beta V} = T - T_o$$

$$\frac{(\Delta V)}{\beta V} - T = -T_o$$

$$\frac{-(\Delta V)}{\beta V} + T = T_o$$

$$\text{or } \frac{-(\Delta V)}{\beta V} + \frac{T\beta V}{\beta V} = T_o$$

$$\frac{T\beta V - (\Delta V)}{\beta V} = T_o$$

89.

$$v = v_o + at$$

$$18\bar{0}m/s = v_o + 9.80m/s^2(10.0)s$$

$$18\bar{0}m/s = v_o + 98.0m/s$$

$$18\bar{0}m/s - 98.0m/s = v_o$$

$$v_o = 82m/s$$

90.

$$Z = \sqrt{R^2 + X_L^{\ 2}}$$

$$Z = \sqrt{(40.0\Omega)^2 + (37.7\Omega)^2}$$

$$Z = 55.0\Omega$$

91.

$$\frac{1}{R} = \frac{1}{R_1} + \frac{1}{R_2} + \frac{1}{R_3}$$

$$\frac{1}{4.5\Omega} = \frac{1}{12.5\Omega} + \frac{1}{8.5\Omega} + \frac{1}{R_3}$$

$$\frac{1}{4.5\Omega} - \frac{1}{12.5\Omega} - \frac{1}{8.5\Omega} = \frac{1}{R_3}$$

$$0.024575/\Omega = \frac{1}{R_3}$$

$$\frac{1}{.024575}\Omega = R_3$$

$$R_3 = 40.7\Omega$$

92. 0.090 in. – 0.018 in.=0.072 in. more needed.
0.072 in. ÷ 0.0040 in./shim=18 shims .

93. Let x = width.
Then $x + 5 =$ length.

$$P = 21 + 2\omega$$

$$90 = 2(x + 5) + 2x$$

$$90 = 2x + 10 + 2x$$

$$90 = 4x + 10$$

$$80 = 4x$$

$$x = 20m$$

$$20m \times 25m$$

94. Let $x =$ amt. 20% silver

$30 - x =$ amt. 15% silver

Pure silver from 20% bar	+	Pure silver from 15% bar	=	Pure silver in 18% bar

$$0.20x + 0.15(30 - x) = 0.18(30)$$
$$0.20x + 4.5 - 0.15x = 5.4$$
$$0.05x + 4.5 = 5.4$$
$$0.05x = 0.9$$
$$x = 18 \text{ oz}$$

18 oz of 20% and 12 oz of 15%.

95. Distance car travels + 10 = Distance police travel

Let t = time to intercept

$$55t + 10 = 75t$$
$$10 = 20t$$
$$t = \frac{10}{20} = \frac{1}{2}h$$

96. $$\frac{360 \text{ ft.}}{441 \text{ ft.}} = \frac{40}{49}$$

97. $$\frac{600\ mm}{1.5\ m} = \frac{600\ mm}{1.5(1000)mm} = \frac{4}{10} = \frac{2}{5}$$

98. $$\frac{(2.8 \text{ in.})^2}{(1.2 \text{ in.})^2} = \frac{2.8^2}{1.2^2} = \frac{49}{9}$$

99. $$\frac{13}{24} = \frac{x}{96}$$
$$24x = 13(96)$$
$$x = \frac{13(96)}{24}$$
$$x = 52$$

100. $$\frac{39}{12} = \frac{3x}{48}$$
$$36x = 39(48)$$
$$x = \frac{39(48)}{36}$$
$$x = 52$$

101. $$\frac{x}{60 - x} = \frac{2}{3}$$
$$3x = 2(60 - x)$$
$$3x = 120 - 2x$$
$$3x + 2x = 120$$
$$5x = 120$$
$$x = 24$$

102. $$\frac{a}{6 + x} = \frac{b}{a}$$
$$b(6 + x) = a^2$$
$$6b + bx = a^2$$
$$bx = a^2 - 6b$$
$$x = \frac{a^2 - 6b}{b}$$

103. $$\frac{x}{c} = \frac{b}{a}$$
$$ax = bc$$
$$x = \frac{bc}{a}$$

104. $$\frac{270}{15} = \frac{x}{3.15}$$
$$15x = 270(3.15)$$
$$x = \frac{270(3.15)}{15}$$
$$x = 56.7$$

105. Let x = water for 20,000 people.

$$\frac{40{,}000 \text{ gal}}{8000 \text{ people}} = \frac{x}{20{,}000}$$
$$\frac{40{,}000(2000)}{8000} = x$$
$$x = 100{,}000 \text{ gal}$$

106. Let x = one part

$2370 - x$ = other part

$$\frac{3}{7} = \frac{x}{2370 - x}$$

$$7x = 3(2370 - x)$$

$$7x = 7110 - 3x$$

$$10x = 7110$$

$$x = 711$$

$$2370 - x = 1659$$

107. $y = k\sqrt{z}$

108. $y = kvu^2$

109. $y = \frac{kp}{q}$

110. $y = \frac{kmn}{p^2}$

111.

$$y = kxz$$

$$576 = k(24)(8)$$

$$\frac{576}{24(8)} = k$$

$$k = 3$$

$$y = 3xz$$

$$y = 3(36)(48)$$

$$y = 5184$$

112.

$$y = \frac{kp^2}{\sqrt{q}}$$

$$6 = \frac{k(4)^2}{\sqrt{16}}$$

$$6 = \frac{16}{4}k$$

$$\frac{6}{4} = k$$

$$\frac{3}{2} = k$$

$$y = \frac{3p^2}{2\sqrt{q}}$$

$$y = \frac{3(8)^2}{2\sqrt{4}}$$

$$y = \frac{3(64)}{2(2)}$$

$$y = 3(16)$$

$$y = 48$$

113.

$$F = kd$$

$$40.0N = k(1.5)m$$

$$\frac{40.0N}{1.5m} = k$$

$$k = \frac{8\bar{0}}{3.0} N/m$$

$$F = \frac{8\bar{0}}{3.0} N/m\, d$$

$$F = \frac{8\bar{0}}{3.0} N/m(6.0m)$$

$$F = 160N$$

114.

$$F = kWv^2$$

$$19.2\text{ lb}=k(20.0\text{-lb})(16.0\text{ ft/s})^2$$

$$19.2\text{ lb}=k(5120)\text{ lb ft}^2/\text{s}^2)$$

$$\frac{19.2}{5120\text{ ft}^2/\text{s}^2} = k$$

$$\frac{0.00375}{\text{ft}^2/\text{s}^2} = k$$

$$F = \frac{0.00375}{\text{ft}^2/\text{s}^2}Wv^2$$

$$F = \frac{0.00375}{\text{ft}^2/\text{s}^2}\cdot(50.0\text{ lb})(10.0\text{ ft/s})^2$$

$$F = 0.0375(50.0)(10\bar{0})\text{ lb}$$

$$F = 18.8\text{ lb}$$

115.

$$F = \frac{kq_1q_2}{d_2}$$

$$6\bar{0}0N = \frac{k(+2\bar{0}\mu C)(+12\mu C)}{(6.0)^2\text{ cm}^2}$$

$$6\bar{0}0N(36)\text{cm}^2 = k(+240)\mu^2C^2$$

$$\frac{6\bar{0}0(36)N\text{cm}^2}{240\mu^2C^2} = k$$

$$k = \frac{9\bar{0}N\text{cm}^2}{\mu^2C^2}$$

$$F = \frac{9\bar{0}N\text{cm}^2 q_1q_2}{\mu^2C^2d^2}$$

$$F = \frac{9\bar{0}N\text{cm}^2}{\mu^2C^2}\frac{(-36\mu C)(+12\mu C)}{12^2\text{ cm}^2}$$

$$F = -270N$$

attractive force

CHAPTER 1 SPECIAL APPLICATION

Part A

1. It converts time in years to time in months. This is necessary for consistency in units since lead time and safety pad are given in months.

2. 4 tires

3. MSL = 4(8)(1)(9 + 12) ÷ (12) = 56 tires.

4. When the inventory level drops to 56 tires a new order for tires should be placed.

Part B

1. a) 4 tires
   b) 8
   c) 12 months
   d) 56 tires
   e) 0 tires
   f) 54 tires
   g) OQ = 4(8)(1 ÷ 12)(12) + 56 - 0 - 54 = 34 tires.

2. Under the given conditions NASA needs to order 34 more tires.

Chapter 2, Exercises 2.1

1. $\angle 2$ and $\angle 3$

3. $\angle 5$ and $\angle 6$

5. $\angle 1$ and $\angle 4$

7. $\angle 1+\angle 2=180°$, so $125°+\angle 2=180°$, thus $\angle 2=55°$

9. $\angle 5$ and $\angle 6$, or $\angle 7$ and $\angle 8$, or $\angle 8$ and $\angle 2$

11. $\angle 5+\angle 6=90°$, so $50°+\angle 6=90°$, thus $\angle 6=40°$

13. $\angle 2=\angle 6$, so $\angle 2=40°$

15. Given $\angle 1=130°$, then
$\angle 3=180°-130°=50°$
$\angle 4=90°-\angle 3=40°$

17. $\angle 5=90°-\angle 7$
$\angle 3=\angle 7=50°$, thus
$\angle 5=40°$

19. $\angle 1+35°=180°$, thus
$\angle 1=145°$

Exercises 2.2

1.
$$c^2=a^2+b^2$$
$$c=\sqrt{a^2+b^2}$$
$$c=\sqrt{(6.00\text{ cm})^2+(8.00\text{ cm})^2}$$
$$c=\sqrt{(36.0+64.0)\text{ cm}^2}$$
$$c=\sqrt{10\bar{0}\text{ cm}^2}.$$
$$c=10.0\text{ cm}$$

3.
$$a^2+b^2=c^2$$
$$b^2=c^2-a^2$$
$$b=\sqrt{c^2-a^2}$$
$$b=\sqrt{(3640\text{ m})^2-(2250\text{ m})^2}$$
$$b=\sqrt{(3640^2-2250^2)\text{ m}^2}$$
$$b=\sqrt{8187100\text{ m}^2}$$
$$b=2860\text{ m}$$

5.
$$A=\frac{1}{2}bh$$
$$A=\frac{1}{2}(28.6\text{ in.})(14.2\text{ in.})$$
$$A=203\text{ in}^2$$
$$P=a+b+c$$
$$P=16.9\text{ in.}+24.0\text{ in.}+28.6\text{ in.}$$
$$P=69.5\text{ in.}$$

7.
$$A=\sqrt{s(s-a)(s-b)(s-c)}$$
$$s=\frac{1}{2}(a+b+c)$$
$$s=\frac{1}{2}(P)=4330\text{ m}$$
$$A=\sqrt{4330\text{ m}(4330\text{ m}-2040\text{ m})(4330\text{ m}-2960\text{ m})(4330\text{ m}-3660\text{ m})}$$
$$A=\sqrt{4330\text{ m}(2290\text{ m})(1370\text{ m})(670\text{ m})}$$
$$A=3,020,000\text{ m}^2$$
$$P=a+b+c$$
$$P=2040\text{ m}+2960\text{ m}+3660\text{ m}$$
$$P=8660\text{ m}$$

9. $A=\sqrt{s(s-a)(s-b)(s-c)}$

$s=\frac{1}{2}(a+b+c)=\frac{1}{2}(3(124\text{ cm}))$

$s=186\text{ cm}$

$A=\sqrt{186\text{ cm}(186\text{ cm}-124\text{ cm})(186\text{ cm}-124\text{ cm})(186\text{ cm}-124\text{ cm})}$

$A=\sqrt{186\text{ cm}(62\text{ cm})(62\text{ cm})(62\text{ cm})}$

$A=6660\text{ cm}^2$

$P=a+b+c=3(124\text{ cm})$

$P=372\text{ cm}$

11. $A=\frac{1}{2}bh$

$A=\frac{1}{2}(275\text{ ft})(275\text{ ft})$

$A=37,800\text{ ft}^2$

$c^2=a^2+b^2$

$c^2=275^2+275^2$

$c=389\text{ ft}$

$P=275\text{ ft}+275\text{ ft}+389\text{ ft}$

$P=939\text{ ft}$

13. The point of intersection of the three medians, the point of intersection of the three altitudes and the point of intersection of the three angle bisectors coincide. This is because in an equilateral triangle, a median is an altitude and bisects the angle.

15. Since the sum of the angles of a triangle is $180°$, the third angle is $180°-45.7°-65.4°=68.9°$

17. $c^2=a^2+b^2$

$c^2=(6.00\text{ ft})^2+(9.50\text{ ft})^2$

$c=11.2\text{ ft}$

19. The guy wire is attached $(275\text{ ft}-45\text{ ft})$ from the ground, ie $23\bar{0}$ ft. Thus $c^2=(23\bar{0}\text{ ft})^2+(60.0\text{ ft})^2$, $c=238\text{ ft}$

21. $A=\sqrt{s(s-a)(s-b)(s-c)}$

$s=\frac{1}{2}(\text{a}+\text{b}+\text{c})$

$s=\frac{1}{2}(925\text{ ft}+624\text{ ft}+835\text{ ft})=1192\text{ ft}$

$A=\sqrt{1192\text{ ft}(1192\text{ ft}-925\text{ ft})(1192\text{ ft}-624\text{ ft})(1192\text{ ft}-835\text{ ft})}$

$A=\sqrt{(1192\text{ ft})(267\text{ ft})(568\text{ ft})(375\text{ ft})}$

$A=254,000\text{ ft}^2$

23. $\frac{\text{Tree}}{45\text{ ft}}=\frac{6.0\text{ ft}}{4.0\text{ ft}}$

$\text{Tree}=\frac{6.0(45)\text{ ft}^2}{4.0\text{ ft}}$

$\text{Tree}=68\text{ ft}$

Section 2.2

Exercises 2.3

1.

$P = 2a + 2b$

$P = 2(245\text{ m}) + 2(143\text{ m})$

$P = 49\bar{0}\text{ m} + 286\text{ m}$

$P = 776\text{ m}$

$A = bh$

$A = (245\text{ m})(143\text{ m})$

$A = 35{,}\bar{0}00\text{ m}^2$

3.

$P = 2a + 2b$

$P = 2(185\text{ in.}) + 2(131\text{ in.})$

$P = 37\bar{0}\text{ in.} + 262\text{ in.}$

$P = 632\text{ in.}$

$A = bh$

$A = 185\text{ in.}(105\text{ in.})$

$A = 19{,}400\text{ in}^2$

5.

$P = 4b$

$P = 5(15.0\text{ m})$

$P = 60.0\text{ m}$

$A = b^2$

$A = (15.0\text{ m})^2$

$A = 225\text{ m}^2$

7.

$P = a + b + c + d$

$P = 15.0\text{ m} + 23.8\text{ m} + 16.4\text{ m} + 17.2\text{ m}$

$P = 72.4\text{ m}$

$A = \frac{1}{2}h(a + b)$

$A = \frac{1}{2}(15.0\text{ m})(23.8\text{ m} + 17.2\text{ m})$

$A = \frac{1}{2}(15.0\text{ m})(41.0\text{ m})$

$A = 308\text{ m}^2$

9.

$A = bh$

$556\text{ m}^2 = b(33.2\text{ m})$

$\frac{556\text{ m}^2}{33.2\text{ m}} = b$

$b = 16.7\text{ m}$

11.

$A = \frac{1}{2}h(a + b)$

$3350\text{ cm}^2 = \frac{1}{2}(36.0\text{ cm})(45.0\text{ cm} + b)$

$\frac{3350\text{ cm}^2}{0.5(36.0)\text{ cm}} = 45.0\text{ cm} + b$

$186\text{ cm} - 45.0\text{ cm} = b$

$141\text{ cm} = b$

13.

$A = bh$

$7560\text{ in}^2 = b(79.2\text{ in.})$

$b = \frac{7560\text{ in}^2}{79.2\text{ in.}}$

$b = 95.5\text{ in.}$

15. The exterior area is the lateral surface area minus the openings.

Paint area = 2 ends + 2 sides $- 375\text{ ft}^2$

$$PA = 2\left[(10.0\text{ ft})(24.0\text{ ft}) \times \frac{1}{2}(24.0\text{ ft})(5.00\text{ ft})\right] + 2(10.0\text{ ft})(45.0\text{ ft}) - 375\text{ ft}^2$$

$PA = 600\text{ ft}^2 \times 900\text{ ft}^2 - 375\text{ ft}^2$

$PA = 1125\text{ ft}^2$

$\text{Total cost} = \frac{\text{price}}{\text{ft}^2} \times \text{area}$

$\text{Total cost} = \frac{\$0.90}{\text{ft}^2} \times 1125\text{ ft}^2$

Total cost = \$1012.50

17. $A = 60.0\text{ ft}(192\text{ ft}) + (293\text{ ft} - 192\text{ ft})(60.0\text{ ft} + 106.0\text{ ft} + 72.0\text{ ft}) + (124\text{ ft})(72.0\text{ ft}) = 44,500\text{ ft}^2$

19. Extend the 30.0 in. line to hit the 60.0 in. line, forming two parallelograms.

$A = 19.0\text{ in.}(30.0\text{ in.}) + 28.0\text{ in.}(50.0\text{ in.})$

$A = 57\bar{0}\text{ in}^2 + 14\bar{0}0\text{ in}^2 = 1970\text{ in}^2$

Exercises 2.4

1.
$C = 2\pi r$
$C = 2\pi(65.2\text{ m})$
$C = 41\bar{0}\text{ m}$
$A = \pi r^2$
$A = \pi(65.2\text{ m})^2$
$A = 13,400\text{ m}^2$

3.
$C = \pi d$
$C = \pi(125\text{ mi})$
$C = 393\text{ mi}$
$A = \frac{\pi d^2}{4}$
$A = \frac{\pi(125\text{ mi})^2}{4}$
$A = 12,300\text{ mi}^2$

5.
$C = 2\pi r$
$C = 2\pi(10.5\text{ cm})$
$C = 66.0\text{ cm}$
$A = \pi r^2$
$A = \pi(10.5\text{ cm})^2$
$A = 346\text{ cm}^2$

7.
$C = \pi d$
$C = \pi(2240\text{ m})$
$C = 7040\text{ m}$
$A = \frac{\pi d^2}{4}$
$A = \frac{\pi(2240\text{ m})^2}{4}$
$A = 3,940,000\text{ m}^2$

9.
$A = \pi r^2$
$48.0\text{ ft}^2 = \pi r^2$
$r^2 = \frac{48.0\text{ ft}^2}{\pi}$
$r = 3.91\text{ ft}$

11.
$C = 2\pi r$
$73.5\text{ in.} = 2\pi r$
$r = \frac{73.5\text{ in.}}{2\pi}$
$r = 11.7\text{ in.}$

13. a) $\angle DEF = \frac{1}{2}(58.0^\circ + 40.0^\circ)$
$\angle DEF = 49.0^\circ$

b) $\angle BDA = \frac{1}{2}(58.0^\circ)$
$\angle BDA = 29.0^\circ$

c) $\angle CBF = \angle DBF$
$\angle DBF = \frac{1}{2}(40.0^\circ) = 20.0^\circ$

d) $\angle BDF = 90^\circ$ since
BF is a diameter

15. a) $\angle BCA = 90^\circ - 30.0^\circ$
$\angle BCA = 60.0^\circ$
thus $\overset{\frown}{BF} = 60.0^\circ$

b) $\overset{\frown}{BD} = 180^\circ - 60.0^\circ = 120.0^\circ$

c) $\overset{\frown}{EF} = \overset{\frown}{BF} = 60.0^\circ$

d) $\angle ACB = 90^\circ - 30.0^\circ = 60.0^\circ$

e) $\angle ADE = \frac{1}{2}(\overset{\frown}{EF}) = \frac{1}{2}(60.0^\circ) = 30.0^\circ$

17. $\angle CBA = 90°$

$BC^2 = AC^2 - AB^2$

$BC^2 = (74.0 \text{ cm})^2 - (68.2 \text{ cm})^2$

$BC = 28.7 \text{ cm}$

19. Central angle between two holes is $\frac{360°}{6} = 60.0°$.

21. $\frac{360°}{60°} = 6$, so $\frac{1}{6}$ of the area is cut out.

$A = \frac{5}{6}(\text{large circular area} - \text{small circular area})$

$A = \frac{5}{6}\left(\pi(36.0 \text{ ft})^2 - \pi(20.0 \text{ ft})^2\right)$

$A = 2350 \text{ ft}^2$

Exercises 2.5

1. a) $V = lwh$

$V = 49.4 \text{ cm}(22.7 \text{ cm})(27.0 \text{ cm})$

$V = 30{,}300 \text{ cm}^3$

b) $A = 2lh + 2wh$

$A = 2(49.4 \text{ cm})(27.0 \text{ cm}) + 2(22.7 \text{ cm})(27.0 \text{ cm})$

$A = 3893.4 \text{ cm}^2$

$A = 3890 \text{ cm}^2$

c) $SA = \text{Lateral area} + \text{Top} + \text{Bottom}$

$SA = 3893.4 \text{ cm}^2 + 2(49.4 \text{ cm})(22.7 \text{ cm})$

$SA = 6140 \text{ cm}^2$

3. $V = \frac{1}{3}Bh$

$V = \frac{1}{3}(18.6 \text{ in.})(10.1 \text{ in.})(24.6 \text{ in.})$

$V = 1540 \text{ in}^3$

5. a) $V = \frac{4}{3}\pi r^3$

$V = \frac{4}{3}\pi\left(\frac{41.8}{2} \text{ ft}\right)^3$

$V = 38{,}200 \text{ ft}^3$

b) $A = 4\pi r^2$

$A = 4\pi\left(\frac{41.8}{2} \text{ ft}\right)^2$

$A = 5490 \text{ ft}^2$

7) a) $V = \pi r^2 h$

$V = \pi(15.0 \text{ ft})^2(30.0 \text{ ft})$

$V = 21{,}200 \text{ ft}^3$

b) $A = 2\pi rh$

$A = 2\pi(15.0 \text{ ft})(30.0 \text{ ft})$

$A = 2830 \text{ ft}^2$

c) $SA = A + 2(\pi r^2)$

$SA = 2827 \text{ ft}^2 + 2\pi(15.0 \text{ ft})^2$

$SA = 4240 \text{ ft}^2$

9. a) $V = \frac{1}{3}\pi r^2 h$

$h^2 = s^2 - r^2$

$h^2 = (18.4 \text{ ft})^2 - (12.5 \text{ ft})^2$

$h = 13.5 \text{ ft}$

$V = \frac{1}{3}\pi(12.5 \text{ ft})^2(13.5 \text{ ft})$

$V = 2210 \text{ ft}^3$

b) $A = \pi rs$

$A = \pi(12.5 \text{ ft})(18.4 \text{ ft})$

$A = 723 \text{ ft}^2$

c) $SA = A + \pi r^2$

$SA = 722.566 \text{ ft}^2 + \pi(12.5 \text{ ft})^2$

$SA = 1213 \text{ ft}^2$

11. **a)** Volume = Outer Vol − Inner Vol

$V = (9.00 \text{ in.})(24.0 \text{ in.})(12.00 \text{ in.}) - (6.00 \text{ in.})(24.0 \text{ in.})(9.00 \text{ in.})$

$V = 1296 \text{ in}^3$

$V = 13\bar{0}0 \text{ in}^3$

b) Recall $1 \text{ ft}^3 = 1728 \text{ in}^3$

$\text{Weight} = 1296 \text{ in}^3 \left(\frac{1 \text{ ft}^3}{1728 \text{ in}^3}\right)\left(\frac{708 \text{ lb}}{\text{ft}^3}\right)$

$W = 531 \text{ lb}$

13. **a)** SA of top $= \pi rs$

$r = 15.0 \text{ ft}$

$s^2 = r^2 + h^2$

$s^2 = (15.0 \text{ ft})^2 + (10.0 \text{ ft})^2$

$s = 18.0 \text{ ft}$

$SA = \pi(15.0 \text{ ft})(18.0 \text{ ft})$

$SA = 848 \text{ ft}^2$

$\text{Weight} = 848 \text{ ft}^2 \left(\frac{17.8 \text{ lb}}{\text{ft}^2}\right)$

$W = 15{,}100 \text{ lb}$

b) SA of sides $= 2\pi rh$

$SA = 2\pi(15.0 \text{ ft})(30.0 \text{ ft})$

$SA = 2830 \text{ ft}^2$

$\text{Weight of sides} = 2830 \text{ ft}^2 \left(\frac{17.8 \text{ lb}}{\text{ft}^2}\right)$

Weight sides $= 50{,}300$ lb

Weight top and sides $= 15{,}100 \text{ lb} + 50{,}300 \text{ lb} = 65{,}400 \text{ lb}$

c) Volume = Vol cylinder + Vol cone

$V = \pi r^2 h_1 + \frac{1}{3}\pi r^2 h_2$

$V = \pi(15.0 \text{ ft})^2(30.0 \text{ ft}) + \frac{1}{3}\pi(15.0 \text{ ft})^2(10.0 \text{ ft})$

$V = 23{,}600 \text{ ft}^3$

15. **a)** $V = \frac{4}{3}\pi r^3$

$r = 30.0 \text{ ft}$

$V = \frac{4}{3}\pi(30.0 \text{ ft})^3$

$V = 113{,}097 \text{ ft}^3$

$\text{Gallons} = 113{,}097 \text{ ft}^3 \left(\frac{7.48 \text{ gal}}{1 \text{ ft}^3}\right)$

Gallons $= 846{,}000$ gal

b) $SA = 4\pi r^2$

$SA = 4\pi(30.0 \text{ ft})^2$

$SA = 11{,}309.7 \text{ ft}^2$

$\text{Paint} = 11{,}309.7 \text{ ft}^2 \left(\frac{1 \text{ gal}}{125 \text{ ft}^2}\right)$

Paint $= 90.5$ gal

17. $r = 3.70 \text{ cm}$

SA one can $= 2\pi rh$

$SA(\text{one}) = 2\pi(3.70 \text{ cm})(10.0 \text{ cm})$

$SA(\text{one}) = 232.47785 \text{ cm}^2$

Paper $= 232.47785 \text{ cm}^2 (1750)$

Paper $= 407{,}000 \text{ cm}^2$

19. $V = \frac{1}{3}\pi r^2 h$

$r = \frac{256 \text{ ft}}{2\pi}$

$s^2 = r^2 + h^2$

$h^2 = (42 \text{ ft})^2 - \left(\frac{256}{2\pi} \text{ ft}\right)^2$

$h = 10.369 \text{ ft}$

$V = \frac{1}{3}\pi\left(\frac{256}{2\pi} \text{ ft}\right)^2 (10.369 \text{ ft})$

$V = 18{,}000 \text{ ft}^3$

Weight gravel $= 18{,}000 \text{ ft}^3 \left(\frac{1 \text{ yd}^3}{27 \text{ ft}^3}\right)\left(\frac{3400 \text{ lb}}{1 \text{ yd}^3}\right)$

Weight gravel $= 2{,}266{,}667 \text{ lb}$

Trucks $= \frac{2{,}266{,}667 \text{ lb}}{(22)(2000 \text{ lb})} = 52$ truck loads

Chapter 2, Review

1. $\angle 4 = 180° - \angle 5$
$\angle 4 = 180° - 30°$
$\angle 4 = 150°$

$\angle 3 = \angle 1 = \angle 4$
$\angle 3 = \angle 1 = 150°$

$\angle 2 = 180° - \angle 1$
$\angle 2 = 180° - 150°$
$\angle 2 = 30°$

2. **a)** $\angle 1$ and $\angle 3$ **b)** $\angle 3$ and $\angle 4$ **c)** $\angle 1$ and $\angle 4$ **d)** $\angle 2$ and $\angle 3$ or $\angle 4$ and $\angle 5$ or $\angle 1$ and $\angle 2$

3. **a)** triangles **b)** quadrilateral **c)** pentagon **d)** hexagon **e)** octagon

4. **a)** $A = bh$
$A = 423 \text{ m}(249 \text{ m})$
$A = 105{,}000 \text{ m}^2$

b) $P = 2(a+b)$
$P = 2(423 \text{ m} + 275 \text{ m})$
$P = 1396 \text{ m}$

5. **a)** $A = bh$
$A = 35.0 \text{ ft}(11.0 \text{ ft})$
$A = 385 \text{ ft}^2$

b) $P = 2(a+b)$
$P = 2(35.0 \text{ ft} + 11.0 \text{ ft})$
$P = 92.0 \text{ ft}$

6. **a)** $A = \frac{1}{2}bh$
$A = \frac{1}{2}(2.61 \text{ mi})(2.26 \text{ mi})$
$A = 2.95 \text{ mi}^2$

b) $P = a+b+c$
$P = 2.61 \text{ mi} + 4.52 \text{ mi} + 2.61 \text{ mi}$
$P = 9.74 \text{ mi}$

7. **a)** Draw a vertical line to cut the figure into a triangle and a rectangle.

$$A = \frac{1}{2}bh + bh$$
$$A = \frac{1}{2}(3.50 \text{ km} - 1.20 \text{ km})(3.00 \text{ km}) + 1.20 \text{ km}(3.00 \text{ km})$$
$$A = 7.05 \text{ km}^2$$

7. **b)** Find the hypotenuse of the triangle formed in part a).

$$c^2 = (3.00 \text{ km})^2 + (2.30 \text{ km})^2$$
$$c = 3.78 \text{ km}$$
$$P = a+b+c+d$$
$$P = 3.78 \text{ km} + 1.20 \text{ km} + 3.00 \text{ km} + 3.50 \text{ km}$$
$$P = 11.48 \text{ km}$$

8. a) $A = b^2$

$A = (36.0\text{ ft})^2$

$A = 13\bar{0}0\text{ ft}^2$

b) $P = 4b$

$P = 4(36.0\text{ ft})$

$P = 144.0\text{ ft}$

9. a) $b^2 = c^2 - a^2$

$b^2 = (46.8\text{ m})^2 - (35.5\text{ m})^2$

$b = 30.5\text{ m}$

$A = \frac{1}{2}bh$

$A = \frac{1}{2}(35.5\text{ m})(30.5\text{ m})$

$A = 541\text{ m}^2$

b) $P = a + b + c$

$P = 30.5\text{ m} + 35.5\text{ m} + 46.8\text{ m}$

$P = 112.8\text{ m}$

10. a) $A = \pi r^2$

$A = \pi(14.5\text{ m})^2$

$A = 661\text{ m}^2$

b) $C = 2\pi r$

$C = 2\pi(14.5\text{ m})$

$C = 91.1\text{ m}$

11. $c^2 = a^2 + b^2$

$c^2 = (126\text{ ft})^2 + (214\text{ ft})^2$

$c = 248\text{ ft}$

12. $A = bh$

$2440\text{ m}^2 = 76.2\text{ m}(h)$

$h = \frac{2440\text{ m}^2}{76.2\text{ m}}$

$h = 32.0\text{ m}$

13. $A = \pi r^2$

$\frac{1230\text{ in}^2}{\pi} = r^2$

$r = 19.8\text{ in.}$

$d = 39.6\text{ in.}$

14. a) The altitude bisects the side it hits, forming a right triangle with one leg of 6.00 cm and a hypotenuse of 12.0 cm.

$h^2 = (12.0\text{ cm})^2 - (6.00\text{ cm})^2$

$h = 10.4\text{ cm}$

$A = \frac{1}{2}bh$

$A = \frac{1}{2}(12.0\text{ cm})(10.4\text{ cm})$

$A = 62.4\text{ cm}^2$

b) $P = 3b$

$P = 3(12.0\text{ cm})$

$P = 36.0\text{ cm}$

15. a) $V = Bh = lwh$

$V = (24.0\text{ ft})(36.0\text{ ft})(18.0\text{ ft})$

$V = 15{,}600\text{ ft}^3$

b) $A = 2(lh) + 2(wh)$

$A = 2(36.0\text{ ft})(18.0\text{ ft}) + 2(24.0\text{ ft})(18.0\text{ ft})$

$A = 2160\text{ ft}^2$

c) $SA = A + \text{Top} + \text{Bottom}$

$SA = 2160\text{ ft}^2 + 2(24.0\text{ ft})(36.0\text{ ft})$

$SA = 3890\text{ ft}^2$

16. a) $V = \pi r^2 h$

$r = 4.75\ m$

$V = \pi(4.75\text{ m})^2(45.0\text{ m})$

$V = 3190\text{ m}^3$

b) $A = 2\pi rh$

$A = 2\pi(4.75\text{ m})(45.0\text{ m})$

$A = 1340\text{ m}^2$

c) $SA = A + \text{Top} + \text{Bottom}$

$SA = 1343\text{ m}^2 + 2(\pi r^2)$

$SA = 1343\text{ m}^2 + 2\pi(4.75\text{ m})^2$

$SA = 1480\text{ m}^2$

17. a) $r = 11.0\text{ in.}$

$V = \frac{1}{3}\pi r^2 h$

$V = \frac{1}{3}\pi(11.0\text{ in.})^2(41.6\text{ in.})$

$V = 5270\text{ in}^3$

b) $A = \pi rs$

$s^2 = r^2 + h^2$

$s^2 = (11.0\text{ in.})^2 + (41.6\text{ in.})^2$

$s = 43.0\text{ in.}$

$A = \pi(11.0\text{ in.})(43.0\text{ in.})$

$A = 1490\text{ in}^2$

c) $SA = A + \text{base}$

$SA = 1485.97\text{ in}^2 + \pi r^2$

$SA = 1485.97\text{ in}^2 + \pi(11.0\text{ in.})^2$

$SA = 1870\text{ in}^2$

18. $V = \frac{1}{3}Bh$

$V = \frac{1}{3}(28.0\text{ cm})(44.0\text{ cm})(16.5\text{ cm})$

$V = 6780\text{ cm}^3$

19. a) $r = 16.0\text{ m}$

$V = \frac{4}{3}\pi r^3$

$V = \frac{4}{3}\pi(16.0\text{ m})^3$

$V = 17{,}200\text{ m}^3$

b) $A = 4\pi r^2$

$A = 4\pi(16.0\text{ m})^2$

$A = 3220\text{ m}^2$

20. a) $V = \text{Vol cylinder} + \text{Vol cone}$

$r = 6.80\text{ ft}$

$V = \pi r^2 h + \frac{1}{3}\pi r^2 h$

$V = \pi(6.80\text{ ft})^2(18.0\text{ ft}) + \frac{1}{3}\pi(6.80\text{ ft})^2(7.50\text{ ft})$

$V = 2980\text{ ft}^3$

b) $A = 2\pi rh + \pi rs$

$s^2 = (6.8\text{ ft})^2 + (7.50\text{ ft})^2$

$s = 10.1\text{ ft}$

$A = 2\pi(6.80\text{ ft})(18.0\text{ ft}) + \pi(6.80\text{ ft})(10.1\text{ ft})$

$A = 985\text{ ft}^2$

c) $SA = A + \text{base}$

$SA = 985\text{ ft}^2 + \pi(6.80\text{ ft})^2$

$SA = 985\text{ ft}^2 + 145\text{ ft}^2$

$SA = 113\bar{0}\text{ ft}^2$

CHAPTER 2 SPECIAL APPLICATION

1. The units cancelled.

2. It would not change.

3. Because the units cancel when the ratio is formed.

Chapter 3, Exercises 3.1

1. See drawing in the textbook answers.

3. See drawing in the textbook answers.

5. $15' = 15\left(\frac{1°}{60}\right)$

$15' = \frac{1°}{4}$

7. $120' = 120\left(\frac{1°}{60}\right)$

$120' = 2°$

9. $\frac{1°}{2} = \frac{1}{2}(60')$

$\frac{1°}{2} = 30'$

11. $0.4° = 0.4(60')$

$0.4° = 24'$

13. $37°12' = 37° + 12'$

$= 37' + 12\left(\frac{1°}{60}\right)$

$= 37° + 0.2°$

$= 37.2°$

15. $75°47' = 75° + 47'$

$= 75° + 47\left(\frac{1°}{60}\right)$

$= 75° + 0.78°$

$= 75.78°$

17. $69\frac{1°}{3} = 69° + \frac{1°}{3}$

$= 69° + \frac{1}{3}(60')$

$= 69° + 20'$

$= 69°20'$

19. $23.3° = 23° + 0.3°$

$= 23° + 0.3(60')$

$= 23° + 18'$

$= 23°18'$

21. $34°24'15'' = 34° + 24\left(\frac{1°}{60}\right) + 15\left(\frac{1°}{3600}\right)$

$= 34° + 0.4° + 0.0042°$

$= 34.4042°$

23. $19°18'27'' = 19° + 18\left(\frac{1°}{60}\right) + 27\left(\frac{1°}{3600}\right)$

$= 19° + 0.3° + 0.0075°$

$= 19.3075°$

25. $18.21° = 18° + 0.21°$

$= 18° + 0.21(60')$

$= 18° + 12.6'$

$= 18°12' + 0.6'$

$= 18°12' + 0.6(60'')$

$= 18°12' + 36''$

$= 18°12'36''$

27. $8.925° = 8° + 0.925°$

$= 8° + 0.925(60')$

$= 8° + 55.5'$

$= 8°55' + 0.5'$

$= 8°55' + 0.5(60'')$

$= 8°55' + 30''$

$= 8°55'30''$

29. a **31.** c **33.** a **35.** B **37.** B

39.

$$a^2+b^2=c^2$$
$$(5.00\,\text{cm})^2+(12.0\,\text{cm})^2=c^2$$
$$25.0\,\text{cm}^2+144\,\text{cm}^2=c^2$$
$$\sqrt{(25.0+144)\,\text{cm}^2}=c^2$$
$$13.0\,\text{cm}=c$$

41.

$$a^2+b^2=c^2$$
$$(155\,\text{mi})^2+b^2=(208\,\text{mi})^2$$
$$b^2=(208\,\text{mi})^2-(115\,\text{mi})^2$$
$$b^2=43264\text{mi}^2-13225\text{mi}^2$$
$$b=\sqrt{(43264-13225)\,\text{mi}^2}$$
$$b=173\,\text{mi}$$

43.

$$a^2+b^2=c^2$$
$$a^2+(377\text{yd})^2=(506\,\text{yd})^2$$
$$a^2=(506\,\text{yd})^2-(377\text{yd})^2$$
$$a=\sqrt{(256036-142129)\,\text{yd}^2}$$
$$a=338\,\text{yd}$$

45.

$$a^2+b^2=c^2$$
$$(35.7\text{m})^2+(16.8\,\text{m})^2=c^2$$
$$1274.49\,\text{m}^2+282.24\,\text{m}^2=c^2$$
$$\sqrt{(1274.49+282.24)\,\text{m}^2}=c$$
$$39.5\,\text{m}=c$$

47.

$$a^2+b^2=c^2$$
$$(2.25\,\text{cm})^2+b^2=(3.75\,\text{cm})^2$$
$$b^2=(3.75\,\text{cm})^2-(2.25\,\text{cm})^2$$
$$b^2=14.0625\text{cm}^2-5.0625\text{cm}^2$$
$$b=\sqrt{(14.0625-5.0625)\,\text{cm}^2}$$
$$b=3.00\,\text{cm}^2$$

49. First find c.

$$a^2+b^2=c^2$$
$$(5.00\,\text{cm})^2+(12.0\,\text{cm})^2=c^2$$
$$\sqrt{(5.00^2+12.0^2)\,\text{cm}^2}=c^2$$
$$13.0\,\text{cm}=c$$
$$\sin A=\frac{a}{c}=\frac{5.00}{13.0\,\text{cm}}=0.385$$
$$\cos A=\frac{b}{c}=\frac{12.0\,\text{cm}}{13.0\,\text{cm}}=0.923$$
$$\tan A=\frac{a}{b}=\frac{5.00\,\text{cm}}{12.0\,\text{cm}}=0.417$$
$$\csc A=\frac{c}{a}=\frac{13.0\,\text{cm}}{5.00\,\text{cm}}=2.60$$
$$\sec A=\frac{c}{b}=\frac{13.0\,\text{cm}}{12.0\,\text{cm}}=1.08$$
$$\cot A=\frac{b}{a}=\frac{12.0\,\text{cm}}{5.00\,\text{cm}}=2.40$$

51. First find b.

$$a^2+b^2=c^2$$
$$(335\,\text{m})^2+b^2=(685\,\text{m})^2$$
$$b^2=(685\,\text{m})^2-(335\,\text{m})^2$$
$$b^2=469{,}225\,\text{m}^2-112225\,\text{m}^2$$
$$b=\sqrt{(469225-112225)\,\text{m}^2}$$
$$b=597\,\text{m}$$
$$\sin A=\frac{a}{c}=\frac{335\,\text{m}}{685\,\text{m}}=0.489$$
$$\cos A=\frac{b}{c}=\frac{597\,\text{m}}{685\,\text{m}}=0.872$$
$$\tan A=\frac{a}{b}=\frac{335\,\text{m}}{597\,\text{m}}=0.561$$
$$\csc A=\frac{c}{a}=\frac{685\,\text{m}}{335\,\text{m}}=2.04$$
$$\sec A=\frac{c}{b}=\frac{685\,\text{m}}{597\,\text{m}}=1.15$$
$$\cot A=\frac{b}{a}=\frac{597\,\text{m}}{335\,\text{m}}=1.78$$

Section 3.1

53. First find a.

$$a^2+b^2=c^2$$
$$a^2+(3.00\,\text{km})^2=(6.00\,\text{km})^2$$
$$a^2=(6.00\,\text{km})^2-(3.00\,\text{km})^2$$
$$a^2=36.0\,\text{km}^2-9.00\,\text{km}^2$$
$$a=\sqrt{(36.0-9.00)\,\text{km}^2}$$
$$a=5.20\,\text{km}$$
$$\sin A=\frac{a}{c}=\frac{5.20\,\text{km}}{6.00\,\text{km}}=0.867$$
$$\cos A=\frac{b}{c}=\frac{3.00\,\text{km}}{6.00\,\text{km}}=0.500$$
$$\tan A=\frac{a}{b}=\frac{5.20\,\text{km}}{3.00\,\text{km}}=1.73$$
$$\csc A=\frac{c}{a}=\frac{6.00\,\text{km}}{5.20\,\text{km}}=1.15$$
$$\sec A=\frac{c}{b}=\frac{6.00\,\text{km}}{3.00\,\text{km}}=2.00$$
$$\cot A=\frac{b}{a}=\frac{3.00\,\text{km}}{5.20\,\text{km}}=0.577$$

55. First find c.

$$a^2+b^2=c^2$$
$$(5.00\,\text{cm})^2+(12.0\,\text{cm})^2=c^2$$
$$25.0\,\text{cm}^2+144\,\text{cm}^2=c^2$$
$$\sqrt{(25.0+144)\,\text{cm}^2}=c^2$$
$$13.0\,\text{cm}=c$$
$$\sin B=\frac{b}{c}=\frac{12.0\,\text{cm}}{13.0\,\text{cm}}=0.923$$
$$\cos B=\frac{a}{c}=\frac{5.00\,\text{cm}}{13.0\,\text{cm}}=0.385$$
$$\tan B=\frac{b}{a}=\frac{12.0\,\text{cm}}{5.00\,\text{cm}}=2.40$$
$$\csc B=\frac{c}{b}=\frac{13.0\,\text{cm}}{12.0\,\text{cm}}=1.08$$
$$\sec B=\frac{c}{a}=\frac{13.0\,\text{cm}}{5.00\,\text{cm}}=2.60$$
$$\cot B=\frac{a}{b}=\frac{5.00\,\text{cm}}{12.0\,\text{cm}}=0.417$$

57. First find b.

$$a^2+b^2=c^2$$
$$(4.60\,\text{m})^2+b^2=(9.25\,\text{m})^2$$
$$b^2=(9.25\,\text{m})^2-(4.60\,\text{m})^2$$
$$b^2=85.5625\,\text{m}^2-21.16\,\text{m}^2$$
$$b=\sqrt{(85.5625-21.16)\,\text{m}^2}$$
$$b=8.03\text{m}$$
$$\sin B=\frac{b}{c}=\frac{8.03\,\text{m}}{9.25\,\text{m}}=0.868$$
$$\cos B=\frac{a}{c}=\frac{4.60\,\text{m}}{9.25\,\text{m}}=0.497$$
$$\tan B=\frac{b}{a}=\frac{8.03\,\text{m}}{4.60\,\text{m}}=1.75$$
$$\csc B=\frac{c}{b}=\frac{9.25\,\text{m}}{8.03\,\text{m}}=1.15$$
$$\sec B=\frac{c}{a}=\frac{9.25\,\text{m}}{4.60\,\text{m}}=2.01$$
$$\cot B=\frac{a}{b}=\frac{4.60\,\text{m}}{8.03\,\text{m}}=0.573$$

59. First find a.

$$a^2+b^2=c^2$$
$$a^2+(4.50\,\text{ft})^2=(9.00\,\text{ft})^2$$
$$a^2=(9.00\,\text{ft})^2-(4.50\,\text{ft})^2$$
$$a^2=81.0\,\text{ft}^2-20.25\,\text{ft}^2$$
$$a=\sqrt{(81.0-20.25)\,\text{ft}^2}$$
$$a=7.79\,\text{ft}$$
$$\sin B=\frac{b}{c}=\frac{4.50\,\text{ft}}{9.00\,\text{ft}}=0.500$$
$$\cos B=\frac{a}{c}=\frac{7.79\,\text{ft}}{9.00\,\text{ft}}=0.866$$
$$\tan B=\frac{b}{a}=\frac{4.50\text{ft}}{7.79\,\text{ft}}=0.578$$
$$\csc B=\frac{c}{b}=\frac{9.00\,\text{ft}}{4.50\,\text{ft}}=2.00$$
$$\sec B=\frac{c}{a}=\frac{9.00\,\text{ft}}{7.79\,\text{ft}}=1.16$$
$$\cot B=\frac{a}{b}=\frac{7.79\,\text{ft}}{4.50\,\text{ft}}=1.73$$

61.

	(a)	(b)
$\sin A$	$\frac{5}{13}=0.385$	$\frac{15}{39}=0.385$
$\cos A$	$\frac{12}{13}=0.923$	$\frac{36}{39}=0.923$
$\tan A$	$\frac{5}{12}=0.417$	$\frac{15}{36}=0.417$
$\csc A$	$\frac{13}{5}=2.60$	$\frac{39}{15}=2.60$
$\sec A$	$\frac{13}{12}=1.08$	$\frac{39}{36}=1.08$
$\cot A$	$\frac{12}{5}=2.40$	$\frac{36}{15}=2.40$

63.

$$\frac{\sin A}{\cos A}=\frac{\frac{a}{c}}{\frac{b}{c}}$$

$$\frac{\sin A}{\cos A}=\frac{a}{c}\cdot\frac{c}{b}$$

$$\frac{\sin A}{\cos A}=\frac{a}{b}$$

$$\frac{\sin A}{\cos A}=\tan A$$

Exercises 3.2

1. $\sin 18.5^\circ = 0.3173$

3. $\tan 41.4^\circ = 0.8816$

5. $\cos 77.2^\circ = 0.2215$

7. $\sec 34.7^\circ = \frac{1}{\cos 34.7^\circ} = 1.216$

9. $\cot 34.0^\circ = \frac{1}{\tan 34.0^\circ} = 1.483$

11. $\csc 49.8^\circ = \frac{1}{\sin 49.8^\circ} = 1.309$

13. $\sin 46.72^\circ = 0.7280$

15. $\tan 73.8035^\circ = 3.443$

17. $\sec 8.3751^\circ = \frac{1}{\cos 8.3751^\circ} = 1.011$

19. $\sin\theta = 0.4305$
$\theta = 25.5^\circ$

21. $\tan\theta = 0.4684$
$\theta = 25.1^\circ$

23. $\cos\theta = 0.1463$
$\theta = 81.6^\circ$

25. $\tan\theta = 3.214$
$\theta = 72.7^\circ$

27. $\sin\theta = 0.1986$
$\theta = 11.5^\circ$

29. $\sec\theta = 2.363$

$$\frac{1}{\cos\theta} = 2.363$$

$$\frac{1}{2.363} = \cos\theta$$

$$\theta = 65.0^\circ$$

31.
$$\cot\theta = 0.5862$$
$$\frac{1}{\tan\theta} = 0.5862$$
$$\frac{1}{0.5862} = \tan\theta$$
$$\theta = 59.6°$$

33.
$$\csc\theta = 2.221$$
$$\frac{1}{\sin\theta} = 2.221$$
$$\frac{1}{2.221} = \sin\theta$$
$$\theta = 26.8°$$

35.
$$\sec\theta = 6.005$$
$$\frac{1}{\cos\theta} = 6.005$$
$$\frac{1}{6.005} = \cos\theta$$
$$\theta = 80.4°$$

37.
$$\cos\theta = 0.4836$$
$$\theta = 61.08°$$

39.
$$\cot\theta = 1.5392$$
$$\frac{1}{\tan\theta} = 1.5392$$
$$\frac{1}{1.5392} = \tan\theta$$
$$\theta = 33.01°$$

41.
$$\csc\theta = 2.4075$$
$$\frac{1}{\sin\theta} = 2.4075$$
$$\frac{1}{2.4075} = \sin\theta$$
$$\theta = 24.54°$$

43. See Example 6.
$$\sin 36°24' = 0.5934$$

45. See Example 6.
$$\tan 52°43'38'' = 1.314$$

47. See Example 6.
$$\cos 9°56'21'' = 0.985C$$

49.
$$\cot 36°15'44'' = \frac{1}{\tan 36°15'44''}$$
$$= 1.363$$

51.
$$\csc 84°35'53'' = \frac{1}{\sin 84°35'53''}$$
$$= 1.004$$

53.
$$\sec 72°27'' = \frac{1}{\cos 72°0'27''}$$
$$= 3.237$$

55. See Example 7.
$$\sin\theta = 0.8556$$
$$\theta = 58°49'34''$$

57. See Example 7.
$$\tan\theta = 6.2662$$
$$\theta = 80°55'58''$$

59. See Example 7.
$$\cos\theta = 0.5966$$
$$\theta = 53°22'24''$$

61.
$$\cot\theta = 0.8678$$
$$\frac{1}{\tan\theta} = 0.8678$$
$$\frac{1}{0.8678} = \tan\theta$$
$$\theta = 49°2'55''$$

63.
$$\csc\theta = 2.3770$$
$$\frac{1}{\sin\theta} = 2.3770$$
$$\frac{1}{2.3770} = \sin\theta$$
$$\theta = 24°52'43''$$

65.
$$\sec\theta = 1.1678$$
$$\frac{1}{\cos\theta} = 1.1678$$
$$\frac{1}{1.1678} = \cos\theta$$
$$\theta = 31°5'42''$$

Exercises 3.3

1.

$$a^2 + b^2 = c^2$$
$$(4.00\text{ ft})^2 + (8.00\text{ ft})^2 = c^2$$
$$16.0\text{ ft}^2 + 64.0\text{ ft}^2 = c^2$$
$$\sqrt{(16.0 + 64.0)\text{ ft}^2} = c$$
$$8.94\text{ ft} = c$$
$$\tan A = \frac{a}{b} = \frac{4.00\text{ ft}}{8.00\text{ ft}}$$
$$A = 26.6°$$
$$B = 90° - A = 63.4°$$

3.

$$\sin A = \frac{a}{c}$$
$$\sin 27.3° \frac{a}{21.0\text{ cm}}$$
$$a = 21.0 \sin 27.3°\text{ cm}$$
$$a = 9.63\text{ cm}$$
$$a^2 + b^2 = c^2$$
$$(9.63\text{ cm})^2 + b^2 = (21.0\text{ cm})^2$$
$$b^2 = (21.0\text{ cm})^2 - (9.63\text{ cm})^2$$
$$b^2 = 441\text{ cm}^2 - 92.7\text{ cm}^2$$
$$b = \sqrt{(441 - 92.7)\text{ cm}^2}$$
$$b = 18.7\text{ cm}$$
$$B = 90 - A = 62.7°$$

5.

$$a^2 + b^2 = c^2$$
$$a^2 + (7.50\text{ m})^2 = (13.4\text{ m})^2$$
$$a^2 = (13.4\text{ m})^2 - (7.50\text{ m})^2$$
$$a^2 = 18\bar{0}\text{ m}^2 - 56.3\text{ m}^2$$
$$a = \sqrt{(18\bar{0} - 56.3)\text{ mi}^2}$$
$$a = 11.1\text{ m}$$
$$\sin B = \frac{b}{c}$$
$$\sin B = \frac{7.50\text{ m}}{13.4\text{ m}}$$
$$B = 34.0°$$
$$A = 90° - B = 56.0°$$

7.

$$a^2 + b^2 = c^2$$
$$(12.4\text{ mi})^2 + (7.70\text{ mi})^2 = c^2$$
$$154\text{ mi}^2 + 59.3\text{ mi}^2 = c^2$$
$$\sqrt{(154 + 59.3)\text{ mi}^2} = c$$
$$c = 14.6\text{ mi}$$
$$\tan A = \frac{a}{b}$$
$$\tan A = \frac{12.4\text{ mi}}{7.70\text{ mi}}$$
$$A = 58.2°$$
$$A = 90° - A = 31.8°$$

9.

$$\sin B = \frac{b}{c}$$
$$\sin 37.0° = \frac{25\bar{0}\text{ km}}{c}$$
$$c = \frac{25\bar{0}\text{ km}}{\sin 37.0°}$$
$$c = 415\text{ km}$$
$$\tan B = \frac{a}{b}$$
$$\tan 37.0° = \frac{25\bar{0}\text{ km}}{a}$$
$$a = \frac{25\bar{0}\text{ km}}{\tan 37.0°}$$
$$a = 332\text{ km}$$
$$A = 90° - B = 53.0°$$

11.

$$\sin A = \frac{a}{c}$$
$$\sin A = \frac{14.21\text{ cm}}{37.42\text{ cm}}$$
$$A = 22.32°$$
$$B = 90° - A = 67.68°$$
$$\sin B = \frac{b}{c}$$
$$\sin 67.68° = \frac{b}{37.42\text{ cm}}$$
$$b = 37.42\text{ cm}\ \sin 67.68°$$
$$b = 34.62\text{ cm}$$

13. $B = 90° - A = 21.25°$

$\sin A = \dfrac{a}{c}$

$\sin 68.75° = \dfrac{6755\text{ mi}}{c}$

$c = \dfrac{6755\text{ mi}}{\sin 68.75°}$

$c = 7248\text{ mi}$

$\sin B = \dfrac{b}{c}$

$\sin 21.25° = \dfrac{b}{7248\text{ mi}}$

$b = 7248\text{ mi}\ \sin 21.25°$

$b = 2627\text{ mi}$

15. $A = 90° - B = 74.20°$

$\sin B = \dfrac{b}{c}$

$\sin 15.80° = \dfrac{b}{45.32\text{ m}}$

$b = 45.32\text{ m}\ \sin 15.80°$

$b = 12.34\text{ m}$

$\sin A = \dfrac{a}{c}$

$\sin 74.20° = \dfrac{a}{45.32\text{ m}}$

$a = 45.32\text{ m}\ \sin 74.20°$

$a = 43.61\text{ m}$

17. $\cos A = \dfrac{b}{c}$

$\cos A = \dfrac{2572\text{ ft}}{4615\text{ ft}}$

$A = 56.13°$

$B = 90° - A = 33.87°$

$\sin A = \dfrac{a}{c}$

$\sin 56.13° = \dfrac{a}{4615\text{ ft}}$

$a = 4615\text{ ft}\ \sin 56.13°$

$a = 3832\text{ ft}$

19. $\sin B = \dfrac{b}{c}$

$\sin B = \dfrac{1500\text{ mi}}{3500\text{ mi}}$

$B = 25°$

$A = 90° - B = 65°$

$\sin A = \dfrac{a}{c}$

$\sin 65° = \dfrac{a}{3500\text{ mi}}$

$a = 3500\text{ mi}\ \sin 65°$

$a = 3200\text{ mi}$

21. $A = 90° - B = 4\bar{0}°$

$\sin B = \dfrac{b}{c}$

$\sin 5\bar{0}° = \dfrac{b}{45\text{ m}}$

$b = 45\text{ m}\ \sin 5\bar{0}°$

$b = 34\text{ m}$

$\sin A = \dfrac{a}{c}$

$\sin 4\bar{0}° = \dfrac{a}{45\text{ m}}$

$a = 45\text{ m}\ \sin 4\bar{0}°$

$a = 29\text{ m}$

23. $B = 90° - A = 53°$

$\sin A = \dfrac{a}{c}$

$\sin 37° = \dfrac{140\text{ ft}}{c}$

$c = \dfrac{140\text{ ft}}{\sin 37°}$

$c = 230\text{ ft}$

$\tan A = \dfrac{a}{b}$

$\tan 37° = \dfrac{140\text{ ft}}{b}$

$b = \dfrac{140\text{ ft}}{\tan 37°}$

$b = 190\text{ ft}$

25.
$$A = 90° - B = 68°$$
$$\tan B = \frac{b}{a}$$
$$\tan 22° = \frac{b}{3.5\text{ mi}}$$
$$b = 3.5\text{ mi } \tan 22°$$
$$b = 1.4\text{ mi}$$
$$\cos B = \frac{a}{c}$$
$$\cos 22° = \frac{3.5\text{ mi}}{c}$$
$$c = \frac{3.5\text{ mi}}{\cos 22°}$$
$$c = 3.8\text{ mi}$$

27.
$$A = 90° - B$$
$$A = 89°59'60'' - 37°41'30''$$
$$A = 52°18'30''$$
$$\tan B = \frac{b}{a}$$
$$b = a \tan B$$
$$b = 1753\text{ m} \tan 37°41'30''$$
$$b = 1354\text{ m}$$
$$\cos B = \frac{a}{c}$$
$$c = \frac{a}{\cos B}$$
$$c = \frac{1753\text{ m}}{\cos 37°41'30''}$$
$$c = 2215\text{ m}$$

29.
$$\sin A = \frac{a}{c}$$
$$\sin A = \frac{495.5\text{ ft}}{617.0\text{ ft}}$$
See Example 7 in 3.2.
$$A = 53°25'31''$$
$$B = 90° - A$$
$$B = 89°59'60'' - 53°25'31''$$
$$B = 36°24'39''$$
$$\cos A = \frac{b}{c}$$
$$b = c \cos A$$
$$b = 617.0\text{ ft } \cos 53°25'31''$$
$$b = 367.7\text{ ft}$$

31.
$$B = 90° - A$$
$$B = 89°59'60'' - 58°11'25''$$
$$B = 31°48'35''$$
$$\sin A = \frac{a}{c}$$
$$c = \frac{a}{\sin A}$$
$$c = \frac{37.52\text{ m}}{\sin 58°11'25''}$$
$$c = 44.15\text{ m}$$
$$\tan A = \frac{a}{b}$$
$$b = \frac{a}{\tan A}$$
$$b = \frac{37.52\text{ m}}{\tan 58°11'25''}$$
$$b = 23.27\text{ m}$$

33.
$$A = 90° - B$$
$$A = 89°59'60'' - 27°5'16''$$
$$A = 62°54'44''$$
$$\sin B = \frac{b}{c}$$
$$b = c \sin B$$
$$b = 6752\text{ ft} \sin 27°5'16''$$
$$b = 3075\text{ ft}$$
$$\cos B = \frac{a}{c}$$
$$a = c \cos B$$
$$a = 6752\text{ ft} \cos 27°5'16''$$
$$a = 6011\text{ ft}$$

Exercises 3.4

1. Let T = tower height.

$\tan 51° = \frac{T}{\text{shadow}}$

$\tan 31° = \frac{T}{42\text{ m}}$

$T = 42\text{ m} \tan 51°$

$T = 52\text{ m}$

3. Let A = angle of elevation.

$\tan A = \frac{\text{rise}}{\text{run}}$

$\tan A = \frac{250\text{ ft}}{3600\text{ ft}}$

$A = 4°$

5. Let x = length of guy wire. Wire is attached $190\text{ ft} - 25\text{ ft} =$ 165 ft from bottom.

$\sin 40.0° = \frac{165\text{ ft}}{x}$

$x = \frac{165\text{ ft}}{\sin 40.0°}$

$x = 257\text{ ft}$

7. Let x = bridge length.

$\sin 75° = \frac{89.1\text{ m}}{x}$

$x = \frac{89.1\text{ m}}{\sin 75°}$

$x = 92.2\text{ m}$

9. Let y = cliff height. Let x = height from road to boulder. Distance down from top to boulder = $y - x$.

$\tan 62° = \frac{y}{275\text{ ft}}$

$y = 275\text{ ft} \tan 62°$

$y = 517\text{ ft}$

$\tan 42° = \frac{x}{275\text{ ft}}$

$x = 275\text{ ft} \tan 42°$

$x = 248\text{ ft}$

Distance from top to boulder = $y - x =$ $517\text{ ft} - 248\text{ ft}$.

$y - x = 269\text{ ft}$

11. (a) See Figure 3.35

$\tan \phi = \frac{x}{R}$

$\tan \phi = \frac{82.6\Omega}{112\Omega}$

$\phi = 36.4°$

$\sin \phi = \frac{x}{Z}$

$Z = \frac{x}{\sin \phi}$

$Z = \frac{82.6\Omega}{\sin 36.4°}$

$Z = 139\Omega$

11. (b) $\tan \phi = \frac{x}{R}$

$\tan 23° = \frac{x}{250\Omega}$

$x = 250\Omega \tan 23°$

$x = 110\Omega$

$\sin \phi = \frac{x}{z}$

$z = \frac{x}{\sin \phi}$

$z = \frac{106\Omega}{\sin 23°}$

$z = 270\Omega$

13. See Figure 3.36.

$\tan a = \frac{1.70\text{ cm}}{2.30\text{ cm}}$

$a = 36.5°$

$\sin a = \frac{1.70\text{ cm}}{x}$

$x = \frac{1.70\text{ cm}}{\sin a}$

$x = \frac{1.70\text{ cm}}{\sin 36.5°}$

$x = 2.86\text{ cm}$

15. See Figure 3.38.

$$\tan a = \frac{1.90\,\text{cm}}{0.750\,\text{cm}}$$

$a = 68.5°$

The unlabeled length adjacent to β is $7.60\,\text{cm-}4.25\,\text{cm} - 0.750\,\text{cm} = 2.60\,\text{cm}$

Then $\tan\beta = \frac{5.25\,\text{cm}}{2.60\,\text{cm}}$

$\beta = 63.7°$

$$\sin\beta = \frac{5.25\,\text{cm}}{y}$$

$$y = \frac{5.25\,\text{cm}}{\sin\beta}$$

$$y = \frac{5.25\,\text{cm}}{\sin 63.7°}$$

$y = 5.86\,\text{cm}$

$$\sin\alpha = \frac{1.90\,\text{cm}}{x}$$

$$x = \frac{1.90\,\text{cm}}{\sin\alpha}$$

$$x = \frac{1.90\,\text{cm}}{\sin 68.5°}$$

$x = 2.04\,\text{cm}$

Draw a line segment through the point where x and z meet and parallel to the line segment measuring 7.60 cm. The right triangle with z as the hypotenuse has a horizontal leg of 4.25 cm and a vertical leg of $5.25\,\text{cm} - 1.90\,\text{cm} = 3.35\,\text{cm}$

$$z^2 = (4.25\,\text{cm})^2 + (3.35\,\text{cm})^2$$

$$z^2 = 29.285\,\text{cm}^2$$

$$z = 5.41\,\text{cm}$$

In the right triangle with z as the hypotenuse, let w be the angle adjacent to the leg of 4.25 cm. Then

$$\tan w = \frac{3.35\,\text{cm}}{4.25\,\text{cm}}$$

$w = 38.2°$

$\phi = \text{complement of } \beta + \text{complement of } w$

$\phi = 26.3° + 51.8°$

$\phi = 78.1°$

$\theta = \text{complement of } \alpha + 90° + w$

$\theta = 21.5° + 90° + 38.2°$

$\theta = 149.7°$

17. See Figure 3.40.

The distance across the top of the gap is $0.6800\text{-}0.2850 = 0.3950$ in.
Bisect the 26° angle with a vertical line, forming two similar right triangles. The hypotenuse of the larger triangle subtract the hypotenuse of the smaller triangle is x.

$$x = \frac{0.1975}{\sin 13°} = \frac{0.1425}{\sin 13°}$$

$x = 0.2445$ in.

19. See Figure 3.42.
Bisect the angle θ with a vertical line forming two similar right triangles. Let x be the side adjacent to $\frac{\theta}{2}$ in the smaller triangle.
Remember that the horizontal lines have also been bisected. Sides are proportional in similar triangles, so

$$\frac{1.000}{0.425} = \frac{5.200 + x}{x}$$

$x = 0.425(5.200 + x)$

$x = 2.210 + 0.425x$

$0.575x = 2.210$

$x = 3.8435$

$$\tan\frac{\theta}{2} = \frac{0.425}{3.8435}$$

$$\frac{\theta}{2} = 6.31°$$

$\theta = 12.6°$

21. See Figure 3.44

Form a right triangle with x as the hypotenuse by drawing a vertical line parallel to the side of 24.50 m and a horizontal line parallel to the side of 32.00 m. Let angle a be the supplement of α and angle b be the supplement of β. The side adjacent to angle a is $24.50\,\text{m} - 4.00\,\text{m} - 2.50\,\text{m} = 18.00\,\text{m}$. The side adjacent to angle b is $32.00\,\text{m} - 3.00\,\text{m} - 3.5\,0\text{m} = 25.50\,\text{m}$.

$$\tan b = \frac{18.00}{25.50}$$

$$b = 35.22°$$

$$\beta = 180° - b = 144.78°$$

$$a = 90° - b = 54.78°$$

$$\alpha = 180° - a = 125.22°$$

$$\sin b = \frac{18.00\,\text{m}}{\text{x}}$$

$$x = \frac{18.00\,\text{m}}{\sin 35.22°} = 31.21\,\text{m}$$

25.

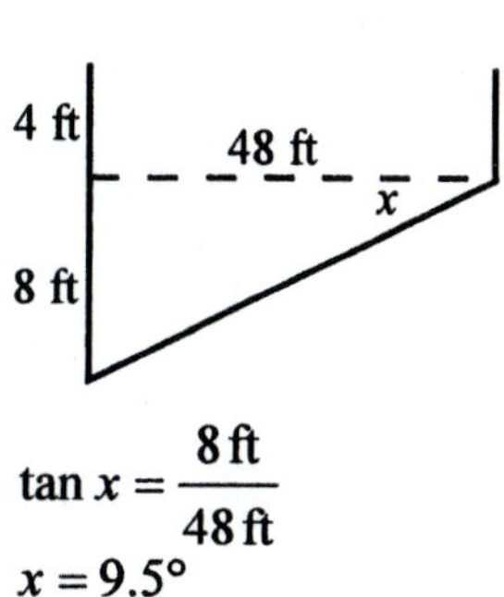

$$\tan x = \frac{8\,\text{ft}}{48\,\text{ft}}$$

$$x = 9.5°$$

29. See Figure 2.49.

Bisect the 30.00° angle forming a right triangle with side r opposite 15.00°.

$$\tan 15.00° = \frac{r}{400.00\,\text{mm}}$$

$$r = 4.00\,\text{mm} \tan 15.00°$$

$$r = 107.2\,\text{mm}$$

23. See Figure 3.46.

$$d^2 = \overline{AC}^2 + \overline{CD}^2$$

$$d^2 = (3.00\,\text{ft})^2 + (5.00\,\text{ft})^2$$

$$d^2 = 9.00\,\text{ft}^2 + 25.0\,\text{ft}^2$$

$$d^2 = 34.0\,\text{ft}^2$$

$$\overline{AB}^2 = d^2 + \overline{BD}^2$$

$$\overline{AB}^2 = 34.0\,\text{ft}^2 + (10.00\,\text{ft})^2$$

$$\overline{AB}^2 = 34.0\,\text{ft}^2 + 100.0\,\text{ft}^2$$

$$\overline{AB}^2 = 34.0\,\text{ft}^2$$

$$\overline{AB} = 11.6\,\text{ft}$$

27. See Figure 3.47.

The distance between the center of the drum and the wall is $8.0\,\text{ft} + 1.0\,\text{ft} = 9.0\,\text{ft}$. Also, the distance between the center of the drum and the pipe is 9.0 ft. Bisect the 45° angle, forming two right triangles.

$$\tan 22.5° = \frac{9.0\,\text{ft}}{x/2}$$

$$\frac{x}{2} = \frac{9.0\,\text{ft}}{\tan 22.5°}$$

$$\frac{x}{2} = 21.7\,\text{ft}$$

$$x = 43\,\text{ft}$$

31. See Figure 3.51.

The two unlabeled angles are equal. Let θ be one of them. Then

$$2\theta + 30.00° = 180.0°$$
$$2\theta = 150.0°$$
$$\theta = 75.00°$$

Bisect the 75.00° angle. The side adjacent to the resulting 37.50° angle is 120.0 mm. Thus

$$\tan 37.50° = \frac{r}{120.0\,\text{mm}}$$
$$r = 120.0\,\text{mm} \tan 37.50°$$
$$r = 92.08\,\text{mm}$$

33. Let x = tower height

$$\tan 71°24'30'' = \frac{x}{125.5\,\text{ft}}$$
$$x = 125.5\,\text{ft} \tan 71°24'30''$$
$$x = 373.1\,\text{ft}$$

35. Let θ be the angle whose adjacent side is the side of the square. Let x be the length of a side of the square.

$$\tan\theta = \frac{x}{2} \div x$$
$$\tan\theta = \frac{1}{2}$$
$$\theta = 26°33'54''$$

The other angle is

$$90° - \theta = 63°26'6''$$

Chapter 3 Review

1. $129°30' = 129° + 30\left(\frac{1°}{60}\right) = 129.5°$

2. $76°12' = 76° + 12\left(\frac{1°}{60}\right) = 76.2°$

3. $35\,{}^{2}\!/_{3}\,° = 35° + \frac{2}{3}(60') = 35°40'$

4. $314.3° = 314° + 0.3(60') = 314°18'$

5. $16°27'45'' = 16° + 27\left(\frac{1°}{60}\right) + 45\left(\frac{1°}{3600}\right) = 16° + 0.45° + 0.0125° = 16.4625°$

6. $38.405° = 38° + 0.405(60') = 38° + 24' + 0.30' = 38°24 + 0.3(60'') = 38°24'18''$

7.
$$a^2 + b^2 = c^2$$
$$(16.0\,\text{m})^2 + b^2 = (36.0\,\text{m})^2$$
$$b^2 = 1296\,\text{m}^2 - 256\,\text{m}^2$$
$$b = \sqrt{(1296 - 256)\,\text{m}^2}$$
$$b = 32.2\,\text{m}$$

8.
$$a^2 + b^2 = c^2$$
$$(18.7\,\text{mi})^2 + (25.5\,\text{mi})^2 = c^2$$
$$(349.69 + 650.25)\,\text{mi}^2 = c^2$$
$$c = \sqrt{999.94\,\text{mi}^2}$$
$$c = 31.6\,\text{mi}$$

9.
$$a^2 + b^2 = c^2$$
$$a = \sqrt{c^2 - b^2}$$
$$a = \sqrt{(235^2 - 127^2)\,\text{cm}^2}$$
$$a = 198\,\text{cm}$$
$$\sin A = \frac{198\,\text{cm}}{235\,\text{cm}} = 0.843 \quad \csc A = \frac{1}{\sin A} = 1.19$$
$$\cos A = \frac{127\,\text{cm}}{235\,\text{cm}} = 0.540 \quad \sec A = \frac{1}{\cos A} = 1.85$$
$$\tan A = \frac{198\,\text{cm}}{127\,\text{cm}} = 1.56 \quad \cot A = \frac{1}{\tan A} = 0.641$$

10. $\cos 14.6° = 0.9677$

11. $\sin 51.7° = 0.7848$

12. $\tan 29.5° = 0.5658$

13. $\sec 16.7° = \frac{1}{\cos 16.7°} = 1.044$

14. $\cot 29.1° = \frac{1}{\tan 29.1°} = 1.797$

15. $\csc 79.2° = \frac{1}{\sin 79.2°} = 1.018$

16. $\sin\theta = 0.6075$
$\theta = 37.4°$

17. $\cos\theta = 0.3522$
$\theta = 69.4°$

18. $\tan\theta = 1.2345$
$\theta = 51.0°$

19.
$$\sec\theta = 1.3290$$
$$\frac{1}{\cos\theta} = 1.3290$$
$$\cos\theta = \frac{1}{1.3290}$$
$$\theta = 41.2°$$

20.
$$\cot\theta = 0.9220$$
$$\frac{1}{\tan\theta} = 0.9220$$
$$\tan\theta = \frac{1}{0.9220}$$
$$\theta = 47.3°$$

21. $\csc\theta = 1.2222$

$\frac{1}{\sin\theta} = 1.2222$

$\sin\theta = \frac{1}{1.2222}$

$\theta = 54.9°$

22. $\sin 41°37'55'' = 0.6643$

23. $\tan 75°9'27'' = 3.774$

24. $\sec 34°14'35'' = \frac{1}{\cos 34°14'35''} = 1.210$

25. See Example 7 in Section 3.2.

$\cos\theta = 0.4470$

$\theta = 63°26'55''$

26. $\tan\theta = 0.2408$

$\theta = 13°32'21''$

27. $\csc\theta = 3.4525$

$\frac{1}{\sin\theta} = 3.4525$

$\sin\theta = \frac{1}{3.4525}$

$\theta = 16°50'12''$

28. $\tan A = \frac{a}{b}$

$\tan A = \frac{7.00\,\text{m}}{9.50\,\text{m}}$

$A = 36.4°$

$B = 90° - A = 53.6°$

$\sin A = \frac{a}{c}$

$\sin A = \frac{7.00\,\text{m}}{c}$

$c = \frac{7.00\,\text{m}}{\sin 36.4°}$

$c = 11.8\,\text{m}$

29. $\sin B = \frac{b}{c}$

$c = \frac{b}{\sin B}$

$c = \frac{17.75\,\text{cm}}{\sin 36.50°}$

$c = 26.48\,\text{cm}$

$A = 90° - B = 53.50°$

$\tan B = \frac{b}{a}$

$a = \frac{b}{\tan B}$

$a = \frac{15.75}{\tan 36.5°}$

$a = 21.28\,\text{cm}$

30. $\sin A = \frac{a}{c}$

$a = c\sin A$

$a = 1700\,\text{km}\sin 2\bar{0}°$

$a = 580\,\text{km}$

$B = 90° - A = 7\bar{0}°$

$\cos A = \frac{b}{c}$

$b = c\cos A$

$b = 1700\,\text{km}\cos 20°$

$b = 1600\,\text{km}$

31. $B = 90° - A$

$B = 90° - 35°14'32''$

$B = 54°45'28''$

$\sin A = \frac{a}{c}$

$c = \frac{a}{\sin A}$

$c = \frac{245.7\,\text{m}}{\sin 35°14'32''}$

$c = 425.8\,\text{m}$

$\cos A = \frac{b}{c}$

$b = c\cos A$

$b = 425.8\,\text{m}\,\cos 35°14'32''$

$b = 347.8\,\text{m}$

32. Let x be the ground distance from the tower to the fire.

$\tan 3° = \frac{250\,\text{ft}}{x}$

$x = \frac{250\,\text{ft}}{\tan 3°}$

$x = 4800\,\text{ft}$

33. Let x be the distance from the sighting to the tower.

$\tan 2° = \frac{175\,\text{ft}}{x}$

$x = \frac{175\,\text{ft}}{\tan 2°}$

$x = 5\bar{0}00\,\text{ft}$

34. Let θ be the angle of elevation.

$$\tan\theta = \frac{175\text{ ft}}{41\bar{0}0\text{ ft}}$$
$$\theta = 2.4°$$

35. See Figure 3.54.

Let θ be the complement of α.

$$\tan\theta = \frac{5.30\text{ m}}{21.7\text{ m}}$$
$$\theta = 13.7°$$
$$\alpha = 90° - \theta = 76.3°$$
$$x^2 = (21.7\text{ m})^2 + (5.30\text{ m})^2$$
$$x = \sqrt{(21.7)^2 + (5.30)^2}\text{ m}$$
$$x = 22.3\text{ m}$$

Form a right triangle with y as the hypotenuse by drawing a vertical line parallel to the line measuring 21.7 m. In that right traingle, the side adjacent to β will be

$$63.5\text{ m} - 47.2\text{ m} - 5.3\text{ m} = 11.0\text{ m}$$
$$\tan\beta = \frac{21.7\text{ m}}{11.0\text{ m}}$$
$$\beta = 63.1°$$
$$\sin\beta = \frac{21.7\text{ m}}{y}$$
$$y = \frac{21.7\text{ m}}{\sin 63.1°}$$
$$y = 24.3\text{ m}$$

36. See Figure 3.55.

$$\tan 20.0° = \frac{x}{\frac{56.0\text{ ft}}{2}}$$
$$\tan 20.0° = \frac{x}{28.0\text{ ft}}$$
$$x = 28.0\text{ ft}\tan 20.0°$$
$$x = 10.2\text{ ft}$$

37. (a) $X = 75\Omega, R = 42\Omega$

$$\tan\phi = \frac{75\Omega}{42\Omega}$$
$$\phi = 61°$$
$$Z = \sqrt{X^2 + R^2}$$
$$Z = \sqrt{(75\Omega)^2 + (42\Omega)^2}$$
$$Z = 86\Omega$$

(b) $X = 94\Omega, \phi = 47°$

$$\sin\phi = \frac{X}{Z}$$
$$Z = \frac{X}{\sin\phi}$$
$$Z = \frac{94\Omega}{\sin 47°}$$
$$Z = 130\Omega$$
$$\tan\phi = \frac{X}{R}$$
$$R = \frac{94\Omega}{\tan 47°}$$
$$R = 88\Omega$$

38. The sum of the measures of the interior angles of a polygon with n sides is given by $S = (n-2)\cdot 180°$. For a hexagon, $S = (6-2)\cdot 180° = 720°$. Thus, each interior angle is 120°. From the center of x, lines drawn to two adjacent vertices will form an equilateral triangle with the included side of the hexagon. Thus

$$\frac{x}{2} = 2.50\text{ cm} \quad \text{or} \quad x = 5.00\text{ cm}$$

From the right triangle,

$$y^2 = 5.00^2 - 2.50^2$$
$$y = 4.33\text{ cm}$$

39.

Using the right triangle containing x and y,

$\cos 65.5° = \dfrac{x}{275}$

$x = 114 \text{ mi}$

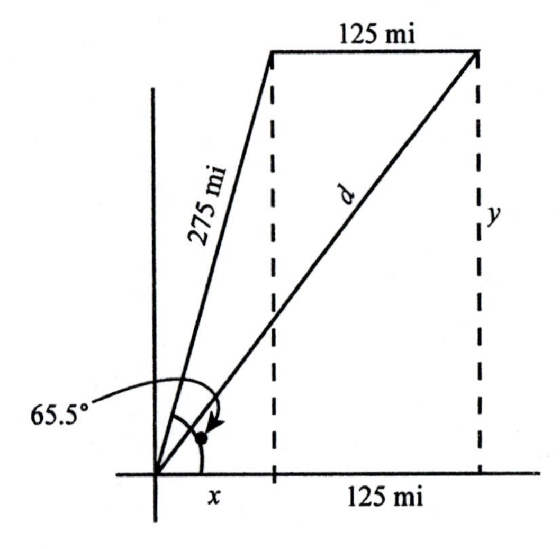

Then to find y use the Pythagorean Theorem.

$275^2 = y^2 + x^2$

$y^2 = 275^2 - 114^2$

$y = 25\overline{0} \text{ mi}$

In the large right triangle the legs are y and $x + 125$, i.e. $25\overline{0}$ and 239. The hypotenuse is the distance we seek.

$d^2 = 25\overline{0}^2 + 239^2$

$d = 346 \text{ mi}$

40.

In the large right triangle the other angle is 39.0°

$\cos 39.0° = \dfrac{y}{28.5}$

$y = 22.1 \text{ m}$

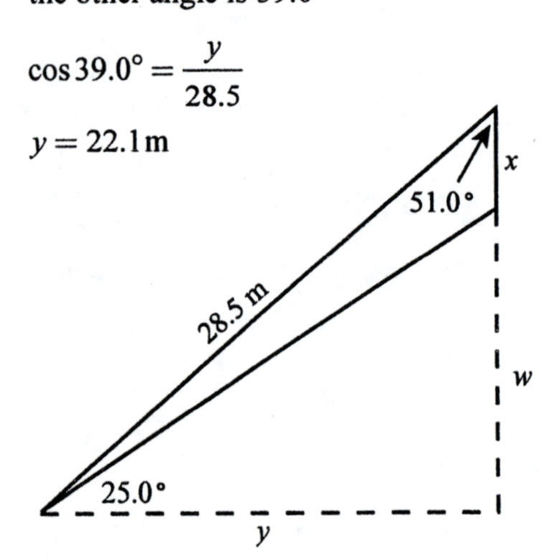

$\tan 25.0° = \dfrac{w}{y}$

$w = 22.1 \tan 25.0°$

$w = 10.3 \text{ m}$

Then $\tan 39.0° = \dfrac{x + w}{y} = \dfrac{x + 10.3}{22.1}$

$x = 22.1 \tan(39.0°) - 10.3$

$x = 7.6 \text{ m}$

Chapter 4, Exercises 4.1

	Function	Domain	Range
1.	Yes	$\{2,3,9\}$	$\{2,4,7\}$
3.	No	$\{1,2,7\}$	$\{1,3,5\}$
5.	Yes	$\{-2,2,3,5\}$	$\{2\}$
7.	Yes	Real Numbers	Real Numbers
9.	Yes	Real Numbers	Real Numbers where $y \ge 1$
11.	$y^2 = x+2$ $y = \pm\sqrt{x+2}$ No	Real Numbers where $x \ge -2$	Real Numbers
13.	Yes	Real Numbers where $x \ge -3$	Real Numbers where $y \ge 0$
15.	Yes	Real Numbers where $x \ge 4$	Real Numbers where $y \ge 6$

17. $f(x) = 8x - 12$

(a) $f(4) = 8(4) - 12$
$f(4) = 20$

(b) $f(0) = 8(0) - 12$
$f(0) = -12$

(c) $f(-2) = 8(-2) - 12$
$f(-2) = -28$

19. $g(x) = 10x + 15$

(a) $g(2) = 10(2) + 15$
$g(2) = 35$

(b) $g(0) = 10(0) + 15$
$g(0) = 15$

(c) $g(-4) = 10(-4) + 15$
$g(-4) = -25$

21. $h(x) = 3x^2 + 4x$

(a) $h(5) = 3(5)^2 + 4(5)$
$h(5) = 3(25) + 20$
$h(5) = 95$

(b) $h(0) = 3(0)^2 + 4(0)$
$h(0) = 0$

(c) $h(-2) = 3(-2)^2 + 4(-2)$
$h(-2) = 3(4) - 8$
$h(-2) = 4$

23. $f(t) = \frac{5-t^2}{2t}$

(a) $f(1) = \frac{5-(1)^2}{2(1)}$

$f(1) = \frac{5-1}{2} = 2$

(b) $f(-3) = \frac{5-(-3)^2}{2(-3)}$

$f(-3) = \frac{5-9}{-6}$

$f(-3) = \frac{-4}{-6} = \frac{2}{3}$

(c) 0 is not in the domain of $f(t)$.

25. $f(x) = 6x + 8$

(a) $f(a) = 6a + 8$

(b) $f(4a) = 6(4a) + 8$

$f(4a) = 24a + 8$

(c) $f(c^2) = 6c^2 + 8$

27. $h(x) = 4x^2 - 12x$

(a) $h(x+2) = 4(x+2)^2 - 12(x+2)$

$h(x+2) = 4(x^2 + 4x + 4) - 12x - 24$

$h(x+2) = 4x^2 + 16x + 16 - 12x - 24$

$h(x+2) = 4x^2 + 4x - 8$

(b) $h(x-3) = 4(x-3)^2 - 12(x-3)$

$h(x-3) = 4(x^2 - 6x + 9) - 12x + 36$

$h(x-3) = 4x^2 - 24x + 36 - 12x + 36$

$h(x-3) = 4x^2 - 36x + 72$

(c) $h(2x+1) = 4(2x+1)^2 - 12(2x+1)$

$h(2x+1) = 4(4x^2 + 4x + 1) - 24x - 12$

$h(2x+1) = 16x^2 + 16x + 4 - 24x - 12$

$h(2x+1) = 16x^2 - 8x - 8$

29. **(a)** $f(x) + g(x) = (3x-1) + (x^2 - 6x + 1)$

$= x^2 - 3x$

(b) $f(x) - g(x) = (3x-1) - (x^2 - 6x + 1)$

$= 3x - 1 - x^2 + 6x - 1$

$= -x^2 + 9x - 2$

(c) $[f(x)][g(x)] = (3x-1)(x^2 - 6x + 1)$

$= 3x^3 - 18x^2 + 3x$

$-x^2 + 6x - 1$

$= 3x^3 - 19x^2 + 9x - 1$

(d) $f(x) = 3x - 1$

$f(x+h) = 3(x+h) - 1$

$f(x+h) = 3x + 3h - 1$

31. Domain is all real numbers x where $x \neq 2$.

33. Domain is all real numbers t where $t \neq 6$ or $t \neq -3$.

35. Domain is all real numbers where $x < 5$.

Exercises 4.2

Graphs appear in the text answer section.

3. $y = 2x + 1$

x	y
0	1
1	3
2	5

5. $-2x - 3y = 6$

x	y
0	–2
3	–4
–3	0

7. $y = x^2 - 9$

x	y	x	y
–3	0	1	–8
–2	–5	2	–5
–1	–8	3	0
0	9	4	7

9. $y = x^2 - 5x + 4$

x	y	x	y
–1	10	3	–2
0	4	4	0
1	0	5	4
2	–2	6	10

11. $y = 2x^2 + 3x - 2$

x	y	x	y
–4	18	0	–2
–3	7	1	3
–2	0	2	12
–1	–3	3	25

13. $y = x^2 + 2x$

x	y	x	y
–4	8	0	0
–3	3	1	3
–2	0	2	8
–1	–1	3	15

15. $y = -2x^2 + 4x$

x	y	x	y
–2	–16	2	0
–1	–6	3	–6
0	0	4	–16
1	2	5	–30

17. $y = x^3 - x^2 - 10x + 8$

x	y	x	y
–4	–32	1	–2
–3	2	2	–8
–2	16	3	–4
–1	16	4	16
0	8	5	58

19. $y = x^3 + 2x^2 - 7x + 4$

x	y	x	y
–5	–36	0	4
–4	0	1	0
–3	16	2	6
–2	18	3	28
–1	12	4	72

21. $y = \sqrt{x + 4}$

x	y	x	y
–4	0	5	3
–3	1	12	4
0	2	21	5

23. $y = \sqrt{12 - 6x}$

x	y	x	y
–13	9.5	–4	6
–10	8.4	0	3.5
–7	7.3	2	0

25. $y = x^2 - 9$

(a) To solve graphically, graph $y_1 = x^2 - 9$ and $y_2 = 0$. Find the values of x for which the graphs intersect.

$0 = x^2 - 9$

$9 = x^2$

$x = 3$ and -3

(b) Graph

$y_1 = x^2 - 9$

$y_2 = -5$

Find x where the graphs intersect.

$-5 = x^2 - 9$

$4 = x^2$

$x = 2$ and -2

(c) Graph

$y_1 = x^2 - 9$

$y_2 = 2$

Find x where the graphs intersect.

$2 = x^2 - 9$

$11 = x^2$

$x = \pm\sqrt{11}$

$x = 3.3$ and -3.3

27. $y = x^2 - 5x + 4$

(a) Graph

$y_1 = x^2 - 5x + 4$

$y_2 = 0$

Find the values of x for which the graphs intersect.

$0 = x^2 - 5x + 4$

$0 = (x - 4)(x - 1)$

$x = 4$ and 1

(b) Graph

$y_1 = x^2 - 5x + 4$

$y_2 = 2$

Find the values of x for which the graphs intersect.

$2 = x^2 - 5x + 4$

$0 = x^2 - 5x + 2$

$x = 4.5$ and 0.5

(c) Graph

$y_1 = x^2 - 5x + 4$

$y_2 = -4$

Find the values of x for which the graphs intersect.
The graphs do not intersect.
No solution.

29. $y = 2x^2 + 3x - 2$

(a) Graph

$y_1 = 2x^2 + 3x - 2$

$y_2 = 0$

Find the values of x for which the graphs intersect.

$0 = 2x^2 + 3x - 2$

$0 = (2x - 1)(x + 2)$

$x = \frac{1}{2}$ and -2

(b) Graph

$y_1 = 2x^2 + 3x - 2$

$y_2 = 3$

Find the values of x for which the graphs intersect.

$3 = 2x^2 + 3x - 2$

$0 = 2x^2 + 3x - 5$

$0 = (2x + 5)(x - 1)$

$x = -\frac{5}{2}$ and 1

(c) Graph

$y_1 = 2x^2 + 3x - 2$

$y_2 = 5$

Find the values of x for which the graphs intersect.

$5 = 2x^2 + 3x - 2$

$0 = 2x^2 + 3x - 7$

By quadratic formula or graph

$x = 1.3$ and -2.8

31.(a) $y_1 = x^2 + 2x$

$y_2 = 0$

Find the values of x for which the graphs intersect.

$0 = x^2 + 2x$

$0 = x(x+2)$

$x = 0$ and -2

(b) $y_1 = x^2 + 2x$

$y_2 = 3$

Find the values of x for which the graphs intersect.

$3 = x^2 + 2x$

$0 = x^2 + 2x - 3$

$0 = (x+3)(x-1)$

$x = -3$ and 1

(c) $y_1 = x^2 + 2x$

$y_2 = 6$

Find the values of x for which the graphs intersect.

$6 = x^2 + 2x$

$0 = x^2 + 2x - 6$

By quadratic formula or graph

$x = 1.6$ and -3.6

33.(a) $y_1 = 2x^2 + 4x$

$y_2 = 0$

Find the values of x for which the graphs intersect.

$0 = -2x^2 + 4x$

$0 = -2x(x-2)$

$x = 0$ and 2

(b) $y_1 = -2x^2 + 4x$

$y_2 = 5$

Find the values of x for which the graphs intersect.

These graphs do not intersect.

No solution.

(c) $y_1 = -2x^2 + 4x$

$y_2 = -4$

Find the values of x for which the graphs intersect.

$-4 = -2x^2 + 4x$

$2x^2 - 4x - 4 = 0$

By quadratic formula or graph

$x = 2.7$ and -0.7

(d) $y_1 = -2x^2 + 4x$

$y_2 = -1.5$

Find the values of x for which the graphs intersect.

$-1.5 = -2x^2 + 4x$

$2x^2 - 4x - \frac{3}{2} = 0$

By the quadratic formula or graph

$x = 2.3$ and -0.3

35.(a) $y_1 = x^3 - x^2 - 10x + 8$

$y_2 = 0$

Find the values of x for which the graphs intersect.

$0 = x^3 - x^2 - 10x + 8$

By graph

$x = -3.1$ and 0.8 and 3.3

(b) $y_1 = x^3 - x^2 - 10x + 8$

$y_2 = 2$

Find the values of x for which the graphs intersect.

$2 = x^3 - x^2 - 10x + 8$

$0 = x^3 - x^2 - 10x + 6$

By graph

$x = -3.0$ and 0.6 and 3.4

(c) $y_1 = x^3 - x^2 - 10x + 8$

$y_2 = -2$

Find the values of x for which the graphs intersect.

$-2 = x^3 - x^2 - 10x + 8$

$0 = x^3 - x^2 - 10x + 10$

$0 = x^2(x-1) - 10(x-1)$

$0 = (x-1)(x^2 - 10)$

$x = 1.0$ and $\sqrt{10}$ and $-\sqrt{10}$

$x = 1.0$ and 3.2 and -3.2

37.(a) $y_1 = x^3 + 2x^2 - 7x + 4$

$y_2 = 0$

Find the values of x for which the graphs intersect.

$0 = x^3 + 2x^2 - 7x + 4$

$0 = (x+4)(x-1)^2$

or by graph

$x = -4$ and 1

37.(b) $y_1 = x^3 + 2x^2 - 7x + 4$

$y_2 = 4$

Find the values of x for which the graphs intersect.

$4 = x^3 + 2x^2 - 7x + 4$

$0 = x^3 + 2x^2 - 7x$

$0 = x(x^2 + 2x - 7)$

By the quadratic formula or graph

$x = 0$ and 1.8 and -3.8

(c) $y_1 = x^3 + 2x^2 - 7x + 4$

$y_2 = 8$

Find the values of x for which the graphs intersect.

$8 = x^3 + 2x^2 - 7x + 4$

$0 = x^3 + 2x^2 - 7x - 4$

By graph

$x = -3.6$ and -0.5 and 2.1

39.(a) $y_1 = x^2 + 3x - 4$

$y_2 = 0$

Find the values of x for which the graphs intersect.

$0 = x^2 + 3x - 4$

$0 = (x+4)(x-1)$

$x = -4$ and 1

(b) $y_1 = x^2 + 3x - 4$

$y_2 = 6$

Find the values of x for which the graphs intersect.

$6 = x^2 + 3x - 4$

$0 = x^2 + 3x - 10$

$0 = (x+5)(x-2)$

$x = -5$ and 2

(c) $y_1 = x^2 + 3x - 4$

$y_2 = -2$

Find the values of x for which the graphs intersect.

$-2 = x^2 + 3x - 4$

$0 = x^2 + 3x - 2$

By the quadratic formula or graph

$x = -3.6$ and 0.6

41.(a) $y_1 = -\frac{1}{2}x^2 + 2$

$y_2 = 0$

Find the values of x for which the graphs intersect.

$0 = -\frac{1}{2}x^2 + 2$

$\frac{1}{2}x^2 = 2$

$x^2 = 4$

$x = 2$ and -2

(b) $y_1 = -\frac{1}{2}x^2 + 2$

$y_2 = 4$

Find the values of x for which the graphs intersect.

These graphs do not intersect.

No solution.

(c) $y_1 = -\frac{1}{2}x^2 + 2$

$y_2 = -4$

Find the values of x for which the graphs intersect.

$-4 = -\frac{1}{2}x^2 + 2$

$\frac{1}{2}x^2 = 6$

$x^2 = 12$

$x = \pm\sqrt{12}$

$x = 3.5$ and -3.5

43.(a)
$y_1 = x^3 - 3x^2 + 1$
$y_2 = 0$
Find the values of x where the graphs intersect.
$0 = x^3 - 3x^2 + 1$
By graph
$x = -0.5$ and 0.7 and 2.9

(b)
$y_1 = x^3 - 3x^2 + 1$
$y_2 = -2$
Find the values of x where the graphs intersect.
$-2 = x^3 - 3x^2 + 1$
$0 = x^3 - 3x^2 + 3$
By graph
$x = -0.9$ and 1.3 and 2.5

(c)
$y_1 = x^3 - 3x^2 + 1$
$y_2 = -0.5$
Find the values of x where the graphs intersect.
$-0.5 = x^3 - 3x^2 + 1$
$0 = x^3 - 3x^2 + 1.5$
By graph
$x = 0.6$ and 0.8 and 2.8

45. $r = 10t^2 + 20$

(a)
$r_1 = 10t^2 + 20$
$r_2 = 90\Omega$
Find the values of t for which the graphs intersect.
$90 = 10t^2 + 20$
$70 = 10t^2$
$7 = t^2$
$t = \sqrt{7}$
By graph or calculator
$t = 2.6$ *ms*

(b)
$r_1 = 10t^2 + 20$
$r_2 = 180\Omega$
Find the values of t for which the graphs intersect.
$180 = 10t^2 + 20$
$160 = 10t^2$
$16 = t^2$
$t = 4$ *ms*

(c)
$r_1 = 10t^2 + 20$
$r_2 = 320$
Find the values of t for which the graphs intersect.
$320 = 10t^2 + 20$
$300 = 10t^2$
$30 = t^2$
$t = \sqrt{30}$
By graph or calculator
$t = 5.5$ *ms*

47. $w = 5t^2 + 6t$

(a)
$w_1 = 5t^2 + 6t$
$w_2 = 2$
Find the values of t for which the graphs intersect.
$2 = 5t^2 + 6t$
$0 = 5t^2 + 6t - 2$
By graph or the quadratic formula
$t = 0.27$ *ms*

(b)
$w_1 = 5t^2 + 6t$
$w_2 = 4$
Find the values of t for which the graphs intersect.
$4 = 5t^2 + 6t$
$0 = 5t^2 + 6t - 4$
By graph or the quadratic formula
$t = 0.48$ *ms*

(c)
$w_1 = 5t^2 + 6t$
$w_2 = 10$
Find the values of t for which the graphs intersect.
$10 = 5t^2 + 6t$
$0 = 5t^2 + 6t - 10$
By graph or the quadratic formula
$t = 0.94$ *ms*

49. $i = t^3 - 15$

(a)
$i_1 = t^3 - 15$
$i_2 = 5A$
Find the values of t for which the graphs intersect.
$5 = t^3 - 15$
$20 = t^3$
$t = \sqrt[3]{20}$
By graph or calculator
$t = 2.7\ s$

(b)
$i_1 = t^3 - 15$
$t_2 = 15A$
Find the values of t for which the graphs intersect.
$15 = t^3 - 15$
$30 = t^3$
$t = \sqrt[3]{30}$
By graph or calculator
$t = 3.1\ s$

51. See Figure 4.14.
Let x = horizontal length
$-y$ = vertical length

(a)
$\sin 36° = \frac{x}{2}$
$x = 1.18$
$\cos 36° = \frac{-y}{2}$
$y = -1.62$
$A(1.18, -1.62)$

(b)
$\sin 36° = \frac{x}{4}$
$x = 2.35$
$\cos 36° = \frac{-y}{4}$
$y = -3.24$
$B(2.35, -3.24)$

(c)
$\sin 36° = \frac{x}{6}$
$x = 3.53$
$\cos 36° = \frac{-y}{6}$
$y = -4.85$
$C(3.53, -4.85)$

Exercises 4.3

1.
$m = \frac{y_2 - y_1}{x_2 - x_1}$
$m = \frac{2-1}{4-3}$
$m = \frac{1}{1} = 1$

3.
$m = \frac{y_2 - y_1}{x_2 - x_1}$
$m = \frac{-5-3}{4-2}$
$m = \frac{-8}{2} = -4$

5.
$m = \frac{y_2 - y_1}{x_2 - x_1}$
$m = \frac{2-2}{-3-6}$
$m = \frac{0}{-9} = 0$

7.
$m = \frac{y_2 - y_1}{x_2 - x_1}$
$m = \frac{7-2}{5-(-3)}$
$m = \frac{5}{8}$

9. $(2, -1)$, $m = 2$
Points on the line go up 2 for every 1 (unit) moved to the right, so another point is (3,1).

11. $(-3, -2)$, $m = \frac{1}{2}$
Points on the line go up 1 for every 2 (units) moved to the right, so another point is (-1,-1).

13. $(4,0)$, $m=-2$

Points on the line drop 2 for every 1 (unit) moved to the right, so another point is (5,-2).

15. $(0,-3)$, $m=-\frac{3}{4}$

Points on the line drop 3 for every 4 (unit) moved to the right, so another point is (4,-6).

17. $(-2,8)$, $m=-3$

$y=mx+b$

where $x=-2$

and $y=8$.

$8=-3(-2)+b$

$8=6+b$

$2=b$

$y=-3x+2$

19. $(-3,-4)$, $m=\frac{1}{2}$

$y=mx+b$

where $x=-3$

and $y=-4$.

$-4=\frac{1}{2}(-3)+b$

$-4=\frac{-3}{2}+b$

$-\frac{5}{2}=b$

$y=\frac{1}{2}x-\frac{5}{2}$

$2y=x-5$

$0=x-2y-5$

21. $(-2,7)$ *and* $(1,4)$

$m=\frac{y_2-y_1}{x_2-x_1}=\frac{7-4}{-2-1}$

$m=\frac{3}{-3}=-1$

Use either point for $y=mx+b$.

Use $x=-2$ and $y=7$.

$7=-1(-2)+b$

$7=2+b$

$5=b$

$y=-1x+5$

$x+y-5=0$

23. $(6,-8)$ *and* $(-4,-3)$

$m=\frac{y_2-y_1}{x_2-x_1}$

$m=\frac{-8-(-3)}{6-(-4)}$

$m=\frac{-5}{10}=\frac{-1}{2}$

Use either point for $y=mx+b$.

Use $x=6$ and $y=-8$.

$-8=\frac{-1}{2}(6)+b$

$-8=-3+b$

$-5=b$

$y=-\frac{1}{2}x-5$

$2y=-x-10$

$x+2y+10=0$

25. $m=-5$ and $y-$intercept is -2.

$y=mx+b$

$y=-5x-2$

27. $m=2$ and $y-$intercept is 7.

$y=mx+b$

$y=2x+7$

29. Horizontal line and 5 units above $x-$axis.

$y=b$ becomes

$y=5$

31. Vertical line through $(-2,0)$.

$x=a$ becomes

$x=-2$

33. Horizontal line through $(2,-3)$.

$y=b$ becomes

$y=-3$

35. Vertical line through $(-7,9)$.

$x=a$ becomes

$x=-7$

37.

$$x+4y=12$$
$$4y=-x+12$$
$$y=-\frac{1}{4}x+3$$
$$m=-\frac{1}{4},\ y-\text{intercept}$$
is 3.

39.

$$4x-2y+14=0$$
$$-2y=-4x-14$$
$$y=2x+7$$
$$m=2,\ y-\text{intercept}$$
is 7.

41.

$$y=6$$
$$y=0x+6$$
$$m=0,\ y-\text{intercept}$$
is 6.

43.

$$y=3x-2$$

x	y
−1	−5
0	−2
1	1
2	4

45.

$$5x-2y+4=0$$
$$-2y=-5x-4$$
$$y=\frac{5}{2}x+2$$

x	y
−2	−3
0	2
2	7
4	12

47.

$$x=7$$
Vertical line.

x	y
7	−1
7	0
7	1
7	2

49.

$$y=-3$$
Horizontal line.

x	y
−1	−3
0	−3
1	−3
2	−3

51.

$$6x+8y=24$$
$$8y=-6x+24$$
$$y=-\frac{3}{4}x+3$$

x	y
−4	6
0	3
4	0
8	−3

53.

$$x-3y=-12$$
$$-3y=-x-12$$
$$y=\frac{1}{3}x+4$$

x	y
−6	2
−3	3
0	4
3	5

55.

$(-15.0, 43.0)$ and $(55.0, 43.2)$

$$\text{m}=\frac{y_2-y_1}{x_2-x_1}$$
$$\text{m}=\frac{43.0-43.2}{-15.0-55.0}$$
$$\text{m}=\frac{-0.2}{-70}=\frac{1}{350}$$

Exercises 4.4

1.

$x+3y-7=0$

$3y=-x+7$

$y=-\frac{1}{3}x+\frac{7}{3}$

and

$-3x+y+2=0$

$y=3x-2$

perpendicular, product of slopes is -1.

3.

$-x+4y+7=0$

$4y=x-7$

$y=\frac{1}{4}x-\frac{7}{4}$

and

$x+4y-5=0$

$4y=-x+5$

$y=-\frac{1}{4}x+\frac{5}{4}$

neither

5.

$y-5x+13=0$

$y=5x-13$

and

$y-5x+9=0$

$y=5x-9$

parallel, slopes are equal.

7.

Parallel to

$-2x+y+13=0$

$y=2x-13$

Desired slope is 2.

For $(-1,5)$, $x=-1$ and $y=5$.

$y=mx+b$

$5=2(-1)+b$

$5=-2+b$

$7=b$

$y=2x+7$

$0=2x-y+7$

9.

Perpendicular to $5y=x$.

$y=\frac{1}{5}x$

$y=2x-13$

Desired slope is -5.

For $(-7,4)$, $x=-7$ and $y=4$.

$y=mx+b$

$4=-5(-7)+b$

$4=35+b$

$-31=b$

$y=-5x-31$

$5x+y+31=0$

11.

Parallel to

$3x-4y=12$

$-4y=-3x+12$

$y=\frac{3}{4}x-3$

Desired slope is $\frac{3}{4}$. Origin is $(0,0)$, so $x=0$ and $y=0$

$y=mx+b$

$0=\frac{3}{4}(0)+b$

$0=b$

$y=\frac{3}{4}x$

$4y=3x$

$0=3x-4y$

13.

Perpendicular to

$4x+6y=9$

$6y=-4x+9$

$y=-\frac{2}{3}x+\frac{3}{2}$

Desired slope is $\frac{3}{2}$.

$x-$intercept 6 is the point $(6,0)$, so $x=6$ and $y=0$.

$y=mx+b$

$0=\frac{3}{2}(6)+b$

$0=9+b$

$b=-9$

$y=\frac{3}{2}x-9$

$2y=3x-18$

$18=3x-2y$

15.

Parallel to $y=2$ means a horizontal line.

$y-$intercept 8 is the point $(0,8)$.

$y=b$

$y=8$

17.

Parallel to $x=-4$ means a vertical line.

$x-$intercept 7 is the point $(7,0)$.

$x=a$

$x=7$

19.(a) Find the slopes of the line segments connecting the points to see if opposite sides have the same slopes.

$m=\frac{y_2-y_1}{x_2-x_1}$

For $\overline{AD}$:

$m=\frac{7-3}{5-(-2)}=\frac{4}{7}$

For $\overline{BC}$:

$m=\frac{6-2}{9-2}=\frac{4}{7}$

Thus $\overline{AD}$ and $\overline{BC}$ are parallel.

For $\overline{AB}$:

$m=\frac{3-2}{-2-2}$

$m=\frac{1}{-4}$

For $\overline{DC}$:

$m=\frac{7-6}{5-9}$

$m=\frac{1}{-4}$

Thus $\overline{AB}$ and $\overline{DC}$ are parallel. Yes, it is a parallelogram.

19.(b) Use the slopes already calculated.

For $\overline{AD}$, $m=\frac{4}{7}$.

For $\overline{AB}$, $m=-\frac{1}{4}$

These sides are adjacent, but the product of their slopes is not -1. The figure is not a rectangle.

Exercises 4.5

1. $d=\sqrt{(x_2-x_1)^2+(y_2-y_1)^2}$

$(4,-7)$ and $(-5,5)$

$d=\sqrt{(-5-4)^2+(5-(-7))^2}$

$d=\sqrt{(-9)^2+(12)^2}$

$d=\sqrt{81+144}=\sqrt{225}$

$d=15$

3. $d=\sqrt{(x_2-x_1)^2+(y_2-y_1)^2}$

$(3,-2)$ and $(10,-2)$

$d=\sqrt{(10-3)^2+(-2-(-2))^2}$

$d=\sqrt{7^2+0^2}$

$d=7$

5. $d=\sqrt{(x_2-x_1)^2+(y_2-y_1)^2}$

$(5,-2)$ and $(1,2)$

$d=\sqrt{(5-1)^2+(-2-2)^2}$

$d=\sqrt{4^2+(-4)^2}=\sqrt{32}$

$d=\sqrt{16}\sqrt{2}=4\sqrt{2}$

7. $d=\sqrt{(x_2-x_1)^2+(y_2-y_1)^2}$

$(3,-5)$ and $(3,2)$

$d=\sqrt{(3-3)^2+(-5-2)^2}$

$d=\sqrt{0^2+(-7)^2}$

$d=\sqrt{49}=7$

9. $x_m=\frac{x_1+x_2}{2},\ y_m=\frac{y_1+y_2}{2}$

$(2,3)$ and $(5,7)$

$x_m=\frac{2+5}{2}=\frac{7}{2}=3.5$

$y_m=\frac{3+7}{2}=5$

$(3.5,5)$

11. $x_m=\frac{x_1+x_2}{2},\ y_m=\frac{y_1+y_2}{2}$

$(3,-2)$ and $(0,0)$

$x_m=\frac{3+0}{2}=1.5$

$y_m=\frac{-2+0}{2}=-1$

$(1.5,-1)$

13. $x_m=\frac{x_1+x_2}{2},\ y_m=\frac{y_1+y_2}{2}$

$(11,4)$ and $(-11,-9)$

$x_m=\frac{11+(-11)}{2}=\frac{0}{2}=0$

$y_m=\frac{4+(-9)}{2}=\frac{-5}{2}=-2.5$

$(0,-2.5)$

15. $A(2,8)$, $B(10,2)$, $C(10,8)$

First, we find the lengths of the sides.

$$d=\sqrt{(x_2-x_1)^2+(y_2-y_1)^2}$$
$$AB=\sqrt{(2-10)^2+(8-2)^2}$$
$$AB=\sqrt{(-8)^2+6^2}$$
$$AB=\sqrt{64+36}=\sqrt{100}$$
$$AB=10$$
$$BC=\sqrt{(10-10)^2+(2-8)^2}$$
$$BC=\sqrt{0^2+(-6)^2}=\sqrt{36}$$
$$BC=6$$
$$AC=\sqrt{(2-10)^2+(8-8)^2}$$
$$AC=\sqrt{(-8)^2+0^2}=\sqrt{64}$$
$$AC=8$$

a. $P=10+6+8=24$

b. Yes, it is a right triangle because

$AB^2=BC^2+AC^2$

i.e. $10^2=6^2+8^2$

$100=36+64$

c. No, it is not isosceles; none of the sides are equal in length.

d.
$$A=\frac{1}{2}bh$$
$$A=\frac{1}{2}(6)(8)$$
$$A=24$$

17. $A(-3,6)$, $B(5,0)$, $C(4,9)$

First find the lengths of the sides.

$$d=\sqrt{(x_2-x_1)^2+(y_2-y_1)^2}$$
$$AB=\sqrt{(-3-5)^2(6-0)^2}$$
$$AB=\sqrt{(-8)^2+6^2}$$
$$AB=\sqrt{64+36}=\sqrt{100}$$
$$AB=10$$
$$BC=\sqrt{(5-4)^2+(0-9)^2}$$
$$BC=\sqrt{1^2+(-9)^2}$$
$$BC=\sqrt{82}$$
$$AC=\sqrt{(-3-4)^2+(6-9)^2}$$
$$AC=\sqrt{(-7)^2+(-3)^2}$$
$$AC=\sqrt{49+9}=\sqrt{58}$$

a. $P=10+\sqrt{82}+\sqrt{58}=26.7$

b. No, it is not a right triangle because

$10^2\neq\left(\sqrt{82}\right)^2+\left(\sqrt{58}\right)^2$

$100\neq 82+58$

$100\neq 140$

c. No, it is not isosceles; none of the sides are equal in length.

19. $A(7,-1)$, $B(9,1)$, $C(-3,5)$

midpoint of BC:

$$x_m = \frac{x_1 + x_2}{2},\ y_m = \frac{y_1 + y_2}{2}$$

$$x_m = \frac{9 + (-3)}{2} = \frac{6}{2} = 3$$

$$y_m = \frac{1 + 5}{2} = \frac{6}{2} = 3$$

midpoint of BC is (3,3)

$$d = \sqrt{(x_2 - x_1)^2 + (y_2 - y_1)^2}$$

$$d = \sqrt{(7-3)^2 + (-1-3)^2}$$

$$d = \sqrt{4^2 + (-4)^2}$$

$$d = \sqrt{16 + 16} = \sqrt{32}$$

$$d = \sqrt{16} = 4\sqrt{2}$$

21. Parallel to

$$3x - 6y = 10$$

$$-6y = -3x + 10$$

$$y = \frac{1}{2}x - \frac{5}{3}$$

Desired slope is $\frac{1}{2}$.

midpoint of $A(4,2)$ and $B(8,-6)$:

$$x_m = \frac{x_1 + x_2}{2} = \frac{4 + 8}{2} = \frac{12}{2} = 6$$

$$y_m = \frac{y_1 + y_2}{2} = \frac{2 + (-6)}{2} = \frac{-4}{2} = -2$$

$(6,-2)$.

Use $y = mx + b$; from the midpoint, $x = 6$ and $y = 2$.

$$-2 = \frac{1}{2}(6) + b$$

$$-2 = 3 + b$$

$$-5 = b$$

$$y = \frac{1}{2}x - 5$$

$$2y = x - 10$$

$$10 = x - 2y$$

23. $A(-8,12)$ and $(6,10)$

midpoint of AB:

$$x_m = \frac{x_1 + x_2}{2} = \frac{-8 + 6}{2}$$

$$x_m = -1$$

$$y_m = \frac{y_1 + y_2}{2} = \frac{12 + 10}{2}$$

$$y_m = 11$$

midpoint $AB = (-1,11)$

Perpendicular to

$$4x + 8y = 16$$

$$8y = -4x + 16$$

$$y = -\frac{1}{2}x + 2$$

Desired slope is 2.

$y = mx + b$; from the midpoint, $x = -1$, $y = 11$.

$$11 = 2(-1) + b$$

$$11 = -2 + b$$

$$13 = b$$

$$y = 2x + 13$$

$$2x - y = -13$$

25. $A(-2,2)$, $B(1,3)$, $C(2,0)$, $D(-1,-1)$.

First find the lengths of the sides of $ABCD$.

$$d = \sqrt{(x_2 - x_1)^2 + (y_2 - y_1)^2}$$

$$AB = \sqrt{(-2-1)^2 + (2-3)^2} = \sqrt{(-3)^2 + (-1)^2}$$

$$AB = \sqrt{9 + 1} = \sqrt{10}$$

$$BC = \sqrt{(1-2)^2 + (3-0)^2} = \sqrt{(-1)^2 + (3)^2}$$

$$BC = \sqrt{1 + 9} = \sqrt{10}$$

$$CD = \sqrt{(2-(-1))^2 + (0-(-1))^2}$$

$$CD = \sqrt{(3)^2 + (1)^2}$$

$$CD = \sqrt{9 + 1} = \sqrt{10}$$

$$DA = \sqrt{(-1-(-2))^2 + (-1-2)^2}$$

$$DA = \sqrt{(1)^2 + (-3)^2}$$

$$DA = \sqrt{1 + 9} = \sqrt{10}$$

Slope of $AB = \frac{y_2 - y_1}{x_2 - x_1}$

$$m = \frac{2-3}{-2-1} = \frac{-1}{-3} = \frac{1}{3}$$

Slope of $BC = \frac{3-0}{1-2} = -3$

The sides are equal and adjacent sides are perpendicular, thus $ABCD$ is a square, a special rectangle.

27. $A(-12,8)$, $B(3,2)$, $C(5,7)$, $D(-5,11)$.

Find slopes to show opposite sides are parallel and two adjacent sides are perpendiculuar.

$$m_{AB} = \frac{y_2 - y_1}{x_2 - x_1} = \frac{8-2}{-12-3}$$

$$m_{AB} = \frac{6}{-15} = \frac{-2}{5}$$

$$m_{BC} = \frac{2-7}{3-5} = \frac{-5}{-2}$$

$$m_{BC} = \frac{5}{2}$$

$$m_{CD} = \frac{7-11}{5-(-5)} = \frac{-4}{10}$$

$$m_{CD} = \frac{-2}{5}$$

$$m_{DA} = \frac{11-8}{-5-(-12)} = \frac{3}{7}$$

Since $m_{AB}(m_{BC}) = -1$,
AB is perpendicular to *BC*.
Since $m_{AB} = m_{CD}$,
AB and *CD* are parallel.
Thus, *ABCD* is a trapezoid with one right angle.

Chapter 4 Review

	Function	Domain	Range
1.	Yes	$\{2,3,4,5\}$	$\{3,4,5,6\}$
2.	No	$\{2,4,6\}$	$\{1,3,4,6\}$
3.	Yes	Real Numbers	Real Numbers
4.	Yes	Real Numbers	Real Numbers where $y \geq -5$
5.	$x = y^2 + 4$ $x - 4 = y^2$ $\pm\sqrt{x-4} = y$ No	Real Numbers where $x \geq 4$	Real Numbers
6.	$y = \sqrt{4-8x}$ Yes	Real Numbers where $x \leq \frac{1}{2}$	Real Numbers where $y \geq 0$

7. $f(x) = 5x + 14$

(a) $f(2) = 5(2) + 14 = 24$

(b) $f(0) = 5(0) + 14 = 14$

(c) $f(-4) = 5(-4) + 14 = -6$

8. $g(t) = 3t^2 + 5t - 12$

(a)
$$g(2) = 3(2)^2 + 5(2) - 12$$
$$g(2) = 3(4) + 10 - 12$$
$$g(2) = 12 + 10 - 12$$
$$g(2) = 10$$

(b)
$$g(0) = 3(0)^2 + 5(0) - 12$$
$$g(0) = 0 + 0 - 12$$
$$g(0) = -12$$

(c)
$$g(-5) = 3(-5)^2 + 5(-5) - 12$$
$$g(-5) = 3(25) - 25 - 12$$
$$g(-5) = 75 - 25 - 12$$
$$g(-5) = 38$$

9. $h(x) = \dfrac{4x^2 - 3x}{2\sqrt{x-1}}$

(a)
$$h(2) = \frac{4(2)^2 - 3(2)}{2\sqrt{2-1}}$$
$$h(2) = \frac{4(4) - 6}{2\sqrt{1}}$$
$$h(2) = \frac{16 - 6}{2}$$
$$h(2) = 5$$

(b)
$$h(5) = \frac{4(5)^2 - 3(5)}{2\sqrt{5-1}}$$
$$h(5) = \frac{4(25) - 15}{2\sqrt{4}}$$
$$h(5) = \frac{100 - 15}{2(2)}$$
$$h(5) = \frac{85}{4}$$

(c) $h(-15)$

-15 is not in the domain of $h(x)$.

10. $g(x) = x^2 - 6x + 4$

(a) $g(a) = a^2 - 6a + 4$ **(b)** $g(2x) = (2x)^2 - 6(2x) + 4$ **(c)**

$g(2x) = 4x^2 - 12x + 4$

$g(z-2) = (z-2)^2 - 6(z-2) + 4$

$g(z-2) = z^2 - 4z + 4 - 6z + 12 + 4$

$g(z-2) = z^2 - 10z + 20$

11. $y = 4x - 5$

x	y
−1	−9
0	−5
1	−1
2	3

12. $y = x^2 + 4$

x	y	x	y
−3	13	1	5
−2	8	2	8
−1	5	3	13
0	4	4	20

13. $y = x^2 + 2x - 8$

x	y	x	y
−4	0	0	−8
−3	−5	1	−5
−2	−8	2	0
−1	−9	3	7

14. $y = 2x^2 + x - 6$

x	y	x	y
−3	9	1	−3
−2	0	2	4
−1	−5	3	15
0	−6	4	30

15. $y = -x^2 - x + 4$

x	y	x	y
−4	−8	0	4
−3	−2	1	2
−2	2	2	−2
−1	4	3	−8

16. $y = \sqrt{2x}$

x	y
0	0
1	1.4
2	2
3	2.4

x	y
4	2.8
5	3.2
6	3.5
7	3.7
8	4

17. $y = \sqrt{-2 - 4x}$

x	y	x	y
−1	−1.4	−9	5.8
−3	3.2	−11	6.5
−5	4.2	−13	7.1
−7	5.1	−15	7.6

18. $y = x^3 - 6x$

x	y	x	y
−3	−9	1	−5
−2	4	2	−4
−1	5	3	−9
0	0	4	40

19. $y = x^2 + 4$

(a) $y_1 = x^2 + 4$

$y_2 = 5$

Find values of x for which the graphs intersect.

$5 = x^2 + 4$

$1 = x^2$

$x = 1$ and -1

(b) $y_1 = x^2 + 4$

$y_2 = 7$

Find values of x for which the graphs intersect.

$7 = x^2 + 4$

$3 = x^2$

$x = \pm\sqrt{3}$

$x = 1.7$ and -1.7

(c) $y_1 = x^2 + 4$

$y_2 = 2$

Find values of x for which the graphs intersect.

These graphs do not intersect.

no solution

20. **(a)** $y_1 = x^2 + 2x - 8$
$y_2 = 0$
Find values of x for which the graphs intersect.
$0 = x^2 + 2x - 8$
$0 = (x+4)(x-2)$
$x = -4$ and 2

(b) $y_1 = x^2 + 2x - 8$
$y_2 = -2$
Find values of x for which the graphs intersect.
$-2 = x^2 + 2x - 8$
$0 = x^2 + 2x - 6$
By graph or quadratic formula,
$x = 1.6$ and -3.6

(c) $y_1 = x^2 + 2x - 8$
$y_2 = 3$
Find values of x for which the graphs intersect.
$3 = x^2 + 2x - 8$
$0 = x^2 + 2x - 11$
By graph or quadratic formula,
$x = 2.5$ and -4.5

21. **(a)** $y_1 = -x^2 - x + 4$
$y_2 = 2$
Find values of x for which the graphs intersect.
$2 = -x^2 - x + 4$
$x^2 + x - 2 = 0$
$(x+2)(x-1) = 0$
$x = -2$ and 1

(b) $y_1 = -x^2 - x + 4$
$y_2 = 0$
Find values of x for which the graphs intersect.
$0 = -x^2 - x + 4$
By graph or quadratic formula,
$x = -2.6$ and 1.6

(c) $y_1 = -x^2 - x + 4$
$y_2 = -2$
Find values of x for which the graphs intersect.
$-2 = -x^2 - x + 4$
$x^2 + x - 6 = 0$
$(x+3)(x-2) = 0$
$x = -3$ and 2

22. **(a)** $y_1 = x^3 - 6x$
$y_2 = 0$
Find values of x for which the graphs intersect.
$0 = x^3 - 6x$
$0 = x(x^2 - 6)$
$x = 0$ and $\pm\sqrt{6}$
$x = 0, 2.4$ and -2.4

(b) $y_1 = x^3 - 6x$
$y_2 = 2$
Find values of x for which the graphs intersect.
$2 = x^3 - 6x$
$0 = x^3 - 6x - 2$
By graph,
$x = -2.3, -0.3$, and 2.6

(c) $y_1 = x^3 - 6x$
$y_2 = -3$
Find values of x for which the graphs intersect.
$-3 = x^3 - 6x$
$0 = x^3 - 6x + 3$
By graph
$x = -2.7, 0.5$, and 2.1

23. **(a)** $i_1 = 2t^2$
$i_2 = 2$
Find values of t for which the graphs intersect.
$2 = 2t^2$
$1 = t^2$
$t = 1$

(b) $i_1 = 2t^2$
$i_2 = 6$
Find values of t for which the graphs intersect.
$6 = 2t^2$
$3 = T^2$
$T = 1.7$

(c) $i_1 = 2t^2$
$i_2 = 8$
Find values of t for which the graphs intersect.
$8 = 2t^2$
$4 = t^2$
$t = 2$

24. **(a)** $V_1 = 4t^3 + t$

$V_2 = 40$

Find values of t for which the graphs intersect.

$0 = 4t^3 + t$

$0 = 4t^3 + t - 40$

By graph,

$t = 2.1$

(b) $V_1 = 4t^3 + t$

$V_2 = 60$

Find values of t for which the graphs intersect.

$60 = 4t^3 + t$

$0 = 4t^3 + t - 60$

By graph,

$t = 2.4$

25. $(3,-4)$ and $(-6,-2)$

$$m = \frac{y_2 - y_1}{x_2 - x_1} = \frac{-4-(-2)}{3-(-6)}$$

$$m = \frac{-2}{9}$$

26. $(3,-4)$ and $(-6,-2)$

$$d = \sqrt{(x_2 - x_1)^2 + (y_2 - y_1)^2}$$

$$d = \sqrt{(3-(-6))^2 + (-4-(-2))^2}$$

$$d = \sqrt{9^2 + (-2)^2} = \sqrt{81+4}$$

$$d = \sqrt{85}$$

27. $(3,-4)$ and $(-6,-2)$

$$x_m = \frac{x_1 + x_2}{2}$$

$$x_m = \frac{3+(-6)}{2} = \frac{-3}{2}$$

$$y_m = \frac{y_1 + y_2}{2}$$

$$y_m = \frac{-4+(-2)}{2}$$

$$y_m = -3$$

$(-1.5,-3)$

28. $(4,7)$ and $(6,-4)$

$$m = \frac{y_2 - y_1}{x_2 - x_1}$$

$$m = \frac{7-(-4)}{4-6} = \frac{11}{-2}$$

Use either point;

for $x = 4$, $y = 7$

$y = mx + b$

$$7 = -\frac{11}{2}(4) + b$$

$7 = -22b$

$29 = b$

$$y = -\frac{11}{2}x + 29$$

$2y = -11x + 58$

$11 + 2y - 58 = 0$

29. $m = \frac{2}{3}$ and $(-3,1)$

$y = mx + b$

$x = -3, y = 1$

$$1 = -3\left(\frac{2}{3}\right) + b$$

$1 = -2 + b$

$3 = b$

$$y = \frac{2}{3}x + 3$$

$3y = 2x + 9$

$0 = 2x - 3y + 9$

30. $m=-\frac{1}{3}$ and

$y-$intercept is $(0,-3)$.

$y=mx+b$

$y=-\frac{1}{3}x-3$

$3y=-x-9$

$x+3y+9=0$

31. Parallel to the $y=$ axis is a vertical line, $x=a$. Three units left of the $y-axis$ means $x=-3$

32.

$3x-2y-6=0$

$-2y=-3x+6$

$y=\frac{3}{2}x-3$

$m=\frac{3}{2}$ and

$y-$intercept is -3.

33.

$3x-4y=12$

$-4y=-3x+12$

$y=\frac{3}{4}x-3$

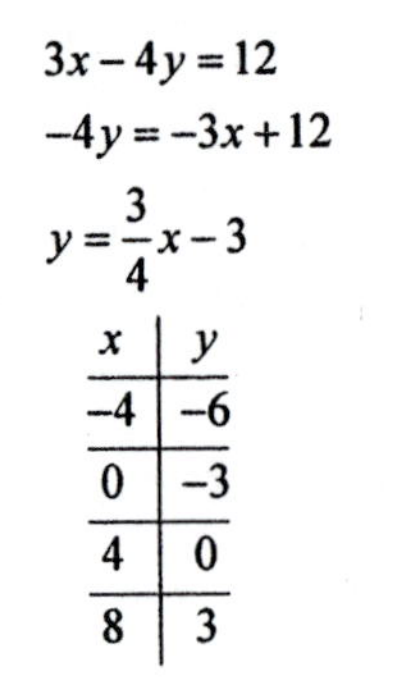

x	y
−4	−6
0	−3
4	0
8	3

34.

$x-2y+3=0$

$-2y=-x-3$

$y=\frac{1}{2}x+\frac{3}{2}$

and $8x+4y-9=0$

$4y=-8x+9$

$y=-2x+\frac{9}{4}$

Lines are perpendicular because $m_1m_2=-1$

35.

$2x-3y+4=0$

$-3y=-2x-4$

$y=\frac{2}{3}x+\frac{4}{3}$

and $-8x+12y=16$

$12y=8x+16$

$y=\frac{2}{3}x+\frac{4}{3}$

Parallel and the same line since the slopes and $y-$intercepts are equal.

36.

$3x-2y+5=0$

$-2y=-3x-5$

$y=\frac{3}{2}x+\frac{5}{2}$

and $2x-3y+9=0$

$-3y=-2x-9$

$y=\frac{2}{3}x+3$

neither

37. $x=2$ and $y=-3$

A vertical line and a horizontal line are perpendicular.

38. $x=4$ and $x=7$

Two vertical lines are parallel.

39.

Parallel to

$2x-y+4=0$

$-y=-2x-4$

$y=2x+4$

Desired slope is 2.

For the point,

$x=5, y=2$

$y=mx+b$

$2=2(5)+b$

$2=10+b$

$-8=b$

$y=2x-8$

$0=2x-y-8$

40.

Perpendicular to

$3x+5y-6=0$

$5y=-3x+6$

$y=-\frac{3}{5}x+\frac{6}{5}$

Desired slope is $\frac{5}{3}$.

For the point,

$x=-4, y=0$

$y=mx+b$

$0=\frac{5}{3}(-4)+b$

$0=\frac{-20}{3}+b$

$\frac{20}{3}=b$

$y=\frac{5}{3}x+\frac{20}{3}$

$3y=5x+20$

$0=5x-3y+20$

Chapter 5, Exercises 5.1

1. $-8x^2\left(5x^3-9x^2+10x\right)=-40x^5+72x^4-80x^3$

3. $6x^3y^5\left(4xy-7x^3y^2+9x^4\right)=24x^4y^6-42x^6y^7+54x^7y^5$

5. $(2x+7)(2x-7)=4x^2-14x+14x-49=4x^2-49$

7. $\left(7x^2+2y\right)^2=\left(7x^2\right)^2+2\left(7x^2\right)(2y)+(2y)^2=49x^4+28x^2y+4y^2$

9. $\left(2a^2-3b\right)^2=\left(2a^2\right)^2-2\left(2a^2\right)(3b)+(3b)^2=4a^4-12a^2b+9b^2$

11. $(x+4)(x+12)=x^2+12x+4x+48=x^2+16x+48$

13. $(2x+5)(-3x-8)=-6x^2-16x-15x-40=-6x^2-31x-40$

15. $(2a-b+3c)^2=(2a)^2+(-b)^2+(3c)^2+2(2a)(-b)+2(2a)(3c)+2(-b)(3c)=4a^2+b^2+9c^2-4ab+12ac-6bc$

17. $(2a+b)^3=(2a)^3+3(2a)^2(b)+3(2a)(b)^2+(b)^3=8a^3+12a^2b+6ab^2+b^3$

19. $\left(5-2x^2\right)^3=(5)^3+3(5)^2\left(-2x^2\right)+3(5)\left(-2x^2\right)^2+\left(-2x^2\right)^3=125-150x^2+60x^4-8x^6$

21. $(6a+5b)(7a-3b)=42a^2-18ab+35ab-15b^2=42a^2+17ab-15b^2$

23. $(4x+7y)^2=(4x)^2+2(4x)(7y)+(7y)^2=16x^2+56xy+49y^2$

25. $(2a+9b)(6a-11b)=12a^2-22ab+54ab-99b^2=12a^2+32ab-99b^2$

27. $\left(3a^3+10b^2\right)^3=\left(3a^3\right)^3+3\left(3a^3\right)^2\left(10b^2\right)+3\left(3a^3\right)\left(10b^2\right)^2+\left(10b^2\right)^3=27a^9+270a^6b^2+900a^3b^4+1000b^6$

29. $-4a^2b^3\left(6a^3b^4-9a^5b^7+20a^5\right)=-24a^5b^7+36a^7b^{10}-80a^7b^3$

31. $\left(9a^3-12b^2\right)^2=\left(9a^3\right)^2-2\left(9a^3\right)\left(12b^2\right)+\left(-12b^2\right)^2=81a^6-216a^3b^2+144b^4$

33. $(x-3y-5z)^2=x^2+(-3y)^2+(-5z)^2+2x(-3y)+2x(-5z)+2(-3y)(-5z)=x^2+9y^2+25z^2-6xy-10xz+30yz$

35. $(9a-4b)(6a-10b)=54a^2-90ab-24ab+40b^2=54a^2-114ab+40b^2$

37. $\left(6x^3+4\right)\left(5x^3-4\right)=30x^6-24x^3+20x^3-16=30x^6-4x^3-16$

39. $(3xy+9yz)(5xy-8yz)=15x^2y^2-24xy^2z+45xy^2z-72y^2z^2=15x^2y^2+21xy^2z-72y^2z^2$

41. $(5abc+6)(5abc-6)=25a^2b^2c^2-36$

43. $\left(3x^2-\frac{1}{2}y\right)\left(x^2+\frac{2}{3}y\right)=3x^4+2x^2y-\frac{1}{2}x^2y-\frac{1}{3}y^2=3x^4+\frac{3}{2}x^2y-\frac{1}{3}y^2$

45. $\left(x-\frac{2}{3}y\right)^2=x^2-2(x)\left(\frac{2}{3}y\right)+\left(\frac{2}{3}y\right)^2=x^2-\frac{4}{3}xy+\frac{4}{9}y^2$

47. $x^2\left(1+x^2\right)^2=x^2\left(1+2(1)\left(x^2\right)+\left(x^2\right)^2\right)=x^2\left(1+2x^2+x^4\right)=x^2+2x^4+x^6$

Exercises 5.2

1. $6x+9y=3(2x+3y)$ **3.** $10x+25y-45z=5(2x+5y-9z)$ **5.** $12x^2-30xy+6xz=6x(2x-5y+z)$

7. $6x^4-12x^2+3x=3x\left(2x^3-4x+1\right)$ **9.** $27x^3+54x=27x\left(x^2+2\right)$

11. $8x^3y^2-6xy^3+12xy^2z=2xy^2\left(4x^2-3y+6z\right)$ **13.** $x^2-16=(x+4)(x-4)$

15. $9x^2-25y^2=(3x+5y)(3x-5y)$ **17.** $2e^2-72=2\left(e^2-36\right)=2(e+6)(e-6)$

19. $16d^2-100=4\left(4d^2-25\right)=4(2d+5)(2d-5)$ **21.** $4R^2-4r^2=4\left(R^2-r^2\right)=4(R+r)(R-r)$

23. $x^2+6x+8=(x+4)(x+2)$ **25.** $b^2+11b+24=(b+3)(b+8)$ **27.** $x^2-9x+18=(x-3)(x-6)$

29. $a^2-18a+32=(a-16)(a-2)$ **31.** $x^2-2x-35=(x-7)(x+5)$ **33.** $a^2+3a-4=(a+4)(a-1)$

35. $2x^2-x-6=(2x+3)(x-2)$ **37.** $15x^2-31x+14=(3x-2)(5x-7)$ **39.** $45y^2+59y+6=(5y+6)(9y+1)$

41. $35a^2+2a-1=(5a+1)(7a-1)$ **43.** $9c^2-24c+16=(3c-4)^2$

45. $25b^2+60b+20=5\left(5b^2+12b+4\right)=5(b+2)(5b+2)$ **47.** $35t^2-4ts-15s^2=(5t+3s)(7t-5s)$

49. $30k^2-95kt+50t^2=5\left(6k^2-19kt+10t^2\right)=5(2k-5t)(3k-2t)$

51. $54x^2+27x-42=3\left(18x^2+9x-14\right)=3(3x-2)(6x+7)$

53. $4a^2+20a+25=(2a+5)^2$ **55.** $9x^2-48xy^2+64y^4=\left(3x-8y^2\right)^2$

57. $25x^2+10xy^3+y^6=\left(5x+y^3\right)^2$ **59.** $x^4-81=\left(x^2+9\right)\left(x^2-9\right)=\left(x^2+9\right)(x+3)(x-3)$

61. $a^4-3a^2-40=\left(a^2-8\right)\left(a^2+5\right)$ **63.** $t^4-13t^2+36=\left(t^2-9\right)\left(t^2-4\right)=(t+3)(t-3)(t+2)(t-2)$

Exercises 5.3

1. $a(m+n)-b(m+n)=(m+n)(a-b)$

3. $mx+my+nx+ny=m(x+y)+n(x+y)=(x+y)(m+n)$

5. $3x+y-6x^2-2xy=1(3x+y)-2x(3x+y)=(3x+y)(1-2x)$

7. $6x^3-4x^2+3x-2=2x^2(3x-2)+1(3x-2)=(3x-2)(2x^2+1)$

9. $(x+y)^2-4z^2=(x+y+2z)(x+y-2z)$

11. $100-49(x-y)^2=(10+7(x-y))(10-7(x-y))=(10+7x-7y)(10-7x+7y)$

13. $x^2-6x+9-4y^2=(x^2-6x+9)-4y^2=(x-3)^2-4y^2=(x-3+2y)(x-3-2y)$

15. $4x^2-4y^2+4y-1=4x^2-(4y^2-4y+1)=4x^2-(2y-1)^2=(2x+(2y-1))(2x-(2y-1))=(2x+2y-1)(2x-2y+1)$

17. $(x+y)^2+13(x+y)+36$ Let $m=x+y$. The problem becomes $m^2+13m+36=(m+9)(m+4)$.
Since $m=x+y$, the factors are $(x+y+9)(x+y+4)$.

19. $24(a+b)^2-14(a+b)-5$. Let $m=a+b$. The problem becomes $24m^2-14m-5=(4m+1)(6m-5)$.
Since $m=a+b$, the factors are $(4(a+b)+1)(6(a+b)-5)=(4a+4b+1)(6a+6b-5)$.

21. $a^3+b^3=(a+b)(a^2-ab+b^2)$

23. $x^3-64=(x-4)(x^2+4x+16)$

25. $a^3b^3+c^3=(ab+c)(a^2b^2+abc+c^2)$

27. $a^3-27b^3=(a-3b)(a^2+3ab+9b^2)$

29. $27a^3+64b^3=(3a+4b)(9a^2-12ab+16b^2)$

31. $27a^6-8b^9=(3a^2)^3-(2b^3)^3=(3a^2-2b^3)(9a^4+6a^2b^3+4b^6)$

33. $(x^6-y^6)=(x^3+y^3)(x^3-y^3)=(x+y)(x^2-xy+y^2)(x-y)(x^2+xy+y^2)$

Exercises 5.4

1. $\dfrac{3x}{18x}=\dfrac{1}{6}$

3. $\dfrac{30x^2y^4}{54x^2y^5}=\dfrac{5}{9y}$

5. $\dfrac{8(x+4)^3}{2(x+4)^5}=\dfrac{4}{(x+4)^2}$

7. $\dfrac{18(x-3)^3(1-5x)^2}{9(1-5x)^2(x-3)}=2(x-3)^2$

9. $\dfrac{5m+15}{6m+18}=\dfrac{5(m+3)}{6(m+3)}=\dfrac{5}{6}$

11. $\dfrac{6m+6}{m^2-1}=\dfrac{6(m+1)}{(m-1)(m+1)}=\dfrac{6}{m-1}$

13. $\dfrac{x^2+x}{x+1}=\dfrac{x(x+1)}{(x+1)}=x$

15. $\dfrac{x^2+5x+4}{x^2-6x-7}=\dfrac{(x+4)(x+1)}{(x-7)(x+1)}=\dfrac{x+4}{x-7}$

17. $\dfrac{6x^2+x-12}{6x^2+19x-36}=\dfrac{(2x+3)(3x-4)}{(2x+9)(3x-4)}=\dfrac{2x+3}{2x+9}$

19. $\dfrac{18-12x}{6x-9}=\dfrac{-6(-3+2x)}{3(2x-3)}=\dfrac{-6(2x-3)}{3(2x-3)}=-2$

21. $\dfrac{x^2-1}{1-x}=\dfrac{(x+1)(x-1)}{-1(-1+x)}=\dfrac{(x+1)(x-1)}{-1(x-1)}=\dfrac{x+1}{-1}=-x-1$

23. $\dfrac{a^2-4}{a^2+4a+4}=\dfrac{(a+2)(a-2)}{(a+2)(a+2)}=\dfrac{a-2}{a+2}$

25. $\dfrac{m^2-16m}{16-m}=\dfrac{m(m-16)}{-1(-16+m)}=\dfrac{m(m-16)}{-1(m-16)}=\dfrac{m}{-1}=-m$

27. $\dfrac{t^2-t^3}{t^2-1}=\dfrac{t^2(1-t)}{(t+1)(t-1)}=\dfrac{t^2(-1)(-1+t)}{(t+1)(t-1)}=\dfrac{-t^2(t-1)}{(t+1)(t-1)}=\dfrac{-t^2}{t+1}$

29. $\dfrac{x^2+xy}{x^2+xy-2x-2y}=\dfrac{x(x+y)}{x(x+y)-2(x+y)}=\dfrac{x(x+y)}{(x+y)(x-2)}=\dfrac{x}{x-2}$

31. $\dfrac{y^3-8}{y^2+2y+4}=\dfrac{(y-2)(y^2+2y+4)}{(y^2+2y+4)}=y-2$

33. $\dfrac{xy-3x-2y+6}{y^3-27}=\dfrac{x(y-3)-2(y-3)}{(y-3)(y^2+3y+4)}=\dfrac{(y-3)(x-2)}{(y-3)(y^2+3y+9)}=\dfrac{x-2}{y^2+3y+9}$

35. $\dfrac{x^2+4x+16}{x^3-64}=\dfrac{1(x^2+4x+16)}{(x-4)(x^2+4x+16)}=\dfrac{1}{x-4}$

37. $\dfrac{3-4x-4x^2}{4x^2-8x+3}=\dfrac{(1-2x)(3+2x)}{(2x-3)(2x-1)}=\dfrac{-1(-1+2x)(2x+3)}{(2x-3)(2x-1)}=\dfrac{-1(2x-1)(2x+3)}{(2x-3)(2x-1)}=\dfrac{-(2x+3)}{2x-3}=\dfrac{-2x-3}{2x-3}$

39. $\dfrac{3a^3+3b^3}{3a^2+6ab+3b^2}=\dfrac{3(a^3+b^3)}{3(a^2+2ab+b^2)}=\dfrac{3(a+b)(a^2-ab+b^2)}{3(a+b)(a+b)}=\dfrac{a^2-ab+b^2}{a+b}$

Exercises 5.5

1. $\dfrac{9}{16}\cdot\dfrac{4}{3}=\dfrac{3}{4}$

3. $\dfrac{15}{32}\div\dfrac{3}{16}=\dfrac{15}{32}\cdot\dfrac{16}{3}=\dfrac{5}{2}$

5. $\dfrac{x^4}{3}\cdot\dfrac{6x}{x^5}=2$

7. $\dfrac{6a^4}{a^2}\div\dfrac{18a^2}{a^5}=\dfrac{6a^4}{a^2}\cdot\dfrac{a^5}{18a^2}=\dfrac{a^5}{3}$

9. $\dfrac{3ab}{4x^2y}\cdot\dfrac{6x^2}{5ab^3}=\dfrac{9}{10yb^2}$

11. $\dfrac{4ax^2}{9b^2y}\div\dfrac{5(ax)^2}{18(by^2)^2}=\dfrac{4ax^2}{9b^2y}\cdot\dfrac{18b^2y^4}{5a^2x^2}=\dfrac{8y^3}{5a}$

13. $\dfrac{(2a^2b^3)^2}{(4ab^2)^3}\cdot\dfrac{(8a^2b)^2}{12a^3b}=\dfrac{4a^4b^6}{64a^3b^6}\cdot\dfrac{64a^4b^2}{12a^3b}=\dfrac{a^2b}{3}$

15. $\dfrac{4x+8}{8xy}\cdot\dfrac{16x}{32x+64}=\dfrac{4(x+2)}{8xy}\cdot\dfrac{16x}{32(x+2)}=\dfrac{1}{4y}$

17. $\dfrac{5k+5}{12}\div\dfrac{9k+9}{4}=\dfrac{5(k+1)}{12}\cdot\dfrac{4}{9(k+1)}=\dfrac{5}{27}$

19. $\dfrac{(y+2)^2}{y}\cdot\dfrac{y^2}{y^2-4}=\dfrac{(y+2)^2}{y}\cdot\dfrac{y^2}{(y+2)(y-2)}=\dfrac{y(y+2)}{y-2}$

21. $(m^2-1)\cdot\dfrac{1+m}{1-m}=\dfrac{(m+1)(m-1)}{1}\cdot\dfrac{1+m}{-1(-1+m)}=\dfrac{(m+1)(m-1)}{1}\cdot\dfrac{(m+1)}{-1(m-1)}=\dfrac{(m+1)^2}{-1}=-(m+1)^2$

23. $\dfrac{x^2-64}{x+2}\div\dfrac{8-x}{x}=\dfrac{(x+8)(x-8)}{x+2}\cdot\dfrac{x}{-1(-8+x)}=\dfrac{(x+8)(x-8)}{x+2}\cdot\dfrac{x}{-1(x-8)}=\dfrac{x(x+8)}{-1(x+2)}=\dfrac{-x(x+8)}{x+2}$

25. $\dfrac{x-2}{x-y}\cdot\dfrac{x^2-y^2}{x^2-4x+4}=\dfrac{(x-2)}{(x-y)}\cdot\dfrac{(x+y)(x-y)}{(x-2)(x-2)}=\dfrac{x+y}{x-2}$

27. $\dfrac{x}{(x-3)^2}\cdot\dfrac{(3-x)(3+x)}{x^4}=\dfrac{x}{(x-3)(x-3)}\cdot\dfrac{-1(-3+x)(x+3)}{x^4}=\dfrac{-(x+3)}{x^3(x-3)}=\dfrac{-x-3}{x^3(x-3)}$

29. $\frac{x^2+6x+9}{x^2-9}\cdot\frac{x^2-6x+9}{x-3}=\frac{(x+3)(x+3)}{(x+3)(x-3)}\cdot\frac{(x-3)(x-3)}{(x-3)}=x+3$

31. $\frac{x^2-x-12}{x^2+2x+1}\cdot\frac{x^2-4x-5}{x^2-9x+20}=\frac{(x-4)(x+3)}{(x+1)(x+1)}\cdot\frac{(x-5)(x+1)}{(x-5)(x-4)}=\frac{x+3}{x+1}$

33. $\frac{9x^2-4}{6x^2-5x-6}\div\frac{9x^2-12x+4}{6x^2-13x+6}=\frac{(3x+2)(3x-2)}{(2x-3)(3x+2)}\cdot\frac{(2x-3)(3x-2)}{(3x-2)(3x-2)}=1$

35. $\frac{15a^2+7ab-2b^2}{6a^2-11ab-10b^2}\cdot\frac{2a^2-13ab+20b^2}{25a^2-b^2}=\frac{(3a+2b)(5a-b)}{(2a-5b)(3a+2b)}\cdot\frac{(a-4b)(2a-5b)}{(5a-b)(5a+b)}=\frac{a-4b}{5a+b}$

37. $\frac{x^2-y^2}{x^2+2xy+y^2}\div\frac{x^2-2xy+y^2}{xy-y^2}=\frac{(x-y)(x+y)}{(x+y)(x+y)}\cdot\frac{y(x-y)}{(x-y)(x-y)}=\frac{y}{x+y}$

39. $\frac{x^2-16}{x^2-9}\div\frac{3x^2+13x+4}{2x^2+x-21}=\frac{(x+4)(x-4)}{(x+3)(x-3)}\cdot\frac{(2x+7)(x-3)}{(3x+1)(x+4)}=\frac{(x-4)(2x+7)}{(x+3)(3x+1)}$

41. $\frac{x^3-8}{x^2-2x+4}\cdot\frac{x^3+8}{x^2+2x+4}=\frac{(x-2)(x^2+2x+4)}{x^2-2x+4}\cdot\frac{(x+2)(x^2-2x+4)}{x^2+2x+4}=(x-2)(x+2)$

43. $\frac{xy+y-x^2-x}{12xy}\div\frac{x^2+x}{4x^2}=\frac{y(x+1)-x(x+1)}{12xy}\cdot\frac{4x^2}{x(x+1)}=\frac{(x+1)(y-x)}{12xy}\cdot\frac{4x^2}{x(x+1)}=\frac{y-x}{3y}$

45. $\frac{x^3+27}{x^2-9}\cdot\frac{5x^2+x-18}{5x^2-9x}\div\frac{x^2-3x+9}{5x^2-6x-27}=\frac{(x+3)(x^2-3x+9)}{(x+3)(x-3)}\cdot\frac{(5x-9)(x+2)}{x(5x-9)}\cdot\frac{(5x+9)(x-3)}{x^2-3x+9}=\frac{(5x+9)(x+2)}{x}$

47. $\frac{18x^2+9x-20}{6x^2-5x-4}\div\left(\frac{2x^2-9x-5}{9x^2-16}\cdot\frac{12x^2-10x}{4x^2+4x+1}\right)=\frac{(3x+4)(6x-5)}{(2x+1)(3x-4)}\div\left(\frac{(2x+1)(x-5)}{(3x-4)(3x+4)}\cdot\frac{2x(6x-5)}{(2x+1)(2x+1)}\right)=$

$\frac{(3x+4)(6x-5)}{(2x+1)(3x-4)}\div\left(\frac{2x(6x-5)(x-5)}{(3x-4)(3x+4)(2x+1)}\right)=\frac{(3x+4)(6x-5)}{(2x+1)(3x-4)}\cdot\frac{(3x-4)(3x+4)(2x+1)}{2x(6x-5)(x-5)}=\frac{(3x+4)^2}{2x(x-5)}$

Exercises 5.6

1. $6x, 8$ L.C.D. $= 24x$

3. $2k, 4k^2$ L.C.D. $= 4k^2$

5. $6ab^2, 5a^2b$ L.C.D $= 30a^2b^2$

7. $5x, x-1$ L.C.D $= 5x(x-1)$

9. $3x+6, 6x+12$ becomes $3(x+2), 6(x+2)$ L.C.D $= 6(x+2)$

11. $x^2-25, (x-5)^2$ becomes $(x+5)(x-5), (x-5)^2$ L.C.D $= (x+5)(x-5)^2$

13. x^2+6x+8, x^2-x-6 becomes $(x+4)(x+2), (x-3)(x+2)$ L.C.D $= (x+4)(x+2)(x-3)$

15. $3x^3-12x^2, 6x^2+12x, 2x^3-4x^2-16x$ becomes $3x^2(x-4), 6x(x+2), 2x(x-4)(x+2)$ L.C.D $= 6x^2(x-4)(x+2)$

17. $6(c+2)(c-4)^2, 12c(c+2)(c-4), 3c(c-4)^2$ L.C.D $= 12c(c+2)(c-4)^2$

19. $\frac{3}{14}+\frac{5}{14}=\frac{3+5}{14}=\frac{8}{14}=\frac{4}{7}$

21. $\frac{16}{a}-\frac{9}{a}=\frac{16-9}{a}=\frac{7}{a}$

23. $\frac{6}{x+1}+\frac{9}{x+1}=\frac{6+9}{x+1}=\frac{15}{x+1}$

25. L.C.D $= 180$ $\frac{7}{36}+\frac{7}{45}=\frac{7\cdot 5}{36\cdot 5}+\frac{7\cdot 4}{45\cdot 4}=\frac{35+28}{180}=\frac{63}{180}=\frac{7}{20}$

27. L.C.D $= 2^2\cdot 3\cdot 5 = 60$ $\frac{2x-1}{12}+\frac{3x+5}{20}-\frac{x}{4}=\frac{(2x-1)\cdot 5}{12\cdot 5}+\frac{(3x+5)\cdot 3}{20\cdot 3}-\frac{x\cdot 15}{4\cdot 15}=$

$$\frac{5(2x-1)+3(3x+5)-15x}{60}=\frac{10x-5+9x+15-15x}{60}=\frac{4x+10}{60}=\frac{2(2x+5)}{60}=\frac{2x+5}{30}$$

29. L.C.D $= 6y$

$\frac{5}{6y}-\frac{7}{2y}=$

$\frac{5}{6y}-\frac{7\cdot 3}{2y\cdot 3}=$

$\frac{5-21}{6y}=\frac{-16}{6y}=$

$-\frac{8}{3y}$

31. L.C.D $= 3p^2$

$\frac{4}{3p}+\frac{2}{p^2}=$

$\frac{4p}{3p^2}+\frac{2\cdot 3}{3p^2}=$

$\frac{4p+6}{3p^2}$

33. L.C.D $= 12x^3$

$\frac{5}{4x^2}+\frac{1}{3x^3}-\frac{3}{2x}=$

$\frac{5\cdot 3x}{4x^2\cdot 3x}+\frac{1\cdot 4}{3x^3\cdot 4}-\frac{3\cdot 6x^2}{2x\cdot 6x^2}=$

$\frac{15x+4-18x^2}{12x^3}$

35.

$$\text{L.C.D} = abc$$

$$\frac{1}{a}+\frac{1}{b}+\frac{1}{c}=$$

$$\frac{1\cdot bc}{a\cdot bc}+\frac{1\cdot ac}{b\cdot ac}+\frac{1\cdot ab}{c\cdot ab}=$$

$$\frac{bc+ac+ab}{abc}$$

37.

$$\text{L.C.D} = 2x$$

$$3x+\frac{4-x}{2x}=$$

$$\frac{3x(2x)}{2x}+\frac{4-x}{2x}=$$

$$\frac{6x^2+4-x}{2x}=$$

$$\frac{6x^2-x+4}{2x}$$

39.

$$\text{L.C.D} = c(c-1)$$

$$\frac{2}{c-1}+\frac{1}{c}=$$

$$\frac{2(c)}{(c-1)(c)}+\frac{1(c-1)}{c(c-1)}=$$

$$\frac{2c+c-1}{c(c-1)}=$$

$$\frac{3c-1}{c(c-1)}$$

41.

$$\text{L.C.D} = (t+2)(t-2)$$

$$\frac{6}{t+2}-\frac{3}{t-2}=$$

$$\frac{6(t-2)}{(t+2)(t-2)}-\frac{3(t+2)}{(t-2)(t+2)}=$$

$$\frac{6t-12-3t-6}{(t+2)(t-2)}=$$

$$\frac{3t-18}{(t+2)(t-2)}$$

43.

$$\text{L.C.D} = a(a+1)$$

$$\frac{2}{a}+\frac{1}{a+1}=$$

$$\frac{2(a+1)}{a(a+1)}+\frac{1(a)}{(a+1)(a)}=$$

$$\frac{2a+2+a}{a(a+1)}=$$

$$\frac{3a+2}{a(a+1)}$$

45.

$$\text{L.C.D} = (d+4)(d-5)$$

$$\frac{d}{d+4}-\frac{2d}{d-5}=$$

$$\frac{d(d-5)}{(d+4)(d-5)}-\frac{2d(d+4)}{(d-5)(d+4)}=$$

$$\frac{d^2-5d-2d^2-8d}{(d+4)(d-5)}=$$

$$\frac{-d^2-13d}{(d+4)(d-5)}$$

47.

$$\text{L.C.D} = x-4$$

$$\frac{8}{x-4}+\frac{2}{4-x}=$$

$$\frac{8}{x-4}+\frac{2}{-1(-4+x)}=$$

$$\frac{8}{x-4}+\frac{-2}{x-4}=$$

$$\frac{8-2}{x-4}=\frac{6}{x-4}$$

49.

$$\text{L.C.D} = 12(a+3)$$

$$\frac{4}{3a+9}-\frac{6}{4a+12}=$$

$$\frac{4}{3(a+3)}-\frac{6}{4(a+3)}=$$

$$\frac{4(4)}{3(a+3)(4)}-\frac{6(3)}{4(a+3)(3)}=$$

$$\frac{16-18}{12(a+3)}=\frac{-2}{12(a+3)}=$$

$$\frac{-1}{6(a+3)}$$

51.

$$\text{L.C.D} = (a+1)(a-1)$$

$$\frac{5a+1}{a^2-1}-\frac{2a-3}{a+1}=$$

$$\frac{5a+1}{(a+1)(a-1)}-\frac{(2a-3)(a-1)}{(a+1)(a-1)}=$$

$$\frac{5a+1-(2a^2-5a+3)}{(a+1)(a-1)}=$$

$$\frac{5a+1-2a^2+5a-3}{(a+1)(a-1)}=$$

$$\frac{-2a^2+10a-2}{(a+1)(a-1)}$$

53.

$$\text{L.C.D} = x-3$$

$$\frac{6}{x-3}-\frac{5}{3-x}=$$

$$\frac{6}{x-3}-\frac{5}{-1(-3+x)}=$$

$$\frac{6}{x-3}-\frac{-5}{(x-3)}=$$

$$\frac{6-(-5)}{x-3}=\frac{11}{x-3}$$

55.

$$\text{L.C.D} = (r-2)^2(r+3)$$

$$\frac{2}{r^2-4r+4}+\frac{1}{r^2+r-6}=$$

$$\frac{2}{(r-2)^2}+\frac{1}{(r+3)(r-2)}=$$

$$\frac{2(r+3)}{(r-2)^2(r+3)}+\frac{1(r-2)}{(r+3)(r-2)(r-2)}=$$

$$\frac{2r+6+r-2}{(r-2)^2(r+3)}=\frac{3r+4}{(r-2)^2(r+3)}$$

57.

$$\text{L.C.D} = (x-2)(x-3)$$

$$\frac{1}{x-2} + \frac{1}{x^2-5x+6} =$$

$$\frac{1(x-3)}{(x-2)(x-3)} + \frac{1}{(x-2)(x-3)} =$$

$$\frac{x-3+1}{(x-2)(x-3)} =$$

$$\frac{x-2}{(x-2)(x-3)} =$$

$$\frac{1}{x-3}$$

59.

$$\text{L.C.D} = 4(t-2)(t+1)$$

$$\frac{t+4}{2t-4} - \frac{2t+5}{t^2-t-2} + \frac{3}{4} =$$

$$\frac{t+4}{2(t-2)} - \frac{2t+5}{(t-2)(t+1)} + \frac{3}{4} =$$

$$\frac{(t+4)(2)(t+1)}{2(t-2)(2)(t+1)} - \frac{(2t+5)(4)}{(t-2)(t+1)(4)} + \frac{3(t-2)(t+1)}{4(t-2)(t+1)} =$$

$$\frac{2(t^2+5t+4)}{4(t-2)(t+1)} - \frac{8t+20}{4(t-2)(t+1)} + \frac{3(t^2-t-2)}{4(t-2)(t+1)} =$$

$$\frac{2t^2+10t+8-8t-20+3t^2-3t-6}{4(t-2)(t+1)} =$$

$$\frac{5t^2-t-18}{4(t-2)(t+1)} = \frac{(t-2)(5t+9)}{4(t-2)(t+1)} =$$

$$\frac{5t+9}{4(t+1)}$$

61.

$$\text{L.C.D} = (x-4)(x+1)(x+3)$$

$$\frac{3x-2}{x^2-3x-4} + \frac{2x-3}{x^2-x-12} =$$

$$\frac{3x-2}{(x-4)(x+1)} + \frac{2x-3}{(x-4)(x+3)} =$$

$$\frac{(3x-2)(x+3)}{(x-4)(x+1)(x+3)} + \frac{(2x-3)(x+1)}{(x-4)(x+3)(x+1)} =$$

$$\frac{3x^2+7x-6+2x^2-x-3}{(x-4)(x+1)(x+3)} =$$

$$\frac{5x^2+6x-9}{(x-4)(x+1)(x+3)}$$

63.

$$\text{L.C.D} = (a+b)(a-b)$$

$$\frac{a^2+b^2}{a^2-b^2} - \frac{a}{a+b} + \frac{b}{b-a} =$$

$$\frac{a^2+b^2}{(a+b)(a-b)} - \frac{a}{a+b} + \frac{b}{-1(-b+a)} =$$

$$\frac{a^2+b^2}{(a+b)(a-b)} - \frac{a}{a+b} + \frac{-b}{a-b} =$$

$$\frac{a^2+b^2}{(a+b)(a-b)} - \frac{a(a-b)}{(a+b)(a-b)} + \frac{-b(a+b)}{(a-b)(a+b)} =$$

$$\frac{a^2+b^2-a^2+ab-ab-b^2}{(a+b)(a-b)} =$$

$$\frac{0}{(a+b)(a-b)} = 0$$

65.

$$\text{L.C.D} = (x+4)(x-4)(x-3)$$

$$\frac{x+3}{x^2-7x+12} + \frac{x+3}{16-x^2} =$$

$$\frac{x+3}{(x-3)(x-4)} + \frac{x+3}{-(x^2-16)} =$$

$$\frac{x+3}{(x-3)(x-4)} + \frac{-1(x+3)}{(x-4)(x+4)} =$$

$$\frac{(x+3)(x+4)}{(x-3)(x-4)(x+4)} + \frac{-1(x+3)(x-3)}{(x-4)(x+4)(x-3)} =$$

$$\frac{x^2+7x+12}{(x-3)(x-4)(x+4)} + \frac{-1(x^2-9)}{(x-4)(x+4)(x-3)} =$$

$$\frac{x^2+7x+12-x^2+9}{(x-3)(x-4)(x+4)} =$$

$$\frac{7x+21}{(x-3)(x-4)(x+4)}$$

67.

$$\text{L.C.D} = 2x(4x+1)$$

$$\frac{5x}{8x+2} - \frac{3}{4x^2+x} + \frac{1}{2x} =$$

$$\frac{5x}{2(4x+1)} - \frac{3}{x(4x+1)} + \frac{1}{2x} =$$

$$\frac{5x(x)}{2(4x+1)(x)} - \frac{3\cdot 2}{x(4x+1)(2)} + \frac{1(4x+1)}{2x(4x+1)} =$$

$$\frac{5x^2-6+4x+1}{2x(4x+1)} =$$

$$\frac{5x^2+4x-5}{2x(4x+1)}$$

Section 5.6

69. L.C.D $=(t-2)(t^2+2t+4)$

$$\frac{1}{t-2}-\frac{6t}{t^3-8}=$$

$$\frac{1}{t-2}-\frac{6t}{(t-2)(t^2+2t+4)}=$$

$$\frac{1(t^2+2t+4)}{(t-2)(t^2+2t+4)}-\frac{6t}{(t-2)(t^2+2t+4)}=$$

$$\frac{t^2+2t+4-6t}{(t-2)(t^2+2t+4)}=$$

$$\frac{t^2-4t+4}{(t-2)(t^2+2t+4)}=$$

$$\frac{(t-2)(t-2)}{(t-2)(t^2+2t+4)}=$$

$$\frac{t-2}{t^2+2t+4}$$

71. L.C.D $=x^2-x+1$

$$\frac{x^2+x}{x^3+1}+1=$$

$$\frac{x(x+1)}{(x+1)(x^2-x+1)}+1=$$

$$\frac{x}{x^2-x+1}+\frac{1(x^2-x+1)}{(x^2-x+1)}=$$

$$\frac{x+x^2-x+1}{x^2-x+1}=$$

$$\frac{x^2+1}{x^2-x+1}$$

Exercises 5.7

1.
$$\frac{\frac{5}{x}}{\frac{15}{x+2}}=\frac{5}{x}\cdot\frac{x+2}{15}=\frac{x+2}{3x}$$

3.
$$\frac{\frac{t-1}{6t}}{\frac{t+1}{9t}}=\frac{t-1}{6t}\cdot\frac{9t}{t+1}=\frac{3(t-1)}{2(t+1)}$$

5.
$$\frac{\frac{s+t}{4k}}{\frac{2s+2t}{8k}}=$$

$$\frac{s+k}{4k}\cdot\frac{8k}{2(s+t)}=$$

$$1$$

7.
$$\frac{2+\frac{7}{16}}{2-\frac{1}{4}}=$$

$$\frac{\frac{32}{16}+\frac{7}{16}}{\frac{8}{4}-\frac{1}{4}}=$$

$$\frac{\frac{39}{16}}{\frac{7}{4}}=$$

$$\frac{39}{16}\cdot\frac{4}{7}=\frac{39}{28}$$

9.
$$\frac{a-4}{2-\frac{8}{a}}=$$

$$\frac{a-4}{\frac{2a}{a}-\frac{8}{a}}=$$

$$\frac{a-4}{\frac{2a-8}{a}}=$$

$$\frac{a-4}{1}\cdot\frac{a}{2(a-4)}=$$

$$\frac{a}{2}$$

11.
$$\frac{a-\frac{25}{a}}{a+5}=$$

$$\frac{\frac{a^2}{a}-\frac{25}{a}}{a+5}=$$

$$\frac{a^2-25}{a}\cdot\frac{1}{a+5}=$$

$$\frac{(a+5)(a-5)}{a}\cdot\frac{1}{a+5}=$$

$$\frac{a-5}{a}$$

13.
$$\frac{\frac{3}{x}+\frac{5}{y}}{\frac{x}{3}+\frac{y}{5}}=$$

$$\frac{\frac{3y+5x}{xy}}{\frac{5x+3y}{15}}=$$

$$\frac{5x+3y}{xy}\cdot\frac{15}{5x+3y}=$$

$$\frac{15}{xy}$$

15.
$$\frac{1-y}{1-\frac{1}{y}}=$$

$$\frac{1-y}{\frac{y-1}{y}}=$$

$$\frac{1-y}{1}\cdot\frac{y}{y-1}=$$

$$\frac{y(-1)(y-1)}{y-1}=$$

$$-y$$

17.

$$\frac{1+\frac{1}{a}}{1-\frac{1}{a}}=$$

$$\frac{\left(1+\frac{1}{a}\right)a}{\left(1-\frac{1}{a}\right)a}=$$

$$\frac{a+1}{a-1}$$

19.

$$\frac{\frac{1}{m}+\frac{1}{n}}{\frac{1}{m}-\frac{1}{n}}=$$

$$\frac{\left(\frac{1}{m}+\frac{1}{n}\right)mn}{\left(\frac{1}{m}-\frac{1}{n}\right)mn}=$$

$$\frac{n+m}{n-m}$$

21.

$$\frac{x-3-\frac{28}{x}}{x+10+\frac{24}{x}}=$$

$$\frac{\left(x-3-\frac{28}{x}\right)x}{\left(x+10+\frac{24}{x}\right)x}=$$

$$\frac{x^2-3x-28}{x^2+10x+24}=$$

$$\frac{(x-7)(x+4)}{(x+6)(x+4)}=$$

$$\frac{x-7}{x+6}$$

23.

$$\frac{\frac{x^2-y^2}{x^3+y^3}}{\frac{x^2+2xy+y^2}{x^2-xy+y^2}}=$$

$$\frac{(x+y)(x-y)}{(x+y)(x^2-xy+y^2)}\cdot\frac{x^2-xy+y^2}{(x+y)^2}=$$

$$\frac{x-y}{(x+y)^2}$$

25.

$$\frac{\frac{x}{x+y}+\frac{y}{x-y}}{\frac{x^2+y^2}{x^2-y^2}}=$$

$$\frac{\frac{x(x-y)+y(x+y)}{(x+y)(x-y)}}{\frac{x^2+y^2}{(x+y)(x-y)}}=$$

$$\frac{x^2-xy+xy+y^2}{(x+y)(x-y)}\cdot\frac{(x+y)(x-y)}{x^2+y^2}=$$

$$\frac{x^2+y^2}{x^2+y^2}=1$$

27.

$$\frac{\frac{6a^2-5ab-4b^2}{a-b}}{\frac{3a^2-7ab+4b^2}{2a+b}}=$$

$$\frac{\frac{(2a+b)(3a-4b)}{a-b}}{\frac{(3a-4b)(a-b)}{2a+b}}=$$

$$\frac{(2a+b)(3a-4b)}{a-b}\cdot\frac{2a+b}{(3a-4b)(a-b)}=$$

$$\frac{(2a+b)^2}{(a-b)^2}$$

29.

$$\frac{1+\frac{1}{x^2-1}}{1+\frac{1}{x-1}}=$$

$$\frac{\frac{x^2-1+1}{x^2-1}}{\frac{x-1+1}{x-1}}=$$

$$\frac{x^2}{(x+1)(x-1)}\cdot\frac{x-1}{x}=$$

$$\frac{x}{x+1}$$

31.

$$1-\frac{1}{2-\frac{1}{x}}=$$

$$1-\frac{1}{\frac{2x-1}{x}}=$$

$$1-\frac{x}{2x-1}=$$

$$\frac{2x-1-x}{2x-1}=$$

$$\frac{x-1}{2x-1}$$

33.

$$1+\frac{1}{x+\frac{1}{x-1}}=$$

$$1+\frac{1}{\frac{x^2-x+1}{x-1}}=$$

$$1+\frac{x-1}{x^2-x+1}=$$

$$\frac{x^2-x+1+x-1}{x^2-x+1}=\frac{x^2}{x^2-x+1}$$

35. $$\frac{\frac{1}{x}-\frac{2}{x^2}-\frac{3}{x^3}}{x+6+\frac{5}{x}}-\frac{\frac{1}{x^2}-\frac{9}{x^4}}{1+\frac{2}{x}-\frac{15}{x^2}}=$$
$$\frac{\left(\frac{1}{x}-\frac{2}{x^2}-\frac{3}{x^3}\right)x^3}{\left(x+6+\frac{5}{x}\right)x^3}-\frac{\left(\frac{1}{x^2}-\frac{9}{x^4}\right)x^4}{\left(1+\frac{2}{x}-\frac{15}{x^2}\right)x^4}=$$
$$\frac{x^2-2x-3}{x^4+6x^3+5x^2}-\frac{x^2-9}{x^4+2x^3-15x^2}=$$
$$\frac{(x-3)(x+1)}{x^2(x+5)(x+1)}-\frac{(x+3)(x-3)}{x^2(x+5)(x-3)}=$$
$$\frac{x-3}{x^2(x+5)}-\frac{x+3}{x^2(x+5)}=$$
$$\frac{x-3-(x+3)}{x^2(x+5)}=$$
$$\frac{x-3-x-3}{x^2(x+5)}=$$
$$\frac{-6}{x^2(x+5)}$$

Exercises 5.8

1. $$\frac{x}{3}-\frac{x}{6}=2$$
L.C.D. = 6
$$6\left(\frac{x}{3}-\frac{x}{6}\right)=6(2)$$
$$2x-x=12$$
$$x=12$$
Check:
$$\frac{12}{3}-\frac{12}{6}=2$$

3. $$\frac{3m}{4}+\frac{2m}{5}=\frac{23}{2}$$
L.C.D. = 20
$$20\left(\frac{3m}{4}+\frac{2m}{5}\right)=20\left(\frac{23}{2}\right)$$
$$15m+8m=230$$
$$23m=230$$
$$m=10$$
Check:
$$\frac{3(10)}{4}+\frac{2(10)}{5}\overset{?}{=}\frac{23}{2}$$
$$\frac{30}{4}+\frac{20}{5}\overset{?}{=}\frac{23}{2}$$
$$7\frac{1}{2}+4\overset{?}{=}\frac{23}{2}$$
$$11\frac{1}{2}=\frac{23}{2}$$

5. $$\frac{x}{6}+2=\frac{x+3}{4}$$
L.C.D. = 12
$$12\left(\frac{x}{6}+2\right)=12\left(\frac{x+3}{4}\right)$$
$$2x+24=3x+9$$
$$15=x$$
Check:
$$\frac{15}{6}+2\overset{?}{=}\frac{15+3}{4}$$
$$2.5+2\overset{?}{=}\frac{18}{4}$$
$$4.5=4.5$$

7. $$\frac{s+4}{8}-1\frac{1}{8}=\frac{3s}{4}$$
$$\frac{s+4}{8}-\frac{9}{8}=\frac{3s}{4}$$
L.C.D. = 8
$$8\left(\frac{s+4}{8}-\frac{9}{8}\right)=8\left(\frac{3s}{4}\right)$$
$$s+4-9-6s$$
$$s-5=6s$$
$$-5=5s$$
$$-1=s$$
Check:
$$\frac{-1+4}{8}-1\frac{1}{8}\overset{?}{=}\frac{3(-1)}{4}$$
$$\frac{3}{8}-\frac{9}{8}\overset{?}{=}\frac{-3}{4}$$
$$\frac{-6}{8}\overset{?}{=}\frac{-3}{4}$$
$$\frac{-3}{4}=\frac{-3}{4}$$

9. $$\frac{x+5}{2}=\frac{x+4}{3}$$
L.C.D. = 6
$$6\left(\frac{x+5}{2}\right)=6\left(\frac{x+4}{3}\right)$$
$$3x+15=2x+8$$
$$x=-7$$
Check:
$$\frac{-7+5}{2}\overset{?}{=}\frac{-7+4}{3}$$
$$\frac{-2}{2}\overset{?}{=}\frac{-3}{3}$$
$$-1=-1$$

11. $$\frac{x}{20}+\frac{1}{4}=\frac{x+2}{5}$$
L.C.D. = 20
$$20\left(\frac{x}{20}+\frac{1}{4}\right)=20\left(\frac{x+2}{5}\right)$$
$$x+5=4x+8$$
$$-3=3x$$
$$-1=x$$
Check:
$$\frac{-1}{20}+\frac{1}{4}\overset{?}{=}\frac{-1+2}{5}$$
$$\frac{-1}{20}+\frac{5}{20}\overset{?}{=}\frac{-1}{5}$$
$$\frac{-4}{20}=\frac{-1}{5}$$

13. $\frac{2x+6}{4}+1=\frac{3x+1}{8}$

L.C.D. = 8

$8\left(\frac{2x+6}{4}+1\right)=8\left(\frac{3x+1}{8}\right)$

$4x+12+8=3x+1$

$4x+20=3x+1$

$x=-19$

Check:

$\frac{2(-19)+6}{4}+1\overset{?}{=}\frac{3(-19)+1}{8}$

$\frac{-38+6}{4}+1\overset{?}{=}\frac{-57+1}{8}$

$\frac{-32}{4}+1\overset{?}{=}\frac{-56}{8}$

$-8+1=-7$

15. $\frac{5-x}{7}-\frac{2x-1}{14}=1$

L.C.D. = 14

$14\left(\frac{5-x}{7}-\frac{2x-1}{14}\right)=14(1)$

$10-2x-(2x-1)=14$

$10-2x-2x+1=14$

$-4x+11=14$

$-4x=3$

$x=\frac{-3}{4}$

Check:

$\frac{5-\left(-\frac{3}{4}\right)}{7}-\frac{2\left(-\frac{3}{4}\right)-1}{14}\overset{?}{=}1$

$\frac{5+\frac{3}{4}}{7}-\frac{\left(\frac{-3}{2}-1\right)}{14}\overset{?}{=}1$

$\left(\frac{20}{4}+\frac{3}{4}\right)\frac{1}{7}-\left(\frac{-3}{2}-\frac{2}{2}\right)\frac{1}{14}\overset{?}{=}1$

$\frac{23}{4}\cdot\frac{1}{7}+\frac{5}{2}\cdot\frac{1}{14}\overset{?}{=}1$

$\frac{23}{28}+\frac{5}{28}\overset{?}{=}1$

$\frac{28}{28}=1$

17. $F=\frac{mv^2}{r}$ for r

$rF=mv^2$

$r=\frac{mv^2}{F}$

19. $I_L=\frac{V}{R+R_L}$ for R

$I_L(R+R_L)=V$

$R+R_L=\frac{V}{I_L}$

$R=\frac{V}{I_L}-R_L$

$R=\frac{V-R_LI_L}{I_L}$

21. $\frac{1}{R}=\frac{1}{R_1}+\frac{1}{R_2}$ for R_1

$\frac{1}{R}-\frac{1}{R_2}=\frac{1}{R_1}$

$\frac{R_2-R}{RR_2}=\frac{1}{R_1}$

The reciprocals are equal:

$\frac{RR_2}{R_2-R}=R_1$

23. $y=\frac{x+y}{x}$ for y

$xy=x+y$

$xy-y=x$

$y(x-1)=x$

$y=\frac{x}{x-1}$

25. $\frac{x}{a+b}=\frac{x+a}{b}$ for x

$xb=(x+a)(a+b)$

$xb=xa+xb+a^2+ab$

$0=xa+a^2+ab$

$-a^2-ab=xa$

$a(-a-b)=xa$

$\frac{a(-a-b)}{a}=x$

$-a-b=x$

27. $V=\frac{Q}{R_1}-\frac{Q}{R_2}$ for Q

$V=Q\left(\frac{1}{R_1}-\frac{1}{R_2}\right)$

$V=Q\left(\frac{R_2-R_1}{R_1R_2}\right)$

$\frac{R_1R_2V}{R_2-R_1}=Q$

29. $\frac{3}{2x}=\frac{5}{x}-7$

L.C.D. = 2x

$2x\left(\frac{3}{2x}\right)=2x\left(\frac{5}{x}-7\right)$

$3=10-14x$

$-7=-14x$

$\frac{1}{2}=x$

Check :

$\frac{3}{2\left(\frac{1}{2}\right)}\overset{?}{=}\frac{5}{\frac{1}{2}}-7$

$\frac{3}{1}\overset{?}{=}5\cdot\frac{2}{1}-7$

$3=10-7$

Section 5.8

31. $\frac{x-3}{x}=\frac{1}{x}+\frac{2}{3}$

L.C.D. $=3x$

$3x\left(\frac{x-3}{x}\right)=3x\left(\frac{1}{x}+\frac{2}{3}\right)$

$3x-9=3+2x$

$x=9+3$

$x=12$

Check:

$\frac{12-3}{12}\overset{?}{=}\frac{1}{12}+\frac{2}{3}$

$\frac{9}{12}\overset{?}{=}\frac{1}{12}+\frac{8}{12}$

$\frac{9}{12}=\frac{9}{12}$

33. $\frac{1}{x-2}=\frac{3}{x+4}$

L.C.D. $=(x-2)(x+4)$

$(x-2)(x+4)\left(\frac{1}{x-2}\right)=(x-2)(x+4)\left(\frac{3}{x+4}\right)$

$x+4=3x-6$

$10=2x$

$x=5$

Check:

$\frac{1}{5-2}\overset{?}{=}\frac{3}{5+4}$

$\frac{1}{3}=\frac{3}{9}$

35. $\frac{x+2}{x-3}=\frac{x+1}{x-1}$

L.C.D. $=(x-3)(x-1)$

$(x-3)(x-1)\left(\frac{x+2}{x-3}\right)=(x-3)(x-1)\left(\frac{x+1}{x-1}\right)$

$(x-1)(x+2)=(x-3)(x+1)$

$x^2+x-2=x^2-2x-3$

$3x=-1$

$x=\frac{-1}{3}$

Check:

$\frac{\frac{-1}{3}+2}{-\frac{1}{3}-3}\overset{?}{=}\frac{\frac{-1}{3}+1}{-\frac{1}{3}-1}$

$\frac{-\frac{1}{3}+\frac{6}{3}}{-\frac{1}{3}-\frac{9}{3}}\overset{?}{=}\frac{\frac{2}{3}}{-\frac{4}{3}}$

$\frac{5}{3}\div\left(\frac{-10}{3}\right)\overset{?}{=}\frac{2}{3}\left(-\frac{3}{4}\right)$

$\frac{5}{3}\left(-\frac{3}{10}\right)\overset{?}{=}\frac{2}{3}\left(-\frac{3}{4}\right)$

$-\frac{1}{2}=\frac{-1}{2}$

37. $\frac{2x^2+8}{x^2-1}-\frac{x-4}{x+1}=\frac{x}{x-1}$

L.C.D. $=(x+1)(x-1)$

$(x+1)(x-1)\left(\frac{2x^2+8}{(x+1)(x-1)}-\frac{x-4}{x+1}\right)=(x+1)(x-1)\left(\frac{x}{x-1}\right)$

$2x^2+8-(x-4)(x-1)=x(x+1)$

$2x^2+8-\left(x^2-5x+4\right)=x^2+x$

$2x^2+8-x^2+5x-4=x^2+x$

$x^2+5x+4=x^2+x$

$4x=-4$

$x=-1$

-1 is an excluded value.

No solution

39. $\dfrac{x+2}{x+7}=\dfrac{2}{x-2}+\dfrac{x^2+3x-28}{x^2+5x-14}$

L.C.D. $=(x+7)(x-2)$

$(x+7)(x-2)\left(\dfrac{x+2}{x+7}\right)=(x+7)(x-2)\left(\dfrac{2}{x-2}+\dfrac{x^2+3x-28}{x^2+5x-14}\right)$

$(x-2)(x+2)=2(x+7)+x^2+3x-28$

$x^2-4=2x+14+x^2+3x-28$

$-4=5x-14$

$10=5x$

$x=2$

2 is an excluded value.

No solution

41. $\dfrac{x}{x+4}+\dfrac{x+1}{x-5}=2$

L.C.D. $=(x+4)(x-5)$

$(x+4)(x-5)\left(\dfrac{x}{x+4}+\dfrac{x+1}{x-5}\right)=(x+4)(x-5)(2)$

$x(x-5)+(x+1)(x+4)=2(x^2-x-20)$

$x^2-5x+x^2+5x+4=2x^2-2x-40$

$4=-2x-40$

$44=-2x$

$x=-22$

Check :

$\dfrac{-22}{-22+4}+\dfrac{-22+1}{-22-5}\stackrel{?}{=}2$

$\dfrac{-22}{-18}+\dfrac{-21}{-27}\stackrel{?}{=}2$

$\dfrac{11}{9}+\dfrac{7}{9}\stackrel{?}{=}2$

$\dfrac{18}{9}=2$

43. $\dfrac{x^2+3x+7}{x^2-x-12}+\dfrac{x+6}{x+3}=\dfrac{2x+1}{x-4}$

L.C.D. $=(x+3)(x-4)$

$(x+3)(x-4)\left(\dfrac{x^2+3x+7}{(x-3)(x-4)}+\dfrac{x+6}{x+3}\right)=(x+3)(x-4)\left(\dfrac{2x+1}{x-4}\right)$

$x^2+3x+7+(x+6)(x-4)=(x+3)(2x+1)$

$x^2+3x+7+x^2+2x-24=2x^2+7x+3$

$2x^2+5x-17=2x^2+7x+3$

$-2x=20$

$x=-10$

Check :

$\dfrac{(-10)^2+3(-10)+7}{(-10)^2-(-10)-12}+\dfrac{(-10)+6}{(-10)+3}\stackrel{?}{=}\dfrac{2(-10)+1}{-10-4}$

$\dfrac{100-30+7}{100+10-12}+\dfrac{-4}{-7}\stackrel{?}{=}\dfrac{-20+1}{-14}$

$\dfrac{77}{98}+\dfrac{4}{7}\stackrel{?}{=}\dfrac{-19}{-14}$

$\dfrac{77}{98}+\dfrac{4(14)}{7(14)}\stackrel{?}{=}\dfrac{19}{14}$

$\dfrac{77+56}{98}\stackrel{?}{=}\dfrac{19}{14}$

$\dfrac{133}{98}\stackrel{?}{=}\dfrac{19}{14}$

$\dfrac{19}{14}=\dfrac{19}{14}$

Chapter 5 Review

1. $-6a^2b^4\left(a^3-12ab^2-9b^5\right)=-6a^5b^4+72a^3b^6+54a^2b^9$

2. $(3x+13)(5x-4)=15x^2-12x+65x-52=15x^2+53x-52$

3. $(4x-7)(4x+7)=16x^2-49$

4. $\left(5a^2-9b^2\right)^2=25a^4-90a^2b^2+81b^4$

5. $\left(4a-b^2\right)^3=(4a)^3-3(4a)^2\left(b^2\right)+3(4a)\left(b^2\right)^2-\left(b^2\right)^3=64a^3-48a^2b^2+12ab^4-b^6$

6. $\left(6x+7yz^2\right)\left(5x-2yz^2\right)=30x^2+23xyz^2-14y^2z^4$

7. $18a^3b-9a^2b^4+9a^3b^2=9a^2b\left(2a-b^3+ab\right)$

8. $16x^2-9y^2=(4x+3y)(4x-3y)$

9. $5x^2+28x+32=(x+4)(5x+8)$

10. $14x^2+59x-18=(2x+9)(7x-2)$

11. $4a^2-20a+25=(2a-5)^2$

12. $m^3-n^3=(m-n)\left(m^2+mn+n^2\right)$

13. $3x^2-18x-21=3\left(x^2-6x-7\right)=3(x-7)(x+1)$

14. $x^4-81=\left(x^2+9\right)\left(x^2-9\right)=\left(x^2+9\right)(x+3)(x-3)$

15. $15x^2+23x-90=(3x+10)(5x-9)$

16. $10x^2+25xy-60y^2=5\left(2x^2+5xy-12y^2\right)=5(x+4y)(2x-3y)$

17. $x^2+6x+9-25y^2=(x+3)^2-25y^2=(x+3+5y)(x+3-5y)$

18. $ax+2bx-4ay-8by=x(a+2b)-4y(a+2b)=(a+2b)(x-4y)$

19. $\dfrac{90x^4y^9}{48x^8y^3}=\dfrac{15y^6}{8x^4}$

20. $\dfrac{16x^2-40x}{8x^2-20x}=\dfrac{8x(2x-5)}{4x(2x-5)}=2$

21. $\dfrac{x^2-4x-32}{2x^2+11x+12}=\dfrac{(x-8)(x+4)}{(2x+3)(x+4)}=\dfrac{x-8}{2x+3}$

22. $\dfrac{-2x^2+7x+4}{x^2-16}=\dfrac{-\left(2x^2-7x-4\right)}{(x+4)(x-4)}=\dfrac{-(2x+1)(x-4)}{(x+4)(x-4)}=\dfrac{-(2x+1)}{x+4}$

23. $\dfrac{2x^2+5x-25}{4x^2+21x+5}\cdot\dfrac{4x^2+x}{4x^2-25}=$

$\dfrac{(2x-5)(x+5)}{(4x+1)(x+5)}\cdot\dfrac{x(4x+1)}{(2x+5)(2x-5)}=$

$\dfrac{x}{2x+5}$

24. $\dfrac{6x^2+11x-7}{4x^2+4x+1}\div\dfrac{3x^2+4x-7}{-2x^2+x+1}=$

$\dfrac{(2x-1)(3x+7)}{(2x+1)^2}\cdot\dfrac{-\left(2x^2-x+1\right)}{(3x+7)(x-1)}=$

$\dfrac{(2x-1)(3x+7)}{(2x+1)^2}\cdot\dfrac{-(2x+1)(x-1)}{(3x+7)(x-1)}=$

$\dfrac{-(2x-1)}{2x+1}$

25. $\dfrac{y^6+125}{4y^2+20} \div \dfrac{y^4-5y^2+25}{12y} =$

$\dfrac{(y^2+5)(y^4-5y^2+25)}{4(y^2+5)} \cdot \dfrac{12y}{y^4-5y^2+25} =$

$3y$

26. $\dfrac{2-x}{x^2+10x+21} \cdot \dfrac{3x^2+21x}{18x+36} \div \dfrac{x^3-4x}{x+3} =$

$\dfrac{-(x-2)}{(x+3)(x+7)} \cdot \dfrac{3x(x+7)}{18(x+2)} \cdot \dfrac{x+3}{x(x^2-4)} =$

$\dfrac{-(x-2)}{(x+3)(x+7)} \cdot \dfrac{3x(x+7)}{18(x+2)} \cdot \dfrac{x+3}{x(x+2)(x-2)} =$

$\dfrac{-1}{6(x+2)^2}$

27. L.C.D. $= 12a^2$

$\dfrac{4}{3a} + \dfrac{9}{4a^2} - \dfrac{5}{2} =$

$\dfrac{4(4a)}{3a(4a)} + \dfrac{9(3)}{4a^2(3)} - \dfrac{5(6a^2)}{2(6a^2)} =$

$\dfrac{16a+27-30a^2}{12a^2}$

28. L.C.D. $= 2x+1$

$4 - \dfrac{3x+4}{2x+1} =$

$\dfrac{4(2x+1)}{2x+1} - \dfrac{3x+4}{2x+1} =$

$\dfrac{8x+4-3x-4}{2x+1} =$

$\dfrac{5x}{2x+1}$

29. L.C.D. $= (x-5)(x-2)(2x-1)$

$\dfrac{x}{x^2-7x+10} + \dfrac{x}{2x^2-5x+2} =$

$\dfrac{x}{(x-5)(x-2)} + \dfrac{x}{(2x-1)(x-2)} =$

$\dfrac{x(2x-1)}{(x-5)(x-2)(2x-1)} + \dfrac{x(x-5)}{(2x-1)(x-2)(x-5)} =$

$\dfrac{2x^2-x+x^2-5x}{(x-5)(x-2)(2x-1)} =$

$\dfrac{3x^2-6x}{(x-5)(x-2)(2x-1)} =$

$\dfrac{3x(x-2)}{(x-5)(x-2)(2x-1)} =$

$\dfrac{3x}{(x-5)(2x-1)}$

30. L.C.D. $= (2x+1)(x+1)(3x-4)$

$\dfrac{x-1}{2x^2+3x+1} + \dfrac{x-3}{3x^2-x-4} =$

$\dfrac{x-1}{(2x+1)(x+1)} + \dfrac{x-3}{(3x-4)(x+1)} =$

$\dfrac{(x-1)(3x-4)}{(2x+1)(x+1)(3x-4)} + \dfrac{(x-3)(2x+1)}{(3x-4)(x+1)(2x+1)} =$

$\dfrac{3x^2-7x+4+2x^2-5x-3}{(2x+1)(x+1)(3x-4)} =$

$\dfrac{5x^2-12x+1}{(2x+1)(x+1)(3x-4)}$

31. $\dfrac{\frac{1}{x}+\frac{1}{x^2}}{x+\frac{1}{x^2}} =$

$\dfrac{\left(\frac{1}{x}+\frac{1}{x^2}\right)x^2}{\left(x+\frac{1}{x^2}\right)x^2} =$

$\dfrac{x+1}{x^3+1} =$

$\dfrac{x+1}{(x+1)(x^2-x+1)} =$

$\dfrac{1}{x^2-x+1}$

32. $4 + \dfrac{2}{3+\frac{1}{y}} =$

$4 + \dfrac{2}{\frac{3y+1}{y}} =$

$4 + 2\left(\dfrac{y}{3y+1}\right) =$

$4 + \dfrac{2y}{3y+1} =$

$\dfrac{4(3y+1)+2y}{3y+1} =$

$\dfrac{12y+4+2y}{3y+1} =$

$\dfrac{14y+4}{3y+1}$

33. $\frac{x+2}{5}+\frac{x+4}{30}=\frac{x}{6}$

L.C.D. = 30

$30\left(\frac{x+2}{5}+\frac{x+4}{30}\right)=30\left(\frac{x}{6}\right)$

$6x+12+x+4=5x$

$7x+16=5x$

$2x=-16$

$x=-8$

Check:

$\frac{-8+2}{5}+\frac{-8+4}{30}\overset{?}{=}\frac{-8}{6}$

$\frac{-6}{5}+\frac{-4}{30}\overset{?}{=}\frac{-4}{3}$

$\frac{-36}{30}+\frac{-4}{30}\overset{?}{=}\frac{-4}{3}$

$\frac{-40}{30}=\frac{-4}{3}$

34. $\frac{ax+b}{b}=\frac{bx}{a}+\frac{a}{b}$

L.C.D. = ab

$ab\left(\frac{ax+b}{b}\right)=ab\left(\frac{bx}{a}+\frac{a}{b}\right)$

$a^2x+ab=b^2x+a^2$

$a^2x-b^2x=a^2-ab$

$x\left(a^2-b^2\right)=a(a-b)$

$x=\frac{a(a-b)}{a^2-b^2}$

$x=\frac{a(a-b)}{(a+b)(a-b)}$

$x=\frac{a}{a+b}$

Check:

$\frac{a\left(\frac{a}{a+b}\right)+b}{b}\overset{?}{=}\frac{b}{a}\left(\frac{a}{a+b}\right)+\frac{a}{b}$

$\frac{\frac{a^2}{a+b}+\frac{b(a+b)}{a+b}}{b}\overset{?}{=}\frac{b}{a+b}+\frac{a}{b}$

$\frac{1}{b}\left(\frac{a^2+ab+b^2}{a+b}\right)=\frac{b^2+a^2+ab}{b(a+b)}$

35. $\frac{x+1}{x-3}+1=\frac{4}{x-3}$

$\frac{x+1}{x-3}-\frac{4}{x-3}=-1$

$\frac{x+1-4}{x-3}=-1$

$\frac{x-3}{x-3}=-1$

Left side is positive one. Contradiction implies no solution.

36. $\frac{4}{x-4}-\frac{x^2}{16-x^2}=1$

$\frac{4}{x-4}-\frac{x^2}{-1\left(x^2-16\right)}=1$

$\frac{4}{x-4}+\frac{x^2}{(x+4)(x-4)}=1$

L.C.D. = $(x-4)(x+4)$

$(x-4)(x+4)\left(\frac{4}{x-4}+\frac{x^2}{(x+4)(x-4)}\right)=(x-4)(x+4)$

$4(x+4)+x^2=x^2-16$

$4x+16+x^2=x^2-16$

$4x=-32$

$x=-8$

Check:

$\frac{4}{-8-4}-\frac{(-8)^2}{16-(-8)^2}\overset{?}{=}1$

$\frac{4}{-12}-\frac{64}{16-64}\overset{?}{=}1$

$-\frac{1}{3}-\frac{64}{-48}\overset{?}{=}1$

$-\frac{1}{3}+\frac{4}{3}\overset{?}{=}1$

$\frac{3}{3}=1$

Chapter 6, Exercises 6.1

1. $2x+3y=5$

x	-2	1	4
y	3	1	-1

$x-2y=6$

x	0	2	4
y	-3	-2	-1

$(4,-1)$

Check:

$2(4)+3(-1)=$ $8-3=5$ and $4-2(-1)=$ $4+2=6$

3. $5x+2y=-5$

x	-3	-1	1
y	5	0	-5

$2x+3y=9$

x	-3	0	3
y	5	3	1

$(-3,5)$

Check:

$5(-3)+2(5)=$ $-15+10=-5$ and $2(-3)+3(5)=$ $-6+15=9$

5. $9x-6y=15$

x	-1	3	5
y	-4	2	5

$-3x+2y=-5$

x	-1	3	5
y	-4	2	5

System is dependent, lines coincide.

7. $4x+y=15$

$3x+y=13$

Subtract the second from the first.

$x=2$

From $4x+y=15$

$4(2)+y=15$

$y=15-8=7$

$(2,7)$

Check: $4(2)+7=8+7=15$

and

$3(2)+7=6+7=13$

9. $-4x+3y=-22$

$4x-5y=34$

Add the equations.

$-2y=12$

$y=-6$

From $4x-5y=34$

$4x-5(-6)=34$

$4x+30=34$

$4x=4$

$x=1$

$(1,-6)$

Check: $-4(1)+3(-6)=$

$-4-18=-22$

and

$4(1)-5(-6)=$

$4+30=34$

11. $4x-5y=-34$

$2x+3y=16$

Multiply the second equation by -2 and add.

$$\begin{array}{r} 4x-5y=-34 \\ \underline{-4x-6y=-32} \\ -11y=-66 \end{array}$$

$y=6$

From $2x+3y=16$

$2x+3(6)=16$

$2x+18=16$

$2x=-2$

$x=-1$

$(-1,6)$

Check: $4(-1)-5(6)=$

$-4-30=-34$

and

$2(-1)+3(6)=$

$-2+18=16$

13. $3x+2y=-15$
$x+5y=-5$
Multiply the second equation by -3 and add.

$$\begin{array}{r} 3x+2y=-15 \\ \underline{-3x-15y=15} \\ -13y=0 \end{array}$$

$y=0$

From $x+5y=-5$
$x+5(0)=-5$
$x=-5$
$(-5,0)$

Check: $3(-5)+2(0)=-15$
and
$-5+5(0)=-5$

15. $4x-5y=7$
$-2x+3y=-3$
Multiply the second equation by 2 and add.

$$\begin{array}{r} 4x-5y=7 \\ \underline{-4x+6y=-6} \\ y=1 \end{array}$$

From $4x-5y=7$
$4x-5(1)=7$
$4x=12$
$x=3$
$(3,1)$

Check: $4(3)-5(1)=$
$12-5=7$
and
$-2(3)+3(1)=$
$-6+3=-3$

17. $4x+3y=-3$
$12x+9y=-12$
Multiply the first equation by -3 and add.

$$\begin{array}{r} -12x-9y=9 \\ \underline{12x+9y=-12} \\ 0x+0y=-3 \end{array}$$

System inconsistent, lines parallel, no solution.

19. $3x+4y=11$
$-2x+3y=21$
Multiply the first equation by 2 and the second by 3 and add.

$$\begin{array}{r} 6x+8y=22 \\ \underline{-6x+9y=63} \\ 17y=85 \end{array}$$

$y=5$

From $3x+4y=11$
$3x+4(5)=11$
$3x=-9$
$x=-3$
$(-3,5)$

Check: $3(-3)+4(5)=11$
$-9+20=11$
and $-2(-3)+3(5)=21$
$6+15=21$

21. $12x+5y=-18$
$8x-7y=-74$
Multiply the first equation by 7 and the second by 5 and add.

$$\begin{array}{r} 84x+35y=-126 \\ \underline{40x-35y=-370} \\ 124x=-496 \end{array}$$

$x=-4$

From $8x-7y=-74$
$8(-4)-7y=-74$
$-32-7y=-74$
$-7y=-42$
$y=6$
$(-4,6)$

Check: $12(-4)+5(6)=$
$-48+30=-18$
and
$8(-4)-7(6)=$
$-32-42=-74$

23.

$-12x+15y=-43$

$9x-12y=34$

Multiply the first equation by 3 and the second by 4 and add.

$-36x+45y=-129$

$\underline{36x-48y=136}$

$-3y=7$

$y=-\frac{7}{3}$

From $9x-12y=34,$

$9x-12\left(-\frac{7}{3}\right)=34$

$9x+28=34$

$9x=6$

$x=\frac{6}{9}=\frac{2}{3}$

$\left(\frac{2}{3},-\frac{7}{3}\right)$

Check: $-12\left(\frac{2}{3}\right)+15\left(-\frac{7}{3}\right)=$

$-8-35=-43$

and

$9\left(\frac{2}{3}\right)-12\left(-\frac{7}{3}\right)=$

$6+28=34$

25.

$2x-5y=-36$

$y=4x$

$2x-5(4x)=-36$

$2x-20x=-36$

$-18x=-36$

$x=2$

From $y=4x$

$y=4(2)=8$

$(2,8)$

Check: $2(2)-5(8)=$

$4-40=-36$

and

$8=4(2)$

27.

$8x+9y=-5$

$x=-3y$

$8(-3y)+9y=-5$

$-24y+9y=-5$

$-15y=-5$

$y=\frac{5}{15}=\frac{1}{3}$

From $x=-3y$

$x=-3\left(\frac{1}{3}\right)=-1$

$\left(-1,\frac{1}{3}\right)$

Check: $8(-1)+9\left(\frac{1}{3}\right)=$

$-8+3=-5$

and

$-1=-3\left(\frac{1}{3}\right)$

29.

$3x-5y=64$

$y=6x-2$

$3x-5(6x-2)=64$

$3x-30x+10=64$

$-27x=54$

$x=-2$

From $y=6x-2$

$y=6(-2)-2$

$y=-12-2=-14$

$(-2,-14)$

Check: $3(-2)-5(-14)=$

$-6+70=64$

and

$-14=6(-2)-2$

31.

$3x+4y=-6$

$2x-y=-15$

The second equation becomes

$2x+15=y$

Substitute into the first equation.

$3x+4(2x+15)=-6$

$3x+8x+60=-6$

$11x=-66$

$x=-6$

From $y=2x+15$

$y=2(-6)+15$

$y=-12+15$

$y=3$

$(-6,3)$

Check: $3(-6)+4(3)=$

$-18+12=-6$

and

$2(-6)-(3)=$

$-12-3=-15$

33.

$5x-7y=23$

$3x+2y=-11$

The second equation becomes

$2y=-3x-11$

$y=-\frac{3}{2}x-\frac{11}{2}$

Substitute into the first equation.

$5x-7\left(-\frac{3}{2}x-\frac{11}{2}\right)=23$

$5x+\frac{21}{2}x+\frac{77}{2}=23$

$\frac{31}{2}x=-\frac{31}{2}$

$x=-1$

From $y=-\frac{3}{2}x-\frac{11}{2}$

$y=-\frac{3}{2}(-1)-\frac{11}{2}$

$y=\frac{3}{2}-\frac{11}{2}=\frac{-8}{2}$

$y=-4$

Check: $5(-1)-7(-4)=$

$-5+28=23$

and

$3(-1)+2(-4)=$

$-3-8=-11$

35.

$\frac{2}{3}x+\frac{1}{2}y=\frac{4}{9}$

$\frac{3}{4}x-\frac{2}{3}y=\frac{-23}{72}$

Multiply the first equation by 18 and the second by 72.

$12x+9y=8$

$54x-48y=-23$

Multiply the first equation by 9 and the second by − 2.

$108x+81y=72$

$-108x+96y=46$

$177y=118$

$y=\frac{118}{177}=\frac{2}{3}$

From $\frac{2}{3}x+\frac{1}{2}y=\frac{4}{9}$

$\frac{2}{3}x+\frac{1}{2}\left(\frac{2}{3}\right)=\frac{4}{9}$

$\frac{2}{3}x+\frac{1}{3}=\frac{4}{9}$

$\frac{2}{3}x=\frac{1}{9}$

$x=\frac{1}{6}$

$\left(\frac{1}{6},\frac{2}{3}\right)$

Check: $\frac{2}{3}\left(\frac{1}{6}\right)+\frac{1}{2}\left(\frac{2}{3}\right)=$

$\frac{1}{9}+\frac{1}{3}=\frac{4}{9}$

and

$\frac{3}{4}\left(\frac{1}{6}\right)-\frac{2}{3}\left(\frac{2}{3}\right)=$

$\frac{1}{8}-\frac{4}{9}=$

$\frac{9}{72}-\frac{32}{72}=\frac{-23}{72}$

37. $\frac{3}{5}x + 2y = \frac{5}{2}$

$\frac{2}{5}x - 4y = 3$

Multiply the first equation by 2 and add.

$$\begin{array}{l} \frac{6}{5}x + 4y = 5 \\ \underline{\frac{2}{5}x - 4y = 3} \\ \quad \frac{8}{5}x = 8 \end{array}$$

$x = 5$

From $\frac{2}{5}x - 4y = 3$

$\frac{2}{5}(5) - 4y = 3$

$2 - 4y = 3$

$-4y = 1$

$y = \frac{-1}{4}$

$\left(5, \frac{-1}{4}\right)$

Check: $\frac{3}{5}(5) + 2\left(\frac{-1}{4}\right) =$

$3 - \frac{1}{2} = \frac{5}{2}$

and

$\frac{2}{5}(5) - 4\left(\frac{-1}{4}\right) =$

$2 + 1 = 3$

39. $1.4x - 2.7y = 5.66$

$0.5x + 2y = 3.8$

Multiply the first equation by 2 and the second by 2.7.

$$\begin{array}{l} 2.8x - 5.4y = 11.32 \\ \underline{1.35x + 5.4y = 10.26} \\ \quad 4.15x = 21.58 \end{array}$$

$x = 5.2$

From $0.5x + 2y = 3.8$

$0.5(5.2) + 2y = 3.8$

$2.6 + 2y = 3.8$

$2y = 1.2$

$y = 0.6$

$(5.2, 0.6)$

Check: $1.4(5.2) - 2.7(0.6) =$

$7.28 - 1.62 = 5.66$

and

$0.5(5.2) + 2(0.6) =$

$2.6 + 1.2 = 3.8$

41. $0.002x + 0.008y = 2.28$

$0.04x + 0.09y = 28.8$

The first equation becomes

$0.002x = -0.008y + 2.28$

$x = -4y + 1140$

Substitute into the second equation.

$0.04(-4y + 1140) + 0.09y = 28.8$

$-0.16y + 45.6 + 0.09y = 28.8$

$-0.07y = -16.8$

$y = 240$

From $x = -4y + 1140$

$x = -4(240) + 1140$

$x = -960 + 1140$

$x = 180$

$(180, 240)$

Check: $0.002(180) + 0.008(240) =$

$.36 + 1.92 = 2.28$

and

$0.04(180) + 0.09(240) =$

$7.2 + 21.6 = 28.8$

43. Let X and Y be the two capacitors.

Then $X + Y = 55$

$X - Y = 25$

Add the equations

$2X = 80$

$X = 40$

Then $Y = 15$

So one is $40\mu F$ and the other is $15\mu F$

45. Let x = amount of cement.

Then $4x$ = amount of gravel.

$x + 4x = 11.5m^3$

$5x = 11.5m^3$

$x = 2.3m^3$

$2.3m^3$ of cement

$9.2m^3$ of gravel

47. Let x = amount of 8%

$2000 - x$ = amount of 3%

$.08x + .03(2000 - x) = .04(2000)$

$.08x + 60 - .03x = 80$

$.05x = 20$

$x = 400$

$400L$ of 8% and

$1600L$ of 3%

49. Let b = boat speed

c = current speed

$b + c$ = rate downstream

$b - c$ = rate upstream

Use $r \cdot t = d$

$5(b - c) = 20$

and $4(b + c) = 20$

Thus
$$\begin{array}{r} b - c = 4 \\ b + c = 5 \\ \hline 2b = 9 \end{array}$$

$b = \frac{9}{2}$ and

$c = \frac{1}{2}$

Boat is 4.5 *mph*

Current is 0.5 *mph*

51. Let x = width

$12 + x$ = length

perimeter = 96m

$2x + 2(x + 12) = 96$

$2x + 2x + 24 = 96$

$4x = 72$

$x = 18\text{m}$

18m by 30m

53. Let X = current in branch 1

Y = current in branch 2

$50X = 300Y$

ie $X = 6Y$

Also $X + Y = 840$

$6Y + Y = 840$

$7Y = 840$

$Y = 120\text{mA}$

$X = 720\text{mA}$

55. $4.3\,I_1 + 2.3(I_1 + I_2) = 27$

$2.8\,I_2 + 2.3(I_1 + I_2) = 35.2$

The first equation becomes

$4.3\,I_1 + 2.3\,I_1 + 2.3\,I_2 = 27$

$6.6\,I_1 + 2.3\,I_2 = 27$

The second equation becomes

$2.8\,I_2 + 2.3\,I_1 + 2.3\,I_2 = 35.2$

$2.3\,I_1 + 5.1\,I_2 = 35.2$

Multiply the first equation by 2.3 and the second by −6.6 and add.

$$\begin{array}{r} 15.18\,I_1 + 5.29\,I_2 = 62.1 \\ -15.18\,I_1 - 33.66\,I_2 = -232.32 \\ \hline -28.37\,I_2 = -170.22 \end{array}$$

$I_2 = 6$

From $6.6\,I_1 + 2.3\,I_2 = 27$

$6.6\,I_1 + 2.3(6) = 27$

$6.6\,I_1 + 13.8 = 27$

$6.6\,I_1 = 13.2$

$I_1 = 2$

57.

$3x+4y=25$

$6x+8y=15$

The first equation is $y=\frac{-3}{4}x+\frac{25}{4}$

Slope of both lines is $\frac{-3}{4}$.

A point on the first line is $(3,4)$. Write the equation of the line through $(3,4)$ and perpendicular to the given lines.

Slope $=\frac{4}{3}$.

$y=\frac{4}{3}x+b$

$4=\frac{4}{3}(3)+b$

$0=b$

$y=\frac{4}{3}x$. Find the point where $y=\frac{4}{3}x$ intersects the second line and use the distance formula.

$6x+8y=15$

and $y=\frac{4}{3}x$

$6x+8\left(\frac{4}{3}x\right)=15$

$6x+\frac{32}{3}x=15$

$\frac{50}{3}x=15$

$x=\frac{9}{10}$

$y=\frac{4}{3}\left(\frac{9}{10}\right)=\frac{6}{5}$

$\left(\frac{9}{10},\frac{6}{5}\right)$.

$d=\sqrt{(x_2-x_1)^2+(y_2-y_1)^2}$

$d=\sqrt{\left(3-\frac{9}{10}\right)^2+\left(4-\frac{6}{5}\right)^2}$

$d=\sqrt{4.41+7.84}$

$d=\sqrt{12.25}$

$d=3.5$

Exercises 6.2

1.

$ax+y=b$

$bx-y=a$

Add

$(a+b)x=a+b$

$x=1$

From $ax+y=b$

$a(1)+y=b$

$y=b-a$

$(1,b-a)$

3.

$x+y=a^2+ab$

$ax=by$

$x=\frac{b}{a}y$

Substitute

$\frac{b}{a}y+y=a^2+ab$

$\frac{b}{a}y+\frac{a}{a}y=a^2+ab$

$\left(\frac{a+b}{a}\right)y=a(a+b)$

$y=a(a+b)\left(\frac{a}{a+b}\right)$

$y=a^2$

From $x=\frac{b}{a}y$

$x=\frac{b}{a}\cdot a^2$

$x=ab$

(ab,a^2)

5.

$ax-by=0$

$bx+ay=1$

The first equation becomes

$ax=by$

$x=\frac{b}{a}y$

Substitute into the second equation.

$b\left(\frac{b}{a}y\right)+ay=1$

$\frac{b^2}{a}y+\frac{a^2}{a}y=1$

$\left(\frac{a^2+b^2}{a}\right)y=1$

$y=\frac{a}{a^2+b^2}$

From $x=\frac{b}{a}y$

$x=\frac{b}{a}\left(\frac{a}{a^2+b^2}\right)=\frac{b}{a^2+b^2}$

$\left(\frac{b}{a^2+b^2},\frac{a}{a^2+b^2}\right)$

7.
$$ax + by = c$$
$$bx + ay = c$$
Multiply the first equation by b and the second by $-a$, then add.
$$abx + b^2y = bc$$
$$\underline{-abx - a^2y = -ac}$$
$$\left(b^2 - a^2\right)y = bc - ac$$
$$y = \frac{c(b-a)}{(b-a)(b+a)}$$
$$y = \frac{c}{a+b}$$
From $ax + by = c$
$$ax + \frac{bc}{a+b} = c$$
$$ax = \frac{c(a+b)}{a+b} - \frac{bc}{a+b}$$
$$ax = \frac{ca + cb - bc}{a+b}$$
$$ax = \frac{ca}{a+b}$$
$$x = \frac{c}{a+b}$$
$$\left(\frac{c}{a+b}, \frac{c}{a+b}\right)$$

9.
$$(a+3b)x - by = a$$
$$(a-b)x - ay = b$$
Multiply the first equation by a and the second by $-b$ and add.
$$a(a+3b)x - aby = a^2$$
$$\underline{-b(a-b)x + aby = -b^2}$$
$$\left(a^2 + 3ab - ab + b^2\right)x = a^2 - b^2$$
$$\left(a^2 + 2ab + b^2\right)x = a^2 - b^2$$
$$x = \frac{(a+b)(a-b)}{(a+b)(a+b)}$$
$$x = \frac{a-b}{a+b}$$
From $(a+3b)x - by = a$
$$(a+3b)\left(\frac{a-b}{a+b}\right) - by = a$$
$$\frac{a^2 + 2ab - 3b^2}{a+b} - a = by$$
$$\frac{a^2 + 2ab - 3b^2}{a+b} - \frac{a^2 + ab}{ab} = by$$
$$\frac{a^2 + 2ab - 3ab^2 - a^2 - ab}{a+b} = by$$
$$\frac{ab - 3b^2}{a+b} = by$$
$$\frac{b(a-3b)}{a+b} = by$$
$$\frac{a-3b}{a+b} = y$$
$$\left(\frac{a-b}{a+b}, \frac{a-3b}{a+b}\right)$$

11.
$$\frac{3}{x} + \frac{4}{y} = 2$$
$$\frac{6}{x} + \frac{12}{y} = 5$$
Multiply the first equation by -2 and add.
$$\frac{-6}{x} - \frac{8}{y} = -4$$
$$\underline{\frac{6}{x} + \frac{12}{y} = 5}$$
$$\frac{4}{y} = 1$$
$$y = 4$$
From $\frac{3}{x} + \frac{4}{y} = 2$
$$\frac{3}{x} + 1 = 2$$
$$\frac{3}{x} = 1$$
$$x = 3$$
$$(3, 4)$$

13.
$$\frac{3}{x} + \frac{2}{y} = 0$$
$$\frac{2}{x} - \frac{5}{y} = 19$$
Multiply the first equation by 5 and the second by 2 and add.
$$\frac{15}{x} + \frac{10}{y} = 0$$
$$\underline{\frac{4}{x} - \frac{10}{y} = 38}$$
$$\frac{19}{x} = 38$$
$$x = \frac{1}{2}$$
From $\frac{3}{x} + \frac{2}{y} = 0$
$$\frac{3}{\frac{1}{2}} + \frac{2}{y} = 0$$
$$6 + \frac{2}{y} = 0$$
$$\frac{2}{y} = -6$$
$$y = -\frac{1}{3}$$
$$\left(\frac{1}{2}, -\frac{1}{3}\right)$$

15.
$$\frac{1}{x} + \frac{1}{y} = \frac{13}{6}$$
$$\frac{1}{x} - \frac{1}{y} = \frac{5}{6}$$
Add the equations.
$$\frac{2}{x} = \frac{18}{6} = 3$$
$$\frac{2}{3} = x$$
From $\frac{1}{x} + \frac{1}{y} = \frac{13}{6}$
$$\frac{1}{\frac{2}{3}} + \frac{1}{y} = \frac{13}{6}$$
$$\frac{3}{2} + \frac{1}{y} = \frac{13}{6}$$
$$\frac{1}{y} = \frac{2}{3}$$
$$y = \frac{3}{2}$$
$$\left(\frac{2}{3}, \frac{3}{2}\right)$$

Section 6.2

17.

$$\frac{6}{s}+\frac{4}{t}=16$$
$$\frac{9}{s}-\frac{5}{t}=2$$

Multiply the first equation by 5 and the second by 4 and add.

$$\frac{30}{s}+\frac{20}{t}=80$$
$$\frac{36}{s}-\frac{20}{t}=8$$
$$\frac{66}{s}=88$$
$$s=\frac{66}{66}=\frac{3}{4}$$

From $\frac{9}{5}-\frac{5}{t}=2$

$$\frac{9}{\frac{3}{4}}-\frac{5}{t}=2$$
$$12-\frac{5}{t}=2$$
$$\frac{-5}{t}=-10$$
$$t=\frac{1}{2}$$

$$\left(\frac{3}{4},\frac{1}{2}\right)$$

19.

$$\frac{1}{x}+\frac{1}{y}=a$$
$$\frac{1}{x}-\frac{1}{y}=b$$

Add the equations.

$$\frac{2}{x}=a+b$$
$$x=\frac{2}{a+b}$$

From $\frac{1}{\text{x}}+\frac{1}{y}=a$

$$\frac{a+b}{2}+\frac{1}{y}=a$$
$$\frac{1}{y}=\frac{2a}{2}-\frac{a+b}{2}$$
$$\frac{1}{y}=\frac{2a-a-b}{2}$$
$$\frac{1}{y}=\frac{a-b}{2}$$
$$y=\frac{2}{a-b}$$

$$\left(\frac{2}{a+b},\frac{2}{a-b}\right)$$

Exercises 6.3

1.

$$x+y+z=4$$
$$x-y+z=0$$
$$x-y-z=2$$

Eliminate y in equations 2 and 3.

Add equations 1 and 2:

$$2x+2z=4$$

Add equations 1 and 3:

$$2x=6$$
$$x=3$$

From $2x+2z=4$

$$2(3)+2z=4$$
$$2z=-2$$
$$z=-1$$

From $x-y+z=0,$

$$3-y-1=0$$
$$-y+2=0$$
$$y=2$$

$(3,2,-1)$

3.

$$3x-5y-6z=-19$$
$$3y+6z=15$$
$$4x-2y-5z=-2$$

Eliminate z in equations 1 and 3.

Add equations 1 and 2.

$$3x-2y=-4$$

Multiply equation 2 by $\frac{5}{3}$.

Multiply equation 3 by 2.

Add the results.

$$5y+10z=25$$
$$8x-4y-10z=-4$$
$$8x+y=21$$

Solve the two results:

$$3x-2y=-4$$
$$8x+y=21$$

$2(8x+y=21)$ becomes $16x+2y=42$. Add to

$$3x-2y=-4$$
$$19x=38$$
$$x=2$$

From $8x+y=21,$

$$8(2)+y=21$$
$$y=5$$

From $3y+6z=15$

$$3(5)+6z=15$$
$$6z=0$$

$z=0$ thus $(2,5,0)$

5.
$$\begin{aligned}2x+3y-5z&=56\\6x-4y+7z&=-42\\x-2y+3z&=-26\end{aligned}$$

Eliminate x in equations 1 and 2. Multiply equation 3 by -2 and add to equation 1.

$$\begin{aligned}2x+3y-5z&=56\\-2x+4y-6z&=52\\\hline 7y-11z&=108\end{aligned}$$

Multiply equation 3 by -6 and add to equation 2.

$$\begin{aligned}6x-4y+7z&=-42\\-6x+12y-18z&=156\\\hline 8y-11z&=114\end{aligned}$$

From the results:

$7y-11z=108$

$8y-11z=114$

Subtract the first from the second:

$y=6.$

From $7y-11z=108,$

$7(6)-11z=108$

$-11z=66$

$z=-6$

From $x-2y+3z=-26$

$x-2(6)+3(-6)=-26$

$x-12-18=-26$

$x=4$

$(4,6,-6)$

7.
$$\begin{aligned}x+y+z&=16\\y+z&=3\\x-z&=11\end{aligned}$$

Subtract equation 2 from equation 1.

$x=13$

From equation 3:

$13-z=11$

$z=2$

From equation 2:

$y+2=3$

$y=1$

$(13,1,2)$

9.
$$\begin{aligned}2x-4y+z&=17\\4x+5y-z&=-8\\x-3y+5z&=16\end{aligned}$$

Eliminate x. Multiply equation 1 by -2 and add to equation 2.

$$\begin{aligned}-4x+8y-2z&=-34\\4x+5y-z&=-8\\\hline 13y-3z&=-42\end{aligned}$$

Multiply equation 3 by -4 and add to equation 2.

$$\begin{aligned}-4x+12y-20z&=-64\\4x+5y-z&=-8\\\hline 17y-21z&=-72\end{aligned}$$

Multiply the first result by -7 and add to the second result.

$$\begin{aligned}-91y+21z&=294\\17y-21z&=-72\\\hline -74y&=222\end{aligned}$$

$y=-3$

From $13y-3z=-42,$

$13(-3)-3z=-42$

$-39-3z=-42$

$-3z=-3$

$z=1$

From $x-3y+5z=16$

$x-3(-3)+5(1)=16$

$x+9+5=16$

$x=2$ Thus $(2,-3,1)$

11.
$$\frac{1}{x}+\frac{1}{y}+\frac{1}{z}=\frac{23}{6}$$
$$\frac{3}{x}-\frac{5}{y}-\frac{2}{z}=\frac{10}{3}$$
$$\frac{2}{x}+\frac{3}{y}+\frac{6}{z}=7$$

Eliminate $\frac{1}{x}$. Multiply equation 1 by -3 and add to equation 2.

$$\frac{-3}{x}-\frac{3}{y}-\frac{3}{z}=\frac{-23}{2}$$
$$\frac{3}{x}-\frac{5}{y}-\frac{2}{z}=\frac{10}{3}$$
$$\frac{-8}{y}-\frac{5}{z}=\frac{-49}{6}$$

Multiply equation 1 by -2 and add to equation 3.

$$\frac{-2}{x}-\frac{2}{y}-\frac{2}{z}=\frac{-23}{3}$$
$$\frac{2}{x}+\frac{3}{y}+\frac{6}{z}=7$$
$$\frac{1}{y}+\frac{4}{z}=\frac{-2}{3}$$

Multiply the second result by 8 and add it to the first result.

$$\frac{8}{y}+\frac{32}{z}=\frac{-16}{3}$$
$$\frac{-8}{y}-\frac{5}{z}=\frac{-49}{6}$$
$$\frac{27}{z}=\frac{-27}{2}$$
$$z=-2$$

From $\frac{1}{y}+\frac{4}{z}=-\frac{2}{3}$

$$\frac{1}{y}+\frac{4}{-2}=-\frac{2}{3}$$
$$\frac{1}{y}=\frac{4}{3}$$
$$y=\frac{3}{4}$$

From $\frac{1}{x}+\frac{1}{y}+\frac{1}{z}=\frac{23}{6}$

$$\frac{1}{x}+\frac{4}{3}-\frac{1}{2}=\frac{23}{6}$$
$$\frac{1}{x}=3$$
$$x=\frac{1}{3}$$
$$\left(\frac{1}{3},\frac{3}{4},-2\right)$$

13. $2x+3y+4z-w=7$
$-x-2y+3z+2w=-3$
$3x+y \quad -3w=-5$
$4x \quad +2z-5w=-19$
Eliminate y from equations 1 and 2 and use those results with equation 4. Multiply equation 3 by -3 and add to equation 1.
$-9x-3y \quad +9w=15$
$2x+3y+4z-w=7$
$* -7x+4z+8w=22$
Multiply equation 3 by 2 and add to equation 2.
$6x+2y \quad -6w=-10$
$-x-2y+3z+2w=-3$
$* \ 5x+3z-4w=-13$
Completely solve the system:
$-7x+4z+8w=22$
$5x+3z-4w=-13$
$4x+2z-5w=-19$
Eliminate z. Multiply equation 3 by -2 and add to equation 1.
$-8x-4z+10w=38$
$-7x+4z+8w=22$
$-15x+18w=60$
which becomes:
$** -5x+6w=20$
Multiply equation 3 by -1.5 and add to equation 2.
$-6x-3z+7.5w=28.5$
$5x+3z-4w=-13$
$** -x+3.5w=15.5$
Using the ** results, multiply the second by -5 and add to the first.
$5x-17.5w=-77.5$
$-5x+6w=20$
$-11.5w=-57.5$
$w=5$
From $-x+3.5w=15.5$
$-x+3.5(5)=15.5$
$-x+17.5=15.5$
$-x=-2$
$x=2$
From $4x+2z-5w=-19$
$4(2)+2z-5(5)=-19$
$8+2z-25=-19$
$2z-17=-19$
$2=-2$
$z=-1$
From $3x+y-3w=-5$
$3(2)+y-3(5)=-5$
$6+y-15=-5$
$y-9=-5$
$y=4$
$(2,4,-1,5)$

15. $R_1+R_2+R_3=1950$
$*R_1 \quad +R_3=1800$
$R_1-8R_2 \quad =0$
Eliminate R_2. Multiply equation 1 by 8 and add to equation 3.
$8R_1+8R_2+8R_3=15,600$
$R_1-8R_2 \quad =0$
$* \ 9R_1 \quad +8R_3=15,600$
$* \ R_1 \quad +R_3 \ =1800$
Multiply the second equation by -8 and add to the first.
$-8R_1-8R_3=-14,400$
$9R_1+8R_3=15,600$
$R_1=1200\ \Omega$
From $R_1+R_3=1800$
$1200+R_3=1800$
$R_3=600\ \Omega$
From $R_1-8R_2=0$
$1200-8R_2=0$
$R_2=150\ \Omega$

17. Let x = length shortest side
$2x$ = length medium
$2x+5$ = length longest
$x+2x+2x+5=65$
$5x=60$
$x=12$ cm
$2x=24$ cm
$2x+5=29$ cm

19. Let x = shortest diameter
y = medium diameter
z = longest diameter
$x+y=21$
$y+z=80$
$x+z=69$
Subtract the second equation from the first and use the result with the ?
$x-z=-59$
$x+z=69$
$2x=10$
$x=5$ mm
From $x+y=21$
$5+y=21$
$y=16$ mm
From $x+z=69$
$5+z=69$
$z=64$ mm

21. $f(x) = ax^2 + bx + c$

$4 = a(1)^2 + b(1) + c$

$* \; 4 = a + b + c$

$10 = a(3)^2 + b(3) + c$

$* \; 10 = 9a + 3b + c$

$14 = a(-1)^2 + b(-1) + c$

$* \; 14 = a - b + c$

The * equations form a system:

$a + b + c = 4$

$9a + 3b + c = 10$

$a - b + c = 14$

Eliminate b. Add equations 1 and 3.

$2a + 2c = 18$

or

$** \; a + c = 9$

Multiply equation 3 by 3 and add to equation 2.

$$\begin{array}{r} 3a - 3b + 3c = 42 \\ \underline{9a + 3b + c = 10} \\ 12a + 4c = 52 \end{array}$$

or

$** \; 3a + c = 13$

The ** equations are:

$a + c = 9$

$3a + c = 13$

Subtract the first from the second.

$2a = 4$

$a = 2$

From $a + c = 9$

$2 + c = 9$

$c = 7$

From $a + b + c = 4$

$2 + b + 7 = 4$

$b + 9 = 4$

$b = -5$

So $f(x) = 2x^2 - 5x + 7$

Exercises 6.4

1. $\begin{vmatrix} 4 & 3 \\ 2 & 5 \end{vmatrix} = 4(5) - 2(3) = 14$

3. $\begin{vmatrix} 2 & -3 \\ -4 & 5 \end{vmatrix} = 2(5) - (-4)(-3) = -2$

5. $\begin{vmatrix} 3 & -7 \\ 6 & -1 \end{vmatrix} = 3(-1) - (6)(-7) = 39$

7. $\begin{vmatrix} -5 & 8 \\ 9 & -6 \end{vmatrix} = (-5)(-6) - 9(8) = -42$

9. $\begin{vmatrix} 4 & 0 \\ 6 & -3 \end{vmatrix} = 4(-3) - 6(0) = -12$

11. $\begin{vmatrix} -4 & -2 \\ -6 & -7 \end{vmatrix} = (-4)(-7) - (-6)(-2) = 16$

13. $\begin{vmatrix} -7 & -2 \\ 4 & 1 \end{vmatrix} = -7(1) - 4(-2) = 1$

15. $\begin{vmatrix} 8 & -9 \\ 7 & 4 \end{vmatrix} = 8(4) - 7(-9) = 95$

17. $\begin{vmatrix} m & -n \\ n^2 & n \end{vmatrix} = mn - (n^2)(-n) = mn + n^3$

19. $$\begin{vmatrix} 1 & 1 & -4 \\ -3 & 7 & 11 \\ 2 & 1 & -5 \end{vmatrix} = 1\begin{vmatrix} 7 & 11 \\ 1 & -5 \end{vmatrix} - (-3)\begin{vmatrix} 1 & -4 \\ 1 & -5 \end{vmatrix} + 2\begin{vmatrix} 1 & -4 \\ 7 & 11 \end{vmatrix}$$

$$= 1[7(-5) - 1(11)] + 3[1(-5) - 1(-4)] + 2[1(11) - 7(-4)]$$
$$= -35 - 11 + 3[-5 + 4] + 2[11 + 28]$$
$$= -46 + 3[-1] + 2[39]$$
$$= -46 - 3 + 78 = 29$$

21. $$\begin{vmatrix} -1 & 3 & 8 \\ 0 & 0 & -6 \\ -5 & 2 & -3 \end{vmatrix} = -0\begin{vmatrix} 3 & 8 \\ 2 & -3 \end{vmatrix} + 0\begin{vmatrix} -1 & 8 \\ -5 & -3 \end{vmatrix} - (-6)\begin{vmatrix} -1 & 3 \\ -5 & 2 \end{vmatrix}$$

$$= 0 + 0 + 6[-1(2) - 3(-5)]$$
$$= 6[-2 + 15] = 6[13] = 78$$

23. $$\begin{vmatrix} 1 & 5 & -3 \\ 2 & -2 & 2 \\ -5 & -6 & 1 \end{vmatrix} = 1\begin{vmatrix} -2 & 2 \\ -6 & 1 \end{vmatrix} - 2\begin{vmatrix} 5 & -3 \\ -6 & 1 \end{vmatrix} + (-5)\begin{vmatrix} 5 & -3 \\ -2 & 2 \end{vmatrix}$$

$$= 1[-2(1) - 2(-6)] - 2[5(1) - (-3)(-6) - 5[5(2) - (-3)(-2)]$$
$$= -2 + 12 - 2[5 - 18] - 5[10 - 6]$$
$$= 10 - 2[-13] - 5[4]$$
$$= 10 + 26 - 20 = 16$$

25. $$\begin{vmatrix} 1 & -3 & 4 \\ 4 & 6 & -2 \\ 1 & -3 & 4 \end{vmatrix} = 1\begin{vmatrix} 6 & -2 \\ -3 & 4 \end{vmatrix} - 4\begin{vmatrix} -3 & 4 \\ -3 & 4 \end{vmatrix} + 1\begin{vmatrix} -3 & 4 \\ 6 & -2 \end{vmatrix}$$

$$= 1[6(4) - (-2)(-3)] - 4[-3(4) - (-3)(4)] + 1[-3(-2) - 4(6)]$$
$$= 24 - 6 - 4[-12 + 12] + 6 - 24 = 0$$

27. $$\begin{vmatrix} 3 & 6 & -9 \\ 1 & -5 & 2 \\ 0 & 0 & 0 \end{vmatrix} = 0\begin{vmatrix} 6 & -9 \\ -5 & 2 \end{vmatrix} - 0\begin{vmatrix} 3 & -9 \\ 1 & 2 \end{vmatrix} + 0\begin{vmatrix} 3 & 6 \\ 1 & -5 \end{vmatrix}$$

$$= 0 + 0 + 0 = 0$$

29. Look for the row or column with the most zeros and expand by that.

$$\begin{vmatrix} 1 & 3 & -7 & 2 \\ -5 & 3 & 0 & -2 \\ 1 & -2 & 5 & 1 \\ 3 & 1 & 0 & 2 \end{vmatrix} = (-7)\begin{vmatrix} -5 & 3 & -2 \\ 1 & -2 & 1 \\ 3 & 1 & 2 \end{vmatrix} - 0\begin{vmatrix} 1 & 3 & 2 \\ 1 & -2 & 1 \\ 3 & 1 & 2 \end{vmatrix}$$

$$+5\begin{vmatrix} 1 & 3 & 2 \\ -5 & 3 & -2 \\ 3 & 1 & 2 \end{vmatrix} - 0\begin{vmatrix} 1 & 3 & 2 \\ -5 & 3 & -2 \\ 1 & -2 & 1 \end{vmatrix} =$$

$$-7\left\{-5\begin{vmatrix} -2 & 1 \\ 1 & 2 \end{vmatrix} - 1\begin{vmatrix} 3 & -2 \\ 1 & 2 \end{vmatrix} + 3\begin{vmatrix} 3 & -2 \\ -2 & 1 \end{vmatrix}\right\} - 0 +$$

$$5\left\{1\begin{vmatrix} 3 & -2 \\ 1 & 2 \end{vmatrix} - (-5)\begin{vmatrix} 3 & 2 \\ 1 & 2 \end{vmatrix} + 3\begin{vmatrix} 3 & 2 \\ 3 & -2 \end{vmatrix}\right\} - 0 =$$

$$-7\{-5[-2(2)-1(1)] - 1[3(2)-1(-2)] + 3[3(1)-(-2)(-2)]\} +$$

$$5\{1[3(2)-1(-2)] + 5[3(2)-1(2)] + 3[3(-2)-2(3)]\} =$$

$$-7\{-5[-4-1] - [6+2] + 3[3-4]\} + 5\{6+2+5[6-2] + 3[-6-6]\} =$$

$$-7\{-5[-5] - 8 + 3[-1]\} + 5\{8 + 5[4] + 3[-12]\} =$$

$$-7\{25-8-3\} + 5\{8+20-36\} =$$

$$-7\{14\} + 5\{-8\} = -98 - 40 = -138$$

31.

$$\begin{vmatrix} 3 & -1 & 6 & 2 \\ -5 & 3 & -8 & 7 \\ 1 & 0 & -5 & 0 \\ 2 & -6 & 3 & 1 \end{vmatrix} = 1\begin{vmatrix} -1 & 6 & 2 \\ 3 & -8 & 7 \\ -6 & 3 & 1 \end{vmatrix} - 0\begin{vmatrix} 3 & 6 & 2 \\ -5 & -8 & 7 \\ 2 & 3 & 1 \end{vmatrix} +$$

$$(-5)\begin{vmatrix} 3 & -1 & 2 \\ -5 & 3 & 7 \\ 2 & -6 & 1 \end{vmatrix} - 0\begin{vmatrix} 3 & -1 & 6 \\ -5 & 3 & -8 \\ 2 & -6 & 3 \end{vmatrix} =$$

$$1\left\{(-1)\begin{vmatrix} -8 & 7 \\ 3 & 1 \end{vmatrix} - 3\begin{vmatrix} 6 & 2 \\ 3 & 1 \end{vmatrix} + (-6)\begin{vmatrix} 6 & 2 \\ -8 & 7 \end{vmatrix}\right\} - 0 +$$

$$-5\left\{3\begin{vmatrix} 3 & 7 \\ -6 & 1 \end{vmatrix} - (-5)\begin{vmatrix} -1 & 2 \\ -6 & 1 \end{vmatrix} + 2\begin{vmatrix} -1 & 2 \\ 3 & 7 \end{vmatrix}\right\} - 0 =$$

$$1\{-1[-8(1)-3(7)] - 3[6(1)-3(2)] - 6[6(7)-2(-8)]\} +$$

$$-5\{3[3(1)-7(-6)] + 5[-1(1)-2(-6)] + 2[-1(7)-2(3)]\} =$$

$$1\{-1[-8-21] - 3[6-6] - 6[42+16]\} +$$

$$-5\{3[3+42] + 5[-1+12] + 2[-7-6]\} =$$

$$1\{-1(-29) - 3(0) - 6(58)\} - 5\{3[45] + 5[11] + 2[-13]\} =$$

$$29 - 348 - 5\{135 + 55 - 26\} =$$

$$-319 - 5\{164\} = -319 - 820 = -1139$$

33. $\begin{vmatrix} 1 & 3 & 6 & -2 \\ 0 & 2 & -5 & 7 \\ 0 & 0 & 3 & 1 \\ 0 & 0 & 0 & 4 \end{vmatrix} = 1\begin{vmatrix} 2 & -5 & 7 \\ 0 & 3 & 1 \\ 0 & 0 & 4 \end{vmatrix} - 0\begin{vmatrix} 3 & 6 & -2 \\ 0 & 3 & 1 \\ 0 & 0 & 4 \end{vmatrix} +$

$0\begin{vmatrix} 3 & 6 & -2 \\ 2 & -5 & 7 \\ 0 & 0 & 4 \end{vmatrix} - 0\begin{vmatrix} 3 & 6 & -2 \\ 2 & -5 & 7 \\ 0 & 3 & 1 \end{vmatrix} =$

$1\left\{2\begin{vmatrix} 3 & 1 \\ 0 & 4 \end{vmatrix} - 0\begin{vmatrix} -5 & 7 \\ 0 & 4 \end{vmatrix} + 0\begin{vmatrix} -5 & 7 \\ 3 & 1 \end{vmatrix}\right\} - 0 + 0 - 0 =$

$2[3(4) - 1(0)] - 0 + 0 = 2[12 - 0] = 24$

35. $\begin{vmatrix} 3 & -2 & 0 & 2 & -1 \\ 1 & 2 & -3 & 1 & 2 \\ 0 & -1 & 0 & 6 & 1 \\ -3 & 2 & 0 & 6 & -7 \\ 0 & 0 & 0 & 5 & 2 \end{vmatrix} = 0\begin{vmatrix} 1 & 2 & 1 & 2 \\ 0 & -1 & 6 & 1 \\ -3 & 2 & 6 & -7 \\ 0 & 0 & 5 & 2 \end{vmatrix} - (-3)\begin{vmatrix} 3 & -2 & 2 & -1 \\ 0 & -1 & 6 & 1 \\ -3 & 2 & 6 & -7 \\ 0 & 0 & 5 & 2 \end{vmatrix} +$

$0\begin{vmatrix} 3 & 2 & 2 & -1 \\ 1 & -2 & 1 & 2 \\ -3 & 2 & 6 & -7 \\ 0 & 0 & 5 & 2 \end{vmatrix} - 0\begin{vmatrix} 3 & -2 & 2 & -1 \\ 1 & 2 & 1 & 2 \\ 0 & -1 & 6 & 1 \\ 0 & 0 & 5 & 2 \end{vmatrix} + 0\begin{vmatrix} 3 & -2 & 2 & -1 \\ 1 & 2 & 1 & 2 \\ 0 & -1 & 6 & 1 \\ -3 & 2 & 6 & -7 \end{vmatrix} =$

$0 + 3\left\{3\begin{vmatrix} -1 & 6 & 1 \\ 2 & 6 & -7 \\ 0 & 5 & 2 \end{vmatrix} - 0\begin{vmatrix} -2 & 2 & -1 \\ 2 & 6 & -7 \\ 0 & 5 & 2 \end{vmatrix} - 3\begin{vmatrix} -2 & 2 & -1 \\ -1 & 6 & 1 \\ 0 & 5 & 2 \end{vmatrix} - 0\begin{vmatrix} -2 & 2 & -1 \\ -1 & 6 & 1 \\ 2 & 6 & -7 \end{vmatrix}\right\} + 0 - 0 + 0 =$

$3\left\{3\left[-1\begin{vmatrix} 6 & -7 \\ 5 & 2 \end{vmatrix} - 2\begin{vmatrix} 6 & 1 \\ 5 & 2 \end{vmatrix} + 0\begin{vmatrix} 6 & 1 \\ 6 & -7 \end{vmatrix}\right] - 0 - 3\left[-2\begin{vmatrix} 6 & 1 \\ 5 & 2 \end{vmatrix} - 1(-1)\begin{vmatrix} 2 & -1 \\ 5 & 2 \end{vmatrix} + 0\begin{vmatrix} 2 & -1 \\ 6 & 1 \end{vmatrix} - 0\right]\right\} =$

$3\{3[-1(6[2] - 5[-7]) - 2(6[2] - 1[5] + 0)] - 3[-2(6[2] - 1[5]) + 1(2[2] - 5[-1]) + 0]\} =$

$3\{3[-1(12 + 35) - 2(12 - 5) + 0] - 3[-2(12 - 5) + (4 + 5) + 0]\} =$

$3\{3[-47 - 2(7)] - 3[-2(7) + 9]\} =$

$3\{3[-61] - 3[-5]\} =$

$3\{-183 + 15\} = 3\{-168\} = -504$

Exercises 6.5

1. $\begin{vmatrix} 4 & 3 & -8 \\ 2 & -7 & 9 \\ 0 & 0 & 0 \end{vmatrix} = 0$ by Property 1.

3. $\begin{vmatrix} 3 & 6 & 0 & -8 \\ 2 & 7 & 5 & 4 \\ 0 & -5 & 8 & 1 \\ 2 & 7 & 5 & 4 \end{vmatrix} = 0$ by Property 2.

5. $\begin{vmatrix} 4 & 6 & -6 \\ 0 & 0 & -9 \\ 1 & -2 & 7 \end{vmatrix} =$ Use Property 5 to multiply row 3 by –4 and add to row 1, replacing row 1 with the answer.

$$\begin{vmatrix} 0 & 14 & -34 \\ 0 & 0 & -9 \\ 1 & -2 & 7 \end{vmatrix} = 0\begin{vmatrix} 0 & -9 \\ -2 & 7 \end{vmatrix} - 0\begin{vmatrix} 14 & -34 \\ -2 & 7 \end{vmatrix} + 1\begin{vmatrix} 14 & -34 \\ 0 & -9 \end{vmatrix}$$

$$= 0 - 0 + 1[14(-9) - 0(-34)] = -126$$

7. $\begin{vmatrix} 3 & 0 & 8 & 0 \\ 0 & 7 & 0 & 0 \\ 0 & 0 & -7 & 5 \\ 0 & 0 & 1 & 0 \end{vmatrix} = 3\begin{vmatrix} 7 & 0 & 0 \\ 0 & -7 & 5 \\ 0 & 1 & 0 \end{vmatrix} - 0 + 0 - 0$ by expanding by the first column.

$$= 3\left[7\begin{vmatrix} -7 & 5 \\ 1 & 0 \end{vmatrix} - 0 + 0\right] = 21(-7(0) - 5(1))$$

$$= 21(-5) = -105$$

9. $\begin{vmatrix} 3 & 0 & 0 & 0 \\ 0 & 1 & 0 & 0 \\ 0 & 0 & -2 & 0 \\ 0 & 0 & 0 & 5 \end{vmatrix} = 3\begin{vmatrix} 1 & 0 & 0 \\ 0 & -2 & 0 \\ 0 & 0 & 5 \end{vmatrix} = 3\left[1\begin{vmatrix} -2 & 0 \\ 0 & 5 \end{vmatrix}\right] =$

$$3(1)[-2(5) - 0(0)] = 3(-10) = -30$$

11. $\begin{vmatrix} 1 & 0 & 4 \\ 6 & 3 & 2 \\ 5 & 4 & 1 \end{vmatrix}$ Use Property 5 to multiply row 1 by –6 and add to row 2.

$\begin{vmatrix} 1 & 0 & 4 \\ 0 & 3 & -22 \\ 5 & 4 & 1 \end{vmatrix}$ Use Property 5 to multiply row 1 by –5 and add to row 3.

$$\begin{vmatrix} 1 & 0 & 4 \\ 0 & 3 & -22 \\ 0 & 4 & -19 \end{vmatrix} = 1\begin{vmatrix} 3 & -22 \\ 4 & -19 \end{vmatrix} = 3(-19) - (4)(-22) = -57 + 88 = 31.$$

13. $\begin{vmatrix} 3 & 1 & -6 \\ -4 & 0 & 7 \\ 5 & 6 & 2 \end{vmatrix}$ Use Property 5 to multiply row 1 by –6 and add to row 3.

$$\begin{vmatrix} 3 & 1 & -6 \\ -4 & 0 & 7 \\ -13 & 0 & 38 \end{vmatrix} = -1\begin{vmatrix} -4 & 7 \\ -13 & 38 \end{vmatrix} = -1[-4(38) - (7)(-13)] =$$

$$-1[-152 + 91] = 61$$

15. $\begin{vmatrix} 5 & 3 & 7 \\ 1 & 2 & 3 \\ 6 & 2 & 5 \end{vmatrix}$ Use Property 5 to multiply row 2 by –5 and add to row 1.

$\begin{vmatrix} 0 & -7 & -8 \\ 1 & 2 & 3 \\ 6 & 2 & 5 \end{vmatrix}$ Use Property 5 to multiply row 2 by –6 and add to row 3.

$$\begin{vmatrix} 0 & -7 & -8 \\ 1 & 2 & 3 \\ 0 & -10 & -13 \end{vmatrix} = -1\begin{vmatrix} -7 & -8 \\ -10 & -13 \end{vmatrix} = -1[(-7)(-13)-(-8)(-10)] =$$

$$-1[91-80] = -11$$

17. $\begin{vmatrix} 3 & -7 & 5 \\ 2 & 4 & 6 \\ -9 & 7 & -2 \end{vmatrix}$ Multiply row 2 by –1 and add to row 1.

$\begin{vmatrix} 1 & -11 & -1 \\ 2 & 4 & 6 \\ -9 & 7 & -2 \end{vmatrix}$ Multiply row 1 by –2 and add to row 2.

$\begin{vmatrix} 1 & -11 & -1 \\ 0 & 26 & 8 \\ -9 & 7 & -2 \end{vmatrix}$ Multiply row 1 by 9 and add to row 3.

$$\begin{vmatrix} 1 & -11 & -1 \\ 0 & 26 & 8 \\ 0 & -92 & -11 \end{vmatrix} = 1\begin{vmatrix} 26 & 8 \\ -92 & -11 \end{vmatrix} = 26(-11)-8(-92) = 450$$

19. $\begin{vmatrix} 1 & 0 & 5 & 0 \\ 6 & 2 & -3 & 7 \\ -1 & 2 & 3 & -4 \\ -3 & 2 & -7 & 1 \end{vmatrix}$ Multiply row 2 by –1 and add to row 3.

Multiply row 2 by –1 and add to row 4.

$$\begin{vmatrix} 1 & 0 & 5 & 0 \\ 6 & 2 & -3 & 7 \\ -7 & 0 & 6 & -11 \\ -9 & 0 & -4 & -6 \end{vmatrix} = 2\begin{vmatrix} 1 & 5 & 0 \\ -7 & 6 & -11 \\ -9 & -4 & -6 \end{vmatrix}$$

Multiply row 1 by 7 and add to row 2.

Multiply row 1 by 9 and add to row 3.

$$\begin{vmatrix} 1 & 5 & 0 \\ 0 & 41 & -11 \\ 0 & 41 & -6 \end{vmatrix} = 2(1)\begin{vmatrix} 41 & -11 \\ 41 & -6 \end{vmatrix} = 2[41(-6)-(41)(-11)] =$$

$$2[-246+451] = 2[205] = 410$$

21. $\begin{vmatrix} 1 & 3 & -7 & 2 \\ 5 & 6 & 1 & 5 \\ 4 & -8 & 9 & -2 \\ -6 & 7 & -4 & 3 \end{vmatrix}$ Multiply row 1 by –5 and add to row 2. Multiply row 1 by –4 and add to row 3. Multiply row 1 by 6 and add to row 4 .

$$\begin{vmatrix} 1 & 3 & -7 & 2 \\ 0 & -9 & 36 & -5 \\ 0 & -20 & 37 & -10 \\ 0 & 25 & -46 & 15 \end{vmatrix} = 1\begin{vmatrix} -9 & 36 & -5 \\ -20 & 37 & -10 \\ 25 & -46 & 15 \end{vmatrix} = 1(5)\begin{vmatrix} -9 & 36 & -1 \\ -20 & 37 & -2 \\ 25 & -46 & 3 \end{vmatrix}$$

$5\begin{vmatrix} -9 & 36 & -1 \\ -20 & 37 & -2 \\ 25 & -46 & 3 \end{vmatrix}$ Multiply row 1 by –2 and add to row 2. Multiply row 1 by 3 and add to row 3.

$$5\begin{vmatrix} -9 & 36 & -1 \\ -2 & -35 & 0 \\ -2 & 62 & 0 \end{vmatrix} = 5(-1)\begin{vmatrix} -2 & -35 \\ -2 & 62 \end{vmatrix} = -5[-2(62)-(-2)(-35)] =$$

$$-5[-124-70] = 970$$

23. $\begin{vmatrix} 1 & 0 & 3 & 2 & 5 \\ 3 & 2 & 3 & -2 & 3 \\ -4 & -2 & 4 & 9 & -6 \\ -3 & 7 & 6 & 2 & 3 \\ 0 & 2 & -2 & -4 & 5 \end{vmatrix}$ Multiply row 1 by –3 and add to row 2. Multiply row 1 by 4 and add to row 3. Multiply row 1 by 3 and add to row 4 .

$$\begin{vmatrix} 1 & 0 & 3 & 2 & 5 \\ 0 & 2 & -6 & -8 & -12 \\ 0 & -2 & 16 & 17 & 14 \\ 0 & 7 & 15 & 8 & 18 \\ 0 & 2 & -2 & -4 & 5 \end{vmatrix} = 1\begin{vmatrix} 2 & -6 & -8 & -12 \\ -2 & 16 & 17 & 14 \\ 7 & 15 & 8 & 18 \\ 2 & -2 & -4 & 5 \end{vmatrix} =$$

$1(2)\begin{vmatrix} 1 & -3 & -4 & -6 \\ -2 & 16 & 17 & 14 \\ 7 & 15 & 8 & 18 \\ 2 & -2 & -4 & 5 \end{vmatrix}$ Multiply row 1 by 2 and add to row 2. Multiply row 1 by –7 and add to row 3. Multiply row 1 by –2 and add to row 4 .

$2\begin{vmatrix} 1 & -3 & -4 & -6 \\ 0 & 10 & 9 & 2 \\ 0 & 36 & 36 & 60 \\ 0 & 4 & 4 & 17 \end{vmatrix} = 2\begin{vmatrix} 10 & 9 & 2 \\ 36 & 36 & 60 \\ 4 & 4 & 17 \end{vmatrix}$ Multiply column 1 by –4 and add to column 3.

$2\begin{vmatrix} 10 & 9 & -38 \\ 36 & 36 & -84 \\ 4 & 4 & 1 \end{vmatrix}$ Multiply column 3 by –4 and add to column 1. Multiply column 3 by –4 and add to column 2.

$$2\begin{vmatrix} 162 & 161 & -38 \\ 372 & 372 & -84 \\ 0 & 0 & 1 \end{vmatrix} = 2(1)\begin{vmatrix} 162 & 161 \\ 372 & 372 \end{vmatrix} = 2(372)\begin{vmatrix} 162 & 161 \\ 1 & 1 \end{vmatrix} =$$

$$2(372)[162-161] = 744$$

Exercises 6.6

In using Cramer's Rule, let D be the determinant of coefficients of the variables. Let D_x be the determinant found when the column of coefficients of x is replaced with the column of constants. Let D_y be the determinant found when the column of coefficients of y is replaced with the column of constants. Let D_z be the determinant found when the column of coefficients of z is replaced with the column of constants.

1. $\begin{aligned} 3x+5y&=-1 \\ 2x-3y&=12 \end{aligned}$ $\quad D=\begin{vmatrix} 3 & 5 \\ 2 & -3 \end{vmatrix}=3(-3)-2(5)=-19$

$$D_x=\begin{vmatrix} -1 & 5 \\ 12 & -3 \end{vmatrix}=-1(-3)-12(5)=3-60=-57$$

$$D_y=\begin{vmatrix} 3 & -1 \\ 2 & 12 \end{vmatrix}=3(12)-2(-1)=36+2=38$$

$$x=\frac{D_x}{D}=\frac{-57}{-19}=3 \text{ and } y=\frac{D_y}{D}=\frac{38}{-19}=-2$$

3. $\begin{aligned} 8x-3y&=-43 \\ 5x-7y&=-73 \end{aligned}$ $\quad D=\begin{vmatrix} 8 & -3 \\ 5 & -7 \end{vmatrix}=8(-7)-5(-3)=-41$

$$D_x=\begin{vmatrix} -43 & -3 \\ -73 & -7 \end{vmatrix}=(-43)(-7)-(-3)(-73)=301-219=82$$

$$D_y=\begin{vmatrix} 8 & -43 \\ 5 & -73 \end{vmatrix}=8(-73)-5(-43)=-584+215=-369$$

$$x=\frac{D_x}{D}=\frac{82}{-41}=-2 \text{ and } y=\frac{D_y}{D}=\frac{-369}{-41}=9$$

5. $\begin{aligned} 6x-7y&=28 \\ -4x+5y&=-20 \end{aligned}$ $\quad D=\begin{vmatrix} 6 & -7 \\ -4 & 5 \end{vmatrix}=6(5)-(-4)(-7)=30-28=2$

$$D_x=\begin{vmatrix} 28 & -7 \\ -20 & 5 \end{vmatrix}=28(5)-(-20)(-7)=140-140=0$$

$$D_y=\begin{vmatrix} 6 & 28 \\ -4 & -20 \end{vmatrix}=6(-20)-(-4)(28)=-120+112=-8$$

$$x=\frac{D_x}{D}=\frac{0}{2}=0 \text{ and } y=\frac{D_y}{D}=\frac{-8}{2}=-4$$

7. $3x+4y=5$ $\quad D=\begin{vmatrix} 3 & 4 \\ 6 & 8 \end{vmatrix}=3(8)-6(4)=0$

$6x+8y=10$

$D_x=\begin{vmatrix} 5 & 4 \\ 10 & 8 \end{vmatrix}=5(8)-10(4)=0$

$D_y=\begin{vmatrix} 3 & 5 \\ 6 & 10 \end{vmatrix}=3(10)-6(5)=0$

The system is dependent.

9. $12x-16y=24$ $\quad D=\begin{vmatrix} 12 & -16 \\ 15 & -20 \end{vmatrix}=12(-20)-15(-16)=0$

$15x-20y=36$

$D_x=\begin{vmatrix} 24 & 12 \\ 36 & 15 \end{vmatrix}=24(15)-36(12)=-72$

$D_y=\begin{vmatrix} 12 & 24 \\ 15 & 36 \end{vmatrix}=12(36)-15(24)=72$

$x=\dfrac{-72}{0}$ and $y=\dfrac{72}{0}$; both are meaningless.

The system is inconsistent.

11. $15x-6y=-15$ $\quad D=\begin{vmatrix} 15 & -6 \\ 9 & 12 \end{vmatrix}=15(12)-9(-6)=234$

$9x+12y=4$

$D_x=\begin{vmatrix} -15 & -6 \\ 4 & 12 \end{vmatrix}=-15(12)-4(-6)=-156$

$D_y=\begin{vmatrix} 15 & -15 \\ 9 & 4 \end{vmatrix}=15(4)-9(-15)=195$

$x=\dfrac{D_x}{D}=\dfrac{-156}{234}=-\dfrac{2}{3}$ and $y=\dfrac{D_y}{D}=\dfrac{195}{234}=\dfrac{5}{6}$

13. $8x+7y=18$ $\quad D=\begin{vmatrix} 8 & 7 \\ -4 & 1 \end{vmatrix}=8(1)-(-4)(7)=36$

$-4x+y=0$

$D_x=\begin{vmatrix} 18 & 7 \\ 0 & 1 \end{vmatrix}=18(1)-0(7)=18$

$D_y=\begin{vmatrix} 8 & 18 \\ -4 & 0 \end{vmatrix}=8(0)-(-4)(18)=72$

$x=\dfrac{D_x}{D}=\dfrac{18}{36}=\dfrac{1}{2}$ and $y=\dfrac{D_y}{D}=\dfrac{72}{36}=2$

15. $\begin{aligned} ax+by&=2 \\ bx+ay&=4 \end{aligned}$ $\quad D=\begin{vmatrix} a & b \\ b & a \end{vmatrix}=a^2-b^2$

$$D_x=\begin{vmatrix} 2 & b \\ 4 & a \end{vmatrix}=2a-4b \text{ and } D_y=\begin{vmatrix} a & 2 \\ b & 4 \end{vmatrix}=4a-2b$$

$$x=\frac{D_x}{D}=\frac{2a-4b}{a^2-b^2} \text{ and } y=\frac{D_y}{D}=\frac{4a-2b}{a^2-b^2}$$

17. $\begin{aligned} 5x+ay&=b \\ 2x-by&=a \end{aligned}$ $\quad D=\begin{vmatrix} 5 & a \\ 2 & -b \end{vmatrix}=-5b-2a$

$$D_x=\begin{vmatrix} b & a \\ a & -b \end{vmatrix}=-b^2-a^2 \text{ and } D_y=\begin{vmatrix} 5 & b \\ 2 & a \end{vmatrix}=5a-2b$$

$$x=\frac{D_x}{D}=\frac{-(a^2+b^2)}{-(2a+5b)}=\frac{a^2+b^2}{2a+5b}$$

$$y=\frac{D_y}{D}=\frac{-(2b-5a)}{-(2a+5b)}=\frac{2b-5a}{2a+5b}$$

19. $\begin{aligned} ax+by&=c \\ -bx+y&=0 \end{aligned}$ $\quad D=\begin{vmatrix} a & b \\ -b & 1 \end{vmatrix}=a-(-b)(b)=a+b^2$

$$D_x=\begin{vmatrix} c & b \\ 0 & 1 \end{vmatrix}=c-0(b)=c \text{ and } D_y=\begin{vmatrix} a & c \\ -b & 0 \end{vmatrix}=a(0)-(-b)(c)=bc$$

$$x=\frac{D_x}{D}=\frac{c}{a+b^2} \text{ and } y=\frac{D_y}{D}=\frac{bc}{a+b^2}$$

21. Let X = the smaller voltage and Y = the larger voltage.

$X+Y=210$ These equations $X+Y=210$

$Y=2X-15$ become: $-2X+Y=-15$

$$D=\begin{vmatrix} 1 & 1 \\ -2 & 1 \end{vmatrix}=1(1)-(-2)(1)=1+2=3$$

$$D_x=\begin{vmatrix} 210 & 1 \\ -15 & 1 \end{vmatrix}=210(1)-(-15)(1)=225$$

$$D_y=\begin{vmatrix} 1 & 210 \\ -2 & -15 \end{vmatrix}=1(-15)-(-2)(210)=405$$

$$X=\frac{D_x}{D}=\frac{225}{3}=75V \text{ and } Y=\frac{D_y}{D}=\frac{405}{3}=135V$$

23.

$$\begin{aligned} 3x - 4y + 7z &= -26 \\ -2x + y - 3z &= 9 \\ 12x \quad + 15z &= -36 \end{aligned} \qquad D = \begin{vmatrix} 3 & -4 & 7 \\ -2 & 1 & -3 \\ 12 & 0 & 15 \end{vmatrix}$$

Multiply row 2 by 4 and add to row 1.

$$D = \begin{vmatrix} -5 & 0 & -5 \\ -2 & 1 & -3 \\ 12 & 0 & 15 \end{vmatrix} = 1\begin{vmatrix} -5 & -5 \\ 12 & 15 \end{vmatrix} = 1[-5(15) - 12(-5)] = -15$$

$$D_x = \begin{vmatrix} -26 & -4 & 7 \\ 9 & 1 & -3 \\ -36 & 0 & 15 \end{vmatrix}$$

Multiply row 2 by 4 and add to row 1.

$$D_x = \begin{vmatrix} 10 & 0 & -5 \\ 9 & 1 & -3 \\ -36 & 0 & 15 \end{vmatrix} = 1\begin{vmatrix} 10 & -5 \\ -36 & 15 \end{vmatrix} = 10(15) - (-36)(-5) = -30$$

$$D_y = \begin{vmatrix} 3 & -26 & 7 \\ -2 & 9 & -3 \\ 12 & -36 & 15 \end{vmatrix}$$

Multiply row 2 by 1 and add to row 1.

$$D_y = \begin{vmatrix} 1 & -17 & 4 \\ -2 & 9 & -3 \\ 12 & -36 & 15 \end{vmatrix}$$

Multiply row 1 by 2 and add to row 2. Multiply row 1 by -12 and add to row 3.

$$D_y = \begin{vmatrix} 1 & -17 & 4 \\ 0 & -25 & 5 \\ 0 & 168 & -33 \end{vmatrix}$$

$$D_y = 1\begin{vmatrix} -25 & 5 \\ 168 & -33 \end{vmatrix} = (-25)(-33) - 5(168) = 825 - 840 = -15$$

$$D_z = \begin{vmatrix} 3 & -4 & -26 \\ -2 & 1 & 9 \\ 12 & 0 & -36 \end{vmatrix}$$

Multiply column 1 by 3 and add to column 3.

$$D_z = \begin{vmatrix} 3 & -4 & -17 \\ -2 & 1 & 3 \\ 12 & 0 & 0 \end{vmatrix}$$

$$D_z = 12\begin{vmatrix} -4 & -17 \\ 1 & 3 \end{vmatrix} = 12[-4(3) - (-17)(1)] = 12[-12 + 17] = 60$$

$$x = \frac{D_x}{D} = \frac{-30}{-15} = 2 \text{ and } y = \frac{D_y}{D} = \frac{-15}{-15} = 1 \text{ and } z = \frac{D_z}{D} = \frac{60}{-15} = -4$$

25. $\begin{aligned} 3x+2y+5z&=-7 \\ 8x-3y+2z&=10 \\ 7x-2y+4z&=1 \end{aligned}$ $\quad D=\begin{vmatrix} 3 & 2 & 5 \\ 8 & -3 & 3 \\ 7 & -2 & 4 \end{vmatrix}$ Multiply row 3 by -1 and add to row 2.

$$D=\begin{vmatrix} 3 & 2 & 5 \\ 1 & -1 & -2 \\ 7 & -2 & 4 \end{vmatrix}$$ Multiply row 2 by -3 and add to row 1. Multiply row 2 by -7 and add to row 3.

$$D=\begin{vmatrix} 0 & 5 & 11 \\ 1 & -1 & -2 \\ 0 & 5 & 18 \end{vmatrix}=-(1)\begin{vmatrix} 5 & 11 \\ 5 & 18 \end{vmatrix}=-[5(18)-5(11)]=-35$$

$$D_x=\begin{vmatrix} -7 & 2 & 5 \\ 10 & -3 & 2 \\ 1 & -2 & 4 \end{vmatrix}$$ Multiply row 3 by 7 and add to row 1. Multiply row 3 by -10 and add to row 2.

$$D_x=\begin{vmatrix} 0 & -12 & 33 \\ 0 & 17 & -38 \\ 1 & -2 & 4 \end{vmatrix}=1\begin{vmatrix} -12 & 33 \\ 17 & -38 \end{vmatrix}=(-12)(-38)-17(33)=-105$$

$$D_y=\begin{vmatrix} 3 & -7 & 5 \\ 8 & 10 & 2 \\ 7 & 1 & 4 \end{vmatrix}$$ Multiply row 3 by 7 and add to row 1. Multiply row 3 by -10 and add to row 2.

$$D_y=\begin{vmatrix} 52 & 0 & 33 \\ -62 & 0 & -38 \\ 7 & 1 & 4 \end{vmatrix}=-(1)\begin{vmatrix} 52 & 33 \\ -62 & -38 \end{vmatrix}=-[52(-38)-(-62)(33)]=-70$$

$$D_z=\begin{vmatrix} 3 & 2 & -7 \\ 8 & -3 & 10 \\ 7 & -2 & 1 \end{vmatrix}$$ Multiply row 3 by 7 and add to row 1. Multiply row 3 by -10 and add to row 2.

$$D_z=\begin{vmatrix} 52 & -12 & 0 \\ -62 & 17 & 0 \\ 7 & -2 & 1 \end{vmatrix}=1\begin{vmatrix} 52 & -12 \\ -62 & 17 \end{vmatrix}=52(17)-(-62)(-12)=140$$

$$x=\frac{D_x}{D}=\frac{-105}{-35}=3 \text{ and } y=\frac{D_y}{D}=\frac{-70}{-35}=2 \text{ and } z=\frac{D_z}{D}=\frac{140}{-35}=-4$$

27.

$$\begin{aligned} 7x-5y-7z&=8 \\ 9x+3y-6z&=33 \\ 4x-2y-8z&=28 \end{aligned}$$

Divide equation 2 by 3. Divide equation 3 by 2.

$$\begin{aligned} 7x-5y-7z&=8 \\ 3x+y-2z&=11 \\ 2x-y-4z&=14 \end{aligned}$$

$$D=\begin{vmatrix} 7 & -5 & -7 \\ 3 & 1 & -2 \\ 2 & -1 & -4 \end{vmatrix}$$

Multiply row 2 by 5 and add to row 1.
Multiply row 2 by 1 and add to row 3.

$$D=\begin{vmatrix} 22 & 0 & -17 \\ 3 & 1 & -2 \\ 5 & 0 & -6 \end{vmatrix}=1\begin{vmatrix} 22 & -17 \\ 5 & -6 \end{vmatrix}=22(-6)-5(-17)=-47$$

$$D_x=\begin{vmatrix} 8 & -5 & -7 \\ 11 & 1 & -2 \\ 14 & -1 & -4 \end{vmatrix}$$

Multiply row 2 by 5 and add to row 1.
Multiply row 2 by 1 and add to row 3.

$$D_x=\begin{vmatrix} 63 & 0 & -17 \\ 11 & 1 & -2 \\ 25 & 0 & -6 \end{vmatrix}=1\begin{vmatrix} 63 & -17 \\ 25 & -6 \end{vmatrix}=63(-6)-25(-17)=47$$

$$D_y=\begin{vmatrix} 7 & 8 & -7 \\ 3 & 11 & -2 \\ 2 & 14 & -4 \end{vmatrix}$$

Multiply row 2 by-2 and add to row 1.

$$D_y=\begin{vmatrix} 1 & -14 & -3 \\ 3 & 11 & -2 \\ 2 & 14 & -4 \end{vmatrix}$$

$$D_y=\begin{vmatrix} 1 & -14 & -3 \\ 3 & 11 & -2 \\ 2 & 14 & -4 \end{vmatrix}$$

Multiply row 1 by -3 and add to row 2.
Multiply row 1 by -2 and add to row 3.

$$D_y=\begin{vmatrix} 1 & -14 & -3 \\ 0 & 53 & 7 \\ 0 & 42 & 2 \end{vmatrix}=1\begin{vmatrix} 53 & 7 \\ 42 & 2 \end{vmatrix}=53(2)-42(7)=-188$$

$$D_z=\begin{vmatrix} 7 & -5 & 8 \\ 3 & 1 & 11 \\ 2 & -1 & 14 \end{vmatrix}$$

Multiply row 2 by 5 and add to row 1.
Multiply row 2 by 1 and add to row 3.

$$D_z=\begin{vmatrix} 22 & 0 & 63 \\ 3 & 1 & 11 \\ 5 & 0 & 25 \end{vmatrix}=1\begin{vmatrix} 22 & 63 \\ 5 & 25 \end{vmatrix}=22(25)-5(63)=235$$

$$x=\frac{D_x}{D}=\frac{47}{-47}=-1 \text{ and } y=\frac{D_y}{D}=\frac{-188}{-47}=4 \text{ and } z=\frac{D_z}{D}=\frac{235}{-47}=-5$$

29.

$$3x-4y-5z=0$$
$$9x+6y+10z=11$$
$$12x+2y-20z=36$$

$$D=\begin{vmatrix}3 & -4 & -5\\ 9 & 6 & 10\\ 12 & 2 & -20\end{vmatrix}$$ Multiply column 1 by -1 and add to column 3.

$$D=\begin{vmatrix}3 & -4 & -8\\ 9 & 6 & 1\\ 12 & 2 & -32\end{vmatrix}$$ Multiply row 2 by 8 and add to row 1. Multiply row 2 by 32 and add to row 3.

$$D=\begin{vmatrix}75 & 44 & 0\\ 9 & 6 & 1\\ 300 & 194 & 0\end{vmatrix}=-(1)\begin{vmatrix}75 & 44\\ 300 & 194\end{vmatrix}=-\left[75(194)-300(44)\right]=-1350$$

$$D_x=\begin{vmatrix}0 & -4 & -5\\ 11 & 6 & 10\\ 36 & 2 & -20\end{vmatrix}$$ Multiply row 1 by 2 and add to row 2. Multiply row 1 by -4 and add to row 3.

$$D_x=\begin{vmatrix}0 & -4 & -5\\ 11 & -2 & 0\\ 36 & 18 & 0\end{vmatrix}=-5\begin{vmatrix}11 & -2\\ 36 & 18\end{vmatrix}=-5\left[11(18)-(-2)(36)\right]=-1350$$

$$D_y=\begin{vmatrix}3 & 0 & -5\\ 9 & 11 & 10\\ 12 & 36 & -20\end{vmatrix}$$ Multiply row 1 by 2 and add to row 2. Multiply row 1 by -4 and add to row 3.

$$D_y=\begin{vmatrix}3 & 0 & -5\\ 15 & 11 & 0\\ 0 & 36 & 0\end{vmatrix}=-5\begin{vmatrix}15 & 11\\ 0 & 36\end{vmatrix}=-5\left[15(36)-0(11)\right]=-2700$$

$$D_z=\begin{vmatrix}3 & -4 & 0\\ 9 & 6 & 11\\ 12 & 2 & 36\end{vmatrix}$$ Multiply row 1 by -3 and add to row 2. Multiply row 1 by -4 and add to row 3.

$$D_z=\begin{vmatrix}3 & -4 & 0\\ 0 & 18 & 11\\ 0 & 18 & 36\end{vmatrix}=3\begin{vmatrix}18 & 11\\ 18 & 36\end{vmatrix}=3\left[18(36)-18(11)\right]=1350$$

$$x=\frac{D_x}{D}=\frac{-1350}{-1350}=1 \text{ and } y=\frac{D_y}{D}=\frac{-2700}{-1350}=2 \text{ and } z=\frac{D_z}{D}=\frac{1350}{-1350}=-1$$

31.

$$\begin{aligned} 2x \quad\; + 5z &= 29 \\ 5y - 7z &= -40 \\ 8x + y \quad\; &= 15 \end{aligned} \qquad D = \begin{vmatrix} 2 & 0 & 5 \\ 0 & 5 & -7 \\ 8 & 1 & 0 \end{vmatrix}$$

Multiply row 1 by -4 and add to row 3.

$$D = \begin{vmatrix} 2 & 0 & 5 \\ 0 & 5 & -7 \\ 0 & 1 & -20 \end{vmatrix} = 2\begin{vmatrix} 5 & -7 \\ 1 & -20 \end{vmatrix} = 2\left[5(-20) - (1)(-7)\right] = -186$$

$$D_x = \begin{vmatrix} 29 & 0 & 5 \\ -40 & 5 & -7 \\ 15 & 1 & 0 \end{vmatrix}$$

Multiply row 3 by -5 and add to row 2.

$$D_x = \begin{vmatrix} 29 & 0 & 5 \\ -115 & 0 & -7 \\ 15 & 1 & 0 \end{vmatrix} = -(1)\begin{vmatrix} 29 & 5 \\ -115 & -7 \end{vmatrix} = -\left[29(-7) - 5(-115)\right] = -372$$

$$D_y = \begin{vmatrix} 2 & 29 & 5 \\ 0 & -40 & -7 \\ 8 & 15 & 0 \end{vmatrix}$$

Multiply row 1 by -4 and add to row 3.

$$D_y = \begin{vmatrix} 2 & 29 & 5 \\ 0 & -40 & -7 \\ 0 & -101 & -20 \end{vmatrix} = 2\begin{vmatrix} -40 & -7 \\ -101 & -20 \end{vmatrix} = 2\left[-40(-20) - (-7)(-101)\right] = 186$$

$$D_z = \begin{vmatrix} 2 & 0 & 29 \\ 0 & 5 & -40 \\ 8 & 1 & 15 \end{vmatrix}$$

Multiply row 3 by -5 and add to row 2.

$$D_z = \begin{vmatrix} 2 & 0 & 29 \\ -40 & 0 & -115 \\ 8 & 1 & 15 \end{vmatrix} = -(1)\begin{vmatrix} 2 & 29 \\ -40 & -115 \end{vmatrix} = -\left[2(-115) - 29(-40)\right] = -930$$

$$x = \frac{D_x}{D} = \frac{-372}{-186} = 2 \text{ and } y = \frac{D_y}{D} = \frac{186}{-186} = -1 \text{ and } z = \frac{D_z}{D} = \frac{-930}{-186} = 5$$

33.

$$\begin{aligned} 4x+6y+8z&=-8 \\ -x \quad\quad +5z&=-19 \\ 5y+7z&=-21 \end{aligned} \qquad D=\begin{vmatrix} 4 & 6 & 8 \\ -1 & 0 & 5 \\ 0 & 5 & 7 \end{vmatrix}$$

Multiply row 2 by 4 and add to row 1.

$$D=\begin{vmatrix} 0 & 6 & 28 \\ -1 & 0 & 5 \\ 0 & 5 & 7 \end{vmatrix} = -(-1)\begin{vmatrix} 6 & 28 \\ 5 & 7 \end{vmatrix} = 6(7)-5(28)=-98$$

$$D_x=\begin{vmatrix} -8 & 6 & 8 \\ -19 & 0 & 5 \\ -21 & 5 & 7 \end{vmatrix}$$

Multiply row 3 by -1 and add to row 1.

$$D_x=\begin{vmatrix} 13 & 1 & 1 \\ -19 & 0 & 5 \\ -21 & 5 & 7 \end{vmatrix}$$

Multiply row 1 by -5 and add to row 3.

$$D_x=\begin{vmatrix} 13 & 1 & 1 \\ -19 & 0 & 5 \\ -86 & 0 & 2 \end{vmatrix}$$

$$D_x=-(1)\begin{vmatrix} -19 & 5 \\ -86 & 2 \end{vmatrix} = -\left[-19(2)-5(-86)\right]=-392$$

$$D_y=\begin{vmatrix} 4 & -8 & 8 \\ -1 & -19 & 5 \\ 0 & -21 & 7 \end{vmatrix}$$

Multiply row 2 by 4 and add to row 1.

$$D_y=\begin{vmatrix} 0 & -84 & 28 \\ -1 & -19 & 5 \\ 0 & -21 & 7 \end{vmatrix} = -(-1)\begin{vmatrix} -84 & 28 \\ -21 & 7 \end{vmatrix} = 7(-84)-(-21)(28)=0$$

$$D_z=\begin{vmatrix} 4 & 6 & -8 \\ -1 & 0 & -19 \\ 0 & 5 & -21 \end{vmatrix}$$

Multiply row 2 by 4 and add to row 1.

$$D_z=\begin{vmatrix} 0 & 6 & -84 \\ -1 & 0 & -19 \\ 0 & 5 & -21 \end{vmatrix} = -(-1)\begin{vmatrix} 6 & -84 \\ 5 & -21 \end{vmatrix} = 6(-21)-5(-84)=294$$

$$x=\frac{D_x}{D}=\frac{-392}{-98}=4 \text{ and } y=\frac{D_y}{D}=\frac{0}{-98}=0 \text{ and } z=\frac{D_z}{D}=\frac{294}{-98}=-3$$

35.

Let $x =$ the shortest side

$x+2 =$ the longest side

$21-(x+x+2) =$ the third side

$x+x+2=2(21-2x-2)$

$2x+2=42-4x-4$

$6x=42-6$

$6x=36$

$x=6$ cm

$x+2=8$ cm

$21-(6+8)=7$ cm

37.
$$\begin{aligned} 3x-2y+z-3w&=-20\\ 2x \quad +5z+w&=-3\\ 5x+y-z+4w&=30\\ 6x-3y+4z \quad &=-11 \end{aligned} \qquad D=\begin{vmatrix} 3 & -2 & 1 & -3\\ 2 & 0 & 5 & 1\\ 5 & 1 & -1 & 4\\ 6 & -3 & 4 & 0 \end{vmatrix}$$

Multiply row 3 by 2 and add to row 1. Multiply row 3 by 3 and add to row 4.
$$D=\begin{vmatrix} 13 & 0 & -1 & 5\\ 2 & 0 & 5 & 1\\ 5 & 1 & -1 & 4\\ 21 & 0 & 1 & 12 \end{vmatrix}$$

$$D=-1\begin{vmatrix} 13 & -1 & 5\\ 2 & 5 & 1\\ 21 & 1 & 12 \end{vmatrix}$$
Multiply row 1 by 5 and add to row 2. Multiply row 1 by 1 and add to row 3.

$$D=-1\begin{vmatrix} 13 & -1 & 5\\ 67 & 0 & 26\\ 34 & 0 & 17 \end{vmatrix}=-1(-)(-1)\begin{vmatrix} 67 & 26\\ 34 & 17 \end{vmatrix}=-[67(17)-34(26)]=-255$$

$$D_x=\begin{vmatrix} -20 & -2 & 1 & -3\\ -3 & 0 & 5 & 1\\ 30 & 1 & -1 & 4\\ -11 & -3 & 4 & 0 \end{vmatrix}$$
Multiply row 3 by 2 and add to row 1. Multiply row 3 by 3 and add to row 4.

$$D_x=\begin{vmatrix} 40 & 0 & -1 & 5\\ -3 & 0 & 5 & 1\\ 30 & 1 & -1 & 4\\ 79 & 0 & 1 & 12 \end{vmatrix}=-1\begin{vmatrix} 40 & -1 & 5\\ -3 & 5 & 1\\ 79 & 1 & 12 \end{vmatrix}$$

Multiply row 1 by 5 and add to row 2. Multiply row 1 by 1 and add to row 3.
$$D_x=-1\begin{vmatrix} 40 & -1 & 5\\ 197 & 0 & 26\\ 119 & 0 & 17 \end{vmatrix}$$

$$D_x=-1(-)(-1)\begin{vmatrix} 197 & 26\\ 119 & 17 \end{vmatrix}=-1[197(17)-119(26)]=-255$$

$$D_y=\begin{vmatrix} 3 & -20 & 1 & -3\\ 2 & -3 & 5 & 1\\ 5 & 30 & -1 & 4\\ 6 & -11 & 4 & 0 \end{vmatrix}$$
Multiply row 2 by 3 and add to row 1. Multiply row 2 by – 4 and add to row 3.

$$D_y=\begin{vmatrix} 9 & -29 & 26 & 0\\ 2 & -3 & 5 & 1\\ -3 & 42 & -21 & 0\\ 6 & -11 & 4 & 0 \end{vmatrix}=1\begin{vmatrix} 9 & -29 & 16\\ -3 & 42 & -21\\ 6 & -11 & 4 \end{vmatrix}$$

37. (con't)

Multiply row 2 by 3 and add to row 1. Multiply row 2 by 2 and add to row 3.

$$D_y = \begin{vmatrix} 0 & 97 & -47 \\ -3 & 42 & -21 \\ 0 & 73 & -38 \end{vmatrix}$$

$$D_y = -(-3)\begin{vmatrix} 97 & -47 \\ 73 & -38 \end{vmatrix} = 3[97(-38) - 73(-47)] = -765$$

$$D_z = \begin{vmatrix} 3 & -2 & -20 & -3 \\ 2 & 0 & -3 & 1 \\ 5 & 1 & 30 & 4 \\ 6 & -3 & -11 & 0 \end{vmatrix}$$

Multiply row 3 by 2 and add to row 1. Multiply row 3 by 3 and add to row 4.

$$D_z = \begin{vmatrix} 13 & 0 & 40 & 5 \\ 2 & 0 & -3 & 1 \\ 5 & 1 & 30 & 4 \\ 21 & 0 & 79 & 12 \end{vmatrix} = -1\begin{vmatrix} 13 & 40 & 5 \\ 2 & -3 & 1 \\ 21 & 79 & 12 \end{vmatrix}$$

Multiply row 2 by -5 and add to row 1. Multiply row 2 by -12 and add to row

$$D_z = -1\begin{vmatrix} 3 & 55 & 0 \\ 2 & -3 & 1 \\ -3 & 115 & 0 \end{vmatrix}$$

$$D_z = -1(-)(1)\begin{vmatrix} 3 & 55 \\ -3 & 115 \end{vmatrix} = 3(115) - (-3)(55) = 510$$

$$D_w = \begin{vmatrix} 3 & -2 & 1 & -20 \\ 2 & 0 & 5 & -3 \\ 5 & 1 & -1 & 30 \\ 6 & -3 & 4 & -11 \end{vmatrix}$$

Multiply row 3 by 2 and add to row 1. Multiply row 3 by 3 and add to row 4.

$$D_w = \begin{vmatrix} 13 & 0 & -1 & 40 \\ 2 & 0 & 5 & -3 \\ 5 & 1 & -1 & 30 \\ 21 & 0 & 1 & 79 \end{vmatrix} = -(1)\begin{vmatrix} 13 & -1 & 40 \\ 2 & 5 & -3 \\ 21 & 1 & 79 \end{vmatrix}$$

Multiply row 1 by 5 and add to row 2. Multiply row 1 by 1 and add to row 3.

$$D_w = -1\begin{vmatrix} 13 & -1 & 40 \\ 67 & 0 & 197 \\ 34 & 0 & 119 \end{vmatrix}$$

$$D_w = -1(-)(-1)\begin{vmatrix} 67 & 197 \\ 34 & 119 \end{vmatrix} = -[67(119) - 34(197)] = -1275$$

$$x = \frac{D_x}{D} = \frac{-255}{-255} = 1 \text{ and } y = \frac{D_y}{D} = \frac{-765}{-255} = 3$$

$$z = \frac{D_z}{D} = \frac{510}{-255} = -2 \text{ and } w = \frac{D_w}{D} = \frac{-1275}{-255} = 5$$

39.

$$\begin{aligned} 2x+2y-z+3w-4v&=13 \\ 3x+7y-z \quad -3v&=-8 \\ 3y+2z+w-v&=5 \\ -2x \quad +3z-w+v&=0 \\ 5x+ \quad 7w \quad &=38 \end{aligned}$$

$$D=\begin{vmatrix} 2 & 2 & -1 & 3 & -4 \\ 3 & 7 & -1 & 0 & -3 \\ 0 & 3 & 2 & 1 & -1 \\ -2 & 0 & 3 & -1 & 1 \\ 5 & 0 & 0 & 7 & 0 \end{vmatrix}$$

Multiply row 3 by -3 and add to row 1. Multiply row 3 by 1 and add to row 4. Multiply row 3 by -7 and add to row 5.

$$D=\begin{vmatrix} 2 & -7 & -7 & 0 & -1 \\ 3 & 7 & -1 & 0 & -3 \\ 0 & 3 & 2 & 1 & -1 \\ -2 & 3 & 5 & 0 & 0 \\ 5 & -21 & -14 & 0 & 7 \end{vmatrix}$$

$$D=-1\begin{vmatrix} 2 & -7 & -7 & -1 \\ 3 & 7 & -1 & -3 \\ -2 & 3 & 5 & 0 \\ 5 & -21 & -14 & 7 \end{vmatrix}$$

Multiply row 1 by -3 and add to row 2. Multiply row 1 by 7 and add to row 4.

$$D=-1\begin{vmatrix} 2 & -7 & -7 & -1 \\ -3 & 28 & 20 & 0 \\ -2 & 3 & 5 & 0 \\ 19 & -70 & -63 & 0 \end{vmatrix}=-1(-)(-1)\begin{vmatrix} -3 & 28 & 20 \\ -2 & 3 & 5 \\ 19 & -70 & -63 \end{vmatrix}$$

Multiply row 2 by -1 and add to row 1.

$$D=-\begin{vmatrix} -1 & 25 & 15 \\ -2 & 3 & 5 \\ 19 & -70 & -63 \end{vmatrix}$$

Multiply row 1 by -2 and add to row 2. Multiply row 1 by 19 and add to row 3.

$$D=-\begin{vmatrix} -1 & 25 & 15 \\ 0 & -47 & -25 \\ 0 & 405 & 222 \end{vmatrix}$$

$$D=-(-1)\begin{vmatrix} -47 & -25 \\ 405 & 222 \end{vmatrix}=(-47)(222)-405(-25)=-309$$

$$D_x=\begin{vmatrix} 13 & 2 & -1 & 3 & -4 \\ -8 & 7 & -1 & 0 & -3 \\ 5 & 3 & 2 & 1 & -1 \\ 0 & 0 & 3 & -1 & 1 \\ 38 & 0 & 0 & 7 & 0 \end{vmatrix}$$

Multiply column 5 by -3 and add to column 3. Multiply column 5 by 1and add to column 4.

$$D_x=\begin{vmatrix} 13 & 2 & 11 & -1 & -4 \\ -8 & 7 & 8 & -3 & -3 \\ 5 & 3 & 5 & 0 & -1 \\ 0 & 0 & 0 & 0 & 1 \\ 38 & 0 & 0 & 7 & 0 \end{vmatrix}=-1\begin{vmatrix} 13 & 2 & 11 & -1 \\ -8 & 7 & 8 & -3 \\ 5 & 3 & 5 & 0 \\ 38 & 0 & 0 & 7 \end{vmatrix}$$

39. (con't)

Multiply row 1 by -3 and add to row 2. Multiply row 1 by 7 and add to row 4.

$$D_x = -1\begin{vmatrix} 13 & 2 & 11 & -1 \\ -47 & 1 & -25 & 0 \\ 5 & 3 & 5 & 0 \\ 129 & 14 & 77 & 0 \end{vmatrix}$$

$$D_x = -1(-)(-1)\begin{vmatrix} -47 & 1 & -25 \\ 5 & 3 & 5 \\ 129 & 14 & 77 \end{vmatrix}$$

Multiply row 1 by -3 and add to row 2. Multiply row 1 by -14 and add to row 3.

$$D_x = -1\begin{vmatrix} -47 & 1 & -25 \\ 146 & 0 & 80 \\ 787 & 0 & 427 \end{vmatrix} = -1(-)(1)\begin{vmatrix} 146 & 80 \\ 787 & 427 \end{vmatrix}$$

$$D_x = 146(427) - 80(787) = -618$$

$$D_y = \begin{vmatrix} 2 & 13 & -1 & 3 & -4 \\ 3 & -8 & -1 & 0 & -3 \\ 0 & 5 & 2 & 1 & -1 \\ -2 & 0 & 3 & -1 & 1 \\ 5 & 38 & 0 & 7 & 0 \end{vmatrix}$$

Multiply row 1 by -1 and add to row 2. Multiply row 1 by 2 and add to row 3. Multiply row 1 by 3 and add to row 4.

$$D_y = \begin{vmatrix} 2 & 13 & -1 & 3 & -4 \\ 1 & -21 & 0 & -3 & 1 \\ 4 & 31 & 0 & 7 & -9 \\ 4 & 39 & 0 & 8 & -11 \\ 5 & 38 & 0 & 7 & 0 \end{vmatrix} = (-1)\begin{vmatrix} 1 & -21 & -3 & 1 \\ 4 & 31 & 7 & -9 \\ 4 & 39 & 8 & -11 \\ 5 & 38 & 7 & 0 \end{vmatrix}$$

Multiply row 1 by 9 and add to row 2. Multiply row 1 by 11 and add to row 3.

$$D_y = -\begin{vmatrix} 1 & -21 & -3 & 1 \\ 13 & -158 & -20 & 0 \\ 15 & -192 & -25 & 0 \\ 5 & 38 & 7 & 0 \end{vmatrix}$$

$$D_y = -(-1)\begin{vmatrix} 13 & -158 & -20 \\ 15 & -192 & -25 \\ 5 & 38 & 7 \end{vmatrix}$$

Multiply row 3 by 3 and add to row 1.

$$D_y = \begin{vmatrix} 28 & -44 & 1 \\ 15 & -192 & -25 \\ 5 & 38 & 7 \end{vmatrix}$$

Multiply row 1 by 25 and add to row 2. Multiply row 1 by -7 and add to row 3.

$$D_y = \begin{vmatrix} 28 & -44 & 1 \\ 715 & -1292 & 0 \\ -191 & 346 & 0 \end{vmatrix} = -1\begin{vmatrix} 715 & -1292 \\ -191 & 346 \end{vmatrix}$$

$$D_y = [715(346) - (-191)(-1292)] = 618$$

39. (con't)

$$D_z = \begin{vmatrix} 2 & 2 & 13 & 3 & -4 \\ 3 & 7 & -8 & 0 & -3 \\ 0 & 3 & 5 & 1 & -1 \\ -2 & 0 & 0 & -1 & 1 \\ 5 & 0 & 38 & 7 & 0 \end{vmatrix}$$

Multiply row 4 by 4 and add to row 1.
Multiply row 4 by 3 and add to row 2.
Multiply row 4 by 1 and add to row 3.

$$D_z = \begin{vmatrix} -6 & 2 & 13 & -1 & 0 \\ -3 & 7 & -8 & -3 & 0 \\ -2 & 3 & 5 & 0 & 0 \\ -2 & 0 & 0 & -1 & 1 \\ 5 & 0 & 38 & 7 & 0 \end{vmatrix} = -1\begin{vmatrix} -6 & 2 & 13 & -1 \\ -3 & 7 & -8 & -3 \\ -2 & 3 & 5 & 0 \\ 5 & 0 & 38 & 7 \end{vmatrix}$$

Multiply row 1 by -3 and add to row 2.
Multiply row 1 by 7 and add to row 4.

$$D_z = -1\begin{vmatrix} -6 & 2 & 13 & -1 \\ 15 & 1 & -47 & 0 \\ -2 & 3 & 5 & 0 \\ -37 & 14 & 129 & 0 \end{vmatrix}$$

$$D_z = -1(-)(-1)\begin{vmatrix} 15 & 1 & -47 \\ -2 & 3 & 5 \\ -37 & 14 & 129 \end{vmatrix}$$

Multiply row 1 by -3 and add to row 2.
Multiply row 1 by -14 and add to row 3.

$$D_z = -\begin{vmatrix} 15 & 1 & -47 \\ -47 & 0 & 146 \\ -247 & 0 & 787 \end{vmatrix} = -1(-1)\begin{vmatrix} -47 & 146 \\ -247 & 787 \end{vmatrix}$$

$$D_z = (-47)(787) - 146(-247) = -927$$

$$D_w = \begin{vmatrix} 2 & 2 & -1 & 13 & -4 \\ 3 & 7 & -1 & -8 & -3 \\ 0 & 3 & 2 & 5 & -1 \\ -2 & 0 & 3 & 0 & 1 \\ 5 & 0 & 0 & 38 & 0 \end{vmatrix}$$

Multiply row 4 by 4 and add to row 1.
Multiply row 4 by 3 and add to row 2.
Multiply row 4 by 1 and add to row 3.

$$D_w = \begin{vmatrix} -6 & 2 & 11 & 13 & 0 \\ -3 & 7 & 8 & -8 & 0 \\ -2 & 3 & 5 & 5 & 0 \\ -2 & 0 & 3 & 0 & 1 \\ 5 & 0 & 0 & 38 & 0 \end{vmatrix} = -\begin{vmatrix} -6 & 2 & 11 & 13 \\ -3 & 7 & 8 & -8 \\ -2 & 3 & 5 & 5 \\ 5 & 0 & 0 & 38 \end{vmatrix}$$

Multiply row 1 by -1 and add to row 3.

$$D_w = -\begin{vmatrix} -6 & 2 & 11 & 13 \\ -3 & 7 & 8 & -8 \\ 4 & 1 & -6 & -8 \\ 5 & 0 & 0 & 38 \end{vmatrix}$$

Multiply row 3 by -2 and add to row 1.
Multiply row 3 by -7 and add to row 2.

$$D_w = -\begin{vmatrix} -14 & 0 & 23 & 29 \\ -31 & 0 & 50 & 48 \\ 4 & 1 & -6 & -8 \\ 5 & 0 & 0 & 38 \end{vmatrix}$$

$$D_w = -(-1)\begin{vmatrix} -14 & 23 & 29 \\ -31 & 50 & 48 \\ 5 & 0 & 38 \end{vmatrix}$$

Multiply row 3 by 3 and add to row 1.

39. (con't)

$$D_w = \begin{vmatrix} 1 & 23 & 143 \\ -31 & 50 & 48 \\ 5 & 0 & 38 \end{vmatrix}$$

Multiply row 1 by 31 and add to row 2.
Multiply row 1 by -5 and add to row 3.

$$D_w = \begin{vmatrix} 1 & 23 & 143 \\ 0 & 763 & 4481 \\ 0 & -115 & -677 \end{vmatrix} = 1\begin{vmatrix} 763 & 4481 \\ -115 & -677 \end{vmatrix}$$

$$D_w = (763)(-677) - (-115)(4481) = -1236$$

$$D_v = \begin{vmatrix} 2 & 2 & -1 & 3 & 13 \\ 3 & 7 & -1 & 0 & -8 \\ 0 & 3 & 2 & 1 & 5 \\ -2 & 0 & 3 & -1 & 0 \\ 5 & 0 & 0 & 7 & 38 \end{vmatrix}$$

Multiply row 1 by -1 and add to row 2.
Multiply row 1 by 2 and add to row 3.
Multiply row 1 by 3 and add to row 4.

$$D_v = \begin{vmatrix} 2 & 2 & -1 & 3 & 13 \\ 1 & 5 & 0 & -3 & -21 \\ 4 & 7 & 0 & 7 & 31 \\ 4 & 6 & 0 & 8 & 39 \\ 5 & 0 & 0 & 7 & 38 \end{vmatrix} = -1\begin{vmatrix} 1 & 5 & -3 & -21 \\ 4 & 7 & 7 & 31 \\ 4 & 6 & 8 & 39 \\ 5 & 0 & 7 & 38 \end{vmatrix}$$

Multiply row 1 by -4 and add to row 2.
Multiply row 1 by -4 and add to row 3.
Multiply row 1 by -5 and add to row 4.

$$D_v = -1\begin{vmatrix} 1 & 5 & -3 & -21 \\ 0 & -13 & 19 & 115 \\ 0 & -14 & 20 & 123 \\ 0 & -25 & 22 & 143 \end{vmatrix}$$

$$D_v = -1\begin{vmatrix} -13 & 19 & 115 \\ -14 & 20 & 123 \\ -25 & 22 & 143 \end{vmatrix}$$

Multiply row 1 by -1 and add to row 2.

$$D_v = -1\begin{vmatrix} -13 & 19 & 115 \\ -1 & 1 & 8 \\ -25 & 22 & 143 \end{vmatrix}$$

Multiply row 2 by -19 and add to row 1.
Multiply row 2 by -22 and add to row 3.

$$D_v = -1\begin{vmatrix} 6 & 0 & -37 \\ -1 & 1 & 8 \\ -3 & 0 & -33 \end{vmatrix} = -1\begin{vmatrix} 6 & -37 \\ -3 & -33 \end{vmatrix}$$

$$D_v = -1[6(-33) - (-3)(-37)] = 309$$

$$x = \frac{D_x}{D} = \frac{-618}{-309} = 2 \text{ and } y = \frac{D_y}{D} = \frac{618}{-309} = -2 \text{ and } z = \frac{D_z}{D} = \frac{-927}{-309} = 3$$

$$w = \frac{D_w}{D} = \frac{-1236}{-309} = 4 \text{ and } v = \frac{D_v}{D} = \frac{309}{-309} = -1$$

Exercises 6.7

1. $\dfrac{8x-29}{(x+2)(x-7)}=\dfrac{A}{x+2}+\dfrac{B}{x-7}$

Multiply each side by L.C.D: $(x+2)(x-7)$.

$8x-29=A(x-7)+B(x+2)$

$8x-29=(A+B)x-7A+2B$

Thus: $A+B=8$

$-7A+2B=-29$

The solution of this system of equations is: $A=5,\ B=3$.

$\dfrac{5}{x+2}+\dfrac{3}{x-7}$

3. $\dfrac{-x-18}{2x^2-5x-12}=$

$\dfrac{-x-18}{(2x+3)(x-4)}=\dfrac{A}{2x+3}+\dfrac{B}{x-4}$

Multiply each side by L.C.D.: $(2x+3)(x-4)$.

$-x-18=A(x-4)+B(2x+3)$

$-x-18=(A+2B)x-4A+3B$

Thus: $A+2B=-1$

$-4A+3B=-18$

The solution of this system is: $A=3,\ B=-2$

$\dfrac{3}{2x+3}+\dfrac{-2}{x-4}$

5. $\dfrac{61x^2-53x-28}{x(3x-4)(2x+1)}=\dfrac{A}{x}+\dfrac{B}{3x-4}+\dfrac{C}{2x+1}$

Multiply each side by the L.C.D.: $x(3x-4)(2x+1)$.

$61x^2-53x-28=A(3x-4)(2x+1)+Bx(2x+1)+Cx(3x-4)$

$61x^2-53x-28=A(6x^2-5x-4)+2Bx^2+Bx+3Cx^2-4Cx$

$61x^2-53x-28=6Ax^2-5Ax-4A+2Bx^2+Bx+3Cx^2-4Cx$

Thus: $6A+2B+3C=61$

$-5A+B-4C=-53$

$-4A\qquad=-28$

The solution to this system is: $A=7,\ B=2,\ C=5$

$\dfrac{7}{x}+\dfrac{2}{3x-4}+\dfrac{5}{2x+1}$

7. $\dfrac{x^2+7x-10}{(x+1)(x+3)^2}=\dfrac{A}{x+1}+\dfrac{B}{x+3}+\dfrac{C}{(x+3)^2}$

Multiply each side by the L.C.D.: $(x+1)(x+3)^2$.

$x^2+7x+10=A(x+3)^2+B(x+1)(x+3)+C(x+1)$

$x^2+7x+10=A(x^2+6x+9)+B(x^2+4x+3)+Cx+C$

$x^2+7x-10=Ax^2+6Ax+9A+Bx^2+4Bx+3B+Cx+C$

Thus: $A+B=1$

$6A+4B+C=7$

$9A+3B+C=10$

The solution to this system is: $A=1,\ B=0,\ C=1$

$\dfrac{1}{x+1}+\dfrac{1}{(x+3)^2}$

9. $$\frac{48x^2-20x-5}{(4x-1)^3}=\frac{A}{4x-1}+\frac{B}{(4x-1)^2}+\frac{C}{(4x-1)^3}$$

Multiply each side by the L.C.D.: $(4x-1)^3$.

$$48x^2-20x-5=A(4x-1)^2+B(4x-1)+C$$
$$48x^2-20x-5=A(16x^2-8x+1)+4Bx-B+C$$
$$48x^2-20x-5=16Ax^2-8Ax+A+4Bx-B+C$$

Thus: $16A=48$, $-8A+4B=-20$, $A-B+C=-5$

The solution to this system is: $A=3,\ B=1,\ C=-7$

$$\frac{3}{4x-1}+\frac{1}{(4x-1)^2}+\frac{-7}{(4x-1)^3}$$

11. $$\frac{11x^2-18x+3}{x(x-1)^2}=\frac{A}{x}+\frac{B}{x-1}+\frac{C}{(x-1)^2}$$

Multiply each side by the L.C.D.: $x(x-1)^2$.

$$11x^2-18x+3=A(x-1)^2+Bx(x-1)+Cx$$
$$11x^2-18x+3=A(x^2-2x+1)+Bx^2-Bx+Cx$$
$$11x^2-18x+3=Ax^2-2Ax+A+Bx^2-Bx+Cx$$

Thus: $A+B=11$, $-2A-B+C=-18$, $A=3$

The solution to this system is: $A=3,\ B=8,\ C=-4$

$$\frac{3}{x}+\frac{8}{x-1}+\frac{-4}{(x-1)^2}$$

13. $$\frac{-x^2-4x+3}{(x^2+1)(x^2-3)}=\frac{Ax+B}{x^2+1}+\frac{Cx+D}{x^2-3}$$

Multiply each side the L.C.D.: $(x^2+1)(x^2-3)$.

$$-x^2-4x+3=(Ax+B)(x^2-3)+(Cx+D)(x^2+1)$$
$$-x^2-4x+3=Ax^3+Bx^2-3Ax-3B+Cx^3+Dx^2+Cx+D$$

Thus: $A+C=0$, $B+D=-1$, $-3A+C=-4$, $-3B+D=3$

The solution to this system is: $A=1,\ B=-1,\ C=-1,\ D=0$

$$\frac{x-1}{x^2+1}+\frac{-x}{x^2-3}$$

15. $$\frac{4x^3-21x-6}{\left(x^2+x+1\right)\left(x^2-5\right)}=\frac{Ax+B}{x^2+x+1}+\frac{Cx+D}{x^2-5}$$

Multiply each side by the L.C.D.:$\left(x^2+x+1\right)\left(x^2-5\right)$.

$4x^3-21x-6=(Ax+B)\left(x^2-5\right)+(Cx+D)\left(x^2+x+1\right)$

$4x^3-21x-6=Ax^3+Bx^2-5Ax-5B+Cx^3+Cx^2+Cx+Dx^2+Dx+D$

Thus:
$$\begin{aligned} A \quad\ +C \qquad &=4\\ B+C+D&=0\\ -5A \qquad +C+D&=-21\\ -5B \qquad +D&=-6 \end{aligned}$$

The solution to this system is: $A=4,\ B=1,\ C=0,\ D=-1$

$$\frac{4x+1}{x^2+x+1}+\frac{-1}{x^2-5}$$

17. $$\frac{4x^3-16x^2-93x-9}{\left(x^2+5x+3\right)(x+3)(x-3)}=\frac{A}{x+3}+\frac{B}{x-3}+\frac{Cx+D}{x^2+5x+3}$$

Multiply each side by the L.C.D.: $(x+3)(x-3)\left(x^2+5x+3\right)$

$4x^3-16x^2-93x-9=A(x-3)\left(x^2+5x+3\right)$
$\qquad +B(x+3)\left(x^2+5x+3\right)+(Cx+D)(x+3)(x-3)$

$4x^3-16x^2-93x-9=A\left(x^3+5x^2+3x-3x^2-15x-9\right)$
$\qquad +B\left(x^3+5x^2+3x+3x^2+15x+9\right)+(Cx+D)\left(x^2-9\right)$

$4x^3-16x^2-93x-9=Ax^3+2Ax^2-12Ax-9A+$
$\qquad Bx^3+8Bx^2+18Bx+9B+Cx^3+Dx^2-9Cx-9D$

Thus:
$$\begin{aligned} A+B+C \qquad &=4\\ 2A+8B \qquad +D&=-16\\ -12A+18B-9C \qquad &=-93\\ -9A+9B \qquad -9D&=-9 \end{aligned}$$

The solution to this system is: $A=1,\ B=-2,\ C=5,\ D=-2$

$$\frac{1}{x+3}+\frac{-2}{x-3}+\frac{5x-2}{x^2+5x+3}$$

19. $$\frac{8x^4-x^3+13x^2-6x+5}{x\left(x^2+1\right)^2}=\frac{A}{x}+\frac{Bx+C}{x^2+1}+\frac{Dx+E}{\left(x^2+1\right)^2}$$

Multiply each side by the L.C.D.: $x\left(x^2+1\right)^2$.

$8x^4-x^3+13x^2-6x+5=A\left(x^2+1\right)^2+(Bx+C)(x)\left(x^2+1\right)+(Dx+E)x$

$8x^4-x^3+13x^2-6x+5=A\left(x^4+2x^2+1\right)+Bx^4+Cx^3+Bx^2+Cx+Dx^2+Ex$

$8x^4-x^3+13x^2-6x+5=Ax^4+2Ax^2+A+Bx^4+Cx^3+Bx^2+Cx+Dx^2+Ex$

Thus:
$$\begin{aligned} A+B \qquad &=8\\ C \qquad &=-1\\ 2A+B \quad +D \qquad &=13\\ C \quad +E&=-6\\ A \qquad &=5 \end{aligned}$$

The solution to this system is: $A=5,\ B=3,\ C=-1,\ D=0,\ E=-5$

$$\frac{5}{x}+\frac{3x-1}{x^2+1}+\frac{-5}{\left(x^2+1\right)^2}$$

21. $$\frac{x^5-2x^4-8x^2+4x-8}{x^2\left(x^2+2\right)^2}=\frac{A}{x}+\frac{B}{x^2}+\frac{Cx+D}{x^2+2}+\frac{Ex+F}{\left(x^2+2\right)^2}$$

Multiply each side by the L.C.D.: $x^2\left(x^2+2\right)^2$.

$$x^5-2x^4-8x^2+4x-8=Ax\left(x^2+2\right)^2+B\left(x^2+2\right)^2+(Cx+D)x^2\left(x^2+2\right)+(Ex+F)x^2$$

$$x^5-2x^4-8x^2+4x-8=Ax\left(x^4+4x^2+4\right)+B\left(x^4+4x^2+4\right)+Cx^5+Dx^4+2Cx^3+2Dx^2+Ex^3+Fx^2$$

$$x^5-2x^4-8x^2+4x-8=Ax^5+4Ax^3+4Ax+Bx^4+4Bx^2+4B+Cx^5+Dx^4+2Cx^3+2Dx^2+Ex^3+Fx^2$$

Thus:

$$\begin{aligned} A+C&=1\\ B+D&=-2\\ 4A+2C+E&=0\\ 4B+2D+F&=-8\\ 4A&=4\\ 4B&=-8 \end{aligned}$$

The solution to this system is: $A=1,\ B=-2,\ C=0,\ D=0,\ E=-4,\ F=0$

$$\frac{1}{x}+\frac{-2}{x^2}+\frac{-4x}{\left(x^2+2\right)^2}$$

23. $$\frac{6x^2+108x+54}{x^4-81}=\frac{6x^2+108x+54}{\left(x^2+9\right)\left(x^2-9\right)}=\frac{6x^2+108x+54}{\left(x^2+9\right)(x+3)(x-3)}$$

$$\frac{6x^2+108x+54}{\left(x^2+9\right)(x+3)(x-3)}=\frac{A}{x+3}+\frac{B}{x-3}+\frac{Cx+D}{x^2+9}$$

Multiply each side by the L.C.D.: $(x+3)(x-3)\left(x^2+9\right)$.

$$6x^2+108x+54=A(x-3)\left(x^2+9\right)+B(x+3)\left(x^2+9\right)+(Cx+D)\left(x^2-9\right)$$

$$6x^2+108x+54=A\left(x^3-3x^2+9x-27\right)+B\left(x^3+3x^2+9x+27\right)+Cx^3+Dx^2-9Cx-9D$$

$$6x^2+108x+54=Ax^3-3Ax^2+9Ax-27A+Bx^3+3Bx^2+9Bx+27B+Cx^3+Dx^2-9Cx-9D$$

Thus:

$$\begin{aligned} A+B+C&=0\\ -3A+3B+D&=6\\ 9A+9B-9C&=108\\ -27A+27B-9D&=54 \end{aligned}$$

The solution to this system is: $A=2,\ B=4,\ C=-6,\ D=0$

$$\frac{2}{x+3}+\frac{4}{x-3}+\frac{-6x}{x^2+9}$$

25. $\dfrac{x^3}{x^2-1}$ Since the degree of the numerator is greater than the degree of the demoninator, divide first.

$$x^2-1\overline{)x^3}\quad\text{quotient } x;\quad \frac{x^3-x}{x}$$

or $x+\dfrac{x}{x^2-1}=x+\dfrac{x}{(x-1)(x+1)}$

$$\frac{x}{(x-1)(x+1)}=\frac{A}{x-1}+\frac{B}{x+1}$$

Multiply each side by the L.C.D.: $(x-1)(x+1)$.

$x=A(x+1)+B(x-1)=Ax+A+Bx-B$

Thus: $A+B=1$ The solution to this system is:

$A-B=0$ $A=\dfrac{1}{2}, B=\dfrac{1}{2}$

$$x+\frac{\frac{1}{2}}{x-1}+\frac{\frac{1}{2}}{x+1}$$

27. $\dfrac{x^3-x^2+8}{x^2-4}$ Since the degree of the numerator is greater than the degree of the demoninator, divide first.

$$x^2-4\overline{)x^3-x^2+0x+8}\quad\text{quotient } x-1$$

$$\begin{array}{l} x^3 \qquad -4x \\ \hline -x^2+4x+8 \\ -x^2 \qquad +4 \\ \hline 4x+4 \end{array}$$

or $x-1+\dfrac{4x+4}{(x+2)(x-2)}$

$$\frac{4x+4}{(x+2)(x-2)}=\frac{A}{x+2}+\frac{B}{x-2}$$

Multiply each side by the L.C.D.: $(x+2)(x-2)$.

$4x+4=A(x-2)+B(x+2)=Ax-2A+Bx+2B$

Thus: $A+B=4$ The solution to this system is:

$-2A+2B=4$ $A=1, B=3$

$$x-1+\frac{1}{x+2}+\frac{3}{x-2}$$

29. $\dfrac{3x^4 - 2x^3 - 2x + 5}{x(x^2+1)}$ Since the degree of the numerator is greater than the degree of the demoninator, divide first.

$$\begin{array}{r}
3x - 2 \\
x^3 + x \overline{)\,3x^4 - 2x^3 + 0x^2 - 2x + 5} \\
\underline{3x^4 \qquad\quad + 3x^2 \qquad\qquad} \\
-2x^3 - 3x^2 - 2x + 5 \\
\underline{-2x^3 \qquad\quad - 2x \qquad} \\
-3x^2 + 5
\end{array}$$

or $3x - 2 + \dfrac{-3x^2 + 5}{x^3 + x}$

$$\frac{-3x^2 + 5}{x(x^2+1)} = \frac{A}{x} + \frac{Bx + C}{x^2 + 1}$$

Multiply each side by the L.C.D.: $x(x^2+1)$.

$$-3x^2 + 5 = A(x^2 + 1) + (Bx + C)x$$

$$-3x^2 + 5 = Ax^2 + A + Bx^2 + Cx$$

Thus: $A + B = -3$ The solution to this system is:

$C = 0$ $A = 5,\ B = -8,\ C = 0$

$A = 5$

$$3x - 2 + \frac{5}{x} + \frac{-8x}{x^2 + 1}$$

Chapter 6 Review

1. $x+y=3$ and $x-y=-1$

x	0	1	2
y	3	2	1

x	0	1	2
y	1	2	3

$(1,2)$

2. $y-3x=2$ and $y-x=2$

x	−1	0	1
y	−1	2	5

x	−1	0	1
y	1	2	3

$(0,2)$

3. $2x-y=-10$ and $4x+y=4$

x	−2	−1	0
y	6	8	10

x	−2	−1	0
y	12	8	4

$(-1,8)$

4. $2x-3y=6$ and $-4x+6y=8$

x	−3	0	3
y	−4	−2	0

x	−2	1	4
y	0	2	4

System is inconsistent; lines parallel.

5.
$3x-4y=5$
$x+7y=10$

Multiply equation 2 by -3 and add.

$$\begin{aligned} -3x-21y&=-30 \\ 3x-4y&=5 \\ \hline -25y&=-25 \\ y&=1 \end{aligned}$$

From $x+7y=10$

$x+7(1)=10$

$x=3$

$(3,1)$

6.
$2x+3y=7$
$4x-6y=11$

Multiply equation 1 by 2 and add.

$$\begin{aligned} 4x+6y&=14 \\ 4x-6y&=11 \\ \hline 8x&=25 \\ x&=\frac{25}{8} \end{aligned}$$

From $4x-6y=11$

$4\left(\frac{25}{8}\right)-6y=11$

$\frac{25}{2}-11=6y$

$\frac{3}{2}=6y$

$y=\frac{1}{4}$

$\left(\frac{25}{8},\frac{1}{4}\right)$

7.
$x-2y=6$
$4x+y=6$

Multiply equation 2 by 2 and add.

$$\begin{aligned} 8x+2y&=12 \\ x-2y&=6 \\ \hline 9x&=18 \\ x&=2 \end{aligned}$$

From $4x+y=6$

$4(2)+y=6$

$y=-2$

$(2,-2)$

8.

$$3x - 2y = 5$$
$$2x + 2y = 15$$

Add.

$$5x = 20$$
$$x = 4$$

From $3x - 2y = 5$

$$3(4) - 2y = 5$$
$$-2y = -7$$
$$y = \frac{7}{2}$$
$$\left(4, \frac{7}{2}\right)$$

9.

$$x = -4y$$
$$3x - 5y = 17$$
$$3(-4y) - 5y = 17$$
$$-12y - 5y = 17$$
$$-17y = 17$$
$$y = -1$$
$$x = -4(-1) = 4$$
$$(4, -1)$$

10.

$$y = 3x$$
$$2x - 15y = 86$$
$$2x - 15(3x) = 86$$
$$2x - 45x = 86$$
$$-43x = 86$$
$$x = -2$$
$$y = 3(-2) = -6$$
$$(-2, -6)$$

11.

$$5x - y = 7$$
$$2x + 3y = 13$$

Equation 1 becomes

$$y = 5x - 7.$$
$$2x + 3(5x - 7) = 13$$
$$2x + 15x - 21 = 13$$
$$17x = 34$$
$$x = 2$$

From $y = 5x - 7$

$$y = 5(2) - 7$$
$$y = 3$$
$$(2, 3)$$

12.

$$3x + 4y = -7$$
$$-5x + 12y = 14$$

Equation 1 becomes

$$y = -\frac{3}{4}x - \frac{7}{4}$$
$$-5x + 12\left(-\frac{3}{4}x - \frac{7}{4}\right) = 14$$
$$-5x - 9x - 21 = 14$$
$$-14x = 35$$
$$x = -\frac{5}{2}$$

From $y = -\frac{3}{4}x - \frac{7}{4}$

$$y = -\frac{3}{4}\left(-\frac{5}{2}\right) - \frac{7}{4}$$
$$y = \frac{15}{8} - \frac{14}{8}$$
$$y = \frac{1}{8}$$
$$\left(-\frac{5}{2}, \frac{1}{8}\right)$$

13. $8x + by = 4$

$16x - ay = 12$

Multiply equation 1 by -2 and add.

$$-16x - 2by = -8$$
$$\underline{16x - ay = 12}$$
$$(-a - 2b)y = 4$$

$$y = \frac{4}{-a-2b} = \frac{-4}{a+2b}$$

From $8x + by = 4$

$$8x + b\left(\frac{-4}{a+2b}\right) = 4$$

$$8x = \frac{4b}{a+2b} + \frac{4(a+2b)}{a+2b}$$

$$8x = \frac{4b+4a+8b}{a+2b}$$

$$8x = \frac{4a+12b}{a+2b}$$

$$x = \frac{1}{8}\cdot 4\left(\frac{a+3b}{a+2b}\right)$$

$$x = \frac{1}{2}\left(\frac{a+3b}{a+2b}\right) = \frac{a+3b}{2a+4b}$$

$$\left(\frac{a+3b}{2a+4b}, \frac{-4}{a+2b}\right)$$

14. $\frac{3}{x} - \frac{4}{y} = -5$

$\frac{6}{x} + \frac{5}{y} = 16$

Multiply equation 1 by -2 and add.

$$\frac{-6}{x} + \frac{8}{y} = 10$$
$$\underline{\frac{6}{x} + \frac{5}{y} = 16}$$
$$\frac{13}{y} = 26$$

$$y = \frac{13}{26} = \frac{1}{2}$$

From $\frac{6}{x} + \frac{5}{y} = 16$

$$\frac{6}{x} + \frac{5}{\frac{1}{2}} = 16$$

$$\frac{6}{x} + 10 = 16$$

$$\frac{6}{x} = 6$$

$$x = 1$$

$$\left(1, \frac{1}{2}\right)$$

15. $x+y+z=2$ Eliminate y from equations 2 and 3 and use the results.

$x-y-2z=3$ Multiply equation 1 by 1 and add to equation 2.

$x+2y-z=\frac{3}{2}$ Multiply equation 1 by –2 and add to equation 3.

$$\begin{array}{r} x+y+z=2 \\ x-y-2z=3 \\ \hline *\ 2x-z=5 \end{array} \quad \text{and} \quad \begin{array}{r} -2x-2y-2z=-4 \\ x+2y-z=\frac{3}{2} \\ \hline *\ -x-3z=-\frac{5}{2} \end{array}$$

Solve the system.

$2x-z=5$ Multiply equation 2 by 2 and add to equation 1.

$-x-3z=-\frac{5}{2}$

$$\begin{array}{r} -2x-6z=-5 \\ 2x-z=5 \\ \hline -7z=0 \\ z=0 \end{array}$$

From $2x-z=5$, we get $2x-0=5$; $x=\frac{5}{2}$.

From $x+y+z=2$, we get $\frac{5}{2}+y+0=2$;

$y=2-\frac{5}{2}$ or $y=-\frac{1}{2}$.

$\left(\frac{5}{2},-\frac{1}{2},0\right)$

16. $2x+2y-z=4$
$2x-y+2z=-2$
$x-5y+z=8$

Eliminate x from equations 1 and 2 and use the results. Multiply equation 3 by -2 and add to equation 1.

$$\begin{array}{r} -2x+10y-2z=-16 \\ 2x+2y-z=4 \\ \hline *\,12y-3z=-12 \end{array}$$

Multiply equation 3 by -2 and add to equation 1.

$$\begin{array}{r} -2x+10y-2z=-16 \\ 2x-y+2z=-2 \\ \hline 9y=-18 \\ y=-2 \end{array}$$

Multiply equation 3 by -2 and add to equation 2.

From $12y-3z=-12$
$12(-2)-3z=-12$
$-3z=12$
$z=-4$

From $x-5y+z=8$
$x-5(-2)+-4=8$
$x+10-4=8$
$x=2$

$(2,-2,-4)$

17. $\begin{vmatrix} 2 & -3 \\ 6 & 2 \end{vmatrix} = 2(2)-6(-3)=4+18=22$

18. $\begin{vmatrix} -1 & 7 \\ 4 & -3 \end{vmatrix} = (-1)(-3)-4(7)=3-28=-25$

19. $\begin{vmatrix} 1 & 3 & 5 \\ 7 & 2 & -1 \\ -4 & 2 & -8 \end{vmatrix}$ Multiply row 1 by -7 and add to row 2. Multiply row 1 by 4 and add to row 3.

$\begin{vmatrix} 1 & 3 & 5 \\ 0 & -19 & -36 \\ 0 & 14 & 12 \end{vmatrix} = 1\begin{vmatrix} -19 & -36 \\ 14 & 12 \end{vmatrix} = (-19)(12)-14(-36)=276$

20. $\begin{vmatrix} 2 & 0 & -3 \\ 1 & 4 & 2 \\ 5 & 1 & -1 \end{vmatrix}$ Multiply row 3 by -4 and add to row 2.

$\begin{vmatrix} 2 & 0 & -3 \\ -19 & 0 & 6 \\ 5 & 1 & -1 \end{vmatrix} = -1\begin{vmatrix} 2 & -3 \\ -19 & 6 \end{vmatrix} = -[2(6)-(-3)(-19)]=45$

See the beginning of the solutions for Section 6.6 for definitions of D, D_x, D_y and D_z.

21. $\begin{aligned} 4x+3y&=46 \\ 2x-3y&=14 \end{aligned}$ $\quad D=\begin{vmatrix} 4 & 3 \\ 2 & -3 \end{vmatrix}=4(-3)-2(3)=-18$

$D_x=\begin{vmatrix} 46 & 3 \\ 14 & -3 \end{vmatrix}=46(-3)-14(3)=-180$

$D_y=\begin{vmatrix} 4 & 46 \\ 2 & 14 \end{vmatrix}=4(14)-2(46)=-36$

$x=\frac{D_x}{D}=\frac{-180}{-18}=10$ and $y=\frac{D_y}{D}=\frac{-36}{-18}=2$

22. $\begin{aligned} 3x-2y&=13 \\ -6x+4y&=-26 \end{aligned}$ $\quad D=\begin{vmatrix} 3 & -2 \\ -6 & 4 \end{vmatrix}=3(4)-(-2)(-6)=0$

$D_x=\begin{vmatrix} 13 & 3 \\ -26 & -6 \end{vmatrix}=13(-6)-(-26)(3)=0$

$D_y=\begin{vmatrix} 3 & 13 \\ -6 & -26 \end{vmatrix}=3(-26)-(-6)(13)=0$

System is dependent; lines coincide.

23. $\begin{aligned} 2x+9y&=4 \\ 5x-9y&=10 \end{aligned}$ $\quad D=\begin{vmatrix} 2 & 9 \\ 5 & -9 \end{vmatrix}=2(-9)-5(9)=-63$

$D_x=\begin{vmatrix} 4 & 9 \\ 10 & -9 \end{vmatrix}=4(-9)-10(9)=-126$

$D_y=\begin{vmatrix} 2 & 4 \\ 5 & 10 \end{vmatrix}=2(10)-5(4)=0$

$x=\frac{D_x}{D}=\frac{-126}{-63}=2$ and $y=\frac{D_y}{D}=\frac{0}{-63}=0$

24. $\begin{aligned} 5x-2y&=2 \\ -3x+3y&=1 \end{aligned}$ $\quad D=\begin{vmatrix} 5 & -2 \\ -3 & 3 \end{vmatrix}=5(3)-(-3)(-2)=9$

$D_x=\begin{vmatrix} 2 & -2 \\ 1 & 3 \end{vmatrix}=2(3)-(1)(-2)=8$

$D_y=\begin{vmatrix} 5 & 2 \\ -3 & 1 \end{vmatrix}=5(1)-(-3)(2)=11$

$x=\frac{D_x}{D}=\frac{8}{9}$ and $y=\frac{D_y}{D}=\frac{11}{9}$

25.

$$\begin{aligned} 2x+2y-z&=11 \\ 3x+4y+z&=-16 \\ 4x-8y+3z&=-113 \end{aligned} \qquad D=\begin{vmatrix} 2 & 2 & -1 \\ 3 & 4 & 1 \\ 4 & -8 & 3 \end{vmatrix}$$

Multiply row 1 by 1 and add to row 2.
Multiply row 1 by 3 and add to row 3.

$$D=\begin{vmatrix} 2 & 2 & -1 \\ 5 & 6 & 0 \\ 10 & -2 & 0 \end{vmatrix}$$

$$D=-1\begin{vmatrix} 5 & 6 \\ 10 & -2 \end{vmatrix}=-1[5(-2)-10(6)]=70$$

$$D_x=\begin{vmatrix} 11 & 2 & -1 \\ -16 & 4 & 1 \\ -113 & -8 & 3 \end{vmatrix}$$

Multiply row 1 by 1 and add to row 2.
Multiply row 1 by 3 and add to row 3.

$$D_x=\begin{vmatrix} 11 & 2 & -1 \\ -5 & 6 & 0 \\ -80 & -2 & 0 \end{vmatrix}=-1\begin{vmatrix} -5 & 6 \\ -80 & -2 \end{vmatrix}=-[-5(-2)-6(-80)]=-490$$

$$D_y=\begin{vmatrix} 2 & 11 & -1 \\ 3 & -16 & 1 \\ 4 & -113 & 3 \end{vmatrix}$$

Multiply row 1 by 1 and add to row 2.
Multiply row 1 by 3 and add to row 3.

$$D_y=\begin{vmatrix} 2 & 11 & -1 \\ 5 & -5 & 0 \\ 10 & -80 & 0 \end{vmatrix}=-1\begin{vmatrix} 5 & -5 \\ 10 & -80 \end{vmatrix}=-[5(-80)-10(-5)]=350$$

$$D_z=\begin{vmatrix} 2 & 2 & 11 \\ 3 & 4 & -16 \\ 4 & -8 & -113 \end{vmatrix}$$

Multiply row 1 by -2 and add to row 2.
Multiply row 1 by 4 and add to row 3.

$$D_z=\begin{vmatrix} 2 & 2 & 11 \\ -1 & 0 & -38 \\ 12 & 0 & -69 \end{vmatrix}=-2\begin{vmatrix} -1 & -38 \\ 12 & -69 \end{vmatrix}=-2[-1(-69)-12(-38)]=-1050$$

$$x=\frac{D_x}{D}=\frac{-490}{70}=-7 \text{ and } y=\frac{D_y}{D}=\frac{350}{70}=5 \text{ and } z=\frac{D_z}{D}=\frac{-1050}{70}=-15$$

26.
$$\begin{aligned} x+y+z&=9 \\ 2x-y+z&=3 \\ x+3y-z&=3 \end{aligned} \qquad D=\begin{vmatrix} 1 & 1 & 1 \\ 2 & -1 & 1 \\ 1 & 3 & -1 \end{vmatrix}$$

Multiply row 1 by -2 and add to row 2.
Multiply row 1 by -1 and add to row 3.

$$D=\begin{vmatrix} 1 & 1 & 1 \\ 0 & -3 & -1 \\ 0 & 2 & -2 \end{vmatrix}$$

$$D=1\begin{vmatrix} -3 & -1 \\ 2 & -2 \end{vmatrix}=-3(-2)-(2)(-1)=8$$

$$D_x=\begin{vmatrix} 9 & 1 & 1 \\ 3 & -1 & 1 \\ 3 & 3 & -1 \end{vmatrix}$$

Multiply row 1 by -1 and add to row 2.
Multiply row 1 by 1 and add to row 3.

$$D_x=\begin{vmatrix} 9 & 1 & 1 \\ -6 & -2 & 0 \\ 12 & 4 & 0 \end{vmatrix}=1\begin{vmatrix} -6 & -2 \\ 12 & 4 \end{vmatrix}=-6(4)-(12)(-2)=0$$

$$D_y=\begin{vmatrix} 1 & 9 & 1 \\ 2 & 3 & 1 \\ 1 & 3 & -1 \end{vmatrix}$$

Multiply row 1 by -1 and add to row 2.
Multiply row 1 by 1 and add to row 3.

$$D_y=\begin{vmatrix} 1 & 9 & 1 \\ 1 & -6 & 0 \\ 2 & 12 & 0 \end{vmatrix}=1\begin{vmatrix} 1 & -6 \\ 2 & 12 \end{vmatrix}=1(12)-2(-6)=24$$

$$D_z=\begin{vmatrix} 1 & 1 & 9 \\ 2 & -1 & 3 \\ 1 & 3 & 3 \end{vmatrix}$$

Multiply row 1 by -2 and add to row 2.
Multiply row 1 by -1 and add to row 3.

$$D_z=\begin{vmatrix} 1 & 1 & 9 \\ 0 & -3 & -15 \\ 0 & 2 & -6 \end{vmatrix}=1\begin{vmatrix} -3 & -15 \\ 2 & -6 \end{vmatrix}=-3(-6)-2(-15)=48$$

$$x=\frac{D_x}{D}=\frac{0}{8}=0 \text{ and } y=\frac{D_y}{D}=\frac{24}{8}=3 \text{ and } z=\frac{D_z}{D}=\frac{48}{8}=6$$

27.
$$\begin{aligned} 2x+7y&=3 \\ 10x+3y&=-1 \end{aligned}$$

Multiply equation 1 by -5 and add to equation 2.

$$\begin{aligned} -10x-35y&=-15 \\ 10x+3y&=-1 \\ \hline -32y&=-16 \\ y&=\frac{1}{2} \end{aligned}$$

From $2x+7y=3$

$$2x+7\left(\frac{1}{2}\right)=3$$

$$2x=3-\frac{7}{2}=-\frac{1}{2}$$

$$x=-\frac{1}{4}$$

$$\left(-\frac{1}{4},\frac{1}{2}\right)$$

28. $\begin{aligned} -5x+2y&=5 \\ -3x+7y&=32 \end{aligned}$ Multiply equation 1 by 3 and equation 2 by -5 and add.

$$\begin{aligned} -15x+6y&=15 \\ 15x-35y&=-160 \\ \hline -29y&=-145 \\ y&=\frac{145}{29}=5 \end{aligned}$$

From
$$\begin{aligned} -5x+2y&=5 \\ -5x+2(5)&=5 \\ -5x&=-5 \\ x&=1 \end{aligned}$$
$(1,5)$

29. $\begin{aligned} 2x-3y&=10 \\ 2x+5y&=2 \end{aligned}$ Multiply equation 1 by -1 and add to equation 2.

$$\begin{aligned} -2x+3y&=-10 \\ 2x+5y&=2 \\ \hline 8y&=-8 \\ y&=-1 \end{aligned}$$

From
$$\begin{aligned} 2x-3y&=10 \\ 2x-3(-1)&=10 \\ 2x&=7 \\ x&=\frac{7}{2} \end{aligned}$$
$\left(\frac{7}{2},-1\right)$

30. $\begin{aligned} x+y&=-25 \\ 2x-3y&=70 \end{aligned}$ Multiply equation 1 by 3 and add to equation 2.

$$\begin{aligned} 3x+3y&=-75 \\ 2x-3y&=70 \\ \hline 5x&=-5 \\ x&=-1 \end{aligned}$$

From
$$\begin{aligned} x+y&=-25 \\ -1+y&=-25 \\ y&=-24 \end{aligned}$$
$(-1,-24)$

31. $\begin{aligned} x+y+z&=7 \\ x\quad +z&=2 \\ y-z&=4 \end{aligned}$ Multiply equation 1 by -1 and add to equation 2.

$$\begin{aligned} -x-y-z&=-7 \\ x\quad +z&=70 \\ \hline -y&=-5 \\ y&=5 \end{aligned}$$

From
$$\begin{aligned} y-z&=4 \\ 5-z&=4 \\ z&=1 \end{aligned}$$

From
$$\begin{aligned} x+z&=2 \\ x+1&=2 \\ x&=1 \end{aligned}$$

$(1,5,1)$

32.

$$7x-2y+5z=9$$
$$3x+4y-9z=27$$
$$10x-3y+7z=7$$

Eliminate y. Multiply equation 1 by 2 and add to equation 2.

$$14x-4y+10z=18$$
$$3x+4y-9z=27$$
$$*\ 17x+z=45$$

Multiply equation 2 by 3 and equation 3 by 4.

$$9x+12y-27z=81$$
$$40x-12y+28z=28$$
$$*\ 49x+z=109$$

Use the two * equations.

$$17x+z=45$$
$$49x+z=109$$

Multiply equation 1 by -1 and add to equation 2.

$$-17x-z=-45$$
$$49x+z=109$$
$$32x=64$$
$$x=2$$

From $17x+z=45$

$$17(2)+z=45$$
$$z=11$$

From $7x-2y+5z=9$

$$7(2)-2y+5(11)=9$$
$$-2y+69=9$$
$$y=30$$
$$(2,30,11)$$

33. Let $x=$ amount of 10%

$y=$ amount of 20%

Multiply equation 1 by -0.1 and add to equation 2.

$$x+y=30$$
$$0.1x+0.2y=0.12(30)$$

$$-0.1x-0.1y=-3$$
$$0.1x+0.2y=3.6$$
$$0.1y=0.6$$
$$y=6\ kg$$

From $x+y=30$

$$x+6=30$$
$$x=24\ kg$$

24 *kg* of 10% and 6 *kg* of 20%.

34. Let $x =$ amount at $5\frac{1}{2}\%$ and $y =$ amount at $6\frac{1}{2}\%$.

$x + y = 3600$
$0.055x + 0.065y = 220$

Multiply equation 1 by -0.055 and add to equation 2.

$$\begin{aligned} -0.055x - 0.055y &= -198 \\ 0.055x + 0.065y &= 220 \\ \hline 0.01y &= 22 \\ y &= 2200 \end{aligned}$$

From
$x + y = 3600$
$x + 2200 = 3600$
$x = 1400$

\$1400 at $5\frac{1}{2}\%$ and \$2200 at $6\frac{1}{2}\%$

35. Let $p =$ plane velocity (speed) and $w =$ wind velocity. Velocity into head wind is $p - w$. Velocity with the wind is $p + w$. Convert 15 minutes to 0.25 hours. Use $R \cdot T = D$.

$1.25(p - w) = 100$
$1(p + w) = 100$

Divide equation 1 by 1.25; add the results to equation 2.

$$\begin{aligned} p - w &= 80 \\ p + w &= 100 \\ \hline 2p &= 180 \\ p &= 90 \end{aligned}$$

From
$p + w = 100$
$90 + w = 100$
$w = 10$

Plane velocity is 90 *km/h*
Wind velocity is 10 *km/h*

36. Let $x =$ number of pieces per hour for Jack.
y = number of pieces per hour for John.
z = number of pieces per hour for Bob.

$x + y = 3z$
$x = 2 + y$
$x + y + z = 32$

Substitute x from equation 2 into equation 1 and equation 3.

$2 + y + y = 3z$ ie $2y = 3z - 2$
$2 + y + y + z = 32$ ie $2y = 32 - 2 - z$ ie $2y = 30 - z$

Set the two expressions for $2y$ equal.

$3z - 2 = 30 - z$
$4z = 32$
$z = 8$

From
$2y = 3z - 2$
$2y = 3(8) - 2 = 22$
$y = 11$

From
$x = 2 + y$
$x = 2 + 11$
$x = 13$

37. $\begin{vmatrix} 1 & 0 & 3 & 6 \\ 4 & -2 & 1 & -5 \\ -3 & 2 & 5 & 2 \\ 1 & 0 & 7 & -9 \end{vmatrix}$ Multiply row 2 by 1 and add to row 3.

$\begin{vmatrix} 1 & 0 & 3 & 6 \\ 4 & -2 & 1 & -5 \\ 1 & 0 & 6 & -3 \\ 1 & 0 & 7 & -9 \end{vmatrix} = -2\begin{vmatrix} 1 & 3 & 6 \\ 1 & 6 & -3 \\ 1 & 7 & -9 \end{vmatrix}$ Multiply row 1 by -1 and add to row 2.
Multiply row 1 by -1 and add to row 3.

$$-2\begin{vmatrix} 1 & 3 & 6 \\ 0 & 3 & -9 \\ 0 & 4 & -15 \end{vmatrix} = -2\begin{vmatrix} 3 & -9 \\ 4 & -15 \end{vmatrix} = -2[3(-15) - 4(-9)] = 18$$

38. $\begin{vmatrix} 5 & 1 & 6 & -3 \\ -2 & 2 & 5 & 1 \\ 3 & -1 & 7 & 5 \\ 3 & 2 & 3 & -2 \end{vmatrix}$ Multiply row 1 by -2 and add to row 2.
Multiply row 1 by 1 and add to row 3.
Multiply row 1 by -2 and add to row 4.

$\begin{vmatrix} 5 & 1 & 6 & -3 \\ -12 & 0 & -7 & 7 \\ 8 & 0 & 13 & 2 \\ -7 & 0 & -9 & 4 \end{vmatrix} = -1\begin{vmatrix} -12 & -7 & 7 \\ 8 & 13 & 2 \\ -7 & -9 & 4 \end{vmatrix}$ Multiply row 2 by 1 and add to row 3.

$-\begin{vmatrix} -12 & -7 & 7 \\ 8 & 13 & 2 \\ 1 & 4 & 6 \end{vmatrix}$ Multiply row 3 by 12 and add to row1.
Multiply row 3 by -8 and add to row 2.

$$-1\begin{vmatrix} 0 & 41 & 79 \\ 0 & -19 & -46 \\ 1 & 4 & 6 \end{vmatrix} = -1\begin{vmatrix} 41 & 79 \\ -19 & -46 \end{vmatrix} = -[41(-46) - (-19)(79)] = 385$$

39. $$\frac{6x+14}{(x-3)(x+5)} = \frac{A}{x-3} + \frac{B}{x+5}$$

Multiply each side by the L.C.D.: $(x-3)(x+5)$

$6x + 14 = A(x+5) + B(x-3) = Ax + 5A + Bx - 3B$

Thus: $A + B = 6$
$5A - 3B = 14$

The solution to this system is: $A = 4$ and $B = 2$

$$\frac{4}{x-3} + \frac{2}{x+5}$$

40. $$\frac{-x^2-6x-1}{x(x-1)(x+1)}=\frac{A}{x}+\frac{B}{x-1}+\frac{C}{x+1}$$

Multiply each side by the L.C.D.: $x(x-1)(x+1)$

$$-x^2-6x-1=A(x-1)(x+1)+Bx(x+1)+Cx(x-1)$$
$$-x^2-6x-1=A(x^2-1)+Bx^2+Bx+Cx^2-Cx$$
$$-x^2-6x-1=Ax^2-A+Bx^2+Bx+Cx^2-Cx$$

$A+B+C=-1$

$B-C=-6$

$-A=-1$

The solution to this system is:

$A=1,\ B=-4,\ c=2$

$$\frac{1}{x}+\frac{-4}{x-1}+\frac{2}{x+1}$$

41. $$\frac{3x^2+5x+1}{x^2(x+1)}=\frac{A}{x}+\frac{B}{x^2}+\frac{C}{x+1}$$

Multiply each side by the L.C.D.: $x^2(x+1)$.

$$3x^2+5x+1=Ax(x+1)+B(x+1)+Cx^2$$
$$3x^2+5x+1=Ax^2+Ax+Bx+B+Cx^2$$

$A+C=3$

$A+B=5$

$B=1$

The solution to this system is:

$A=4,\ B=1,\ C=-1$

$$\frac{4}{x}+\frac{1}{x^2}+\frac{-1}{x+1}$$

42. $$\frac{10x+4}{(x+1)^2}=\frac{A}{x+1}+\frac{B}{(x+1)^2}$$

Multiply each side by the L.C.D.: $(x+1)^2$.

$$10x+4=A(x+1)+B=Ax+A+B$$

$A=10$

$A+B=4$

The solution to this system is:

$A=10,\ B=-6$

$$\frac{10}{x+1}+\frac{-6}{(x+1)^2}$$

43. $$\frac{7x^2 - x + 2}{(x^2+1)(x-1)} = \frac{Ax+B}{x^2+1} + \frac{C}{x-1}$$

Multiply each side by the L.C.D.: $(x^2+1)(x-1)$.

$$7x^2 - x + 2 = (Ax+B)(x-1) + C(x^2+1)$$

$$7x^2 - x + 2 = Ax^2 + Bx - Ax - B + Cx^2 + C$$

$$\begin{aligned} A \quad\quad + C &= 7 \\ -A + B \quad\quad &= -1 \\ -B + C &= 2 \end{aligned}$$

The solution to this system is: $A = 3,\ B = 2,\ C = 4$

$$\frac{3x+2}{x^2+1} + \frac{4}{x-1}$$

44. $$\frac{5x^2 + 2x + 21}{(x^2+4)^2} = \frac{Ax+B}{x^2+4} + \frac{Cx+D}{(x^2+4)^2}$$

Multiply each side by the L.C.D.: $(x^2+4)^2$.

$$5x^2 + 2x + 21 = (Ax+B)(x^2+4) + Cx + D$$

$$5x^2 + 2x + 21 = Ax^3 + Bx^2 + 4Ax + 4B + Cx + D$$

$$\begin{aligned} A \quad\quad\quad\quad &= 0 \\ B \quad\quad\quad &= 5 \\ 4A \quad + C \quad &= 2 \\ 4B \quad\quad + D &= 21 \end{aligned}$$

The solution to this system is: $A = 0,\ B = 5,\ C = 2,\ D = 1$

$$\frac{5}{x^2+4} + \frac{2x+1}{(x^2+4)^2}$$

CHAPTER 6 SPECIAL APPLICATION

1. junctions: $I_1 = I_2 + I_4$ (1)

$I_2 = I_3$ (2) Current through each resister is the same.

$I_5 = I_3 + I_4$ (3)

outer loop: $80.0\,I_1 + 80.0\,I_2 + 40.0\,I_3 + 60.0\,I_5 = 40.0$ (4)

right loop: $80.0\,I_1 + 60.0\,I_4 + 60.0\,I_5 = 40.0$ (5)

Solve equations (1) through (5) simultaneously using a graphing calculator.

$I_1 = I_5 = 0.222\text{A}$

$I_2 = I_3 = 0.0741\text{A}$

$I_4 = 0.148\text{A}$

2. junctions: $I_1 = I_2 + I_3$ (1)

$I_4 = I_2 + I_3$ (2)

right loop: $75.0\,I_2 - 125\,I_3 = 0$ (3)

inside loop: $175\,I_1 + 75.0\,I_2 + 60.0\,I_4 = 90.0$ (4)

Solve equations (1) through (4) simultaneously using a graphing calculator.

$I_1 = I_4 = 0.319\text{A}$

$I_2 = 0.200\text{A}$

$I_3 = 0.120\text{A}$

3. $I_1 = I_2 + I_3$ (1)

Also, $I_4 = I_2 + I_3$, thus $I_1 = I_4$

right loop: $75.0\,I_2 - 125\,I_3 = 0$ (2)

inside loop: $175\,I_1 + 75.0\,I_2 + 60.0\,I_4 = 90.0$

since $I_1 = I_4 : 235\,I_1 + 75.0\,I_2 = 90.0$ (3)

Solve equations (1), (2), and (3) simultaneously:

$I_1 = I_4 = 0.319290$ or 0.319A

$I_2 = 0.199556$ or 0.200A

$I_3 = 0.119733$ or 0.120A

4. $I_1 = I_3$ (1) and $I_5 = I_6$ (2)

$I_2 = I_4 + I_6$ (3) and $I_3 = I_4 + I_5$ (4)

outer loop: $20.0\,I_1 + 40.0\,I_3 + 50.0\,I_5 + 75.0\,I_6 + 80.0\,I_2 = 48.0$ (5)

top loop: $20.0\,I_1 + 40.0\,I_2 + 200.0\,I_4 + 80.0\,I_2 = 48.0$ (6)

Solving equations (1) through (6) we have

$$I_1 = I_2 = I_3 = 0.221$$
$$I_4 = 0.0851$$
$$I_5 = I_6 = 0.136$$

Chapter 7, Exercises 7.1

1.
$$(x+4)(x-7)=0$$
$$x+4=0 \text{ or } x-7=0$$
$$x=-4 \text{ or } x=7$$

3.
$$x^2-6x+8=0$$
$$(x-2)(x-4)=0$$
$$x-2=0 \text{ or } x-4=0$$
$$x=2 \text{ or } x=4$$

5.
$$x^2+3x=10$$
$$x^2+3x-10=0$$
$$(x+5)(x-2)=0$$
$$x+5=0 \text{ or } x-2=0$$
$$x=-5 \text{ or } x=2$$

7.
$$2x^2=3x+9$$
$$2x^2-3x-9=0$$
$$(x-3)(2x+3)=0$$
$$x-3=0 \text{ or } 2x+3=0$$
$$x=3 \text{ or } x=\frac{-3}{2}$$

9.
$$14x^2+17x+5=0$$
$$(2x+1)(7x+5)=0$$
$$2x+1=0 \text{ or } 7x+5=0$$
$$x=\frac{-1}{2} \text{ or } x=\frac{-5}{7}$$

11.
$$18x^2+56=69x$$
$$18x^2-69x+56=0$$
$$(3x-8)(6x-7)=0$$
$$3x-8=0 \text{ or } 6x-7=0$$
$$x=\frac{8}{3} \text{ or } x=\frac{7}{6}$$

13.
$$8x^2+x=0$$
$$x(8x+1)=0$$
$$x=0 \text{ or } 8x+1=0$$
$$x=0 \text{ or } x=\frac{-1}{8}$$

15.
$$-7x^2+21x=0$$
$$-7x(x-3)=0$$
$$-7x=0 \text{ or } x-3=0$$
$$x=0 \text{ or } x=3$$

17.
$$x^2-36=0$$
$$(x+6)(x-6)=0$$
$$x+6=0 \text{ or } x-6=0$$
$$x=-6 \text{ or } x=6$$

19.
$$16x^2-25=0$$
$$16x^2=25$$
$$x^2=\frac{25}{16}$$
$$x=\frac{5}{4} \text{ or } \frac{-5}{4}$$

21.
$$5x^2-12=0$$
$$5x^2=12$$
$$x^2=\frac{12}{5}$$
$$x=\pm\sqrt{\frac{12}{5}}=\pm\frac{\sqrt{12}}{\sqrt{5}}$$
$$x=\pm\frac{2\sqrt{3}\sqrt{5}}{\sqrt{5}\sqrt{5}}=\pm\frac{2\sqrt{15}}{5}$$

23.
$$4x^2=21$$
$$x^2=\frac{21}{4}$$
$$x=\pm\sqrt{\frac{21}{4}}=\pm\frac{\sqrt{21}}{2}$$

25.
$$5a^2-40=0$$
$$5a^2=40$$
$$a^2=8$$
$$a=\pm\sqrt{8}=\pm2\sqrt{2}$$

27.
$$2=\frac{1}{3}x^2$$
$$6=x^2$$
$$\pm\sqrt{6}=x$$

29.
$$30x^2+16x=24$$
$$15x^2+8x=12$$
$$15x^2+8x-12=0$$
$$(3x-2)(5x+6)=0$$
$$3x-2=0 \text{ or } 5x+6=0$$
$$x=\frac{2}{3} \text{ or } \frac{-6}{5}$$

31.

$$40x^2 + 100x + 40 = 0$$
$$2x^2 + 5x + 2 = 0$$
$$(2x+1)(x+2) = 0$$
$$2x + 1 = 0 \text{ or } x + 2 = 0$$
$$x = \frac{-1}{2} \text{ or } x = -2$$

33.

$$\frac{x-1}{2x} = \frac{5}{x+12}$$

Multiply each side
by the L.C.D.:
$2x(x+12)$

$$(x-1)(x+12) = 5(2x)$$
$$x^2 + 11x - 12 - 10x = 0$$
$$x^2 + x - 12 = 0$$
$$(x+4)(x-3) = 0$$
$$x + 4 = 0 \text{ or } x - 3 = 0$$
$$x = -4 \text{ or } x = 3$$

35.

$$\frac{2x-3}{x-4} = x + 2$$

Multiply each side
by the L.C.D.: $x - 4$

$$2x - 3 = (x+2)(x-4)$$
$$2x - 3 = x^2 - 2x - 8$$
$$0 = x^2 - 4x - 5$$
$$0 = (x-5)(x+1)$$
$$x - 5 = 0 \text{ or } x + 1 = 0$$
$$x = 5 \text{ or } x = -1$$

37.

$$\frac{x+1}{x-2} + \frac{x-1}{x+1} = \frac{9}{2}$$

Multiply each side
by the L.C.D.:
$2(x-2)(x+1)$

$$(x+1)(2)(x+1) + (x-1)(2)(x-2) = 9(x-2)(x+1)$$
$$2(x^2 + 2x + 1) + 2(x^2 - 3x + 2) = 9(x^2 - x - 2)$$
$$2x^2 + 4x + 2 + 2x^2 - 6x + 4 = 9x^2 - 9x - 18$$
$$4x^2 - 2x + 6 = 9x^2 - 9x - 18$$
$$0 = 5x^2 - 7x - 24$$
$$0 = (x-3)(5x+8)$$
$$x - 3 = 0 \text{ or } 5x + 8 = 0$$
$$x = 3 \text{ or } x = \frac{-8}{5}$$

Exercises 7.2

1.

$$x^2+6x-16=0$$
$$x^2+6x=16$$
$$\text{Find } \left[\frac{1}{2}(6)\right]^2=9$$

Add 9 to both sides.

$$x^2+6x+9=16+9$$
$$(x+3)^2=25$$
$$x+3=\pm\sqrt{25}=\pm 5$$
$$x=-3\pm 5$$
$$x=-3+5=2 \text{ or}$$
$$x=-3-5=-8$$

3.

$$x^2+35=12x$$
$$x^2-12x=-35$$
$$\text{Find } \left[\frac{1}{2}(-12)\right]^2=36$$
$$x^2-12x+36=-35+36$$
$$(x-6)^2=1$$
$$x-6=\pm\sqrt{1}=\pm 1$$
$$x=6\pm 1$$
$$x=7 \text{ or } 5$$

5.

$$2x^2+x=1$$
$$x^2+\frac{1}{2}x=\frac{1}{2}$$
$$\text{Find } \left[\frac{1}{2}\left(\frac{1}{2}\right)\right]^2=\frac{1}{16}$$
$$x^2+\frac{1}{2}x+\frac{1}{16}=\frac{1}{2}+\frac{1}{16}$$
$$\left(x+\frac{1}{4}\right)^2=\frac{9}{16}$$
$$x+\frac{1}{4}=\pm\sqrt{\frac{9}{16}}=\pm\frac{3}{4}$$
$$x=-\frac{1}{4}\pm\frac{3}{4}$$
$$x=\frac{1}{2} \text{ or } -1$$

7.

$$25x^2=15x+18$$
$$x^2=\frac{15x}{25}+\frac{18}{25}$$
$$x^2-\frac{3}{5}x=\frac{18}{25}$$
$$\text{Find } \left[\frac{1}{2}\left(-\frac{3}{5}\right)\right]^2=\frac{9}{100}$$
$$x^2-\frac{3}{5}x+\frac{9}{100}=\frac{18}{25}+\frac{9}{100}$$
$$\left(x-\frac{3}{10}\right)^2=\frac{81}{100}$$
$$x-\frac{3}{10}=\pm\sqrt{\frac{81}{100}}=\pm\frac{9}{10}$$
$$x=\frac{3}{10}\pm\frac{9}{10}$$
$$x=\frac{6}{5} \text{ or } -\frac{3}{5}$$

9.

$$x^2+4x-7=0$$
$$x^2+4x=7$$
$$\text{Find } \left[\frac{1}{2}(4)\right]^2=4$$
$$x^2+4x+4=7+4$$
$$(x+2)^2=11$$
$$x+2=\pm\sqrt{11}$$
$$x=-2\pm\sqrt{11}$$

11.

$$2x^2 + 2x - 9 = 0$$
$$x^2 + x - \frac{9}{2} = 0$$
$$x^2 + x = \frac{9}{2}$$
$$\text{Find } \left[\frac{1}{2}(1)\right]^2 = \frac{1}{4}$$
$$x^2 + x + \frac{1}{4} = \frac{9}{2} + \frac{1}{4}$$
$$\left(x + \frac{1}{2}\right)^2 = \frac{19}{4}$$
$$x + \frac{1}{2} = \pm\sqrt{\frac{19}{4}} = \pm\frac{\sqrt{19}}{2}$$
$$x = -\frac{1}{2} \pm \frac{\sqrt{19}}{2} = \frac{-1 \pm \sqrt{19}}{2}$$

13.

$$3x^2 + 9x + 5 = 0$$
$$x^2 + 3x + \frac{5}{3} = 0$$
$$x^2 + 3x = -\frac{5}{3}$$
$$\text{Find } \left[\frac{1}{2}(3)\right]^2 = \frac{9}{4}$$
$$x^2 + 3x + \frac{9}{4} = -\frac{5}{3} + \frac{9}{4}$$
$$\left(x + \frac{3}{2}\right)^2 = \frac{7}{12}$$
$$x + \frac{3}{2} = \frac{\pm\sqrt{7}}{\sqrt{12}} = \frac{\pm\sqrt{7}}{2\sqrt{3}}$$
$$x + \frac{3}{2} = \frac{\pm\sqrt{7}\sqrt{3}}{2\sqrt{3}\sqrt{3}} = \frac{\pm\sqrt{21}}{6}$$
$$x = \frac{-3}{2} \pm \frac{\sqrt{21}}{6} = \frac{-9 \pm \sqrt{21}}{6}$$

15.

$$-2x^2 + 3x + 9 = 0$$
$$x^2 - \frac{3}{2}x - \frac{9}{2} = 0$$
$$x^2 - \frac{3}{2}x = \frac{9}{2}$$
$$\text{Find } \left[\frac{1}{2}\left(-\frac{3}{2}\right)\right]^2 = \frac{9}{16}$$
$$x^2 - \frac{3}{2}x + \frac{9}{16} = \frac{9}{2} + \frac{9}{16}$$
$$\left(x - \frac{3}{4}\right)^2 = \frac{81}{16}$$
$$x - \frac{3}{4} = \pm\sqrt{\frac{81}{16}} = \pm\frac{9}{4}$$
$$x = \frac{3}{4} \pm \frac{9}{4}$$
$$x = 3 \text{ or } \frac{-3}{2}$$

17.

$$4x^2 + 11x = 3$$
$$x^2 + \frac{11}{4}x = \frac{3}{4}$$
$$\text{Find } \left[\frac{1}{2}\left(\frac{11}{4}\right)\right]^2 = \frac{121}{64}$$
$$x^2 + \frac{11}{4}x + \frac{121}{64} = \frac{3}{4} + \frac{121}{64}$$
$$\left(x + \frac{11}{8}\right)^2 = \frac{169}{64}$$
$$x + \frac{11}{8} = \pm\sqrt{\frac{169}{64}} = \pm\frac{13}{8}$$
$$x = \frac{-11}{8} \pm \frac{13}{8}$$
$$x = \frac{1}{4} \text{ or } -3$$

19.

$$6m^2 + 15m + 3 = 0$$

$$m^2 + \frac{15}{6}m + \frac{3}{6} = 0$$

$$m^2 + \frac{5}{2}m = -\frac{1}{2}$$

$$\text{Find } \left[\frac{1}{2}\left(\frac{5}{2}\right)\right]^2 = \frac{25}{16}$$

$$m^2 + \frac{5}{2}m + \frac{25}{16} = \frac{-1}{2} + \frac{25}{16}$$

$$\left(m + \frac{5}{4}\right)^2 = \frac{17}{16}$$

$$m + \frac{5}{4} = \pm\sqrt{\frac{17}{16}} = \frac{\pm\sqrt{17}}{4}$$

$$m = \frac{-5}{4} \pm \frac{\sqrt{17}}{4} = \frac{-5 \pm \sqrt{17}}{4}$$

Exercises 7.3

For these problems, we will use $x = \dfrac{-b \pm \sqrt{b^2 - 4ac}}{2a}$.

1.

$$x^2 - 4x - 32 = 0$$

$$a = 1,\ b = -4,\ c = -32$$

$$x = \frac{-(-4) \pm \sqrt{(-4)^2 - 4(1)(-32)}}{2(1)}$$

$$x = \frac{4 \pm \sqrt{16 + 128}}{2}$$

$$x = \frac{4 \pm \sqrt{144}}{2} = \frac{4 \pm 12}{2}$$

$$x = 8 \text{ or } -4$$

3.

$$2x^2 = 3x + 9$$

$$2x^2 - 3x - 9 = 0$$

$$a = 2,\ b = -3,\ c = -9$$

$$x = \frac{-(-3) \pm \sqrt{(-3)^2 - 4(2)(-9)}}{2(2)}$$

$$x = \frac{3 \pm \sqrt{9 + 72}}{4}$$

$$x = \frac{3 \pm \sqrt{81}}{4} = \frac{3 \pm 9}{4}$$

$$x = 3 \text{ or } -\frac{3}{2}$$

5.

$$12x^2 + 5x = 28$$

$$12x^2 + 5x - 28 = 0$$

$$a = 12,\ b = 5,\ c = -28$$

$$x = \frac{-5 \pm \sqrt{5^2 - 4(12)(-28)}}{2(12)}$$

$$x = \frac{-5 \pm \sqrt{25 + 1344}}{24}$$

$$x = \frac{-5 \pm 37}{24}$$

$$x = \frac{4}{3} \text{ or } \frac{-7}{4}$$

7.

$$15x^2 + 75x + 90 = 0$$

$$a = 15,\ b = 75,\ c = 90$$

$$x = \frac{-75 \pm \sqrt{75^2 - 3(15)(90)}}{2(15)}$$

$$x = \frac{-75 \pm \sqrt{5625 - 5400}}{30}$$

$$x = \frac{-75 \pm 15}{30}$$

$$x = -2 \text{ or } -3$$

9.

$$8x^2 + 27 = 30x$$
$$8x^2 - 30x + 27 = 0$$
$$a = 8,\ b = -30,\ c = 27$$
$$x = \frac{-(-30) \pm \sqrt{(-30)^2 - 4(8)(27)}}{2(8)}$$
$$x = \frac{30 \pm \sqrt{900 - 864}}{16}$$
$$x = \frac{30 \pm \sqrt{36}}{16} = \frac{30 \pm 6}{16}$$
$$x = \frac{9}{4} \text{ or } \frac{3}{2}$$

11.

$$6x^2 + 15x = 0$$
$$a = 6,\ b = 15,\ c = 0$$
$$x = \frac{-15 \pm \sqrt{15^2 - 4(6)(0)}}{2(6)}$$
$$x = \frac{-15 \pm \sqrt{15^2}}{12}$$
$$x = \frac{-15 \pm 15}{12}$$
$$x = 0 \text{ or } -\frac{5}{2}$$

13.

$$x^2 + 5x = 7$$
$$x^2 + 5x - 7 = 0$$
$$a = 1,\ b = 5,\ c = -7$$
$$x = \frac{-5 \pm \sqrt{5^2 - 4(1)(-7)}}{2(1)}$$
$$x = \frac{-5 \pm \sqrt{25 + 28}}{2}$$
$$x = \frac{-5 \pm \sqrt{53}}{2}$$

15.

$$x^2 = 4x + 14$$
$$x^2 - 4x - 14 = 0$$
$$a = 1,\ b = -4,\ c = -14$$
$$x = \frac{-(-4) \pm \sqrt{(-4)^2 - 4(1)(-14)}}{2(1)}$$
$$x = \frac{4 \pm \sqrt{16 + 56}}{2}$$
$$x = \frac{4 \pm \sqrt{72}}{2} = \frac{4 \pm 6\sqrt{2}}{2}$$
$$x = \frac{2\left(2 \pm 3\sqrt{2}\right)}{2}$$
$$x = 2 \pm 3\sqrt{2}$$

17.

$$4x^2 = 8x + 23$$
$$4x^2 - 8x - 23 = 0$$
$$a = 4,\ b = -8,\ c = -23$$
$$x = \frac{-(-8) \pm \sqrt{(-8)^2 - 4(4)(-23)}}{2(4)}$$
$$x = \frac{8 \pm \sqrt{64 + 368}}{8}$$
$$x = \frac{8 \pm \sqrt{432}}{8}$$
$$x = \frac{8 \pm \sqrt{144(3)}}{8}$$
$$x = \frac{8 \pm 12\sqrt{3}}{8} = \frac{4\left(2 \pm 3\sqrt{3}\right)}{8}$$
$$x = \frac{2 \pm 3\sqrt{3}}{2}$$

19.
$$9x^2 + 25 = 42x$$
$$9x^2 - 42x + 25 = 0$$
$$a = 9,\ b = -42,\ c = 25$$
$$x = \frac{-(-42) \pm \sqrt{(-42)^2 - 4(9)(25)}}{2(9)}$$
$$x = \frac{42 \pm \sqrt{1764 - 900}}{18}$$
$$x = \frac{42 \pm \sqrt{864}}{18}$$
$$x = \frac{42 \pm \sqrt{144(6)}}{18}$$
$$x = \frac{42 \pm 12\sqrt{6}}{18}$$
$$x = \frac{6\left(7 \pm 2\sqrt{6}\right)}{18}$$
$$x = \frac{7 \pm 2\sqrt{6}}{3}$$

21.
$$x^2 - 4x + 1 = 0$$
$$a = 1,\ b = -4,\ c = 1$$
$$x = \frac{-(-4) \pm \sqrt{(-4)^2 - 4(1)(1)}}{2(1)}$$
$$x = \frac{4 \pm \sqrt{16 - 4}}{2}$$
$$x = \frac{4 \pm \sqrt{12}}{2} = \frac{4 \pm 2\sqrt{3}}{2}$$
$$x = \frac{2\left(2 \pm \sqrt{3}\right)}{2} = 2 \pm \sqrt{3}$$

23.
$$-3x^2 + 7x = 2$$
$$-3x^2 + 7x - 2 = 0$$
$$a = -3,\ b = 7,\ c = -2$$
$$x = \frac{-7 \pm \sqrt{7^2 - 4(-3)(-2)}}{2(-3)}$$
$$x = \frac{-7 \pm \sqrt{49 - 24}}{-6}$$
$$x = \frac{-7 \pm \sqrt{25}}{-6} = \frac{-7 \pm 5}{-6}$$
$$x = \frac{1}{3} \text{ or } 2$$

25.
$$x^2 + 10x = 24$$
$$x^2 + 10x - 24 = 0$$
$$a = 1,\ b = 10,\ c = -24$$
$$x = \frac{-10 \pm \sqrt{10^2 - 4(1)(-24)}}{2(1)}$$
$$x = \frac{-10 \pm \sqrt{100 + 96}}{2}$$
$$x = \frac{-10 \pm \sqrt{196}}{2} = \frac{-10 \pm 14}{2}$$
$$x = 2,\ -12$$

27.
$$4x^2 + 12x + 9 = 0$$
$$a = 4,\ b = 12,\ c = 9$$
$$x = \frac{-12 \pm \sqrt{12^2 - 4(4)(9)}}{2(4)}$$
$$x = \frac{-12 \pm \sqrt{144 - 144}}{8}$$
$$x = \frac{-12}{8} = \frac{-3}{2}$$

29.
$$2x^2 - 5x + 1 = 0$$
$$a = 2,\ b = -5,\ c = 1$$
$$x = \frac{-(-5) \pm \sqrt{(-5)^2 - 4(2)(1)}}{2(2)}$$
$$x = \frac{5 \pm \sqrt{25 - 8}}{4}$$
$$x = \frac{5 \pm \sqrt{17}}{4}$$

31.
$$5x^2 + 6x + 4 = 0$$
$$a = 5,\ b = 6,\ c = 4$$
$$x = \frac{-6 \pm \sqrt{6^2 - 4(5)(4)}}{2(5)}$$
$$x = \frac{-6 \pm \sqrt{36 - 80}}{10}$$
$$x = \frac{-6 \pm \sqrt{-44}}{10}$$
No real solution.

33.
$$21.4x^2 - 16.4x = 12.4$$
$$21.4x^2 - 16.4x - 12.4 = 0$$
$$a = 21.4,\ b = -16.4,\ c = -12.4$$
$$x = \frac{16.4 \pm \sqrt{(-16.4)^2 - 4(21.4)(-12.4)}}{2(21.4)}$$
$$x = \frac{16.4 \pm \sqrt{1330.4}}{42.8}$$
$$x = 1.24 \text{ or } -0.469$$

35.
$$208.4x^2 + 187.2 = 444.5x$$
$$208.4x^2 - 444.5x + 187.2 = 0$$
$$a = 208.4,\ b = -444.5,\ c = 187.2$$
$$x = \frac{444.5 \pm \sqrt{(-444.5)^2 - 4(208.4)(187.2)}}{2(208.4)}$$
$$x = \frac{444.5 \pm \sqrt{41530.33}}{416.8}$$
$$x = 1.56 \text{ or } 0.578$$

37.
$$3/4\,x^2 + 2/3\,x = 1/2$$
$$3/4\,x^2 + 2/3\,x - 1/2 = 0$$
$$a = 0.75,\ b = 2/3,\ c = -0.5$$
$$x = \frac{-2/3 \pm \sqrt{(2/3)^2 - 4(0.75)(-0.5)}}{2(0.75)}$$
$$x = \frac{-2/3 \pm \sqrt{1.94444}}{1.5}$$
$$x = 0.485 \text{ or } -1.37$$

39.
$$3\ 5/6\,x^2 + 5\ 6/7\,x = 2\ 9/16$$
$$3\ 5/6\,x^2 + 5\ 6/7\,x - 2\ 9/16 = 0$$
$$a = 3\ 5/6,\ b = 5\ 6/7,\ c = -2\ 9/16$$
$$x = \frac{-41/7 \pm \sqrt{(-41/7)^2 - 4(23/6)(-41/16)}}{2(23/6)}$$
$$x = \frac{-41/7 \pm \sqrt{(73.59778912)}}{23/3}$$
$$x = 0.355 \text{ or } -1.88$$

41.
$$y = -2x^2 + 2x$$

(a) For $y = 0$:
$$0 = -2x^2 + 2x$$
$$0 = -2x(x-1)$$
$$-2x = 0 \text{ or } x - 1 = 0$$
$$x = 0 \text{ or } 1$$

(b) For $y = -12$:
$$-12 = -2x^2 + 2x$$
$$2x^2 - 2x - 12 = 0$$
$$x^2 - x - 6 = 0$$
$$(x-3)(x+2) = 0$$
$$x - 3 = 0 \text{ or } x + 2 = 0$$
$$x = 3 \text{ or } -2$$

(c) For $y = 7$:
$$7 = -2x^2 + 2x$$
$$2x^2 - 2x + 7 = 0$$
$$a = 2,\ b = -2,\ c = 7$$
$$b^2 - 4ac =$$
$$(-2)^2 - 4(2)(7) =$$
$$4 - 56 =$$
$$-52$$
No real solution.

43. $y = 3x^2 + x$

(a) For $y = 0$:

$0 = 3x^2 + x$

$0 = x(3x + 1)$

$x = 0$ or $3x + 1 = 0$

$x = 0$ or $-\frac{1}{3}$

(b) For $y = 4$:

$4 = 3x^2 + x$

$0 = 3x^2 + x - 4$

$0 = (3x + 4)(x - 1)$

$3x + 4 = 0$ or $x - 1 = 0$

$x = -\frac{4}{3}$ or 1

(c) For $y = 1$:

$1 = 3x^2 + x$

$0 = 3x^2 + x - 1$

$a = 3, b = 1, c = -1$

$$x = \frac{-1 \pm \sqrt{1^2 - 4(3)(-1)}}{2(3)}$$

$$x = \frac{-1 \pm \sqrt{1 + 12}}{6}$$

$$x = \frac{-1 \pm \sqrt{13}}{6}$$

45. $m^2x^2 - 2mx + 1 = 0$

$(mx - 1)^2 = 0$

$mx - 1 = 0$

$x = \frac{1}{m}$

47. $n^2x^2 + 4nx + 4 = 0$

$(nx + 2)^2 = 0$

$nx + 2 = 0$

$x = -\frac{2}{n}$

49. $x^2 + 5x = 16a^2 - 20a$

$x^2 + 5x - 16a^2 + 20a = 0$

$a = 1, b = 5, c = -16a^2 + 20a$

$$x = \frac{-5 \pm \sqrt{5^2 - 4(1)(-16a^2 + 20a)}}{2(1)}$$

$$x = \frac{-5 \pm \sqrt{25 + 64a^2 - 80a}}{2}$$

$$x = \frac{-5 \pm \sqrt{(8a - 5)^2}}{2}$$

$$x = \frac{-5 \pm (8a - 5)}{2} \text{ for } a \geq \frac{5}{8}$$

$$x = \frac{-5 + 8a - 5}{2} = 4a - 5$$

$$\text{or } x = \frac{-5 - 8a + 5}{2} = -4a$$

51. $A = \pi r^2$

$\frac{A}{\pi} = r^2$

$\pm\sqrt{\frac{A}{\pi}} = r$

Since $r \geq 0$,

$r = \sqrt{\frac{A}{\pi}}$

Exercises 7.4

1. Let x = one part

$13 - x$ = other part

$x(13-x) = 40$

$13x - x^2 = 40$

$0 = x^2 - 13x + 40$

$0 = (x-8)(x-5)$

$x - 8 = 0$ or $x - 5 = 0$

$x = 8$ or 5

3. Let x = width

$3.00 + 2x$ = length

$x(2x + 3.00) = 35.0$

$2x^2 + 3.00x - 35.0 = 0$

$(2x - 7.00)(x + 5.00) = 0$

$2x - 7.00 = 0$

$x = 3.50$ cm

$2x + 3.00 = 10.0$ cm

5. Let x = height

$x + 1$ = base

$\frac{1}{2}x(x+1) = 66.0$

$x^2 + x = 132$

$x^2 + x - 132 = 0$

$(x + 12.0)(x - 11.0) = 0$

$x - 11.0 = 0$

$x = 11.0$ m

base = 12.0 m

7. $i = t^2 - 7t + 12$

(a) $i = 2.00A$

$2.00 = t^2 - 7t + 12$

$0 = t^2 - 7t + 10$

$0 = (t - 5.00)(t - 2.00)$

$t - 5.00 = 0$ or $t - 2.00 = 0$

$t = 5.00s$ or $2.00s$

(b) $i = 0A$

$0 = t^2 - 7t + 12$

$(t - 4.00)(t - 3.00) = 0$

$t - 4.00 = 0$ or $t - 3.00 = 0$

$t = 4.00s$ or $3.00s$

(c) $i = 4.00A$

$4.00 = t^2 - 7t + 12$

$0 = t^2 - 7t + 8$

$a = 1, b = -7, c = 8$

$$t = \frac{-(-7) \pm \sqrt{(-7)^2 - 4(1)(8)}}{2(1)}$$

$$t = \frac{7 \pm \sqrt{49 - 32}}{2}$$

$$t = \frac{7 \pm \sqrt{17}}{2}s$$

$t = 5.56s$ or $1.44s$

9. $q = 2t^2 - 4t + 4$

(a) $q = 2.00$

$2.00 = 2t^2 - 4t + 4$

$0 = 2t^2 - 4t + 2$

$0 = t^2 - 2t + 1$

$(t-1)^2 = 0$

$t = 1.00\mu s$

(b) $q = 3.00$

$3.00 = 2t^2 - 4t + 4$

$0 = 2t^2 - 4t + 1$

$a = 2, b = -4, c = 1$

$$t = \frac{-(-4) \pm \sqrt{(-4)^2 - 4(2)(1)}}{2(2)}$$

$$t = \frac{4 \pm \sqrt{8}}{4} = \frac{4 \pm 2\sqrt{2}}{4}$$

$$t = \frac{2(2 \pm \sqrt{2})}{4} = \frac{2 \pm \sqrt{2}}{2}$$

$t = 1.71\mu s$ or $0.293\mu s$

11.

$\phi = 0.4t - 2t^2$

when $\phi = 0.0200$

$0.0200 = 0.4t - 2t^2$

$2t^2 - 0.4t + 0.0200 = 0$

$t^2 - 0.2t + 0.0100 = 0$

$(t - 0.100)^2 = 0$

$t = 0.100ms$

13.

Given $2w + 2l = 80.0$

$w + l = 40.0$

Thus $l = 40.0 - w$

$w(40.0 - w) = 375$

$40.0w - w^2 = 375$

$0 = w^2 - 40.0w + 375$

$0 = (w - 15.0)(w - 25.0)$

$w = 15.0m$ or 25.0m

Since width is less than length, width = $15.0m$ and length = $25.0m$

15.

Let x = side of the new container.

Thus $4xx = 400$

$x^2 = 100$

$x = 10.0\text{in.}$

original side = $10.0 + 4.00 + 4.00$

original side = 18.0in.

17.

Let x = width

y = length

Then $2x + y = 25\bar{0}0$

$y = 25\bar{0}0 - 2x$

$x(25\bar{0}0 - 2x) = 72\bar{0},000$

$25\bar{0}0x - 2x^2 = 72\bar{0},000$

$0 = 2x^2 - 25\bar{0}0x + 72\bar{0},000$

$0 = x^2 - 1250x + 36\bar{0},000$

$0 = (x - 45\bar{0})(x - 80\bar{0})$

$x - 45\bar{0} = 0$ or $x - 80\bar{0} = 0$

$x = 45\bar{0}m$ or $80\bar{0}m$

If $x = 45\bar{0}m$, then

$y = 25\bar{0}0m - 2(45\bar{0})m$

$y = 16\bar{0}0m$

So $45\bar{0}m \times 16\bar{0}0m$

If $x = 80\bar{0}m$, then

$y = 25\bar{0}0m - 2(80\bar{0})m$

$y = 90\bar{0}m$

So $80\bar{0}m \times 90\bar{0}m$

19.

Let x = width of the strip

Area mowed = $2(40.0)x + 2(60 - 2x)(x)$

Also, area mowed = $\frac{1}{2}(40.0)(60.0) = 12\bar{0}0$

Thus $80.0x + 12\bar{0}x - 4x^2 = 12\bar{0}0$

$0 = 4x^2 - 20\bar{0}x + 1200$

$0 = x^2 - 50.0x + 30\bar{0}$

$a = 1, b = -50.0, c = 30\bar{0}$

$$x = \frac{50.0 \pm \sqrt{(-50.0)^2 - 4(30\bar{0})}}{2}$$

$$x = \frac{50.0 + \sqrt{13\bar{0}0}}{2} = 43.0m$$

This is impossible.

$$\text{or } x = \frac{50.0 - \sqrt{1300}}{2} = 6.97$$

The mowed width is $6.97m$

Section 7.4

21.

Let x = width of closed tube

y = length of closed tube

$2x + 2y = 48.0$

$x + y = 24.0$

$y = 24.0 - x$

$x(24.0 - x) = 108$

$24.0x - x^2 = 108$

$0 = x^2 - 24.0x + 108$

$0 = (x - 6.00)(x - 18.0)$

$x - 6.00 = 0$ or $x - 18.0 = 0$

$x = 6.00$ or 18.0

width $= 6.00$ in.

length $= 18.0$ in.

23.

Let x = amount increase in width

Then x = amount increase in length

New width $= 80.0 + x$

New length $= 10\bar{0} + x$

$(80.0 + x)(10\bar{0} + x) = (80.0)(100.0) + 40\bar{0}0$

$80\bar{0}0 + 18\bar{0}x + x^2 = 12\bar{0}00$

$x^2 + 18\bar{0}x - 40\bar{0}0 = 0$

$(x - 20)(x + 200) = 0$

$x - 20 = 0$ *ie*

$x = 20m$

$10\bar{0}m \times 12\bar{0}m$

25.

$R_1 = 3R_2$

$$R = \frac{R_1 R_2}{R_1 + R_2}$$

$$45.0 = \frac{3R_2 R_2}{3R_2 + R_2} = \frac{3R_2^{\,2}}{4R_2}$$

$45.0 = 0.75R_2$

$R_2 = 60.0\Omega$ and $R_1 = 18\bar{0}\Omega$

27.

New width is $10\bar{0} + 2x$

New length is $20\bar{0} + 2x$

$(2x + 10\bar{0})(2x + 20\bar{0}) = 3(10\bar{0})(20\bar{0})$

$4x^2 + 60\bar{0}x + 20\bar{0}00 = 60\bar{0}00$

$4x^2 + 60\bar{0}x - 40\bar{0}00 = 0$

$x^2 + 15\bar{0}x - 10\bar{0}00 = 0$

$(x + 20\bar{0})(x - 50.0) = 0$

$x = 50.0m$

Chapter 7 Review

1.

$x^2 + 4x = 21$

$x^2 + 4x - 21 = 0$

$(x + 7)(x - 3) = 0$

$x + 7 = 0$ or $x - 3 = 0$

$x = -7$ or 3

2.

$6x^2 + 40 = 31x$

$6x^2 - 31x + 40 = 0$

$(2x - 5)(3x - 8) = 0$

$2x - 5 = 0$ or $3x - 8 = 0$

$$x = \frac{5}{2} \text{ or } \frac{8}{3}$$

3.

$20x^2 + 21x + 4 = 0$

$(4x + 1)(5x + 4) = 0$

$4x + 1 = 0$ or $5x + 4 = 0$

$$x = \frac{-1}{4} \text{ or } \frac{-4}{5}$$

4.

$36x^2 - 1 = 0$

$36x^2 = 1$

$$x^2 = \frac{1}{36}$$

$$x = \frac{1}{6} \text{ or } -\frac{1}{6}$$

5.

$18x^2 + 45x + 18 = 0$

$2x^2 + 5x + 2 = 0$

$(2x + 1)(x + 2) = 0$

$2x + 1 = 0$ or $x + 2 = 0$

$$x = -\frac{1}{2} \text{ or } -2$$

6.

$6x^2 - 36x = 0$

$6x(x - 6) = 0$

$6x = 0$ or $x - 6 = 0$

$x = 0$ or 6

7.

$$-8x^2+6x+9=0$$
$$8x^2-6x-9=0$$
$$(2x-3)(4x+3)=0$$
$$2x-3=0 \text{ or } 4x+3=0$$
$$x=\frac{3}{2} \text{ or } x=-\frac{3}{4}$$

8.

$$36-4x^2=0$$
$$-4x^2=-36$$
$$x^2=9$$
$$x=3 \text{ or } -3$$

9.

$$x^2+10x+25=0$$
$$(x+5)^2=0$$
$$x+5=0$$
$$x=-5$$

10.

$$9x^2+16=24x$$
$$9x^2-24x+16=0$$
$$(3x-4)^2=0$$
$$3x-4=0$$
$$x=\frac{4}{3}$$

11.

$$2x^2+13x+20=0$$
$$x^2+\frac{13}{2}x+10=0$$
$$x^2+\frac{13}{2}x=-10$$
$$\text{Find }\left[\frac{1}{2}\left(\frac{13}{2}\right)\right]^2=\frac{169}{16}$$
$$x^2+\frac{13}{2}x+\frac{169}{16}=-10+\frac{169}{16}$$
$$\left(x+\frac{13}{4}\right)^2=\frac{9}{16}$$
$$x+\frac{13}{4}=\pm\sqrt{\frac{9}{16}}=\pm\frac{3}{4}$$
$$x=\frac{-13}{4}\pm\frac{3}{4}$$
$$x=\frac{-5}{2} \text{ or } -4$$

12.

$$3x^2=4x+7$$
$$3x^2-4x=7$$
$$x^2-\frac{4}{3}x=\frac{7}{3}$$
$$\text{Find }\left[\frac{1}{2}\left(-\frac{4}{3}\right)\right]^2=\frac{4}{9}$$
$$x^2-\frac{4}{3}x+\frac{4}{9}=\frac{7}{3}+\frac{4}{9}$$
$$\left(x-\frac{2}{3}\right)^2=\frac{25}{9}$$
$$x-\frac{2}{3}=\pm\sqrt{\frac{25}{9}}$$
$$x-\frac{2}{3}=\pm\frac{5}{3}$$
$$x=\frac{2}{3}\pm\frac{5}{3}$$
$$x=\frac{7}{3} \text{ or } -1$$

13.

$$x^2-8x+9=0$$
$$x^2-8x=-9$$
$$\text{Find }\left[\frac{1}{2}(-8)\right]^2=16$$
$$x^2-8x+16=-9+16$$
$$(x-4)^2=7$$
$$x-4=\pm\sqrt{7}$$
$$x=4\pm\sqrt{7}$$

14.

$$-5x^2+6x+1=0$$
$$x^2-\frac{6}{5}x-\frac{1}{5}=0$$
$$x^2-\frac{6}{5}x=\frac{1}{5}$$
$$\text{Find }\left[\frac{1}{2}\left(-\frac{6}{5}\right)\right]^2=\frac{9}{25}$$
$$x^2-\frac{6}{5}x+\frac{9}{25}=\frac{1}{5}+\frac{9}{25}$$
$$\left(x-\frac{3}{5}\right)^2=\frac{14}{25}$$
$$x-\frac{3}{5}=\pm\sqrt{\frac{14}{25}}=\pm\frac{\sqrt{14}}{5}$$
$$x=\frac{3}{5}\pm\frac{\sqrt{14}}{5}=\frac{3\pm\sqrt{14}}{5}$$

For problems 15 – 21 use: $x = \dfrac{-b \pm \sqrt{b^2 - 4ac}}{2a}$

15.
$$3x^2 - 16x + 20 = 0$$
$$a = 3,\ b = -16,\ c = 20$$
$$x = \frac{-1(-16) \pm \sqrt{(-16)^2 - 4(3)(20)}}{2(3)}$$
$$x = \frac{16 \pm \sqrt{256 - 240}}{6}$$
$$x = \frac{16 \pm \sqrt{16}}{6} = \frac{16 \pm 4}{6}$$
$$x = \frac{10}{3} \text{ or } 2$$

16.
$$4x^2 = 6x + 3$$
$$4x^2 - 6x - 3 = 0$$
$$a = 4,\ b = -6,\ c = -3$$
$$x = \frac{-(-6) \pm \sqrt{(-6)^2 - 4(4)(-3)}}{2(4)}$$
$$x = \frac{6 \pm \sqrt{84}}{8} = \frac{6 \pm 2\sqrt{21}}{8}$$
$$x = \frac{2\left(3 \pm \sqrt{21}\right)}{8}$$
$$x = \frac{3 \pm \sqrt{21}}{4}$$

17.
$$-2x^2 + 3x + 1 = 0$$
$$a = -2,\ b = 3,\ c = 1$$
$$x = \frac{-3 \pm \sqrt{3^2 - 4(-2)(1)}}{2(-2)}$$
$$x = \frac{-3 \pm \sqrt{17}}{-4}$$
$$x = \frac{3 \pm \sqrt{17}}{4}$$

18.
$$6x^2 + 8x - 5 = 0$$
$$a = 6,\ b = 8,\ c = -5$$
$$x = \frac{-8 \pm \sqrt{8^2 - 4(6)(-5)}}{2(6)}$$
$$x = \frac{-8 \pm \sqrt{184}}{12} = \frac{-8 \pm 2\sqrt{46}}{12}$$
$$x = \frac{2\left(-4 \pm \sqrt{46}\right)}{12} = \frac{-4 \pm \sqrt{46}}{6}$$

19.
$$mnx^2 - mx + n = 0$$
$$a = mn,\ b = -m,\ c = n$$
$$x = \frac{-(-m) \pm \sqrt{(-m)^2 - 4(mn)(n)}}{2mn}$$
$$x = \frac{m \pm \sqrt{m^2 - 4mn^2}}{2mn}$$

20.
$$n^4x^2 - 2m^2n^2x + m^4 = 0$$
$$\left(n^2x - m^2\right)^2 = 0$$
$$n^2x - m^2 = 0$$
$$n^2x = m^2$$
$$x = \frac{m^2}{n^2}$$

21.

$$s = vt + \frac{1}{2}at^2$$

solve for t

$$0 = \frac{1}{2}at^2 + vt - s$$

coefficient of t^2 is $\frac{1}{2}a$

$b = v$, $c = -s$

$$t = \frac{-v \pm \sqrt{v^2 - 4\left(\frac{1}{2}a\right)(-s)}}{2\left(\frac{1}{2}a\right)}$$

$$t = \frac{-v \pm \sqrt{v^2 + 2as}}{a}$$

22.

$x =$ one part

$15 - x =$ other part

$x(15 - x) = 36$

$15x - x^2 = 36$

$0 = x^2 - 15x + 36$

$0 = (x - 12)(x - 3)$

$x - 12 = 0$ or $x - 3 = 0$

$x = 12$ or 3

23.

Let $x =$ width

$y =$ length

$2x + 2y = 30\bar{0}0$

$x + y = 15\bar{0}0$

$y = 15\bar{0}0 - x$

$x(15\bar{0}0 - x) = 54\bar{0},000$

$15\bar{0}0x - x^2 = 54\bar{0},000$

$0 = x^2 - 15\bar{0}0x + 54\bar{0}000$

$0 = (x - 60\bar{0})(x - 90\bar{0})$

$x - 60\bar{0} = 0$ or $x - 90\bar{0} = 0$

$x = 60\bar{0}$ or $90\bar{0}$

$60\bar{0}$ ft $\times$ $90\bar{0}$ ft

24.

Let $x =$ one side of cutout

Bottom area $= 90\bar{0}$ cm^2.

$(50.0 - 2x)^2 = 90\bar{0}$

$50.0 - 2x = 30.0$

$-2x = -20.0$

$x = 10.0$ cm

$V = lwh$

$V = (50.0 - 20.0)^2(10.0)$ cm^3

$V = 90\bar{0}0$ cm^3

25. $h = 64t - 16t^2$

(a) Find t when $h = 48.0$

$48.0 = 64t - 16t^2$

$16t^2 - 64t + 48.0 = 0$

$t^2 - 4t + 3.00 = 0$

$(t - 3.00)(t - 1.00) = 0$

$t = 3.00s$ or $t = 1.00s$

(b) Can $h = 80.0$ ft?

$80.0 = 64t - 16t^2$

$16t^2 - 64t + 80.0 = 0$

$t^2 - 4t + 5.00 = 0$

$b^2 - 4ac = (-4)^2 - 4(1)(5)$

$= 16 - 20$

$= -4$

No real solution. So h cannot be 80.0 ft.

(c) Find the time for $h = 0$.

$0 = 64t - 16t^2$

$0 = 16t(4 - t)$

$16t = 0$ or $4 - t = 0$

$t = 0$ or $t = 4s$

Air time is $4 - 0 = 4s$

(d) Maximum height will occur half way between initial time and end time, i.e. half way between $t = 0$ and $t = 4$, i.e. at $t = 2$.

$h = 64(2) - 16(2)^2$

$h = 128 - 16(4)$

$h = 128 - 64.0$

$h = 64.0$ ft

26.

$$\frac{1}{R} + \frac{1}{55\bar{0} - R} = \frac{1}{12\bar{0}}$$

Multiply by the L.C.D.: $12\bar{0}R(55\bar{0} - R)$

$12\bar{0}(55\bar{0} - R) + 12\bar{0}R = R(55\bar{0} - R)$

$66\bar{0}00 - 12\bar{0}R + 12\bar{0}R = 55\bar{0}R - R^2$

$R^2 - 55\bar{0}R + 66\bar{0}00 = 0$

$a = 1, b = -55\bar{0}, c = 66\bar{0}00$

$$R = \frac{-(-55\bar{0}) \pm \sqrt{(-55\bar{0})^2 - 4(1)(66\bar{0}00)}}{2(1)}$$

$$R = \frac{55\bar{0} \pm \sqrt{38500}}{2} = \frac{55\bar{0} \pm 10\sqrt{385}}{2} = 275 \pm 5\sqrt{385}$$

$R = (275 + 5\sqrt{385})\Omega$ or $(275 - 5\sqrt{385})\Omega$

$R = 373\Omega$ or 177Ω

Chapter 8, Exercises 8.1

1. $x^2 \cdot x^3 = x^{2+3} = x^5$

3. $\dfrac{5^6}{5^3} = 5^{6-3} = 5^3$

5. $\left(7^2\right)^3 = 7^{2(3)} = 7^6$

7. $m^{-2} \cdot m^5 = m^{-2+5} = m^3$

9. $\dfrac{d^{-3}}{d^4} = d^{-3-4} = d^{-7} = \dfrac{1}{d^7}$

11. $\left(2^{-3}\right)^2 = 2^{-3(2)} = 2^{-6} = \dfrac{1}{2^6}$

13. $y^{-3} \cdot y^{-5} = y^{-3+(-5)} = y^{-8} = \dfrac{1}{y^8}$

15. $\left(5^{-2}\right)^{-5} = 5^{-2(-5)} = 5^{10}$

17. $\dfrac{t^{-2}}{t^{-5}} = t^{-2-(-5)} = t^3$

19. $\left(\dfrac{4}{7}\right)^2 = \dfrac{16}{49}$

21. $\left(\dfrac{1}{3}\right)^{-1} = \dfrac{1}{\frac{1}{3}} = 3$

23. $\left(\dfrac{5}{6}\right)^{-1} = \dfrac{1}{\frac{5}{6}} = \dfrac{6}{5}$

25. $\left(\dfrac{1}{2}\right)^{-3} = \dfrac{1^{-3}}{2^{-3}} = \dfrac{1}{2^{-3}} = 2^3 = 8$

27. $\left(\dfrac{2}{3}\right)^{-2} = \dfrac{2^{-2}}{3^{-2}} = \dfrac{3^2}{2^2} = \dfrac{9}{4}$

29. $\dfrac{a^5}{a^{-5}} = a^{5-(-5)} = a^{10}$

31. $a^6 \cdot a^{-4} = a^{6+(-4)} = a^2$

33. $\left(a^3\right)^{-4} = a^{3(-4)} = a^{-12} = \dfrac{1}{a^{12}}$

35. $\left(3a^2\right)^2 = 3^2 a^{2(2)} = 9a^4$

37. $\left(2a^{-2}b\right)^{-2} = 2^{-2} a^{-2(-2)} b^{-2} = \dfrac{a^4}{4b^2}$

39. $6k^2(-2k)\left(4k^{-5}\right) = 6(-2)(4)k^{2+1+(-5)} = -48k^{-2} = \dfrac{-48}{k^2}$

41. $\left(3a^{-2}b\right)\left(5a^3b^{-3}\right) = 3(5)a^{-2+3}b^{1-3} = 15ab^{-2} = \dfrac{15a}{b^2}$

43. $w^{-4}w^2w^{-3} = w^{-4+2-3} = w^{-5} = \dfrac{1}{w^5}$

45. $\dfrac{x^3 \cdot x^{-4}}{x^{-2} \cdot x^{-5}} = \dfrac{x^{3-4}}{x^{-2-5}} = \dfrac{x^{-1}}{x^{-7}} = x^{-1-(-7)} = x^6$

47. $\dfrac{x^{-2}y^3}{x^3y^{-2}} = x^{-2-3}y^{3-(-2)} = x^{-5}y^5 = \dfrac{y^5}{x^5}$

49. $\left(\dfrac{1}{t^4}\right)^{-2} = \dfrac{1^{-2}}{t^{4(-2)}} = \dfrac{1}{t^{-8}} = t^8$

51. $\left(\dfrac{1}{b^{-2}}\right)^{-1} = \dfrac{1^{-1}}{b^{-2(-1)}} = \dfrac{1}{b^2}$

53. $\dfrac{ab^{-4}c^5}{a^2b^{-2}c} = a^{1-2}b^{-4-(-2)}c^{5-1} = a^{-1}b^{-2}c^4 = \dfrac{c^4}{ab^2}$

55. $\left(\dfrac{a^2}{a^{-4}}\right)^3 = \left(a^{2-(-4)}\right)^3 = \left(a^6\right)^3 = a^{6(3)} = a^{18}$

57.
$$\frac{(3a^{-2})^3}{(2a^{-4})^{-4}} =$$
$$\frac{3^3 a^{-2(3)}}{2^{-2} a^{-4(-2)}} =$$
$$\frac{27a^{-6}}{2^{-2}a^{8}} =$$
$$27(2^2)a^{-6-8} =$$
$$27(4)a^{-14} =$$
$$\frac{108}{a^{14}}$$

59.
$$(2a^4b^0c^{-2})^3 =$$
$$2^3 a^{4(3)} b^{0(3)} c^{-2(3)} =$$
$$8a^{12}b^0c^{-6} =$$
$$\frac{8a^{12}}{c^6}$$

61.
$$\left(\frac{14a^3b^{-8}}{2a^{-2}b^{-4}}\right)^2 =$$
$$\left(7a^{3-(-2)}b^{-8-(-4)}\right)^2 =$$
$$(7a^5b^{-4})^2 =$$
$$7^2 a^{5(2)} b^{-4(2)} =$$
$$49a^{10}b^{-8} =$$
$$\frac{49a^{10}}{b^8}$$

63.
$$a^{-1} + b^{-1} =$$
$$\frac{1}{a} + \frac{1}{b} =$$
$$\frac{1\cdot b}{ab} + \frac{1\cdot a}{b\cdot a} =$$
$$\frac{b+a}{ab} = \frac{a+b}{ab}$$

65.
$$a^{-2}b + a^{-1}b =$$
$$\frac{b}{a^2} + \frac{b}{a} =$$
$$\frac{b}{a^2} + \frac{ba}{a^2} =$$
$$\frac{b+ab}{a^2}$$

67.
$$3^{-2} + 9^{-1} =$$
$$\frac{1}{3^2} + \frac{1}{9} =$$
$$\frac{1}{9} + \frac{1}{9} =$$
$$\frac{2}{9}$$

69.
$$(2^{-1} + 3^{-1})^{-1} =$$
$$\left(\frac{1}{2} + \frac{1}{3}\right)^{-1} =$$
$$\left(\frac{3}{6} + \frac{2}{6}\right)^{-1} =$$
$$\left(\frac{5}{6}\right)^{-1} =$$
$$\frac{1}{\frac{5}{6}} =$$
$$\frac{6}{5}$$

71.
$$\frac{1}{a^0 + b^0} = \frac{1}{1+1} = \frac{1}{2}$$

Exercises 8.2

1. $36^{1/2} = \sqrt{36} = 6$

3. $64^{1/3} = \sqrt[3]{64} = 4$

5. $64^{2/3} = \left(\sqrt[3]{64}\right)^2 = (4)^2 = 16$

7. $8^{5/3} = \left(\sqrt[3]{8}\right)^5 = (2)^5 = 32$

9. $16^{1/2} = \dfrac{1}{16^{1/2}} = \dfrac{1}{\sqrt{16}} = \dfrac{1}{4}$

11. $27^{-1/3} = \dfrac{1}{27^{1/3}} = \dfrac{1}{\sqrt[3]{27}} = \dfrac{1}{3}$

13. $27^{2/3} = \left(\sqrt[3]{27}\right)^2 = (3)^2 = 9$

15. $9^{-3/2} = \dfrac{1}{9^{3/2}} = \dfrac{1}{\left(\sqrt{9}\right)^3} = \dfrac{1}{3^3} = \dfrac{1}{27}$

17. $\left(\dfrac{16}{25}\right)^{1/2} = \sqrt{\dfrac{16}{25}} = \dfrac{4}{5}$

19. $\left(\dfrac{4}{9}\right)^{3/2} = \left(\sqrt{\dfrac{4}{9}}\right)^3 = \left(\dfrac{2}{3}\right)^3 = \dfrac{8}{27}$

21. $\left(\dfrac{25}{16}\right)^{-3/2} = \dfrac{1}{\left(\dfrac{25}{16}\right)^{3/2}} = \dfrac{1}{\left(\sqrt{\dfrac{25}{16}}\right)^3} = \dfrac{1}{\left(\dfrac{5}{4}\right)^3} = \dfrac{1}{\dfrac{125}{64}} = \dfrac{64}{125}$

23. $(-8)^{-1/3} = \dfrac{1}{(-8)^{1/3}} = \dfrac{1}{\sqrt[3]{-8}} = \dfrac{1}{-2} = -\dfrac{1}{2}$

25. $3^{3/5} \cdot 3^{7/5} = 3^{3/5+7/5} = 3^{10/5} = 3^2 = 9$

27. $\dfrac{9^{7/4}}{9^{5/4}} = 9^{7/4-5/4} = 9^{2/4} = 9^{1/2} = \sqrt{9} = 3$

29. $\left(16^{1/2}\right)^{1/2} = 16^{1/2(1/2)} = 16^{1/4} = 2$

31. $x^{4/3} \cdot x^{2/3} = x^{4/3+2/3} = x^{6/3} = x^2$

33. $x^{3/4} \cdot x^{-1/2} = x^{3/4-1/2} = x^{1/4}$

35. $\dfrac{x^{2/3}}{x^{1/6}} = x^{2/3-1/6} = x^{3/6} = x^{1/2}$

37. $\left(x^{2/3}\right)^{-1/2} = x^{2/3(-1/2)} = x^{-1/3} = \dfrac{1}{x^{1/3}}$

39. $\left(x^{2/5} \cdot x^{-4/5}\right)^{5/6} = \left(x^{2/5-4/5}\right)^{5/6} = \left(x^{-2/5}\right)^{5/6} = x^{-2/5(5/6)} = x^{-1/3} = \dfrac{1}{x^{1/3}}$

41. $\left(\dfrac{x^{3/2}}{x^{3/4}}\right)^{1/3} = \left(x^{3/2-3/4}\right)^{1/3} = \left(x^{3/4}\right)^{1/3} = x^{1/4}$

43. $\left(\dfrac{a^4 b^{-8}}{c^{16}}\right)^{3/4} =$

$\dfrac{a^{4(3/4)} b^{-8(3/4)}}{c^{16(3/4)}} =$

$\dfrac{a^3 b^{-6}}{c^{12}} =$

$\dfrac{a^3}{b^6 c^{12}}$

45. $\left(a^{-12} b^9 c^{-6}\right)^{-4/3} =$

$a^{-12(-4/3)} b^{9(-4/3)} c^{-6(-4/3)} =$

$a^{16} b^{-12} c^8 =$

$\dfrac{a^{16} c^8}{b^{12}}$

47.

$$x^{2/3}\left(x^{1/3}+4x^{4/3}\right)=$$
$$x^{2/3+1/3}+4x^{2/3+4/3}=$$
$$x+4x^{6/3}=$$
$$x+4x^2$$

49.

$$4t^{1/2}\left(2t^{1/2}+\frac{1}{2}t^{-1/2}\right)=$$
$$8t^{1/2+1/2}+2t^{1/2-1/2}=$$
$$8t+2t^0=$$
$$8t+2$$

51.

$$2c^{-1/3}\left(3c^{2/3}+4c^{-2/3}\right)=$$
$$6c^{-1/3+2/3}+8c^{-1/3-2/3}=$$
$$6c^{1/3}+8c^{-3/3}=$$
$$6c^{1/3}+8c^{-1}=$$
$$6c^{1/3}+\frac{8}{c}$$

53.

$$\left(x^{1/2}+y^{1/2}\right)\left(x^{1/2}-y^{1/2}\right)=$$
$$x^{1/2+1/2}-x^{1/2}y^{1/2}+x^{1/2}y^{1/2}-y^{1/2+1/2}=$$
$$x-y$$

55.

$$\left(x^{1/2}+y^{1/2}\right)^2=$$
$$\left(x^{1/2}\right)^2+2x^{1/2}y^{1/2}+\left(y^{1/2}\right)^2=$$
$$x+2x^{1/2}y^{1/2}+y$$

57.

$$v=14t^{2/5} \text{ for } t=32$$
$$v=14(32)^{2/5}$$
$$v=14[(32)^{1/5}]^2$$
$$v=14[2]^2$$
$$v=14(4)=56$$

59.

$$y=4^{2t} \text{ for } t=5/4$$
$$y=4^{2(5/4)}=4^{5/2}$$
$$y=\left(4^{1/2}\right)^5=2^5$$
$$y=32$$

61.

$$Q=\frac{bH^{3/2}}{3} \text{ for } b=12,\ H=16$$
$$Q=\frac{12(16)^{3/2}}{3}=\frac{12[16^{1/2}]^3}{3}$$
$$Q=\frac{12[4]^3}{3}=\frac{12(64)}{3}$$
$$Q=256$$

63.

$$H=\frac{4M^{3/4}N^{1/3}}{5Q}$$
$$H=\frac{4(16)^{3/4}64^{1/3}}{5(0.04)}$$
$$H=\frac{4\left(16^{1/4}\right)^3(4)}{0.2}$$
$$H=\frac{4(2)^3(4)}{0.2}$$
$$H=\frac{4(8)(4)}{0.2}$$
$$H=640$$

65.

$$f=\frac{2}{\pi}\left(\frac{3EIg}{w}\right)^{1/2}l^{-3/2}$$
$$f=\frac{2}{\pi}\left(\frac{3\cdot4\cdot1\cdot32}{6}\right)^{1/2}4^{-3/2}$$
$$f=\frac{2}{\pi}(64)^{1/2}4^{-3/2}$$
$$f=\frac{\frac{2}{\pi}(8)}{4^{3/2}}=\frac{\frac{16}{\pi}}{\left(4^{1/2}\right)^3}$$
$$f=\frac{\frac{16}{\pi}}{2^3}=\frac{16}{\pi}\cdot\frac{1}{8}=\frac{2}{\pi}$$

Exercises 8.3

1. $\sqrt{49} = 7$

3. $\sqrt{75} = \sqrt{25 \cdot 3} = \sqrt{25}\sqrt{3} = 5\sqrt{3}$

5. $\sqrt{180} = \sqrt{36}\sqrt{5} = 6\sqrt{5}$

7. $\sqrt{8a^2} = \sqrt{4a^2}\sqrt{2} = 2a\sqrt{2}$

9. $\sqrt{72b^2} = \sqrt{36b^2}\sqrt{2} = 6b\sqrt{2}$

11. $\sqrt{80a^5b^2} = \sqrt{16a^4b^2}\sqrt{5a} = 4a^2b\sqrt{5a}$

13. $\sqrt{32a^2b^4c^4} = \sqrt{16a^2b^4c^8}\sqrt{2c} = 4ab^2c^4\sqrt{2c}$

15. $\sqrt{\frac{3}{4}} = \frac{\sqrt{3}}{\sqrt{4}} = \frac{\sqrt{3}}{2}$

17. $\sqrt{\frac{5}{8}} = \sqrt{\frac{5 \cdot 2}{8 \cdot 2}} = \sqrt{\frac{10}{16}} = \frac{\sqrt{10}}{\sqrt{16}} = \frac{\sqrt{10}}{4}$

19. $\frac{\sqrt{6}}{\sqrt{10}} = \sqrt{\frac{6}{10}} = \sqrt{\frac{3}{5} \cdot \frac{5}{5}} = \frac{\sqrt{15}}{\sqrt{25}} = \frac{\sqrt{15}}{5}$

21. $\frac{5}{\sqrt{24}} = \frac{5}{\sqrt{4}\sqrt{6}} = \frac{5\sqrt{6}}{2\sqrt{6}\sqrt{6}} = \frac{5\sqrt{6}}{2(6)} = \frac{5\sqrt{6}}{12}$

23. $\sqrt{\frac{4a}{15b^2}} = \sqrt{\frac{4a}{15b^2} \cdot \frac{15}{15}} = \frac{\sqrt{4}\sqrt{15a}}{\sqrt{15^2b^2}} = \frac{2\sqrt{15a}}{15b}$

25. $\frac{2}{\sqrt{8b}} = \frac{2}{\sqrt{4}\sqrt{2b}} = \frac{2\sqrt{2b}}{2\sqrt{2b}\sqrt{2b}} = \frac{\sqrt{2b}}{2b}$

27. $\frac{\sqrt{5a^4b}}{\sqrt{20a^2b^3}} = \sqrt{\frac{5a^4b}{20a^2b^3}} = \sqrt{\frac{a^2}{4b^2}} = \frac{a}{2b}$

29. $\sqrt[3]{125} = \sqrt[3]{5^3} = 5$

31. $\sqrt[3]{16a^4} = \sqrt[3]{2^3a^3 \cdot 2a} = 2a\sqrt[3]{2a}$

33. $\sqrt[3]{40a^8} = \sqrt[3]{2^3a^6 5a^2} = 2a^2\sqrt[3]{5a^2}$

35. $\sqrt[3]{54x^5} = \sqrt[3]{27x^3 \cdot 2x^2} = 3x\sqrt[3]{2x^2}$

37. $\sqrt[3]{56x^7y^5z^3} = \sqrt[3]{8x^6y^3z^3 \cdot 7xy^2} = 2x^2yz\sqrt[3]{7xy^2}$

39. $\sqrt[3]{\frac{3}{8}} = \frac{\sqrt[3]{3}}{\sqrt[3]{8}} = \frac{\sqrt[3]{3}}{2}$

41. $\sqrt[3]{\frac{5}{4}} = \sqrt[3]{\frac{5 \cdot 2}{4 \cdot 2}} = \frac{\sqrt[3]{10}}{\sqrt[3]{8}} = \frac{\sqrt[3]{10}}{2}$

43. $\sqrt[3]{\frac{5}{12}} = \sqrt[3]{\frac{5}{2^2 \cdot 3} \cdot \frac{2 \cdot 3^2}{2 \cdot 3^2}} = \sqrt[3]{\frac{90}{2^3 \cdot 3^3}} = \frac{\sqrt[3]{90}}{6}$

45. $\frac{1}{\sqrt[3]{2}} = \frac{1}{\sqrt[3]{2}} \cdot \frac{\sqrt[3]{4}}{\sqrt[3]{4}} = \frac{\sqrt[3]{4}}{\sqrt[3]{8}} = \frac{\sqrt[3]{4}}{2}$

47. $\sqrt[3]{\frac{4a}{50b^2}} = \sqrt[3]{\frac{2a}{25b^2} \cdot \frac{5b}{5b}} = \sqrt[3]{\frac{10ab}{5^3b^3}} = \frac{\sqrt[3]{10ab}}{5b}$

49. $\sqrt[3]{\frac{8a}{63a^2b^3}} = \sqrt[3]{\frac{8a}{63a^2b^3} \cdot \frac{3 \cdot 7^2a}{3 \cdot 7^2a}} = \sqrt[3]{\frac{8 \cdot 147a^2}{7^3 \cdot 3^3a^3 \cdot b^3}} = \frac{2\sqrt[3]{147a^2}}{21ab}$

51. $\dfrac{\sqrt[3]{5a^4b}}{\sqrt[3]{20a^2b^3}} =$

$\sqrt[3]{\dfrac{5a^4b}{20a^2b^3}} =$

$\sqrt[3]{\dfrac{a^2b}{4b^3} \cdot \dfrac{2}{2}} =$

$\sqrt[3]{\dfrac{2a^2b}{8b^3}} =$

$\dfrac{\sqrt[3]{2a^2b}}{2b}$

53. $\sqrt[4]{80x^6} = \sqrt[4]{2^4x^4 \cdot 5x^2} = 2x\sqrt[4]{5x^2}$

55. $\sqrt[4]{25a^5b^4} = \sqrt[4]{a^4b^4 \cdot 25a} = ab\sqrt[4]{25a}$

57. $\sqrt[5]{\dfrac{32a^8b^{12}}{c^6}} = \sqrt[5]{\dfrac{32a^5b^{10}}{c^5} \cdot \dfrac{a^3b^2}{c} \cdot \dfrac{c^4}{c^4}} =$

$\sqrt[5]{\dfrac{32a^5b^{10}}{c^{10}}}\sqrt[5]{a^3b^2c^4} =$

$\dfrac{2ab^2\sqrt[5]{a^3b^2c^4}}{c^2}$

59. $\sqrt[4]{a^2} = a^{2/4} = a^{1/2} = \sqrt{a}$

61. $\sqrt[6]{27a^3b^9} = \left(3^3a^3b^9\right)^{1/6} = 3^{3/6}a^{3/6}b^{9/6} = 3^{1/2}a^{1/2}b^{3/2} = b\sqrt{3ab}$

63. $\sqrt{\sqrt{3}} = \sqrt[2(2)]{3} = \sqrt[4]{3}$

65. $\sqrt[3]{\sqrt{64}} = \sqrt[3(2)]{64} = \sqrt[6]{64} = 2$

67. $\sqrt[4]{64x^6} = \sqrt[4]{16x^3 \cdot 4x^2} = 2x\sqrt[4]{4x^2} = 2x\left(2^2x^2\right)^{1/4} = 2x \cdot \left(2^{1/2}x^{1/2}\right) = 2x\sqrt{2x}$

69. $\sqrt{4a^2 - 4b^2} = \sqrt{4\left(a^2 - b^2\right)} = 2\sqrt{a^2 - b^2}$

71. $\sqrt{4(a-b)^2} = 2(a-b)$ for $a \geq b$

Exercises 8.4

1. $3\sqrt{2} - 5\sqrt{2} + \sqrt{2} = (3 - 5 + 1)\sqrt{2} = -\sqrt{2}$

3. $\sqrt{8} - \sqrt{2} = \sqrt{4}\sqrt{2} - \sqrt{2} = 2\sqrt{2} - \sqrt{2} = \sqrt{2}$

5. $3\sqrt{48} + 4\sqrt{75} =$

$3\sqrt{16}\sqrt{3} + 4\sqrt{25}\sqrt{3} =$

$3(4)\sqrt{3} + 4(5)\sqrt{3} =$

$12\sqrt{3} + 20\sqrt{3} =$

$32\sqrt{3}$

7. $5\sqrt{80} - 6\sqrt{45} =$

$5\sqrt{16}\sqrt{5} - 6\sqrt{9}\sqrt{5} =$

$5(4)\sqrt{5} - 6(3)\sqrt{5} =$

$20\sqrt{5} - 18\sqrt{5} =$

$2\sqrt{5}$

9. $4\sqrt{12} - \sqrt{27} - 3\sqrt{48} =$

$4\sqrt{4}\sqrt{3} - \sqrt{9}\sqrt{3} - 3\sqrt{16}\sqrt{3} =$

$4(2)\sqrt{3} - 3\sqrt{3} - 3(4)\sqrt{3} =$

$8\sqrt{3} - 3\sqrt{3} - 12\sqrt{3} =$

$-7\sqrt{3}$

11. $\sqrt{32} + 3\sqrt{50} - 6\sqrt{18} =$

$\sqrt{16}\sqrt{2} + 3\sqrt{25}\sqrt{2} - 6\sqrt{9}\sqrt{2} =$

$4\sqrt{2} + 3(5)\sqrt{2} - 6(3)\sqrt{2} =$

$4\sqrt{2} + 15\sqrt{2} - 18\sqrt{2} =$

$\sqrt{2}$

13. $3\sqrt{27}-9\sqrt{48}+\sqrt{75}=$
$3\sqrt{9}\sqrt{3}-9\sqrt{16}\sqrt{3}+\sqrt{25}\sqrt{3}=$
$3(3)\sqrt{3}-9(4)\sqrt{3}+5\sqrt{3}=$
$9\sqrt{3}-36\sqrt{3}+5\sqrt{3}=$
$-22\sqrt{3}$

15. $2\sqrt{54}+3\sqrt{24}-3\sqrt{96}=$
$2\sqrt{9}\sqrt{6}+3\sqrt{4}\sqrt{6}-3\sqrt{16}\sqrt{6}=$
$2(3)\sqrt{6}+3(2)\sqrt{6}-3(4)\sqrt{6}=$
$6\sqrt{6}+6\sqrt{6}-12\sqrt{6}=$
0

17. $5\sqrt{2x}-\sqrt{18x}+\sqrt{8x}=$
$5\sqrt{2x}-\sqrt{9}\sqrt{2x}+\sqrt{4}\sqrt{2x}=$
$5\sqrt{2x}-3\sqrt{2x}+2\sqrt{2x}=$
$4\sqrt{2x}$

19. $4\sqrt{32x^2}-\sqrt{72x^2}+3\sqrt{50x^2}=$
$4\sqrt{16x^2}\sqrt{2}-\sqrt{36x^2}\sqrt{2}+3\sqrt{25x^2}\sqrt{2}=$
$4(4x)\sqrt{2}-6x\sqrt{2}+3(5x)\sqrt{2}=$
$16x\sqrt{2}-6x\sqrt{2}+15x\sqrt{2}=$
$25x\sqrt{2}$

21. $3\sqrt[3]{6}-7\sqrt[3]{6}+\sqrt[3]{6}=$
$-3\sqrt[3]{6}$

23. $2\sqrt[3]{54}-2\sqrt[3]{16}=$
$2\sqrt[3]{27}\sqrt[3]{2}-2\sqrt[3]{8}\sqrt[3]{2}=$
$2(3)\sqrt[3]{2}-2(2)\sqrt[3]{2}=$
$6\sqrt[3]{2}-4\sqrt[3]{2}=$
$2\sqrt[3]{2}$

25. $\dfrac{2\sqrt[3]{40}+8\sqrt[3]{5}}{4}=$
$\dfrac{2\sqrt[3]{8}\sqrt[3]{5}+8\sqrt[3]{5}}{4}=$
$\dfrac{2(2)\sqrt[3]{5}+8\sqrt[3]{5}}{4}=$
$\dfrac{4\sqrt[3]{5}+8\sqrt[3]{5}}{4}=$
$\dfrac{12\sqrt[3]{5}}{4}=3\sqrt[3]{5}$

27. $\sqrt[3]{81}+3\sqrt[3]{3}-\sqrt{3}=$
$\sqrt[3]{27}\sqrt[3]{3}+3\sqrt[3]{3}-\sqrt{3}=$
$3\sqrt[3]{3}+3\sqrt[3]{3}-\sqrt{3}=$
$6\sqrt[3]{3}-\sqrt{3}$

29. $\sqrt[3]{54}-4\sqrt[3]{16}+\sqrt[3]{128}=$
$\sqrt[3]{27}\sqrt[3]{2}-4\sqrt[3]{8}\sqrt[3]{2}+\sqrt[3]{64}\sqrt[3]{2}=$
$3\sqrt[3]{2}-4(2)\sqrt[3]{2}+4\sqrt[3]{2}=$
$3\sqrt[3]{2}-8\sqrt[3]{2}+4\sqrt[3]{2}=$
$-\sqrt[3]{2}$

31. $\dfrac{5}{\sqrt{2}}-\dfrac{\sqrt{2}}{2}=$
$\dfrac{5\sqrt{2}}{\sqrt{2}\sqrt{2}}-\dfrac{\sqrt{2}}{2}=$
$\dfrac{5\sqrt{2}}{2}-\dfrac{\sqrt{2}}{2}=$
$\dfrac{4\sqrt{2}}{2}=2\sqrt{2}$

33. $\dfrac{3\sqrt{3}}{2}+\dfrac{2}{\sqrt{3}}=$
$\dfrac{3\sqrt{3}}{2}+\dfrac{2}{\sqrt{3}}\cdot\dfrac{\sqrt{3}}{\sqrt{3}}=$
$\dfrac{3\sqrt{3}}{2}+\dfrac{2\sqrt{3}}{3}=$
$\dfrac{9\sqrt{3}}{6}+\dfrac{4\sqrt{3}}{6}=$
$\dfrac{13\sqrt{3}}{6}$

35.
$$\sqrt{\frac{2}{3}}+\sqrt{\frac{1}{6}}-\sqrt{24}=$$
$$\sqrt{\frac{2}{3}\cdot\frac{3}{3}}+\sqrt{\frac{1}{6}\cdot\frac{6}{6}}-\sqrt{4}\sqrt{6}=$$
$$\frac{\sqrt{6}}{3}+\frac{\sqrt{6}}{6}-2\sqrt{6}=$$
$$\frac{2\sqrt{6}}{6}+\frac{\sqrt{6}}{6}-\frac{12\sqrt{6}}{6}=$$
$$\frac{-9\sqrt{6}}{6}=\frac{-3\sqrt{6}}{2}$$

37.
$$\sqrt{3}-\frac{1}{\sqrt{3}}=$$
$$\sqrt{3}-\frac{1}{\sqrt{3}}\cdot\frac{\sqrt{3}}{\sqrt{3}}=$$
$$\sqrt{3}-\frac{\sqrt{3}}{3}=$$
$$\frac{3\sqrt{3}-\sqrt{3}}{3}=\frac{2\sqrt{3}}{3}$$

39.
$$\frac{\sqrt{18}}{6}+\frac{\sqrt{2}}{2}+\sqrt{\frac{1}{2}}=$$
$$\frac{\sqrt{9}\sqrt{2}}{6}+\frac{\sqrt{2}}{2}+\sqrt{\frac{1}{2}\cdot\frac{2}{2}}=$$
$$\frac{3\sqrt{2}}{6}+\frac{\sqrt{2}}{2}\cdot\frac{3}{3}+\frac{\sqrt{2}}{2}\cdot\frac{3}{3}=$$
$$\frac{3\sqrt{2}}{6}+\frac{3\sqrt{2}}{6}+\frac{3\sqrt{2}}{6}=$$
$$\frac{9\sqrt{2}}{6}=\frac{3\sqrt{2}}{2}$$

41.
$$\sqrt{10}-\sqrt{\frac{2}{5}}=$$
$$\sqrt{10}-\sqrt{\frac{2}{5}\cdot\frac{5}{5}}=$$
$$\sqrt{10}-\frac{\sqrt{10}}{5}=$$
$$\frac{4\sqrt{10}}{5}$$

43.
$$4\sqrt[3]{3}-\frac{6}{\sqrt[3]{9}}=$$
$$4\sqrt[3]{3}-\frac{6}{\sqrt[3]{9}}\cdot\frac{\sqrt[3]{3}}{\sqrt[3]{3}}=$$
$$4\sqrt[3]{3}-\frac{6\sqrt[3]{3}}{3}=$$
$$4\sqrt[3]{3}-2\sqrt[3]{3}=2\sqrt[3]{3}$$

45.
$$\sqrt[3]{\frac{3}{4}}-\sqrt[3]{\frac{2}{9}}-\frac{1}{\sqrt[3]{6}}=$$
$$\sqrt[3]{\frac{3}{4}\cdot\frac{2}{2}}-\sqrt[3]{\frac{2}{9}\cdot\frac{3}{3}}-\frac{1}{\sqrt[3]{6}}\cdot\frac{\sqrt[3]{36}}{\sqrt[3]{36}}=$$
$$\frac{\sqrt[3]{6}}{2}-\frac{\sqrt[3]{6}}{3}-\frac{\sqrt[3]{36}}{6}=$$
$$\frac{3\sqrt[3]{6}}{6}-\frac{2\sqrt[3]{6}}{6}-\frac{\sqrt[3]{36}}{6}=$$
$$\frac{\sqrt[3]{6}}{6}-\frac{\sqrt[3]{36}}{6}$$

47.
$$\sqrt{\frac{1}{8}}-\sqrt{50}+\sqrt[4]{\frac{1}{4}}=$$
$$\sqrt{\frac{1}{8}\cdot\frac{2}{2}}-\sqrt{25}\sqrt{2}+\sqrt[4]{\frac{4}{16}}=$$
$$\frac{\sqrt{2}}{4}-5\sqrt{2}+\frac{\sqrt[4]{4}}{2}=$$
$$\frac{\sqrt{2}}{4}-5\sqrt{2}+\frac{\sqrt[4]{2^2}}{2}=$$
$$\frac{-19\sqrt{2}}{4}+\frac{\sqrt{2}}{2}=$$
$$\frac{-19\sqrt{2}}{4}+\frac{2\sqrt{2}}{4}=$$
$$\frac{-17\sqrt{2}}{4}$$

49.
$$\sqrt{50ax^3}+\sqrt{72a^3x^3}-\sqrt{8a^5x}=$$
$$\sqrt{25x^2}\sqrt{2ax}+\sqrt{36a^2x^2}\sqrt{2ax}-\sqrt{4a^4}\sqrt{2ax}=$$
$$5x\sqrt{2ax}+6ax\sqrt{2ax}-2a^2\sqrt{2ax}=$$
$$\left(5x+6ax-2a^2\right)\sqrt{2ax}$$

51.
$$\sqrt[3]{16x^4}+\sqrt[3]{54x^7}+\sqrt[3]{250x}=$$
$$\sqrt[3]{8x^3}\sqrt[3]{2x}+\sqrt[3]{27x^6}\sqrt[3]{2x}+\sqrt[3]{125}\sqrt[3]{2x}=$$
$$2x\sqrt[3]{2x}+3x^2\sqrt[3]{2x}+5\sqrt[3]{2x}=$$
$$\left(3x^2+2x+5\right)\sqrt[3]{2x}$$

Section 8.4

53. $\sqrt{\frac{a}{6b}}+\sqrt{\frac{6b}{a}}-\sqrt{\frac{2}{3ab}}=$

$\sqrt{\frac{a}{6b}\cdot\frac{6b}{6b}}+\sqrt{\frac{6ba}{a^2}}-\sqrt{\frac{2(3ab)}{3ab(eab)}}=$

$\frac{\sqrt{6ab}}{6b}+\frac{\sqrt{6ab}}{a}-\frac{\sqrt{6ab}}{3ab}=$

$\frac{a\sqrt{6ab}}{6ab}+\frac{6b\sqrt{6ab}}{6ab}-\frac{2\sqrt{6ab}}{6ab}=$

$\left(\frac{a+6b-2}{6ab}\right)\sqrt{6ab}$

55. $\sqrt[3]{\frac{a}{b^2}}+\sqrt[3]{\frac{b}{a^2}}-\frac{1}{\sqrt[3]{a^2b^2}}=$

$\sqrt[3]{\frac{ab}{b^3}}+\sqrt[3]{\frac{ab}{a^3}}-\frac{\sqrt[3]{ab}}{\sqrt[3]{a^3b^3}}=$

$\frac{\sqrt[3]{ab}}{b}+\frac{\sqrt[3]{ab}}{a}-\frac{\sqrt[3]{ab}}{ab}=$

$\frac{a\sqrt[3]{ab}}{ab}+\frac{b\sqrt[3]{ab}}{ab}-\frac{\sqrt[3]{ab}}{ab}=$

$\left(\frac{a+b-1}{ab}\right)\sqrt[3]{ab}$

Exercises 8.5

1. $\sqrt{3}\sqrt{6}=\sqrt{3\cdot 3\cdot 2}=3\sqrt{2}$

3. $\sqrt[3]{4}\sqrt[3]{18}=\sqrt[3]{2^2\cdot 2\cdot 9}=2\sqrt[3]{9}$

5. $\frac{\sqrt[4]{36}}{\sqrt[4]{9}}=\sqrt[4]{\frac{36}{9}}=\sqrt[4]{4}$

7. $\sqrt[3]{2}\cdot\sqrt{5}=2^{1/3}\cdot 5^{1/2}=2^{2/6}\cdot 5^{3/6}=\sqrt[6]{2^2\cdot 5^3}=\sqrt[6]{500}$

9. $\sqrt{2}\cdot\sqrt[3]{4}=2^{1/2}\cdot 4^{1/3}=2^{3/6}\cdot 4^{2/6}=\sqrt[6]{2^3\cdot 4^2}=\sqrt[6]{2^3\cdot 2^4}=2\sqrt[6]{2}$

11. $\sqrt{27}\cdot\sqrt[3]{9}=27^{1/2}\cdot 9^{1/3}=3^{3/2}\cdot 3^{2/3}=3^{9/6}\cdot 3^{4/6}=3^{13/6}=3^2\cdot 3^{1/6}=9\sqrt[6]{3}$

13. $\frac{\sqrt{3}}{\sqrt[3]{3}}=\frac{3^{1/2}}{3^{1/3}}=3^{1/2-1/3}=3^{1/6}=\sqrt[6]{3}$

15. $\frac{2\sqrt[3]{3}}{3\sqrt{2}}=\frac{2\cdot 3^{1/3}}{3\cdot 2^{1/2}}=\frac{2\cdot 3^{2/6}}{3\cdot 2^{3/6}}=\frac{2^{6/6-3/6}\cdot 3^{2/6}}{3}=\frac{2^{3/6}\cdot 3^{2/6}}{3}=\frac{\sqrt[6]{2^3\cdot 3^2}}{3}=\frac{\sqrt[6]{72}}{3}$

17. $\left(2+\sqrt{5}\right)\left(1-\sqrt{5}\right)=2-\sqrt{5}-5=-3-\sqrt{5}$

19. $\left(5-\sqrt{3}\right)\left(5+\sqrt{3}\right)=25-\sqrt{3}\sqrt{3}=25-3=22$

21. $\left(3\sqrt{3}+2\right)\left(4\sqrt{3}+3\right)=12\sqrt{3}\sqrt{3}+17\sqrt{3}+6=12(3)+6+17\sqrt{3}=42+17\sqrt{3}$

23. $\left(a+\sqrt{b}\right)\left(2a-4\sqrt{b}\right)=2a^2-2a\sqrt{b}-4\sqrt{b}\sqrt{b}=2a^2-2a\sqrt{b}-4b$

25. $\left(2\sqrt{7}+2\sqrt{3}\right)\left(-2\sqrt{7}-\sqrt{3}\right)=-4\sqrt{7}\sqrt{7}-6\sqrt{7}\sqrt{3}-2\sqrt{3}\sqrt{3}=-28-6\sqrt{21}-6=-34-6\sqrt{21}$

27. $\left(2\sqrt{a}-3\sqrt{b}\right)\left(\sqrt{a}+2\sqrt{b}\right)=2\sqrt{a}\sqrt{a}+\sqrt{a}\sqrt{b}-6\sqrt{b}\sqrt{b}=2a+\sqrt{ab}-6b$

29. $\left(2-\sqrt{3}\right)^2=2^2-2(2)\left(\sqrt{3}\right)+\left(\sqrt{3}\right)^2=4-4\sqrt{3}+3=7-4\sqrt{3}$

31. $\left(2\sqrt{3}-\sqrt{5}\right)^2 = 2^2\left(\sqrt{3}\right)^2 - 2\left(2\sqrt{3}\right)\left(\sqrt{5}\right)+\left(\sqrt{5}\right)^2 = 12-4\sqrt{15}+5 = 17-4\sqrt{15}$

33. $\left(a+\sqrt{b}\right)^2 = a^2+2a\sqrt{b}+\left(\sqrt{b}\right)^2 = a^2+2a\sqrt{b}+b$

35. $\left(2\sqrt{a}+3\sqrt{b}\right)^2 = 2^2\left(\sqrt{a}\right)^2+2\left(2\sqrt{a}\right)\left(3\sqrt{b}\right)+3^2\left(\sqrt{b}\right)^2 = 4a+12\sqrt{ab}+9b$

37. $\frac{\sqrt{2}}{3+\sqrt{2}} = \frac{\sqrt{2}}{3+\sqrt{2}}\cdot\frac{3-\sqrt{2}}{3-\sqrt{2}} = \frac{3\sqrt{2}-2}{9-2} = \frac{3\sqrt{2}-2}{7}$

39. $\frac{2+\sqrt{5}}{3-2\sqrt{5}} = \frac{2+\sqrt{5}}{3-2\sqrt{5}}\cdot\frac{3+2\sqrt{5}}{3+2\sqrt{5}} = \frac{6+7\sqrt{5}+2(5)}{9-4(5)} = \frac{16+7\sqrt{5}}{-11}$

41. $\frac{4\sqrt{6}+2\sqrt{3}}{2\sqrt{6}+5\sqrt{3}} = \frac{4\sqrt{6}+2\sqrt{3}}{2\sqrt{6}+5\sqrt{3}}\cdot\frac{2\sqrt{6}-5\sqrt{3}}{2\sqrt{6}-5\sqrt{3}} = \frac{8(6)-16\sqrt{18}-10(3)}{4(6)-25(3)} =$

$\frac{48-16\sqrt{9}\sqrt{2}-30}{24-75} = \frac{18-16(3)\sqrt{2}}{-51} = \frac{3\left(6-16\sqrt{2}\right)}{3(-17)} = \frac{6-16\sqrt{2}}{-17}$

43. $\sqrt{3+\sqrt{5}}\sqrt{3-\sqrt{5}} = \sqrt{\left(3+\sqrt{5}\right)\left(3-\sqrt{5}\right)} = \sqrt{9-5} = \sqrt{4} = 2$

45. $\frac{\sqrt{3+\sqrt{5}}}{\sqrt{3-\sqrt{5}}} = \frac{\sqrt{3+\sqrt{5}}}{\sqrt{3-\sqrt{5}}}\cdot\frac{\sqrt{3+\sqrt{5}}}{\sqrt{3+\sqrt{5}}} = \frac{\sqrt{\left(3+\sqrt{5}\right)^2}}{\sqrt{9-5}} = \frac{3+\sqrt{5}}{4}$

47. $\frac{a}{a+\sqrt{b}} = \frac{a}{a+\sqrt{b}}\cdot\frac{a-\sqrt{b}}{a-\sqrt{b}} = \frac{a^2-a\sqrt{b}}{a^2-b}$

Exercises 8.6

1.
$\sqrt{x-4} = 7$
$x-4 = 7^2 = 49$
$x = 53$
Check:
$\sqrt{53-4} = ?7$
$\sqrt{49} = 7$ Yes
Solution is 53.

3.
$\sqrt{3-x} = 2x-3$
$3-x = (2x-3)^2$
$3-x = 4x^2-12x+9$
$0 = 4x^2-11x+6$
$(x-2)(4x-3) = 0$
$x = 2$ or $\frac{3}{4}$

Check $x = 2$:
$\sqrt{3-2} = ?2(2)-3$
$1 = 1$ Yes
Check $x = \frac{3}{4}$:
$\sqrt{3-\frac{3}{4}} = ?2\left(\frac{3}{4}\right)-3$
$\sqrt{\frac{9}{4}} = ?\frac{3}{2}-3$
$\frac{3}{2} = -\frac{3}{2}$ No
Solution is 2.

5.

$\sqrt[3]{3-x}=5$

$3-x=5^3$

$3-x=125$

$x=-122$

Check:

$\sqrt[3]{3-(-122)}=?5$

$\sqrt[3]{125}=5$ Yes

Solution is -122.

7.

$\sqrt[4]{2x-3}=3$

$2x-3=3^4=81$

$2x=84$

$x=42$

Check:

$\sqrt[4]{2(42)-3}=?3$

$\sqrt[4]{84-3}=?3$

$\sqrt[4]{81}=3$ Yes

Solution is 42.

9.

$\sqrt{x}+6=x$

$\sqrt{x}=x-6$

$x=(x-6)^2$

$x=x^2-12x+36$

$0=x^2-13x+36$

$0=(x-9)(x-4)$

$x=9$ or 4

Check $x=9$:

$\sqrt{9}+6=?9$

$3+6=9$ Yes

Check 4:

$\sqrt{4}+6=?\ 4$

$2+6=?\ 4$ No

Solution is 9.

11.

$\sqrt{x^2+2x+6}=x+2$

$x^2+2x+6=(x+2)^2$

$x^2+2x+6=x^2+4x+4$

$-2x=-2$

$x=1$

Check: $\sqrt{1+2+6}=?1+2$

$\sqrt{9}=3$ Yes

Solution is 1.

13.

$\sqrt{a^2-5a+20}-4a=0$

$\sqrt{a^2-5a+20}=4a$

$a^2-5a+20=16a^2$

$0=15a^2+5a-20$

$0=3a^2+a-4$

$(3a+4)(a-1)=0$

$a=-\frac{4}{3}$ or 1

Check $a=-\frac{4}{3}$:

$\sqrt{\frac{16}{9}-5\left(-\frac{4}{3}\right)+20}-4\left(-\frac{4}{3}\right)=?0$

$\sqrt{\frac{256}{9}}+\frac{16}{3}=?0$

$\frac{16}{3}+\frac{16}{3}\neq 0$ No

Check $a=1$:

$\sqrt{1-5+20}-4(1)=?0$

$\sqrt{16}-4=0$ Yes

Solution is 1.

15.

$\sqrt{m^2+3m}-3m+1=0$

$\sqrt{m^2+3m}=3m-1$

$m^2+3m=(3m-1)^2$

$m^2+3m=9m^2-6m+1$

$0=8m^2-9m+1$

$0=(8m-1)(m-1)$

$m=\frac{1}{8}$ or 1

Check $m=1$:

$\sqrt{1+3}-3+1=?0$

$\sqrt{4}-2=?0$ Yes

Check $m=\frac{1}{8}$:

$\sqrt{\left(\frac{1}{8}\right)^2+3\left(\frac{1}{8}\right)}-3\left(\frac{1}{8}\right)+1=?0$

$\sqrt{\frac{1}{64}+\frac{24}{64}}-\frac{3}{8}+1=?0$

$\frac{5}{8}-\frac{3}{8}+1=?0$

$\frac{1}{4}+1\neq 0$ No

Solution is 1.

17. $\sqrt{3x^2+4x+2}=\sqrt{x^2+x+11}$

$3x^2+4x+2=x^2+x+11$

$2x^2+3x-9=0$

$(x+3)(2x-3)=0$

$x=-3 \text{ or } \frac{3}{2}$

Check $x=-3$:

$\sqrt{3(-3)^2+4(-3)+2}=?\sqrt{(-3)^2+(-3)+11}$

$\sqrt{17}=?\sqrt{17}$ Yes

Check $x=\frac{3}{2}$:

$\sqrt{3\left(\frac{3}{2}\right)^2+4\left(\frac{3}{2}\right)+2}=?\sqrt{\left(\frac{3}{2}\right)^2+\frac{3}{2}+11}$

$\sqrt{\frac{59}{4}}\stackrel{?}{=}\sqrt{\frac{59}{4}}$ Yes

Solutions are -3 or $\frac{3}{2}$.

19. $\sqrt[3]{3x-4}=\sqrt[3]{5x+8}$

$3x-4=5x+8$

$-2x=12$

$x=-6$

Check:

$\sqrt[3]{(3)(-6)-4}=?\sqrt[3]{5(-6)+8}$

$\sqrt[3]{-22}=\sqrt[3]{-22}$ Yes

Solution is -6.

21. $\sqrt{x+6}-\sqrt{x}=4$

$\sqrt{x+6}=\sqrt{x}+4$

$x+6=\left(\sqrt{x}+4\right)^2$

$x+6=x+8\sqrt{x}+16$

$-10=8\sqrt{x}$

$-5=4\sqrt{x}$

The left side is negative; the right side is positive – no solution.

23. $\sqrt{x+9}-\sqrt{x+2}=\sqrt{4x-27}$

$\left(\sqrt{x+9}-\sqrt{x+2}\right)^2=4x-27$

$x+9-2\sqrt{x+9}\sqrt{x+2}+x+2=4x-27$

$2x+11-2\sqrt{(x+9)(x+2)}=4x-27$

$-2\sqrt{(x+9)(x+2)}=2x-38$

$-\sqrt{(x+9)(x+2)}=x-19$

$x^2+11x+18=(x-19)^2$

$x^2+11x+18=x^2-38x+361$

$49x=343$

$x=7$

Check $x=7$:

$\sqrt{7+9}-\sqrt{7+2}=?\sqrt{4(7)-27}$

$\sqrt{16}-\sqrt{9}=?\sqrt{1}$

$4-3=1$ yes

Solution is 7.

25. $\sqrt{13+\sqrt{x}}=\sqrt{x}+1$

$13+\sqrt{x}=\left(\sqrt{x}+1\right)^2$

$13+\sqrt{x}=x+2\sqrt{x}+1$

$-x+12=\sqrt{x}$

$(-x+12)^2=x$

$x^2-24x+144=x$

$x^2-25x+144=0$

$(x-9)(x-16)=0$

$x=9$ or 16

Check $x=9$:

$\sqrt{13+\sqrt{9}}=?\sqrt{9}+1$

$\sqrt{13+3}=?3+1$

$\sqrt{16}=4$ Yes

Check $x=16$:

$\sqrt{13+\sqrt{16}}=?\sqrt{16}+1$

$\sqrt{13+4}=?4+1$

$\sqrt{17}=?5$ No

Solution is 9.

27. $v=\sqrt{v_o^{\,2}-2gh}$ for h

$v^2=v_o^{\,2}-2gh$

$2gh=v_o^{\,2}-v^2$

$h=\frac{v_o^{\,2}-v^2}{2g}$

29. $P = 2\pi\sqrt{\frac{1}{g}}$ for 1

$$P^2 = 4\pi^2\left(\frac{1}{g}\right)$$

$$gP^2 = 4\pi^2 1$$

$$\frac{gP^2}{4\pi^2} = 1$$

31. $f = \frac{1}{2\pi\sqrt{LC}}$ for C

$$f\sqrt{LC} = \frac{1}{2\pi}$$

$$\sqrt{LC} = \frac{1}{2\pi f}$$

$$LC = \left(\frac{1}{2\pi f}\right)^2$$

$$C = \left(\frac{1}{2\pi f}\right)^2 \div L$$

$$C = \frac{1}{4\pi^2 f^2} \cdot \frac{1}{L}$$

$$C = \frac{1}{4\pi^2 f^2 L}$$

33. $Z = \sqrt{R^2 + (2\pi fL)^2}$ for L

$$Z^2 = R^2 + 4\pi^2 f^2 L^2$$

$$Z^2 - R^2 = 4\pi^2 f^2 L^2$$

$$\frac{Z^2 - R^2}{4\pi^2 f^2} = L^2$$

$$L = \sqrt{\frac{Z^2 - R^2}{4\pi^2 f^2}}$$

$$L = \frac{\sqrt{Z^2 - R^2}}{2\pi f}$$

35. $T = 2\pi\sqrt{\frac{R_2 C}{R_1 + R_2}}$ for C

$$T^2 = 4\pi^2\left(\frac{R_2 C}{R_1 + R_2}\right)$$

$$(R_1 + R_2)T^2 = 4\pi^2 R_2 C$$

$$\frac{(R_1 + R_2)T^2}{4\pi^2 R_2} = C$$

37. $v = \sqrt{Gm\left(\frac{1}{D} - \frac{1}{r}\right)}$ for m

$$v^2 = Gm\left(\frac{1}{D} - \frac{1}{r}\right)$$

$$v^2 = Gm\left(\frac{r - D}{Dr}\right)$$

$$Drv^2 = Gm(r - D)$$

$$\frac{Drv^2}{G(r - D)} = m$$

39. $(x-3)^2 = \left(\sqrt{x+6}\right)^2 + \left(\sqrt{x+12}\right)^2$

$$x^2 - 6x + 9 = x + 6 + x + 12$$

$$x^2 - 8x - 9 = 0$$

$$(x-9)(x+1) = 0$$

$x = 9$ (x is not negative)

Sides: $x - 3 = 6$

$$\sqrt{x+6} = \sqrt{15}$$

$$\sqrt{x+12} = \sqrt{21}$$

Exercises 8.7

1.

$x^4 - 11x^2 + 18 = 0$

$(x^2 - 9)(x^2 - 2) = 0$

$x^2 - 9 = 0$ or $x^2 - 2 = 0$

$x^2 = 9$ or $x^2 = 2$

$x = \pm 3$ or $\pm\sqrt{2}$

3.

$x^{-4} - 17x^{-2} + 16 = 0$

$(x^{-2} - 1)(x^{-2} - 16) = 0$

$x^{-2} = 1$ or $x^{-2} = 16$

$\frac{1}{x^2} = 1$ or $\frac{1}{x^2} = 16$

$x^2 = 1$ or $x^2 = \frac{1}{16}$

$x = \pm 1$ or $x = \pm\frac{1}{4}$

5.

$(x+2)^2 + 3(x+2) + 2 = 0$

Let $m = x + 2$

$m^2 + 3m + 2 = 0$

$(m+2)(m+1) = 0$

$m = -2$ or $m = -1$

$x + 2 = -2$ or $x + 2 = -1$

$x = -4$ or $x = -3$

7.

$(x-1)^4 - 5(x-1)^2 + 4 = 0$

Let $m = (x-1)^2$

$m^2 - 5m + 4 = 0$

$(m-4)(m-1) = 0$

$m = 4$ or $m = 1$

$(x-1)^2 = 4$ or $(x-1)^2 = 1$

$x - 1 = \pm 2$ or $x - 1 = \pm 1$

$x = 1 \pm 2$ or $x = 1 \pm 1$

$x = 3, -1, 2, 0$

9.

$x - 2\sqrt{x} - 3 = 0$

Let $m = \sqrt{x}$.

Then $m^2 = x$.

$m^2 - 2m - 3 = 0$

$(m-3)(m+1) = 0$

$m = 3$ or $m = -1$

$\sqrt{x} = 3$ or $\sqrt{x} = -1$

$x = 9$ or no real number.

Check $x = 9$:

$9 - 2\sqrt{9} - 3 = ?0$

$9 - 2(3) - 3 = ?0$

$9 - 6 - 3 = 0$ Yes

Solution is 9.

11.

$(3x+2)^{-4} - 1 = 0$

$[(3x+2)^{-2} + 1][(3x+2)^{-2} - 1] = 0$

$(3x+2)^{-2} = -1$ or $(3x+2)^{-2} = 1$

$\frac{1}{(3x+2)^2} = -1$ or $\frac{1}{(3x+2)^2} = 1$

No real solution. $\quad 1 = (3x+2)^2$

For $1 = (3x+2)^2$

$3x + 2 = \pm 1$

$3x = -2 \pm 1$

$3x = -1$ or $3x = -3$

$x = \frac{-1}{3}$ or $x = -1$

13. $x^{2/3} + 2x^{1/3} - 8 = 0$

Let $m = x^{1/3}$.

Then $m^2 = x^{2/3}$.

$m^2 + 2m - 8 = 0$

$(m+4)(m-2) = 0$

$m = -4$ or $m = 2$

From $m = x^{1/3}$:

$x^{1/3} = -4$ or $x^{1/3} = 2$

$x = (-4)^3$ or $x = (2)^3$

$x = -64$ or $x = 8$

Check $x = -64$:

$(-64)^{2/3} + 2(-64)^{1/3} - 8 = ?0$

$(-4)^2 + 2(-4) - 8 = 0$ Yes

Check $x = 8$:

$8^{2/3} + 2(8)^{1/3} - 8 = ?0$

$(2)^2 + 2(2) - 8 = 0$ Yes

Solutions are: $-64, 8$

15. $(3x+1)^{4/3} - 2(3x+1)^{2/3} - 8 = 0$

let $m = (3x+1)^{2/3}$;

the equation becomes:

$m^2 - 2m - 8 = 0$

$(m-4)(m+2) = 0$

$m - 4 = 0$ or $m + 2 = 0$

$m = 4$ or $m = -2$

i.e. $(3x+1)^{2/3} = 4$

$3x + 1 = 4^{3/2} = 8$

$3x = 7$

$x = \dfrac{7}{3}$

or

$(3x+1)^{2/3} = -2$

$(3x+1) = (-2)^{3/2}$

The right side is not a Real Number.

17. $x^4 - 3x^2 + 1 = 0$

$a = 1, b = -3, c = 1$

$$x^2 = \frac{-(-3) \pm \sqrt{(-3)^2 - 4(1)(1)}}{2(1)}$$

$$x^2 = \frac{3 \pm \sqrt{9-4}}{2}$$

$$x = \pm\sqrt{\frac{3 \pm \sqrt{5}}{2}} \text{ or}$$

$$x^2 = \frac{3+\sqrt{5}}{2} \text{ or } x^2 = \frac{3-\sqrt{5}}{2}$$

$x^2 = 2.6180$ or $x^2 = 0.38197$

$x = \pm 1.62$ or $x = \pm 0.618$

19. $\sqrt[3]{x} - 3\sqrt[6]{x} + 2 = 0$

Let $m = \sqrt[6]{x}$; then $m^2 = \sqrt[3]{x}$.

$m^2 - 3m + 2 = 0$

$(m-2)(m-1) = 0$

$m - 2 = 0$ or $m - 1 = 0$

$m = 2$ or $m = 1$

$\sqrt[6]{x} = 2$ $\quad$ $\sqrt[6]{x} = 1$

$x = 2^6 = 64$ $\quad$ $x = 1^6 = 1$

Chapter 8 Review

1. $3a^{-2}=\dfrac{3}{a^2}$ **2.** $(2a)^{-4}=\dfrac{1}{(2a)^4}=\dfrac{1}{2^4a^4}=\dfrac{1}{16a^4}$ **3.** $a^{-5}a^{10}=\dfrac{a^{10}}{a^5}=a^5$

4. $\left(a^{-2}\right)^{-4}=a^{-2(-4)}=a^8$ **5.** $\dfrac{a^3b^0c^{-3}}{a^5b^{-2}c^{-9}}=a^{3-5}b^{0-(-2)}c^{-3-(-9)}=a^{-2}b^2c^6=\dfrac{b^2c^6}{a^2}$

6. $a^{-2}+a^{-1}=\dfrac{1}{a^2}+\dfrac{1}{a}\cdot\dfrac{a}{a}=\dfrac{1+a}{a^2}$ **7.** $49^{1/2}=7$ **8.** $16^{3/2}=\left(16^{1/2}\right)^3=4^3=64$

9. $8^{-2/3}=\dfrac{1}{8^{2/3}}=\dfrac{1}{\left(8^{1/3}\right)^2}=\dfrac{1}{(2)^2}=\dfrac{1}{4}$ **10.** $\left(x^{2/3}y^{-1/3}\right)^{-3/5}=x^{2/3(-3/5)}y^{-1/3(-3/5)}=x^{-2/5}y^{1/5}=\dfrac{y^{1/5}}{x^{2/5}}$

11. $\dfrac{x^{-2/3}}{x^{-1/4}}=x^{-2/3-(-1/4)}=x^{-5/12}=\dfrac{1}{x^{5/12}}$ **12.** $\left(x^{-3/4}\right)^{-2/3}=x^{-3/4(-2/3)}=x^{1/2}$

13. $w=2t^{-2/3}$ for $t=27$; $w=2(27)^{-2/3}=\dfrac{2}{27^{2/3}}=\dfrac{2}{\left(27^{1/3}\right)^2}=\dfrac{2}{3^2}=\dfrac{2}{9}$

14. $\sqrt{80}=\sqrt{16}\sqrt{5}=4\sqrt{5}$ **15.** $\sqrt{72a^2b^3}=\sqrt{36a^2b^2}\sqrt{2b}=6ab\sqrt{2b}$

16. $\sqrt{80x^5y^6}=\sqrt{16x^4y^6}\sqrt{5x}=4x^2y^3\sqrt{5x}$ **17.** $\sqrt{48x^3y}=\sqrt{16x^2}\sqrt{3xy}=4x\sqrt{3xy}$

18. $\sqrt{\dfrac{5}{54}}=\sqrt{\dfrac{5}{9\cdot 6}\cdot\dfrac{6}{6}}=\dfrac{\sqrt{30}}{\sqrt{9\cdot 36}}=\dfrac{\sqrt{30}}{18}$

19. $\sqrt{\dfrac{45a^2}{7b^4}}=\sqrt{\dfrac{9a^2\cdot 5}{b^4}\cdot\dfrac{7}{7}\cdot\dfrac{7}{7}}=\dfrac{\sqrt{9a^2}\sqrt{35}}{\sqrt{49b^4}}=\dfrac{3a\sqrt{35}}{7b^2}$ **20.** $\sqrt[3]{250}=\sqrt[3]{125}\sqrt[3]{2}=5\sqrt[3]{2}$

21. $\sqrt[3]{108a^4b^2}=\sqrt[3]{27a^3}\sqrt[3]{4ab^2}=3a\sqrt[3]{4ab^2}$

22. $\sqrt[3]{256a^5b^{10}}=\sqrt[3]{64a^3b^9}\sqrt[3]{4a^2b}=4ab^3\sqrt[3]{4a^2b}$

23. $\sqrt[3]{\dfrac{9a}{20b^2}}=\sqrt[3]{\dfrac{9a}{4\cdot 5b^2}\cdot\dfrac{2(25)b}{2(25)b}}=\dfrac{\sqrt[3]{450ab}}{\sqrt[3]{8\cdot 125b^3}}=\dfrac{\sqrt[3]{450ab}}{10b}$

24. $\frac{8}{\sqrt[3]{40}}=\frac{8\sqrt[3]{25}}{\sqrt[3]{8\cdot5}\sqrt[3]{25}}=\frac{8\sqrt[3]{25}}{\sqrt[3]{8}\sqrt[3]{125}}=\frac{8\sqrt[3]{25}}{10}=\frac{4\sqrt[3]{25}}{5}$

25. $\sqrt[4]{9a^2}=\sqrt[4]{3^2a^2}=\sqrt{3a}$

26. $\frac{6}{\sqrt{12a}}=\frac{6}{\sqrt{4}\sqrt{3a}}\cdot\frac{\sqrt{3a}}{\sqrt{3a}}=\frac{6\sqrt{3a}}{2(3a)}=\frac{\sqrt{3a}}{a}$

27. $\frac{6}{\sqrt[3]{12a}}=\frac{6}{\sqrt[3]{4\cdot3a}}\cdot\frac{\sqrt[3]{18a^2}}{\sqrt[3]{18a^2}}=\frac{6\sqrt[3]{18a^2}}{\sqrt[3]{216a^3}}=\frac{6\sqrt[3]{18a^2}}{6a}=\frac{\sqrt[3]{18a^2}}{a}$

28. $\sqrt{\sqrt{5}}=\sqrt[4]{5}$

29. $\sqrt[4]{\sqrt[3]{10}}=\sqrt[12]{10}$

30.
$\sqrt{12}+\sqrt{27}-\sqrt[4]{9}=$
$\sqrt{4}\sqrt{3}+\sqrt{9}\sqrt{3}-\sqrt{3}=$
$2\sqrt{3}+3\sqrt{3}-\sqrt{3}=$
$4\sqrt{3}$

31.
$\sqrt{\frac{2}{5}}+\frac{1}{\sqrt{10}}-\sqrt{40}=$
$\sqrt{\frac{2}{5}\cdot\frac{5}{5}}+\frac{1}{\sqrt{10}}\cdot\frac{\sqrt{10}}{\sqrt{10}}-\sqrt{4}\sqrt{10}=$
$\frac{\sqrt{10}}{5}+\frac{\sqrt{10}}{10}-2\sqrt{10}=$
$\frac{-17\sqrt{10}}{10}$

32.
$\sqrt[3]{54}-\sqrt[3]{250}+\sqrt[3]{16}=$
$\sqrt[3]{27}\sqrt[3]{2}-\sqrt[3]{125}\sqrt[3]{2}+\sqrt[3]{8}\sqrt[3]{2}=$
$3\sqrt[3]{2}-5\sqrt[3]{2}+2\sqrt[3]{2}=$
0

33.
$2\sqrt[3]{\frac{3}{4}}-\frac{3}{\sqrt[3]{36}}+5\sqrt[3]{\frac{2}{9}}=$
$2\sqrt[3]{\frac{3}{4}\cdot\frac{2}{2}}-\frac{3}{\sqrt[3]{6^2}}\cdot\frac{\sqrt[3]{6}}{\sqrt[3]{6}}+5\sqrt[3]{\frac{2}{9}\cdot\frac{3}{3}}=$
$2\frac{\sqrt[3]{6}}{2}-\frac{3\sqrt[3]{6}}{6}+\frac{5\sqrt[3]{6}}{3}=$
$\sqrt[3]{6}-\frac{\sqrt[3]{6}}{2}+\frac{5\sqrt[3]{6}}{3}=$
$\frac{13\sqrt[3]{6}}{6}$

34.
$\sqrt{8}\cdot\sqrt{20}=$
$\sqrt{2^3\cdot2^2\cdot5}=$
$\sqrt{2^4\cdot10}=$
$4\sqrt{10}$

35.
$\sqrt{3}\sqrt[4]{9}=$
$\sqrt{3}\sqrt{3}=$
3

36.
$\sqrt{2}\sqrt[3]{4}=$
$\sqrt[6]{2^3}\sqrt[6]{4^2}=$
$\sqrt[6]{2^3\cdot2^4}=$
$\sqrt[6]{2^6}\sqrt[6]{2}=$
$2\sqrt[6]{2}$

37. $\left(2+\sqrt{3}\right)\left(-4-\sqrt{3}\right)=$

$-8-2\sqrt{3}-4\sqrt{3}-3=$

$-11-6\sqrt{3}$

38. $\left(4-3\sqrt{5}\right)^2=$

$16-24\sqrt{5}+9(5)=$

$16+45-24\sqrt{5}=$

$61-24\sqrt{5}$

39. $\dfrac{\sqrt{3}}{2-3\sqrt{3}}\cdot\dfrac{\left(2+3\sqrt{3}\right)}{\left(2+3\sqrt{3}\right)}=$

$\dfrac{2\sqrt{3}+3(3)}{4-9(3)}=$

$\dfrac{9+2\sqrt{3}}{-23}$

40. $\sqrt{x+2}=8$

$x+2=8^2$

$x=62$

Check:

$\sqrt{62+2}=8$ Yes

Solution is 62.

41. $\sqrt[3]{x}+2=-1$

$\sqrt[3]{x}=-3$

$x=(-3)^3=-27$

Check:

$\sqrt[3]{-27}+2?-1$ Yes

Solution is -27.

42. $\sqrt{x-5}+\sqrt{x}=5$

$\sqrt{x-5}=-\sqrt{x}+5$

$x-5=\left(-\sqrt{x}+5\right)^2$

$x-5=x-10\sqrt{x}+25$

$-30=-10\sqrt{x}$

$3=\sqrt{x}$

$3^2=x$

$x=9$

Check:

$\sqrt{9-5}+\sqrt{9}=?5$

$\sqrt{4}+\sqrt{9}=?5$

$2+3=5$ Yes

Solution is 9.

43. $\sqrt{x+9}=\sqrt[4]{x^2+9x}$

$(x+9)^2=x^2+9x$

$x^2+18x+81=x^2+9x$

$9x=-81$

$x=-9$

Check:

$\sqrt{-9+9}=?\sqrt[4]{(-9)^2+9(-9)}$

$\sqrt{0}=?\sqrt[4]{81-81}$

$\sqrt{0}=\sqrt[4]{0}$ yes

Solution is -9.

44. $4x^4-41x^2+45=0$

$\left(x^2-9\right)\left(4x^2-5\right)=0$

$x^2=9$ or $4x^2=5$

$x=\pm 3$ or $x^2=\dfrac{5}{4}$

$x=\pm 3$ or $x=\pm\dfrac{\sqrt{5}}{2}$

45. $x^{2/3}-2x^{1/3}-8=0$

$\left(x^{1/3}-4\right)\left(x^{1/3}+2\right)=0$

$x^{1/3}=4$ or $x^{1/3}=-2$

$x=64$ or $x=-8$

Check $x=64$:

$(64)^{2/3}-2(64)^{1/3}-8=?0$

$4^2-2(4)-8=0$ Yes

Check $x=-8$:

$(-8)^{2/3}-2(-8)^{1/3}-8=?0$

$(-2)^2-2(-2)-8=0$ Yes

Solutions are 64, -8.

Chapter 9, Exercises 9.1

The graphs are on pages 731 and 732.

1. $y = 4^x$

x	-1	0	1	2
y	0.25	1	4	16

3. $y = 10^x$

x	-1	0	1	2
y	0.1	1	10	100

5. $y = \left(\frac{1}{3}\right)^x$

x	-2	-1	0	1
y	9	3	1	$\frac{1}{3}$

7. $y = \left(\frac{1}{10}\right)^x$

x	-2	-1	0	1
y	100	10	1	0.1

9. $y = \left(\frac{3}{4}\right)^x$

x	-1	0	1	2
y	$\frac{4}{3}$	1	0.75	0.56

11. $y = \left(\frac{2}{5}\right)^x$

x	-1	0	1	2
y	2.5	1	0.4	0.16

13. $y = 4^{-x}$

x	-2	-1	0	1
y	16	4	1	0.25

15. $y = \left(\frac{4}{3}\right)^{-x}$

x	-2	-1	0	1
y	1.8	1.3	1	0.75

17. $y = (1.2)^x$

x	-2	-1	0	1
y	0.69	0.83	1	1.2

19. $y = 3^{-x+2}$

x	-1	0	1	2
y	27	9	3	1

21. $y = 2^{x-3}$

x	1	2	3	4
y	0.25	0.5	1	2

23. $y = 2^{x^2}$

x	-2	-1	0	1	2
y	16	2	1	2	16

25. $y = 3^x + 2$

x	-1	0	1	2
y	2.3	3	5	11

27. $5^{0.2} = 1.38$

29. $10^{5.5} = 316{,}000$

31. $5^{2\pi} = 24{,}600$

33. $8^{\pi/3} = 8.82$

35. $9^{3/4} = 5.20$

37. $\sqrt[4]{6} = 1.57$

39. $\sqrt[6]{140} = 2.28$

41.
$y = A(1+r)^n$
$A = 5.0 \times 10^{11}\ kwh$
$r = 0.08$
$n = 6$
$y = 5.0(10^{11})(1.08)^6$
$y = 7.93 \times 10^{11}\ kwh$

43. $A = P\left(1+\frac{r}{x}\right)^{xn}$

$P = 1000,\ r = 0.08,\ n = 3$

(a) $x = 1$

$$A = 1000\left(1+\frac{0.08}{1}\right)^{3}$$

$A = \$1260$

(b) $x = 2$

$$A = 1000\left(1+\frac{0.08}{2}\right)^{2(3)}$$

$A = \$1265$

(c) $x = 4$

$$A = 1000\left(1+\frac{0.08}{4}\right)^{4(3)}$$

$A = \$1268$

(d) $x = 365$

$$A = 1000\left(1+\frac{0.08}{365}\right)^{365(3)}$$

$A = \$1271$

45.

$$\frac{T_2}{T_1} = \left(\frac{P_1}{P_2}\right)^{(1-\gamma)/\gamma}$$

$$T_2 = T_1\left(\frac{P_1}{P_2}\right)^{(1-\gamma)/\gamma}$$

$$T_2 = 575\left(\frac{25.0}{65\bar{0}}\right)^{(1-1.50)/1.50}$$

$$T_2 = 575\left(\frac{25.0}{65\bar{0}}\right)^{-1/3}$$

$$T_2 = 17\bar{0}0R$$

Exercises 9.2

1. $3^2 = 9$

$\log_3 9 = 2$

3. $5^3 = 125$

$\log_5 125 = 3$

5. $2^5 = 32$

$\log_2 32 = 5$

7. $9^{1/2} = 3$

$\log_9 3 = \frac{1}{2}$

9. $5^{-2} = \frac{1}{25}$

$\log_5\left(\frac{1}{25}\right) = -2$

11. $10^{-5} = 0.00001$

$\log_{10} 0.00001 = -5$

13. $\log_5 25 = 2$

$5^2 = 25$

15. $\log_2 16 = 4$

$2^4 = 16$

17. $\log_{25} 5 = \frac{1}{2}$

$25^{1/2} = 5$

19. $\log_8 2 = \frac{1}{3}$

$8^{1/3} = 2$

21. $\log_2\left(\frac{1}{4}\right) = -2$

$2^{-2} = \frac{1}{4}$

23. $\log_{10} 0.01 = -2$

$10^{-2} = 0.01$

The graphs for 25, 27, and 29 appear in the text answer section.

25. $y = \log_4 x$

x	0.25	1	4	16
y	−1	0	1	2

27. $y = \log_{10} x$

x	0.1	1	10	100
y	−1	0	1	2

29. $y = \log_{1/4} x$

x	4	1	0.25	0.06
y	−1	0	1	2

31. $\log_4 x = 3$

$4^3 = x$

$x = 64$

33. $\log_9 3 = x$

$9^x = 3$

$(3^2)^x = 3^1$

$2x = 1$

$x = \frac{1}{2}$

35. $\log_2 8 = x$

$2^x = 8$

$2^x = 2^3$

$x = 3$

37. $\log_{25} 5 = x$

$25^x = 5$

$(5^2)^x = 5^1$

$2x = 1$

$x = \frac{1}{2}$

39. $\log_x 25 = 2$

$x^2 = 25$

$x = 5$

(Base cannot be negative.)

41. $\log_{1/2}\left(\frac{1}{8}\right) = x$

$\left(\frac{1}{2}\right)^x = \frac{1}{8}$

$\left(\frac{1}{2}\right)^x = \left(\frac{1}{2}\right)^3$

$x = 3$

43. $\log_{12} x = 2$

$12^2 = x$

$x = 144$

45. $\log_x 9 = \frac{2}{3}$

$x^{2/3} = 9$

$x = 9^{3/2}$

$x = 27$

47. $\log_x\left(\frac{1}{8}\right) = -\frac{3}{2}$

$x^{-3/2} = \frac{1}{8}$

$x = \left(\frac{1}{8}\right)^{-2/3}$

$x = \left[\left(\frac{1}{8}\right)^{1/3}\right]^{-2}$

$x = \left(\frac{1}{2}\right)^{-2}$

$x = \frac{1^{-2}}{2^{-2}}$

$x = 4$

Exercises 9.3

1. $\log_2 5x^3y = \log_2 5 + \log_2 x^3 + \log_2 y = \log_2 5 + 3\log_2 x + \log_2 y$

3. $\log_{10} \dfrac{2x^2}{y^3z} = \log_{10} 2x^2 - \log_{10} y^3z = \log_{10} 2 + \log_{10} x^2 -$

$\left(\log_{10} y^3 + \log_{10} z\right) = \log_{10} 2 + 2\log_{10} x - 3\log_{10} y - \log_{10} z$

5. $\log_b \dfrac{y^3\sqrt{x}}{z^2} = \log_b y^3\sqrt{x} - \log_b z^2 = \log_b y^3 + \log_b \sqrt{x} -$

$\left(\log_b z^2\right) = 3\log_b y + \dfrac{1}{2}\log_b x - 2\log_b z$

7. $\log_b \sqrt[3]{\dfrac{x^2}{y}} = \dfrac{1}{3}\log_b \dfrac{x^2}{y} = \dfrac{1}{3}\left(\log_b x^2 - \log_b y\right) =$

$\dfrac{1}{3}(2\log_b x - \log_b y) = \dfrac{2}{3}\log_b x - \dfrac{1}{3}\log_b y$

9. $\log_2 \dfrac{1}{x}\sqrt{\dfrac{y}{z}} = \log_2 \dfrac{1}{x} + \log_2 \sqrt{\dfrac{y}{z}} = -\log_2 x + \dfrac{1}{2}\log_2 \dfrac{y}{z} =$

$-\log_2 x + \dfrac{1}{2}\left(\log_2 y - \log_2 z\right) = -\log_2 x + \dfrac{1}{2}\log_2 y - \dfrac{1}{2}\log_2 z$

11. $\log_b \dfrac{z^3\sqrt{x}}{\sqrt[3]{y}} = \log_b z^3\sqrt{x} - \log_b \sqrt[3]{y} = \log_b z^3 + \log_b \sqrt{x} -$

$\left(\dfrac{1}{3}\log_b y\right) = 3\log_b z + \dfrac{1}{2}\log_b x - \dfrac{1}{3}\log_b y$

13. $\log_b \dfrac{x^2(x+1)}{\sqrt{x+2}} = \log_b x^2(x+1) - \log_b \sqrt{x+2} =$

$\log_b x^2 + \log_b (x+1) - \dfrac{1}{2}\log_b (x+2) = 2\log_b x + \log_b (x+1) - \dfrac{1}{2}\log_b (x+2)$

15. $\log_b x + 2\log_b y = \log_b x + \log_b y^2 = \log_b xy^2$

17. $\log_b x + 2\log_b y - 3\log_b z = \log_b x + \log_b y^2 - \log_b z^3 = \log_b \dfrac{xy^2}{z^3}$

19. $\log_3 x + \dfrac{1}{3}\log_3 y - \dfrac{1}{2}\log_3 z = \log_3 x + \log_3 \sqrt[3]{y} - \log_3 \sqrt{z} = \log_3 \dfrac{x\sqrt[3]{y}}{\sqrt{z}}$

21. $2\log_{10} x - \frac{1}{2}\log_{10}(x-3) - \log_{10}(x+1) =$

$\log_{10} x^2 - \log_{10}\sqrt{x-3} - \log_{10}(x+1) = \log_{10}\dfrac{x^2}{(x+1)\sqrt{x-3}}$

23. $5\log_b x + \frac{1}{3}\log_b(x-1) - \log_b(x+2) =$

$\log_b x^5 + \log_b \sqrt[3]{x-1} - \log_b(x+2) = \log_b \dfrac{x^5\sqrt[3]{x-1}}{x+2}$

25. $\log_{10} x + 2\log_{10}(x-1) - \frac{1}{3}[\log_{10}(x+2) + \log_{10}(x-5)] =$

$\log_{10} x + \log_{10}(x-1)^2 - \frac{1}{3}[\log_{10}(x+2)(x-5)] = \log_{10}\dfrac{x(x-1)^2}{\sqrt[3]{(x+2)(x-5)}}$

27. $\log_b b^3 = 3$

29. $\log_3 9 = \log_3 3^2 = 2$

31. $\log_5 125 = \log_5 5^3 = 3$

33. $\log_2 \frac{1}{4} = \log_2(2^{-2}) = -2$

35. $\log_{10} 0.001 = \log_{10} 10^{-3} = -3$

37. $\log_3 1 = \log_3 3^0 = 0$

39. $6^{\log_6 5} = 5$

41. $25^{\log_5 6} = (5^2)^{\log_5 6} = 5^{2\log_5 6} = 5^{\log_5 6^2} = 36$

43. $4^{\log_2\left(\frac{1}{5}\right)} = 2^{2\log_2\left(\frac{1}{5}\right)} = 2^{\log_2\left(\frac{1}{5}\right)^2} = \left(\frac{1}{5}\right)^2 = \frac{1}{25}$

45. $\log_3 9^{\log_3 27} = \log_3 9^{\log_3 3^3} = \log_3 9^3 = \log_3(3^2)^3 = \log_3 3^6 = 6$

47. $\log_2 16^{\log_4 16} = \log_2 16^{\log_4 4^2} = \log_2 16^2 = \log_2(2^4)^2 = \log_2 2^8 = 8$

Exercises 9.4

1. $\log 68.1 = 1.833$

3. $\log 45 = 1.653$

5. $\log 0.142 = -0.8477$

7. $\log 0.00621 = -2.207$

9. $\log 805 = 2.906$

11. $\log 9.25 = 0.9661$

13. $\log N = 1.4048$
$N = 25.4$

15. $\log N = 2.8484$
$N = 705$

17. $\log N = 0.2782$
$N = 1.90$

19. $\log N = -1.6050$
$N = 0.0248$

21. $\log N = -3.6345$
$N = 2.32 \times 10^{-4}$

23. $\log N = -4.8145$
$N = 1.53 \times 10^{-5}$

25. $\dfrac{\log 685}{\log 6} =$ $\dfrac{2.8357}{0.77815} =$ 3.64

27. $\dfrac{\log 1675}{\log 12.5} =$ $\dfrac{3.2240}{1.0969} =$ 2.94

29. $\dfrac{\log 16.5}{\log 1350} =$ $\dfrac{1.2175}{3.1303} =$ 0.389

31. $\log\left(\dfrac{596}{45}\right) =$ $\log(13.244) =$ 1.12

33. $\log\left(\dfrac{4}{15}\right) =$ $\log(0.2667) =$ -0.574

35. $\log 14.6 - \log 3.75 =$ $1.1644 - 0.5740 =$ 0.590

37. $\log 145 + \log 25 =$ $2.1614 + 1.3979 =$ 3.56

39. $\dfrac{\log 486 + \log 680}{\log 14} =$ $\dfrac{2.6866 + 2.8325}{1.1461} =$ 4.82

41. $pH = -\log(H^+)$ $pH = -\log(10^{-6})$ $pH = 6.0$

43. $pH = -\log(H^+)$ $pH = -\log(3.2 \times 10^{-7})$ $pH = 6.5$

45. $pH = -\log(H^+)$ $pH = -\log(5.5 \times 10^{-8})$ $pH = 7.3$

47. $\beta = 10 \log \dfrac{I}{10^{-16}}$ $\beta = 10 \log \dfrac{10^{-6}}{10^{-16}}$ $\beta = 100 dB$

49. $\beta = 10 \log \dfrac{I}{10^{-16}}$ $\beta = 10 \log \dfrac{10^{-10}}{10^{-16}}$ $\beta = 60 dB$

51. $n = 10 \log\left(\dfrac{P_o}{P_i}\right)$ $n = 10 \log\left(\dfrac{12}{0.50}\right)$ $n = 14 dB$

53. $n = 10 \log\left(\dfrac{P_o}{P_i}\right)$ $n = 10 \log\left(\dfrac{0.60}{0.80}\right)$ $n = -1.2 dB$

Exercises 9.5

1. $e^2 = 7.39$

3. $e^6 = 403$

5. $e^{-2} = 0.135$

7. $e^{-6} = 0.00248$

9. $e^{3.5} = 33.1$

11. $e^{0.15} = 1.16$

13. $e^{-2.5} = 0.0821$

15. $e^{-0.08} = 0.923$

17. $e^{2/3} = 1.95$

19. $e^{-1/3} = 0.717$

21. $\ln 56 = 4.025$

23. $\ln 406 = 6.006$

25. $\ln 4.3 = 1.459$

27. $\ln 0.00582 = -5.146$

29. $\ln 1 = 0$

31. $\ln x = 0.475$

$x = 1.61$

33. $\ln x = -1.445$

$x = 0.236$

35. $\ln x = 14.75$

$x = 2{,}550{,}000$

37. $y = Ae^{rn}$

$r = 0.08$

$A = 5.0 \times 10^{11}$

$n = 6$

$y = 5.0 \times 10^{11} e^{0.08(6)}$

$y = 8.08 \times 10^{11} kwh$

This formula yields a greater answer.

39. $y = Ae^{rn}$

$A = 3500$

$r = 0.0625$

$n = 6$

$y = 3500e^{0.0625(6)}$

$y = \$5092$

41. $y = Ae^{rn}$

$A = 95{,}000$

$r = 0.031$

$n = 10$

$y = 95{,}000e^{0.031(10)}$

$y = 130{,}000$

43. $N = N_o e^{0.04t}$

$N = 3\bar{0}00e^{0.04(5.0)}$

$N = 3700$

45. $3\% - 3.5\% = -0.5\%$,

will loose 0.5% per year.

$A = 10{,}000,\ r = -0.005,\ t = 20.$

$y = Ae^{rt}$

$y = 10{,}000e^{-0.005(20)}$

$y = \$9048$

47. $N = N_o e^{rt}$

$N_o = 27.0$

$r = -0.0250$

$t = 60.0$ sec.

$N = 27.0e^{-0.0250(60.0)}$

$N = 6.02g$

49. $\ln A - \ln 8\bar{0} = -4.10 \times 10^{-4} t$

$\ln 79 - \ln 80 = -4.10 \times 10^{-4} t$

$t = \dfrac{-0.012579}{-4.10 \times 10^{-4}}$

$t = 31$ years

51. $\ln 6\bar{0} - \ln 8\bar{0} = -4.10 \times 10^{-4} t$

$t = \dfrac{-0.28768}{-4.10 \times 10^{-4}}$

$t = 7\bar{0}0$ years

53. $\ln 2\bar{0} - \ln 8\bar{0} = -4.10 \times 10^{-4} t$

$t = \dfrac{-1.38629}{-4.10 \times 10^{-4}}$

$t = 3400$ years

$$V_O = E_{AS} - (E_{AS} - E_O)e^{-t/TC}$$

55. $V_O = 10.0 - (10.0 - 0)e^{-50.0/20.0}$

$V_O = 9.18V$

57. $V_O = -40.0 - (-40.0 - 60.0)e^{-10.0/18.0}$

$V_O = -40.0 + 10\bar{0}e^{-10.0/18.0}$

$V_O = 17.4V$

59. $k=\frac{q}{\pi\left(H_2^{\,2}-H_1^{\,2}\right)}\ln\left(\frac{R_2}{R_1}\right)$

From Appendix E, p. 714, 1 gallon $=0.134\,\text{ft}^3$.

$$k=\frac{\frac{51(0.134)\text{ft}^3}{\text{min}}}{\pi\left(48.5^2-44.5^2\right)\text{ft}^2}\ln\left(\frac{24\,\text{ft}}{11\,\text{ft}}\right)$$

$$k=\frac{\frac{6.834\,\text{ft}^3}{\text{min}}}{\pi(372)\text{ft}^2}(0.7801586)$$

$k=0.0046\,\text{ft/min}$

61. $\log_3 84.1=$

$\frac{\log 84.1}{\log 3}=$

4.03

63. $\log_4 2360=$

$\frac{\log 2360}{\log 4}=$

5.60

65. $\log_5 374=$

$\frac{\ln 374}{\ln 5}=$

3.68

67. $\log_6 9600=$

$\frac{\ln 9600}{\ln 6}=$

5.12

Exercises 9.6

1. $3^x=12$

$\log 3^x=\log 12$

$x\log 3=\log 12$

$x=\frac{\log 12}{\log 3}$

$x=2.26$

3. $2^{-x}=43.7$

$\log 2^{-x}=\log 43.7$

$-x\log 2=\log 43.7$

$-x=\frac{\log 43.7}{\log 2}$

$x=-5.45$

5. $3^{2x}=0.21$

$\log 3^{2x}=\log 0.21$

$2x\log 3=\log 0.21$

$x=\frac{\log 0.21}{2\log 3}$

$x=-0.710$

7. $5^{x+1}=3^x$

$\log 5^{x+1}=\log 3^x$

$(x+1)\log 5=x\log 3$

$x\log 5+\log 5=x\log 3$

$x\log 5-x\log 3=-\log 5$

$x(\log 5-\log 3)=-\log 5$

$x=\frac{-\log 5}{\log 5-\log 3}$

$x=-3.15$

9. $4^{-20x}=50$

$\log 4^{-20x}=\log 50$

$-20x\log 4=\log 50$

$x=\frac{\log 50}{-20\log 4}$

$x=-0.1411$

11. $3^{2x+1}=5^{-x}$

$\log 3^{2x+1}=\log 5^{-x}$

$(2x+1)\log 3=-x\log 5$

$2x\log 3+\log 3+x\log 5=0$

$x(2\log 3+\log 5)=-\log 3$

$x=\frac{-\log 3}{2\log 3+\log 5}$

$x=-0.2886$

13.
$$e^x = 23$$
$$\ln e^x = \ln 23$$
$$x \ln e = \ln 23$$
$$x = \ln 23$$
$$x = 3.135$$

15.
$$10^{-x} = 0.146$$
$$\log 10^{-x} = \log 0.146$$
$$-x \log 10 = \log 0.146$$
$$x = -\log 0.146$$
$$x = 0.8356$$

17.
$$e^{2x} = 40.5$$
$$\ln e^{2x} = \ln 40.5$$
$$2x \ln e = \ln 40.5$$
$$x = \frac{\ln 40.5}{2}$$
$$x = 1.851$$

19.
$$e^{-3x} = 850$$
$$\ln e^{-3x} = \ln 850$$
$$-3x \ln e = \ln 850$$
$$x = \frac{\ln 850}{-3}$$
$$x = -2.248$$

21.
$$10^{x/2} = 0.45$$
$$\log 10^{x/2} = \log 0.45$$
$$\frac{x}{2} \log 10 = \log 0.45$$
$$x = 2 \log 0.45$$
$$x = -0.6936$$

23.
$$4e^x = 94.7$$
$$\ln\left(4e^x\right) = \ln 94.7$$
$$\ln 4 + \ln e^x = \ln 94.7$$
$$x \ln e = \ln 94.7 - \ln 4$$
$$x = \ln 94.7 - \ln 4$$
$$x = 3.164$$

25.
$$e^x = 5^{x-1}$$
$$\ln e^x = \ln 5^{x-1}$$
$$x = (x-1)\ln 5$$
$$x = x \ln 5 - \ln 5$$
$$x - x \ln 5 = -\ln 5$$
$$x(1 - \ln 5) = -\ln 5$$
$$x = \frac{-\ln 5}{1 - \ln 5}$$
$$x = 2.641$$

27.
$$e^x = 4^{2x+1}$$
$$\ln e^x = \ln 4^{2x+1}$$
$$x = (2x+1)\ln 4$$
$$x = 2x \ln 4 + \ln 4$$
$$x - 2x \ln 4 = \ln 4$$
$$x(1 - 2\ln 4) = \ln 4$$
$$x = \frac{\ln 4}{1 - 2\ln 4}$$
$$x = -0.7821$$

29.
$$5e^{-2x} = (12)3^{x-1}$$
$$\ln 5e^{-2x} = \ln(12)(3)^{x-1}$$
$$\ln 5 + \ln e^{-2x} = \ln 12 + \ln 3^{x-1}$$
$$\ln 5 - 2x \ln e = \ln 12 + (x-1)\ln 3$$
$$\ln 5 - 2x = \ln 12 + x \ln 3 - \ln 3$$
$$-2x - x \ln 3 = \ln 12 - \ln 3 - \ln 5$$
$$x(-2 - \ln 3) = \ln\left(\frac{12}{15}\right)$$
$$x = \frac{\ln\left(\frac{4}{5}\right)}{-2 - \ln 3}$$
$$x = 0.07201$$

31.
$$e^{x^2} = 600$$
$$\ln e^{x^2} = \ln 600$$
$$x^2 = \ln 600$$
$$x = \pm\sqrt{\ln 600}$$
$$x = \pm 2.529$$

33.
$$3 = 4\left(1 - e^x\right)$$
$$\frac{3}{4} = 1 - e^x$$
$$0.75 - 1 = -e^x$$
$$-0.25 = -e^x$$
$$0.25 = e^x$$
$$\ln 0.25 = \ln e^x$$
$$x = -1.386$$

35.
$$35 = 80\left(1 - e^{-x}\right)$$
$$\frac{35}{80} = 1 - e^{-x}$$
$$0.4375 - 1 = -e^{-x}$$
$$-0.5625 = -e^{-x}$$
$$e^{-x} = 0.5625$$
$$\ln e^{-x} = \ln 0.5625$$
$$-x = \ln 0.5625$$
$$x = 0.5754$$

37.

$$8=10\left(1-e^{x/2}\right)$$
$$\frac{8}{10}=1-e^{x/2}$$
$$0.8+e^{x/2}=1$$
$$e^{x/2}=1-0.8=0.2$$
$$\ln e^{x/2}=\ln 0.2$$
$$\frac{x}{2}=\ln 0.2$$
$$x=-3.219$$

39.

$$175=225\left(1-e^{-x/3}\right)$$
$$\frac{175}{225}=1-e^{-x/3}$$
$$0.77778+e^{-x/3}=1$$
$$e^{-x/3}=0.22222$$
$$\ln e^{-x/3}=\ln 0.22222$$
$$\frac{-x}{3}=\ln 0.22222$$
$$x=4.512$$

41.

$$48=64\left(1-e^{-3x}\right)$$
$$\frac{48}{64}=1-e^{-3x}$$
$$0.75+e^{-3x}=1$$
$$e^{-3x}=0.25$$
$$\ln e^{-3x}=\ln 0.25$$
$$-3x=\ln 0.25$$
$$x=0.4621$$

43.

$$135=145\left(1-e^{-4x/3}\right)$$
$$\frac{135}{145}=1-e^{-4x/3}$$
$$\frac{27}{29}+e^{-4x/3}=1$$
$$e^{-4x/3}=\frac{2}{29}$$
$$\ln e^{-4x/3}=\ln\left(\frac{2}{29}\right)$$
$$\frac{-4x}{3}=\ln 2-\ln 29$$
$$x=\frac{-3}{4}(\ln 2-\ln 29)$$
$$x=2.006$$

45.

$$50=75(1+e)^{x}$$
$$\frac{50}{75}=(1+e)^{x}$$
$$\ln\left(\frac{2}{3}\right)=\ln(1+e)^{x}$$
$$\ln 2-\ln 3=x\ln(1+e)$$
$$\frac{\ln 2-\ln 3}{\ln(1+e)}=x$$
$$x=-0.3087$$

47.

$$9=15(1+e)^{-x}$$
$$\frac{9}{15}=(1+e)^{-x}$$
$$\ln(0.6)=\ln(1+e)^{-x}$$
$$\ln(0.6)=-x\ln(1+e)$$
$$\frac{\ln(0.6)}{-\ln(1+e)}=x$$
$$x=0.3890$$

49.

$$7=10(1+e)^{x/2}$$
$$0.7=(1+e)^{x/2}$$
$$\ln(0.7)=\ln(1+e)^{x/2}$$
$$\ln(0.7)=\frac{x}{2}\ln(1+e)$$
$$2\ln(0.7)=x\ln(1+e)$$
$$\frac{2\ln(0.7)}{\ln(1+e)}=x$$
$$x=-0.5432$$

51.

$$V_O=E_{AS}-(E_{AS}-E_O)e^{-t/TC}$$
$$0=20.0-(20.0-(-20.0))e^{-t/50.0}$$
$$0=20.0-(40.0)e^{-t/50.0}$$
$$40.0e^{-t/50.0}=20.0$$
$$e^{-t/50.0}=\frac{20.0}{40.0}=0.500$$
$$\ln e^{-t/50.0}=\ln(0.500)$$
$$\frac{-t}{50.0}=\ln(0.500)$$
$$t=-50.0\ln(0.500)$$
$$t=34.7\mu s$$

53.

$$28.5 = 27.0 - (27.0 - 30.0)e^{-145/TC}$$
$$1.50 = -(-3.00)e^{-145/TC}$$
$$\frac{1.50}{3.00} = e^{-145/TC}$$
$$\ln 0.500 = \ln e^{-145/TC}$$
$$\ln 0.500 = \frac{-145}{TC}$$
$$TC = \frac{-145}{\ln 0.500}$$
$$TC = 209\mu s$$

55.

$$50.0 = E_{AS} - (E_{AS} - 15.0)e^{-32.0/10.0}$$
$$50.0 = E_{AS} - (E_{AS} - 15.0)e^{-3.2}$$
$$50.0 = E_{AS} - E_{AS}e^{-3.2} + 15.0e^{-3.2}$$
$$50.0 - 15.0e^{-3.2} = E_{AS}(1 - e^{-3.2})$$
$$E_{AS} = \frac{50.0 - 15.0e^{-3.2}}{1 - e^{-3.2}}$$
$$E_{AS} = 51.5V$$

Exercises 9.7

1.

$$\log(3x - 4) = \log(x + 6)$$
$$3x - 4 = x + 6$$
$$2x = 10$$
$$x = 5$$

3.

$$\ln(x - 4) = \ln(2x + 5)$$
$$x - 4 = 2x + 5$$
$$-x = 9$$
$$x = -9$$

This results in $\ln(-9-4) = \ln(-13)$. Since we cannot take the log of a negative number, no solution.

5.

$$\log 2 + \log(x - 3) = \log(x + 1)$$
$$\log 2(x - 3) = \log(x + 1)$$
$$2(x - 3) = x + 1$$
$$2x - 6 = x + 1$$
$$x = 7$$

7.

$$2\ln x = \ln 49$$
$$\ln x^2 = \ln 49$$
$$x^2 = 49$$
$$x = 7$$

9.

$$\log x + \log(2x) = \log 72$$
$$\log x(2x) = \log 72$$
$$2x^2 = 72$$
$$x^2 = 36$$
$$x = 6$$

11.

$$\log(x + 4) - \log(x - 2) = \log 2$$
$$\log\left(\frac{x + 4}{x - 2}\right) = \log 2$$
$$\frac{x + 4}{x - 2} = 2$$
$$x + 4 = 2x - 4$$
$$8 = x$$

13.

$$\ln x + \ln(x - 2) = \ln 3$$
$$\ln x(x - 2) = \ln 3$$
$$x^2 - 2x = 3$$
$$x^2 - 2x - 3 = 0$$
$$(x - 3)(x + 1) = 0$$
$$x = 3 \text{ or } -1$$

But $x \neq -1$, so solution is 3.

15.
$$\ln x + \ln(x-4) = \ln(x+6)$$
$$\ln x(x-4) = \ln(x+6)$$
$$x^2 - 4x = x + 6$$
$$x^2 - 5x - 6 = 0$$
$$(x-6)(x+1) = 0$$
$$x = 6 \text{ or } -1$$
But $x \neq -1$, so
solution is 6.

17.
$$2\log x = 2$$
$$\log x^2 = 2$$
$$10^2 = x^2$$
$$x = 10$$

19.
$$\log(x+1) = 1$$
$$10^1 = x + 1$$
$$9 = x$$

21.
$$2\log(x-1) = 1$$
$$\log(x-1)^2 = 1$$
$$10^1 = (x-1)^2$$
$$x - 1 = \pm\sqrt{10}$$
$$x = 1 \pm \sqrt{10}$$
x cannot be
negative, so
$$x = 1 + \sqrt{10}$$
$$x = 4.162$$

23.
$$\log(x+1) - 2\log 3 = 2$$
$$\log(x+1) - \log 3^2 = 2$$
$$\log\frac{x+1}{9} = 2$$
$$10^2 = \frac{x+1}{9}$$
$$900 = x + 1$$
$$899 = x$$

25.
$$\ln(x-3) - \ln 2 = 1$$
$$\ln\frac{x-3}{2} = 1$$
$$e^1 = \frac{x-3}{2}$$
$$2e = x - 3$$
$$2e + 3 = x$$
$$x = 8.437$$

27.
$$\ln(2x+1) + \ln 3 = 2$$
$$\ln 3(2x+1) = 2$$
$$e^2 = 6x + 3$$
$$e^2 - 3 = 6x$$
$$x = \frac{e^2 - 3}{6}$$
$$x = 0.7315$$

29.
$$\log(x+3) + 2\log 4 = 2$$
$$\log(x+3) + \log 4^2 = 2$$
$$\log 16(x+3) = 2$$
$$10^2 = 16x + 48$$
$$52 = 16x$$
$$x = 3.25$$

31.
$$\log(2x+1) + \log(x-1) = 2$$
$$\log(2x+1)(x-1) = 2$$
$$10^2 = (2x+1)(x-1)$$
$$100 = 2x^2 - x - 1$$
$$0 = 2x^2 - x - 101$$
$$x = \frac{-b \pm \sqrt{b^2 - 4ac}}{2a}$$
$a = 2$, $b = -1$, $c = -101$
$x = 7.361$ or -6.861
x cannot be negative
$$x = 7.361$$

33.

$$\ln x + \ln(x+2) = 1$$
$$\ln x(x+2) = 1$$
$$e^1 = x^2 + 2x$$
$$0 = x^2 + 2x - e$$
$$x = \frac{-b \pm \sqrt{b^2 - 4ac}}{2a}$$
$$a = 1,\ b = 2,\ c = -e$$
$$x = 0.9293 \text{ or } -2.928$$
x cannot be negative
$$x = 0.9283$$

35.

$$\ln x + \ln(2x - 3e) - 2 = \ln 2$$
$$\ln x(2x - 3e) = 2 + \ln 2$$
$$x(2x - 3e) = e^{2+\ln 2}$$
$$2x^2 - 3ex = e^2 \cdot e^{\ln 2}$$
$$2x^2 - 3ex = 2e^2$$
$$2x^2 - 3e - 2e^2 = 0$$
$$(2x + e)(x - 2e) = 0$$
x cannot be negative
$$x = 2e \text{ or } 5.437$$

Exercises 9.8

1. $A = P\left(1 + \frac{r}{4}\right)^{4n}$

(a)

$$P = 2500$$
$$r = 0.075$$
$$A = 4500$$
$$4500 = 2500\left(1 + \frac{0.075}{4}\right)^{4n}$$
$$\frac{4500}{2500} = (1.01875)^{4n}$$
$$\log 1.8 = \log(1.01875)^{4n}$$
$$\log 1.8 = 4n \log 1.01875$$
$$n = \frac{\log 1.8}{4 \log 1.01875}$$
$$n = 7.9 \text{ years}$$

(b)

$$P = 2500$$
$$A = 3(2500)$$
$$r = 0.075$$
$$7500 = 2500\left(1 + \frac{0.075}{4}\right)^{4n}$$
$$\frac{7500}{2500} = (1.01875)^{4n}$$
$$\log 3 = \log(1.01875)^{4n}$$
$$\log 3 = 4n \log(1.01875)$$
$$n = \frac{\log 3}{4 \log(1.01875)}$$
$$n = 14.8 \text{ years}$$

3.

$$y = y_0 e^{-0.4t}$$
$$y_0 = 175g$$
$$y = 25g$$
$$25 = 175e^{-0.4t}$$
$$\frac{25}{175} = e^{-0.4t}$$
$$\ln\left(\frac{1}{7}\right) = -0.4t$$
$$\frac{\ln 1 - \ln 7}{-0.4} = t$$
$$t = 4.9\ s$$

5.

$$pH = -\log(H^+)$$
$$3.5 = -\log(H^+)$$
$$-3.5 = \log(H^+)$$
$$10^{-3.5} = H^+$$
$$H^+ = 3.2 \times 10^{-4} M/L$$

7.

$$pH = -\log(H^+)$$
$$9.5 = -\log(H^+)$$
$$-9.5 = \log(H^+)$$
$$10^{-9.5} = H^+$$
$$H^+ = 3.2 \times 10^{-10} M/L$$

9.

$$pH = -\log(H^+)$$
$$2.7 = -\log(H^+)$$
$$-2.7 = \log(H^+)$$
$$10^{-2.7} = H^+$$
$$H^+ = 2.0 \times 10^{-3} M/L$$

For problems 11-15, use $\beta = 10\log\frac{I}{I_0}$ to find I.

11.

$$15 = 10\log\frac{I}{10^{-16}}$$
$$\frac{15}{10} = \log I - \log 10^{-16}$$
$$\frac{3}{2} = \log I - (-16)$$
$$1.5 - 16 = \log I$$
$$-14.5 = \log I$$
$$10^{-14.5} = I$$
$$I = 3.2 \times 10^{-15}\ W/\text{cm}^2$$

13.

$$120 = 10\log\frac{I}{10^{-16}}$$
$$12 = \log I - \log 10^{-16}$$
$$12 = \log I - (-16)$$
$$-4 = \log I$$
$$10^{-4} = I$$
$$I = 1.0 \times 10^{-4}\ W/\text{cm}^2$$

15.

$$3\bar{0} = 10\log\frac{I}{10^{-16}}$$
$$3.0 = \log I - \log 10^{-16}$$
$$3.0 = \log I - (-16)$$
$$-13 = \log I$$
$$10^{-13} = I$$
$$I = 1.0 \times 10^{-13}\ W/\text{cm}^2$$

17.

$$P = P_0 e^{rn}$$
$$P_0 = 75{,}000$$
$$r = 0.04$$

(a)

$$100{,}000 = 75{,}000e^{0.04n}$$
$$\frac{4}{3} = e^{0.04n}$$
$$\ln\left(\frac{4}{3}\right) = 0.04n$$
$$\frac{\ln 4 - \ln 3}{0.04} = n$$
$$n = 7.2 \text{ years}$$

(b)

$$2(75{,}000) = 75{,}000e^{0.04n}$$
$$2 = e^{0.04n}$$
$$\ln 2 = 0.04n$$
$$\frac{\ln 2}{0.04} = n$$
$$n = 17 \text{ years}$$

19.

$$i = 2.7e^{-0.2t}$$
$$0.35A = 2.7e^{-0.2t}$$
$$\frac{0.35}{2.7} = e^{-0.2t}$$
$$\ln\left(\frac{0.35}{2.7}\right) = -0.2t$$
$$\frac{\ln 0.35 - \ln 2.7}{-0.2} = t$$
$$t = 1\bar{0}s$$

21.

$$i = \frac{E}{R}e^{-t/RC}$$
$$3.91 \times 10^{-5} = \frac{10\bar{0}}{2.00 \times 10^4}e^{-t/2.00\times10^4(2.00\times10^{-8})}$$
$$3.91 \times 10^{-5} = .00500e^{-t/4.00\times10^{-4}}$$
$$\frac{3.91 \times 10^{-5}}{.00500} = e^{-t/4.00\times10^{-4}}$$
$$0.00782 = e^{-t/4.00\times10^{-4}}$$
$$\ln(0.00782) = \frac{-t}{4.00 \times 10^{-4}}$$
$$t = -\ln(0.00782)(4.00 \times 10^{-4})$$
$$t = 1.94 \times 10^{-3} ms$$

23. $V_C = V\left(1 - e^{-t/RC}\right)$

Note: $4.00\mu F = 4.00 \times 10^{-6} F$

$75.0 = 10\bar{0}\left(1 - e^{-t/2.00\times10^6(4.00\times10^{-6})}\right)$

$0.750 = 1 - e^{-t/8.00}$

$e^{-t/8.00} = 0.250$

$\frac{-t}{8.00} = \ln 0.250$

$t = -\ln(0.25)(8.00)$

$t = 11.1\mu s$

25. $\ln P = C - \gamma \ln V$

$\ln P = 2.5 - 1.6 \ln 2.5$

$\ln P = 1.0339$

$P = 2.8$

For problems **27** and **29** use $\ln A - \ln 50.0 = \left(-4.10\times10^{-4}\right)t$

27. $\ln A - \ln 50.0 = \left(-4.10\times10^{-4}\right)(1.00)$

$\ln A = 3.9116$

$A = 50.0\,\text{mg}$

29. $\ln A - \ln 50.0 = \left(-4.10\times10^{-4}\right)(10\bar{0})$

$\ln A = 3.8710$

$A = 48.0\,\text{mg}$

Exercises 9.9

1. Choose points $(10,50)$ and $(14,64)$.

$m = \frac{64-50}{14-10} = \frac{14}{4}$

$m = 3.5$

At $x = 0$, $y = 15$, so

$y = mx + b$ becomes

$y = 3.5x + 15$

Using the calculator routine

$y = 3.8x + 14$

3. Choose points $(60,1125)$ and $(100,1854)$.

$m = \frac{1854-1125}{100-60} = 18.2$

At $x = 0$, $y = -60$, so

$y = mx + b$ becomes

$y = 18.2x - 60$

Using the calculator routine

$y = 21.0x - 186$

5. See text answer section.

7. See text answer section.

9. Choose points $(1,1.7)$ and $(4,32)$.

$m = \frac{\log 32 - \log 1.7}{\log 4 - \log 1} = 2.11$

Round m (which is k) to 2.

The line crosses the $\log y$ – axis at 1.7 rounded to 2, so $a = 2$.
From $y = ax^k$, the equation is $y = 2x^2$.

11. Choose points $(1,12)$ and $(5,175)$.

$m = \frac{\log 175 - \log 12}{\log 5 - \log 1} = 1.67$

Round m (which is k) to 2.

The line crosses the $\log y$ – axis at 12, so $a = 12$. From $y = ax^k$, the equation is $y = 12x^2$.

For problems **13, 15, 17** see the text answer section.

19. Choose points $(3,20)$ and $(4.4,81)$.

$\log k = m = \dfrac{\log 81 - \log 20}{4.4 - 3} = 0.433896$

Thus, $k = 10^{0.433896} = 2.7.$

Use the point $(6.6, 700)$ in $y = a(2.7)^x$

$700 = a(2.7)^{6.6}$

$a = 0.996;\ a = 1$

So: $y = 1(2.7^x)$

21. Choose points $(3,0.72)$ and $(12,0.29)$.

$\log k = m = \dfrac{\log 0.29 - \log 0.72}{12 - 3} = -0.04388$

Thus, $k = 10^{-0.04388} = 0.9.$

Use the point $(4, 0.66)$ in $y = a(0.9)^x$

$0.66 = a(0.9)^4$

$a = 1.0$

So: $y = 1(0.9)^x$

23. Choose points $(3.2,69)$ and $(4.4,255)$.

$\log k = m = \dfrac{\log 255 - \log 69}{4.4 - 3.2}$

$\log k = m = \dfrac{0.56769}{1.2}$

$\log k = 0.473075$

$k = 10^{0.473075}$

$k = 2.97$

$k = 3$

Use the point $(1, 6.2)$ in $y = a(3)^x$

$6.2 = a(3)^1$

$a = 2.1$

So: $y = 2.1(3)^x$

Chapter 9 Review

Graphs appear in the text answer section.

1. $y = 3^x$

x	-1	0	1	2
y	$\frac{1}{3}$	1	3	9

2. $y = \log_3 x$

x	$\frac{1}{3}$	1	3	9
y	-1	0	1	2

3. $2^4 = 16$

$\log_2 16 = 4$

4. $10^{-3} = 0.001$

$\log_{10} 0.001 = -3$

5. $\log_{10} 7.389 = 0.8686$

$10^{0.8686} = 7.389$

6. $\log_4\left(\frac{1}{16}\right) = -2$

$4^{-2} = \frac{1}{16}$

7. $\log_9 x = 2$

$9^2 = x$

$x = 81$

8. $\log_x 8 = 3$

$x^3 = 8$

$x = 2$

9. $\log_2 32 = x$

$2^x = 32$

$2^x = 2^5$

$x = 5$

10. $\log_4 6x^2 y =$

$\log_4 6 + \log_4 x^2 + \log_4 y =$

$\log_4 6 + 2\log_4 x + \log_4 y$

11. $\log_3 \frac{5x\sqrt{y}}{z^3} =$

$\log_3 5 + \log_3 x + \log_3 \sqrt{y} - \log_3 z^3 =$

$\log_3 5 + \log_3 x + \frac{1}{2}\log_3 y - 3\log_3 z$

12. $\log \frac{x^2 (x+1)^3}{\sqrt{x-4}} =$

$\log x^2 + \log(x+1)^3 - \log\sqrt{x-4} =$

$2\log x + 3\log(x+1) - \frac{1}{2}\log(x-4)$

13. $\ln \frac{[x(x-1)]^3}{\sqrt{x+1}} =$

$\ln[x(x-1)]^3 - \ln\sqrt{x+1} =$

$3\ln x(x-1) - \frac{1}{2}\ln(x+1) =$

$3\ln x + 3\ln(x-1) - \frac{1}{2}\ln(x+1)$

14. $\log_2 x + 3\log_2 y - 2\log_2 z =$

$\log_2 x + \log_2 y^3 - \log_2 z^2 =$

$\log_2 \frac{xy^3}{z^2}$

15. $\frac{1}{2}\log(x+1) - 3\log(x-2) =$

$\log\sqrt{x+1} - \log(x-2)^3 =$

$\log \frac{\sqrt{x+1}}{(x-2)^3}$

16.
$$4\ln x - 5\ln(x+1) - \ln(x+2) =$$
$$\ln x^4 - \ln(x+1)^5 - \ln(x+2) =$$
$$\ln\frac{x^4}{(x+1)^5(x+2)}$$

17.
$$\frac{1}{2}[\ln x + \ln(x+2)] - 2\ln(x-5) =$$
$$\frac{1}{2}\ln x + \frac{1}{2}\ln(x+2) - \ln(x-5)^2 =$$
$$\ln\sqrt{x} + \ln\sqrt{x+2} - \ln(x-5)^2 =$$
$$\ln\sqrt{x}\sqrt{x+2} - \ln(x-5)^2 =$$
$$\ln\frac{\sqrt{x(x+2)}}{(x-5)^2}$$

18. $\log 1000 = 3$

19. $\log 10^{x^2} = x^2$

20. $\ln e^2 = 2$

21. $\ln e^x = x$

22. $\log 664.8 = 2.823$

23. $\log 0.04046 = -1.393$

24. $\log 14,420 = 4.159$

25. $\log N = 3.0737$
$N = 1180$

26. $\log N = -2.4289$
$N = 0.00372$

27. $\log N = -1.7522$
$N = 0.0177$

28. $\ln 72 = 4.277$

29. $\ln 421 = 6.043$

30. $\ln 0.00185 = -6.293$

31. $\ln x = 1.315$
$x = e^{1.315}$
$x = 3.72$

32. $\ln x = 3.45$
$x = e^{3.45}$
$x = 31.5$

33. $\ln x = -0.24$
$x = e^{-0.24}$
$x = 0.787$

34.
$$\log_4 20 =$$
$$\frac{\log 20}{\log 4} =$$
$$2.16$$

35.
$$6^{-2x} = 48.1$$
$$\log 6^{-2x} = \log 48.1$$
$$-2x\log 6 = \log 48.1$$
$$x = \frac{\log 48.1}{-2\log 6}$$
$$x = -1.08$$

36.
$$3^{4x-1} = 14^x$$
$$\log 3^{4x-1} = \log 14^x$$
$$(4x-1)\log 3 = x\log 14$$
$$4x\log 3 - \log 3 = x\log 14$$
$$4x\log 3 - x\log 14 = \log 3$$
$$x(4\log 3 - \log 14) = \log 3$$
$$x = \frac{\log 3}{4\log 3 - \log 14}$$
$$x = 0.626$$

37.
$$26.5 = 3.81e^{4x}$$
$$\frac{26.5}{3.81} = e^{4x}$$
$$\ln\left(\frac{26.5}{3.81}\right) = 4x$$
$$x = \frac{1}{4}\ln\left(\frac{26.5}{3.81}\right)$$
$$x = 0.485$$

38.

$$48 = 72\left(1-e^{-x/2}\right)$$
$$\frac{48}{72} = 1-e^{-x/2}$$
$$e^{-x/2} = 1-\frac{48}{72}$$
$$e^{-x/2} = \frac{1}{3}$$
$$\frac{-x}{2} = \ln\left(\frac{1}{3}\right)$$
$$x = -2\ln\left(\frac{1}{3}\right)$$
$$x = 2.20$$

39.

$$\log(x+4) = 2$$
$$10^2 = x+4$$
$$100-4 = x$$
$$x = 96$$

40.

$$\log(2x+3)-3\log 2 = 2\log 2$$
$$\log(2x+3)-\log 2^3 = \log 2^2$$
$$\log\frac{2x+3}{8} = \log 4$$
$$\frac{2x+3}{8} = 4$$
$$2x+3 = 32$$
$$2x = 29$$
$$x = 14.5$$

41.

$$\log(x+1)+\log(x-2) = 1$$
$$\log(x+1)(x-2) = 1$$
$$10^1 = (x+1)(x-2)$$
$$10 = x^2-x-2$$
$$0 = x^2-x-12$$
$$0 = (x-4)(x+3)$$
$x = 4$ or -3

x cannot be -3

Solution is 4.

42.

$$\ln x = \ln(3x-2)$$
$$x = 3x-2$$
$$-2x = -2$$
$$x = 1$$

43.

$$\ln(x+1)-\ln x = \ln 3$$
$$\ln\frac{x+1}{x} = \ln 3$$
$$\frac{x+1}{x} = 3$$
$$x+1 = 3x$$
$$1 = 2x$$
$$\frac{1}{2} = x$$

44.

$$2\ln x = 3$$
$$\ln x^2 = 3$$
$$e^3 = x^2$$
$$x = \sqrt{e^3}$$
$$x = 4.48$$

x cannot be negative.

45. $y = 125{,}000e^{-0.03t}$

$y = 125{,}000e^{-0.03(5.0)}$

$y = 110{,}000$

46. If P = price level

$2P$ = double price level

$y = Ae^{rn}$

$2P = Pe^{0.08n}$

$\ln 2 = 0.08n$

$n = \dfrac{\ln 2}{0.08}$

$n = 8.7$ *years*

47. $E = P_0V_0 \ln\left(\dfrac{V_1}{V_0}\right)$

$15{,}100 = 85.0(265)\ln\left(\dfrac{V_1}{265}\right)$

$\dfrac{15{,}100}{85.0(265)} = \ln\left(\dfrac{V_1}{265}\right)$

$0.670366 = \ln\left(\dfrac{V_1}{265}\right)$

$e^{0.670366} = \dfrac{V_1}{265}$

$V_1 = 265e^{0.670366}$

$V_1 = 518$

48. Choose points $(10, 51)$ and $(30, 34.5)$.

$m = \dfrac{34.5 - 51}{30 - 10} = -0.825$

$m = -0.83$

At $x = 0$, $y = 61$

So: $y = mx + b$ becomes

$y = -0.83x + 61.$

Using the calculator routine on page 295, $y = -0.80x + 59$

49. Choose points $(55, 215)$ and $(111, 390)$.

$m = \dfrac{390 - 215}{111 - 55} = 3.125$

$m = 3.1$

At $x = 0$, $y = 60$

So: $y = mx + b$ becomes

$y = 3.1x + 60.$

Using the calculator routine on page 295, $y = 3.3x + 45$

50. See text answer section.

51. See text answer section.

52. Choose points $(2,16.5)$ and $(3,35)$.

$$k = m = \frac{\log 35 - \log 16.5}{\log 3 - \log 2}$$

$k = 1.85$ i.e. $k \approx 2$

The line crosses the $\log y-$ axis at 3.6, which is a. So, $y = ax^k$ *is* $y = 3.6x^2$.

53. Choose points $(2.15,10)$ and $(5.5,70)$.

$$k = m = \frac{\log 70 - \log 10}{\log 5.5 - \log 2.15}$$

$k = m = 2.07$ i.e. $k \approx 2$

The line crosses the $\log y-$ axis at 2.1, which is a. So, $y = ax^k$ *is* $y = 2.1x^2$.

54. See text answer section.

55. See text answer section.

56. Choose points $(1,3.9)$ and $(3.6,130)$.

$$\log k = m = \frac{\log 130 - \log 3.9}{3.6 - 1}$$

$\log k = 0.585723$

$k = 3.9$

At $x = 1$, $y = 3.9$

So : $y = ak^x$ becomes

$3.9 = a(3.9)^1$

$$a = \frac{3.9}{3.9} = 1$$

So : $y = 3.9^x$

57. Choose points $(2,9.9)$ and $(3.6,30)$.

$$\log k = m = \frac{\log 30 - \log 9.9}{3.6 - 2}$$

$\log k = 0.300929$

$k = 1.999;\ k = 2$

At $x = 2$, $y = 9.9$

So : $y = ak^x$ becomes

$9.9 = a(2)^2$

$$a = \frac{9.9}{4} = 2.475$$

$a = 2.5$

So : $y = 2.5(2^x)$

Chapter 10, Exercises 10.1

For problems **1, 3, 5** the graphs appear in the text answer section.

7. $60°:(60°+360°)$ ie $420°$ and $(60°-360°)$ ie $-300°$

9. $-86°:(-86°+360°)$ ie $274°$ and $(-86°-360°)$ ie $-446°$

11. $225°:(225°+360°)$ ie $585°$ and $(225°-360°)$ ie $-135°$

13. $412°:(412°-360°)$ ie $52°$ and $(52°-360°)$ ie $-308°$

15. See text answer section. **17.** See text answer section. **19.** See text answer section.

	r	$\sin\theta$	$\cos\theta$	$\tan\theta$	$\cot\theta$	$\sec\theta$	$\csc\theta$
21.	$\sqrt{3^2+(-4)^2}$	$\frac{-4}{5}$	$\frac{3}{5}$	$\frac{-4}{3}$	$\frac{3}{-4}$	$\frac{5}{3}$	$\frac{5}{-4}$
23.	$\sqrt{1^2+1^2}$	$\frac{1}{\sqrt{2}}$	$\frac{1}{\sqrt{2}}$	1	1	$\sqrt{2}$	$\sqrt{2}$
25.	$\sqrt{(-1)^2+(-\sqrt{3})^2}$	$\frac{-\sqrt{3}}{2}$	$\frac{-1}{2}$	$\sqrt{3}$	$\frac{1}{\sqrt{3}}$	-2	$\frac{-2}{\sqrt{3}}$
27.	$\sqrt{(-4)^2+5^2}$	$\frac{5}{\sqrt{41}}$	$\frac{-4}{\sqrt{41}}$	$\frac{-5}{4}$	$\frac{-4}{5}$	$\frac{-\sqrt{41}}{4}$	$\frac{\sqrt{41}}{5}$
29.	$\sqrt{0^2+(-3)^2}$	-1	0	undefined	0	undefined	-1

31. For P, $r=\sqrt{8^2+6^2}=10$; for R, $r=\sqrt{12^2+9^2}=15$

$\sin\theta$	$\cos\theta$	$\tan\theta$	$\cot\theta$	$\sec\theta$	$\csc\theta$
P R	P R	P R	P R	P R	P R
$\frac{6}{10}=\frac{9}{15}$	$\frac{8}{10}=\frac{12}{15}$	$\frac{6}{8}=\frac{9}{12}$	$\frac{8}{6}=\frac{12}{9}$	$\frac{10}{8}=\frac{15}{12}$	$\frac{10}{6}=\frac{15}{9}$
$\frac{3}{5}$	$\frac{4}{5}$	$\frac{3}{4}$	$\frac{4}{3}$	$\frac{5}{4}$	$\frac{5}{3}$

33. $\sec\theta = \dfrac{1}{\cos\theta}$

$\sec\theta = \dfrac{r}{x} =$

$\dfrac{1}{\frac{x}{r}} = \dfrac{1}{\cos\theta}$

35. $\sin\theta = \dfrac{1}{\csc\theta}$

$\sin\theta = \dfrac{y}{r} =$

$\dfrac{1}{\frac{r}{y}} = \dfrac{1}{\csc\theta}$

37. $\tan\theta = \dfrac{\sin\theta}{\cos\theta}$

$\tan\theta = \dfrac{y}{x} =$

$\dfrac{\frac{y}{r}}{\frac{x}{r}} = \dfrac{\sin\theta}{\cos\theta}$

Exercises 10.2

1. $120°: \alpha = 180° - 120° = 60°$

3. $253°: \alpha = 253° - 180° = 73°$

5. $293.4°: \alpha = 360° - 293.4° = 66.6°$

7. $-116.7°: \alpha = -116.7° + 180° = 63.3°$

9. $462°4'$: coterminal is $462°4' - 360° = 102°4'$

$\alpha = 180° - 102°4' = 77°56'$

11. $1920°$: coterminal is $1920° - 5(360°) = 120°$

$\alpha = 180° - 120° = 60°$

13. $\sin 125.7° = 0.8121$

15. $\tan 349.7° = -0.1817$

17. $\cos 265.7° = -0.07498$

19. $\cos(-143.5°) = -0.8039$

21. $\sec 192.0° = -1.022$

23. $\cot(-36.5°) = -1.351$

25. $\sin\theta = 0.3684$

Quadrants I and II

$\theta = 21.6°$ or $180° - 21.6°$

$\theta = 21.6°$ or $158.4°$

27. $\tan\theta = 0.7250$

Quadrants I and II

$\theta = 35.9°$ or $180° + 35.9°$

$\theta = 35.9°$ or $215.9°$

29. $\cos\theta = -0.1050$

Quadrants II and III

$\theta = 96.0°$; $\alpha = 180° - 96.0° = 84.0°$

or

$\theta = 180° + 84.0° = 264.0°$

31. $\sin\theta = -0.9111$

Quadrants III and IV

$\theta = -65.7°$, coterminal is

$-65.7° + 360° = 294.3°$, or

$\theta =$ reference angle $+ 180°$

$\theta = 65.7° + 180° = 245.7°$

33. $\sec\theta = -1.7632$

Quadrants II and III

$$\frac{1}{\cos\theta} = -1.7632$$

$$\frac{1}{-1.7632} = \cos\theta$$

$\theta = 124.6°$. Reference angle is $180° - 124.6° = 55.4°$.

Or θ = reference angle $+180°$.

Or $\theta = 55.4° + 180° = 235.4°$.

35. $\cot\theta = 3.6994$

Quadrants I and III.

$$\frac{1}{\tan\theta} = 3.6994$$

$$\frac{1}{3.6994} = \tan\theta$$

$\theta = 15.1°$ or

$\theta = 15.1° + 180° = 195.1°$

37. $\csc\theta = -1.3250$

Quadrants III and IV

$$\frac{1}{\sin\theta} = -1.3250$$

$$\frac{1}{-1.3250} = \sin\theta$$

$\theta = -49.0°$, coterminal angle is $-49.0° + 360° = 311.0°$.

Reference angle is $49.0°$.

Also, $\theta = 49.0° + 180° = 229.0°$.

39. $\sec\theta = 2.3766$

Quadrants I and IV

$$\frac{1}{\cos\theta} = 2.3766$$

$$\frac{1}{2.3766} = \cos\theta$$

$\theta = 65.1°$ or

$\theta = 360° - 65.1° = 294.9°$

41. $\cos\theta = 0.7140$

Quadrants I and IV

$\theta = 44.4°$ or

$\theta = 360° - 44.4° = 315.6°$

Exercises 10.3

1. $$135° = 135°\left(\frac{\pi rad}{180°}\right)$$

$$135° = \frac{3\pi}{4} rad$$

3. $$90° = 90°\left(\frac{\pi rad}{180°}\right)$$

$$90° = \frac{\pi}{2} rad$$

5. $$-75° = -75°\left(\frac{\pi rad}{180°}\right)$$

$$-75° = -\frac{5\pi}{12} rad$$

7. $$1260° = 1260°\left(\frac{\pi rad}{180°}\right)$$

$$120° = 7\pi rad$$

9. $$\frac{7\pi}{4} = \frac{7\pi}{4}\left(\frac{180°}{\pi rad}\right)$$

$$\frac{7\pi}{4} = 315°$$

11. $$\frac{19\pi}{4} = \frac{19\pi}{4}\left(\frac{180°}{\pi rad}\right)$$

$$\frac{19\pi}{4} = 855°$$

13. $9\pi = 9\pi\left(\frac{180°}{\pi rad}\right)$

$9\pi = 1620°$

15. $3.7 = 3.7\left(\frac{180°}{\pi rad}\right)$

$3.7 = 212°$

17. $\frac{7\pi}{4}: \alpha = 2\pi - \frac{7\pi}{4} = \frac{\pi}{4}$

19. $\frac{9\pi}{4}: \alpha = \frac{9\pi}{4} - 2\pi = \frac{\pi}{4}$

21. $\frac{11\pi}{12}: \alpha = \pi - \frac{11\pi}{12} = \frac{\pi}{12}$

23. $\frac{-8\pi}{5}$: coterminal angle is $2\pi + \frac{-8\pi}{5} = \frac{2\pi}{5}: \alpha = \frac{2\pi}{5}$

25. $\sin\frac{3\pi}{4} = \frac{\sqrt{2}}{2}$

27. $\tan\frac{4\pi}{3} = \sqrt{3}$

29. $\csc\left(-\frac{3\pi}{4}\right) = \csc\left(-\frac{3\pi}{4} + \frac{8\pi}{4}\right) = \csc\frac{5\pi}{4} = -\sqrt{2}$

31. $\cos\theta = -\frac{1}{2}$

Quadrants II and III

reference angle is $\frac{\pi}{3}$.

$\theta = \pi - \frac{\pi}{3} = \frac{2\pi}{3}$ or

$\theta = \pi + \frac{\pi}{3} = \frac{4\pi}{3}$

33. $\sec\theta = 2$

Quadrants I and IV

reference angle is $\frac{\pi}{3}$.

$\theta = \frac{\pi}{3}$ or

$\theta = 2\pi - \frac{\pi}{3} = \frac{5\pi}{3}$

35. $\cot\theta = 1$

Quadrants I and III

reference angle is $\frac{\pi}{4}$.

$\theta = \frac{\pi}{4}$ or $\pi + \frac{\pi}{4} = \frac{5\pi}{4}$

37. $\sin 0.8 = 0.7174$

39. $\tan 1.2 = 2.572$

41. $\cos 1.0 = 0.5403$

43. $\sin(-1.65) = -0.9969$

45. $\tan 18.7 = -0.1507$

47. $\sec(-5.6) = \frac{1}{\cos(-5.6)} = 1.289$

49. $\sin\left(\frac{3\pi}{4}\right) = 0.7071$

51. $\tan\left(\frac{3\pi}{5}\right) = -3.078$

53. $\cos\left(\frac{-\pi}{24}\right) = 0.9914$

55. $\sin\theta = 0.9845$

Quadrants I and II

$\theta = 1.394$ or $\pi - \sin^{-1}(0.9845)$

$\theta = 1.394$ or 1.747

57. $\tan\theta = 1.685$

Quadrants I and III

$\theta = 1.035$ or $\pi + 1.035$

$\theta = 1.035$ or 4.177

59. $\cos\theta = -0.7540$

Quadrants II and III

$\theta = 2.425$ (in Quad II).

reference angle is $\pi - 2.425$.

reference angle is 0.7166.

Also, $\theta = \pi + 0.7166$

$\theta = 3.858$

61. $\cos\theta = 0.6924$

Quadrants I and IV

$\theta = 0.8060$ or

$\theta = 2\pi - 0.8060 = 5.477$

63. $\tan\theta = -4.672$

Quadrants II and IV

$\theta = -1.360$. Reference angle is 1.360.

$\theta = 2\pi - 1.360 = 4.923$ or

$\theta = \pi - 1.360 = 1.782$

65. $\csc\theta = -2.140$

Quadrants III and IV

$$\frac{1}{\sin\theta} = -2.140$$

$$\frac{1}{-2.140} = \sin\theta$$

$\theta = -0.4862$ Reference angle is 0.4862.

$\theta = 2\pi - 0.4862 = 5.797$ or

$\theta = \pi + 0.4862 = 3.628$

67. $\cos\theta = \dfrac{-\sqrt{3}}{2}$

Quadrants II and III

reference angle is $\dfrac{\pi}{6}$.

$\theta = \pi - \dfrac{\pi}{6} = \dfrac{5\pi}{6}$ or

$\theta = \pi + \dfrac{\pi}{6} = \dfrac{7\pi}{6}$

ie $\theta = 2.618$ or

$\theta = 3.665$

69. $\sin\theta = \dfrac{1}{2}$

Quadrants I and II

$\theta = \dfrac{\pi}{6}$ or

$\theta = \pi - \dfrac{\pi}{6} = \dfrac{5\pi}{6}$

ie $\theta = 0.5236$ or

$\theta = 2.618$

71. $\sec\theta = \dfrac{-2}{\sqrt{3}}$

Quadrants II and III

reference angle is $\dfrac{\pi}{6}$.

$\theta = \pi - \dfrac{\pi}{6} = \dfrac{5\pi}{6}$ or

$\theta = \pi + \dfrac{\pi}{6} = \dfrac{7\pi}{6}$

ie $\theta = 2.618$ or

$\theta = 3.665$

Exercises 10.4

1. $s = r\theta$

$s = 12.0\left(\dfrac{2\pi}{3}\right)$ in.

$s = 8\pi$ in. or

$s = 25.1$ in.

3. $A = \dfrac{1}{2}r^2\theta$

$A = \dfrac{1}{2}(12.0)^2\left(\dfrac{2\pi}{3}\right)$

$A = 48\pi$ or

$A = 151$ in.^{2}

5. $s = r\theta$

$5.00 = 6.00\theta$

$\dfrac{5.00}{6.00} = \theta$

or $\dfrac{5.00}{6.00}\left(\dfrac{180°}{\pi}\right)$

$\theta = 47.7°$

7. Area shaded is

$A_2 - A_1 =$

$\frac{1}{2}r_2^2\theta - \frac{1}{2}r_1^2\theta =$

$\frac{1}{2}\theta\left(r_2^2 - r_1^2\right) =$

$\frac{1}{2}\left(\frac{\pi}{6}\right)\left(5.00^2 - 3.00^2\right) =$

$4.19m^2$

9. $s = r\theta$

$s = 5.00\left(6°\left(\frac{\pi}{180°}\right)\right)$

$s = 0.524m$

11. $G = \frac{P}{\omega}$

$P = 275\,\text{hp}\left(\frac{55\bar{0}\,\frac{\text{ft-lb}}{\text{s}}}{1\,\text{hp}}\right)$

$P = 151{,}250\,\frac{\text{ft-lb}}{\text{s}}$

$\omega = 42\bar{0}0\,\frac{\text{rev}}{\text{min}} \times \frac{2\pi\,\text{rad}}{\text{rev}} \times \frac{1\,\text{min}}{60\,\text{s}}$

$\omega = 44\bar{0}\,\frac{\text{rad}}{\text{s}}$

$G = \frac{151{,}250}{44\bar{0}} = 344\,\text{ft-lb}$

13. **(a)** $42\bar{0}\,\text{rpm} =$

$42\bar{0}\,\frac{\text{rev}}{\text{min}}\left(\frac{2\pi\,\text{rad}}{\text{rev}}\right)\left(\frac{1\,\text{min}}{60\,\text{s}}\right) =$

$44.0\,\text{rad/s}$

(b) $\theta = \omega t$

$\omega = 44.0\,\text{rad/s}$

$\theta = 44.0\,\frac{\text{rad}}{\text{s}}.\,10.0\,\text{s}$

$\theta = 44\bar{0}\,\text{rad}.$

or $\theta = 44\bar{0}\,\text{rad}\left(\frac{1\,\text{rev}}{2\pi\,\text{rad}}\right)$

$\theta = 70.0\,\text{rev}$

(c) $V = \omega r$

$\omega = 44.0\,\frac{\text{rad}}{\text{s}}$

$r = 1.75\,\text{ft}$

$V = 44.0(1.75)$

$V = 77.0\,\text{ft/s}$

15. $18\bar{0}0\,\frac{\text{rev}}{\text{min}} = 18\bar{0}0(2\pi)\left(\frac{1}{60}\right)$

$18\bar{0}0\,\text{rpm} = 188\,\text{rad/s}$

$22\bar{0}0\,\frac{\text{rev}}{\text{min}} = 22\bar{0}0(2\pi)\left(\frac{1}{60}\right)$

$22\bar{0}0\,\frac{\text{rev}}{\text{min}} = 23\bar{0}\,\text{rad/s}$

$\alpha = \frac{\Delta\omega}{\Delta t}$

$\alpha = \frac{(23\bar{0} - 188)\,\text{rad/s}}{10.0\,\text{s}}$

$\alpha = \frac{42\,\text{rad/s}}{10.0\,\text{s}}$

$\alpha = 4.2\,\text{rad/s}^2$

17.

$$V = \omega r$$

$$V = \frac{1\,\text{rev}}{24\,\text{h}}\left(\frac{2\pi\,\text{rad}}{1\,\text{rev}}\right)(3960\,\text{mi})$$

$$V = 1040\,\text{mi/h}$$

19. (a)

$$V = \omega r$$

$$\omega = \frac{1250\,\text{rev}}{\text{min}}\left(\frac{2\pi\,\text{rad}}{1\,\text{rev}}\right)\left(\frac{1\,\text{min}}{60\,\text{s}}\right)$$

$$\omega = \frac{131\,\text{rad}}{\text{s}}$$

$$V = \frac{131\,\text{rad}}{\text{s}}(2.00\,\text{cm})\left(\frac{1\,\text{m}}{100\,\text{cm}}\right)$$

$$V = 2.62\,\text{m/s}$$

19. (b) $V = 2.62\,\text{m/s}$

19. (c)

$$V = \omega r$$

$$2.62\frac{\text{m}}{\text{s}} = \omega(8.00\,\text{cm})\left(\frac{1\,\text{m}}{100\,\text{cm}}\right)$$

$$\frac{2.62(100)\,\text{rad}}{8.00\quad\text{s}} = \omega$$

$$\omega = 32.8\,\text{rad/s}$$

Section 10.4

Chapter 10 Review

	$r=\sqrt{x^2+y^2}$	$\sin\theta$	$\cos\theta$	$\tan\theta$	$\cot\theta$	$\sec\theta$	$\csc\theta$
1.	5	$\frac{3}{5}$	$\frac{4}{5}$	$\frac{3}{4}$	$\frac{4}{3}$	$\frac{5}{4}$	$\frac{5}{3}$
2.	2	$\frac{-1}{2}$	$\frac{-\sqrt{3}}{2}$	$\frac{1}{\sqrt{3}}$	$\sqrt{3}$	$\frac{2}{-\sqrt{3}}$	-2
3.	4	0	-1	0	undefined	-1	undefined

4. $135°: \alpha = 180° - 135° = 45°$

5. $208°20': \alpha = 208°20' - 180° = 28°20'$

6. $-125°$: coterminal is $-125° + 360° = 235°$; $\alpha = 235° - 180° = 55°$

7. $1250°$: coterminal is $1250° - 3(360°) = 170°, \alpha = 180° - 170° = 10°$

8. $\sin 244.3° = -0.9011$

9. $\tan 337.5° = -0.4142$

10. $\sec 98.7° = \frac{1}{\cos 98.7°} = -6.611$

11. $\cos(-297.4°) = 0.4602$

12. $\cot 402.1° = \frac{1}{\tan 402.1°} = 1.107$

13. $\csc(-168.0°) = \frac{1}{\sin(-168.0°)} = -4.810$

14. $\sin\theta = 0.3448$
Quadrants I and II
$\theta = 20.2°$ or $\theta = 180° - 20.2°$
$\theta = 20.2°$ or $159.8°$

15. $\cos\theta = -0.5495$
Quadrants II and III
$\theta = 123.3°; \alpha = 56.7°$
or $\theta = 180° + 56.7° = 236.7°$

16. $\tan\theta = -1.050$

Quadrants II and IV

$\theta = -46.4°$, coterminal is

$-46.4° + 360° = 313.6°$

or $\theta = 180° - 46.6° = 133.6°$

17. $\sec\theta = 1.956$

Quadrants I and IV

$$\frac{1}{\cos\theta} = 1.956$$

$$\frac{1}{1.956} = \cos\theta$$

$\theta = 59.3°$ or

$\theta = 360° - 59.3° = 300.7°$

18. $\cot\theta = -1.855$

Quadrants II and IV

$$\frac{1}{\tan\theta} = -1.855$$

$$\frac{1}{-1.855} = \tan\theta$$

$\theta = -28.3°$, coterminal is

$-28.3° + 360° = 331.7°$

or $\theta = 180° - 28.3°$

$\theta = 151.7°$

19. $\csc\theta = 1.353$

Quadrants I and II

$$\frac{1}{\sin\theta} = 1.353$$

$$\frac{1}{1.353} = \sin\theta$$

$\theta = 47.7°$ or

$\theta = 180° - 47.7° = 132.3°$

20. $72° = 72°\left(\frac{\pi}{180°}\right)$

$72° = \frac{2\pi}{5}$

21. $315° = 315°\left(\frac{\pi}{180°}\right)$

$315° = \frac{7\pi}{4}$

22. $\frac{5\pi}{6} = \frac{5\pi}{6}\left(\frac{180°}{\pi}\right)$

$\frac{5\pi}{6} = 150°$

23. $\frac{3\pi}{4} = \frac{3\pi}{4}\left(\frac{180°}{\pi}\right)$

$\frac{3\pi}{4} = 135°$

24. $\frac{5\pi}{3}: \alpha = 2\pi - \frac{5\pi}{3}$

$\alpha = \frac{\pi}{3}$

25. $\frac{3\pi}{5}: \alpha = \pi - \frac{3\pi}{5}$

$\alpha = \frac{2\pi}{5}$

26. $\cos\frac{5\pi}{6} = -\cos\frac{\pi}{6}$

$\cos\frac{5\pi}{6} = -\frac{\sqrt{3}}{2}$

27. $\tan\frac{2\pi}{3} = -\tan\frac{\pi}{3}$

$\tan\frac{2\pi}{3} = -\sqrt{3}$

28. $\sec\frac{7\pi}{4} = \sec\frac{\pi}{4}$

$\sec\frac{7\pi}{4} = \sqrt{2}$

29. $\sin\frac{7\pi}{6} = -\sin\frac{\pi}{6}$

$\sin\frac{7\pi}{6} = -\frac{1}{2}$

30. $\cot\left(-\frac{\pi}{4}\right) = \cot\frac{7\pi}{4} = -\cot\frac{\pi}{4}$

$\cot\left(-\frac{\pi}{4}\right) = -1$

31. $\cos\left(-\frac{11\pi}{3}\right) = \cos\left(\frac{12\pi}{3} + \frac{-11\pi}{3}\right) = \cos\left(\frac{\pi}{3}\right)$

$\cos\left(-\frac{11\pi}{3}\right) = \frac{1}{2}$

32. $\cos\theta = -\frac{\sqrt{2}}{2}$

Quadrants II and III

$\alpha = \frac{\pi}{4}$

$\theta = \pi - \frac{\pi}{4} = \frac{3\pi}{4}$ or

$\theta = \pi + \frac{\pi}{4} = \frac{5\pi}{4}$

33. $\tan\theta = -1$

Quadrants II and IV

$\alpha = \frac{\pi}{4}$

$\theta = \pi - \frac{\pi}{4} = \frac{3\pi}{4}$ or

$\theta = 2\pi - \frac{\pi}{4} = \frac{7\pi}{4}$

34. $\csc\theta = \frac{2\sqrt{3}}{3}$

Quadrants I and II

$\alpha = \frac{\pi}{3}$

$\theta = \frac{\pi}{3}$ or $\pi - \frac{\pi}{3}$

$\theta = \frac{\pi}{3}$ or $\frac{2\pi}{3}$

35. $\sin\theta = -\frac{\sqrt{3}}{2}$

Quadrants III and IV

$\alpha = \frac{\pi}{3}$

$\theta = \pi + \frac{\pi}{3} = \frac{4\pi}{3}$ or

$\theta = 2\pi - \frac{\pi}{3} = \frac{5\pi}{3}$

36. $\sec\theta = -2$

Quadrants II and III

$\alpha = \frac{\pi}{3}$

$\theta = \pi - \frac{\pi}{3} = \frac{2\pi}{3}$ or

$\theta = \pi + \frac{\pi}{3} = \frac{4\pi}{3}$

37. $\tan\theta = \frac{\sqrt{3}}{3}$

Quadrants I and III

$\alpha = \frac{\pi}{6}$

$\theta = \frac{\pi}{6}$ or

$\theta = \pi + \frac{\pi}{6} = \frac{7\pi}{6}$

38. $\sin 1.5 = 0.9975$

39. $\cos(0.25) = 0.9689$

40. $\tan\frac{\pi}{6} = 0.5774$

41. $\sin\left(-\frac{2\pi}{3}\right) = -0.8660$

42. $\cos\theta = 0.1981$

Quadrants 1 and IV

$\theta = 1.371$ or

$\theta = 2\pi - 1.371 = 4.912$

43. $\sin\theta = -0.6472$

Quadrants III and IV

$\theta = -0.7039$, coterminal

is $2\pi - 0.7039 = 5.579$

or $\theta = \pi + 0.7039 = 3.845$

44. $\tan\theta = 1.6182$

Quadrants I and III

$\theta = 1.017$ or

$\theta = \pi + 1.017 = 4.159$

45.

$$\sec\theta = -2.8061$$

Quadrants II and III

$$\frac{1}{\cos\theta} = -2.8061$$

$$\frac{1}{-2.8061} = \cos\theta$$

$$\theta = 1.935$$

$$\alpha = \pi - 1.935 = 1.207$$

$$\text{or } \theta = \pi + 1.207 = 4.348$$

46.

$$s = r\theta$$

$$\theta = 56° = 0.9774\,\text{rad}$$

$$s = 9.00(0.9774)\,\text{in.}$$

$$s = 8.80\,\text{in.}$$

$$A = \frac{1}{2}r^2\theta$$

$$A = \frac{1}{2}(9.00)^2(0.9774)\,\text{in.}^2$$

$$A = 39.6\,\text{in.}^2$$

47.

$$G = \frac{P}{\omega}$$

$$325\,\text{hp} = 325\,\text{hp}\frac{\left(55\bar{0}\,\frac{\text{ft-lb}}{\text{s}}\right)}{1\,\text{hp}}$$

$$325\,\text{hp} = 179{,}000\,\text{ft-lb/s}$$

$$\omega = 40\bar{0}0\,\frac{\text{rev}}{\text{min}}\left(\frac{2\pi\,\text{rad}}{\text{rev}}\right)\left(\frac{1\,\text{min}}{60\,\text{s}}\right)$$

$$\omega = 419\,\text{rad/s}$$

$$G = \frac{179{,}000\,\text{ft-lb/s}}{419\,\text{rad/s}}$$

$$G = 427\,\text{ft-lb}$$

48.

(a)

$$63\bar{0}\,\frac{\text{rev}}{\text{min}} = 63\bar{0}\,\frac{\text{rev}}{\text{min}}\left(\frac{2\pi\,\text{rad}}{\text{rev}}\right)\left(\frac{1\,\text{min}}{60\,\text{s}}\right)$$

$$63\bar{0}\,\frac{\text{rev}}{\text{min}} = 66.0\,\text{rad/s}$$

(b)

$$\theta = \omega t$$

$$\theta = 66.0\,\frac{\text{rad}}{\text{s}}(5.00\,\text{s})$$

$$\theta = 33\bar{0}\,\text{rad}$$

(c)

$$V = \omega r$$

$$V = 66.0\,\frac{\text{rad}}{\text{s}}(1.50\,\text{ft})$$

$$V = 99.0\,\text{ft/s}$$

Chapter 11, Exercises 11.1

1. $A = 69.0°$, $a = 25.0\text{m}$, $b = 16.5\text{m}$

$$\frac{a}{\sin A} = \frac{b}{\sin B}$$
$$\frac{25.0\text{m}}{\sin 69.0°} = \frac{16.5\text{m}}{\sin B}$$
$$\sin B = \frac{16.5\text{m} \sin 69.0°}{25.0\text{m}}$$
$$B = 38.0°$$

$$C = 180° - 69.0° - 38.0°$$
$$C = 73.0°$$
$$\frac{c}{\sin C} = \frac{25.0\text{m}}{\sin 69.0°}$$
$$c = \frac{25.0\text{m} \sin 73.0°}{\sin 69.0°}$$
$$c = 25.6\text{m}$$

3. $B = 61.4°$, $b = 124\text{cm}$, $c = 112\text{cm}$

$$\frac{b}{\sin B} = \frac{c}{\sin C}$$
$$\frac{124\text{cm}}{\sin 61.4°} = \frac{112\text{cm}}{\sin C}$$
$$\sin C = \frac{112\text{cm} \sin 61.4°}{124\text{cm}}$$
$$C = 52.5°$$

$$A = 180° - 61.4° - 52.5°$$
$$A = 66.1°$$
$$\frac{a}{\sin A} = \frac{124\text{cm}}{\sin 61.4°}$$
$$a = \frac{124\text{cm} \sin 66.1°}{\sin 61.4°}$$
$$a = 129\text{cm}$$

5. $B = 75.3°$, $A = 57.1°$, $b = 257\text{ft}$

$$\frac{b}{\sin B} = \frac{a}{\sin A}$$
$$\frac{257\text{ft}}{\sin 75.3°} = \frac{a}{\sin 57.1°}$$
$$a = \frac{257\text{ft} \sin 57.1°}{\sin 75.3°}$$
$$a = 223\text{ft}$$

$$C = 180° - 75.3° - 57.1°$$
$$C = 47.6°$$
$$\frac{c}{\sin C} = \frac{257\text{ft}}{\sin 75.3°}$$
$$c = \frac{257\text{ft} \sin 47.6°}{\sin 75.3°}$$
$$c = 196\text{ft}$$

7. $A = 115.0°$, $a = 5870\text{m}$, $b = 4850\text{m}$

$$\frac{a}{\sin A} = \frac{b}{\sin B}$$
$$\frac{5870\text{m}}{\sin 115°} = \frac{4850\text{m}}{\sin B}$$
$$\sin B = \frac{4850\text{m} \sin 115°}{5870\text{m}}$$
$$B = 48.5°$$

$$C = 180° - 115.0° - 48.5°$$
$$C = 16.5°$$
$$\frac{c}{\sin C} = \frac{5870\text{m}}{\sin 115°}$$
$$c = \frac{5870\text{m} \sin 16.5°}{\sin 115°}$$
$$c = 1840\text{m}$$

9. $C = 72.58°, b = 28.63\,\text{cm}, c = 42.19\,\text{cm}$

$$\frac{c}{\sin C} = \frac{b}{\sin B}$$

$$\frac{42.19\,\text{cm}}{\sin 72.58°} = \frac{28.63\,\text{cm}}{\sin B}$$

$$\sin B = \frac{28.63\,\text{cm} \sin 72.58°}{42.19\,\text{cm}}$$

$$B = 40.35°$$

$$A = 180° - 72.58° - 40.35°$$

$$A = 67.07°$$

$$\frac{a}{\sin A} = \frac{42.19\,\text{cm}}{\sin 72.58°}$$

$$a = \frac{42.19\,\text{cm} \sin 67.07°}{\sin 72.58°}$$

$$a = 40.72\,\text{cm}$$

11. $B = 28.76°, C = 19.30°, c = 39{,}750\,\text{mi}$

$$\frac{b}{\sin B} = \frac{c}{\sin C}$$

$$\frac{b}{\sin 28.76°} = \frac{39{,}750\,\text{mi}}{\sin 19.30°}$$

$$b = \frac{39{,}750\,\text{mi} \sin 28.76°}{\sin 19.30°}$$

$$b = 57{,}870\,\text{mi}$$

$$A = 180° - 28.76° - 19.30°$$

$$A = 131.94°$$

$$\frac{a}{\sin A} = \frac{39{,}750\,\text{mi}}{\sin 19.30°}$$

$$a = \frac{39{,}750\,\text{mi} \sin 131.94°}{\sin 19.30°}$$

$$a = 89{,}460\,\text{mi}$$

13. $A = 25°, a = 5\bar{0}\,\text{cm}, b = 4\bar{0}\,\text{cm}$

$$\frac{a}{\sin A} = \frac{b}{\sin B}$$

$$\frac{5\bar{0}\,\text{cm}}{\sin 25°} = \frac{4\bar{0}\,\text{cm}}{\sin B}$$

$$\sin B = \frac{4\bar{0}\,\text{cm} \sin 25°}{5\bar{0}\,\text{cm}}$$

$$B = 2\bar{0}°$$

$$C = 180° - 25° - 2\bar{0}°$$

$$C = 135°$$

$$\frac{c}{\sin C} = \frac{5\bar{0}\,\text{cm}}{\sin 25°}$$

$$c = \frac{5\bar{0}\,\text{cm} \sin 135°}{\sin 25°}$$

$$c = 84\,\text{cm}$$

15. $C = 8°, c = 16\,\text{m}, a = 12\,\text{m}$

$$\frac{c}{\sin C} = \frac{a}{\sin A}$$

$$\frac{16\,\text{m}}{\sin 8°} = \frac{12\,\text{m}}{\sin A}$$

$$\sin A = \frac{12\,\text{m} \sin 8°}{16\,\text{m}}$$

$$A = 6°$$

$$B = 180° - 8° - 6°$$

$$B = 166°$$

$$\frac{b}{\sin B} = \frac{16\,\text{m}}{\sin 8°}$$

$$b = \frac{16\,\text{m} \sin 166°}{\sin 8°}$$

$$b = 28\,\text{m}$$

17. $B = 51°17'', b = 1948\,\text{ft}, c = 1525\,\text{ft}$

$$\frac{b}{\sin B} = \frac{c}{\sin C}$$

$$\frac{1948}{\sin 51°17''} = \frac{1525\,\text{ft}}{\sin C}$$

$$\sin C = \frac{1525\,\text{ft} \sin 51°17''}{1948}$$

$$C = 37.476255°$$

$$C = 37°28'35''$$

On calculator use: angle, DMS

$$A = 180° - 51°17'' - 37°28'35''$$

$$A = 91°31'8''$$

$$\frac{a}{\sin A} = \frac{1948\,\text{ft}}{\sin 51°17''}$$

$$a = \frac{1948\,\text{ft} \sin 91°31'8''}{\sin 51°17''}$$

$$a = 2506\,\text{ft}$$

19. $A = 31°14'35'', B = 85°45'15'', c = 4.575\,\text{mi}$

$$C = 180° - 31°14'35'' - 85°45'15''$$

$$C = 63°10''$$

$$\frac{c}{\sin C} = \frac{b}{\sin B}$$

$$\frac{4.575\,\text{mi}}{\sin 63°10''} = \frac{b}{\sin 85°45'15''}$$

$$b = \frac{4.575\,\text{mi} \sin 85°45'15''}{\sin 63°10''}$$

$$b = 5.120\,\text{mi}$$

$$\frac{a}{\sin A} = \frac{4.575\,\text{mi}}{\sin 63°10''}$$

$$a = \frac{4.575\,\text{mi} \sin 31°14'35''}{\sin 63°10''}$$

$$a = 2.663\,\text{mi}$$

21. $\angle P = 180° - 112.0° - 24.5° = 43.5°$

$$\frac{87.0\,in.}{\sin 24.5°} = \frac{PQ}{\sin 112.0°}$$

$$PQ = \frac{87.0\,in. \sin 112.0°}{\sin 24.5°}$$

PQ = 195 in.

$$\frac{QR}{\sin 43.5°} = \frac{87.0\,in.}{\sin 24.5°}$$

$$QR = \frac{87.0\,in. \sin 43.5°}{\sin 24.5°}$$

QR = 144 in.

23. $\angle F = 180° - 64.0° - 33.0° = 83.0°$

$$\frac{EF}{\sin 64.0°} = \frac{18.0\,ft}{\sin 83.0°}$$

$$EF = \frac{18.0\,ft \sin 64.0°}{\sin 83.0°}$$

EF = 16.3 ft

$$\frac{DF}{\sin 33.0°} = \frac{18.0\,ft}{\sin 83.0°}$$

$$DF = \frac{18.0\,ft \sin 33.0°}{\sin 83.0°}$$

DF = 9.88 ft

Exercises 11.2

1. $A = 37.0°$, $a = 21.5\text{cm}$, $b = 16.4\text{cm}$

$$\frac{b}{\sin B} = \frac{a}{\sin A}$$
$$\frac{16.4\text{cm}}{\sin B} = \frac{21.5\text{cm}}{\sin 37.0°}$$
$$\sin B = \frac{16.4\text{cm} \sin 37.0°}{21.5\text{cm}}$$
$$B = 27.3° \quad (\text{or } 152.7°)$$

For $B = 27.3°$, $C = 180° - 27.3° - 37.0°$

$$C = 115.7°$$
$$\frac{c}{\sin C} = \frac{21.5\text{cm}}{\sin 37.0°}$$
$$c = \frac{21.5\text{cm} \sin 115.7°}{\sin 37.0°}$$
$$c = 32.2\text{cm}$$

If $B = 152.7°$, then $C = 180° - 152.7° - 37.0° = -9.7°$. This is impossible, so there is only one triangle.

3. $C = 26.5°, c = 42.7\text{km}, a = 47.2\text{km}$

$$\frac{a}{\sin A} = \frac{c}{\sin C}$$
$$\frac{47.2\text{km}}{\sin A} = \frac{42.7\text{km}}{\sin 26.5°}$$
$$\sin A = \frac{47.2\text{km} \sin 26.5°}{42.7\text{km}}$$
$$A = 29.6° \text{ or } (150.4°)$$

For $A = 29.6°$, $B = 180° - 29.6° - 26.5°$

$$B = 123.9°$$
$$\frac{b}{\sin 123.9°} = \frac{42.7\text{km}}{\sin 26.5°}$$
$$b = \frac{42.7\text{km} \sin 123.9°}{\sin 26.5°}$$
$$b = 79.4\text{km}$$

If $A = 150.4°$, then $B = 180° - 150.4° - 26.5° = 3.1°$. So, there is a second triangle.

For $A = 150.4°$, $B = 3.1°$,
$C = 26.5°$, $c = 42.7\text{km}$, $a = 47.2\text{km}$.

$$\frac{b}{\sin 3.1°} = \frac{42.7\text{km}}{\sin 26.5°}$$
$$b = \frac{42.7\text{km} \sin 3.1°}{\sin 26.5°}$$
$$b = 5.18\text{km}$$

5. $A = 71.5°$, $a = 3.45\text{m}$, $c = 3.50\text{m}$

$$\frac{a}{\sin A} = \frac{c}{\sin C}$$
$$\frac{3.45\text{m}}{\sin 71.5°} = \frac{3.50\text{m}}{\sin C}$$
$$\sin C = \frac{3.50\text{m} \sin 71.5°}{3.45\text{m}}$$
$$C = 74.2° \text{ or } (105.8°)$$

For $C = 74.2°$, $B = 180° - 74.2° - 71.5°$

$$B = 34.3°$$
$$\frac{b}{\sin B} = \frac{3.45\text{m}}{\sin 71.5°}$$
$$b = \frac{3.45\text{m} \sin 34.3°}{\sin 71.5°}$$
$$b = 2.05\text{m}$$

For $C = 105.8°$, $B = 180° - 105.8° - 71.5°$, $B = 2.7°$, so there is a second triangle.

$$\frac{b}{\sin 2.7°} = \frac{3.45\text{m}}{\sin 71.5°}$$
$$b = \frac{3.45 \sin 2.7°}{\sin 71.5°}$$
$$b = 0.171\text{m}$$

7. $B = 105.0°, b = 16.5\,\text{mi}, a = 12.0\,\text{mi}$

$$\frac{a}{\sin A} = \frac{b}{\sin B}$$

$$\frac{12.0\,\text{mi}}{\sin A} = \frac{16.5\,\text{mi}}{\sin 105.0°}$$

$$\sin A = \frac{12.0\,\text{mi}\,\sin 105.0°}{16.5\,\text{mi}}$$

$A = 44.6°$ or $(135.4°)$

For $A = 44.6°, C = 180° - 44.6° - 105.0° = 30.4°$

$$\frac{c}{\sin 30.4°} = \frac{16.5\,\text{mi}}{\sin 105.0°}$$

$$c = \frac{16.5\,\text{mi}\,\sin 30.4°}{\sin 105.0°}$$

$c = 8.64\,\text{mi}$

For $A = 135.4°$, $C = 180° - 135.4° - 105.0° = -60.4°$, which is impossible.
There is only one triangle.

9. $C = 18°, c = 24\,\text{mi}, a = 45\,\text{mi}$

$$\frac{a}{\sin A} = \frac{c}{\sin C}$$

$$\frac{45\,\text{mi}}{\sin A} = \frac{24\,\text{mi}}{\sin 18°}$$

$$\sin A = \frac{45\,\text{mi}\,\sin 18°}{24\,\text{mi}}$$

$A = 35°$ or $(145°)$

For $A = 35°, B = 180° - 35° - 18°$

$B = 127°$

$$\frac{b}{\sin 127°} = \frac{24\,\text{mi}}{\sin 18°}$$

$$b = \frac{24\,\text{mi}\,\sin 127°}{\sin 18°}$$

$b = 62\,\text{mi}$

For $A = 145°, B = 180° - 145° - 18°, B = 17°,$ so there is a second triangle.

$$\frac{b}{\sin 17°} = \frac{24\,\text{mi}}{\sin 18°}$$

$$b = \frac{24\,\text{mi}\,\sin 17°}{\sin 18°}$$

$b = 23\,\text{mi}$

11. $C = 60°, c = 150\,\text{m}, b = 180\,\text{m}$

$$\frac{b}{\sin B} = \frac{c}{\sin C}$$

$$\frac{180\,\text{m}}{\sin B} = \frac{150\,\text{m}}{\sin 60°}$$

$$\sin B = \frac{180\,\text{m}\,\sin 60°}{150\,\text{m}}$$

$\sin B = 1.0392$, which is impossible. No triangle.

13. $B=8°$, $b=450\,\text{m}$, $c=850\,\text{m}$

$$\frac{c}{\sin C}=\frac{b}{\sin B}$$

$$\frac{850\,\text{m}}{\sin C}=\frac{450\,\text{m}}{\sin 8°}$$

$$\sin C=\frac{850\,\text{m}\sin 8°}{450\,\text{m}}$$

$C=15°$ or $(165°)$

For $C=15°$, $A=180°-15°-8°$

$A=157°$

$$\frac{a}{\sin A}=\frac{450\,\text{m}}{\sin 8°}$$

$$a=\frac{450\,\text{m}\sin 157°}{\sin 8°}$$

$a=1300\,\text{m}$

For $C=165°$, $A=180°-165°-8°=7°$, so there are two triangles.

For $C=165°$, $A=7°$

$$\frac{a}{\sin A}=\frac{450\,\text{m}}{\sin 8°}$$

$$a=\frac{450\,\text{m}\sin 7°}{\sin 8°}$$

$a=390\,\text{m}$

15. $B=41.50°$, $b=14.25\,\text{km}$, $a=18.50\,\text{km}$

$$\frac{a}{\sin A}=\frac{b}{\sin B}$$

$$\frac{18.50\,\text{km}}{\sin A}=\frac{14.25\,\text{km}}{\sin 41.50°}$$

$$\sin A=\frac{18.50\,\text{km}\sin 41.50°}{14.25\,\text{km}}$$

$A=59.34°$ or $(120.66°)$

For $A=59.34°$, $C=180°-59.34°-41.50°$

$C=79.16°$

$$\frac{c}{\sin C}=\frac{14.25\,\text{km}}{\sin 41.50°}$$

$$c=\frac{14.25\,\text{km}\sin 79.16°}{\sin 41.50°}$$

$c=21.12\,\text{km}$

For $A=120.66°$,

$C=180°-120.66°-41.50°=17.84°$

$$\frac{c}{\sin C}=\frac{14.25\,\text{km}}{\sin 41.50°}$$

$$c=\frac{14.25\,\text{km}\sin 17.84°}{\sin 41.50°}$$

$c=6.588\,\text{km}$

17. $C=63.85°$, $c=29.50\,\text{cm}$, $b=38.75\,\text{cm}$

$$\frac{b}{\sin B}=\frac{c}{\sin C}$$

$$\frac{38.75\,\text{cm}}{\sin B}=\frac{29.50\,\text{cm}}{\sin 63.85°}$$

$$\sin B=\frac{38.75\sin 63.85°}{29.50\,\text{cm}}$$

$\sin B=1.179$

This is impossible; so there is no triangle.

19. $C = 8.75°$, $c = 89.30\,\text{m}$, $a = 61.93\,\text{m}$

$$\frac{a}{\sin A} = \frac{c}{\sin C}$$

$$\frac{61.93\,\text{m}}{\sin A} = \frac{89.30\,\text{m}}{\sin 8.75°}$$

$$\sin A = \frac{61.93\,\text{m} \sin 8.75°}{89.30\,\text{m}}$$

$$A = 6.06° \text{ or } (173.94°)$$

For $A = 6.06°$,

$$B = 180° - 6.06° - 8.75°$$

$$B = 165.19°$$

$$\frac{b}{\sin B} = \frac{89.30\,\text{m}}{\sin 8.75°}$$

$$b = \frac{89.30\,\text{m} \sin 165.19°}{\sin 8.75°}$$

$$b = 150.1\,\text{m}$$

For $A = 173.94°$, $B = 180° - 173.94° - 8.75° = -2.69°$

Thus, there is only one triangle.

21. $B = 29°16'37''$, $b = 215.6\,\text{m}$, $c = 304.5\,\text{m}$

$$\frac{c}{\sin C} = \frac{b}{\sin B}$$

$$\frac{304.5\,\text{m}}{\sin C} = \frac{215.6\,\text{m}}{\sin 29°16'37''}$$

$$\sin C = \frac{304.5\,\text{m} \sin 29°16'37''}{215.6\,\text{m}}$$

$$C = 43.68377°$$

$$C = 43°41'2'' \text{ or } (136°18'58'')$$

For $C = 136°18'58''$

$$A = 180° - 136°18'58'' - 29°16'37''$$

$$A = 14°24'25''$$

$$\frac{a}{\sin A} = \frac{215.6\,\text{m}}{\sin 29°16'37''}$$

$$a = \frac{215.6\,\text{m} \sin 14°24'25''}{\sin 29°16'37''}$$

$$a = 109.7\,\text{m}$$

For $C = 43°41'2''$

$$A = 180° - 43°41'2'' - 29°16'37''$$

$$A = 107°2'21''$$

$$\frac{a}{\sin A} = \frac{215.6\,\text{m}}{\sin 29°16'37''}$$

$$a = \frac{215.6\,\text{m} \sin 107°2'21''}{\sin 29°16'37''}$$

$$a = 421.5\,\text{m}$$

23. $C = 25°45''$, $a = 524.5\,\text{ft}$, $c = 485.6\,\text{ft}$

$$\frac{a}{\sin A} = \frac{c}{\sin C}$$

$$\frac{524.5\,\text{ft}}{\sin A} = \frac{485.6\,\text{ft}}{\sin 25°45''}$$

$$\sin A = \frac{524.5\,\text{ft} \sin 25°45''}{485.6\,\text{ft}}$$

$$A = 27.17350$$

$$A = 27°10'25'' \text{ or}$$

$$A = 152°49'35''$$

For $A = 152°49'35''$,

$$B = 180° - 152°49'35'' - 25°45''$$

$$B = 2°9'40''$$

$$\frac{b}{\sin B} = \frac{485.6\,\text{ft}}{\sin 25°45''}$$

$$b = \frac{485.6\,\text{ft} \sin 2°9'40''}{\sin 25°45''}$$

$$b = 43.31\,\text{ft}$$

For $A = 27°10'25''$,

$$B = 180° - 27°10'25'' - 25°45''$$

$$B = 127°48'50''$$

$$\frac{b}{\sin B} = \frac{485.6\,\text{ft}}{\sin 25°45''}$$

$$b = \frac{485.6\,\text{ft} \sin 127°48'50''}{\sin 25°45''}$$

$$b = 907.3\,\text{ft}$$

Exercises 11.3

1. $A = 60.0°,\ b = 19.5\text{ m},\ c = 25.0\text{ m}$

$a^2 = b^2 + c^2 - 2bc\cos A$

$a^2 = 19.5^2 + 25.0^2 - 2(19.5)(25.0)\cos 60.0°$

$a^2 = 517.75$

$a = 22.8\text{ m}$

$\frac{b}{\sin B} = \frac{a}{\sin A}$

$\frac{19.5\text{ m}}{\sin B} = \frac{22.8\text{ m}}{\sin 60.0°}$

$\sin B = \frac{19.5\text{ m}\ \sin 60.0°}{22.8\text{ m}}$

$B = 47.8°$

$C = 180° - 47.8° - 60.0° = 72.2°$

3. $C = 109.0°,\ a = 14\bar{0}\text{ km},\ b = 215\text{ km}$

$c^2 = a^2 + b^2 - 2ab\cos C$

$c^2 = (14\bar{0})^2 + 215^2 - 2(14\bar{0})(215)\cos 109.0°$

$c^2 = 85{,}424$

$c = 292\text{ km}$

$\frac{a}{\sin A} = \frac{c}{\sin C}$

$\frac{14\bar{0}\text{ km}}{\sin A} = \frac{292\text{ km}}{\sin 109.0°}$

$\sin A = \frac{14\bar{0}\text{ km}\ \sin 109.0°}{292\text{ km}}$

$A = 27.0°$

$B = 180° - 27.0° - 109.0°$

$B = 44.0°$

5. $a = 19.2\text{ m},\ b = 21.3\text{ m},\ c = 27.2\text{ m}$

$a^2 = b^2 + c^2 - 2bc\cos A$

$19.2^2 = 21.3^2 + 27.2^2 - 2(21.3)(27.2)\cos A$

$19.2^2 - 21.3^2 - 27.2^2 = -2(21.3)(27.2)\cos A$

$\frac{19.2^2 - 21.3^2 - 27.2^2}{-2(21.3)(27.2)} = \cos A$

$0.7118976 = \cos A$

$A = 44.6°$

$\frac{b}{\sin B} = \frac{a}{\sin A}$

$\frac{21.3\text{ m}}{\sin B} = \frac{19.2\text{ m}}{\sin 44.6°}$

$\sin B = \frac{21.3\text{ m}\ \sin 44.6°}{19.2\text{ m}}$

$B = 51.2°$

$C = 180° - 51.2° - 44.6°$

$C = 84.2°$

7. $a = 4.25\text{ ft},\ b = 7.75\text{ ft},\ c = 5.50\text{ ft}$

$a^2 = b^2 + c^2 - 2bc\cos A$

$4.25^2 = 7.75^2 + 5.50^2 - 2(7.75)(5.50)\cos A$

$4.25^2 - 7.75^2 - 5.50^2 = -2(7.75)(5.50)\cos A$

$\frac{4.25^2 - 7.75^2 - 5.50^2}{-2(7.75)(5.50)} = \cos A$

$0.84750733 = \cos A$

$A = 32.1°$

$b^2 = a^2 + c^2 - 2ac\cos B$

$7.75^2 = 4.25^2 + 5.50^2 - 2(4.25)(5.50)\cos B$

$7.75^2 - 4.25^2 - 5.50^2 = -2(4.25)(5.50)\cos B$

$\frac{7.75^2 - 4.25^2 - 5.50^2}{-2(4.25)(5.50)} = \cos B$

$B = 104.6°$

$C = 180° - 104.6° - 32.1° = 43.3°$

9. $A = 45°,\ b = 51\text{ m},\ c = 39\text{ m}$

$a^2 = b^2 + c^2 - 2bc\cos A$

$a^2 = 51^2 + 39^2 - 2(51)(39)\cos 45°$

$a^2 = 1309$

$a = 36\text{ m}$

$\frac{c}{\sin C} = \frac{36\text{ m}}{\sin 45°}$

$\frac{39\text{ m}}{\sin C} = \frac{36\text{ m}}{\sin 45°}$

$\sin C = \frac{39\text{ m}\ \sin 45°}{36\text{ m}}$

$C = 5\bar{0}°$

$B = 180° - 5\bar{0}° - 45° = 85°$

11. $a = 7\bar{0}00\,\text{m}$, $b = 5600\,\text{m}$, $c = 4800\,\text{m}$

$$a^2 = b^2 + c^2 - 2bc\cos A$$
$$a^2 - b^2 - c^2 = -2bc\cos A$$
$$\frac{a^2 - b^2 - c^2}{-2bc} = \cos A$$
$$\frac{7\bar{0}00^2 - 5600^2 - 4800^2}{-2(5600)(4800)} = \cos A$$
$$0.1004464 = \cos A$$
$$A = 84°$$

$$\frac{c}{\sin C} = \frac{a}{\sin A}$$
$$\frac{4800\,\text{m}}{\sin C} = \frac{7\bar{0}00\,\text{m}}{\sin 84°}$$
$$\sin C = \frac{4800\sin 84°}{7\bar{0}00}$$
$$C = 43°$$
$$B = 180° - 43° - 84°$$
$$B = 53°$$

13. $C = 135°$, a $= 36\,\text{ft}$, $b = 48\,\text{ft}$

$$c^2 = a^2 + b^2 - 2ab\cos C$$
$$c^2 = 36^2 + 48^2 - 2(36)(48)\cos 135°$$
$$c^2 = 6044$$
$$c = 78\,\text{ft}$$

$$\frac{a}{\sin A} = \frac{c}{\sin C}$$
$$\frac{36\,\text{ft}}{\sin A} = \frac{78\,\text{ft}}{\sin 135°}$$
$$\sin A = \frac{36\sin 135°}{78}$$
$$A = 19°$$
$$B = 180° - 19° - 135° = 26°$$

15. $B = 19.25°$, $a = 4815\,\text{m}$

$$c = 1925\,\text{m}$$
$$b^2 = a^2 + c^2 - 2ac\cos B$$
$$b^2 = 4815^2 + 1925^2 - 2(4815)(1925)\cos 19.25°$$
$$b^2 = 9388564$$
$$b = 3064\,\text{m}$$

$$\frac{c}{\sin C} = \frac{b}{\sin B}$$
$$\frac{1925\,\text{m}}{\sin C} = \frac{3064\,\text{m}}{\sin 19.25°}$$
$$\sin C = \frac{1925\sin 19.25°}{3064}$$
$$C = 11.95°$$
$$A = 180° - 11.95° - 19.25°$$
$$A = 148.80°$$

17. $C = 108.75°$, $a = 405.0\,\text{mm}$,

$$b = 325.0\,\text{mm}$$
$$c^2 = a^2 + b^2 - 2ab\cos C$$
$$c^2 = 405.0^2 + 325.0^2 - 2(405.0)(325.0)\cos 108.75°$$
$$c^2 = 354269$$
$$c = 595.2\,\text{mm}$$

$$\frac{b}{\sin B} = \frac{c}{\sin C}$$
$$\frac{325.0\,\text{mm}}{\sin B} = \frac{595.2\,\text{mm}}{\sin 108.75°}$$
$$\sin B = \frac{325.0\,\text{mm}\,\sin 108.75°}{595.2\,\text{mm}}$$
$$B = 31.14°$$
$$A = 180° - 31.14° - 108.75°$$
$$A = 40.11°$$

19. $a = 207.5\,\text{km},\ b = 105.6\,\text{km}$

$c = 141.5\,\text{km}$

$a^2 = b^2 + c^2 - 2bc\cos A$

$a^2 - b^2 - c^2 = -2bc\cos A$

$\dfrac{a^2 - b^2 - c^2}{-2bc} = \cos A$

$\dfrac{207.5^2 - 105.6^2 - 141.5^2}{-2(105.6)(141.5)} = \cos A$

$-0.3976148 = \cos A$

$A = 113.43°$

$\dfrac{b}{\sin B} = \dfrac{a}{\sin A}$

$\dfrac{105.6\,\text{km}}{\sin B} = \dfrac{207.5\,\text{km}}{\sin 113.43°}$

$\sin B = \dfrac{105.6 \sin 113.43°}{207.5}$

$B = 27.84°$

$C = 180° - 27.84° - 113.43°$

$C = 38.73°$

21. $A = 72°18'0'',\ b = 1074\,\text{m},$

$c = 1375\,\text{m}$

$a^2 = b^2 + c^2 - 2bc\cos A$

$a^2 = 1074^2 + 1375^2 - 2(1074)(1375)\cos 72°18'0''$

$a^2 = 2146139$

$a = 1465\,\text{m}$

$\dfrac{b}{\sin B} = \dfrac{a}{\sin A}$

$\dfrac{1074}{\sin B} = \dfrac{1465}{\sin 72°18'0''}$

$\sin B = \dfrac{1074 \sin 72°18'0''}{1465}$

$B = 44.2989°$

$B = 44°17'56''$

$\dfrac{c}{\sin C} = \dfrac{a}{\sin A}$

$\dfrac{1375\,\text{m}}{\sin C} = \dfrac{1465\,\text{m}}{\sin 72°18'0''}$

$\sin C = \dfrac{1375 \sin 72°18'0''}{1465}$

$C = 63°23'52''$

Note: If C is calculated by subtracting A and B from 180°, then $C = 63°24'4''$. These differences are due to rounding.

23. $a = 1.250\,\text{mi},\ b = 1.975\,\text{mi}$

$c = 1.250\,\text{mi}$

$a^2 = b^2 + c^2 - 2bc\cos A$

$a^2 - b^2 - c^2 = -2bc\cos A$

$\dfrac{a^2 - b^2 - c^2}{-2bc} = \cos A$

$\dfrac{1.250^2 - 1.975^2 - 1.250^2}{-2(1.975)(1.250)} = \cos A$

$0.7900 = \cos A$

$A = 37.814488°$

$A = 37°48'52''$

Since $a = c$, $A = C$

so, $C = 37°48'52''$

$B = 180° - 2(37°48'52'')$

$B = 104°22'16''$

25. a)

$$(146\,\text{m})^2 = (127\,\text{m})^2 + (109\,\text{m})^2 - 2(109\,\text{m})(127\,\text{m})\cos Q$$
$$(146^2 - 127^2 - 109^2)\,\text{m}^2 = -2(109)(127)\,\text{m}^2 \cos Q$$
$$-6694 = -27686\cos Q$$
$$\cos Q = \frac{6694}{27686}$$
$$Q = 76.0°$$

b)

$$(127\,\text{m})^2 = (146\,\text{m})^2 + (109\,\text{m})^2 - 2(109)(146)\,\text{m}^2 \cos S$$
$$(127^2 - 146^2 - 109^2)\,\text{m}^2 - 2(109)(146)\,\text{m}^2 \cos S$$
$$-17068 = -31828\cos S$$
$$\cos S = \frac{17068}{31828}$$
$$S = 57.6°$$

27. a)

$$SM^2 = (10.0\,\text{ft})^2 + (24.0\,\text{ft})^2 - 2(10.0)(24.0)\,\text{ft}^2 \cos 36.5°$$
$$SM^2 = 676\,\text{ft}^2 - 480\,\text{ft}^2 \cos 36.5°$$
$$SM = 17.0\,\text{ft}$$

b)

$$(10.0\,\text{ft})^2 = (24.0\,\text{ft})^2 + (17.0\,\text{ft})^2 - 2(24.0)(17.0)\,\text{ft}^2 \cos S$$
$$-765\,\text{ft}^2 = -816\,\text{ft}^2 \cos S$$
$$\cos S = \frac{765}{816}$$
$$S = 20.4°$$

Exercises 11.4

1.

$$C = 180° - 30.0° - 65.0°$$
$$C = 85.0°$$
$$\frac{AC}{\sin B} = \frac{AB}{\sin C}$$
$$\frac{AC}{\sin 65.0°} = \frac{12.0\,\text{m}}{\sin 85.0°}$$
$$AC = \frac{12.0\,\text{m} \sin 65.0°}{\sin 85.0°}$$
$$AC = 10.9\,\text{m}$$

$$\frac{BC}{\sin A} = \frac{AB}{\sin C}$$
$$\frac{BC}{\sin 30.0°} = \frac{12.0\,\text{m}}{\sin 85.0°}$$
$$BC = \frac{12.0\,\text{m} \sin 30.0°}{\sin 85.0°}$$
$$BC = 6.02\,\text{m}$$

3.

$$c^2 = 81.3^2 + 67.5^2 - 2(81.3)(67.5)\cos 49.0°$$
$$c^2 = 3965.364$$
$$c = 63.0\,\text{m}$$

5.

$$\angle ABC = 180° - 58° - 49°$$
$$\angle ABC = 73°$$
$$\frac{AB}{\sin C} = \frac{AC}{\sin B}$$
$$\frac{AB}{\sin 49°} = \frac{3\bar{0}0\,\text{m}}{\sin 73°}$$
$$AB = \frac{3\bar{0}0\,\text{m} \sin 49°}{\sin 73°}$$
$$AB = 240\,\text{m}$$

7. Draw a triangle with tower B east of tower A and the fire, F, north of them. $\angle A = 90° - 51.5° = 38.5°$.

$\angle B = 90° - 17.2° = 72.8°$.

$\angle F = 180° - 38.5° - 72.8° = 68.7°$

$AB = 5.00\,\text{mi}$

$$\frac{AF}{\sin B} = \frac{AB}{\sin F}$$

$$\frac{AF}{\sin 72.8°} = \frac{5.00\,\text{mi}}{\sin 68.7°}$$

$AF = 5.13\,\text{mi}$

$$\frac{BF}{\sin A} = \frac{AB}{\sin F}$$

$$\frac{BF}{\sin 38.5°} = \frac{5.00\,\text{m}}{\sin 68.7°}$$

$$BF = \frac{5.00\,\text{m} \sin 38.5°}{\sin 68.7°}$$

$BF = 3.34\,\text{mi}$

9. Draw line segments intersecting at $75°$. Since distance = rate (time), the length of one line segment will be $6\bar{0}\,\text{mph}\ (1.5\,\text{h}) = 9\bar{0}\,\text{mi}$. The length of the other will be $45\,\text{mph}\ (1.5\,\text{h}) = 68\,\text{mi}$. The length, x, of the third side of the triangle will be the distance between the cars after 1.5h.

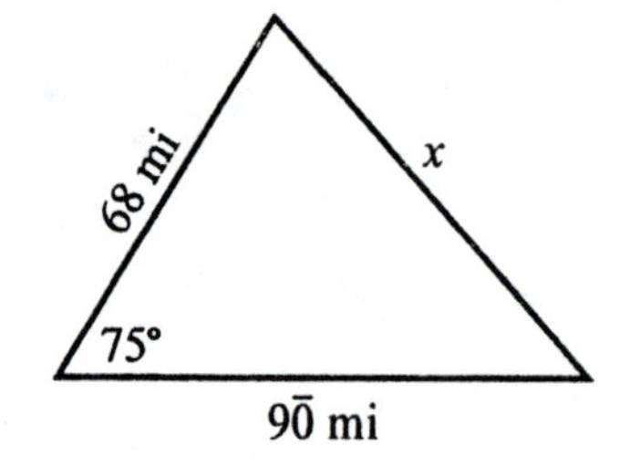

$$x^2 = 9\bar{0}^2 + 68^2 - 2(9\bar{0})(68)\cos 75°$$

$$x^2 = 9556$$

$$x = 98\,\text{mi}$$

11. Label the distance between A and the base of the cliff as y and the vertical distance from the top of the cliff to its base as x. In the smaller right triangle, the other angle is $90° - 37° = 53°$. At point A, the other angle is $180° - 37° = 143°$. In the triangle containing AB, the angle opposite AB is $180° - 24° - 143° = 13°$.

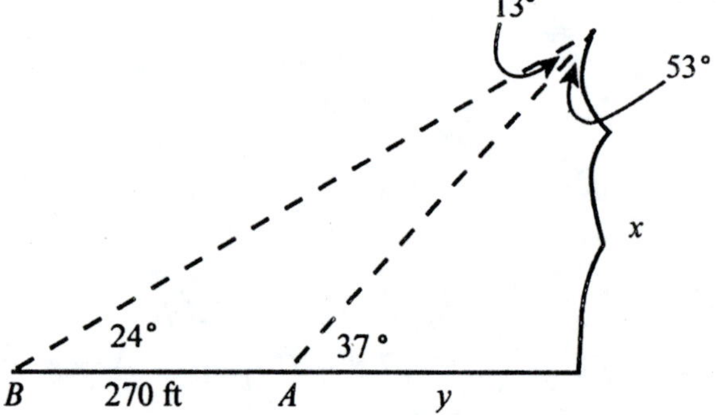

From the small right triangle,

$$\frac{x}{\sin 37°} = \frac{y}{\sin 53°}, \text{ ie } y = \frac{\sin 53°}{\sin 37°}x.$$

From the large right triangle,

$$\frac{x}{\sin 24°} = \frac{270\,\text{ft} + y}{\sin(13° + 53°)}$$

$$\frac{\sin 66°}{\sin 24°}x = 270 + y$$

$$\frac{\sin 66°}{\sin 24°}x = 270 + \frac{\sin 53°}{\sin 37°}x$$

$$2.246037x = 270 + 1.32704x$$

$$0.918997x = 270$$

$$x = 290\,\text{ft}$$

13. 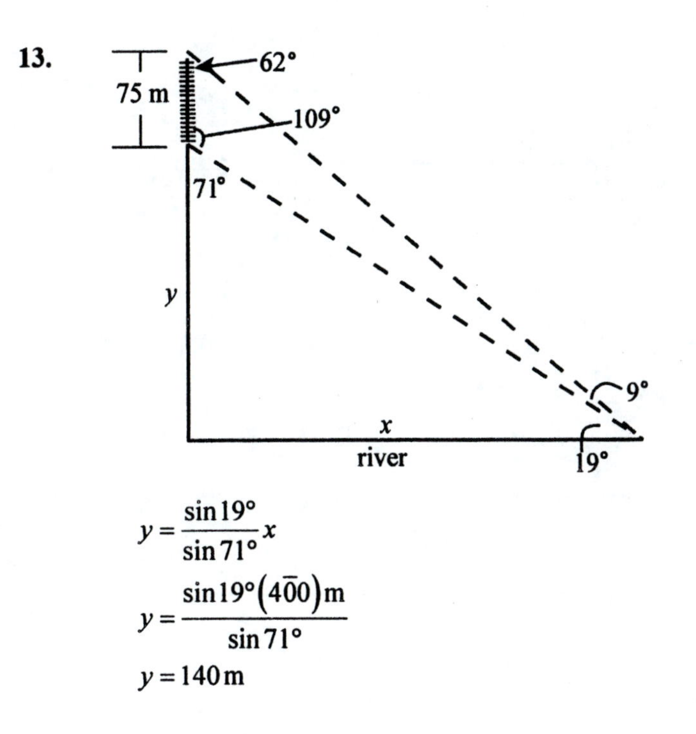

$$\frac{x}{\sin 71^\circ} = \frac{y}{\sin 19^\circ}$$

$$\frac{\sin 19^\circ}{\sin 71^\circ}x = y$$

Also: $\frac{x}{\sin 62^\circ} = \frac{y+75}{\sin(19^\circ + 9^\circ)}$

$$\frac{\sin 28^\circ}{\sin 62^\circ}x = y + 75$$

Substitue for y

$$\frac{\sin 28^\circ}{\sin 62^\circ}x = \frac{\sin 19^\circ}{\sin 71^\circ}x + 75$$

$$\left(\frac{\sin 28^\circ}{\sin 62^\circ} - \frac{\sin 19^\circ}{\sin 71^\circ}\right)x = 75$$

$$0.1873818x = 75$$

$$x = 4\bar{0}0\,\text{m}$$

$$y = \frac{\sin 19^\circ}{\sin 71^\circ}x$$

$$y = \frac{\sin 19^\circ (4\bar{0}0)\,\text{m}}{\sin 71^\circ}$$

$$y = 140\,\text{m}$$

15.

$$\frac{BC}{\sin A} = \frac{AB}{\sin C}$$

$$\frac{185\,\text{ft}}{\sin 35.0^\circ} = \frac{245\,\text{ft}}{\sin C}$$

$$\sin C = \frac{245\,\text{ft}\,\sin 35.0^\circ}{185\,\text{ft}}$$

$C = 49.4^\circ$ or 130.6°

For $C = 49.4^\circ$,

$$B = 180^\circ - 49.4^\circ - 35.0^\circ$$

$$B = 95.6^\circ$$

$$\frac{AC}{\sin B} = \frac{BC}{\sin A}$$

$$AC = \frac{185\,\text{ft}\,\sin 95.6^\circ}{\sin 35.0^\circ}$$

$$AC = 321\,\text{ft}$$

For $C = 130.6^\circ$

$$B = 180^\circ - 130.6^\circ - 35.0^\circ$$

$$B = 14.4^\circ$$

$$\frac{AC}{\sin B} = \frac{BC}{\sin A}$$

$$AC = \frac{185\,\text{ft}\,\sin 14.4^\circ}{\sin 35.0^\circ}$$

$$AC = 80.2\,\text{ft}$$

so $AC = 321\,\text{ft}$ or $80.2\,\text{ft}$

17.

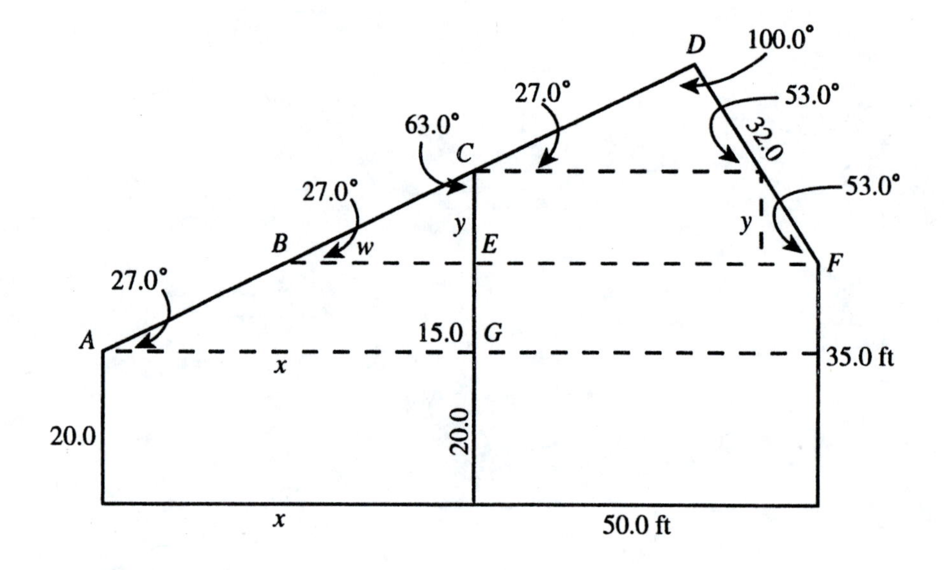

Use triangle BDF:

$$\frac{50.0\,\text{ft} + w}{\sin 100.0°} = \frac{32.0\,\text{ft}}{\sin 27.0°}$$

$$50.0 + w = \frac{32.0(\sin 100.0°)}{\sin 27.0°}$$

$$w = 19.4152\,\text{ft}$$

Use triangle BCE:

$$\tan 27.0° = \frac{y}{w}$$

$$y = w \tan 27.0°$$

$$y = 19.4152 \tan 27.0°$$

$$y = 9.8925\,\text{ft}$$

Use triangle ACG:

$$\tan 27.0° = \frac{y + 15.0}{x}$$

$$x = \frac{y + 15.0}{\tan 27.0°}$$

$$x = \frac{9.8925 + 15.0}{\tan 27.0°}$$

$$x = 48.9\,\text{ft}$$

19.(a)

$$BG^2 = AG^2 + AB^2 - 2(AG)(AB)\cos A$$

$$BG^2 = 7.50^2 + 9.00^2 - 2(7.50)(9.00)\cos 25.0°$$

$$BG^2 = 14.898449$$

$$BG = 3.86\,\text{m}$$

(b)

$$\frac{7.50\,\text{m}}{\sin \angle ABG} = \frac{3.86}{\sin 25.0°}$$

$$\sin \angle ABG = \frac{7.50 \sin 25.0°}{3.86}$$

$$\angle ABG = 55.2°$$

19.(c) $\angle AGB = 180° - 55.2° - 25.0°$

$\angle AGB = 99.8°$. Thus,

$\angle BGF = 180° - 99.8°$

$\angle BGF = 80.2°$

$BF^2 = BG^2 + FG^2 - 2(BG)(FG)\cos\angle BGF$

$BF^2 = 3.86^2 + 6.00^2 - 2(3.86)(6.00)\cos 80.2°$

$BF^2 = 43.0155$

$BF = 6.56\,\text{m}$

(d) $\frac{BG}{\sin\angle GFB} = \frac{BF}{\sin\angle BGF}$

$\frac{3.86\,\text{m}}{\sin\angle GFB} = \frac{6.56\,\text{m}}{\sin 80.2°}$

$\sin\angle GFB = \frac{3.86\sin 80.2°}{6.56}$

$\angle GFB = 35.4°$

(e) $\angle BGF + \angle GFB + \angle FBG = 180°$

$80.2° + 35.4° + \angle FBG = 180°$

$\angle FBG = 64.4°$

Also $\angle ABG + \angle FBG + \angle FBC = 180°$

$55.2° + 64.4° + \angle FBC = 180°$

$\angle FBC = 60.4°\angle$

21. $AC = 16.0\,\text{ft} - 4.00\,\text{ft}$

$AC = 12.0\,\text{ft}$

$\tan 25.0° = \frac{BC}{AC}$

$\tan 25.0° = \frac{BC}{12.0}$

$BC = 5.6\,\text{ft}$

$BE = BC + 8.00\,\text{ft}$

$BE = 5.6\,\text{ft} + 8.00\,\text{ft}$

$BE = 13.6\,\text{ft}$

$\angle ABC = 90.0° - 25.0°$

$\angle ABC = 65.0°$

$\angle EBD = 180° - \angle ABC$

$\angle EBD = 180° - 65.0°$

$\angle EBD = 115.0°$

$\angle BDE = 180° - 115.0° - 35.0°$

$\angle BDE = 30.0°$

$\frac{1}{\sin 115°} = \frac{BE}{\sin 30°}$

$1 = \frac{13.6\ \sin 115°}{\sin 30°}$

$1 = 24.7\,\text{ft}$

23. Label the dotted line y.

$y^2 = 58.0^2 = 10.0^2 - 2(58.0)(10.0)\cos 68.0°$

$y^2 = 3029.46$

$y = 55.0\,\text{ft}$

Let θ be the angle formed by

y and the side measuring 58.0ft.

$\frac{10.0\text{ft}}{\sin\theta} = \frac{55.0\text{ft}}{\sin 68.0°}$

$\sin\theta = \frac{10.0\sin 68.0°}{55.0}$

$\theta = 9.7°$

The angle opposite x

in the large triangle is

$94.0° - 9.7° = 84.3°$

$x^2 = 55.0^2 + 63.0^2 - 2(55.0)(63.0)\cos 84.3°$

$x^2 = 6305.7$

$x = 79.4\,\text{ft}$

25. $58.7^2 = 46.5^2 + (x+16.2)^2 - 2(46.5)(x+16.2)\cos 64.0°$

$3445.69 = 2162.25 + x^2 + 32.4x + 262.44 - 93(x+16.2)\cos 64.0°$

$3445.69 = x^2 + 32.4x + 2424.69 - 40.769x - 660.45$

$0 = x^2 - 8.369x - 1681.45$

Using the quadratic formula

$x = 45.4\,\text{m}$ (The negative formula is discarded.)

Exercises 11.5

1.

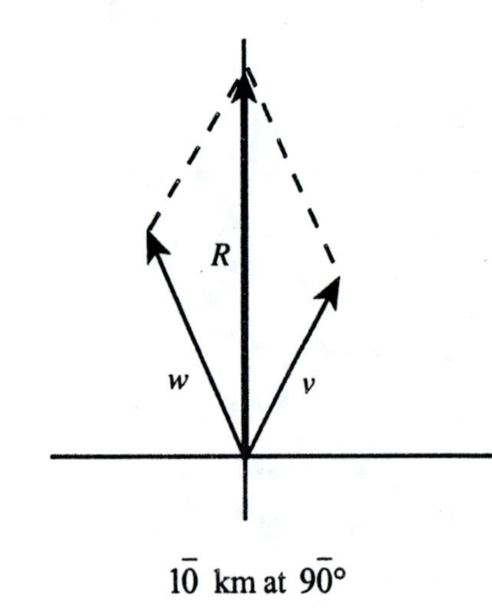

$1\bar{0}$ km at $9\bar{0}$°

3.

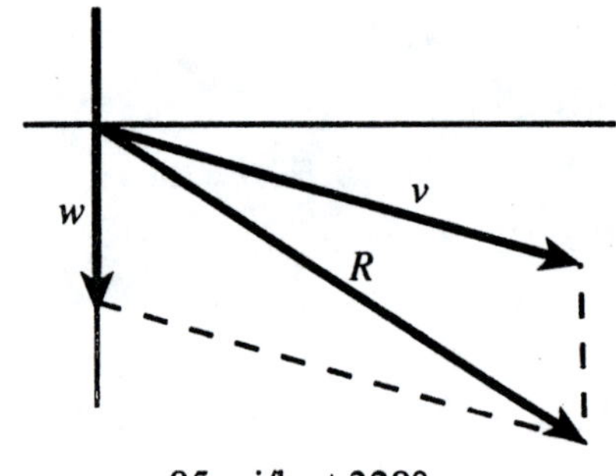

85 mi/h at 328°

5.

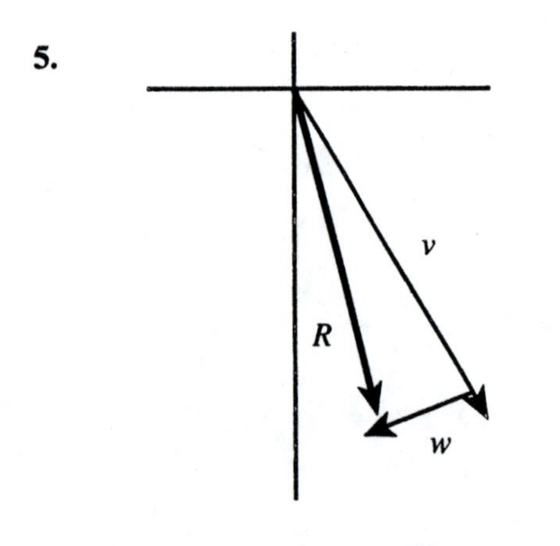

73 mi at 281°

7.

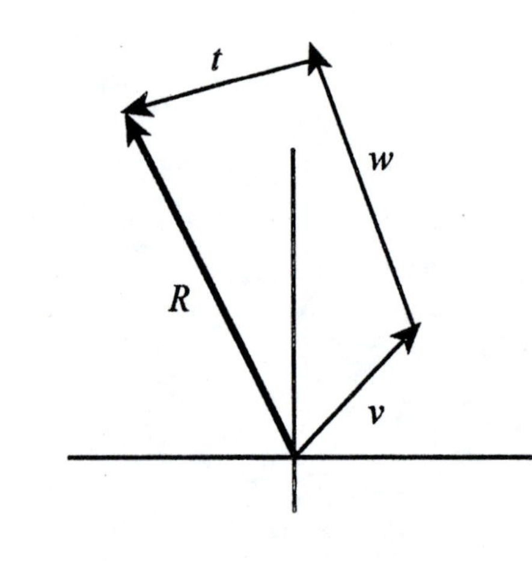

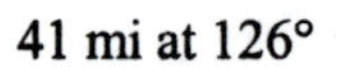

41 mi at 126°

9.

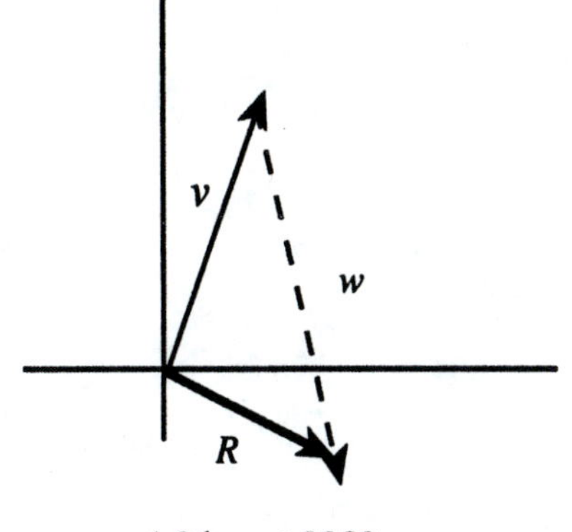

4.0 km at 328°

11.

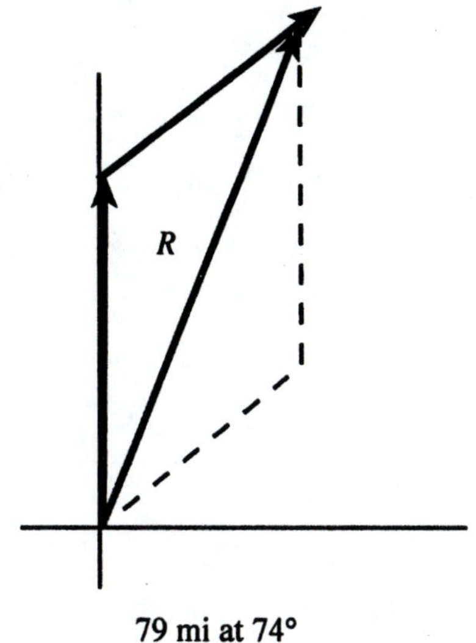

79 mi at 74°
(16° east of north)

13.

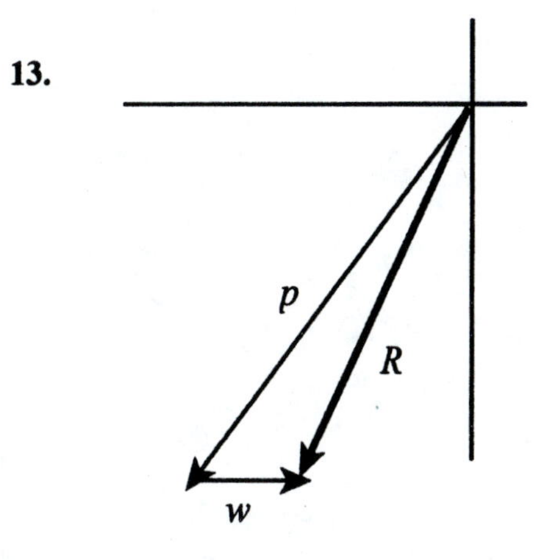

270 mi/h at $24\bar{0}$
(30° west of south)

Exercises 11.6

1. $v = 65.3$ km/h at 270.0°

$w = 40.5$ km/h at 180.0°

$|R| = \sqrt{|v|^2 + |w|^2}$

$|R| = \sqrt{65.3^2 + 40.5^2}$

$|R| = 76.8$ km/h

$\tan\alpha = \frac{|v|}{|w|}$

$\tan\alpha = \frac{65.3}{40.5}$

$\alpha = 58.2°$

$\theta = 180° + \alpha = 238.2°$

3. $v = 4.50\ km$ at 67.0°

$w = 6.50\ km$ at 105.0°

Draw the vectors with the initial point of w starting at the end point v. The angle between v and w is $67.0°+75.0° = 142.0°$.

$|R|^2 = |v|^2 + |w|^2 - 2|v||w|\cos 142.0°$

$|R|^2 = 4.50^2 + 6.50^2 - 2(4.50)(6.50)\cos 142.0°$

$|R|^2 = 108.5986$

$|R| = 10.4$ km

Let ϕ be the angle between v and R.

$\frac{6.50}{\sin\phi} = \frac{10.4}{\sin 142.0°}$

$\sin\phi = \frac{6.50 \sin 142.0°\theta}{10.4}$

$\phi = 22.6°$

$\theta = 67.0° + 22.6°$

$\theta = 89.6°$

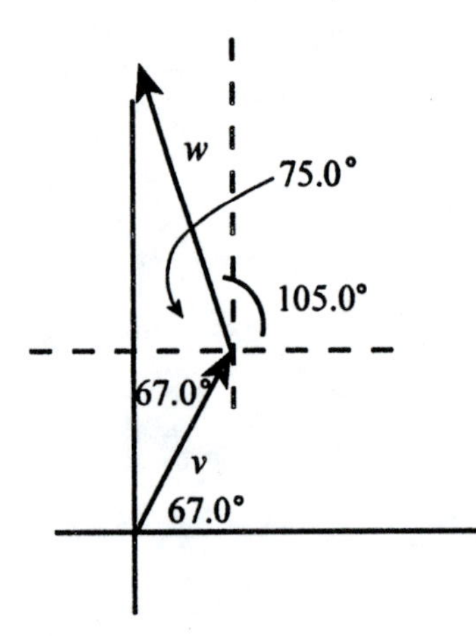

5. $v = 87.1$ mi/h at $130.5°$
$w = 46.7$ mi/h at $207.0°$
Draw the vectors with the initial point of w starting at the end point of v. The angle between v and w is
$360.0° - (49.5° + 207.0°) = 103.5°$.

$$|R|^2 = |v|^2 + |w|^2 - 2|v||w|\cos 103.5° w$$

$$|R|^2 = 87.1^2 + 46.7^2 - 2(87.1)(46.7)\cos 103.5° R$$

$$|R|^2 = 11666.41$$

$$|R| = 108 \text{ mi/h}$$

Let ϕ be the angle between v and R.

$$\frac{46.7}{\sin\phi} = \frac{108}{\sin 103.5°}$$

$$\sin\phi = \frac{46.7 \sin 103.5°}{108}$$

$$\phi = 24.9°$$

$$\theta = 130.5° + 24.9°$$

$$\theta = 155.4°$$

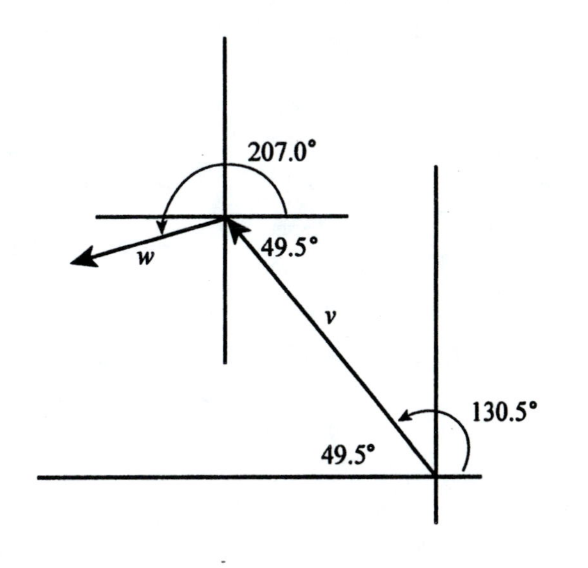

7. $v = 605\text{ m at } 60.0°$

$w = 415\text{ m at } 120.0°$

$t = 295\text{ m at } 90.0°$

First we find the sum of v and w, then add t to it. Draw the vectors with the initial point of w at the end point of v. The angle between v and w is $60.0° + 60.0° = 120.0°$.

$|R|^2 = |v|^2 + |w|^2 - 2|v||w|\cos 120.0°$

$|R|^2 = 605^2 + 415^2 - 2(605)(415)\cos 120.0°$

$|R|^2 = 789325$

$|R| = 888\text{ m}$

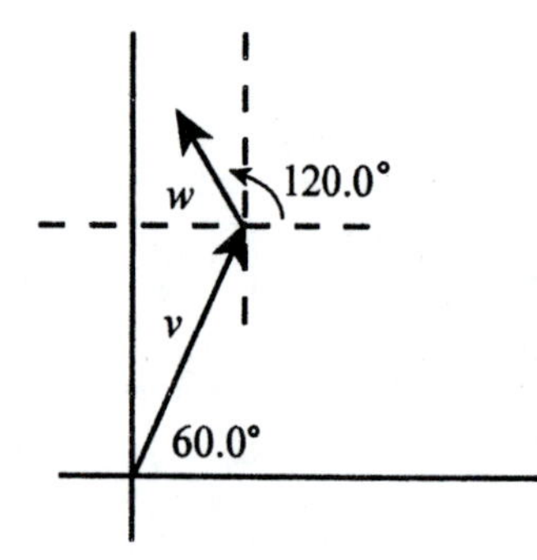

Let ϕ be the angle between R and v.

$$\frac{415\text{ m}}{\sin\phi} = \frac{888\text{ m}}{\sin 120.0°}$$

$$\sin\phi = \frac{415 \sin 120.0°}{888}$$

$\phi = 23.9°$

$\theta = 60.0° + 23.9°$

$\theta = 83.9°$

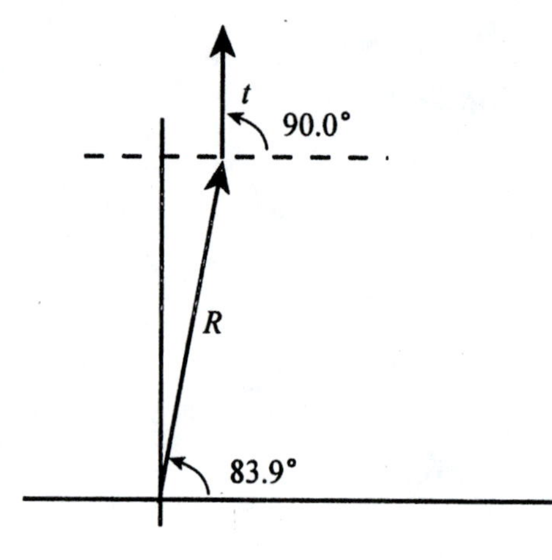

$R = 888\text{ m at } 83.9°$

$t = 295\text{ m at } 90.0°$

Draw the vectors with the initial point of t at the end point of R. The angle between t and R is $90.0° + 83.9° = 173.9°$.

Let the final resultant be z.

$|z|^2 = |R|^2 + |t|^2 - 2|R||t|\cos 173.9°$

$|z|^2 = 888^2 + 295^2 - 2(888)(295)\cos 173.9°$

$|z|^2 = 1396523$

$|z| = 1180\text{ m}$

Let β be the angle between z and R.

$$\frac{|t|}{\sin\beta} = \frac{|z|}{\sin 173.9°}$$

$$\frac{295\text{ m}}{\sin\beta} = \frac{1180\text{ m}}{\sin 173.9°}$$

$$\sin\beta = \frac{295 \sin 173.9°}{1180}$$

$\beta = 1.5°$

So $\theta = 83.9° + 1.5°$

$\theta = 85.4°$

Thus the sum of the three vectors is 1180 m at 85.4°.

9. $v = 4.50\,\text{km}$ at $67.0°$

$w = 6.50\,\text{km}$ at $105.0°$

$-w = 6.50\,\text{km}$ at $105.0° + 180.0° = 285.0°$

Draw v and $-w$ with the initial point of $-w$ starting at the end point of v. The angle between v and $-w$ is $180.0° - 67.0° - 75.0° = 38.0°$

$$|R|^2 = |v|^2 + |-w|^2 - 2|v||-w|\cos 38.0°$$

$$|R|^2 = 4.50^2 + 6.50^2 - 2(4.50)(6.50)\cos 38.0°$$

$$|R|^2 = 16.4014$$

$$|R| = 4.05\,\text{km}$$

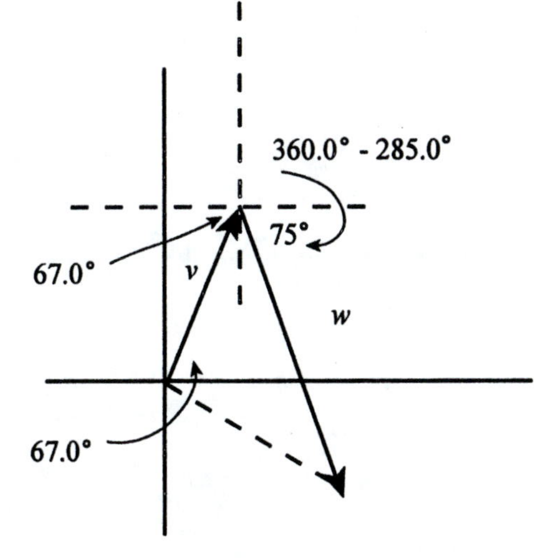

Let ϕ be the angle across from v the smaller magnitude of v and w.

$$\frac{|v|}{\sin\phi} = \frac{|R|}{\sin 38.0°}$$

$$\frac{4.50}{\sin\phi} = \frac{4.05}{\sin 38.0°}$$

$$\sin\phi = \frac{4.50 \sin 38.0°}{4.05}$$

$\phi = 43.2°$. The angle opposite $-w$ is $180.0° - 43.2° - 38.0° = 98.8°$.

The reference angle of R is $98.8° - 67.0°$. ie $\alpha = 31.8°$. R is in quadrant IV, so

$$\theta = 360.0° - 31.8°$$

$$\theta = 328.2°$$

Note: If we use components rounded to three significant digits (3.44, -2.14) before finding the angle, we will get $\theta = 328.1°$.

11. $v_1 = 175\,\text{km at } 105.0°$

$v_2 = 105\,\text{km at } 90.0°$

Draw the vectors with the initial point of v_2 starting at the end point of v_1. The angle between v_1 and v_2 is $90° + 75° = 165°$.

$$|R|^2 = |v_1|^2 + |v_2|^2 - 2|v_1||v_2|\cos 165°$$

$$|R|^2 = 175^2 + 105^2 - 2(175)(105)\cos 165°$$

$$|R|^2 = 77147.77$$

$$|R| = 278\,\text{km}$$

Let ϕ be the angle opposite v_2 (ie between R and v_1.

$$\frac{|v_2|}{\sin\phi} = \frac{|R|}{\sin 165°}$$

$$\frac{105}{\sin\phi} = \frac{278}{\sin 165°}$$

$$\sin\phi = \frac{105\sin 165°}{278}$$

$$\phi = 5.6°$$

$$\theta = 105° - 5.6°$$

$$\theta = 99.4°$$

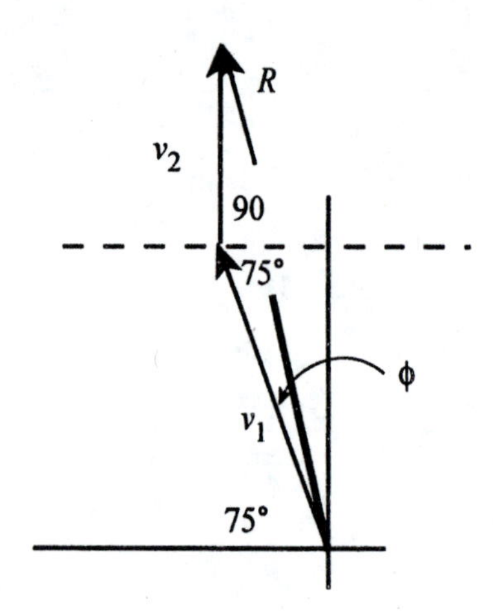

13.

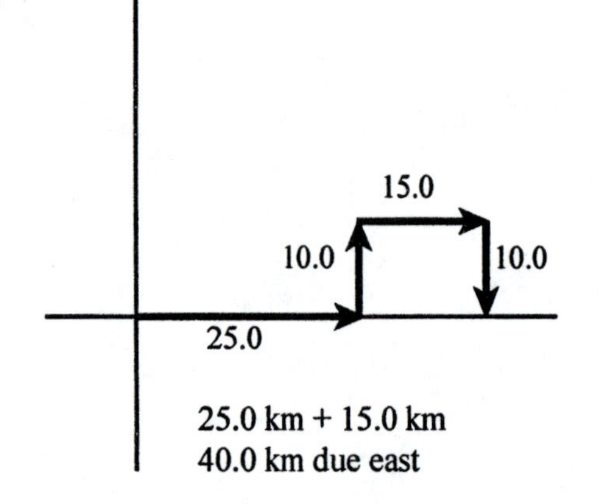

25.0 km + 15.0 km
40.0 km due east

15. Plane $= 175\,\text{mi/h}$ at $270°$
Wind $= 65.0\,\text{mi/h}$ at $180°$
The angle between the plane and wind is $90.0°$.

$|\boldsymbol{R}|^2 = |\boldsymbol{p}|^2 + |\boldsymbol{w}|^2$

$|\boldsymbol{R}|^2 = 175^2 + 65.0^2$

$|\boldsymbol{R}|^2 = 34850$

$|\boldsymbol{R}| = 187\,\text{mi/h}$

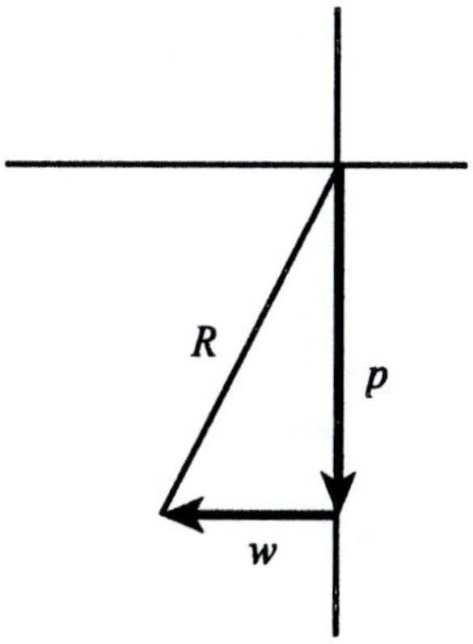

Let ϕ be the angle between $\boldsymbol{R}$ and $\boldsymbol{w}$.

$\tan\phi = \dfrac{|\boldsymbol{p}|}{|\boldsymbol{w}|}$

$\tan\phi = \dfrac{175}{65.0}$

$\phi = 69.6°$

$\theta = 180° + 69.6°$

$\theta = 249.6°$

Exercises 11.7

1. $\boldsymbol{v} = 18.2\,\text{km}$ at $85.0°$

$\boldsymbol{v}_x = 18.2\cos 85.0°$

$\boldsymbol{v}_x = 1.59\,\text{km}$

$\boldsymbol{v}_y = 18.2\sin 85.0°$

$\boldsymbol{v}_y = 18.1\,\text{km}$

3. $\boldsymbol{v} = 135\,\text{mi/h}$ at $270.0°$

$\boldsymbol{v}_x = 135\cos 270.0°$

$\boldsymbol{v}_x = 0$

$\boldsymbol{v}_y = 135\sin 270.0°$

$\boldsymbol{v}_y = -135\,\text{mi/h}$

5. $\boldsymbol{v} = 2680\,\text{ft}$ at $152.5°$

$\boldsymbol{v}_x = 2680\cos 152.5°$

$\boldsymbol{v}_x = -2380\,\text{ft}$

$\boldsymbol{v}_y = 2680\sin 152.5°$

$\boldsymbol{v}_y = 1240\,\text{ft}$

7. $\boldsymbol{v}_x = 8.70\,\text{m}$

$\boldsymbol{v}_y = 6.40\,\text{m}$

$|\boldsymbol{v}| = \sqrt{|\boldsymbol{v}_x|^2 + |\boldsymbol{v}_y|^2}$

$|\boldsymbol{v}| = \sqrt{8.70^2 + 6.40^2}$

$|\boldsymbol{v}| = 10.8\,\text{m}$

$\tan\alpha = \dfrac{|\boldsymbol{v}_y|}{|\boldsymbol{v}_x|}$

$\tan\alpha = \dfrac{6.40}{8.70}$

$\alpha = 36.3°$

Since $\boldsymbol{v}$ is in quandrant I,

$\theta = 36.3°$

9.

$$v_x = 4.70\,\text{m/s}$$
$$v_y = -6.60\,\text{m/s}$$
$$|v| = \sqrt{|v_x|^2 + |v_y|^2}$$
$$|v| = \sqrt{4.70^2 + (-6.60)^2}$$
$$|v| = 8.10\,\text{m/s}$$

$$\tan\alpha = \frac{|v_y|}{|v_x|}$$
$$\tan\alpha = \frac{6.60}{4.70}$$
$$\alpha = 54.5°$$

Since v is in quandrant IV,

$$\theta = 360.0° - 54.5°$$
$$\theta = 305.5°$$

11.

$$v_x = -14.7\,\text{km}$$
$$v_y = 0$$
$$|v| = \sqrt{|v_x|^2 + |v_y|^2}$$
$$|v| = \sqrt{(-14.7)^2 + 0^2}$$
$$|v| = 14.7$$

$$\tan\alpha = \frac{|v_y|}{|v_x|}$$
$$\tan\alpha = \frac{0}{14.7} = 0$$
$$\alpha = 0°$$
$$\theta = 180.0° - 0° = 180.0°$$

13.

$\phi = 2\bar{0}°$ and $|z| = 9\bar{0}\Omega$

$$|R| = 9\bar{0}\cos 2\bar{0}°$$
$$|R| = 85\Omega$$
$$|X_L| = 9\bar{0}\sin 2\bar{0}°$$
$$|X_L| = 31\Omega$$

15.

$$|z| = \sqrt{240^2 + 140^2}$$
$$|z| = 280\Omega$$
$$\tan\phi = \frac{140\Omega}{240\Omega}$$
$$\phi = 3\bar{0}°$$

17.

$v = 324\,\text{ft at } 0°$

$w = 576\,\text{ft at } 90.0°$

$$v_x = 324\cos 0° = 324$$
$$v_y = 324\sin 0° = 0$$
$$w_x = 576\cos 90.0° = 0$$
$$w_y = 576\sin 90.0° = 576$$
$$R_x = v_x + w_x = 324 + 0 = 324$$
$$R_y = v_y + w_y = 0 + 576 = 576$$

$$|R| = \sqrt{324^2 + 576^2}$$
$$|R| = 661\,\text{ft}$$
$$\tan\alpha = \frac{|R_y|}{|R_x|} = \frac{576}{324}$$
$$\alpha = 60.6°$$

Since R is in quadrant I,

$$\theta = 60.6°$$

19.

$v = 28.9\,\text{mi/h at } 52.0°$
$w = 16.2\,\text{mi/h at } 310.0°$
$v_x = 28.9\cos 52.0°$
$v_x = 17.8\,\text{mi/h}$
$v_y = 28.9\sin 52.0°$
$v_y = 22.8\,\text{mi/h}$
$w_x = 16.2\cos 310.0°$
$w_x = 10.4\,\text{mi/h}$
$w_y = 16.2\sin 310.0°$
$w_y = -12.4\,\text{mi/h}$

$R_x = v_x + w_x = 17.8 + 10.4$
$R_x = 28.2\,\text{mi/h}$
$R_y = v_y + w_y = 22.8 + (-12.4)$
$R_y = 10.4\,\text{mi/h}$
$|R| = \sqrt{28.2^2 + 10.4^2}$
$|R| = 30.1\,\text{mi/h}$
$\tan\alpha = \dfrac{10.4}{28.2}$
$\alpha = 20.2°$
Quadrant I, $\theta = 20.2°$

21.

$v = 655\,\text{km at } 108.0°$
$w = 655\,\text{km at } 27.0°$
$u = 655\,\text{km at } 270.0°$
$v_x = 655\cos 108.0° = -202.4$
$v_y = 655\sin 108.0° = 622.9$
$w_x = 655\cos 27.0° = 583.6$
$w_y = 655\sin 27.0° = 297.4$
$u_x = 655\cos 270.0° = 0$
$u_y = 655\sin 270.0° = -655$

$R_x = v_x + w_x + u_x$
$R_x = -202.4 + 583.6 + 0 = 381.2$
$R_y = v_y + w_y + u_y$
$R_y = 622.9 + 297.4 - 655 = 265.3$
$|R| = \sqrt{381.2^2 + 265.3^2}$
$|R| = 464\,\text{km}$
$\tan\alpha = \dfrac{265.3}{381.2}$
$\alpha = 34.8°$
Quadrant I, $\theta = 34.8°$

23.

$v = 5020\,\text{m at } 0°$
$w = 3130\,\text{m at } 148.0°$
$u = 6250\,\text{m at } 65.0°$
$t = 4620\,\text{m at } 335.0°$
$v_x = 5020\cos 0° = 5020$
$v_y = 5020\sin 0° = 0$
$w_x = 3130\cos 148.0° = -2654.4$
$w_y = 3130\sin 148.0° = 1658.6$
$u_x = 6250\cos 65.0° = 2641.4$
$u_y = 6250\sin 65.0° = 5664.4$
$t_x = 4620\cos 335.0° = 4187.1$
$t_y = 4620\sin 335.0° = -1952.5$

$R_x = v_x + w_x + u_x + t_x$
$R_x = 5020 - 2654.4 + 2641.4 + 4187.1$
$R_x = 9190\,\text{m}$
$R_y = v_y + w_y + u_y + t_y$
$R_y = 0 + 1658.6 + 5664.4 - 1952.5$
$R_y = 5370\,\text{m}$
$|R| = \sqrt{9190^2 + 5370^2}$
$|R| = 10{,}600\,\text{m}$
$\tan\alpha = \dfrac{5370}{9190}$
$\alpha = 30.3°$
Quadrant I; $\theta = 30.3°$

25.

$v = 28.9\,\text{mi/h at } 52.0°$

$-w = 16.2\,\text{mi/h at } 310.0° - 180.0°$

$-w = 16.2\,\text{mi/h at } 130.0°$

$v_x = 28.9\cos 52.0° = 17.79$

$v_y = 28.9\sin 52.0° = 22.77$

$-w_x = 16.2\cos 130.0° = -10.41$

$-w_y = 16.2\sin 130.0° = 12.41$

$R_x = v_x + (-w_x)$

$R_x = 17.79 - 10.41$

$R_x = 7.38$

$R_y = v_y + (-w_y)$

$R_y = 22.77 + 12.41$

$R_y = 35.18$

$|R| = \sqrt{7.38^2 + 35.18^2}$

$|R| = 36.0\,\text{mi/h}$

$\tan\alpha = \dfrac{35.18}{7.38}$

$\alpha = 78.2°$, Quadrant I

$\theta = 78.2°$

27.

$p_1 = 165\,\text{mi at } 245°$

$p_2 = 125\,\text{mi at } 165°$

$p_{1x} = 165\cos 245° = -69.73$

$p_{1y} = 165\sin 245° = -149.54$

$p_{2x} = 125\cos 165° = -120.74$

$p_{2y} = 125\sin 165° = 32.35$

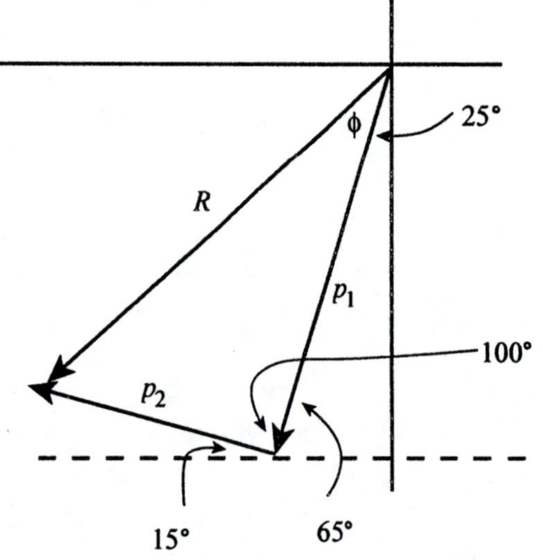

$R_x = p_{1x} + p_{2x} = -69.73 - 120.74$

$R_x = -190.47$

$R_y = p_{1y} + p_{2y} = -149.54 + 32.35$

$R_y = -117.19$

In quadrant III.

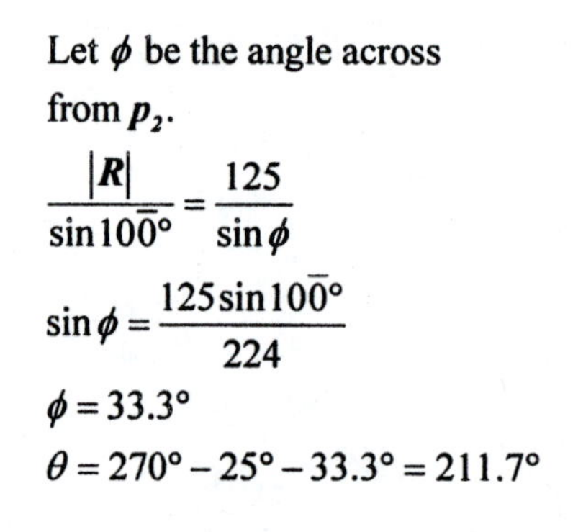

$|R| = \sqrt{(-190.47)^2 + (-117.19)^2}$

$|R| = 223.6$

$|R| = 224\,\text{mi}$

Let ϕ be the angle across from p_2.

$\dfrac{|R|}{\sin 10\bar{0}°} = \dfrac{125}{\sin\phi}$

$\sin\phi = \dfrac{125\sin 10\bar{0}°}{224}$

$\phi = 33.3°$

$\theta = 270° - 25° - 33.3° = 211.7°$

Exercises 11.8

1. $p_1 = 125\,\text{mi}$ at $241.5°$ at

$p_2 = 185\,\text{mi}$ at $270.0°$

$p_{1x} = 125\cos 241.5° = -59.64$

$p_{1y} = 125\sin 241.5° = -109.85$

$p_{2x} = 185\cos 270.0° = 0$

$p_{2y} = 185\sin 270.0° = -185$

$R_x = p_{1x} + p_{2x}$

$R_x = -59.64 + 0 = -59.64$

$R_y = p_{1y} + p_{2y}$

$R_y = -109.85 - 185 = -294.85$

$|R| = \sqrt{|-59.64|^2 + |-294.85|^2}$

$|R| = 301\,\text{mi}$

$\tan\alpha = \dfrac{|R_y|}{|R_x|} = \dfrac{294.85}{59.64}$

$\alpha = 78.6°$

Quadrant III, $\theta = 180.0° + 78.6° = 258.6°$

3. $a = 115\,\text{km}$ at $270.0°$

$p = 195\,\text{km}$ at $225.0°$

$b = 45.0\,\text{km}$ at $180.0°$

$a_x = 115\cos 270.0° = 0$

$a_y = 115\sin 270.0° = -115$

$p_x = 195\cos 225.0° = -137.89$

$p_y = 195\sin 225.0° = -137.89$

$b_x = 45.0\cos 180.0° = -45.0$

$b_y = 45.0\sin 180.0° = 0$

$R_x = a_x + p_x + b_x$

$R_x = 0 - 137.89 - 45.0 = -182.89$

$R_y = a_y + p_y + b_y$

$R_y = -115 - 137.89 + 0 = -252.89$

$|R| = \sqrt{|-182.89|^2 + |-252.89|^2}$

$|R| = 312\,\text{km}$

$\tan\alpha = \dfrac{|R_y|}{|R_x|}$

$\tan\alpha = \dfrac{252.89}{182.89}$

$\alpha = 54.1°$

Quadrant III, so

$\theta = 180.0° + 54.1°$

$\theta = 234.1°$

5. The maximum displacement will occur when the vectors are going in the same direction. Thus the maximum displacement is $10.0\,\text{mi} + 15.0\,\text{mi} = 25.0\,\text{mi}$

7. Let ϕ be the angle opposite the resultant. The supplement of ϕ will be the angle between the vectors.

$12.0^2 = 10.0^2 + 15.0^2 - 2(12.0)(15.0)\cos\phi$

$12.0^2 - 10.0^2 - 15.0^2 = -2(12.0)(15.0)\cos\phi$

$\cos\phi = \dfrac{12.0^2 - 10.0^2 - 15.0^2}{-2(12.0)(15.0)}$

$\phi = 52.9°$

The angle is $180.0° - 52.9° = 127.1°$

9. The maximum displacement will occur when the vectors are going in the same direction. Thus the maximum displacement is $10.0\,\text{km}+15.0\,\text{km}+20.0\,\text{km}=45.0\,\text{km}$

11. **(a)** With the current the speed will be the sum of the boat and current speed, ie 15mi/h.
(b) Against the current the speed will be the difference of the boat and current speed, ie $12\,\text{mi/h}-3\,\text{mi/h}=9\,\text{mi/h}$.

13.

$$p=175\,\text{mi/h at }180.0^\circ$$
$$w=25.0\,\text{mi/h at }90.0^\circ$$
$$p_x=175\cos 180.0^\circ=-175$$
$$p_y=175\sin 180.0^\circ=0$$
$$w_x=25.0\cos 90.0^\circ=0$$
$$w_y=25.0\sin 90.0^\circ=25.0$$
$$R_x=p_x+w_x=-175$$
$$R_y=p_y+w_y=25.0$$
$$|R|=\sqrt{|-175|^2+|25.0|^2}$$
$$|R|=177\,\text{mi/h}$$
$$\tan a=\frac{|R_y|}{|R_x|}=\frac{25.0}{175},\text{ Quadrant II}$$
$$a=8.1^\circ,\ \theta=180.0^\circ-8.1^\circ$$
$$\theta=171.9^\circ$$

15.

$$p=315\,\text{km/h at }125.0^\circ$$
$$w=50.0\,\text{km/h at }0^\circ$$
$$p_x=315\cos 125.0^\circ=-180.68$$
$$p_y=315\sin 125.0^\circ=258.03$$
$$w_x=50.0\cos 0^\circ=50.0$$
$$w_y=50.0\sin 0^\circ=0$$
$$R_x=p_x+w_x=-130.68$$
$$R_y=p_y+w_y=258.03$$
$$|R|=\sqrt{|-130.68|^2+|258.03|^2}$$
$$|R|=289\,\text{km/h}$$
$$\tan\alpha=\frac{|R_y|}{|R_x|}=\frac{258.03}{130.68}$$
$$\alpha=63.1^\circ$$
Quadrant II, thus $\theta=180.0^\circ-63.1^\circ$
$$\theta=116.9^\circ$$

17.

Given: $R=215\,\text{mi/h at }154.5^\circ$

$w=45.6\,\text{mi/h at }180.0^\circ$

Find air speed and heading of plane.

$$R_x=215\cos 154.5^\circ=-194.06$$
$$R_y=215\sin 154.5^\circ=92.56$$
$$w_x=45.6\cos 180.0^\circ=-45.6$$
$$w_y=45.6\sin 180.0^\circ=0$$
$$R_x=p_x+w_x\text{ and }R_y=p_y+w_y$$
$$R_x-w_x=p_x\text{ and }R_y-w_y=p_y$$
$$p_x=-194.06-(-45.6)$$
$$p_x=-148$$
$$p_y=92.56-0=92.6$$
$$|p|=\sqrt{|-148.46|^2+|92.56|^2}$$
$$|p|=175\,\text{mi/h}$$
$$\tan\alpha=\frac{|p_y|}{|p_x|}=\frac{92.6}{148}$$
$\alpha=32.0^\circ$, Quadrant II
$$\theta=180.0^\circ-32.0^\circ=148.0^\circ$$

19. Given: 25.0-lb force at $-35.0°$.

horizontal force $= 25.0\cos(-35.0°)$

$$F_x = 20.5\text{-lb}$$

vertical force $= 25.0\sin(-35.0°)$

$$F_y = -14.3\text{-lb}$$

21. $F = 3550$ lb at $0°$

$G = 1750$ lb at $270°$

$F_x = 3550\cos 0° = 3550$

$F_y = 3550\sin 0° = 0$

$G_x = 1750\cos 270° = 0$

$G_y = 1750\sin 270° = -1750$

$R_x = F_x + G_x = 3550$

$R_y = F_y + G_y = -1750$

$|R| = \sqrt{|3550|^2 + |-1750|^2}$

$|R| = 3960$ lb

$\tan\alpha = \dfrac{|R_y|}{|R_x|} = \dfrac{1750}{3550}$

$\alpha = 26.2°$

Quadrant IV,

$\theta = 360.0° - 26.2°$

$\theta = 333.8°$

23. $F = 275$ lb at $180.0°$

$G = 225$ lb at $215.0°$

$H = 175$ lb at $290.0°$

$F_x = 275\cos 180.0° = -275$

$F_y = 275\sin 180.0° = 0$

$G_x = 225\cos 215.0° = -184.31$

$G_y = 225\sin 215.0° = -129.05$

$H_x = 175\cos 290.0° = 59.85$

$H_y = 175\sin 290.0° = -164.45$

$R_x = F_x + G_x + H_x$

$R_x = -399$

$R_y = F_y + G_y + H_y$

$R_y = -294$

$|R| = \sqrt{|-399|^2 + |-294|^2}$

$|R| = 496$ lb

$\tan\alpha = \dfrac{|R_y|}{|R_x|} = \dfrac{294}{399}$

$\alpha = 36.4°$

Quadrant III, so

$\theta = 180.0° + 36.4° = 216.4°$

25.

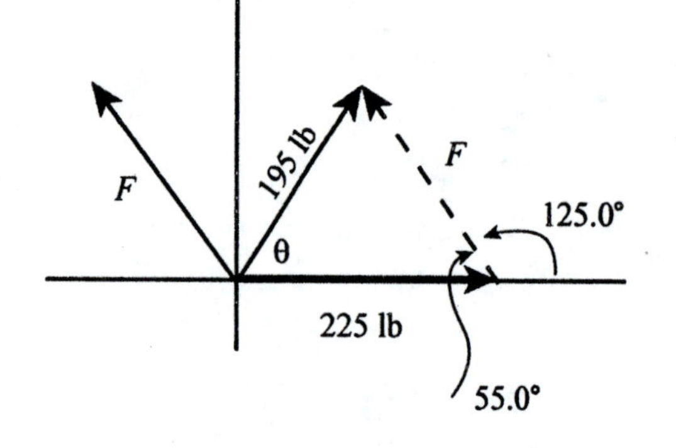

Draw the vectors with the second vector, $\boldsymbol{F}$, starting at the initial point of the first. The angle opposite the resultant is $55.0°$.

$195^2 = 225^2 + |\boldsymbol{F}|^2 - 2(225)(|\boldsymbol{F}|)\cos 55.0°$

$195^2 - 225^2 = |\boldsymbol{F}|^2 - 450(\cos 55.0°)\boldsymbol{F}$

$-12,600 = |\boldsymbol{F}|^2 - 258.109\boldsymbol{F}$

$0 = |\boldsymbol{F}|^2 - 258.109|\boldsymbol{F}| + 12,600$

By the quadratic formula,

$|\boldsymbol{F}| = 192.734$ or 65.375

Let θ be the angle opposite $\boldsymbol{F}$.

If $|\boldsymbol{F}| = 192.734$ lb

$\frac{192.734}{\sin\theta} = \frac{195}{\sin 55.0°}$

$\sin\theta = \frac{192.734 \sin 55.0°}{195}$

$\theta = 54.06°$ or $125.94°$

θ can not be $125.94°$ since $125.94° + 55.0° = 180.95°$, impossible.

The angle between $\boldsymbol{F}$ and the resultant is $180.0° - 54.1° - 55.0°$, ie $70.9°$. The magnitude of $\boldsymbol{F}$; rounded is 193 lb.

If $\boldsymbol{F} = 65.375$ lb, then

$\frac{65.375}{\sin\theta} = \frac{195}{\sin 55.0°}$

$\sin\theta = \frac{65.375 \sin 55.0°}{195}$

$\theta = 15.94°$ or $164.06°$,

θ can not be $164.06°$, since $164.06° + 55.0° > 180.0°$.

The angle between $\boldsymbol{F}$ and the resultant is $180.0° - 15.9° - 55.0° = 109.1°$.

The magnitude of $\boldsymbol{F}$ rounded is 65.4 lb.

27.

$F = 80.0$ lb at $0°$

$G = 30.0$ lb at $270.0°$

$F_x = 80.0\cos 0° = 80.0$

$F_y = 80.0° \sin 0° = 0$

$G_x = 30.0\cos 270.0° = 0$

$G_y = 30.0\sin 270.0° = -30.0$

$R_x = F_x + G_x = 80.0$ lb

$R_y = F_y + G_y = -30.0$ lb

$|R| = \sqrt{|80.0|^2 + |-30.0|^2}$

$|R| = 85.4$ lb

$$\tan\alpha = \frac{|R_y|}{|R_x|} = \frac{30.0}{80.0}$$

$\alpha = 20.6°$

Quadrant IV, $\theta = 360.0° - 20.6°$

$\theta = 339.4°$

The equilibrant force has the same magnitude and opposite direction, so

85.4 lb at $339.4° - 180° = 159.4°$

29.

$F = 375$ lb at $210.0°$

$G = 375$ lb at $240.0°$

$F_x = 375\cos 210.0° = -324.76$

$F_y = 375\sin 210.0° = -187.50$

$G_x = 375\cos 240.0° = -187.50$

$G_y = 375\sin 240.0° = -324.76$

$R_x = -324.76 - 187.50 = -512.26$

$R_y = -187.50 - 324.76 = -512.26$

$|R| = \sqrt{|-512.26|^2 + |-512.26|^2}$

$|R| = 724$ lb

$$\tan\alpha = \frac{|R_y|}{|R_x|} = \frac{512.26}{512.26}$$

$\alpha = 45.0°$, θ in

Quadrant III, $\theta = 225°$.

Equilibrant force is

724 lb at $45.0°$ ie at $225° - 180° = 45.0°$

31.

The tension, T, in the cable must be equal and opposite to the sign's weight, ie the sum of the components is zero. The vertical tension in the cable is

$T_y = |T|\sin 142.0°$

$T_y = |T|(0.61566)$

The vertical tension of the sign is

$S_y = 615\sin 270.0°$

$S_y = -615$

$T_y + S_y = 0$

$0.61566|T| - 615 = 0$

$0.61566|T| = 615$

$|T| = 999$ lb

33.

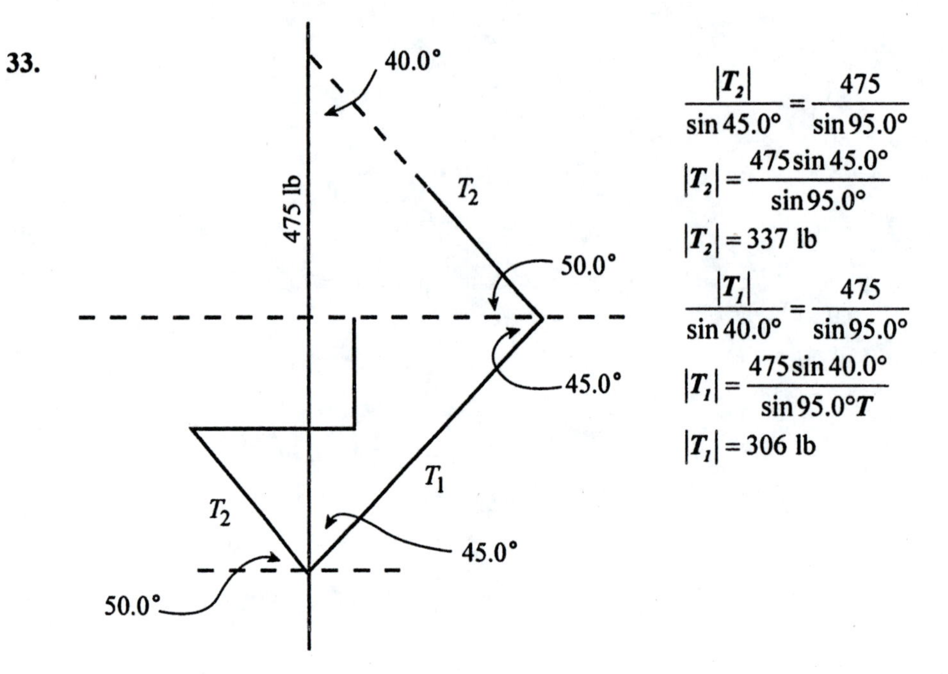

$$\frac{|T_2|}{\sin 45.0°} = \frac{475}{\sin 95.0°}$$
$$|T_2| = \frac{475 \sin 45.0°}{\sin 95.0°}$$
$$|T_2| = 337 \text{ lb}$$
$$\frac{|T_1|}{\sin 40.0°} = \frac{475}{\sin 95.0°}$$
$$|T_1| = \frac{475 \sin 40.0°}{\sin 95.0° T}$$
$$|T_1| = 306 \text{ lb}$$

33. Alternate Solution

Writing the vectors T_1, T_2 and S in components we have:

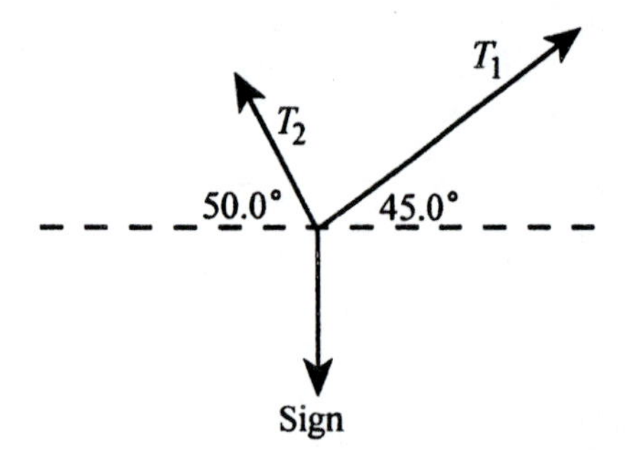

$$S_x = 475 \cos 270° = 0$$
$$S_y = 475 \sin 270° = -475$$
$$T_{1x} = |T_1| \cos 45.0° = 0.707|T_1|$$
$$T_{1y} = |T_1| \cos 45.0° = 0.707|T_1|$$
$$T_{2x} = |T_2| \cos 130.0° = -0.643|T_2|$$
$$T_{2y} = |T_2| \sin 130.0° = 0.766|T_2|$$

We know that

$$T_{1x} + T_{2x} + S_x = 0$$
$$\text{i.e. } 0.707|T_1| - 0.643|T_2| + 0 = 0 \quad (1)$$
$$T_{1y} + T_{2y} + S_y = 0$$
$$\text{i.e. } 0.707|T_1| + 0.766|T_2| - 475 = 0 \quad (2)$$

Solving equations (1) and (2) simultaneously yields:

$$|T_1| = 306.5 \approx 307 \text{ lb}$$
$$|T_2| = 337 \text{ lb}$$

Chapter 11 Review

1. $B = 71.4°, b = 409\text{ ft}, c = 327\text{ ft}.$

$$\frac{c}{\sin C} = \frac{b}{\sin B}$$
$$\frac{327\text{ ft}}{\sin C} = \frac{409}{\sin 71.4°}$$
$$\sin C = \frac{327 \sin 71.4°}{409}$$
$$C = 49.3°$$
$(C \neq 130.7°, \text{ since } 130.7° + 71.4° > 180°.)$

$$A = 180.0° - 49.3° - 71.4°$$
$$A = 59.3°$$
$$\frac{a}{\sin A} = \frac{409}{\sin 71.4°}$$
$$a = \frac{409 \sin 59.3°}{\sin 71.4°}$$
$$a = 371\text{ ft}$$

2. $A = 25.1°, C = 37.7°\ a = 15.7\text{ m}$

$$B = 180.0° - 25.1° - 37.7°$$
$$B = 117.2°$$
$$\frac{c}{\sin C} = \frac{a}{\sin A}$$
$$c = \frac{15.7 \sin 37.7°}{\sin 25.1°}$$
$$c = 22.6\text{ m}$$

$$\frac{b}{\sin B} = \frac{a}{\sin A}$$
$$b = \frac{15.7 \sin 117.2°}{\sin 25.1°}$$
$$b = 32.9\text{ m}$$

3. $A = 15.5°, b = 236\text{ cm}, c = 209\text{ cm}$

$$a^2 = b^2 + c^2 - 2bc\cos A$$
$$a^2 = 236^2 + 209^2 - 2(236)(209)\cos 15.5°$$
$$a^2 = 4316.783$$
$$a = 65.7\text{ cm}$$

$$\frac{c}{\sin C} = \frac{a}{\sin A}$$
$$\sin C = \frac{c \sin A}{a}$$
$$\sin C = \frac{209 \sin 15.5°}{65.7}$$
$$C = 58.2°$$
$$B = 180° - 58.2° - 15.5° = 106.3°$$

4. $a = 25.6\text{ m}, b = 42.2\text{ m}, c = 35.2\text{ m}$

$$a^2 = b^2 + c^2 - 2bc\cos A$$
$$25.6^2 = 42.2^2 + 35.2^2 - 2(42.2)(35.2)\cos A$$
$$\frac{25.6^2 - 42.2^2 - 35.2^2}{-2(42.2)(35.2)} = \cos A$$
$$A = 37.3°$$

$$\frac{a}{\sin A} = \frac{c}{\sin C}$$
$$\sin C = \frac{c \sin A}{a}$$
$$\sin C = \frac{35.2 \sin 37.3°}{25.6}$$
$$C = 56.4°$$
$$B = 180° - 56.4° - 37.3°$$
$$B = 86.3°$$

5. $B = 44°$, $b = 150\,\text{mi}$, $c = 240\,\text{mi}$

$$\frac{c}{\sin C} = \frac{b}{\sin B}$$

$$\frac{240}{\sin C} = \frac{150}{\sin 44°}$$

$$\sin C = \frac{240 \sin 44°}{150}$$

$$\sin C = 1.1115$$

impossible, no triangle

6. $A = 29°$, $a = 41\,\text{cm}$, $b = 49\,\text{cm}$

$$\frac{b}{\sin B} = \frac{a}{\sin A}$$

$$\frac{49}{\sin B} = \frac{41}{\sin 29°}$$

$$\sin B = \frac{49 \sin 29°}{41}$$

$B = 35°$ or $145°$

If $B = 35°$, then

$$C = 180° - 35° - 29° = 116°$$

$$\frac{c}{\sin 116°} = \frac{41}{\sin 29°}$$

$$c = \frac{41 \sin 116°}{\sin 29°}$$

$$c = 76\,\text{cm}$$

If $B = 145°$, then

$$C = 180° - 145° - 29°$$

$$C = 6°$$

$$\frac{c}{\sin 6°} = \frac{41}{\sin 29°}$$

$$c = \frac{41 \sin 6°}{\sin 29°}$$

$$c = 8.8\,\text{cm}$$

7. $C = 36°$ $a = 2100\,\text{ft}$, $b = 3600\,\text{ft}$

$$c^2 = a^2 + b^2 - 2ab\cos C$$

$$c^2 = 2100^2 + 3600^2 - 2(2100)(3600)\cos 36°$$

$$c^2 = 5137663$$

$$c = 2266.6 \approx 2300\,\text{ft}$$

$$\frac{a}{\sin A} = \frac{c}{\sin C}$$

$$\frac{2100}{\sin A} = \frac{2300}{\sin 36°}$$

$$\sin A = \frac{2100 \sin 36°}{2300}$$

$$A = 32°$$

$$B = 180° - 32° - 36°$$

$$B = 112°$$

8. $C = 58°$, $a = 450\text{ m}$, $c = 410\text{ m}$

$$\frac{a}{\sin A} = \frac{c}{\sin C}$$
$$\frac{450}{\sin A} = \frac{410}{\sin 58°}$$
$$\sin A = \frac{450 \sin 58°}{410}$$
$$A = 69° \text{ or } 111°$$

If $A = 69°$, then

$$B = 180° - 69° - 58°$$
$$B = 53°$$
$$\frac{b}{\sin 53°} = \frac{410}{\sin 58°}$$
$$b = \frac{410 \sin 53°}{\sin 58°}$$
$$b = 390\text{ m}$$

If $A = 111°$, then

$$B = 180° - 111° - 58°$$
$$B = 11°$$
$$\frac{b}{\sin 11°} = \frac{410}{\sin 58°}$$
$$b = \frac{410 \sin 11°}{\sin 58°}$$
$$b = 92\text{ m}$$

9. $B = 105.15°$, $a = 231.1\text{m}$, $c = 190.7\text{ m}$

$$b^2 = a^2 + c^2 - 2ac\cos B$$
$$b^2 = 231.1^2 + 190.7^2 - 2(231.1)(190.7)\cos 105.15°$$
$$b^2 = 112809$$
$$b = 335.8708 \approx 335.9\text{ m}$$
$$\frac{c}{\sin C} = \frac{335.9}{\sin 105.15°}$$
$$\sin C = \frac{190.7 \sin 105.15°}{335.9}$$
$$C = 33.23°$$
$$A = 180.00° - 105.15° - 33.23°$$
$$A = 41.62°$$

10. $A = 74.75°$, $a = 22.19\text{ cm}$, $c = 15.28\text{ cm}$

$$\frac{c}{\sin C} = \frac{a}{\sin A}$$
$$\frac{15.28}{\sin C} = \frac{22.19}{\sin 74.75°}$$
$$\sin C = \frac{15.28 \sin 74.75°}{22.19}$$
$$C = 41.63° \text{ or } 138.37°$$

note, 138.37° is impossible

$$B = 180.00° - 41.63° - 74.75°$$
$$B = 63.62°$$
$$\frac{b}{\sin B} = \frac{22.19}{\sin 74.75°}$$
$$b = \frac{22.19 \sin 63.62°}{\sin 74.75°}$$
$$b = 20.60\text{ cm}$$

11. $B = 18.25°$ $a = 1675\text{ ft}$, $b = 1525\text{ ft}$

$$\frac{a}{\sin A} = \frac{b}{\sin B}$$

$$\frac{1675}{\sin A} = \frac{1525}{\sin 18.25°}$$

$$\sin A = \frac{1675 \sin 18.25°}{1525}$$

$A = 20.12°$ or $159.88°$

If $A = 20.12°$, then

$$C = 180.00° - 20.12° - 18.25°$$

$$C = 141.63°$$

$$\frac{c}{\sin C} = \frac{b}{\sin B}$$

$$c = \frac{1525 \sin 141.63°}{\sin 18.25°}$$

$$c = 3023\text{ ft}$$

If $A = 159.88°$, then

$$C = 180.00° - 159.88° - 18.25°$$

$$C = 1.87°$$

$$\frac{c}{\sin 1.87°} = \frac{1525}{\sin 18.25°}$$

$$c = 158.9\text{ ft}$$

12. $C = 40.16°$, $b = 25{,}870\text{ ft}$, $c = 10{,}250\text{ ft}$

$$\frac{b}{\sin B} = \frac{c}{\sin C}$$

$$\frac{25{,}870}{\sin B} = \frac{10{,}250}{\sin 40.16°}$$

$$\sin B = \frac{25{,}870 \sin 40.16°}{10{,}250}$$

$\sin B = 1.628$, impossible;

thus, no triangle

13. $A = 48°15'35''$, $B = 68°7'18''$,

$a = 2755\text{ ft}$

$$C = 180° - 48°15'35'' - 68°7'18''$$

$$C = 63°37'7''$$

$$\frac{b}{\sin B} = \frac{a}{\sin A}$$

$$b = \frac{2755 \sin 68°7'18''}{\sin 48°15'35''}$$

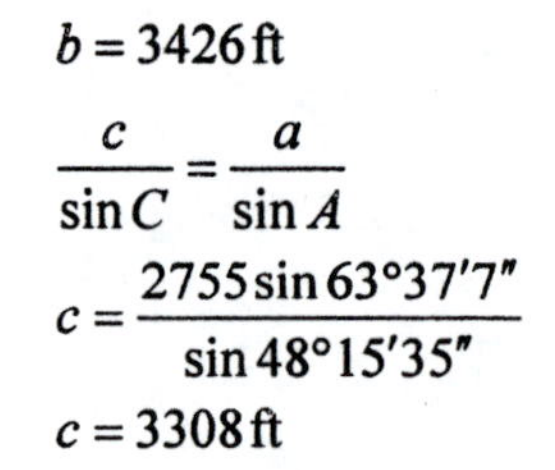

$$b = 3426\text{ ft}$$

$$\frac{c}{\sin C} = \frac{a}{\sin A}$$

$$c = \frac{2755 \sin 63°37'7''}{\sin 48°15'35''}$$

$$c = 3308\text{ ft}$$

14. $C = 29°25'16''$, $a = 13{,}560\,\text{ft}$,

$b = 24{,}140\,\text{ft}$

$c^2 = a^2 + b^2 - 2ab\cos C$

$c^2 = 13{,}560^2 + 24{,}140^2 - 2(13{,}560)(24{,}140)\cos 29°25'16''$

$c^2 = 196368185$

$c = 14013 \approx 14010\,\text{ft}$

$$\frac{a}{\sin A} = \frac{c}{\sin C}$$

$$\frac{13{,}560}{\sin A} = \frac{14010}{\sin 29°25'16''}$$

$$\sin A = \frac{13560 \sin 29°25'16''}{14010}$$

$A = 28°23'18''$

$B = 180° - 28°23'18'' - 29°25'16''$

$B = 122°11'26''$

15. $AC^2 = AB^2 + BC^2 - 2(AB)(BC)\cos 43.25°$

$AC^2 = 165\bar{0}^2 + 1475^2 - 2(165\bar{0})(1475)\cos 43.25°$

$AC^2 = 1352779$

$AC = 1163\,\text{ft}$

$\text{Fence} = 165\bar{0} + 1475 + 1163 = 4288\,\text{ft}$

16. **(a)** $\angle ACB = 180.00° - 79.55° - 85.00°$

$\angle ACB = 15.45°$

$$\frac{AC}{\sin 85.00°} = \frac{300.0}{\sin 15.45°}$$

$$AC = \frac{300.0 \sin 85.00°}{\sin 15.45°}$$

$AC = 1122\,\text{ft}$

Let x be the perpendicular distance. Then

$$\sin 79.55° = \frac{x}{1122}$$

$x = 1103\,\text{ft}$

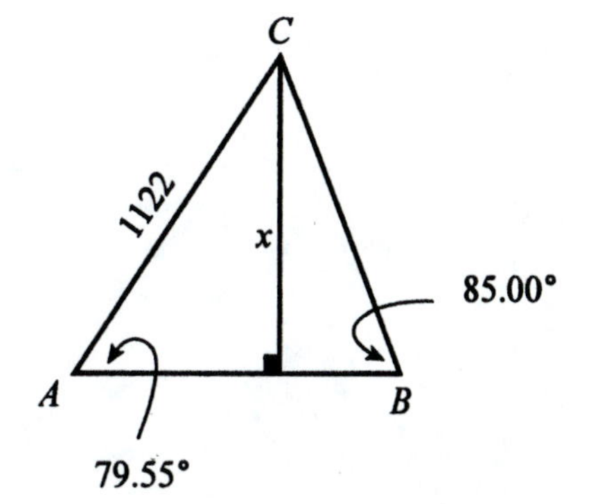

17. $\overline{AC}^2 = AB^2 + BC^2 - 2(AB)(BC)\cos 120.0°$

$AC^2 = 4.350^2 + 5.149^2 - 2(4.350)(5.149)\cos 120.0°$

$AC^2 = 67.832851$

$AC = 8.236\,\text{in.}$

18. Let W be the center of the circle. Then

$\angle AWB = \dfrac{360°}{5} = 72.00°$

$WA = WB = \dfrac{5.000}{2} = 2.500\text{-in.}$

(a) $AB^2 = WA^2 + WB^2 - 2(WA)(WB)\cos 72.00°$

$AB^2 = 2.500^2 + 2.500^2 - 2(2.500)(2.500)\cos 72.00°$

$AB^2 = 8.63728757$

$AB = 2.939\text{-in.}$

(b) $\angle AWC = 2(72.00) = 144.00°$

$AC^2 = WA^2 + WB^2 - 2(WA)(WB)\cos 144.00°$

$AC^2 = 2.500^2 + 2.500^2 - 2(2.500)(2.500)\cos 144.00°$

$AC^2 = 22.61271243$

$AC = 4.755\text{-in.}$

19. From the given information we can determine the following.

$\angle ABF = 180.0° - 42.0° = 138.0°$

$\angle AFB = 180.0° - 138.0° - 30.0° = 12.0°$

$\angle BFC = 90.0° - 42.0° = 48.0°$

$\Delta ACF \cong \Delta ECF$ (side, angle, side)

$\angle AFC = \angle CFE = 12.0° + 48.0° = 60.0°$

(or $\angle AFC = 90.0° - 30.0° = 60.0°$)

Since $AC = CE$, $AC = \dfrac{1}{2}AE = 10.0\,\text{m}$

(a) $\cos 30.0° = \dfrac{AC}{AF}$

$AF = \dfrac{10.0\,\text{m}}{\cos 30.0°}$

$AF = 11.54700538$

$AF = 11.5\,\text{m}$

(b) $\dfrac{BF}{\sin 30.0°} = \dfrac{AF}{\sin 138.0°}$

$BF = \dfrac{11.547\sin 30.0°}{\sin 138.0°}$

$BF = 8.63\,\text{m}$

Note: Using a rounded answer from part (a) will give a slightly different answer.

(c) $\tan 30.0° = \dfrac{CF}{10.0\,\text{m}}$

$CF = 5.77\,\text{m}$

(d) $\dfrac{AB}{\sin 12.0°} = \dfrac{AF}{\sin 138.0°}$

$AB = \dfrac{11.547\sin 12.0°}{\sin 138.0°}$

$AB = 3.59\,\text{m}$

$BC = 10.0\,\text{m} - 3.59\,\text{m}$

$BC = 6.41\,\text{m}$

20.

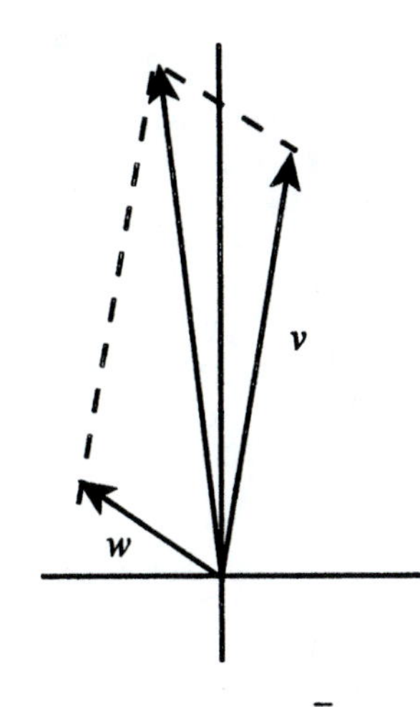

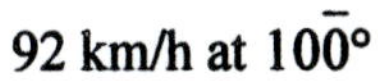

92 km/h at $10\bar{0}°$

21. $v = 126\,\text{mi at } 35.0°$

$w = 306\,\text{mi at } 180.0°$

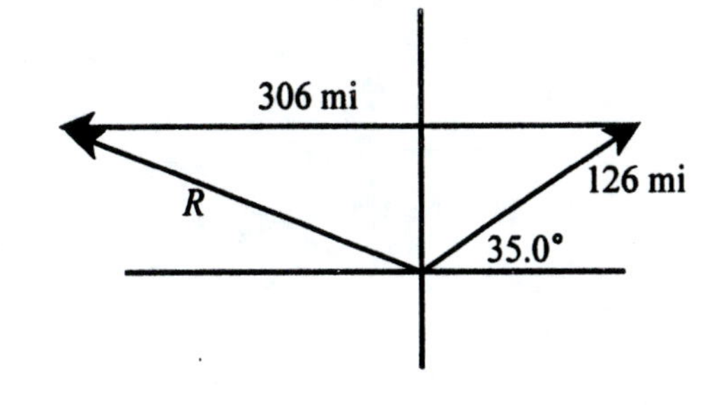

$|R|^2 = 306^2 + 126^2 - 2(306)(126)\cos 35.0°$

$|R|^2 = 46345.54$

$|R| = 215\,\text{mi}$

Let ϕ be the angle across from v.

$$\frac{126}{\sin\phi} = \frac{215}{\sin 35°}$$

$\phi = 19.6°$. The angle across from w is

$180° - 19.6° - 35.0° = 125.4°$

$\theta = 35.0° + 125.4° = 160.4°$

22. $v = 89.4\,\text{mi/h at } 142.0°$

$w = 44.7\,\text{mi/h at } 322.0°$

These vectors differ by 180° and thus have opposite directions.

The magnitude will be $89.4 - 44.7 = 44.7$.

Since $|v| > |w|$, the resultant has the same direction as v, ie 142.0°.

23. $258\,\text{km at } 135.0°$

horizontal: $258\cos 135.0° = -182\,\text{km}$

vertical: $258\sin 135.0° = 182\,\text{km}$

24. $42.2\,\text{mi/h at } 303.0°$

horizontal: $42.2\cos 303.0° = 23.0\,\text{mi/h}$

vertical: $42.2\sin 303.0° = -35.4\,\text{mi/h}$

25. $160N$ at $-45°$

horizontal: $160\cos(-45°) = 110N$

vertical: $160\sin(-45°) = -110N$

26. $110\,\text{lb}$ at $23°$

horizontal : $110\cos 23° = 1\bar{0}0\,\text{lb}$

vertical : $110\sin 23° = 43\,\text{lb}$

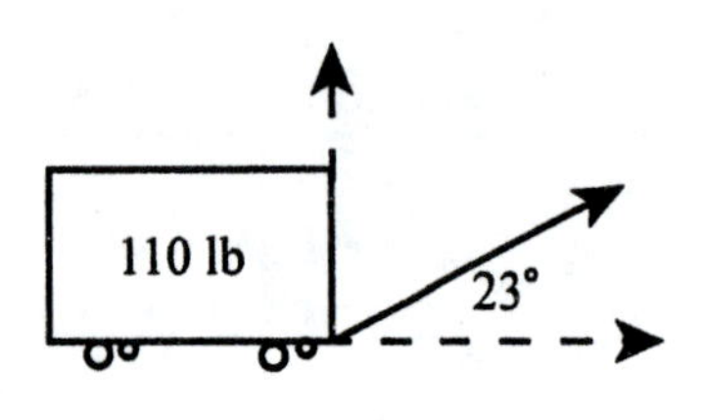

27. $v_x = 18.5N$ and $v_y = -31.0N$

$|v| = \sqrt{v_x^{\,2} + v_y^{\,2}}$

$|v| = \sqrt{18.5^2 + (-31.0)^2}$

$|v| = 36.1N$

$\tan\alpha = \dfrac{|v_y|}{|v_x|} = \dfrac{31.0}{18.5}$

$\alpha = 59.2°$, in IV, so

$\theta = 360° - 59.2°$

$\theta = 300.8°$

28. $v = 87.1\,\text{mi/h}$ at $120.0°$

$w = 25.6\,\text{mi/h}$ at $247.0°$

$v_x = 87.1\cos 120.0° = -43.55$

$v_y = 87.1\sin 120.0° = 75.43$

$w_x = 25.6$ at $\cos 247.0° = -10.00$

$w_y = 25.6$ at $\sin 247.0° = -23.56$

$R_x = -53.55$

$R_y = 51.87$

$|R| = \sqrt{|-53.55|^2 + |51.87|^2}$

$|R| = 74.6\,\text{mi/h}$

$\tan\alpha = \dfrac{|R_y|}{|R_x|} = \dfrac{51.87}{53.55}$

$\alpha = 44.1°$

Quadrant II, $\theta = 135.9°$

29. $v = 2560N$ at $237.1°$

$w = 3890N$ at $346.7°$

$v_x = 2560\cos 237.1° = -1390.53N$

$v_y = 2560\sin 237.1° = -2149.43N$

$w_x = 3890\cos 346.7° = 3785.67N$

$w_y = 3890\sin 346.7° = -894.89N$

$R_x = -1390.53 + 3785.67 = 2395.14$

$R_y = -2149.43 - 894.89 = -3044.32$

$|R| = \sqrt{|2395.14|^2 + |-3044.32|^2}$

$|R| = 3870N$

$\tan\alpha = \dfrac{|R_y|}{|R_x|}$

$\tan\alpha = \dfrac{3044.32}{2395.14}$

$\alpha = 51.8°$; in Quadrant IV,

$\theta = 308.2°$

30.

$u = 325N$ at $90.0°$

$v = 325N$ at $162.0°$

$w = 325N$ at $270.0°$

Notice that u and w cancel each other out - the answer is $v = 325N$ at $162.0°$.

Using components:

$u_x = 325\cos 90.0° = 0$

$u_y = 325\sin 90.0° = 325$

$v_x = 325\cos 162.0° = 309.1$

$v_y = 325\sin 162.0° = 100.4$

$w_x = 325\cos 270.0° = 0$

$w_y = 325\sin 270.0° = -325$

$R_x = 0 - 309.1 + 0 = -309.1$

$R_y = 325 + 100.4 - 325 = 100.4$

$|R| = \sqrt{|-309.1|^2 + |100.4|^2}$

$|R| = 325N$

$\tan\alpha = \dfrac{|R_y|}{|R_x|} = \dfrac{100.4}{309.1}$

$\alpha = 18.0°$, in Quadrant II,

$\theta = 162.0°$

31.

$v = 19.7\text{km}$ at $144.5°$

$w = 28.5\text{km}$ at $180.0°$

$u = 10.3\text{km}$ at $225.5°$

$t = 31.7\text{km}$ at $90.0°$

$v_x = 19.7\cos 144.5° = -16.0381$

$v_y = 19.7\sin 144.5° = 11.4398$

$w_x = 28.5\cos 180.0° = -28.5$

$w_y = 28.5\sin 180.0° = 0$

$u_x = 10.3\cos 225.5° = -7.2194$

$u_y = 10.3\sin 225.5° = -7.3465$

$t_x = 31.7\cos 90.0° = 0$

$t_y = 31.7\sin 90.0° = 31.7$

$R_x = v_x + w_x + u_x + t_x = -51.8$

$R_y = v_y + w_y + u_y + t_y = 35.8$

$|R| = \sqrt{|-51.8|^2 + |35.8|^2}$

$|R| = 63.0\text{km}$

$\tan\alpha = \dfrac{|R_y|}{|R_x|} = \dfrac{35.8}{51.8}$

$\alpha = 34.6°$, in Quadrant II,

$\theta = 145.4°$

32.

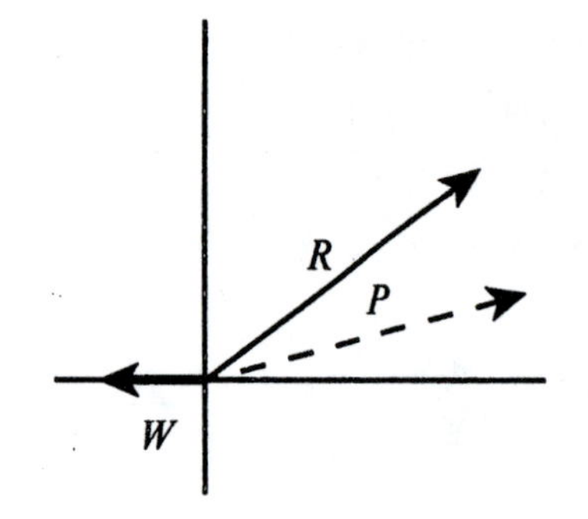

Let $\boldsymbol{R}$ be the resultant of the plane, $\boldsymbol{P}$, and the wind, $\boldsymbol{W}$.

$\boldsymbol{R} = 550\,\text{km/h at } 3\bar{0}°.$

$\boldsymbol{W} = 9\bar{0}\,\text{km/hr at } 180°$

$\boldsymbol{R}_x = 550\cos 3\bar{0}° = 476.314$

$\boldsymbol{R}_y = 550\sin 3\bar{0}° = 275$

$\boldsymbol{w}_x = 9\bar{0}\cos 180° = -9\bar{0}$

$\boldsymbol{w}_y = 9\bar{0}\sin 180° = 0$

$\boldsymbol{P} + \boldsymbol{W} = \boldsymbol{R}$ so

$\boldsymbol{P} = \boldsymbol{R} - \boldsymbol{W}$

$\boldsymbol{P}_x = 476.314 - (-9\bar{0})$

$\boldsymbol{P}_x = 566.314\,\text{km/h}$

$\boldsymbol{P}_y = 275 - 0 = 275\,\text{km/h}$

$|\boldsymbol{P}| = \sqrt{|566.314|^2 + |275|^2}$

$|\boldsymbol{P}| = 630\,\text{km/h}$

$\tan\alpha = \dfrac{|\boldsymbol{P}_y|}{|\boldsymbol{P}_x|} = \dfrac{275}{566.314}$

$\alpha = 26°$; Quadrant I, $\theta = 26°$

33.

$\boldsymbol{u} = 16.5\,\text{mi at } 76.5°$

$\boldsymbol{v} = 24.7\,\text{mi at } 124.5°$

$\boldsymbol{w} = 30.5\,\text{mi at } 90.0°$

$\boldsymbol{u}_x = 16.5\cos 76.5° = 3.85185$

$\boldsymbol{u}_y = 16.5\sin 76.5° = 16.04410$

$\boldsymbol{v}_x = 24.7\cos 124.5° = -13.99023$

$\boldsymbol{v}_y = 24.7\sin 124.5° = 20.3559$

$\boldsymbol{w}_x = 30.5\cos 90.0° = 0$

$\boldsymbol{w}_y = 30.5\sin 90.0° = 30.5$

$\boldsymbol{R}_x = \boldsymbol{u}_x + \boldsymbol{v}_x + \boldsymbol{w}_x = -10.13838$

$\boldsymbol{R}_y = \boldsymbol{u}_y + \boldsymbol{v}_y + \boldsymbol{w}_y = 66.9$

$|\boldsymbol{R}| = \sqrt{|-10.1|^2 + |66.9|^2}$

$|\boldsymbol{R}| = 67.7\,\text{mi}$

$\tan\alpha = \dfrac{|\boldsymbol{R}_y|}{|\boldsymbol{R}_x|} = \dfrac{66.9}{10.1}$

$\alpha = 81.4°$; in Quadrant II,

$\theta = 98.6°$

34.

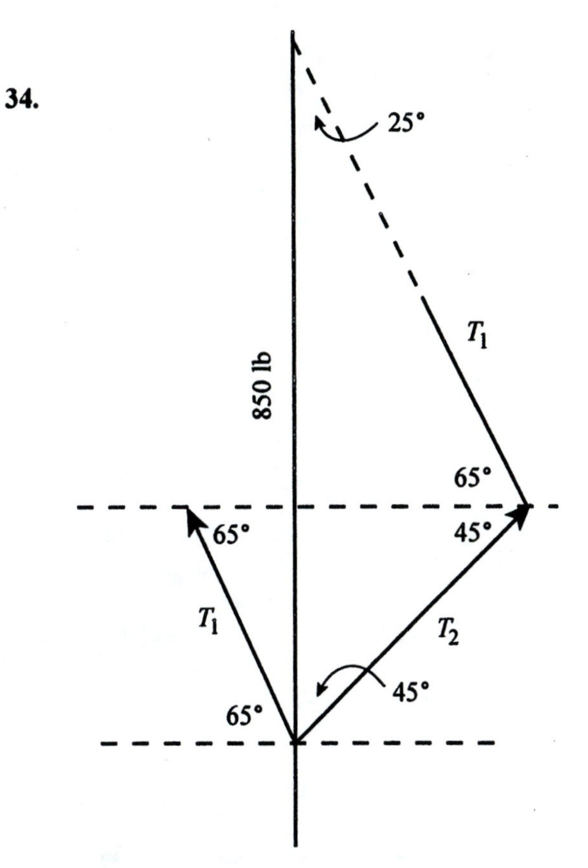

E = equilibrant force

$$\frac{850}{\sin(65° + 45°)} = \frac{|\boldsymbol{T_1}|}{\sin 45°}$$

$$|\boldsymbol{T_1}| = \frac{850 \sin 45°}{\sin 110°}$$

$$|\boldsymbol{T_1}| = 640 \text{ lb}$$

$$\frac{850}{\sin 110°} = \frac{|\boldsymbol{T_2}|}{\sin 25°}$$

$$|\boldsymbol{T_2}| = \frac{850 \sin 25°}{\sin 110°}$$

$$|\boldsymbol{T_2}| = 380 \text{ lb}$$

34. Alternate Solution

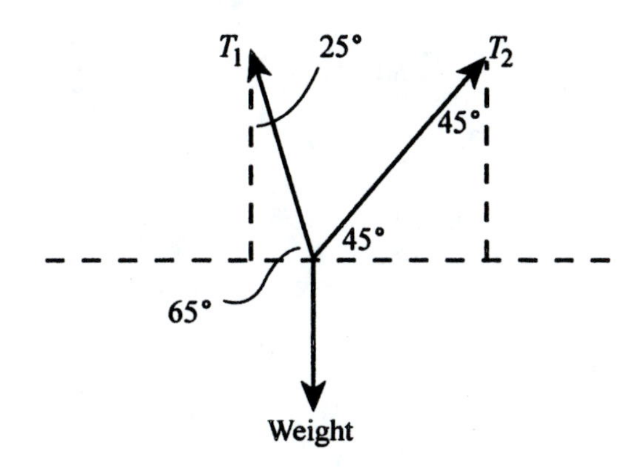

Writing the vectors $\boldsymbol{T_1}$, $\boldsymbol{T_2}$, and $\boldsymbol{W}$ in components we have

$$\boldsymbol{W_x} = 850 \cos 270° = 0$$

$$\boldsymbol{W_y} = 850 \sin 270° = -850$$

$$\boldsymbol{T_{1x}} = |\boldsymbol{T_1}| \cos 115° = -0.4226 |\boldsymbol{T_1}|$$

$$\boldsymbol{T_{1y}} = |\boldsymbol{T_1}| \sin 115° = -0.9063 |\boldsymbol{T_1}|$$

$$\boldsymbol{T_{2x}} = |\boldsymbol{T_2}| \cos 45° = 0.7071 |\boldsymbol{T_2}|$$

$$\boldsymbol{T_{2y}} = |\boldsymbol{T_2}| \sin 45° = 0.7071 |\boldsymbol{T_2}|$$

We know that $\boldsymbol{T_{1x}} + \boldsymbol{T_{2x}} + \boldsymbol{W_x} = 0$ and $\boldsymbol{T_{1y}} + \boldsymbol{T_{2y}} + \boldsymbol{W_y} = 0$

i.e. $-0.4226|\boldsymbol{T_1}| + 0.7071|\boldsymbol{T_2}| + 0 = 0$ (1)

and $0.9063|\boldsymbol{T_1}| + 0.7071|\boldsymbol{T_2}| - 850 = 0$ (2)

Solving equations (1) and (2) simultaneously,

$\boldsymbol{T_1} = 640$ lb

$\boldsymbol{T_2} = 380$ lb

Chapter 12, Exercises 12.1

Graphs appear in the text answer section.

1. $y = 2\cos x$
amp = 2
period = 2π

3. $y = -3\sin x$
amp = $|-3| = 3$
period = 2π

5. $y = \sin 3x$
amp = 1
period = $\frac{2\pi}{3}$

7. $y = \cos 2x$
amp = 1
period = $\frac{2\pi}{2} = \pi$

9. $y = 2\sin 4x$
amp = 2
period = $\frac{2\pi}{4} = \frac{\pi}{2}$

11. $y = \frac{5}{2}\cos\frac{1}{2}x$
amp = $\frac{5}{2}$
period = $\frac{2\pi}{\frac{1}{2}} = 4\pi$

13. $y = \frac{-1}{2}\sin\frac{2}{3}x$
amp = $\left|-\frac{1}{2}\right| = \frac{1}{2}$
period = $\frac{2\pi}{\frac{2}{3}} = 3\pi$

15. $y = 2\sin 3\pi x$
amp = 2
period = $\frac{2\pi}{3\pi} = \frac{2}{3}$

17. $y = -3\cos \pi x$
amp = $|-3| = 3$
period = $\frac{2\pi}{\pi} = 2$

19. $y = 6.5\sin 120\pi x$
amp = 6.5
period = $\frac{2\pi}{120\pi} = \frac{1}{60}$

21. $y = 40\cos 60x$
amp = 40
period = $\frac{2\pi}{60} = \frac{\pi}{30}$

23. $y = -60\sin 80\pi x$
amp = $|-60| = 60$
period = $\frac{2\pi}{80\pi} = \frac{1}{40}$

Exercises 12.2

Graphs appear in the text answer section.

1. $y = \sin\left(x + \frac{\pi}{3}\right)$
amp = 1
period = 2π
phase shift = $\frac{\pi}{3}$, left

3. $y = 2\cos\left(x - \frac{\pi}{6}\right)$
amp = 2
period = 2π
phase shift = $\frac{-\pi}{6}$, right

5. $y = \sin(3x - \pi)$
amp = 1
period = $\frac{2\pi}{3}$
phase shift = $\frac{-\pi}{3}$, right

7. $y = -\cos(4x + \pi)$
amp = $|-1| = 1$
period = $\frac{2\pi}{4} = \frac{\pi}{2}$
phase shift = $\frac{\pi}{4}$, left

9. $y = 3\sin\left(\frac{1}{2}x - \frac{\pi}{4}\right)$

amp = 3

period $= \frac{2\pi}{\frac{1}{2}} = 4\pi$

phase shift $= \frac{-\pi}{4} \div \frac{1}{2}$

phase shift $= \frac{-\pi}{2}$

right

11. $y = 2\sin\left(\frac{4}{3}x + \frac{\pi}{3}\right)$

amp = 2

period $= \frac{2\pi}{\frac{4}{3}} = \frac{3\pi}{2}$

phase shift $= \frac{\pi}{3} \div \frac{4}{3}$

phase shift $= \frac{\pi}{4}$

left

13. $y = 3\sin(\pi x + \pi)$

amp = 3

period $= \frac{2\pi}{\pi} = 2$

phase shift $= \frac{\pi}{\pi} = 1$

left

15. $y = 2\cos\left(\frac{\pi x}{2} - \frac{\pi}{2}\right)$

amp = 2

period $= \frac{2\pi}{\frac{\pi}{2}} = 4$

phase shift $= \frac{-\frac{\pi}{2}}{\frac{\pi}{2}} = -1$

right

17. $y = 40\cos\left(60x - \frac{\pi}{3}\right)$

amp = 40

period $= \frac{2\pi}{60} = \frac{\pi}{30}$

phase shift $= \frac{\frac{-\pi}{3}}{60} = \frac{-\pi}{180}$

right

19. $y = 120\cos\left(40\pi x - \frac{\pi}{2}\right)$

amp = 120

period $= \frac{2\pi}{40\pi} = \frac{1}{20}$

phase shift $= \frac{-\frac{\pi}{2}}{40\pi} = \frac{-1}{80}$

right

21. $y = 7.5\sin(220x + \pi)$

amp = 7.5

period $= \frac{2\pi}{220} = \frac{\pi}{110}$

phase shift $= \frac{\pi}{220}$

left

23. $y = 20\sin(120\pi x + 4\pi)$

amp = 20

period $= \frac{2\pi}{120\pi} = \frac{1}{60}$

phase shift $= \frac{4\pi}{120\pi} = \frac{1}{30}$

left

Exercises 12.3

Graphs appear in the text answer section.

1. $y = \tan 3x$

period $= \dfrac{\pi}{3}$

3. $y = -2\tan\dfrac{1}{2}x$

period $= \dfrac{\pi}{\frac{1}{2}} = 2\pi$

5. $y = \cot 6x$

period $= \dfrac{\pi}{6}$

7. $y = 3\sec x$

period $= 2\pi$

9. $y = -2\csc x$

period $= 2\pi$

11. $y = 2\sec 6x$

period $= \dfrac{2\pi}{6} = \dfrac{\pi}{3}$

13. $y = -3\sec\dfrac{1}{2}x$

period $= \dfrac{2\pi}{\frac{1}{2}} = 4\pi$

15. $y = \csc 2x$

period $= \dfrac{2\pi}{2} = \pi$

17. $y = \tan\left(x + \dfrac{\pi}{2}\right)$

period $= \pi$

phase shift$= \dfrac{\pi}{2}$

left

19. $y = 2\sec\left(2x - \dfrac{\pi}{2}\right)$

period $= \dfrac{2\pi}{2} = \pi$

phase shift $= \dfrac{\frac{-\pi}{2}}{2} = \dfrac{-\pi}{4}$

right

Exercises 12.4

Graphs for these exercises are in the answer section of the text.

Exercises 12.5

Graphs appear in the text answer section.

1. From $y = r\sin(\omega t + \theta)$
$r = 8\text{ in.}, \omega = 10\pi \text{ rad/s}$
There is no phase shift.
$y = 8\sin(10\pi t)$

3. $i = I\sin(\omega t + \theta)$
$I = 15A$
Note: 60 cycles/s is
$\frac{60 \text{ cycles}}{\text{s}} \cdot \frac{2\pi \text{ rad}}{\text{cycle}}$ which
is 120π rad/s.

$\omega = 120\pi, \theta = \frac{\pi}{4}$

$i = 15\sin\left(120\pi t + \frac{\pi}{4}\right)$

5. $e = E\cos(\omega t + \theta)$
$E = 110V, \omega = 60$ cycles/s,
ie $\omega = 120\pi$ rad/s, $\theta = 0°$.
$e = 110\cos(120\pi t)$

7. From $y = r\sin(\omega t + \theta)$,

$r = 10\text{ cm}, \theta = \frac{-\pi}{2}$.

1 complete cycle in 0.80 s is
2π rad in 0.80 s, which is

$\frac{2\pi \text{ rad}}{0.80s} = \frac{5\pi}{2} \text{ rad/s}$

$y = 10\sin\left(\frac{5\pi}{2}t - \frac{\pi}{2}\right)$

9. $y = r\sin(\omega t + \theta)$
$r = 0.25\text{ cm}$

$\frac{200 \text{ vibrations}}{\text{s}} = \frac{200(2\pi)\text{ rad}}{\text{s}}$

$y = 0.25\sin 400\pi t$

Chapter 12 Review

Graphs appear in the text answer section.

1. $y = 4\cos 6x$

amp $= 4$

period $= \frac{2\pi}{6} = \frac{\pi}{3}$

2. $y = -2\sin\frac{1}{3}x$

amp $= |-2| = 2$

period $= \frac{2\pi}{\frac{1}{3}} = 6\pi$

3. $y = 3\cos 2\pi x$

amp $= 3$

period $= \frac{2\pi}{2\pi} = 1$

4. $y = 3\sin\left(x - \frac{\pi}{4}\right)$

amp $= 3$

period $= 2\pi$

phase shift $= \frac{-\pi}{4}$

right

5. $y = \cos\left(2x + \frac{2\pi}{3}\right)$

amp $= 1$

period $= \frac{2\pi}{2} = \pi$

phase shift $= \frac{2\pi}{3} \div 2 = \frac{\pi}{3}$

left

6. $y = 4\sin\left(\pi x + \frac{\pi}{2}\right)$

amp $= 4$

period $= \frac{2\pi}{\pi} = 2$

phase shift $= \frac{2\pi}{\pi} \div \pi = \frac{1}{2}$

left

7. $y = \tan 5x$

period $= \frac{\pi}{5}$

8. $y = -\cot 3x$

period $= \frac{\pi}{3}$

9. $y = 2\sec 4x$

period $= \frac{2\pi}{4} = \frac{\pi}{2}$

10. $y = 3\sin x + 2\cos 2x$

11. $y = 2\sin x + \cos\frac{x}{2}$

12. $y = \sin 2x - 2\cos\frac{x}{2}$

13. $y = 6\cos 12t$

amplitude 6,

period $\frac{\pi}{6}$,

no phase shift.

14. $e = 220\sin\left(120\pi t + \frac{\pi}{4}\right)$

Chapter 13, Exercises 13.1

1.
$$\frac{1}{\csc\theta} = \sin\theta$$
$$\frac{1}{\csc\theta} = \frac{1}{\frac{r}{y}} = \frac{y}{r} = \sin\theta$$

3.
$$\frac{\cos\theta}{\sin\theta} = \cot\theta$$
$$\frac{\cos\theta}{\sin\theta} = \frac{\frac{x}{r}}{\frac{y}{r}} = \frac{x}{y} = \cot\theta$$

5.
$$\cos\theta\sec\theta = 1$$
$$\cos\theta\left(\frac{1}{\cos\theta}\right) = 1$$

7.
$$\cos x\tan x = \sin x$$
$$\cos x\left(\frac{\sin x}{\cos x}\right) = \sin x$$

9.
$$\frac{\csc\theta}{\sec\theta} = \cot\theta$$
$$\frac{\csc\theta}{\sec\theta} = \frac{\frac{1}{\sin\theta}}{\frac{1}{\cos\theta}} = \frac{\cos\theta}{\sin\theta} = \cot\theta$$

11.
$$\frac{\tan\theta}{\sin\theta} = \sec\theta$$
$$\frac{\tan\theta}{\sin\theta} = \frac{\frac{\sin\theta}{\cos\theta}}{\sin\theta} = \frac{1}{\cos\theta} = \sec\theta$$

13.
$$\sec\theta\cot\theta = \csc\theta$$
$$\sec\theta\cot\theta = \frac{1}{\cos\theta}\cdot\frac{\cos\theta}{\sin\theta} = \frac{1}{\sin\theta} = \csc\theta$$

15.
$$\left(1-\cos^2 x\right)\csc^2 x = 1$$
$$\left(1-\cos^2 x\right)\csc^2 x = \sin^2 x\csc^2 x = \sin^2 x\frac{1}{\sin^2 x} = 1$$

17.
$$\left(1-\sin^2\theta\right)\cos^2\theta = \cos^4\theta$$
$$\left(1-\sin^2\theta\right)\cos^2\theta = \cos^2\theta\cos^2\theta = \cos^4\theta$$

19.
$$\frac{\cos^2 x-1}{\sin x} = -\sin x$$
$$\frac{\cos^2 x-1}{\sin x} = \frac{-1\left(1-\cos^2 x\right)}{\sin x} = \frac{-1\left(\sin^2 x\right)}{\sin x} = -\sin x$$

21.
$$\cos\theta(\csc\theta-\sec\theta)=\cot\theta-1$$
$$\begin{aligned}\cos\theta(\csc\theta-\sec\theta)&=\cos\theta\left(\frac{1}{\sin\theta}-\frac{1}{\cos\theta}\right)\\&=\frac{\cos\theta}{\sin\theta}-\frac{\cos\theta}{\cos\theta}\\&=\cot\theta-1\end{aligned}$$

23.
$$\tan^2\theta-\tan^2\theta\sin^2\theta=\sin^2\theta$$
$$\begin{aligned}\tan^2\theta-\tan^2\theta\sin^2\theta&=\tan^2\theta\left(1-\sin^2\theta\right)\\&=\tan^2\theta\cos^2\theta\\&=\frac{\sin^2\theta}{\cos^2\theta}\cdot\frac{\cos^2\theta}{1}\\&=\sin^2\theta\end{aligned}$$

25.
$$\frac{\sec x-\cos x}{\sin x}=\tan x$$
$$\begin{aligned}\frac{\sec x-\cos x}{\sin x}&=\frac{\frac{1}{\cos x}-\frac{\cos x}{1}}{\sin x}\\&=\frac{\frac{1-\cos^2 x}{\cos x}}{\sin x}\\&=\frac{\sin^2 x}{\cos x}\cdot\frac{1}{\sin x}\\&=\frac{\sin x}{\cos x}\\&=\tan x\end{aligned}$$

27.
$$\frac{\sec^2\theta-1}{\sec^2\theta}=\sin^2\theta$$
$$\begin{aligned}\frac{\sec^2\theta-1}{\sec^2\theta}&=\frac{\sec^2\theta}{\sec^2\theta}-\frac{1}{\sec^2\theta}\\&=1-\cos^2\theta\\&=\sin^2\theta\end{aligned}$$

29.
$$\frac{\csc^2 x}{1+\tan^2 x}=\cot^2 x$$
$$\begin{aligned}\frac{\csc^2 x}{1+\tan^2 x}&=\frac{\csc^2 x}{\sec^2 x}\\&=\frac{\frac{1}{\sin^2 x}}{\frac{1}{\cos^2 x}}\\&=\frac{\cos^2 x}{\sin^2 x}\\&=\cot^2 x\end{aligned}$$

31.
$$\frac{(1+\sin\theta)(1-\sin\theta)}{\sin^2\theta}=\cot^2\theta$$
$$\begin{aligned}\frac{(1+\sin\theta)(1-\sin\theta)}{\sin^2\theta}&=\frac{1-\sin^2\theta}{\sin^2\theta}\\&=\frac{\cos^2\theta}{\sin^2\theta}\\&=\cot^2\theta\end{aligned}$$

33.
$$\frac{\sin^2 x}{1+\cos x}=1-\cos x$$
$$\begin{aligned}\frac{\sin^2 x}{1+\cos x}&=\frac{1-\cos^2 x}{1+\cos x}\\&=\frac{(1+\cos x)(1-\cos x)}{1+\cos x}\\&=1-\cos x\end{aligned}$$

35.
$$\cos^4\theta-\sin^4\theta=2\cos^2\theta-1$$
$$\begin{aligned}\cos^4\theta-\sin^4\theta&=\left(\cos^2\theta+\sin^2\theta\right)\left(\cos^2\theta-\sin^2\theta\right)\\&=1\left(\cos^2\theta-\sin^2\theta\right)\\&=\cos^2\theta-\left(1-\cos^2\theta\right)\\&=\cos^2\theta-1+\cos^2\theta\\&=2\cos^2\theta-1\end{aligned}$$

37. $$\frac{1+\sec x}{\csc x}=\sin x+\tan x$$

$$\frac{1+\sec x}{\csc x}=\frac{1+\dfrac{1}{\cos x}}{\dfrac{1}{\sin x}}$$
$$=\left(1+\frac{1}{\cos x}\right)\sin x$$
$$=\sin x+\frac{\sin x}{\cos x}$$
$$=\sin x+\tan x$$

39. $$(\sec x-\tan x)(\csc x+1)=\cot x$$

$$(\sec x-\tan x)(\csc x+1)=\left(\frac{1}{\cos x}-\frac{\sin x}{\cos x}\right)\left(\frac{1}{\sin x}+1\right)$$
$$=\frac{1}{\cos x\sin x}+\frac{1}{\cos x}-\frac{1}{\cos x}-\frac{\sin x}{\cos x}$$
$$=\frac{1}{\cos x\sin x}-\frac{\sin x^2}{\sin x\cos x}$$
$$=\frac{1-\sin^2 x}{\cos x\sin x}$$
$$=\frac{\cos^2 x}{\cos x\sin x}$$
$$=\frac{\cos x}{\sin x}=\cot x$$

41. $$\frac{1-\sin^2 x}{1-\cos^2 x}=\cot^2 x$$

$$\frac{1-\sin^2 x}{1-\cos^2 x}=\frac{\cos^2 x}{\sin^2 x}$$
$$=\cot^2 x$$

43. $$\frac{\sin x}{1+\cos x}=\frac{1-\cos x}{\sin x}$$

$$\frac{\sin x}{1+\cos x}=\frac{\sin x}{(1+\cos x)}\frac{(1-\cos x)}{(1-\cos x)}$$
$$=\frac{\sin x(1-\cos x)}{1-\cos^2 x}$$
$$=\frac{\sin x(1-\cos x)}{\sin^2 x}$$
$$=\frac{1-\cos x}{\sin x}$$

45. $$\frac{\cos^2 x}{1-\sin x}=1+\sin x$$

$$\frac{\cos^2 x}{1-\sin x}=\frac{\cos^2 x}{1-\sin x}\cdot\frac{1+\sin x}{1+\sin x}$$
$$=\frac{\cos^2 x(1+\sin x)}{1-\sin^2 x}$$
$$=\frac{\cos^2 x(1+\sin x)}{\cos^2 x}$$
$$=1+\sin x$$

47. $$\frac{1}{\sec x-1}-\frac{1}{\sec x+1}=2\cot^2 x$$

$$\frac{1}{\sec x-1}-\frac{1}{\sec x+1}=\frac{\sec x+1-(\sec x-1)}{(\sec x-1)(\sec x+1)}$$
$$=\frac{\sec x+1-\sec x+1}{\sec^2 x-1}$$
$$=\frac{2}{\tan^2 x}$$
$$=2\cot^2 x$$

49. $$\cos^4 x - \sin^4 x = 1 - 2\sin^2 x$$

$$\begin{aligned}\cos^4 x - \sin^4 x &= \left(\cos^2 x + \sin^2 x\right)\left(\cos^2 x - \sin^2 x\right)\\ &= 1\left(1 - \sin^2 x - \sin^2 x\right)\\ &= 1 - 2\sin^2 x\end{aligned}$$

51. $$\frac{\tan^2 x - 1}{1 - \cot^2 x} = \tan^2 x$$

$$\begin{aligned}\frac{\tan^2 x - 1}{1 - \cot^2 x} &= \frac{\frac{\sin^2 x}{\cos^2 x} - 1}{1 - \frac{\cos^2 x}{\sin^2 x}}\\ &= \frac{\frac{\sin^2 x - \cos^2 x}{\cos^2 x}}{\frac{\sin^2 x - \cos^2 x}{\sin^2 x}}\\ &= \frac{\sin^2 x - \cos^2 x}{\cos^2 x} \cdot \frac{\sin^2 x}{\sin^2 x - \cos^2 x}\\ &= \frac{\sin^2 x}{\cos^2 x}\\ &= \tan^2 x\end{aligned}$$

53. $$\frac{\tan x + \tan y}{\cot x + \cot y} = \tan x \tan y$$

$$\begin{aligned}\frac{\tan x + \tan y}{\cot x + \cot y} &= \frac{\tan x + \tan y}{\cot x + \cot y} \cdot \frac{\cot x - \cot y}{\cot x - \cot y}\\ &= \frac{\tan x \cot x - \tan x \cot y + \tan y \cot x - \tan y \cot y}{\cot^2 x - \cot^2 y}\\ &= \frac{1 - \frac{\tan x}{\tan y} + \frac{\tan y}{\tan x} - 1}{\frac{1}{\tan^2 x} - \frac{1}{\tan^2 y}}\\ &= \frac{\frac{\tan^2 y - \tan^2 x}{\tan x \tan y}}{\frac{\tan^2 y - \tan^2 x}{\tan^2 x \tan^2 y}}\\ &= \frac{\tan^2 y - \tan^2 x}{\tan x \tan y} \cdot \frac{\tan^2 x \tan^2 y}{\tan^2 y - \tan^2 x}\\ &= \tan x \tan y\end{aligned}$$

Exercises 13.2

1.

$$\sin(x+\pi) = -\sin x$$
$$\sin(x+\pi) = \sin x\cos\pi + \cos x\sin\pi$$
$$= \sin x(-1) + \cos x(0)$$
$$= -\sin x$$

3.

$$\sin(x+2\pi) = \sin x$$
$$\sin(x+2\pi) = \sin x\cos 2\pi + \cos x\sin 2\pi$$
$$= \sin x(1) + \cos x(0)$$
$$= \sin x$$

5.

$$\tan(x+\pi) = \tan x$$
$$\tan(x+\pi) = \frac{\tan x + \tan\pi}{1 - \tan x\tan\pi}$$
$$= \frac{\tan x + 0}{1 - \tan x(0)}$$
$$= \tan x$$

7.

$$\sin(90° - \theta) = \cos\theta$$
$$\sin(90° - \theta) = \sin 90°\cos\theta - \cos 90°\sin\theta$$
$$= \cos\theta - 0(\sin\theta)$$
$$= \cos\theta$$

9.

$$\cos\left(\frac{\pi}{2} + \theta\right) = -\sin\theta$$
$$\cos\left(\frac{\pi}{2} + \theta\right) = \cos\frac{\pi}{2}\cos\theta - \sin\frac{\pi}{2}\sin\theta$$
$$= 0(\cos\theta) - 1(\sin\theta)$$
$$= -\sin\theta$$

11.

$$\tan(180° - \theta) = -\tan\theta$$
$$\tan(180° - \theta) = \frac{\tan 180° - \tan\theta}{1 + \tan 180°\tan\theta}$$
$$= \frac{0 - \tan\theta}{1 + 0\tan\theta}$$
$$= -\tan\theta$$

13.

$$\cos\left(\frac{\pi}{4} + \theta\right) = \frac{\cos\theta - \sin\theta}{\sqrt{2}}$$
$$\cos\left(\frac{\pi}{4} + \theta\right) = \cos\frac{\pi}{4}\cos\theta - \sin\frac{\pi}{4}\sin\theta$$
$$= \frac{1}{\sqrt{2}}\cos\theta - \frac{1}{\sqrt{2}}\sin\theta$$
$$= \frac{\cos\theta - \sin\theta}{\sqrt{2}}$$

15.

$$\tan(90° - x) = \cot x$$
$$\tan(90° - x) = \frac{\sin(90° - x)}{\cos(90° - x)}$$
$$= \frac{\sin 90°\cos x - \cos 90°\sin x}{\cos 90°\cos x + \sin 90°\sin x}$$
$$= \frac{\cos x - 0}{0 + \sin x}$$
$$= \cot x$$

17.

$$\tan(x + 45°) = \frac{1 + \tan x}{1 - \tan x}$$
$$\tan(x + 45°) = \frac{\tan x + \tan 45°}{1 - \tan x\tan 45°}$$
$$= \frac{\tan x + 1}{1 - \tan x}$$

19. $\cos(x+y)\cos(x-y)=\cos^2 x-\sin^2 y$

$$\begin{aligned}\cos(x+y)\cos(x-y)&=(\cos x\cos y-\sin x\sin y)(\cos x\cos y+\sin x\sin y)\\&=\cos^2 x\cos^2 y-\sin^2 x\sin^2 y\\&=\cos^2 x\left(1-\sin^2 y\right)-\left(1-\cos^2 x\right)\sin^2 y\\&=\cos^2 x-\cos^2 x\sin^2 y-\sin^2 y+\cos^2 x\sin^2 y\\&=\cos^2 x-\sin^2 y\end{aligned}$$

21. $\cos M\cos N-\sin M\sin N=$
$\cos(M+N)$

23. $\sin\theta\cos 3\theta+\cos\theta\sin 3\theta=$
$\sin(\theta+3\theta)=\sin 4\theta$

25. $\cos 4\theta\cos 3\theta+\sin 4\theta\sin 3\theta=$
$\cos(4\theta-3\theta)=\cos\theta$

27. $\dfrac{\tan 3\theta+\tan 2\theta}{1-\tan 3\theta\tan 2\theta}=$
$\tan(3\theta+2\theta)=\tan 5\theta$

29. $\sin(\theta+\phi)+\sin(\theta-\phi)=$
$\sin\theta\cos\phi+\cos\theta\sin\phi+\sin\theta\cos\phi-\cos\theta\sin\phi=2\sin\theta\cos\phi$

31. $\sin(A+B)\cos B+\cos(A+B)\sin B=$
$(\sin A\cos B+\cos A\sin B)\cos B+(\cos A\cos B-\sin A\sin B)\sin B=$
$\sin A\cos^2 B+\underline{\sin B\cos A\cos B+\cos A\sin B\cos B}-\sin A\sin^2 B=$
$2\sin B\cos B\cos A+\sin A\left(\cos^2 B-\sin^2 B\right)=$
$\cos A(\sin(B+B))+\sin A(\cos(B+B))=$
$\sin A\cos(2B)+\cos A\sin(2B)=$
$\sin(A+2B)$

Exercises 13.3

1. $2\sin\dfrac{x}{4}\cos\dfrac{x}{4}=$
$\sin 2\left(\dfrac{x}{4}\right)=$
$\sin\dfrac{x}{2}$

3. $1-2\sin^2 3x=$
$\cos 2(3x)=$
$\cos 6x$

5. $\sqrt{\dfrac{1+\cos\dfrac{\theta}{4}}{2}}=$
$\cos\dfrac{1}{2}\left(\dfrac{\theta}{4}\right)=\cos\dfrac{\theta}{8}$

7.

$$\cos^2\frac{x}{6}-\sin^2\frac{x}{6}=$$
$$\cos 2\left(\frac{x}{6}\right)=\cos\frac{x}{3}$$

9.

$$20\sin 4\theta\cos 4\theta=$$
$$10(2\sin 4\theta\cos 4\theta)=$$
$$10(\sin 2(4\theta))=$$
$$10\sin 8\theta$$

11.

$$-\sqrt{\frac{1+\cos 250°}{2}}=$$
$$\cos\frac{1}{2}(250°)=\cos 125°$$

13.

$$4-8\sin^2\theta=$$
$$4(1-2\sin^2\theta)=$$
$$4\cos 2\theta$$

15.

$$100\sin 30t\cos 30t=$$
$$50(2\sin 30t\cos 30t)=$$
$$50\sin 2(30t)=$$
$$50\sin 60t$$

17.

$\cos\theta=\frac{3}{5}$ means

$$\cos^2\theta=\frac{9}{25}.$$
$$\sin^2\theta=1-\cos^2\theta$$
$$=1-\frac{9}{25}$$
$$=\frac{16}{25}$$
$$\sin\theta=\frac{4}{5}(\theta\text{ in I}).$$
$$\sin 2\theta=2\sin\theta\cos\theta$$
$$=2\left(\frac{4}{5}\right)\left(\frac{3}{5}\right)$$
$$=\frac{24}{25}$$

19.

$\cos\theta=\frac{-12}{13}$ means

$$\cos^2\theta=\frac{144}{169}.$$
$$\sin^2\theta=1-\cos^2\theta$$
$$=1-\frac{144}{169}$$
$$=\frac{25}{169}$$
$$\sin\theta=\frac{-5}{13}(\theta\text{ in III}).$$
$$\tan\theta=\frac{\sin\theta}{\cos\theta}=\frac{-5}{13}\div\frac{-12}{13}$$
$$\tan\theta=\frac{5}{12}.$$
$$\tan 2\theta=\frac{2\tan\theta}{1-\tan^2\theta}$$
$$=\frac{2\left(\frac{5}{12}\right)}{1-\frac{25}{144}}=\frac{\frac{5}{6}}{\frac{119}{144}}$$
$$=\frac{120}{119}$$

21.

$\sin\theta=-\frac{2}{3}$ means

$$\sin^2\theta=\frac{4}{9}.$$
$$\cos^2\theta=1-\sin^2\theta$$
$$=1-\frac{4}{9}=\frac{5}{9}$$
$$\cos\theta=\frac{-\sqrt{5}}{3}(\theta\text{ in III})$$

since $\pi<\theta<\frac{3\pi}{2}$

$\frac{\pi}{2}<\frac{\theta}{2}<\frac{3\pi}{4}$, so

$\cos\frac{\theta}{2}$ is negative

$$\cos\frac{\theta}{2}=-\sqrt{\frac{1-\frac{\sqrt{5}}{3}}{2}}$$
$$\cos\frac{\theta}{2}=-\sqrt{\frac{3-\sqrt{5}}{6}}$$

23. $(\sin x+\cos x)^2=1+\sin 2x$

$$\begin{aligned}(\sin x+\cos x)^2&=\sin^2 x+2\sin x\cos x+\cos^2 x\\&=\sin^2 x+\cos^2 x+2\sin x\cos x\\&=1+\sin 2x\end{aligned}$$

25. $\cos^4 x-\sin^4 x=\cos 2x$

$$\begin{aligned}\cos^4 x-\sin^4 x&=\left(\cos^2 x+\sin^2 x\right)\left(\cos^2 x-\sin^2 x\right)\\&=1(\cos 2x)\\&=\cos 2x\end{aligned}$$

27. $\dfrac{1-\tan^2 x}{1+\tan^2 x}=\cos 2x$

$$\begin{aligned}\frac{1-\tan^2 x}{1+\tan^2 x}&=\frac{1-\dfrac{\sin^2 x}{\cos^2 x}}{\sec^2 x}\\&=\frac{\dfrac{\cos^2 x-\sin^2 x}{\cos^2 x}}{\sec^2 x}\\&=\frac{\cos^2 x-\sin^2 x}{\cos^2 x}\div\frac{1}{\cos^2 x}\\&=\cos^2 x-\sin^2 x\\&=\cos 2x\end{aligned}$$

29. $\cot 2x=\dfrac{\cot^2 x-1}{2\cot x}$

$$\begin{aligned}\frac{\cot^2 x-1}{2\cot x}&=\frac{\dfrac{\cos^2 x}{\sin^2 x}-1}{2\dfrac{\cos x}{\sin x}}\\&=\frac{\cos^2 x-\sin^2 x}{\sin^2 x}\cdot\frac{\sin x}{2\cos x}\\&=\frac{\cos 2x}{2\sin x\cos x}\\&=\frac{\cos 2x}{\sin 2x}\\&=\cot 2x\end{aligned}$$

31. $\tan x+\cot 2x=\csc 2x$

ie $\tan x=\csc 2x-\cot 2x$

$$\begin{aligned}\csc 2x-\cot 2x&=\frac{1}{\sin 2x}-\frac{\cos 2x}{\sin 2x}\\&=\frac{1-\cos 2x}{\sin 2x}\\&=\frac{1-\left(1-2\sin^2 x\right)}{2\sin x\cos x}\\&=\frac{1-1+2\sin^2 x}{2\sin x\cos x}\\&=\frac{\sin x}{\cos x}=\tan x\end{aligned}$$

33. $\sin^2\dfrac{x}{2}=\dfrac{\sec x-1}{2\sec x}$

$$\begin{aligned}\frac{\sec x-1}{2\sec x}&=\frac{\dfrac{1}{\cos x}-1}{\dfrac{2}{\cos x}}\\&=\frac{1-\cos x}{\cos x}\cdot\frac{\cos x}{2}\\&=\frac{1-\cos x}{2}\\&=\sin^2\frac{x}{2}\end{aligned}$$

35.
$$\sec^2\frac{x}{2}=\frac{2}{1+\cos x}$$
$$\sec^2\frac{x}{2}=\frac{1}{\cos^2\frac{x}{2}}=\frac{1}{\frac{1+\cos x}{2}}=\frac{2}{1+\cos x}$$

37.
$$\tan\frac{x}{2}=\frac{\sin x}{1+\cos x}$$
$$\frac{\sin x}{1+\cos x}=\frac{\sin 2\left(\frac{x}{2}\right)}{1+\cos 2\left(\frac{x}{2}\right)}=\frac{2\sin\frac{x}{2}\cos\frac{x}{2}}{1+2\cos^2\frac{x}{2}-1}=\frac{2\sin\frac{x}{2}\cos\frac{x}{2}}{2\cos\frac{x}{2}\cos\frac{x}{2}}=\tan\frac{x}{2}$$

39.
$$2\cos\frac{x}{2}=(1+\cos x)\sec\frac{x}{2}$$
$$(1+\cos x)\sec\frac{x}{2}=\frac{\left(1+\cos 2\left(\frac{x}{2}\right)\right)}{\cos\frac{x}{2}}=\frac{1+2\cos^2\frac{x}{2}-1}{\cos\frac{x}{2}}=2\cos\frac{x}{2}$$

41.
$$\tan\frac{x}{2}+\cot\frac{x}{2}=2\csc x$$
$$\tan\frac{x}{2}+\cot\frac{x}{2}=\tan\frac{x}{2}+\frac{1}{\tan\frac{x}{2}}=\frac{\tan^2\frac{x}{2}+1}{\tan\frac{x}{2}}=\frac{\sec^2\frac{x}{2}}{\tan\frac{x}{2}}=\frac{1}{\cos^2\frac{x}{2}}\cdot\frac{\cos\frac{x}{2}}{\sin\frac{x}{2}}=\frac{1}{\sin\frac{x}{2}\cos\frac{x}{2}}=\frac{1}{\frac{1}{2}\left(2\sin\frac{x}{2}\cos\frac{x}{2}\right)}=\frac{1}{\frac{1}{2}\sin x}=\frac{2}{\sin x}=2\csc x$$

43.
$$\left(\sin\frac{\theta}{2}-\cos\frac{\theta}{2}\right)^2=1-\sin\theta$$
$$\left(\sin\frac{\theta}{2}-\cos\frac{\theta}{2}\right)^2=\sin^2\frac{\theta}{2}-2\sin\frac{\theta}{2}\cos\frac{\theta}{2}+\cos^2\frac{\theta}{2}=1-2\sin\frac{\theta}{2}\cos\frac{\theta}{2}=1-\sin 2\left(\frac{\theta}{2}\right)=1-\sin\theta$$

Section 13.3

45.

$$\sin 3x = 3\sin x - 4\sin^3 x$$

$$\begin{aligned}\sin 3x &= \sin(2x + x)\\ &= \sin 2x\cos x + \cos 2x\sin x\\ &= 2\sin x\cos x\cos x + \left(1 - 2\sin^2 x\right)\sin x\\ &= 2\sin x\cos^2 x + \sin x - 2\sin^3 x\\ &= 2\sin x\left(1 - \sin^2 x\right) + \sin x - 2\sin^3 x\\ &= 2\sin x - 2\sin^3 x + \sin x - 2\sin^3 x\\ &= 3\sin x - 4\sin^3 x\end{aligned}$$

Exercises 13.4

1.

$\cos x - 1 = 0$

$\cos x = 1$

$x = 0°$

3.

$\tan x - 1 = 0$

$\tan x = 1$

$x = 45°, 225°$

5.

$4\cos^2 x - 3 = 0$

$\cos^2 x = \dfrac{3}{4}$

$\cos x = \pm\dfrac{\sqrt{3}}{2}$

$x = 30°, 150°, 210°, 330°$

7.

$\sin 2x = 1$

$2x = 90°, 450°$

$x = 45°, 225°$

9.

$3\tan^2 3x - 1 = 0$

$\tan^2 3x = \dfrac{1}{3}$

$\tan 3x = \pm\dfrac{1}{\sqrt{3}}$

$3x = 30°, 150°, 210°, 330°, 390°, 510°, 570°, 690°, 750°, 870°, 930°, 1050°$

$x = 10°, 50°, 70°, 110°, 130°, 170°, 190°, 230°, 250°, 290°, 310°, 350°$

11.

$\sin 2x + \cos x = 0$

$2\sin x\cos x + \cos x = 0$

$\cos x(2\sin x + 1) = 0$

$\cos x = 0$ or $\sin x = \dfrac{-1}{2}$

$90°, 270°$ or $210°, 330°$

13.

$\sin^2 x + \sin x = 0$

$\sin x(\sin x + 1) = 0$

$\sin x = 0$ or $\sin x = -1$

$x = 0°, 180°$, or $270°$

15.

$2\sin^2 x = 1 - 2\sin x$

$2\sin^2 x + 2\sin x - 1 = 0$

Use quadratic formula

$\sin x = \dfrac{-2 \pm \sqrt{4 - 4(2)(-1)}}{4}$

$\sin x = 0.366025$ or -1.366 (impossible)

$x = 21.5°$ or $158.5°$

17.

$$2\cos^2 x + \sin x = 1$$
$$2(1-\sin^2 x) + \sin x - 1 = 0$$
$$2 - 2\sin^2 x + \sin x - 1 = 0$$
$$-2\sin^2 x + \sin x + 1 = 0$$
$$2\sin^2 x - \sin x - 1 = 0$$
$$(2\sin x + 1)(\sin x - 1) = 0$$
$$\sin x = \frac{-1}{2} \text{ or } \sin x = 1$$
$$x = \frac{7\pi}{6}, \frac{11\pi}{6}, \frac{\pi}{2}$$

19.

$$\cos^2 x - \cos x \sec x = 0$$
$$\cos^2 x - 1 = 0$$
$$\cos^2 x = 1$$
$$\cos x = \pm 1$$
$$x = 0, \pi$$

21.

$$2\sin^2 x + 2\cos 2x = 1$$
$$2\sin^2 x + 2(1 - 2\sin^2 x) = 1$$
$$2\sin^2 x + 2 - 4\sin^2 x = 1$$
$$0 = 2\sin^2 x - 1$$
$$\sin^2 x = \frac{1}{2}$$
$$\sin x = \pm\frac{1}{\sqrt{2}}$$
$$x = \frac{\pi}{4}, \frac{3\pi}{4}, \frac{5\pi}{4}, \frac{7\pi}{4}$$

23.

$$2\sin^2 2x - \sin 2x - 1 = 0$$
$$(2\sin 2x + 1)(\sin 2x - 1) = 0$$
$$\sin 2x = \frac{-1}{2} \text{ or } \sin 2x = 1$$
$$2x = \frac{7\pi}{6}, \frac{11\pi}{6}, \frac{19\pi}{6}, \frac{23\pi}{6}, \frac{\pi}{2}, \frac{5\pi}{2}$$
$$x = \frac{7\pi}{12}, \frac{11\pi}{12}, \frac{19\pi}{12}, \frac{23\pi}{12}, \frac{\pi}{4}, \frac{5\pi}{4}$$

25.

$$4\tan^2 x = 3\sec^2 x$$
$$\frac{4\sin^2 x}{\cos^2 x} = \frac{3}{\cos^2 x}$$
$$\sin^2 x = \frac{3}{4}$$
$$\sin x = \pm\frac{\sqrt{3}}{2}$$
$$x = \frac{\pi}{3}, \frac{2\pi}{3}, \frac{4\pi}{3}, \frac{5\pi}{3}$$

27.

$$4\sin 2x \cos 2x = 1$$
$$2(2\sin 2x \cos 2x) = 1$$
$$2\sin 4x = 1$$
$$\sin 4x = \frac{1}{2}$$
$$4x = \frac{\pi}{6}, \frac{5\pi}{6}, \frac{13\pi}{6}, \frac{17\pi}{6},$$
$$\frac{25\pi}{6}, \frac{29\pi}{6}, \frac{37\pi}{6}, \frac{41\pi}{6}$$
$$x = \frac{\pi}{24}, \frac{5\pi}{24}, \frac{13\pi}{24}, \frac{17\pi}{24},$$
$$\frac{25\pi}{24}, \frac{29\pi}{24}, \frac{37\pi}{24}, \frac{41\pi}{24}$$

29.

$$\cos x = \cos\frac{x}{2}$$
$$\cos x = \pm\sqrt{\frac{1+\cos x}{2}}$$
$$\cos^2 x = \frac{1+\cos x}{2}$$
$$2\cos^2 x = 1+\cos x$$
$$2\cos^2 x - \cos x - 1 = 0$$
$$(2\cos x+1)(\cos x-1) = 0$$
$$\cos x = \frac{-1}{2} \text{ or } \cos x = 1$$
$$x = \frac{2\pi}{3}, \frac{4\pi}{3}, 0$$

Must check in original equation.

For $x = 0$: $\cos 0 = \cos\frac{0}{2}$ yes

For $\frac{2\pi}{3}$: $\cos\frac{2\pi}{3} = \cos\frac{\pi}{3}$ no

For $\frac{4\pi}{3}$: $\cos\frac{4\pi}{3} = \cos\frac{2\pi}{3}$ yes

Solution: $x = 0, \frac{4\pi}{3}$

31.

$$\cos\frac{x}{2} = 1+\cos x$$
$$\pm\sqrt{\frac{1+\cos x}{2}} = 1+\cos x$$
$$\frac{1+\cos x}{2} = (1+\cos x)^2$$
$$\frac{1+\cos}{2} = 1+2\cos x+\cos^2 x$$
$$1+\cos x = 2+4\cos x+2\cos^2 x$$
$$0 = 2\cos^2 x+3\cos x+1$$
$$0 = (2\cos x+1)(\cos x+1)$$
$$\cos x = \frac{-1}{2} \text{ or } \cos x = -1$$

$$x = \frac{2\pi}{3}, \frac{4\pi}{3}, \pi$$

Must check:

For $x = \frac{2\pi}{3}$: $\cos\frac{\pi}{3} = 1+\cos\frac{2\pi}{3}$

$\frac{1}{2} = \frac{1}{2}$ yes

For $x = \frac{4\pi}{3}$: $\cos\frac{2\pi}{3} = 1+\cos\frac{4\pi}{3}$

$\frac{-1}{2} \stackrel{?}{=} 1-\frac{1}{2}$ no

For $x = \pi$: $\cos\frac{\pi}{2} = 1+\cos\pi$

$0 = 1-1$ yes

Solutions: $x = \frac{2\pi}{3}, \pi$

33.

$$1+\cos^2\frac{x}{2} = 2\cos x$$
$$1+\frac{1+\cos x}{2} = 2\cos x$$
$$2+1+\cos x = 4\cos x$$
$$3 = 3\cos x$$
$$1 = \cos x$$
$$x = 0$$

Exercises 13.5

1. $y = \arcsin x$

y is the angle whose sine is x

3. $y = \cot^{-1} 4x$

y is the angle whose cotangent is $4x$

5. $y = 3\,\text{arc}\csc \frac{1}{2}x$

y is three times the angle whose cosecant is $\frac{x}{2}$

7.
$$y = \sin 3x$$
$$\arcsin y = 3x$$
$$\frac{1}{3}\arcsin y = x$$

9.
$$y = 4\cos x$$
$$\frac{y}{4} = \cos x$$
$$\arccos \frac{y}{4} = x$$

11.
$$y = 5\tan \frac{x}{2}$$
$$\frac{y}{5} = \tan \frac{x}{2}$$
$$\arctan \frac{y}{5} = \frac{x}{2}$$
$$2\arctan \frac{y}{5} = x$$

13.
$$y = \frac{3}{2}\cot \frac{x}{4}$$
$$\frac{2}{3}y = \cot \frac{x}{4}$$
$$\text{arc}\cot \frac{2}{3}y = \frac{x}{4}$$
$$4\,\text{arc}\cot \frac{2}{3}y = x$$

15.
$$y = 3\sin(x-1)$$
$$\frac{y}{3} = \sin(x-1)$$
$$\arcsin \frac{y}{3} = x-1$$
$$1 + \arcsin \frac{y}{3} = x$$

17.
$$y = \frac{1}{2}\cos(3x+1)$$
$$2y = \cos(3x+1)$$
$$\arccos 2y = 3x+1$$
$$-1 + \arccos 2y = 3x$$
$$\frac{-1}{3} + \frac{1}{3}\arccos 2y = x$$

19. $\arcsin\left(\frac{\sqrt{3}}{2}\right) =$

$\frac{\pi}{3}$

21. $\tan^{-1}\left(-\frac{1}{\sqrt{3}}\right) =$

$\frac{-\pi}{6}$

23. $\arccos\left(-\frac{\sqrt{3}}{2}\right) =$

$\frac{5\pi}{6}$

25. $\text{arc}\sec\sqrt{2} =$

$\frac{\pi}{4}$

27. $\arctan\sqrt{3} =$

$\frac{\pi}{3}$

29. $\cos^{-1}\left(\frac{1}{\sqrt{2}}\right) =$

$\frac{\pi}{4}$

31. $\sin^{-1}\left(-\frac{\sqrt{3}}{2}\right)=$

$-\frac{\pi}{3}$

33. $\operatorname{arc}\cot(-1.5)$

Reference angle is

$\arctan\left(\frac{-1}{1.5}\right)$

$\alpha=-0.5880$

$\pi-0.5880=$

2.554

35. $\operatorname{arc}\sec(-3.2)=$

$\arccos\left(\frac{-1}{3.2}\right)$

$\alpha=1.889$

This is in

quadrant II,

1.889

37. $\csc^{-1}(-1.15)=$

$\sin^{-1}\left(\frac{-1}{1.15}\right)=$

-1.054

39. $\cos\left(\arctan\sqrt{3}\right)=$

$\cos\left(\frac{\pi}{3}\right)=\frac{1}{2}$

41. $\sin\left[\arccos\left(-\frac{1}{\sqrt{2}}\right)\right]=$

$\sin\left[\frac{3\pi}{4}\right]=\frac{1}{\sqrt{2}}$

43. $\tan[\cos^{-1}(-1)]=$

$\tan\pi=0$

45. $\sin\left[\arcsin\frac{\sqrt{3}}{2}\right]=$

$\sin\frac{\pi}{3}=\frac{\sqrt{3}}{2}$

47. $\cos\left[\sin^{-1}\left(\frac{3}{5}\right)\right]$

Let $\theta=\sin^{-1}\frac{3}{5}$

$\cos\theta=\frac{4}{5}$

49. $\tan\left[\arcsin(-0.1560)\right]$

Let $\theta=\arcsin(-0.1560)$

$\theta=-0.1566$

$\tan(-0.1566)=-0.1579$

51. $\cos(\arcsin x)$

Let $\theta=\arcsin x$

$\sin\theta=x$

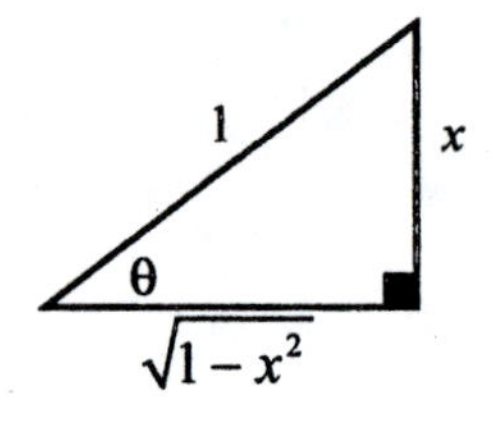

$\cos\theta=\sqrt{1-x^2}$

53. $\sin(\operatorname{arcsec} x)$

Let $\theta = \operatorname{arcsec} x$

$\sec\theta = x$

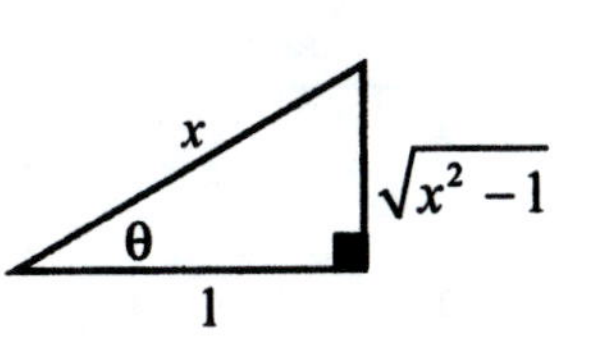

$$\sin\theta = \frac{\sqrt{x^2 - 1}}{x}$$

55. $\sec(\cos^{-1} x)$

Let $\theta = \cos^{-1} x$

$\cos\theta = x$

$$\frac{1}{\sec\theta} = x$$

$$\frac{1}{x} = \sec\theta$$

57. $\tan(\arctan x)$

Let $\theta = \arctan x$

$\tan\theta = x$

59. $\cos(\arcsin 2x)$

Let $\theta = \arcsin 2x$

$\sin\theta = 2x$

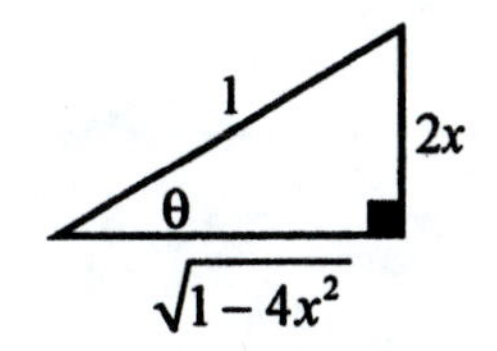

$$\cos\theta = \sqrt{1 - 4x^2}$$

$\sin(2\arcsin x)$

61. Let $\theta = 2\arcsin x$

$\frac{\theta}{2} = \arcsin x$

$\sin\frac{\theta}{2} = x$

$\pm\sqrt{\frac{1-\cos\theta}{2}} = x$

$\frac{1-\cos\theta}{2} = x^2$

$1-\cos\theta = 2x^2$

$1-2x^2 = \cos\theta$

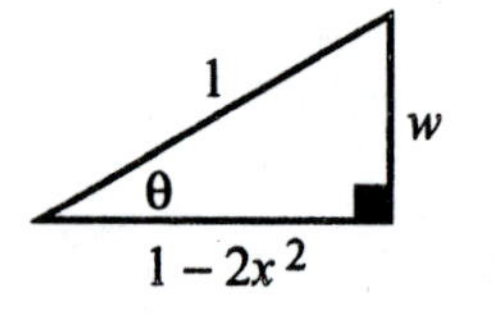

$w^2 = 1-(1-2x^2)^2$

$w^2 = 1-(1-4x^2+4x^4)$

$w^2 = 4x^2-4x^4$

$w^2 = 4x^2(1-x^2)$

$w = 2x\sqrt{1-x^2}$

$\sin(2\arcsin x) = \sin\left(2\left[\frac{\theta}{2}\right]\right)$

$= \sin\theta$

Thus from the triangle

$\sin\theta = \frac{w}{1}$

$\sin\theta = 2x\sqrt{1-x^2}$

63. $y = \arcsin 2x$

$\sin y = 2x$

$\frac{\sin y}{2} = x$

Pick y and solve for x.

x	-0.5	-0.43	-0.35	-0.25	0
y	$\frac{-\pi}{2}$	$\frac{-\pi}{3}$	$\frac{-\pi}{4}$	$\frac{-\pi}{6}$	0

x	0.25	0.35	0.43	0.5
y	$\frac{\pi}{6}$	$\frac{\pi}{4}$	$\frac{\pi}{3}$	$\frac{\pi}{2}$

The graph appears in the text answer section. The x limits on the graph are

$-1 \le 2x \le 1$

$-0.5 \le x \le 0.5$

The y limits on the graph are $\frac{-\pi}{2} \le y \le \frac{\pi}{2}$

65. $y = 2\arctan 3x$

$\frac{y}{2} = \arctan 3x$

$\tan\frac{y}{2} = 3x$

$\frac{1}{3}\tan\frac{y}{2} = x$

Pick y and solve for x.

The graph is in the text answer section.

The y limits on the graph are:

$\frac{-\pi}{2} < \frac{y}{2} < \frac{\pi}{2}$

$-\pi < y < \pi$

x	-4.7	-1	-0.52	-0.31	-0.18
y	-3	-2.5	-2	-1.5	-1

x	0	0.18	0.31	0.52	1	4.7
y	0	1	1.5	2	2.5	3

Section 13.5

Chapter 13 Review

1.

$$\sec x \cot x = \csc x$$

$$\begin{aligned}\sec x \cot x &= \frac{1}{\cot x} \cdot \frac{\cos x}{\sin x} \\ &= \frac{1}{\sin x} \\ &= \csc x\end{aligned}$$

2.

$$\sec^2\theta + \tan^2\theta + 1 = \frac{2}{\cos^2\theta}$$

$$\begin{aligned}\sec^2\theta + \tan^2\theta + 1 &= \sec^2\theta + \sec^2\theta \\ &= 2\sec^2\theta \\ &= \frac{2}{\cos^2\theta}\end{aligned}$$

3.

$$\frac{\cos\theta}{\cos\theta + \sin\theta} = \frac{\cot\theta}{1+\cot\theta}$$

$$\begin{aligned}\frac{\cot\theta}{1+\cot\theta} &= \frac{\frac{\cos\theta}{\sin\theta}}{1+\frac{\cos\theta}{\sin\theta}} \\ &= \frac{\frac{\cos\theta}{\sin\theta}}{\frac{\sin\theta+\cos\theta}{\sin\theta}} \\ &= \frac{\cos\theta}{\sin\theta+\cos\theta}\end{aligned}$$

4.

$$\cos\left(\theta - \frac{3\pi}{2}\right) = -\sin\theta$$

$$\begin{aligned}\cos\left(\theta - \frac{3\pi}{2}\right) &= \cos\theta\cos\frac{3\pi}{2} + \sin\theta\sin\frac{3\pi}{2} \\ &= \cos\theta(0) + \sin\theta(-1) \\ &= -\sin\theta\end{aligned}$$

5.

$$\left(\sin\frac{1}{2}x + \cos\frac{1}{2}x\right)^2 = 1 + \sin x$$

$$\begin{aligned}\left(\sin\frac{1}{2}x + \cos\frac{1}{2}x\right)^2 &= \sin^2\frac{1}{2}x + 2\sin\frac{1}{2}x\cos\frac{1}{2}x + \cos^2\frac{1}{2}x \\ &= 1 + \sin 2\left(\frac{1}{2}x\right) \\ &= 1 + \sin x\end{aligned}$$

6.

$$2\cos^2\frac{\theta}{2} = \frac{1+\sec\theta}{\sec\theta}$$

$$\begin{aligned}2\cos^2\frac{\theta}{2} &= \cos\theta + 1 \\ &= \frac{1}{\sec\theta} + \frac{\sec\theta}{\sec\theta} \\ &= \frac{1+\sec\theta}{\sec\theta}\end{aligned}$$

7.

$$\frac{2\cot\theta}{1+\cot^2\theta} = \sin 2\theta$$

$$\begin{aligned}\frac{2\cot\theta}{1+\cot^2\theta} &= \frac{2\cot\theta}{\csc^2\theta} \\ &= 2\frac{\cos\theta}{\sin\theta} \div \frac{1}{\sin^2\theta} \\ &= 2\sin\theta\cos\theta \\ &= \sin 2\theta\end{aligned}$$

8.

$$\csc x - \cot x = \tan\frac{1}{2}x$$

$$\begin{aligned}\csc x - \cot x &= \frac{1}{\sin x} - \frac{\cos x}{\sin x} \\ &= \frac{1-\cos x}{\sin x} \\ &= \tan\frac{x}{2}\end{aligned}$$

9. $$\tan 2x = \frac{2\cos x}{\csc x - 2\sin x}$$

$$\frac{2\cos x}{\csc x - 2\sin x} = \frac{2\cos x}{\frac{1}{\sin x} - 2\sin x}$$
$$= \frac{2\cos x}{\frac{1-2\sin^2 x}{\sin x}}$$
$$= \frac{2\cos x}{\frac{\cos 2x}{\sin x}}$$
$$= \frac{2\cos x}{1}\cdot\frac{\sin x}{\cos 2x}$$
$$= \frac{\sin 2x}{\cos 2x}$$
$$= \tan 2x$$

10. $$\tan^2\frac{x}{2} + 1 = 2\tan\frac{x}{2}\csc x$$

$$\tan^2\frac{x}{2} + 1 = \sec^2\frac{x}{2}$$
$$= \frac{1}{\cos^2\frac{x}{2}}$$
$$= \frac{1}{\frac{1+\cos x}{2}}$$
$$= \frac{2}{1+\cos x}$$
$$= \frac{2(1-\cos x)}{(1+\cos x)(1-\cos x)}$$
$$= \frac{2(1-\cos x)}{1-\cos^2 x}$$
$$= \frac{2(1-\cos x)}{\sin^2 x}$$
$$= 2\left(\frac{1-\cos x}{\sin x}\right)\frac{1}{\sin x}$$
$$= 2\tan\frac{x}{2}\csc x$$

11. $\sin\theta\cos\theta =$

$\frac{1}{2}\cdot 2\sin\theta\cos\theta =$

$\frac{1}{2}\sin 2\theta$

12. $\cos^2 3\theta - \sin^2 3\theta =$

$\cos 2(3\theta) =$

$\cos 6\theta$

13. $\frac{1+\cos 4\theta}{2} =$

$\cos^2\frac{4\theta}{2} =$

$\cos^2 2\theta$

14. $1 - 2\sin^2\frac{\theta}{3} =$

$\cos 2\left(\frac{\theta}{3}\right) =$

$\cos\frac{2\theta}{3}$

15. $\cos 2x\cos 3x - \sin 2x\sin 3x =$

$\cos(2x+3x) = \cos 5x$

16. $\sin 2x\cos x - \cos 2x\sin x =$

$\sin(2x - x) = \sin x$

17. $$\sin 2\theta = 2\sin\theta\cos\theta$$
$$= 2\left(\frac{12}{13}\right)\left(\frac{-5}{13}\right)$$
$$= \frac{-120}{169}$$

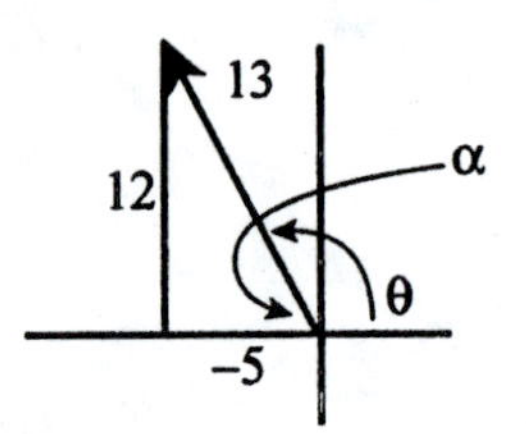

18.

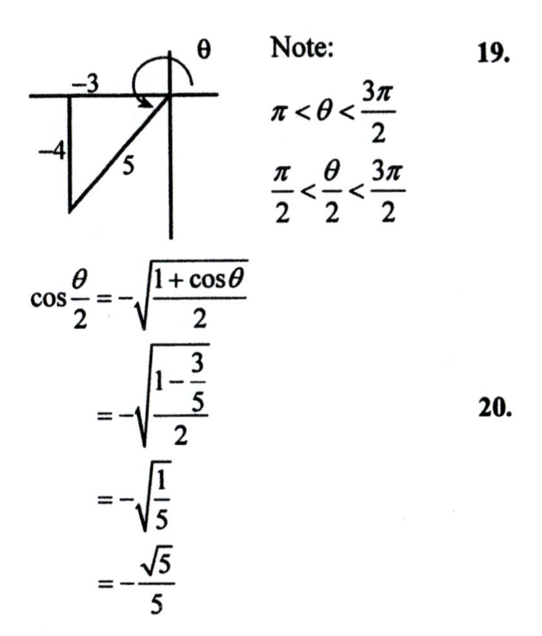

Note:

$$\pi < \theta < \frac{3\pi}{2}$$

$$\frac{\pi}{2} < \frac{\theta}{2} < \frac{3\pi}{2}$$

$$\cos\frac{\theta}{2} = -\sqrt{\frac{1+\cos\theta}{2}}$$

$$= -\sqrt{\frac{1-\frac{3}{5}}{2}}$$

$$= -\sqrt{\frac{1}{5}}$$

$$= -\frac{\sqrt{5}}{5}$$

19.

$$2\cos^2 x = \cos x$$

$$2\cos^2 x - \cos x = 0$$

$$\cos x(2\cos x - 1) = 0$$

$$\cos x = 0 \text{ or } \cos x = \frac{1}{2}$$

$$x = \frac{\pi}{2}, \frac{3\pi}{2} \text{ or } \frac{\pi}{3}, \frac{5\pi}{3}$$

20.

$$4\sin^2 x - 1 = 0$$

$$(2\sin x + 1)(2\sin x - 1) = 0$$

$$\sin x = \frac{-1}{2} \text{ or } \sin x = \frac{1}{2}$$

$$x = \frac{\pi}{6}, \frac{5\pi}{6}, \frac{7\pi}{6}, \frac{11\pi}{6}$$

21.

$$2\cos^2 3x + \sin 3x - 1 = 0$$

$$2(1 - \sin^2 3x) + \sin 3x - 1 = 0$$

$$2 - 2\sin^2 3x + \sin 3x - 1 = 0$$

$$0 = 2\sin^2 3x - \sin 3x$$

$$0 = (2\sin 3x + 1)(\sin 3x - 1)$$

$$\sin 3x = \frac{-1}{2} \text{ or } \sin 3x = 1$$

$$3x = \frac{7\pi}{6}, \frac{11\pi}{6}, \frac{19\pi}{6}, \frac{23\pi}{6}, \frac{31\pi}{6}, \frac{35\pi}{6}$$

$$\text{or } 3x = \frac{\pi}{2}, \frac{5\pi}{2}, \frac{9\pi}{2}$$

$$\text{So } x = \frac{7\pi}{18}, \frac{11\pi}{18}, \frac{19\pi}{18}, \frac{23\pi}{18}, \frac{31\pi}{18}, \frac{35\pi}{18}, \frac{\pi}{6}, \frac{5\pi}{6}, \frac{3\pi}{2}$$

22.

$$\tan^2 x = \sin^2 x$$

$$\frac{\sin^2 x}{\cos^2 x} = \sin^2 x$$

$$\frac{1}{\cos^2 x} = 1$$

$$\cos^2 x = 1$$

$$\cos x = \pm 1$$

$$x = 0, \pi$$

23.

$$\sin\frac{x}{2} + \cos\frac{x}{2} = 0$$

$$\left(\sin\frac{x}{2} + \cos\frac{x}{2}\right)^2 = 0^2$$

$$\sin^2\frac{x}{2} + \cos^2\frac{x}{2} + 2\sin\frac{x}{2}\cos\frac{x}{2} = 0$$

$$1 + \sin 2\left(\frac{x}{2}\right) = 0$$

$$\sin x = -1$$

$$x = \frac{3\pi}{2}$$

Check:

$$\sin\left(\frac{3\pi}{2}\cdot\frac{1}{2}\right) + \cos\left(\frac{3\pi}{2}\cdot\frac{1}{2}\right) =$$

$$\sin\left(\frac{3\pi}{4}\right) + \cos\left(\frac{3\pi}{4}\right) =$$

$$\frac{1}{\sqrt{2}} - \frac{1}{\sqrt{2}} = 0$$

24.
$$\sin 2x = \cos^2 x - \sin^2 x$$
$$\sin 2x = \cos 2x$$
$$\frac{\sin 2x}{\cos 2x} = 1$$
$$\tan 2x = 1$$
$$2x = \frac{\pi}{4} \text{ or } \frac{5\pi}{4} \text{ or } \frac{9\pi}{4} \text{ or } \frac{13\pi}{4}$$
$$x = \frac{\pi}{8}, \frac{5\pi}{8}, \frac{9\pi}{8}, \frac{13\pi}{8}$$

25.
$$y = \frac{1}{2}\sin\frac{3x}{4}$$
$$2y = \sin\frac{3x}{4}$$
$$\arcsin 2y = \frac{3x}{4}$$
$$\frac{4}{3}\arcsin 2y = x$$

26.
$$\arcsin\left(\frac{1}{\sqrt{2}}\right)$$
Quadrant I
$$\frac{\pi}{4}$$

27.
$$\arctan\left(-\frac{1}{\sqrt{3}}\right)$$
Quadrant IV
$$-\frac{\pi}{6}$$

28.
$$\sec^{-1}(-1)$$
$$\pi$$

29.
$$\cos^{-1}\left(-\frac{1}{2}\right)$$
Quadrant II
$$\frac{2\pi}{3}$$

30.
$$\sin\left[\arccos\left(-\frac{1}{2}\right)\right] =$$
$$\sin\left[\frac{2\pi}{3}\right] = \frac{\sqrt{3}}{2}$$

31.
$$\tan\left(\tan^{-1}\sqrt{3}\right) =$$
$$\tan\left(\frac{\pi}{3}\right) = \sqrt{3}$$

32.
$$\sin(\operatorname{arc}\cot x)$$
Let $\theta = \operatorname{arc}\cot x$

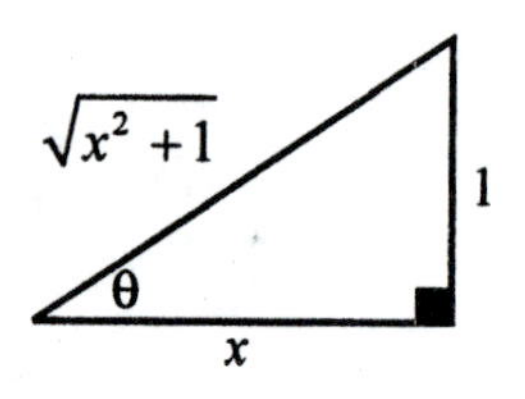

$$\sin\theta = \frac{1}{\sqrt{x^2+1}}$$
$$= \frac{\sqrt{x^2+1}}{\sqrt{x^2+1}\sqrt{x^2+1}}$$
$$= \frac{\sqrt{x^2+1}}{x^2+1}$$

33.
$$y = 1.5\arccos 2x$$
$$y = \frac{3}{2}\arccos 2x$$
$$\frac{2}{3}y = \arccos 2x$$
$$\cos\frac{2y}{3} = 2x$$
$$\frac{1}{2}\cos\frac{2y}{3} = x$$
Pick y and solve for x.
Note: The x limits on the graph are:
$$-1 \le 2x \le 1$$
$$-0.5 \le x \le 0.5$$
Also, the y intercept is $\left(0, \frac{3\pi}{4}\right)$

The y limits on the graph are:
$$0 \le \frac{2y}{3} \le \pi$$
$$0 \le 2y \le 3\pi$$
$$0 \le y \le \frac{3\pi}{2}$$

x	0.5	0.38	0.09	−0.25
y	0	$\frac{\pi}{3}$	$\frac{2\pi}{3}$	π

x	−0.47	−0.5
y	$\frac{4\pi}{3}$	$\frac{3\pi}{2}$

Chapter 14, Exercises 14.1

1. $\sqrt{-49} = 7j$

3. $\sqrt{-64} = 8j$

5. $\sqrt{-12} = \sqrt{12}j = \sqrt{4}\sqrt{3}j = 2\sqrt{3}j$

7. $\sqrt{-54} = \sqrt{54}j = \sqrt{9}\sqrt{6}j = 3\sqrt{6}j$

9. $j^{19} = j^{16} \cdot j^{3} = 1(-j) = -j$

11. $j^{22} = j^{20} \cdot j^{2} = 1(-1) = -1$

13. $j^{81} = j^{80} \cdot j^{1} = 1(j) = j$

15. $j^{246} = j^{244} \cdot j^{2} = j^{4(61)} \cdot j^{2} = 1(-1) = -1$

17. $(3+4j)+(9+2j) = (3+9)+(4+2)j = 12+6j$

19. $(4-9j)-(2-j) = 4-9j-2+j = 2-8j$

21. $(4+2j)+(-4-3j) = 4+2j-4-3j = -j$

23. $(2+j)(8-3j) = 16-6j+8j-3j^2 = 16+2j-3(-1) = 19+2j$

25. $(-4+5j)(3+2j) = -12+7j+10j^2 = -12+7j-10 = -22+7j$

27. $(2+5j)(2-5j) = 4+0j-25j^2 = 4+25 = 29$

29. $(-3+4j)^2 = 9-24j+16j^2 = 9-24j-16 = -7-24j$

31. $\dfrac{3+7j}{4-j} = \dfrac{(3+7j)}{(4-j)} \cdot \dfrac{(4+j)}{(4+j)} = \dfrac{12+31j+7j^2}{16-j^2} = \dfrac{5+31j}{17}$

33. $\dfrac{6-3j}{4+8j} = \dfrac{(6-3j)}{(4+8j)} \cdot \dfrac{(4-8j)}{(4-8j)} = \dfrac{24-60j+24j^2}{16-64j^2} = \dfrac{-60j}{80} = \dfrac{-3j}{4}$

35. $\dfrac{-9+8j}{6-2j}=\dfrac{(-9+8j)}{(6-2j)}\cdot\dfrac{(6+2j)}{(6+2j)}=\dfrac{-54+30j+16j^2}{36-4j^2}=\dfrac{-70+30j}{40}=\dfrac{-7+3j}{4}$

37.
$$x^2+4=0$$
$$x^2=-4$$
$$x=\pm\sqrt{-4}$$
$$x=\pm 2j$$

39.
$$3x^2+4x+9=0$$
$$a=3,\ b=4,\ c=9$$
$$x=\frac{-b\pm\sqrt{b^2-4ac}}{2a}$$
$$x=\frac{-4\pm\sqrt{4^2-4(3)(9)}}{2(3)}$$
$$x=\frac{-4\pm\sqrt{-92}}{6}$$
$$x=\frac{-4\pm\sqrt{4}\sqrt{23}j}{6}$$
$$x=\frac{-4\pm 2\sqrt{23}j}{6}$$
$$x=\frac{-2\pm\sqrt{23}j}{3}$$

41.
$$5x^2-2x+5=0$$
$$a=5,\ b=-2,\ c=5$$
$$x=\frac{-(-2)\pm\sqrt{(-2)^2-4(5)(5)}}{2(5)}$$
$$x=\frac{2\pm\sqrt{-96}}{10}$$
$$x=\frac{2\pm\sqrt{16}\sqrt{6}j}{10}$$
$$x=\frac{2\pm 4\sqrt{6}j}{10}$$
$$x=\frac{1\pm 2\sqrt{6}j}{5}$$

43.
$$3x^2-2x+1=0$$
$$a=3,\ b=-2,\ c=1$$
$$x=\frac{-(-2)\pm\sqrt{(-2)^2-4(3)(1)}}{2(3)}$$
$$x=\frac{2\pm\sqrt{-8}}{6}$$
$$x=\frac{2\pm 2\sqrt{2}j}{6}$$
$$x=\frac{1\pm\sqrt{2}j}{3}$$

45.
$$x^3+1=0$$
$$(x+1)(x^2-x+1)=0$$
$x+1=0$ or $x^2-x+1=0$

$x=-1$ or use quadratic formula

$$a=1,\ b=-1,\ c=1$$
$$x=\frac{-(-1)\pm\sqrt{(-1)^2-4(1)(1)}}{2(1)}$$
$$x=\frac{1\pm\sqrt{-3}}{2}$$
$x=\dfrac{1\pm\sqrt{3}j}{2}$ or -1

47. $x^4 - 1 = 0$

$(x^2+1)(x^2-1)=0$

$(x^2+1)(x+1)(x-1)=0$

$x^2+1=0$ or $x+1=0$ or $x-1=0$

$x^2=-1$ or $x=-1$ or $x=1$

$x=\pm\sqrt{-1}$ or -1 or 1

$x=\pm j$ or -1 or 1

49. $x^4+80x^2=0$

$x^2(x^2+80)=0$

$x^2=0$ or $x^2+80=0$

$x=0$ or $x^2=-80$

$x=0$ or $x=\pm\sqrt{-80}$

$x=0$ or $x=\pm\sqrt{16}\sqrt{5}j$

$x=0$ or $x=\pm4\sqrt{5}j$

51. $2x^4+54x=0$

$2x(x^3+27)=0$

$2x(x+3)(x^2-3x+9)=0$

$2x=0$ or $x+3=0$ or $x^2-3x+9=0$

$x=0$ or -3. Use quadratic formula

$a=1,\ b=-3,\ c=9$

$$x=\frac{-(-3)\pm\sqrt{(-3)^2-4(1)(9)}}{2(1)}$$

$$x=\frac{3\pm\sqrt{-27}}{2}$$

$$x=\frac{3\pm3\sqrt{3}j}{2} \text{ or } 0 \text{ or } -3$$

53. $x^5=x^2$

$x^5-x^2=0$

$x^2(x^3-1)=0$

$x^2(x-1)(x^2+x+1)=0$

$x^2=0$ or $x-1=0$ or

$x^2+x+1=0$

$x=0,\ 1,$ or use

$a=1,\ b=1,\ c=1$

$$x=\frac{-(1)\pm\sqrt{(1)^2-4(1)(1)}}{2(1)}$$

$$x=\frac{-1\pm\sqrt{-3}}{2}$$

$$x=\frac{-1\pm\sqrt{3}j}{2} \text{ or } 0 \text{ or } 1$$

For 55, 57, 59 see the text answer section.

61. $(3+2j)+(-2+j)$

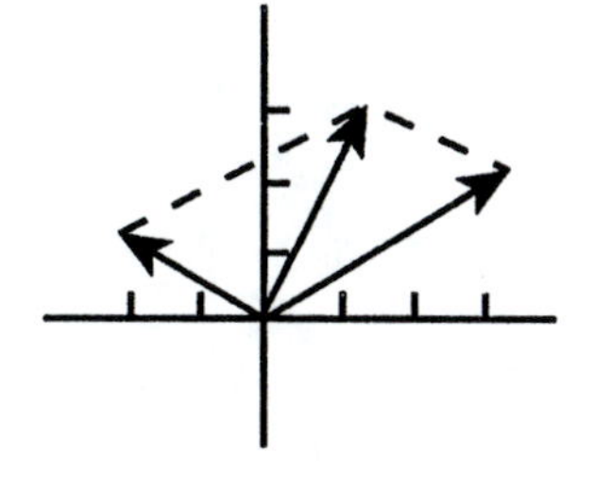

$1+3j$

63. $(5-3j)-(1+5j)=$

$(5-3j)+(-1-5j)$

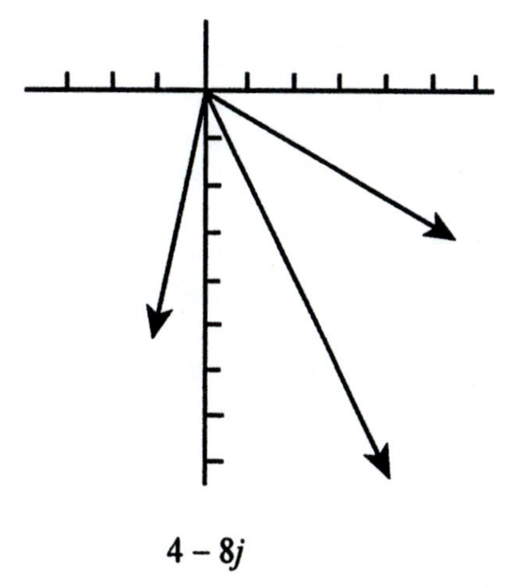

$4-8j$

65. $(-3-4j)-(-6+2j)=$
$(-3-4j)+(6-2j)$

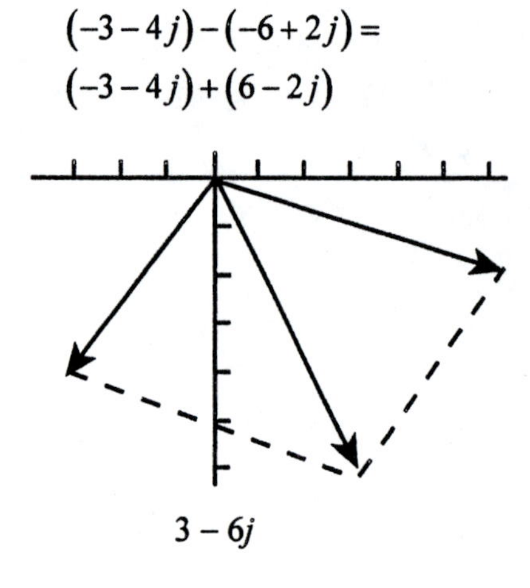

$3-6j$

67. $(2+4j)+(-2+3j)$

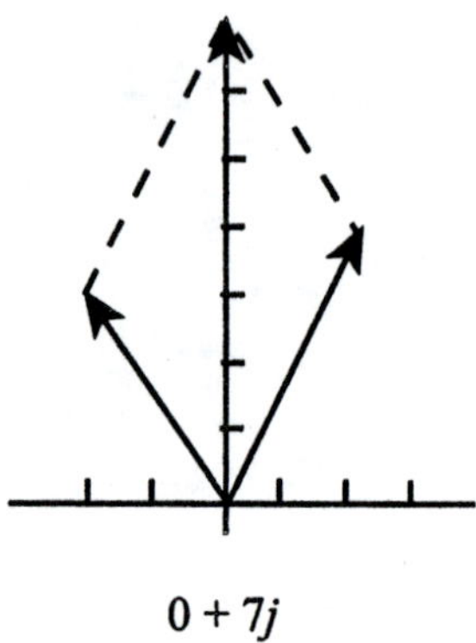

$0+7j$

Exercises 14.2

1. $2+2j$
$a=2,\ b=2$
$r=\sqrt{a^2+b^2}$
$r=\sqrt{2^2+2^2}$
$r=\sqrt{8}=2\sqrt{2}$
θ is in quadrant I
$\tan\theta=\dfrac{b}{a}$
$\tan\theta=\dfrac{2}{2}=1$
$\theta=45°$
$2\sqrt{2}(\cos 45°+j\sin 45°)$

3. $-1-\sqrt{3}j$
$a=-1,\ b=-\sqrt{3}$
$r=\sqrt{a^2+b^2}$
$r=\sqrt{(-1)^2+(-\sqrt{3})^2}$
$r=\sqrt{4}=2$
θ is in quadrant III
$\tan\theta=\dfrac{b}{a}=\dfrac{-\sqrt{3}}{-1}=\sqrt{3}$
$\theta=180°+60°=240°$
$2(\cos 240°+j\sin 240°)$

5. $4j$
$r=\sqrt{0^2+4^2}$
$r=\sqrt{16}=4$
The number is on the positive y axis, thus,
$\theta=90°$
$4(\cos 90°+j\sin 90°)$

7. $-6-6j$
$r=\sqrt{(-6)^2+(-6)^2}$
$r=\sqrt{72}=6\sqrt{2}$
θ is in quadrant III
$\tan\theta=\dfrac{-6}{-6}=1$
$\theta=180°+45°=225°$
$6\sqrt{2}(\cos 225°+j\sin 225°)$

9. $-2+3j$
$r=\sqrt{(-2)^2+3^2}$
$r=\sqrt{13}$
θ is in quadrant II
$\tan\theta=\dfrac{3}{-2}=-1.5$
$\theta=180°-56°$
$\theta=124°$
$\sqrt{13}(\cos 124°+j\sin 124°)$

11. $4(\cos 60°+j\sin 60°)=$
$4\left(\dfrac{1}{2}+\dfrac{\sqrt{3}}{2}j\right)=$
$2+2\sqrt{3}j$

13. $2(\cos 330° + j\sin 330°) =$

$2\left(\frac{\sqrt{3}}{2} - \frac{1}{2}j\right) =$

$\sqrt{3} - j$

15. $3\sqrt{2}(\cos 135° + j\sin 135°) =$

$3\sqrt{2}\left(\frac{-\sqrt{2}}{2} + \frac{\sqrt{2}}{2}j\right) =$

$-3 + 3j$

17. $3(\cos 270° + j\sin 270°) =$

$3(0 + j(-1)) =$

$-3j$

19. $\sqrt{53}(\cos 344° + j\sin 344°) =$

$\sqrt{53}(0.9613 - 0.2756j) =$

$7.00 - 2.01j$

21. $\sqrt{3} - j$

$r = \sqrt{(\sqrt{3})^2 + (-1)^2}$

$r = \sqrt{4} = 2$

θ is in quadrant IV

$\tan\theta = \frac{-1}{\sqrt{3}}$

$\theta = 360° - 30° = 330°$

$\theta = 330°\left(\frac{\pi}{180°}\right) = 5.76 \text{ rad}$

$2e^{5.76j}$

23. $-\sqrt{2} - \sqrt{2}j$

$r = \sqrt{(-\sqrt{2})^2 + (-\sqrt{2})^2}$

$r = \sqrt{4} = 2$

θ is in quadrant III

$\tan\theta = \frac{-\sqrt{2}}{-\sqrt{2}} = 1$

$\theta = 180° + 45° = 225°$

$\theta = 225°\left(\frac{\pi}{180°}\right) = 3.93 \text{ rad}$

$2e^{3.93j}$

25. $4 + 6j$

$r = \sqrt{4^2 + 6^2}$

$r = \sqrt{52} = \sqrt{4}\sqrt{13}$

$r = 2\sqrt{13}$

θ is in quadrant I

$\tan\theta = \frac{6}{4} = \frac{3}{2}$

$\theta = 0.983 \text{ rad}$

$2\sqrt{13}e^{0.983j}$ or $7.21e^{0.983j}$

27. $3e^{1.35j}$

$r = 3, \theta = 1.35 \text{ rad}$

$\theta = 1.35\left(\frac{180°}{\pi}\right) = 77°$

$3e^{1.35j} =$

$3(\cos 77° + j\sin 77°) =$

$3\cos 77° + 3\sin 77° j =$

$0.67 + 2.92j$

29. $4e^{5.76j}$

$r = 4, \theta = 5.76\,\text{rad}$

$\theta = 5.76\left(\dfrac{180°}{\pi}\right) = 330°$

$4e^{5.76j} =$

$4(\cos 330° + j\sin 330°) =$

$4\cos 330° + 4\sin 330° j =$

$4\left(\dfrac{\sqrt{3}}{2}\right) + 4\left(\dfrac{-1}{2}\right)j =$

$2\sqrt{3} - 2j$ or

$3.46 - 2j$

31. $2e^{j}$

$r = 2, \theta = 1\,\text{rad}$

$\theta = 1\left(\dfrac{180°}{\pi}\right) = 57.3°$

$2e^{j} =$

$2(\cos 57° + j\sin 57°) =$

$2\cos 57° + 2\sin 57° j =$

$1.09 + 1.68j$

Exercises 14.3

1. $(4e^{j})(7e^{3j}) =$

$4(7)e^{j+3j} =$

$28e^{4j}$

3. $(9e^{5j})(3e^{-3j}) =$

$9(3)e^{5j-3j} =$

$27e^{2j}$

5. $(6e^{5.6j})(4e^{3.7j}) =$

$6(4)e^{5.6j+3.7j} =$

$24e^{9.3j}$ or $24e^{3.0j}$

NOTE: $9.3\,\text{rad}$ has the same trig values as $9.3 - 2\pi = 3.0\,\text{rad}$

7. $(3e^{\pi j})\left(8e^{\pi j/3}\right) =$

$24e^{4\pi j/3}$

NOTE: $\dfrac{4\pi}{3} = 240°$

$24(\cos 240° + j\sin 240°)$

$24\left(\dfrac{-1}{2} - \dfrac{\sqrt{3}}{2}j\right) =$

$-12 - 12\sqrt{3}j$

9. $(4e^{3.7j})(20e^{6.1j}) =$

$80e^{9.8j}$

NOTE: $9.8 = 561°$

$561° - 360° = 201°$

$80(\cos 201° + j\sin 201°) =$

$-74.7 - 28.7j$

11. $(6e^{-1.4j})(9e^{-2.5j}) =$

$54e^{-3.9j}$

NOTE: $-3.9 = -223°$

$-223° + 360° = 137°$

$54(\cos 137° + j\sin 137°) =$

$-39.5 + 36.8j$

13. $[3(\cos 75° + j\sin 75°)][4(\cos 38° + j\sin 38°)] =$

$3(4)[\cos(75° + 38°) + j\sin(75° + 38°)] =$

$12(\cos 113° + j\sin 113°)$

15. $[3(\cos 150° + j\sin 150°)][3(\cos 150° + j\sin 150°)] =$
$3(3)[\cos(150° + 150°) + j\sin(150° + 150°)] =$
$9(\cos 300° + j\sin 300°)$

17. $[1(\cos 180° + j\sin 180°)][7(\cos 315° + j\sin 315°)] =$
$7[\cos(180° + 315°) + j\sin(180° + 315°)] =$
$7(\cos 495° + j\sin 495°) = 7(\cos 135° + j\sin 135°)$

19. $[5(\cos 50° + j\sin 50°)][5(\cos 10° + j\sin 10°)] =$
$25(\cos 60° + j\sin 60°) =$
$25\left(\frac{1}{2} + \frac{\sqrt{3}}{2}j\right) =$
$\frac{25}{2} + \frac{25\sqrt{3}}{2}j$

21. $[2(\cos 120° + j\sin 120°)][6(\cos 60° + j\sin 60°)] =$
$12(\cos 180° + j\sin 180°) =$
$12(-1 + 0j) = -12$

23. $[7(\cos 162° + j\sin 162°)][8(\cos 213° + j\sin 213°)] =$
$56(\cos 375° + j\sin 375°) = 56(\cos 15° + j\sin 15°) =$
$54.1 + 14.5j$

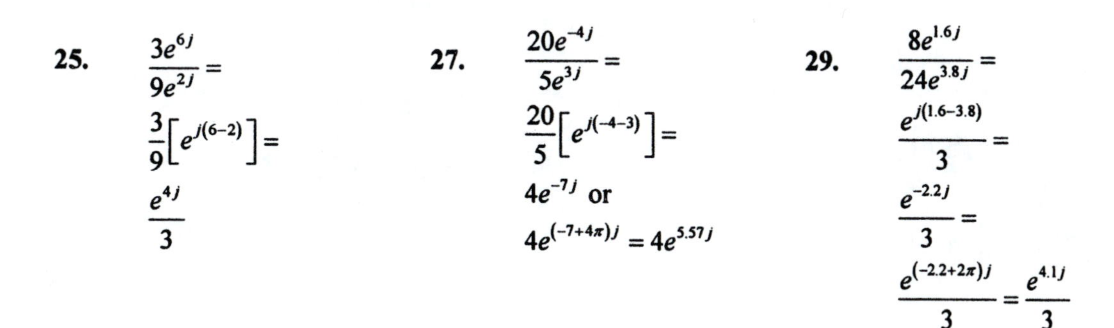

25. $\frac{3e^{6j}}{9e^{2j}} =$
$\frac{3}{9}\left[e^{j(6-2)}\right] =$
$\frac{e^{4j}}{3}$

27. $\frac{20e^{-4j}}{5e^{3j}} =$
$\frac{20}{5}\left[e^{j(-4-3)}\right] =$
$4e^{-7j}$ or
$4e^{(-7+4\pi)j} = 4e^{5.57j}$

29. $\frac{8e^{1.6j}}{24e^{3.8j}} =$
$\frac{e^{j(1.6-3.8)}}{3} =$
$\frac{e^{-2.2j}}{3} =$
$\frac{e^{(-2.2+2\pi)j}}{3} = \frac{e^{4.1j}}{3}$

31. $\dfrac{10e^{\pi j/6}}{5e^{\pi j}} = 2e^{\pi j/6 - \pi j} =$

$2e^{-5\pi j/6} = 2e^{7\pi j/6}$

NOTE: $\dfrac{7\pi}{6} = 180° + 30° = 210°$

$2e^{7\pi j/6} = 2(\cos 210° + j\sin 210°)$

$= 2\left(-\dfrac{\sqrt{3}}{2} - \dfrac{1}{2}j\right)$

$= -\sqrt{3} - j$

33. $\dfrac{14e^{4.6j}}{2e^{1.3j}} =$

$7e^{3.3j}$

NOTE: $3.3 = 189°$

$7e^{3.3j} =$

$7(\cos 189° + j\sin 189°) =$

$-6.91 - 1.10j$

35. $\dfrac{35e^{6.7j}}{7e^{-5.2j}} =$

$5e^{11.9j} = 5e^{5.62j}$

NOTE: $5.62 = 322°$

$5e^{5.62j} = 5(\cos 322° + j\sin 322°) =$

$3.94 - 3.08j$

37. $\dfrac{25(\cos 120° + j\sin 120°)}{5(\cos 50° + j\sin 50°)} =$

$\dfrac{25}{5}(\cos(120° - 50°) + j\sin(120° - 50°)) =$

$5(\cos 70° + j\sin 70°)$

39. $\dfrac{42(\cos 275° + j\sin 275°)}{7(\cos 156° + j\sin 156°)} =$

$6(\cos 119° + j\sin 119°)$

41. $\dfrac{40(\cos 86° + j\sin 86°)}{5(\cos 215° + j\sin 215°)} =$

$8(\cos(-129°) + j\sin(-129°)) =$

$8(\cos 231° + j\sin 231°)$

43. $\dfrac{72(\cos 240° + j\sin 240°)}{8(\cos 120° + j\sin 120°)} =$

$9(\cos 120° + j\sin 120°) =$

$9\left(-\dfrac{1}{2} + \dfrac{\sqrt{3}}{2}j\right) =$

$\dfrac{-9}{2} + \dfrac{9\sqrt{3}}{2}j$

45. $\dfrac{8(\cos 185° + j\sin 185°)}{40(\cos 35° + j\sin 35°)} =$

$\dfrac{1}{5}(\cos 150° + j\sin 150°) =$

$\dfrac{1}{5}\left(-\dfrac{\sqrt{3}}{2} + \dfrac{1}{2}j\right) =$

$\dfrac{-\sqrt{3}}{10} + \dfrac{1}{10}j$

47. $\dfrac{96(\cos 85° + j\sin 85°)}{16(\cos 145° + j\sin 145°)} =$

$6(\cos(-60°) + j\sin(-60°)) =$

$6(\cos 300° + j\sin 300°) =$

$6\left(\dfrac{1}{2} - \dfrac{\sqrt{3}}{2}j\right) = 3 - 3\sqrt{3}j$

49. $[2(\cos 60° + j\sin 60°)]^3 =$

$2^3(\cos(3\cdot 60°) + j\sin(3\cdot 60°)) =$

$2^3(\cos 180° + j\sin 180°) =$

$8(-1 + 0j) = -8$

51. $[3(\cos 157.5° + j\sin 157.5°)]^4 =$

$3^4(\cos(4\cdot 157.5°) + j\sin(4\cdot 157.5°)) =$

$3^4(\cos 630° + j\sin 630°) =$

$81(\cos 270° + j\sin 270°) =$

$81(0 - 1j) = -81j$

Exercises 14.4

1. $\left(3e^{1.4j}\right)^5 =$
$3^5 e^{5(1.4)j} =$
$243e^{7j}$ or
$243e^{0.72j}$

3. $\left(5e^{4.6j}\right)^2 =$
$5^2 e^{2(4.6j)} =$
$25e^{9.2j}$ or
$25e^{2.9j}$

5. $[3(\cos 20° + j\sin 20°)]^4 =$
$3^4\left(\cos(4 \cdot 20°) + j\sin(4 \cdot 20°)\right) =$
$81(\cos 80° + j\sin 80°)$

7. $[2(\cos 150° + j\sin 150°)]^5 =$
$2^5(\cos 750° + j\sin 750°) =$
$32(\cos 30° + j\sin 30°)$

9. $[2(\cos 240° + j\sin 240°)]^{-3} =$
$2^{-3}\left(\cos(-720°) + j\sin(-720°)\right) =$
$\frac{1}{8}(\cos 0° + j\sin 0°) = \frac{1}{8}$

11. $(1-j)^8$
$r = \sqrt{1^2 + (-1)^2} = \sqrt{2}$
θ is in quadrant IV
$\tan\theta = \frac{-1}{1} = -1$
$\theta = 360° - 45°$
$\theta = 315°$
$[\sqrt{2}(\cos 315° + j\sin 315°)]^8 =$
$\left(\sqrt{2}\right)^8(\cos 2520° + j\sin 2520°) =$
$16(\cos 0° + j\sin 0°) =$
$16(1 + 0j) = 16$

13. $\left(-2\sqrt{3} - 2j\right)^3$
$r = \sqrt{\left(-2\sqrt{3}\right)^2 + (-2)^2}$
$r = \sqrt{16} = 4$
θ is in quadrant III
$\tan\theta = \frac{-2}{-2\sqrt{3}} = \frac{1}{\sqrt{3}}$
$\theta = 180° + 30° = 210°$
$[4(\cos 210° + j\sin 210°)]^3 =$
$4^3(\cos 630° + j\sin 630°) =$
$64(0 - 1j) = -64j$

15. $\left(1 + \sqrt{3}j\right)^5$
$r = \sqrt{1^2 + \left(\sqrt{3}\right)^2} = 2$
θ is in quadrant I
$\tan\theta = \frac{\sqrt{3}}{1}$
$\theta = 60°$
$[2(\cos 60° + j\sin 60°)]^5 =$
$2^5(\cos 300° + j\sin 300°) =$
$32\left(\frac{1}{2} - \frac{\sqrt{3}}{2}j\right) =$
$16 - 16\sqrt{3}j$

17. $(-3 + 3j)^{-4}$
$r = \sqrt{(-3)^2 + (3)^2} = 3\sqrt{2}$
θ is in quadrant II
$\tan\theta = \frac{3}{-3} = -1$
$\theta = 180° - 45° = 135°$
$[3\sqrt{2}(\cos 135° + j\sin 135°)]^{-4} =$
$\left(3\sqrt{2}\right)^{-4}\left(\cos(-540°) + j\sin(-540°)\right) =$
$\frac{1}{324}(-1 + 0j) = -\frac{1}{324}$

19. $1 = 1(\cos 0° + j\sin 0°)$

Thus the cube roots are:

$$[1(\cos 0° + j\sin 0°)]^{1/3} =$$

$$1^{1/3}\left[\cos\left(\frac{0° + k(360°)}{3}\right) + j\sin\left(\frac{0° + k(360°)}{3}\right)\right]$$

for $k = 0, 1, 2$

$$k = 0:\ 1(\cos 0° + j\sin 0°) =$$

$$1(1 + 0j) = 1$$

$$k = 1:\ 1\left(\cos\left(\frac{360°}{3}\right) + j\sin\left(\frac{360°}{3}\right)\right) =$$

$$1(\cos 120° + j\sin 120°) =$$

$$1\left(-\frac{1}{2} + \frac{\sqrt{3}}{2}j\right) =$$

$$-\frac{1}{2} + \frac{\sqrt{3}}{2}j$$

$$k = 2:\ 1\left(\cos\left(\frac{720°}{3}\right) + j\sin\left(\frac{720°}{3}\right)\right) =$$

$$1(\cos 240° + j\sin 240°) =$$

$$-\frac{1}{2} - \frac{\sqrt{3}}{2}j$$

21. $j = 1(\cos 90° + j\sin 90°)$

The fifth roots are:

$$[1(\cos 90° + j\sin 90°)]^{1/5} =$$

$$1^{1/5}\left(\cos\left(\frac{90° + k(360°)}{5}\right) + j\sin\left(\frac{90° + k(360°)}{5}\right)\right)$$

$k = 0, 1, 2, 3, 4$

$$k = 0:\ 1\left(\cos\frac{90°}{5} + j\sin\frac{90°}{5}\right) =$$

$$1(\cos 18° + j\sin 18°) =$$

$$0.951 + 0.309j$$

$$k = 1:\ 1\left(\cos\left(\frac{90° + 360°}{5}\right) + j\sin\left(\frac{90° + 360°}{5}\right)\right) =$$

$$1(\cos 90° + j\sin 90°) = j$$

$$k = 2:\ 1\left(\cos\left(\frac{90° + 720°}{5}\right) + j\sin\left(\frac{90° + 720°}{5}\right)\right) =$$

$$1(\cos 162° + j\sin 162°) = -0.951 + 0.309j$$

$$k = 3:\ 1\left(\cos\left(\frac{90° + 1080°}{5}\right) + j\sin\left(\frac{90° + 1080°}{5}\right)\right)$$

$$1(\cos 234° + j\sin 234°) = -0.588 - 0.809j$$

$$k = 4:\ 1\left(\cos\left(\frac{90° + 1440°}{5}\right) + j\sin\left(\frac{90° + 1440°}{5}\right)\right)$$

$$1(\cos 306° + j\sin 306°) = 0.588 - 0.809j$$

23.

$$x^3 = 27(\cos 405° + j\sin 405°)$$

$$x = 27^{1/3}\left(\cos\frac{405° + k(360°)}{3} + j\sin\frac{405° + k(360°)}{3}\right)$$

for $k = 0, 1, 2$

$$k = 0:\ 3\left(\cos\frac{405°}{3} + j\sin\frac{405°}{3}\right) = 3(\cos 135° + j\sin 135°) =$$

$$-2.12 + 2.12j$$

$$k = 1:\ 3\left(\cos\frac{405° + 360°}{3} + j\sin\frac{405° + 360°}{3}\right) =$$

$$3(\cos 255° + j\sin 255°) = -0.776 - 2.90j$$

$$k = 2:\ 3\left(\cos\frac{405° + 720°}{3} + j\sin\frac{405° + 720°}{3}\right) =$$

$$3(\cos 375° + j\sin 375°) =$$

$$3(\cos 15° + j\sin 15°) = 2.90 + 0.776j$$

25.

$$x^2 = j = 1(\cos 90° + j\sin 90°)$$

$$x = 1^{1/2}\left(\cos\frac{90° + k(360°)}{2} + j\sin\frac{90° + k(360°)}{2}\right)$$

for $k = 0, 1$

$$k = 0:\ 1\left(\cos\frac{90°}{2} + j\sin\frac{90°}{2}\right) =$$

$$1(\cos 45° + j\sin 45°) = \frac{1}{\sqrt{2}} + \frac{1}{\sqrt{2}}j =$$

$$\frac{\sqrt{2}}{2} + \frac{\sqrt{2}j}{2}$$

$$k = 1:\ 1\left(\cos\frac{90° + 360°}{2} + j\sin\frac{90° + 360°}{2}\right) =$$

$$1(\cos 225° + j\sin 225°) = -\frac{\sqrt{2}}{2} - \frac{\sqrt{2}}{2}j$$

27.

$$x^5 = 1$$

$$x^5 = 1(\cos 0° + j\sin 0°)$$

$$x = 1^{1/5}\left(\cos\frac{0° + k(360°)}{5} + j\sin\frac{0° + k(360°)}{5}\right)$$

for $k = 0, 1, 2, 3, 4$

$$k = 0:\ 1(\cos 0° + j\sin 0°) = 1$$

$$k = 1:\ 1(\cos 72° + j\sin 72°) = 0.309 + 0.951j$$

$$k = 2:\ 1\left(\cos\frac{2(360°)}{5} + j\sin\frac{2(360°)}{5}\right) =$$

$$1(\cos 144° + j\sin 144°) = -0.809 + 0.588j$$

$$k = 3:\ 1\left(\cos\frac{3(360°)}{5} + j\sin\frac{3(360°)}{5}\right) =$$

$$1(\cos 216° + j\sin 216°) = -0.809 - 0.588j$$

$$k = 4:\ 1\left(\cos\frac{4(360°)}{5} + j\sin\frac{4(360°)}{5}\right) =$$

$$1(\cos 288° + j\sin 288°) = 0.309 - 0.951j$$

29.

$$x^5 = -1$$

$$x^5 = 1(\cos 180° + j\sin 180°)$$

$$x = 1^{1/5}\left(\cos\frac{180° + k(360°)}{5} + j\sin\frac{180° + k(360°)}{5}\right)$$

for $k = 0, 1, 2, 3, 4$

$$k = 0:\ 1\left(\cos\frac{180°}{5} + j\sin\frac{180°}{5}\right) =$$

$$1(\cos 36° + j\sin 36°) = 0.809 + 0.588j$$

$$k = 1:\ 1\left(\cos\frac{180° + 360°}{5} + j\sin\frac{180° + 360°}{5}\right) =$$

$$\cos 108° + j\sin 108° = -0.309 + 0.951j$$

$$k = 2:\ 1\left(\cos\frac{180° + 720°}{5} + j\sin\frac{180° + 720°}{5}\right) =$$

$$\cos 180° + j\sin 180° = -1$$

$$k = 3:\ 1\left(\cos\frac{180° + 1080°}{5} + j\sin\frac{180° + 1080°}{5}\right) =$$

$$\cos 252° + j\sin 252° = -0.309 - 0.951j$$

$$k = 4:\ 1\left(\cos\frac{180° + 1440°}{5} + j\sin\frac{180° + 1440°}{5}\right) =$$

$$1(\cos 324° + j\sin 324°) = 0.809 - 0.588j$$

31.

$$x^4 = -16$$

$$x^4 = 16(\cos 180° + j\sin 180°)$$

$$x = 16^{1/4}\left(\cos\frac{180° + k(360°)}{4} + j\sin\frac{180° + k(360°)}{4}\right)$$

for $k = 0, 1, 2, 3$

$$k = 0:\ 2\left(\cos\frac{180°}{4} + j\sin\frac{180°}{4}\right) =$$

$$2(\cos 45° + j\sin 45°) = 2\left(\frac{\sqrt{2}}{2} + j\frac{\sqrt{2}}{2}\right) = \sqrt{2} + \sqrt{2}j$$

$$k = 1:\ 2\left(\cos\frac{180° + 360°}{4} + j\sin\frac{180° + 360°}{4}\right) =$$

$$2(\cos 135° + j\sin 135°) = 2\left(\frac{-\sqrt{2}}{2} + j\frac{\sqrt{2}}{2}\right) = -\sqrt{2} + \sqrt{2}j$$

$$k = 2:\ 2\left(\cos\frac{180° + 720°}{4} + j\sin\frac{180° + 720°}{4}\right) =$$

$$2(\cos 225° + j\sin 225°) = 2\left(\frac{-\sqrt{2}}{2} - \frac{\sqrt{2}}{2}j\right) = -\sqrt{2} - \sqrt{2}j$$

$$k = 3:\ 2\left(\cos\frac{180° + 1080°}{4} + j\sin\frac{180° + 1080°}{4}\right) =$$

$$2(\cos 315° + j\sin 315°) = 2\left(\frac{\sqrt{2}}{2} - \frac{\sqrt{2}}{2}j\right) = \sqrt{2} - \sqrt{2}j$$

33.

$$x^3 = -2 + 11j$$

$$r = \sqrt{(-2)^2 + 11^2} = \sqrt{125}$$

$$r = 5\sqrt{5}$$

θ is in quadrant II

$$\tan\theta = \frac{11}{-2}$$

$$\theta = 180° - 80° = 100°$$

$$x^3 = 5\sqrt{5}\left(\cos 100° + j\sin 100°\right)$$

NOTE: $\left(125^{1/2}\right)^{1/3} = \left(125^{1/3}\right)^{1/2} = 5^{1/2}$

$$x = \sqrt{5}\left(\cos\frac{100° + k(360°)}{3} + j\sin\frac{100° + k(360°)}{3}\right)$$

for $k = 0, 1, 2$

$$k = 0: \sqrt{5}\left(\cos\frac{100°}{3} + j\sin\frac{100°}{3}\right) =$$

$$\sqrt{5}\left(\cos 33° + j\sin 33°\right) = 1.88 + 1.22j$$

$$k = 1: \sqrt{5}\left(\cos\frac{100° + 360°}{3} + j\sin\frac{100° + 360°}{3}\right) =$$

$$\sqrt{5}\left(\cos 153° + j\sin 153°\right) = -1.99 + 1.02j$$

$$k = 2: \sqrt{5}\left(\cos\frac{100° + 720°}{3} + j\sin\frac{100° + 720°}{3}\right) =$$

$$\sqrt{5}\left(\cos 273° + j\sin 273°\right) = 0.117 - 2.23j$$

Chapter 14 Review

1. $\sqrt{-81} = 9j$

2. $\sqrt{-18} = \sqrt{9}\sqrt{2}j = 3\sqrt{2}j$

3. $j^{18} = \left(j^4\right)^4 \cdot j^2 = -1$

4. $j^{23} = \left(j^4\right)^5 \cdot j^3 = -j$

5. $j^{48} = \left(j^4\right)^{12} = 1$

6. $j^{145} = \left(j^4\right)^{36} \cdot j^1 = j$

7. $(9+3j)+(-4+7j) = (9+-4)+(3j+7j) = 5+10j$

8. $(-1+j)-(-4+5j) = -1+j+4-5j = 3-4j$

9. $(5+2j)(6-7j) = 5(6)+5(-7j)+6(2j)+(-7)(2)j^2 =$
$30-35j+12j-14j^2 = 30-23j-14(-1) = 44-23j$

10. $(3-2j)^2 = 3^2+2(3)(-2)j+(-2j)^2 = 9-12j+4j^2 = 9-12j+4(-1) = 5-12j$

11. $\dfrac{1-2j}{4+j} = \dfrac{(1-2j)}{(4+j)} \cdot \dfrac{(4-j)}{(4-j)} = \dfrac{4-9j+2j^2}{16-j^2} = \dfrac{2-9j}{17}$

12. $\dfrac{5+j}{3-7j} = \dfrac{(5+j)(3+7j)}{(3-7j)(3+7j)} = \dfrac{15+38j+7j^2}{9-49j^2} = \dfrac{8+38j}{58} = \dfrac{4+19j}{29}$

13.
$$\frac{4+3j}{1-2j} = \frac{4+3j}{1-2j} \cdot \frac{1+2j}{1+2j}$$
$$= \frac{4+11j+6j^2}{1-4j^2}$$
$$= \frac{-2+11j}{5}$$
$$= \frac{-2}{5} + \frac{11j}{5}$$

14.
$$\frac{7-2j}{3+j} = \frac{7-2j}{3+j} \cdot \frac{3-j}{3-j}$$
$$= \frac{21-13j+2j^2}{9-j^2}$$
$$= \frac{19-13j}{10}$$
$$= \frac{19}{10} - \frac{13j}{10}$$

15.

$$x^2+36=0$$
$$x^2=-36$$
$$x=\pm\sqrt{-36}$$
$$x=\pm 6j$$

16.

$$2x^2+3x+2=0$$
$$x=\frac{-b\pm\sqrt{b^2-4ac}}{2a}$$
$$x=\frac{-3\pm\sqrt{3^2-4(2)(2)}}{2(2)}$$
$$x=\frac{-3\pm\sqrt{-7}}{4}$$
$$x=\frac{-3\pm\sqrt{7}j}{4}$$

17.

$$3x^2-5x+4=0$$
$a=3,\ b=-5,\ c=4$ in.
$$x=\frac{-b\pm\sqrt{b^2-4ac}}{2a}$$
$$x=\frac{5\pm\sqrt{25-4(3)(4)}}{2(3)}$$
$$x=\frac{5\pm\sqrt{-23}}{6}$$
$$x=\frac{5}{6}\pm\frac{\sqrt{23}}{6}j$$

18.

$$2x^2+50=0$$
$$2x^2=-50$$
$$x^2=-25$$
$$x=\pm\sqrt{-25}$$
$$x=\pm 5j$$

19.

$$-1+j$$
$$r=\sqrt{(-1)^2+(1)^2}$$
$$r=\sqrt{2}$$
θ is in quadrant II
$$\tan\theta=\frac{1}{-1}=-1$$
$$\theta=180°-45°=135°$$
$$-1+j=\sqrt{2}(\cos 135°+j\sin 135°)$$
$$-1+j=\sqrt{2}e^{3\pi j/4} \text{ or } \sqrt{2}e^{2.36j}$$

20.

$$1-\sqrt{3}j$$
$$r=\sqrt{1^2+\left(-\sqrt{3}\right)^2}$$
$$r=\sqrt{4}=2$$
θ is in quadrant IV
$$\tan\theta=\frac{-\sqrt{3}}{1}=-\sqrt{3}$$
$$\theta=360°-60°=300°$$
$$1-\sqrt{3}j=2(\cos 300°+j\sin 300°)$$
$$1-\sqrt{3}j=2e^{5\pi j/3} \text{ or } 2e^{5.24j}$$

21. $3-5j$

$r=\sqrt{(3)^2+(-5)^2}$

$r=\sqrt{34}$

θ is in quadrant IV

$\tan\theta=\dfrac{-5}{3}$

$\theta=360°-59°=301°$

$3-5j=\sqrt{34}\left(\cos 301°+j\sin 301°\right)$

$3-5j=\sqrt{34}e^{5.25j}$

22. $-2-8j$

$r=\sqrt{(-2)^2+(-8)^2}$

$r=\sqrt{68}=2\sqrt{17}$

θ is in quadrant III

$\tan\theta=\dfrac{-8}{-2}=4$

$\theta=180°+76°=256°$

$-2-8j=2\sqrt{17}\left(\cos 256°+j\sin 256°\right)$

$-2-8j=2\sqrt{17}e^{4.47j}$

23. $6\left(\cos 315°+j\sin 315°\right)=$

$6\left(\dfrac{\sqrt{2}}{2}-\dfrac{\sqrt{2}}{2}j\right)=$

$3\sqrt{2}-3\sqrt{2}j$

NOTE: $315°=\dfrac{7\pi}{4}$ rad

$6\left(\cos 315°+j\sin 315°\right)=$

$6e^{7\pi j/4}$ or $6e^{5.50j}$

24. $4\left(\cos 210°+j\sin 210°\right)=$

$4\left(-\dfrac{\sqrt{3}}{2}-\dfrac{1}{2}j\right)=$

$-2\sqrt{3}-2j$

NOTE: $210°=\dfrac{7\pi}{6}$ rad

$4\left(\cos 210°+j\sin 210°\right)=$

$4e^{7\pi j/6}$ or $4e^{3.67j}$

25. $5.52\left(\cos 105°+j\sin 105°\right)=$

$5.52\left(-0.2588+j\left(0.9659\right)\right)=$

$-1.43+5.33j$

NOTE: $105°=1.83$ rad

$5.52\left(\cos 105°+j\sin 105°\right)=$

$5.52e^{1.83j}$

26. $6.50\left(\cos 253°+j\sin 253°\right)=$

$6.50\left(-0.2924-0.9563j\right)=$

$-1.90-6.22j$

NOTE: $253°=4.42$ rad

$6.50\left(\cos 253°+j\sin 253°\right)=$

$6.50e^{4.42j}$

27. $2e^{0.489j}$

$0.489\text{ rad}=28°$

$2\left(\cos 28°+j\sin 28°\right)=$

$1.77+0.939j$

28. $3e^{8.75j}$

$8.75\text{ rad}=501°$

$501°-360°=141°$

$3\left(\cos 141°+j\sin 141°\right)=$

$-2.33+1.89j$

29. $7e^{1.46j}$

$1.46 \text{ rad} = 84°$

$7(\cos 84° + j\sin 84°)$

$0.732 + 6.96j$

30. $8e^{8.59j}$

$8.59 \text{ rad} = 492°$

use $492° - 360° = 132°$

$8(\cos 132° + j\sin 132°)$

$-5.35 + 5.95j$

31. $(5e^{2j})(3e^{3j}) =$

$15e^{5j}$

32. $[2(\cos 150° + j\sin 150°)][4(\cos 300° + j\sin 300°)] =$

$8[\cos(150° + 300°) + j\sin(150° + 300°)] =$

$8[\cos 450° + j\sin 450°] =$

$8[\cos 90° + j\sin 90°]$

33. $\dfrac{12e^{3j}}{3e^{-2j}} =$

$4e^{5j}$

34. $\dfrac{24(\cos 150° + j\sin 150°)}{8(\cos 275° + j\sin 275°)} =$

$3(\cos(-125°) + j\sin(-125°)) =$

$3(\cos 235° + j\sin 235°)$

35. $(4e^{2j})^3 =$

$4^3 e^{6j} =$

$64e^{6j}$

36. $[2(\cos 60° + j\sin 60°)]^7 =$

$2^7(\cos 420° + j\sin 420°) =$

$128(\cos 60° + j\sin 60°)$

37. $(-2 + 2j)^4$

$r = \sqrt{(-2)^2 + 2^2} =$

$\sqrt{8} = 2\sqrt{2}$

θ is in quadrant II

$\tan\theta = \dfrac{2}{-2} = -1$

$\theta = 180° - 45° = 135°$

$[2\sqrt{2}(\cos 135° + j\sin 135°)]^4 =$

$(2\sqrt{2})^4(\cos 540° + j\sin 540°) =$

$64(\cos 180° + j\sin 180°) =$

$64(-1 + 0j) = -64$

38. $(1 + j\sqrt{3})^6$

$r = \sqrt{1^2 + (\sqrt{3})^2} =$

$\sqrt{4} = 2$

θ is in quadrant I

$\tan\theta = \dfrac{\sqrt{3}}{1}$

$\theta = 60°$

$[2(\cos 60° + j\sin 60°)]^6 =$

$2^6(\cos 360° + j\sin 360°) =$

$64(1 + 0j) = 64$

39. $(1+j)^{-4}$

$r=\sqrt{1^2+1^2}=$

$\sqrt{2}$

θ is in quadrant I

$\tan\theta=\frac{1}{1}=1$

$\theta=45°$

$[\sqrt{2}(\cos 45°+j\sin 45°)]^{-4}=$

$(\sqrt{2})^{-4}(\cos(-180°)+j\sin(-180°))=$

$\frac{1}{4}(\cos 180°+j\sin(180°))=$

$\frac{1}{4}(-1+0j)=-\frac{1}{4}$

40. $x^3=j$

$x^3=1(\cos 90°+j\sin 90°)$

$x=1^{1/3}\left(\cos\frac{90°+k(360°)}{3}+j\sin\frac{90°+k(360°)}{3}\right)$

for $k=0, 1, 2$

$k=0:\ 1(\cos 30°+j\sin 30°)=\frac{\sqrt{3}}{2}+\frac{1}{2}j$

$k=1:\ 1\left(\cos\frac{450°}{3}+j\sin\frac{450°}{3}\right)=$

$\cos 150°+j\sin 150°=\frac{-\sqrt{3}}{2}+\frac{1}{2}j$

$k=2:\ 1\left(\cos\frac{810°}{3}+j\sin\frac{810°}{3}\right)=$

$\cos 270°+j\sin 270°=0-j=-j$

41. $x^4=-1$

$x^4=1(\cos 180°+j\sin 180°)$

$x=1^{1/4}\left(\cos\frac{180°+k(360°)}{4}+j\sin\frac{180°+k(360°)}{4}\right)$

for $k=0, 1, 2, 3$

$k=0:\ 1(\cos 45°+j\sin 45°)=\frac{\sqrt{2}}{2}+\frac{\sqrt{2}}{2}j$

$k=1:\ 1(\cos 135°+j\sin 135°)=\frac{-\sqrt{2}}{2}+\frac{\sqrt{2}}{2}j$

$k=2:\ 1(\cos 225°+j\sin 225°)=-\frac{\sqrt{2}}{2}-\frac{\sqrt{2}}{2}j$

$k=3:\ 1(\cos 315°+j\sin 315°)=\frac{\sqrt{2}}{2}-\frac{\sqrt{2}}{2}j$

42. $x^4=16$

$x^4=16(\cos 0°+j\sin 0°)$

$x=16^{1/4}\left(\cos\frac{0°+k(360°)}{4}+j\sin\frac{0°+k(360°)}{4}\right)$

for $k=0, 1, 2, 3$

$k=0:\ 2(\cos 0°+j\sin 0°)=2(1+0j)=2$

$k=1:\ 2(\cos 90°+j\sin 90°)=2(0+j)=2j$

$k=2:\ 2(\cos 180°+j\sin 180°)=1(-1+0j)=-2$

$k=3:\ 2(\cos 270°+j\sin 270°)=2(0-j)=-2j$

Chapter 14 Review

43. $x^5 = 4 - 4j$

$r = \sqrt{4^2 + (-4)^2} =$

$\sqrt{32} = 4\sqrt{2}$

θ is in quadrant IV

$\tan\theta = \frac{-4}{4} = -1$

$\theta = 360° - 45° = 315°$

$x^5 = 4\sqrt{2}(\cos 315° + j\sin 315°)$

$$x = \left(4\sqrt{2}\right)^{1/5}\left(\cos\frac{315° + k(360°)}{5} + j\sin\frac{315° + (360°)}{5}\right)$$

for $k = 0, 1, 2, 3, 4$

NOTE: $\left(\sqrt{32}\right)^{1/5} = \left(32^{1/2}\right)^{1/5} = \left(32^{1/5}\right)^{1/2} = 2^{1/2} = \sqrt{2}$

$k = 0: \sqrt{2}(\cos 63° + j\sin 63°) = 0.642 + 1.26j$

$k = 1: \sqrt{2}(\cos 135° + j\sin 135°) = \sqrt{2}\left(\frac{-1}{\sqrt{2}} + \frac{1}{\sqrt{2}}j\right) = -1 + j$

$k = 2: \sqrt{2}(\cos 207° + j\sin 207°) = -1.26 - 0.642j$

$k = 3: \sqrt{2}(\cos 279° + j\sin 279°) = 0.221 - 1.40j$

$k = 4: \sqrt{2}(\cos 351° + j\sin 351°) = 1.40 - 0.221j$

Chapter 15, Exercises 15.1

1. 2×2
(2 rows, 2 columns)

3. 2×4
(2 rows, 4 columns)

5. 4×1
(4 rows, 1 column)

7. 1×2
(1 row, 2 columns)

9. $\begin{bmatrix} x & 4 \\ 3 & -5 \end{bmatrix} = \begin{bmatrix} 2 & y \\ 3 & z \end{bmatrix}$

Corresponding entries are equal.

$x = 2,\ 4 = y,\ -5 = z$

11. $\begin{bmatrix} x & 2y & 3z \\ a+1 & b-3 & \frac{c}{4} \end{bmatrix} = \begin{bmatrix} 1 & 6 & -12 \\ 5 & 9 & -3 \end{bmatrix}$

$x = 1,\ 2y = 6,$ so $y = 3,$

$3z = -12$ so $z = -4,$

$a+1 = 5$ so $a = 4,$

$b-3 = 9$ so $b = 12,$

$\frac{c}{4} = -3$ so $c = -12$

13. $\begin{bmatrix} 2x+y \\ x-y \end{bmatrix} = \begin{bmatrix} 2 \\ 10 \end{bmatrix}$

$2x + y = 2$

$x - y = 10$

From the first equation, $y = -2x + 2.$

The second equation becomes:

$x - (-2x+2) = 10$

$x + 2x - 2 = 10$

$3x = 12$

$x = 4.$ Then

$y = -2(4) + 2$

$y = -6$

15. $\begin{bmatrix} x & x-y \\ x+z & y-2z \\ x-2w & y+w \end{bmatrix} = \begin{bmatrix} 3 & -2 \\ 13 & -15 \\ -9 & 11 \end{bmatrix}$

$x = 3 \qquad x - y = -2$

$x + z = 13 \qquad y - 2z = -15$

$x - 2w = -9 \qquad y + w = 11$

Use substitution:

$x = 3$

$x - y = -2$, $3 - y = -2$, $y = 5$

$x + z = 13$, $3 + z = 13$, $z = 10$

$x - 2w = -9$, $3 - 2w = -9$, $w = 6$

17. $\begin{bmatrix} 4 & 7 & -2 \\ -9 & 0 & 6 \end{bmatrix} + \begin{bmatrix} 4 & -3 & -1 \\ 5 & 8 & 10 \end{bmatrix} =$

$\begin{bmatrix} 4+4 & 7-3 & -2-1 \\ -9+5 & 0+8 & 6+10 \end{bmatrix} = \begin{bmatrix} 8 & 4 & -3 \\ -4 & 8 & 16 \end{bmatrix}$

19. $\begin{bmatrix} 1 & 0 & 5 \\ 11 & -9 & 1 \\ 4 & -2 & -8 \end{bmatrix} + \begin{bmatrix} 4 & -10 & 3 \\ -7 & 2 & 0 \\ 6 & 3 & -6 \end{bmatrix} = \begin{bmatrix} 5 & -10 & 8 \\ 4 & -7 & 1 \\ 10 & 1 & -14 \end{bmatrix}$

21. $-A = -\begin{bmatrix} 1 & 10 & 6 \\ 5 & -2 & 0 \end{bmatrix} = \begin{bmatrix} -1 & -10 & -6 \\ -5 & 2 & 0 \end{bmatrix}$

23. $3A = 3\begin{bmatrix} 1 & 10 & 6 \\ 5 & -2 & 0 \end{bmatrix} = \begin{bmatrix} 3 & 30 & 18 \\ 15 & -6 & 0 \end{bmatrix}$

25. $-2C = -2\begin{bmatrix} 1 & -9 \\ 4 & 3 \end{bmatrix} = \begin{bmatrix} -2 & 18 \\ -8 & -6 \end{bmatrix}$

27. $A + B = \begin{bmatrix} 1 & 10 & 6 \\ 5 & -2 & 0 \end{bmatrix} + \begin{bmatrix} 0 & -2 & -8 \\ 4 & 7 & 5 \end{bmatrix} = \begin{bmatrix} 1 & 8 & -2 \\ 9 & 5 & 5 \end{bmatrix}$

29. $B - A = B + -A = \begin{bmatrix} 0 & -2 & -8 \\ 4 & 7 & 5 \end{bmatrix} + \begin{bmatrix} -1 & -10 & -6 \\ -5 & 2 & 0 \end{bmatrix} = \begin{bmatrix} -1 & -12 & -14 \\ -1 & 9 & 5 \end{bmatrix}$

31. $B + C$, not defined since B and C do not have the same dimensions.

33. $3A - 4B = 3A + (-4)B =$

$3\begin{bmatrix} 1 & 10 & 6 \\ 5 & -2 & 0 \end{bmatrix} + (-4)\begin{bmatrix} 0 & -2 & -8 \\ 4 & 7 & 5 \end{bmatrix} =$

$\begin{bmatrix} 3 & 30 & 18 \\ 15 & -6 & 0 \end{bmatrix} + \begin{bmatrix} 0 & 8 & 32 \\ -16 & -28 & -20 \end{bmatrix} = \begin{bmatrix} 3 & 38 & 50 \\ -1 & -34 & -20 \end{bmatrix}$

35. $2A + 3C$, not defined since A and C do not have the same dimensions.

37. $D + E = E + D$

$\begin{bmatrix} 3 & -4 & 5 \\ -2 & 0 & 10 \end{bmatrix} + \begin{bmatrix} 4 & -6 & 2 \\ 1 & 5 & -8 \end{bmatrix} = \begin{bmatrix} 3+4 & -4+-6 & 5+2 \\ -2+1 & 0+5 & 10+-8 \end{bmatrix} =$

$\begin{bmatrix} 4+3 & -6+-4 & 2+5 \\ 1+-2 & 5+0 & -8+10 \end{bmatrix} = \begin{bmatrix} 4 & -6 & -2 \\ 1 & 5 & -8 \end{bmatrix} + \begin{bmatrix} 3 & -4 & 5 \\ -2 & 0 & 10 \end{bmatrix}$

39. $E + (-E) = O$

$\begin{bmatrix} 4 & -6 & 2 \\ 1 & 5 & -8 \end{bmatrix} + \begin{bmatrix} -4 & 6 & -2 \\ -1 & -5 & 8 \end{bmatrix} = \begin{bmatrix} 0 & 0 & 0 \\ 0 & 0 & 0 \end{bmatrix} = O$

41. $D+O=D$

$$\begin{bmatrix} 3 & -4 & 5 \\ -2 & 0 & 10 \end{bmatrix} + \begin{bmatrix} 0 & 0 & 0 \\ 0 & 0 & 0 \end{bmatrix} = \begin{bmatrix} 3 & -4 & 5 \\ -2 & 0 & 10 \end{bmatrix}$$

43. Double B ie

$$2B = 2\begin{bmatrix} 85 & 65 & 48 \\ 70 & 60 & 25 \\ 34 & 25 & 15 \end{bmatrix} = \begin{bmatrix} 170 & 130 & 96 \\ 140 & 120 & 50 \\ 68 & 50 & 30 \end{bmatrix}$$

45. The inventory for July 1 is: $A+B-C=$

$$\begin{bmatrix} 6 & 3 & 4 & 5 \\ 7 & 4 & 3 & 2 \\ 1 & 3 & 0 & 2 \\ 2 & 3 & 4 & 0 \end{bmatrix} + \begin{bmatrix} 24 & 30 & 14 & 10 \\ 20 & 23 & 12 & 8 \\ 15 & 19 & 11 & 4 \\ 15 & 17 & 16 & 5 \end{bmatrix} - \begin{bmatrix} 27 & 25 & 12 & 9 \\ 25 & 22 & 14 & 0 \\ 12 & 6 & 3 & 4 \\ 10 & 15 & 18 & 6 \end{bmatrix} = \begin{bmatrix} 3 & 8 & 6 & 6 \\ 2 & 5 & 1 & 10 \\ 4 & 16 & 8 & 2 \\ 7 & 5 & 2 & 1 \end{bmatrix}$$

Exercises 15.2

1. 1×3 by $3\times 1 = 1\times 1$

$$\begin{bmatrix} 5 & 2 & 4 \end{bmatrix}\begin{bmatrix} 1 \\ -1 \\ 3 \end{bmatrix} = [5(1)+2(-1)+4(3)] = [15]$$

3. 1×3 by $3\times 2 = 1\times 2$

$$\begin{bmatrix} 3 & 1 & -1 \end{bmatrix}\begin{bmatrix} -2 & 4 \\ 6 & 5 \\ 0 & 2 \end{bmatrix} = [3(-2)+1(6)+(-1)(0) \quad 3(4)+1(5)+(-1)(2)] = \begin{bmatrix} 0 & 15 \end{bmatrix}$$

5. 2×2 by $2\times 2 = 2\times 2$

$$\begin{bmatrix} 3 & 0 \\ 2 & 1 \end{bmatrix}\begin{bmatrix} -1 & 2 \\ 0 & 5 \end{bmatrix} = \begin{bmatrix} 3(-1)+0(0) & 3(2)+0(5) \\ 2(-1)+1(0) & 2(2)+1(5) \end{bmatrix} = \begin{bmatrix} -3 & 6 \\ -2 & 9 \end{bmatrix}$$

7. 2×2 by $2\times 3 = 2\times 3$

$$\begin{bmatrix} 3 & -1 \\ 0 & 4 \end{bmatrix}\begin{bmatrix} 4 & 3 & -1 \\ 5 & 0 & 2 \end{bmatrix} = \begin{bmatrix} 7 & 9 & -5 \\ 20 & 0 & 8 \end{bmatrix}$$

9. 2×3 by $3\times1=2\times1$

$$\begin{bmatrix}4&1&2\\6&0&-3\end{bmatrix}\begin{bmatrix}8\\-1\\-3\end{bmatrix}=\begin{bmatrix}25\\57\end{bmatrix}$$

11. 2×3 by 2×2 not defined.

13. 3×2 by $2\times3=3\times3$

$$\begin{bmatrix}1&2\\4&-3\\0&5\end{bmatrix}\begin{bmatrix}2&-5&0\\2&-3&4\end{bmatrix}=\begin{bmatrix}6&-11&8\\2&-11&-12\\10&-15&20\end{bmatrix}$$

15. 4×2 by $2\times4=4\times4$

$$\begin{bmatrix}1&7\\6&-3\\0&4\\-2&5\end{bmatrix}\begin{bmatrix}4&6&-4&0\\-2&5&1&3\end{bmatrix}=\begin{bmatrix}-10&41&3&21\\30&21&-27&-9\\-8&20&4&12\\-18&13&13&15\end{bmatrix}$$

17. 2×4 by 2×4 is not defined.

19. 3×3 by $3\times3=3\times3$

$$\begin{bmatrix}1&2&-2\\-5&3&0\\6&4&-1\end{bmatrix}\begin{bmatrix}-3&0&2\\4&1&-4\\7&5&6\end{bmatrix}=\begin{bmatrix}-9&-8&-18\\27&3&-22\\-9&-1&-10\end{bmatrix}$$

21. 2×5 by $5\times1=2\times1$

$$\begin{bmatrix}6&2&1&10&-4\\-5&0&3&-7&6\end{bmatrix}\begin{bmatrix}3\\-2\\5\\-1\\4\end{bmatrix}=\begin{bmatrix}-7\\31\end{bmatrix}$$

23. **(a)** $AB=$

$$\begin{bmatrix}3&2\\-4&1\end{bmatrix}\begin{bmatrix}0&1\\5&2\end{bmatrix}=\begin{bmatrix}10&7\\5&-2\end{bmatrix}$$

(b) $BA=$

$$\begin{bmatrix}0&1\\5&2\end{bmatrix}\begin{bmatrix}3&2\\-4&1\end{bmatrix}=\begin{bmatrix}-4&1\\7&12\end{bmatrix}$$

(c) $A^2=$

$$\begin{bmatrix}3&2\\-4&1\end{bmatrix}\begin{bmatrix}3&2\\-4&1\end{bmatrix}=\begin{bmatrix}1&8\\-16&-7\end{bmatrix}$$

25. **(a)** $AB =$

$$\begin{bmatrix} 1 & 2 & 3 \\ 0 & 4 & 1 \\ 5 & -2 & 6 \end{bmatrix}\begin{bmatrix} 2 & 0 & 1 \\ -1 & 3 & 4 \\ 0 & -5 & 8 \end{bmatrix} = \begin{bmatrix} 0 & -9 & 33 \\ -4 & 7 & 24 \\ 12 & -36 & 45 \end{bmatrix}$$

(b) $BA =$

$$\begin{bmatrix} 2 & 0 & 1 \\ -1 & 3 & 4 \\ 0 & -5 & 8 \end{bmatrix}\begin{bmatrix} 1 & 2 & 3 \\ 0 & 4 & 1 \\ 5 & -2 & 6 \end{bmatrix} = \begin{bmatrix} 7 & 2 & 12 \\ 19 & 2 & 24 \\ 40 & -36 & 43 \end{bmatrix}$$

(c) $A^2 = AA =$

$$\begin{bmatrix} 1 & 2 & 3 \\ 0 & 4 & 1 \\ 5 & -2 & 6 \end{bmatrix}\begin{bmatrix} 1 & 2 & 3 \\ 0 & 4 & 1 \\ 5 & -2 & 6 \end{bmatrix} = \begin{bmatrix} 16 & 4 & 23 \\ 5 & 14 & 10 \\ 35 & -10 & 49 \end{bmatrix}$$

27. **(a)** $AB =$

$$\begin{bmatrix} 1 & 0 & 5 \\ -3 & 2 & 6 \end{bmatrix}\begin{bmatrix} 3 & 2 \\ -5 & 0 \\ 1 & 1 \end{bmatrix} =$$

$$\begin{bmatrix} 8 & 7 \\ -13 & 0 \end{bmatrix}$$

(b) BA not defined

(c) A^2 not defined

29.

$$\begin{bmatrix} 5 & 3 \\ 2 & -6 \end{bmatrix}\begin{bmatrix} 1 & 0 \\ 0 & 1 \end{bmatrix} = \begin{bmatrix} 5 & 3 \\ 2 & -6 \end{bmatrix}$$

$$\begin{bmatrix} 1 & 0 \\ 0 & 1 \end{bmatrix}\begin{bmatrix} 5 & 3 \\ 2 & -6 \end{bmatrix} = \begin{bmatrix} 5 & 3 \\ 2 & -6 \end{bmatrix}$$

31.

$$\begin{bmatrix} 1 & -4 & 1 \\ 5 & 0 & 3 \\ 3 & 1 & 2 \end{bmatrix}\begin{bmatrix} 1 & 0 & 0 \\ 0 & 1 & 0 \\ 0 & 0 & 1 \end{bmatrix} = \begin{bmatrix} 1 & -4 & 1 \\ 5 & 0 & 3 \\ 3 & 1 & 2 \end{bmatrix}$$

$$\begin{bmatrix} 1 & 0 & 0 \\ 0 & 1 & 0 \\ 0 & 0 & 1 \end{bmatrix}\begin{bmatrix} 1 & -4 & 1 \\ 5 & 0 & 3 \\ 3 & 1 & 2 \end{bmatrix} = \begin{bmatrix} 1 & -4 & 1 \\ 5 & 0 & 3 \\ 3 & 1 & 2 \end{bmatrix}$$

Exercises 15.3

1. Let $A = \begin{bmatrix} 2 & 6 \\ 1 & 4 \end{bmatrix}$

$$\begin{vmatrix} 2 & 6 \\ 1 & 4 \end{vmatrix} = 2$$

$$A^{-1} = \frac{1}{2}\begin{bmatrix} 4 & -6 \\ -1 & 2 \end{bmatrix}$$

$$A^{-1} = \begin{bmatrix} 2 & -3 \\ -\frac{1}{2} & 1 \end{bmatrix}$$

3. Let $A = \begin{bmatrix} 5 & -2 \\ 11 & -4 \end{bmatrix}$

$$\begin{vmatrix} 5 & -2 \\ 11 & -4 \end{vmatrix} = 2$$

$$A^{-1} = \frac{1}{2}\begin{bmatrix} -4 & 2 \\ -11 & 5 \end{bmatrix}$$

$$A^{-1} = \begin{bmatrix} -2 & 1 \\ -\frac{11}{2} & \frac{5}{2} \end{bmatrix}$$

5. Let $A = \begin{bmatrix} -6 & -4 \\ 3 & 5 \end{bmatrix}$

$$\begin{vmatrix} -6 & -4 \\ 3 & 5 \end{vmatrix} = -18$$

$$A^{-1} = \frac{-1}{18}\begin{bmatrix} 5 & 4 \\ -3 & -6 \end{bmatrix}$$

$$A^{-1} = \begin{bmatrix} \frac{-5}{18} & \frac{-2}{9} \\ \frac{1}{6} & \frac{1}{3} \end{bmatrix}$$

7. Let $A=\begin{bmatrix}-4 & 2\\ 5 & -2\end{bmatrix}$

$$\begin{vmatrix}-4 & 2\\ 5 & -2\end{vmatrix}=-2$$

$$A^{-1}=\frac{-1}{2}\begin{bmatrix}-2 & -2\\ -5 & -4\end{bmatrix}$$

$$A^{-1}=\begin{bmatrix}1 & 1\\ \frac{5}{2} & 2\end{bmatrix}$$

9. Let $A=\begin{bmatrix}1 & 4\\ 1 & 3\end{bmatrix}$

$$\left[\begin{array}{cc|cc}1 & 4 & 1 & 0\\ 1 & 3 & 0 & 1\end{array}\right]$$

$$\left[\begin{array}{cc|cc}1 & 4 & 1 & 0\\ 0 & -1 & -1 & 1\end{array}\right](-\boldsymbol{R_1}+\boldsymbol{R_2})$$

$$\left[\begin{array}{cc|cc}1 & 0 & -3 & 4\\ 0 & -1 & -1 & 1\end{array}\right](4\boldsymbol{R_2}+\boldsymbol{R_1})$$

$$\left[\begin{array}{cc|cc}1 & 0 & -3 & 4\\ 0 & 1 & 1 & -1\end{array}\right](-1\boldsymbol{R_2})$$

$$A^{-1}=\begin{bmatrix}-3 & 4\\ 1 & -1\end{bmatrix}$$

11. Let $A=\begin{bmatrix}1 & 3\\ -2 & 6\end{bmatrix}$

$$\left[\begin{array}{cc|cc}1 & 3 & 1 & 0\\ -2 & 6 & 0 & 1\end{array}\right]$$

$$\left[\begin{array}{cc|cc}1 & 3 & 1 & 0\\ 0 & 12 & 2 & 1\end{array}\right](2\boldsymbol{R_1}+\boldsymbol{R_2})$$

$$\left[\begin{array}{cc|cc}1 & 3 & 1 & 0\\ 0 & 1 & \frac{1}{6} & \frac{1}{12}\end{array}\right]\left(\frac{1}{12}\boldsymbol{R_2}\right)$$

$$\left[\begin{array}{cc|cc}1 & 0 & \frac{1}{2} & \frac{-1}{4}\\ 0 & 1 & \frac{1}{6} & \frac{1}{12}\end{array}\right](-3\boldsymbol{R_2}+\boldsymbol{R_1})$$

$$A^{-1}=\begin{bmatrix}\frac{1}{2} & -\frac{1}{4}\\ \frac{1}{6} & \frac{1}{12}\end{bmatrix}$$

13. Let $A=\begin{bmatrix}-2 & 4\\ 3 & -5\end{bmatrix}$

$$\left[\begin{array}{cc|cc}-2 & 4 & 1 & 0\\ 3 & -5 & 0 & 1\end{array}\right]$$

$$\left[\begin{array}{cc|cc}1 & -1 & 1 & 1\\ 3 & -5 & 0 & 1\end{array}\right](\boldsymbol{R_2}+\boldsymbol{R_1})$$

$$\left[\begin{array}{cc|cc}1 & -1 & 1 & 1\\ 0 & -2 & -3 & -2\end{array}\right](-1\boldsymbol{R_1}+\boldsymbol{R_2})$$

$$\left[\begin{array}{cc|cc}1 & -1 & 1 & 1\\ 0 & 1 & \frac{3}{2} & 1\end{array}\right]\left(-\frac{1}{2}\boldsymbol{R_2}\right)$$

$$\left[\begin{array}{cc|cc}1 & 0 & \frac{5}{2} & 2\\ 0 & 1 & \frac{3}{2} & 1\end{array}\right](\boldsymbol{R_2}+\boldsymbol{R_1})$$

$$A^{-1}=\begin{bmatrix}\frac{5}{2} & 2\\ \frac{3}{2} & 1\end{bmatrix}$$

15. Let $A = \begin{bmatrix} 3 & 6 \\ 4 & 7 \end{bmatrix}$

$$\left[\begin{array}{cc|cc} 3 & 6 & 1 & 0 \\ 4 & 7 & 0 & 1 \end{array}\right]$$

$$\left[\begin{array}{cc|cc} -1 & -1 & 0 & -1 \\ 4 & 7 & 0 & 1 \end{array}\right] (-R_2 + R_1)$$

$$\left[\begin{array}{cc|cc} 1 & 1 & -1 & 1 \\ 4 & 7 & 0 & 1 \end{array}\right] (-R_1)$$

$$\left[\begin{array}{cc|cc} 1 & 1 & -1 & 1 \\ 0 & 3 & 4 & -3 \end{array}\right] (-4R_1 + R_2)$$

$$\left[\begin{array}{cc|cc} 1 & 1 & -1 & 1 \\ 0 & 1 & \frac{4}{3} & -1 \end{array}\right] \left(\frac{1}{3}R_2\right)$$

$$\left[\begin{array}{cc|cc} 1 & 0 & \frac{-7}{3} & 2 \\ 0 & 1 & \frac{4}{3} & -1 \end{array}\right] (-1R_2 + R_1)$$

$$A^{-1} = \begin{bmatrix} -\frac{7}{3} & 2 \\ \frac{4}{3} & -1 \end{bmatrix}$$

17. Let $A = \begin{bmatrix} 1 & 0 & 4 \\ 0 & 2 & 4 \\ 2 & 3 & 6 \end{bmatrix}$

$$\left[\begin{array}{ccc|ccc} 1 & 0 & 4 & 1 & 0 & 0 \\ 0 & 2 & 4 & 0 & 1 & 0 \\ 2 & 3 & 6 & 0 & 0 & 1 \end{array}\right]$$

$$\left[\begin{array}{ccc|ccc} 1 & 0 & 4 & 1 & 0 & 0 \\ 0 & 2 & 4 & 0 & 1 & 0 \\ 0 & 3 & -2 & -2 & 0 & 1 \end{array}\right] (-2R_1 + R_3)$$

$$\left[\begin{array}{ccc|ccc} 1 & 0 & 4 & 1 & 0 & 0 \\ 0 & 1 & 2 & 0 & \frac{1}{2} & 0 \\ 0 & 3 & -2 & -2 & 0 & 1 \end{array}\right] \left(\frac{1}{2}R_2\right)$$

$$\left[\begin{array}{ccc|ccc} 1 & 0 & 4 & 1 & 0 & 0 \\ 0 & 1 & 2 & 0 & \frac{1}{2} & 0 \\ 0 & 0 & -8 & -2 & -\frac{3}{2} & 1 \end{array}\right] (-3R_2 + R_3)$$

$$\left[\begin{array}{ccc|ccc} 1 & 0 & 4 & 1 & 0 & 0 \\ 0 & 1 & 2 & 0 & \frac{1}{2} & 0 \\ 0 & 0 & 1 & \frac{1}{4} & \frac{3}{16} & -\frac{1}{8} \end{array}\right] \left(-\frac{1}{8}R_3\right)$$

$$\left[\begin{array}{ccc|ccc} 1 & 0 & 0 & 0 & -\frac{3}{4} & \frac{1}{2} \\ 0 & 1 & 0 & -\frac{1}{2} & \frac{1}{8} & \frac{1}{4} \\ 0 & 0 & 1 & \frac{1}{4} & \frac{3}{16} & -\frac{1}{8} \end{array}\right] \begin{array}{l} (-4R_3 + R_1) \\ (-2R_3 + R_2) \end{array}$$

$$A^{-1} = \begin{bmatrix} 0 & -\frac{3}{4} & \frac{1}{2} \\ -\frac{1}{2} & \frac{1}{8} & \frac{1}{4} \\ \frac{1}{4} & \frac{3}{16} & -\frac{1}{8} \end{bmatrix}$$

19. Let $A = \begin{bmatrix} -1 & -2 & 3 \\ 0 & -1 & 2 \\ 3 & 1 & 0 \end{bmatrix}$

$$\left[\begin{array}{ccc|ccc} -1 & -2 & 3 & 1 & 0 & 0 \\ 0 & -1 & 2 & 0 & 1 & 0 \\ 3 & 1 & 0 & 0 & 0 & 1 \end{array}\right]$$

$$\left[\begin{array}{ccc|ccc} -1 & -2 & 3 & 1 & 0 & 0 \\ 0 & -1 & 2 & 0 & 1 & 0 \\ 0 & -5 & 9 & 3 & 0 & 1 \end{array}\right] \begin{matrix} \\ \\ (3\boldsymbol{R_1} + \boldsymbol{R_3}) \end{matrix}$$

$$\left[\begin{array}{ccc|ccc} 1 & 2 & -3 & -1 & 0 & 0 \\ 0 & 1 & -2 & 0 & -1 & 0 \\ 0 & -5 & 9 & 3 & 0 & 1 \end{array}\right] \begin{matrix} (-1\boldsymbol{R_1}) \\ (-1\boldsymbol{R_2}) \\ \\ \end{matrix}$$

$$\left[\begin{array}{ccc|ccc} 1 & 0 & 1 & -1 & 2 & 0 \\ 0 & 1 & -2 & 0 & -1 & 0 \\ 0 & 0 & -1 & 3 & -5 & 1 \end{array}\right] \begin{matrix} (-2\boldsymbol{R_2} + \boldsymbol{R_1}) \\ \\ (5\boldsymbol{R_2} + \boldsymbol{R_3}) \end{matrix}$$

$$\left[\begin{array}{ccc|ccc} 1 & 0 & 1 & -1 & 2 & 0 \\ 0 & 1 & -2 & 0 & -1 & 0 \\ 0 & 0 & 1 & -3 & 5 & -1 \end{array}\right] \begin{matrix} \\ \\ (-1\boldsymbol{R_3}) \end{matrix}$$

$$\left[\begin{array}{ccc|ccc} 1 & 0 & 0 & 2 & -3 & 1 \\ 0 & 1 & 2 & -6 & 9 & -2 \\ 0 & 0 & 1 & -3 & 5 & -1 \end{array}\right] \begin{matrix} (-1\boldsymbol{R_3} + \boldsymbol{R_1}) \\ (2\boldsymbol{R_3} + \boldsymbol{R_2}) \\ \\ \end{matrix}$$

$$A^{-1} = \begin{bmatrix} 2 & -3 & 1 \\ -6 & 9 & -2 \\ -3 & 5 & -1 \end{bmatrix}$$

21. Let $A = \begin{bmatrix} 1 & 1 & 2 \\ 4 & 0 & 1 \\ 2 & 2 & 5 \end{bmatrix}$

$$\left[\begin{array}{ccc|ccc} 1 & 1 & 2 & 1 & 0 & 0 \\ 4 & 0 & 1 & 0 & 1 & 0 \\ 2 & 2 & 5 & 0 & 0 & 1 \end{array}\right]$$

$$\left[\begin{array}{ccc|ccc} 1 & 1 & 2 & 1 & 0 & 0 \\ 0 & -4 & -7 & -4 & 1 & 0 \\ 0 & 0 & 1 & -2 & 0 & 1 \end{array}\right] \begin{matrix} \\ (-4\boldsymbol{R_1} + \boldsymbol{R_2}) \\ (-2\boldsymbol{R_1} + \boldsymbol{R_3}) \end{matrix}$$

$$\left[\begin{array}{ccc|ccc} 1 & 1 & 2 & 1 & 0 & 0 \\ 0 & 1 & \frac{7}{4} & 1 & -\frac{1}{4} & 0 \\ 0 & 0 & 1 & -2 & 0 & 1 \end{array}\right] \begin{matrix} \\ \left(-\frac{1}{4}\boldsymbol{R_2}\right) \\ \\ \end{matrix}$$

$$\left[\begin{array}{ccc|ccc} 1 & 0 & \frac{1}{4} & 0 & \frac{1}{4} & 0 \\ 0 & 1 & \frac{7}{4} & 1 & -\frac{1}{4} & 0 \\ 0 & 0 & 1 & -2 & 0 & 1 \end{array}\right] (-1\boldsymbol{R_2} + \boldsymbol{R_1})$$

$$\left[\begin{array}{ccc|ccc} 1 & 0 & 0 & \frac{1}{2} & \frac{1}{4} & -\frac{1}{4} \\ 0 & 1 & 0 & \frac{9}{2} & -\frac{1}{4} & -\frac{7}{4} \\ 0 & 0 & 1 & -2 & 0 & 1 \end{array}\right] \begin{matrix} \left(-\frac{1}{4}\boldsymbol{R_3} + \boldsymbol{R_1}\right) \\ \left(-\frac{7}{4}\boldsymbol{R_3} + \boldsymbol{R_2}\right) \\ \\ \end{matrix}$$

$$A^{-1} = \begin{bmatrix} \frac{1}{2} & \frac{1}{4} & -\frac{1}{4} \\ \frac{9}{2} & -\frac{1}{4} & -\frac{7}{4} \\ -2 & 0 & 1 \end{bmatrix}$$

23. Let $A = \begin{bmatrix} 9 & -5 & 1 \\ -5 & 1 & 1 \\ 1 & 1 & -1 \end{bmatrix}$

$$\left[\begin{array}{ccc|ccc} 9 & -5 & 1 & 1 & 0 & 0 \\ -5 & 1 & 1 & 0 & 1 & 0 \\ 1 & 1 & -1 & 0 & 0 & 1 \end{array}\right]$$

$$\left[\begin{array}{ccc|ccc} 1 & 1 & -1 & 0 & 0 & 1 \\ -5 & 1 & 1 & 0 & 1 & 0 \\ 9 & -5 & 1 & 1 & 0 & 0 \end{array}\right] \text{Interchange rows 1 and 3}$$

$$\left[\begin{array}{ccc|ccc} 1 & 1 & -1 & 0 & 0 & 1 \\ 0 & 6 & -4 & 0 & 1 & 5 \\ 0 & -14 & 10 & 1 & 0 & -9 \end{array}\right] \begin{array}{l} \\ (5\boldsymbol{R_1} + \boldsymbol{R_2}) \\ (-9\boldsymbol{R_1} + \boldsymbol{R_3}) \end{array}$$

$$\left[\begin{array}{ccc|ccc} 1 & 1 & -1 & 0 & 0 & 1 \\ 0 & 1 & -\frac{2}{3} & 0 & \frac{1}{6} & \frac{5}{6} \\ 0 & -14 & 10 & 1 & 0 & -9 \end{array}\right] \left(\frac{1}{6}\boldsymbol{R_2}\right)$$

$$\left[\begin{array}{ccc|ccc} 1 & 0 & -\frac{1}{3} & 0 & -\frac{1}{6} & \frac{1}{6} \\ 0 & 1 & -\frac{2}{3} & 0 & \frac{1}{6} & \frac{5}{6} \\ 0 & 0 & \frac{2}{3} & 1 & \frac{7}{3} & \frac{8}{3} \end{array}\right] \begin{array}{l} (-1\boldsymbol{R_2} + \boldsymbol{R_1}) \\ \\ (14\boldsymbol{R_2} + \boldsymbol{R_3}) \end{array}$$

$$\left[\begin{array}{ccc|ccc} 1 & 0 & -\frac{1}{3} & 0 & -\frac{1}{6} & \frac{1}{6} \\ 0 & 1 & -\frac{2}{3} & 0 & \frac{1}{6} & \frac{5}{6} \\ 0 & 0 & 1 & \frac{3}{2} & \frac{7}{2} & 4 \end{array}\right] \left(\frac{3}{2}\boldsymbol{R_3}\right)$$

$$\left[\begin{array}{ccc|ccc} 1 & 0 & 0 & \frac{1}{2} & 1 & \frac{3}{2} \\ 0 & 1 & 0 & 1 & \frac{5}{2} & \frac{7}{2} \\ 0 & 0 & 1 & \frac{3}{2} & \frac{7}{2} & 4 \end{array}\right] \begin{array}{l} \left(\frac{1}{3}\boldsymbol{R_3} + \boldsymbol{R_1}\right) \\ \left(\frac{2}{3}\boldsymbol{R_3} + \boldsymbol{R_2}\right) \\ \\ \end{array}$$

$$A^{-1} = \begin{bmatrix} \frac{1}{2} & 1 & \frac{3}{2} \\ 1 & \frac{5}{2} & \frac{7}{2} \\ \frac{3}{2} & \frac{7}{2} & 4 \end{bmatrix}$$

25. Let $A = \begin{bmatrix} 1 & 0 & 2 & 0 \\ 2 & 0 & 0 & 2 \\ -1 & 0 & -1 & 1 \\ 0 & 1 & 4 & 1 \end{bmatrix}$

$$\left[\begin{array}{cccc|cccc} 1 & 0 & 2 & 0 & 1 & 0 & 0 & 0 \\ 2 & 0 & 0 & 2 & 0 & 1 & 0 & 0 \\ -1 & 0 & -1 & 1 & 0 & 0 & 1 & 0 \\ 0 & 1 & 4 & 1 & 0 & 0 & 0 & 1 \end{array}\right]$$

$$\left[\begin{array}{cccc|cccc} 1 & 0 & 2 & 0 & 1 & 0 & 0 & 0 \\ 0 & 0 & -4 & 2 & -2 & 1 & 0 & 0 \\ 0 & 0 & 1 & 1 & 1 & 0 & 1 & 0 \\ 0 & 1 & 4 & 1 & 0 & 0 & 0 & 1 \end{array}\right] \begin{array}{l} \\ (-2\boldsymbol{R_1} + \boldsymbol{R_2}) \\ (1\boldsymbol{R_1} + \boldsymbol{R_3}) \\ \\ \end{array}$$

$$\left[\begin{array}{cccc|cccc} 1 & 0 & 2 & 0 & 1 & 0 & 0 & 0 \\ 0 & 1 & 4 & 1 & 0 & 0 & 0 & 1 \\ 0 & 0 & 1 & 1 & 1 & 0 & 1 & 0 \\ 0 & 0 & -4 & 2 & -2 & 1 & 0 & 0 \end{array}\right] \text{Interchange rows 4 and 2}$$

$$\left[\begin{array}{cccc|cccc} 1 & 0 & 0 & -2 & -1 & 0 & -2 & 0 \\ 0 & 1 & 0 & -3 & -4 & 0 & -4 & 1 \\ 0 & 0 & 1 & 1 & 1 & 0 & 1 & 0 \\ 0 & 0 & 0 & 6 & 2 & 1 & 4 & 0 \end{array}\right] \begin{array}{l} (-2\boldsymbol{R_3} + \boldsymbol{R_1}) \\ (-4\boldsymbol{R_3} + \boldsymbol{R_2}) \\ \\ (4\boldsymbol{R_3} + \boldsymbol{R_4}) \end{array}$$

$$\left[\begin{array}{cccc|cccc} 1 & 0 & 0 & -2 & -1 & 0 & -2 & 0 \\ 0 & 1 & 0 & -3 & -4 & 0 & -4 & 1 \\ 0 & 0 & 1 & 1 & 1 & 0 & 1 & 0 \\ 0 & 0 & 0 & 1 & \frac{1}{3} & \frac{1}{6} & \frac{2}{3} & 0 \end{array}\right] \left(\frac{1}{6}\boldsymbol{R_4}\right)$$

$$\left[\begin{array}{cccc|cccc} 1 & 0 & 0 & 0 & \frac{-1}{3} & \frac{1}{3} & \frac{-2}{3} & 0 \\ 0 & 1 & 0 & 0 & -3 & \frac{1}{2} & -2 & 1 \\ 0 & 0 & 1 & 0 & \frac{2}{3} & \frac{-1}{6} & \frac{1}{3} & 0 \\ 0 & 0 & 0 & 1 & \frac{1}{3} & \frac{1}{6} & \frac{2}{3} & 0 \end{array}\right] \begin{array}{l} (2\boldsymbol{R_4} + \boldsymbol{R_1}) \\ (3\boldsymbol{R_4} + \boldsymbol{R_2}) \\ (-1\boldsymbol{R_4} + \boldsymbol{R_3}) \\ \\ \end{array}$$

$$A^{-1} = \begin{bmatrix} \frac{-1}{3} & \frac{1}{3} & \frac{-2}{3} & 0 \\ -3 & \frac{1}{2} & -2 & 1 \\ \frac{2}{3} & -\frac{1}{6} & \frac{1}{3} & 0 \\ \frac{1}{3} & \frac{1}{6} & \frac{2}{3} & 0 \end{bmatrix}$$

27. Let $A=\begin{bmatrix}2&1&-2&2\\1&-2&4&1\\0&1&2&2\\2&2&-4&4\end{bmatrix}$

$$\left[\begin{array}{cccc|cccc}2&1&-2&2&1&0&0&0\\1&-2&4&1&0&1&0&0\\0&1&2&2&0&0&1&0\\2&2&-4&4&0&0&0&1\end{array}\right]$$

$$\left[\begin{array}{cccc|cccc}1&-2&4&1&0&1&0&0\\2&1&-2&2&1&0&0&0\\0&1&2&2&0&0&1&0\\2&2&-4&4&0&0&0&1\end{array}\right]\begin{matrix}\text{Interchange rows}\\\text{1 and 2}\\ \\ \end{matrix}$$

$$\left[\begin{array}{cccc|cccc}1&-2&4&1&0&1&0&0\\0&5&-10&0&1&-2&0&0\\0&1&2&2&0&0&1&0\\0&6&-12&2&0&-2&0&1\end{array}\right]\begin{matrix} \\(-2\boldsymbol{R_1}+\boldsymbol{R_2})\\ \\(-2\boldsymbol{R_1}+\boldsymbol{R_4})\end{matrix}$$

$$\left[\begin{array}{cccc|cccc}1&-2&4&1&0&1&0&0\\0&1&2&2&0&0&1&0\\0&5&-10&0&1&-2&0&0\\0&6&-12&2&0&-2&0&1\end{array}\right]\begin{matrix}\text{Interchange rows}\\\text{2 and 3}\\ \\ \end{matrix}$$

$$\left[\begin{array}{cccc|cccc}1&0&8&5&0&1&2&0\\0&1&2&2&0&0&1&0\\0&0&-20&-10&1&-2&-5&0\\0&0&-24&-10&0&-2&-6&1\end{array}\right]\begin{matrix}(2\boldsymbol{R_2}+\boldsymbol{R_1})\\ \\(-5\boldsymbol{R_2}+\boldsymbol{R_3})\\(-6\boldsymbol{R_2}+\boldsymbol{R_4})\end{matrix}$$

$$\left[\begin{array}{cccc|cccc}1&0&8&5&0&1&2&0\\0&1&2&2&0&0&1&0\\0&0&1&\frac{1}{2}&\frac{-1}{20}&\frac{1}{10}&\frac{1}{4}&0\\0&0&-24&-10&0&-2&-6&1\end{array}\right]\begin{matrix} \\ \\\left(\frac{-1}{20}\boldsymbol{R_3}\right)\\ \end{matrix}$$

$$\left[\begin{array}{cccc|cccc}1&0&0&1&\frac{2}{5}&\frac{1}{5}&0&0\\0&1&0&1&\frac{1}{10}&\frac{-1}{5}&\frac{1}{2}&0\\0&0&1&\frac{1}{2}&\frac{-1}{20}&\frac{1}{10}&\frac{1}{4}&0\\0&0&0&2&\frac{-6}{5}&\frac{2}{5}&0&1\end{array}\right]\begin{matrix}(-8\boldsymbol{R_3}+\boldsymbol{R_1})\\(-2\boldsymbol{R_3}+\boldsymbol{R_2})\\ \\(24\boldsymbol{R_3}+\boldsymbol{R_4})\end{matrix}$$

$$\left[\begin{array}{cccc|cccc}1&0&0&1&\frac{2}{5}&\frac{1}{5}&0&0\\0&1&0&1&\frac{1}{10}&\frac{-1}{5}&\frac{1}{2}&0\\0&0&1&\frac{1}{2}&\frac{-1}{20}&\frac{1}{10}&\frac{1}{4}&0\\0&0&0&1&\frac{-3}{5}&\frac{1}{5}&0&\frac{1}{2}\end{array}\right]\begin{matrix} \\ \\ \\\left(\frac{1}{2}\boldsymbol{R_4}\right)\end{matrix}$$

$$\left[\begin{array}{cccc|cccc}1&0&0&0&1&0&0&\frac{-1}{2}\\0&1&0&0&\frac{7}{10}&\frac{-2}{5}&\frac{1}{2}&\frac{-1}{2}\\0&0&1&0&\frac{1}{4}&0&\frac{1}{4}&\frac{-1}{4}\\0&0&0&1&\frac{-3}{5}&\frac{1}{5}&0&\frac{1}{2}\end{array}\right]\begin{matrix}(-1\boldsymbol{R_4}+\boldsymbol{R_1})\\(-1\boldsymbol{R_4}+\boldsymbol{R_2})\\\left(-\frac{1}{2}\boldsymbol{R_4}+\boldsymbol{R_3}\right)\\ \end{matrix}$$

$$A^{-1}=\begin{bmatrix}1&0&0&\frac{-1}{2}\\\frac{7}{10}&\frac{-2}{5}&\frac{1}{2}&\frac{-1}{2}\\\frac{1}{4}&0&\frac{1}{4}&\frac{-1}{4}\\\frac{-3}{5}&\frac{1}{5}&0&\frac{1}{2}\end{bmatrix}$$

29. $2x+6y=34$

$x+4y=22$

$$A=\begin{bmatrix}2 & 6\\1 & 4\end{bmatrix},\ X=\begin{bmatrix}x\\y\end{bmatrix},\ B=\begin{bmatrix}34\\22\end{bmatrix}$$

From #1, $A^{-1}=\begin{bmatrix}2 & -3\\ \frac{-1}{2} & 1\end{bmatrix}$

$$X=A^{-1}B=\begin{bmatrix}2 & -3\\ \frac{-1}{2} & 1\end{bmatrix}\begin{bmatrix}34\\22\end{bmatrix}$$

$X=\begin{bmatrix}2\\5\end{bmatrix}$. $\quad x=2,\ y=5$

31. $5x-2y=-18$

$11x-4y=-40$

$$A=\begin{bmatrix}5 & -2\\11 & -4\end{bmatrix},\ X=\begin{bmatrix}x\\y\end{bmatrix},\ B=\begin{bmatrix}-18\\-40\end{bmatrix}$$

Use A^{-1} from #3.

$$X=A^{-1}B=\begin{bmatrix}-2 & 1\\ -\frac{11}{2} & \frac{5}{2}\end{bmatrix}\begin{bmatrix}-18\\-40\end{bmatrix}$$

$X=\begin{bmatrix}-4\\-1\end{bmatrix}$. $\quad x=-4,\ y=-1$

33. $x+4y=33$

$x+3y=25$

$$A=\begin{bmatrix}1 & 4\\1 & 3\end{bmatrix},\ X=\begin{bmatrix}x\\y\end{bmatrix},\ B=\begin{bmatrix}33\\25\end{bmatrix}$$

Use A^{-1} from #9.

$$X=A^{-1}B=\begin{bmatrix}-3 & 4\\1 & -1\end{bmatrix}\begin{bmatrix}33\\25\end{bmatrix}$$

$X=\begin{bmatrix}1\\8\end{bmatrix}$. $\quad x=1,\ y=8$

35. $-2x+4y=26$

$3x-5y=-34$

$$A=\begin{bmatrix}-2 & 4\\3 & -5\end{bmatrix},\ X=\begin{bmatrix}x\\y\end{bmatrix},\ B=\begin{bmatrix}26\\-34\end{bmatrix}$$

Use A^{-1} from #13.

$$X=A^{-1}B=\begin{bmatrix}\frac{5}{2} & 2\\ \frac{3}{2} & 1\end{bmatrix}\begin{bmatrix}26\\-34\end{bmatrix}$$

$X=\begin{bmatrix}-3\\5\end{bmatrix}$. $\quad x=-3,\ y=5$

37. $x+\quad +4z=-27$

$2y+4z=-16$

$2x+3y+6z=-30$

$$A=\begin{bmatrix}1 & 0 & 4\\0 & 2 & 4\\2 & 3 & 6\end{bmatrix},\ X=\begin{bmatrix}x\\y\\z\end{bmatrix},\ B=\begin{bmatrix}-27\\-16\\-30\end{bmatrix}$$

Use A^{-1} from #17.

$$X=A^{-1}B=\begin{bmatrix}0 & \frac{-3}{4} & \frac{1}{2}\\ \frac{-1}{2} & \frac{1}{8} & \frac{1}{4}\\ \frac{1}{4} & \frac{3}{16} & \frac{-1}{8}\end{bmatrix}\begin{bmatrix}-27\\-16\\-30\end{bmatrix}$$

$X=\begin{bmatrix}-3\\4\\-6\end{bmatrix}$. $\quad x=-3,\ y=4,\ z=-6$

39. $-x-2y+3z=37$

$-y+2z=26$

$3x+y\quad =7$

$$A=\begin{bmatrix}-1 & -2 & 3\\0 & -1 & 2\\3 & 1 & 0\end{bmatrix},\ X=\begin{bmatrix}x\\y\\z\end{bmatrix},\ B=\begin{bmatrix}37\\26\\7\end{bmatrix}$$

Use A^{-1} from #19.

$$X=A^{-1}B=\begin{bmatrix}2 & -3 & 1\\-6 & 9 & -2\\-3 & 5 & -1\end{bmatrix}\begin{bmatrix}37\\26\\7\end{bmatrix}$$

$X=\begin{bmatrix}3\\-2\\12\end{bmatrix}$. $\quad x=3,\ y=-2,\ z=12$

41.
$$\begin{aligned} 2x+y-2z+2w&=-3 \\ x-2y+4z+w&=61 \\ y+2z+2w&=21 \\ 2x+2y-4z+4w&=-22 \end{aligned}$$

$$A=\begin{bmatrix} 2 & 1 & -2 & 2 \\ 1 & -2 & 4 & 1 \\ 0 & 1 & 2 & 2 \\ 2 & 2 & -4 & 4 \end{bmatrix},\ X=\begin{bmatrix} x \\ y \\ z \\ w \end{bmatrix},\ B=\begin{bmatrix} -3 \\ 61 \\ 21 \\ -22 \end{bmatrix}$$

Use A^{-1} from #27.

$$X=A^{-1}B=\begin{bmatrix} 1 & 0 & 0 & \frac{-1}{2} \\ \frac{7}{10} & \frac{-2}{5} & \frac{1}{2} & -\frac{1}{2} \\ \frac{1}{4} & 0 & \frac{1}{4} & -\frac{1}{4} \\ -\frac{3}{5} & \frac{1}{5} & 0 & \frac{1}{2} \end{bmatrix}\begin{bmatrix} -3 \\ 61 \\ 21 \\ -22 \end{bmatrix}$$

$$X=\begin{bmatrix} 8 \\ -5 \\ 10 \\ 3 \end{bmatrix}. \qquad x=8,\ y=-5,\ z=10,\ w=3$$

Exercises 15.4

1. $(3,-4)$ **3.** $(17,-2,5)$ **5.** $(3,15,-6,10)$

7.
$$\begin{aligned} 3x-y&=3 \\ -4x+2y&=-2 \end{aligned}$$

$$\begin{bmatrix} 3 & -1 & 3 \\ -4 & 2 & -2 \end{bmatrix}$$

$$\begin{bmatrix} -1 & 1 & 1 \\ -4 & 2 & -2 \end{bmatrix}(1\boldsymbol{R_2}+\boldsymbol{R_1})$$

$$\begin{bmatrix} 1 & -1 & -1 \\ -4 & 2 & -2 \end{bmatrix}(-1\boldsymbol{R_1})$$

$$\begin{bmatrix} 1 & -1 & -1 \\ 0 & -2 & -6 \end{bmatrix}(4\boldsymbol{R_1}+\boldsymbol{R_2})$$

$$\begin{bmatrix} 1 & -1 & -1 \\ 0 & 1 & 3 \end{bmatrix}\left(-\frac{1}{2}\boldsymbol{R_2}\right)$$

$$\begin{bmatrix} 1 & 0 & 2 \\ 0 & 1 & 3 \end{bmatrix}(1\boldsymbol{R_2}+\boldsymbol{R_1})$$

$x=2,\ y=3$

9.
$$\begin{aligned} 4x-6y&=-64 \\ x+5y&=36 \end{aligned}$$

$$\begin{bmatrix} 4 & -6 & -64 \\ 1 & 5 & 36 \end{bmatrix}$$

$$\begin{bmatrix} 1 & 5 & 36 \\ 4 & -6 & -64 \end{bmatrix}\text{Interchange rows 1 and 2}$$

$$\begin{bmatrix} 1 & 5 & 36 \\ 0 & -26 & -208 \end{bmatrix}(-4\boldsymbol{R_2}+\boldsymbol{R_1})$$

$$\begin{bmatrix} 1 & 5 & 36 \\ 0 & 1 & 8 \end{bmatrix}\left(-\frac{1}{26}\boldsymbol{R_2}\right)$$

$$\begin{bmatrix} 1 & 0 & -4 \\ 0 & 1 & 8 \end{bmatrix}(-5\boldsymbol{R_2}+\boldsymbol{R_1})$$

$x=-4,\ y=8$

11.

$$2x-8y=46$$
$$3x+2y=-1$$

$$\begin{bmatrix}2 & -8 & 46\\3 & 2 & -1\end{bmatrix}$$

$$\begin{bmatrix}-1 & -10 & 47\\3 & 2 & -1\end{bmatrix}(-3\boldsymbol{R_2}+\boldsymbol{R_1})$$

$$\begin{bmatrix}1 & 10 & -47\\3 & 2 & -1\end{bmatrix}(-\boldsymbol{R_1})$$

$$\begin{bmatrix}1 & 10 & -47\\0 & -28 & 140\end{bmatrix}(-3\boldsymbol{R_1}+\boldsymbol{R_2})$$

$$\begin{bmatrix}1 & 10 & -47\\0 & 1 & -5\end{bmatrix}\left(-\frac{1}{28}\boldsymbol{R_2}\right)$$

$$\begin{bmatrix}1 & 0 & 3\\0 & 1 & -5\end{bmatrix}(-10\boldsymbol{R_2}+\boldsymbol{R_1})$$

$x=3,\ y=-5$

13.

$$6x+8y=98$$
$$2x+5y=49$$

$$\begin{bmatrix}6 & 8 & 98\\2 & 5 & 49\end{bmatrix}$$

$$\begin{bmatrix}1 & \frac{4}{3} & \frac{49}{3}\\2 & 5 & 49\end{bmatrix}\left(\frac{1}{6}\boldsymbol{R_1}\right)$$

$$\begin{bmatrix}1 & \frac{4}{3} & \frac{49}{3}\\0 & \frac{7}{3} & \frac{49}{3}\end{bmatrix}(-2\boldsymbol{R_1}+\boldsymbol{R_2})$$

$$\begin{bmatrix}1 & \frac{4}{3} & \frac{49}{3}\\0 & 1 & 7\end{bmatrix}\left(\frac{3}{7}\boldsymbol{R_2}\right)$$

$$\begin{bmatrix}1 & 0 & 7\\0 & 1 & 7\end{bmatrix}\left(-\frac{4}{3}\boldsymbol{R_2}+\boldsymbol{R_1}\right)$$

$x=7,\ y=7$

15.

$$x+2y-z=9$$
$$3x+y-2z=13$$
$$y+z=-2$$

$$\begin{bmatrix}1 & 2 & -1 & 9\\3 & 1 & -2 & 13\\0 & 1 & 1 & -2\end{bmatrix}$$

$$\begin{bmatrix}1 & 2 & -1 & 9\\0 & -5 & 1 & -14\\0 & 1 & 1 & -2\end{bmatrix}(-3\boldsymbol{R_1}+\boldsymbol{R_2})$$

$$\begin{bmatrix}1 & 2 & -1 & 9\\0 & 1 & 1 & -2\\0 & -5 & 1 & -14\end{bmatrix}\text{Interchange rows 2 and 3}$$

$$\begin{bmatrix}1 & 0 & -3 & 13\\0 & 1 & 1 & -2\\0 & 0 & 6 & -24\end{bmatrix}\begin{matrix}(-2\boldsymbol{R_2}+\boldsymbol{R_1})\\ \\(5\boldsymbol{R_2}+\boldsymbol{R_3})\end{matrix}$$

$$\begin{bmatrix}1 & 0 & -3 & 13\\0 & 1 & 1 & -2\\0 & 0 & 1 & -4\end{bmatrix}\left(\frac{1}{6}\boldsymbol{R_3}\right)$$

$$\begin{bmatrix}1 & 0 & 0 & 1\\0 & 1 & 0 & 2\\0 & 0 & 1 & -4\end{bmatrix}\begin{matrix}(3\boldsymbol{R_3}+\boldsymbol{R_1})\\(-\boldsymbol{R_3}+\boldsymbol{R_2})\\ \end{matrix}$$

$x=1,\ y=2,\ z=-4$

17.

$$x+y-z=-4$$
$$x+2y+3z=14$$
$$4x+3y-2z=-4$$

$$\begin{bmatrix}1 & 1 & -1 & -4\\1 & 2 & 3 & 14\\4 & 3 & -2 & -4\end{bmatrix}$$

$$\begin{bmatrix}1 & 1 & -1 & -4\\0 & 1 & 4 & 18\\0 & -1 & 2 & 12\end{bmatrix}\begin{matrix}\\(-\boldsymbol{R_1}+\boldsymbol{R_2})\\(-4\boldsymbol{R_1}+\boldsymbol{R_3})\end{matrix}$$

$$\begin{bmatrix}1 & 0 & -5 & -22\\0 & 1 & 4 & 18\\0 & 0 & 6 & 30\end{bmatrix}\begin{matrix}(-\boldsymbol{R_2}+\boldsymbol{R_1})\\ \\(\boldsymbol{R_2}+\boldsymbol{R_3})\end{matrix}$$

$$\begin{bmatrix}1 & 0 & -5 & -22\\0 & 1 & 4 & 18\\0 & 0 & 1 & 5\end{bmatrix}\begin{matrix}\\ \\(\boldsymbol{R_3}/6)\end{matrix}$$

$$\begin{bmatrix}1 & 0 & 0 & 3\\0 & 1 & 0 & -2\\0 & 0 & 1 & 5\end{bmatrix}\begin{matrix}(5\boldsymbol{R_3}+\boldsymbol{R_1})\\(-4\boldsymbol{R_3}+\boldsymbol{R_2})\\ \end{matrix}$$

$x=3,\ y=-2,\ z=5$

19.

$$2x-3y+z=33$$
$$x-4y-2x=29$$
$$x-6y-4z=39$$

$$\begin{bmatrix} 2 & -3 & 1 & 33 \\ 1 & -4 & -2 & 29 \\ 1 & -6 & -4 & 39 \end{bmatrix}$$

$$\begin{bmatrix} 1 & 1 & 3 & 3 \\ 1 & -4 & -2 & 29 \\ 1 & -6 & -4 & 39 \end{bmatrix} \begin{matrix} (-R_2+R_1) \\ \\ \\ \end{matrix}$$

$$\begin{bmatrix} 1 & 1 & 3 & 4 \\ 0 & -5 & -5 & 25 \\ 0 & -7 & -7 & 35 \end{bmatrix} \begin{matrix} \\ (-1R_1+R_2) \\ (-R_1+R_3) \end{matrix}$$

$$\begin{bmatrix} 1 & 1 & 3 & 4 \\ 0 & 1 & 1 & -5 \\ 0 & 1 & 1 & -5 \end{bmatrix} \begin{matrix} \left(-\frac{1}{5}R_2\right) \\ \\ \left(-\frac{1}{7}R_3\right) \end{matrix}$$

$$\begin{bmatrix} 1 & 0 & 2 & 9 \\ 0 & 1 & 1 & -5 \\ 0 & 0 & 0 & 0 \end{bmatrix} \begin{matrix} (-1R_2+R_1) \\ \\ (-1R_2+R_3) \end{matrix}$$

This system is dependent. $(5,-7,2)$ is a solution, but it is not unique. Note: $(7,-6,1)$ is also a solution.

21.

$$x+2y-2z+3w=8$$
$$2x-3y-4z+w=-7$$
$$-x+4y-z+2w=11$$
$$4x+5y-2z+4w=28$$

$$\begin{bmatrix} 1 & 2 & -2 & 3 & 8 \\ 2 & -3 & -4 & 1 & -7 \\ -1 & 4 & -1 & 2 & 11 \\ 4 & 5 & -2 & 4 & 28 \end{bmatrix}$$

$$\begin{bmatrix} 1 & 2 & -2 & 3 & 8 \\ 0 & -7 & 0 & -5 & -23 \\ 0 & 6 & -3 & 5 & 19 \\ 0 & -3 & 6 & -8 & -4 \end{bmatrix} \begin{matrix} \\ (-2R_1+R_2) \\ (1R_1+R_3) \\ (-4R_1+R_4) \end{matrix}$$

$$\begin{bmatrix} 1 & 2 & -2 & 3 & 8 \\ 0 & -1 & -3 & 0 & -4 \\ 0 & 6 & -3 & 5 & 19 \\ 0 & -3 & 6 & -8 & -4 \end{bmatrix} \begin{matrix} \\ (-1R_3+R_2) \\ \\ \\ \end{matrix}$$

$$\begin{bmatrix} 1 & 0 & -8 & 3 & 0 \\ 0 & -1 & -3 & 0 & -4 \\ 0 & 0 & -21 & 5 & -5 \\ 0 & 0 & 15 & -8 & 8 \end{bmatrix} \begin{matrix} (2R_2+R_1) \\ \\ (6R_2+R_3) \\ (-3R_2+R_4) \end{matrix}$$

$$\begin{bmatrix} 1 & 0 & -8 & 3 & 0 \\ 0 & 1 & 3 & 0 & 4 \\ 0 & 0 & 1 & \frac{-5}{21} & \frac{5}{21} \\ 0 & 0 & 15 & -8 & 8 \end{bmatrix} \begin{matrix} \\ (-1R_2) \\ \left(\frac{-1}{21}R_3\right) \\ \\ \end{matrix}$$

$$\begin{bmatrix} 1 & 0 & 0 & \frac{23}{21} & \frac{40}{21} \\ 0 & 1 & 0 & \frac{5}{7} & \frac{23}{7} \\ 0 & 0 & 1 & \frac{-5}{21} & \frac{5}{21} \\ 0 & 0 & 0 & \frac{-31}{7} & \frac{31}{7} \end{bmatrix} \begin{matrix} (8R_3+R_1) \\ (-3R_3+R_2) \\ \\ (-15R_3+R_4) \end{matrix}$$

$$\begin{bmatrix} 1 & 0 & 0 & \frac{23}{21} & \frac{40}{21} \\ 0 & 1 & 0 & \frac{5}{7} & \frac{23}{7} \\ 0 & 0 & 1 & \frac{-5}{21} & \frac{5}{21} \\ 0 & 0 & 0 & 1 & -1 \end{bmatrix} \begin{matrix} \\ \\ \left(\frac{-7}{31}R_4\right) \\ \\ \end{matrix}$$

$$\begin{bmatrix} 1 & 0 & 0 & 0 & 3 \\ 0 & 1 & 0 & 0 & 4 \\ 0 & 0 & 1 & 0 & 0 \\ 0 & 0 & 0 & 1 & -1 \end{bmatrix} \begin{matrix} \left(\frac{-23}{21}R_4+R_1\right) \\ \left(\frac{-5}{7}R_4+R_2\right) \\ \left(\frac{5}{21}R_4+R_3\right) \end{matrix}$$

$x=3,\ y=4,\ z=0,\ w=-1$

23.

$$2x+4y+6z+2w=32$$
$$x-y+2z-3w=-2$$
$$-3x-2y+4z+w=16$$
$$3x+5y+z-2w=-7$$

$$\begin{bmatrix} 2 & 4 & 6 & 2 & 32 \\ 1 & -1 & 2 & -3 & -2 \\ -3 & -2 & 4 & 1 & 16 \\ 3 & 5 & 1 & -2 & -7 \end{bmatrix}$$

$$\begin{bmatrix} 1 & 5 & 4 & 5 & 34 \\ 1 & -1 & 2 & -3 & -2 \\ -3 & -2 & 4 & 1 & 16 \\ 3 & 5 & 1 & -2 & -7 \end{bmatrix} \begin{matrix} (-1\boldsymbol{R_2}+\boldsymbol{R_1}) \\ \\ \\ \\ \end{matrix}$$

$$\begin{bmatrix} 1 & 5 & 4 & 5 & 34 \\ 0 & -6 & -2 & -8 & -36 \\ 0 & 13 & 16 & 16 & 118 \\ 0 & -10 & -11 & -17 & -109 \end{bmatrix} \begin{matrix} \\ (-1\boldsymbol{R_1}+\boldsymbol{R_2}) \\ (3\boldsymbol{R_1}+\boldsymbol{R_3}) \\ (-3\boldsymbol{R_1}+\boldsymbol{R_4}) \end{matrix}$$

$$\begin{bmatrix} 1 & 5 & 4 & 5 & 34 \\ 0 & -6 & -2 & -8 & -36 \\ 0 & 1 & 12 & 0 & 46 \\ 0 & -10 & -11 & -17 & -109 \end{bmatrix} \begin{matrix} \\ \\ (2\boldsymbol{R_2}+\boldsymbol{R_3}) \\ \\ \end{matrix}$$

$$\begin{bmatrix} 1 & 5 & 4 & 5 & 34 \\ 0 & 1 & 12 & 0 & 46 \\ 0 & -6 & -2 & -8 & -36 \\ 0 & -10 & -11 & -17 & -109 \end{bmatrix} \begin{matrix} \\ \text{Interchange rows} \\ \text{2 and 3} \\ \\ \end{matrix}$$

$$\begin{bmatrix} 1 & 0 & -56 & 5 & -196 \\ 0 & 1 & 12 & 0 & 46 \\ 0 & 0 & 70 & -8 & 240 \\ 0 & 0 & 109 & -17 & 351 \end{bmatrix} \begin{matrix} (-5\boldsymbol{R_2}+\boldsymbol{R_1}) \\ \\ (6\boldsymbol{R_2}+\boldsymbol{R_3}) \\ (10\boldsymbol{R_2}+\boldsymbol{R_3}) \end{matrix}$$

$$\begin{bmatrix} 1 & 0 & -56 & 5 & -196 \\ 0 & 1 & 12 & 0 & 46 \\ 0 & 0 & 70 & -8 & 240 \\ 0 & 0 & -31 & -1 & -129 \end{bmatrix} \begin{matrix} \\ \\ \\ (-2\boldsymbol{R_3}+\boldsymbol{R_4}) \end{matrix}$$

$$\begin{bmatrix} 1 & 0 & -56 & 5 & -196 \\ 0 & 1 & 12 & 0 & 46 \\ 0 & 0 & 8 & -10 & -18 \\ 0 & 0 & -31 & -1 & -129 \end{bmatrix} \begin{matrix} \\ \\ (2\boldsymbol{R_4}+\boldsymbol{R_3}) \\ \\ \end{matrix}$$

$$\begin{bmatrix} 1 & 0 & -56 & 5 & -196 \\ 0 & 1 & 12 & 0 & 46 \\ 0 & 0 & 1 & \frac{-5}{4} & \frac{-9}{4} \\ 0 & 0 & -31 & -1 & -129 \end{bmatrix} \begin{matrix} \\ \\ \left(\frac{1}{8}\boldsymbol{R_3}\right) \\ \\ \end{matrix}$$

$$\begin{bmatrix} 1 & 0 & 0 & -65 & -322 \\ 0 & 1 & 0 & 15 & 73 \\ 0 & 0 & 1 & \frac{-5}{4} & \frac{-9}{4} \\ 0 & 0 & 0 & \frac{-159}{4} & \frac{-795}{4} \end{bmatrix} \begin{matrix} (56\boldsymbol{R_3}+\boldsymbol{R_1}) \\ (-12\boldsymbol{R_3}+\boldsymbol{R_2}) \\ \\ (31\boldsymbol{R_3}+\boldsymbol{R_4}) \end{matrix}$$

$$\begin{bmatrix} 1 & 0 & 0 & -65 & -322 \\ 0 & 1 & 0 & 15 & 73 \\ 0 & 0 & 1 & \frac{-5}{4} & \frac{-9}{4} \\ 0 & 0 & 0 & 1 & 5 \end{bmatrix} \begin{matrix} \\ \\ \\ \left(\frac{-4}{159}\boldsymbol{R_4}\right) \end{matrix}$$

$$\begin{bmatrix} 1 & 0 & 0 & 0 & 3 \\ 0 & 1 & 0 & 0 & -2 \\ 0 & 0 & 1 & 0 & 4 \\ 0 & 0 & 0 & 1 & 5 \end{bmatrix} \begin{matrix} (65\boldsymbol{R_4}+\boldsymbol{R_1}) \\ (-15\boldsymbol{R_4}+\boldsymbol{R_2}) \\ \left(\frac{-5}{4}\boldsymbol{R_4}+\boldsymbol{R_3}\right) \\ \end{matrix}$$

$x=3,\ y=-2,\ z=4,\ w=5$

25. Let x = the shortest side

$2x$ = another side

$x+6$ = third side

$x+2x+x+6=34$

$4x+6=34$

$4x=28$

$x=7$ inches

$2x=14$ inches

$x+6=13$ inches

Chapter 15 Review

1. 2×3 (2 rows, 3 columns) **2.** 3×1 (3 rows, 1 column) **3.** 3×4 (3 rows, 4 columns)

4. $$\begin{bmatrix} x & 9 & 10 \\ -1 & y & 12 \end{bmatrix} = \begin{bmatrix} 12 & 9 & z \\ -1 & 3 & 12 \end{bmatrix},$$
$x = 12,\ y = 3,\ z = 10$

5. $$\begin{bmatrix} x + y - z \\ x - y + z \\ x \quad + z \end{bmatrix} = \begin{bmatrix} 0 \\ 2 \\ 5 \end{bmatrix},\quad \begin{matrix} x + y - z = 0 \\ x - y + z = 2 \\ x \quad + z = 5 \end{matrix}$$
Add equations 1 and 2 to yield $2x = 2$. Then $x = 1$. Use equation 3 to find z. $1 + z = 5$, so $z = 4$. Use equation 1 to find y. $1 + y - 4 = 0$, so $y = 3$.

6. $$\begin{bmatrix} 5 & 0 & -3 \\ 2 & 6 & 4 \end{bmatrix} + \begin{bmatrix} 2 & 7 & 10 \\ -2 & 9 & -5 \end{bmatrix} = \begin{bmatrix} 7 & 7 & 7 \\ 0 & 15 & -1 \end{bmatrix}$$

7. $$\begin{bmatrix} 2 & 4 \\ -1 & -5 \\ -3 & 7 \end{bmatrix} + \begin{bmatrix} -2 & 5 \\ 0 & -8 \\ 5 & 10 \end{bmatrix} = \begin{bmatrix} 0 & 9 \\ -1 & -13 \\ 2 & 17 \end{bmatrix}$$

8. $$-C = -\begin{bmatrix} 2 & -4 \\ 8 & 11 \end{bmatrix} = \begin{bmatrix} -2 & 4 \\ -8 & -11 \end{bmatrix}$$

9. $$3A = 3\begin{bmatrix} 2 & -3 & 5 \\ 7 & 0 & -1 \end{bmatrix} = \begin{bmatrix} 6 & -9 & 15 \\ 21 & 0 & -3 \end{bmatrix}$$

10. $$A + B = \begin{bmatrix} 2 & -3 & 5 \\ 7 & 0 & -1 \end{bmatrix} + \begin{bmatrix} 6 & -2 & -5 \\ 10 & 4 & 8 \end{bmatrix} = \begin{bmatrix} 8 & -5 & 0 \\ 17 & 4 & 7 \end{bmatrix}$$

11. A is a 2×3, while C is a 2×2. The sum is not defined.

12. $$B - A = B + -A = \begin{bmatrix} 6 & -2 & -5 \\ 10 & 4 & 8 \end{bmatrix} + \begin{bmatrix} -2 & 3 & -5 \\ -7 & 0 & 1 \end{bmatrix} = \begin{bmatrix} 4 & 1 & -10 \\ 3 & 4 & 9 \end{bmatrix}$$

13. $$2A + 3B = 2\begin{bmatrix} 2 & -3 & 5 \\ 7 & 0 & -1 \end{bmatrix} + 3\begin{bmatrix} 6 & -2 & -5 \\ 10 & 4 & 8 \end{bmatrix} =$$
$$\begin{bmatrix} 4 & -6 & 10 \\ 14 & 0 & -2 \end{bmatrix} + \begin{bmatrix} 18 & -6 & -15 \\ 30 & 12 & 24 \end{bmatrix} = \begin{bmatrix} 22 & -12 & -5 \\ 44 & 12 & 22 \end{bmatrix}$$

14. $5A-2B=5A+-2B=5\begin{bmatrix}2 & -3 & 5\\ 7 & 0 & -1\end{bmatrix}+-2\begin{bmatrix}6 & -2 & -5\\ 10 & 4 & 8\end{bmatrix}=$

$\begin{bmatrix}10 & -15 & 25\\ 35 & 0 & -5\end{bmatrix}+\begin{bmatrix}-12 & 4 & 10\\ -20 & -8 & -16\end{bmatrix}=\begin{bmatrix}-2 & -11 & 35\\ 15 & -8 & -21\end{bmatrix}$

15. $2C-B$ The dimensions are different, thus the difference is not defined.

16. $\begin{bmatrix}1 & 2 & 3\\ -2 & 4 & 0\end{bmatrix}\begin{bmatrix}4 & -5\\ -6 & 2\\ 1 & 1\end{bmatrix}=\begin{bmatrix}-5 & 2\\ -32 & 18\end{bmatrix}$

17. $\begin{bmatrix}1 & 2 & -3\\ -4 & 5 & 2\\ 0 & 1 & -1\end{bmatrix}\begin{bmatrix}0 & 1 & 2\\ -5 & 2 & 0\\ 1 & 2 & -2\end{bmatrix}=\begin{bmatrix}-13 & -1 & 8\\ -23 & 10 & -12\\ -6 & 0 & 2\end{bmatrix}$

18. 3×2 by 3×2, product is not defined.

19. $\begin{bmatrix}2 & 0 & 1 & -1\\ 3 & 1 & 2 & 5\end{bmatrix}\begin{bmatrix}1 & 2 & -3\\ 4 & 1 & 6\\ 1 & 0 & 1\\ -2 & 1 & 2\end{bmatrix}=\begin{bmatrix}5 & 3 & -7\\ -1 & 12 & 9\end{bmatrix}$

20. $A^2=\begin{bmatrix}0 & -2\\ 1 & -3\end{bmatrix}\begin{bmatrix}0 & -2\\ 1 & -3\end{bmatrix}=\begin{bmatrix}-2 & 6\\ -3 & 7\end{bmatrix}$

21. Let $A=\begin{bmatrix}-3 & -2\\ 8 & 5\end{bmatrix}$; $|A|=1\cdot A^{-1}=\frac{1}{1}\begin{bmatrix}5 & 2\\ -8 & -3\end{bmatrix}$

22. Let $A=\begin{bmatrix}8 & -5\\ 4 & -3\end{bmatrix}$; $|A|=-4$; $A^{-1}=\frac{1}{-4}\begin{bmatrix}-3 & 5\\ -4 & 8\end{bmatrix}=\begin{bmatrix}\frac{3}{4} & \frac{-5}{4}\\ 1 & -2\end{bmatrix}$

23.

Let $A=\begin{bmatrix}2 & 4 & -2\\ 1 & 2 & 0\\ 4 & 2 & 2\end{bmatrix}$

$$\left[\begin{array}{ccc|ccc}2 & 4 & -2 & 1 & 0 & 0\\ 1 & 2 & 0 & 0 & 1 & 0\\ 4 & 2 & 2 & 0 & 0 & 1\end{array}\right]$$

$$\left[\begin{array}{ccc|ccc}1 & 2 & 0 & 0 & 1 & 0\\ 2 & 4 & -2 & 1 & 0 & 0\\ 4 & 2 & 2 & 0 & 0 & 1\end{array}\right] \text{Interchange rows 1 and 2.}$$

$$\left[\begin{array}{ccc|ccc}1 & 2 & 0 & 0 & 1 & 0\\ 0 & 0 & -2 & 1 & -2 & 0\\ 0 & -6 & 2 & 0 & -4 & 1\end{array}\right]\begin{array}{l}\\(-2\boldsymbol{R_1}+\boldsymbol{R_2})\\(-4\boldsymbol{R_1}+\boldsymbol{R_3})\end{array}$$

$$\left[\begin{array}{ccc|ccc}1 & 2 & 0 & 0 & 1 & 0\\ 0 & -6 & 2 & 0 & -4 & 1\\ 0 & 0 & -2 & 1 & -2 & 0\end{array}\right] \text{Interchange rows 2 and 3.}$$

$$\left[\begin{array}{ccc|ccc}1 & 2 & 0 & 0 & 1 & 0\\ 0 & 1 & \frac{-1}{3} & 0 & \frac{2}{3} & \frac{-1}{6}\\ 0 & 0 & -2 & 1 & -2 & 0\end{array}\right]\left(\frac{-1}{6}\boldsymbol{R_2}\right)$$

$$\left[\begin{array}{ccc|ccc}1 & 0 & \frac{2}{3} & 0 & \frac{-1}{3} & \frac{1}{3}\\ 0 & 1 & \frac{-1}{3} & 0 & \frac{2}{3} & \frac{-1}{6}\\ 0 & 0 & -2 & 1 & -2 & 0\end{array}\right](-2\boldsymbol{R_2}+\boldsymbol{R_1})$$

$$\left[\begin{array}{ccc|ccc}1 & 0 & \frac{2}{3} & 0 & \frac{-1}{3} & \frac{1}{3}\\ 0 & 1 & \frac{-1}{3} & 0 & \frac{2}{3} & \frac{-1}{6}\\ 0 & 0 & 1 & \frac{-1}{2} & 1 & 0\end{array}\right]\left(\frac{-1}{2}\boldsymbol{R_3}\right)$$

$$\left[\begin{array}{ccc|ccc}1 & 0 & 0 & \frac{1}{3} & -1 & \frac{1}{3}\\ 0 & 1 & 0 & \frac{-1}{6} & 1 & \frac{-1}{6}\\ 0 & 0 & 1 & \frac{-1}{2} & 1 & 0\end{array}\right]\begin{array}{l}\left(\frac{-2}{3}\boldsymbol{R_3}+\boldsymbol{R_1}\right)\\\left(\frac{1}{3}\boldsymbol{R_3}+\boldsymbol{R_2}\right)\\ \end{array}$$

$$A^{-1}=\begin{bmatrix}\frac{1}{3} & -1 & \frac{1}{3}\\ \frac{-1}{6} & 1 & \frac{-1}{6}\\ \frac{-1}{2} & 1 & 0\end{bmatrix}$$

24.

Let $A=\begin{bmatrix}2 & 3 & -1\\ 1 & 2 & 0\\ 4 & 5 & 0\end{bmatrix}$

$$\left[\begin{array}{ccc|ccc}2 & 3 & -1 & 1 & 0 & 0\\ 1 & 2 & 0 & 0 & 1 & 0\\ 4 & 5 & 0 & 0 & 0 & 1\end{array}\right]$$

$$\left[\begin{array}{ccc|ccc}1 & 2 & 0 & 0 & 1 & 0\\ 2 & 3 & -1 & 1 & 0 & 0\\ 4 & 5 & 0 & 0 & 0 & 1\end{array}\right] \text{Interchange rows 1 and 2.}$$

$$\left[\begin{array}{ccc|ccc}1 & 2 & 0 & 0 & 1 & 0\\ 0 & -1 & -1 & 1 & -2 & 0\\ 0 & -3 & 0 & 0 & -4 & 1\end{array}\right]\begin{array}{l}\\(-2\boldsymbol{R_1}+\boldsymbol{R_2})\\(-4\boldsymbol{R_1}+\boldsymbol{R_3})\end{array}$$

$$\left[\begin{array}{ccc|ccc}1 & 2 & 0 & 0 & 1 & 0\\ 0 & 1 & 1 & -1 & 2 & 0\\ 0 & -3 & 0 & 0 & -4 & 1\end{array}\right](-1\boldsymbol{R_2})$$

$$\left[\begin{array}{ccc|ccc}1 & 0 & -2 & 2 & -3 & 0\\ 0 & 1 & 1 & -1 & 2 & 0\\ 0 & 0 & 3 & -3 & 2 & 1\end{array}\right]\begin{array}{l}(-2\boldsymbol{R_2}+\boldsymbol{R_1})\\ \\(3\boldsymbol{R_2}+\boldsymbol{R_3})\end{array}$$

$$\left[\begin{array}{ccc|ccc}1 & 0 & -2 & 2 & -3 & 0\\ 0 & 1 & 1 & -1 & 2 & 0\\ 0 & 0 & 1 & -1 & \frac{2}{3} & \frac{1}{3}\end{array}\right]\left(\frac{1}{3}\boldsymbol{R_3}\right)$$

$$\left[\begin{array}{ccc|ccc}1 & 0 & 0 & 0 & \frac{-5}{3} & \frac{2}{3}\\ 0 & 1 & 0 & 0 & \frac{4}{3} & \frac{-1}{3}\\ 0 & 0 & 1 & -1 & \frac{2}{3} & \frac{1}{3}\end{array}\right]\begin{array}{l}(2\boldsymbol{R_3}+\boldsymbol{R_1})\\(-\boldsymbol{R_3}+\boldsymbol{R_2})\\ \end{array}$$

$$A^{-1}=\begin{bmatrix}0 & \frac{-5}{3} & \frac{2}{3}\\ 0 & \frac{4}{3} & \frac{-1}{3}\\ -1 & \frac{2}{3} & \frac{1}{3}\end{bmatrix}$$

25.

$$\text{Let } A = \begin{bmatrix} 1 & 0 & -1 & 0 \\ 2 & 1 & 4 & 0 \\ -1 & 1 & 0 & -2 \\ 2 & 0 & 1 & 1 \end{bmatrix}$$

$$\left[\begin{array}{cccc|cccc} 1 & 0 & -1 & 0 & 1 & 0 & 0 & 0 \\ 2 & 1 & 4 & 0 & 0 & 1 & 0 & 0 \\ -1 & 1 & 0 & -2 & 0 & 0 & 1 & 0 \\ 2 & 0 & 1 & 1 & 0 & 0 & 0 & 1 \end{array}\right]$$

$$\left[\begin{array}{cccc|cccc} 1 & 0 & -1 & 0 & 1 & 0 & 0 & 0 \\ 0 & 1 & 6 & 0 & -2 & 1 & 0 & 0 \\ 0 & 1 & -1 & -2 & 1 & 0 & 1 & 0 \\ 0 & 0 & 3 & 1 & -2 & 0 & 0 & 1 \end{array}\right] \begin{matrix} \\ (-2\boldsymbol{R_1} + \boldsymbol{R_2}) \\ (1\boldsymbol{R_1} + \boldsymbol{R_3}) \\ (-2\boldsymbol{R_1} + \boldsymbol{R_4}) \end{matrix}$$

$$\left[\begin{array}{cccc|cccc} 1 & 0 & -1 & 0 & 1 & 0 & 0 & 0 \\ 0 & 1 & 6 & 0 & -2 & 1 & 0 & 0 \\ 0 & 0 & -7 & -2 & 3 & -1 & 1 & 0 \\ 0 & 0 & 3 & 1 & -2 & 0 & 0 & 1 \end{array}\right] \begin{matrix} \\ \\ (-1\boldsymbol{R_2} + \boldsymbol{R_3}) \\ \\ \end{matrix}$$

$$\left[\begin{array}{cccc|cccc} 1 & 0 & -1 & 0 & 1 & 0 & 0 & 0 \\ 0 & 1 & 6 & 0 & -2 & 1 & 0 & 0 \\ 0 & 0 & -1 & 0 & -1 & -1 & 1 & 2 \\ 0 & 0 & 3 & 1 & -2 & 0 & 0 & 1 \end{array}\right] \begin{matrix} \\ \\ (2\boldsymbol{R_4} + \boldsymbol{R_3}) \\ \\ \end{matrix}$$

$$\left[\begin{array}{cccc|cccc} 1 & 0 & -1 & 0 & 1 & 0 & 0 & 0 \\ 0 & 1 & 6 & 0 & -2 & 1 & 0 & 0 \\ 0 & 0 & 1 & 0 & 1 & 1 & -1 & -2 \\ 0 & 0 & 3 & 1 & -2 & 0 & 0 & 1 \end{array}\right] \begin{matrix} \\ \\ (-1\boldsymbol{R_3}) \\ \\ \end{matrix}$$

$$\left[\begin{array}{cccc|cccc} 1 & 0 & 0 & 0 & 2 & 1 & -1 & -2 \\ 0 & 1 & 0 & 0 & -8 & -5 & 6 & 12 \\ 0 & 0 & 1 & 0 & 1 & 1 & -1 & -2 \\ 0 & 0 & 0 & 1 & -5 & -3 & 3 & 7 \end{array}\right] \begin{matrix} (1\boldsymbol{R_3} + \boldsymbol{R_1}) \\ (-6\boldsymbol{R_3} + \boldsymbol{R_2}) \\ \\ (-3\boldsymbol{R_3} + \boldsymbol{R_4}) \end{matrix}$$

$$A^{-1} = \begin{bmatrix} 2 & 1 & -1 & -2 \\ -8 & -5 & 6 & 12 \\ 1 & 1 & -1 & -2 \\ -5 & -3 & 3 & 7 \end{bmatrix}$$

26. $\begin{aligned} -3x - 2y &= -17 \\ 8x + 5y &= 44 \end{aligned}$ Let $A = \begin{bmatrix} -3 & -2 \\ 8 & 5 \end{bmatrix}, B = \begin{bmatrix} -17 \\ 44 \end{bmatrix}, X = \begin{bmatrix} x \\ y \end{bmatrix}.$

Use A^{-1} from #21. $A^{-1}AX = A^{-1}B$, so $X = A^{-1}B$

$$X = \begin{bmatrix} 5 & 2 \\ -8 & -3 \end{bmatrix} \begin{bmatrix} -17 \\ 44 \end{bmatrix} = \begin{bmatrix} 3 \\ 4 \end{bmatrix} \text{ so } x = 3,\ y = 4$$

27. $\begin{aligned} 8x - 5y &= 50 \\ 4x - 3y &= 26 \end{aligned}$ Let $A = \begin{bmatrix} 8 & -5 \\ 4 & -3 \end{bmatrix}, B = \begin{bmatrix} 50 \\ 26 \end{bmatrix}, X = \begin{bmatrix} x \\ y \end{bmatrix}.$

Use A^{-1} from #22. $A^{-1}AX = A^{-1}B$, so $X = A^{-1}B$

$$X = \begin{bmatrix} \frac{3}{4} & \frac{-5}{4} \\ 1 & -2 \end{bmatrix} \begin{bmatrix} 50 \\ 26 \end{bmatrix} = \begin{bmatrix} 5 \\ -2 \end{bmatrix} \text{ so } x = 5,\ y = -2$$

28. $\begin{aligned} 2x + 4y - 2z &= -14 \\ x + 2y \quad\quad &= -3 \\ 4x + 2y + 2z &= 8 \end{aligned}$ Let $A = \begin{bmatrix} 2 & 4 & -2 \\ 1 & 2 & 0 \\ 4 & 2 & 2 \end{bmatrix}, B = \begin{bmatrix} -14 \\ -3 \\ 8 \end{bmatrix}, X = \begin{bmatrix} x \\ y \\ z \end{bmatrix}.$

Use A^{-1} from #23.

$$X = A^{-1}B = \begin{bmatrix} \frac{1}{3} & -1 & \frac{1}{3} \\ \frac{-1}{6} & 1 & \frac{-1}{6} \\ \frac{-1}{2} & 1 & 0 \end{bmatrix} \begin{bmatrix} -14 \\ -3 \\ 8 \end{bmatrix} = \begin{bmatrix} 1 \\ -2 \\ 4 \end{bmatrix} \text{ so } x = 1,\ y = -2,\ z = 4$$

29. $\begin{aligned} 2x + 3y - z &= 10 \\ x + 2y \quad &= 7 \\ 4x + 5y \quad &= 22 \end{aligned}$ Let $A = \begin{bmatrix} 2 & 3 & -1 \\ 1 & 2 & 0 \\ 4 & 5 & 0 \end{bmatrix}, B = \begin{bmatrix} 10 \\ 7 \\ 22 \end{bmatrix}, X = \begin{bmatrix} x \\ y \\ z \end{bmatrix}.$

Use A^{-1} from #24.

$$X = A^{-1}B = \begin{bmatrix} 0 & \frac{-5}{3} & \frac{2}{3} \\ 0 & \frac{4}{3} & \frac{-1}{3} \\ -1 & \frac{2}{3} & \frac{1}{3} \end{bmatrix} \begin{bmatrix} 10 \\ 7 \\ 22 \end{bmatrix} = \begin{bmatrix} 3 \\ 2 \\ 2 \end{bmatrix} \text{ so } x = 3,\ y = 2,\ z = 2$$

30.
$$\begin{aligned} x \quad\quad\quad - z \quad &= 5 \\ 2x + y + 4z \quad &= -1 \\ -x + y \quad\quad - 2w &= -10 \\ 2x \quad + z \quad + w &= 8 \end{aligned}$$

Let $A = \begin{bmatrix} 1 & 0 & -1 & 0 \\ 2 & 1 & 4 & 0 \\ -1 & 1 & 0 & -2 \\ 2 & 0 & 1 & 1 \end{bmatrix}$, $B = \begin{bmatrix} 5 \\ -1 \\ -10 \\ 8 \end{bmatrix}$, $X = \begin{bmatrix} x \\ y \\ z \\ w \end{bmatrix}$.

Use A^{-1} from #25.

$$X = A^{-1}B = \begin{bmatrix} 2 & 1 & -1 & -2 \\ -8 & -5 & 6 & 12 \\ 1 & 1 & -1 & -2 \\ -5 & -3 & 3 & 7 \end{bmatrix} \begin{bmatrix} 5 \\ -1 \\ -10 \\ 8 \end{bmatrix} = \begin{bmatrix} 3 \\ 1 \\ -2 \\ 4 \end{bmatrix}$$ so $x = 3,\ y = 1,\ z = -2,\ w = 4$

31. $\left[\begin{array}{cc|c} 1 & 0 & -6 \\ 0 & 1 & 10 \end{array}\right]$ then $x = -6,\ y = 10$

32. $\left[\begin{array}{ccc|c} 1 & 0 & 0 & 4 \\ 0 & 1 & 0 & -3 \\ 0 & 0 & 1 & 15 \end{array}\right]$ then $\begin{aligned} x &= 4 \\ y &= -3 \\ z &= 15 \end{aligned}$

33. $\left[\begin{array}{cccc|c} 1 & 0 & 0 & 0 & 4 \\ 0 & 1 & 0 & 0 & -11 \\ 0 & 0 & 1 & 0 & 8 \\ 0 & 0 & 0 & 1 & 0 \end{array}\right]$ then $\begin{aligned} x &= 4 \\ y &= -11 \\ z &= 8 \\ w &= 0 \end{aligned}$

34. $\begin{aligned} x + 4y &= 15 \\ -2x + 3y &= 3 \end{aligned}$ $\left[\begin{array}{cc|c} 1 & 4 & 15 \\ -2 & 3 & 3 \end{array}\right]$

$\left[\begin{array}{cc|c} 1 & 4 & 15 \\ 0 & 11 & 33 \end{array}\right] (2\boldsymbol{R_1} + \boldsymbol{R_2})$

$\left[\begin{array}{cc|c} 1 & 4 & 15 \\ 0 & 1 & 3 \end{array}\right] \left(\frac{1}{11}\boldsymbol{R_2}\right)$

$\left[\begin{array}{cc|c} 1 & 0 & 3 \\ 0 & 1 & 3 \end{array}\right] (-4\boldsymbol{R_2} + \boldsymbol{R_1})$

So $x = 3,\ y = 3$

35. $\begin{aligned} 3x + 8y &= -4 \\ 6x + y &= 22 \end{aligned}$ $\left[\begin{array}{cc|c} 3 & 8 & -4 \\ 6 & 1 & 22 \end{array}\right]$

$\left[\begin{array}{cc|c} 1 & \frac{8}{3} & \frac{-4}{3} \\ 6 & 1 & 22 \end{array}\right] \left(\frac{1}{3}\boldsymbol{R_1}\right)$

$\left[\begin{array}{cc|c} 1 & \frac{8}{3} & \frac{-4}{3} \\ 0 & -15 & 30 \end{array}\right] (-6\boldsymbol{R_1} + \boldsymbol{R_2})$

$\left[\begin{array}{cc|c} 1 & \frac{8}{3} & \frac{-4}{3} \\ 0 & 1 & -2 \end{array}\right] \left(\frac{-1}{15}\boldsymbol{R_2}\right)$

$\left[\begin{array}{cc|c} 1 & 0 & 4 \\ 0 & 1 & -2 \end{array}\right] \left(\frac{-8}{3}\boldsymbol{R_2} + \boldsymbol{R_1}\right)$

So $x = 4,\ y = -2$

36.

$$x+3y-z=8$$
$$3x-4y+2z=-9$$
$$-2x+y-4z=15$$

$$\left[\begin{array}{ccc|c} 1 & 3 & -1 & 8 \\ 3 & -4 & 2 & -9 \\ -2 & 1 & -4 & 15 \end{array}\right]$$

$$\left[\begin{array}{ccc|c} 1 & 3 & -1 & 8 \\ 0 & -13 & 5 & -33 \\ 0 & 7 & -6 & 31 \end{array}\right] \begin{array}{l} \\ (-3\boldsymbol{R_1}+\boldsymbol{R_2}) \\ (2\boldsymbol{R_1}+\boldsymbol{R_3}) \end{array}$$

$$\left[\begin{array}{ccc|c} 1 & 3 & -1 & 8 \\ 0 & 1 & -7 & 29 \\ 0 & 7 & -6 & 31 \end{array}\right] \begin{array}{l} \\ (2\boldsymbol{R_3}+\boldsymbol{R_2}) \\ \\ \end{array}$$

$$\left[\begin{array}{ccc|c} 1 & 0 & 20 & -79 \\ 0 & 1 & -7 & 29 \\ 0 & 0 & 43 & -172 \end{array}\right] \begin{array}{l} (-3\boldsymbol{R_2}+\boldsymbol{R_1}) \\ \\ (-7\boldsymbol{R_2}+\boldsymbol{R_3}) \end{array}$$

$$\left[\begin{array}{ccc|c} 1 & 0 & 20 & -79 \\ 0 & 1 & -7 & 29 \\ 0 & 0 & 1 & -4 \end{array}\right] \left(\frac{1}{43}\boldsymbol{R_3}\right)$$

$$\left[\begin{array}{ccc|c} 1 & 0 & 0 & 1 \\ 0 & 1 & 0 & 1 \\ 0 & 0 & 1 & -4 \end{array}\right] \begin{array}{l} (-20\boldsymbol{R_3}+\boldsymbol{R_1}) \\ (7\boldsymbol{R_3}+\boldsymbol{R_2}) \\ \\ \end{array}$$

So $x=1,\ y=1,\ z=-4$

37.

$$3x-4y+2z=49$$
$$-x+y+5z=31$$
$$2x-3y+2z=40$$

$$\left[\begin{array}{ccc|c} 3 & -4 & 2 & 49 \\ -1 & 1 & 5 & 31 \\ 2 & -3 & 2 & 40 \end{array}\right]$$

$$\left[\begin{array}{ccc|c} 1 & -1 & 0 & 9 \\ -1 & 1 & 5 & 31 \\ 2 & -3 & 2 & 40 \end{array}\right] \begin{array}{l} (-1\boldsymbol{R_3}+\boldsymbol{R_1}) \\ \\ \\ \end{array}$$

$$\left[\begin{array}{ccc|c} 1 & -1 & 0 & 9 \\ 0 & 0 & 5 & 40 \\ 0 & -1 & 2 & 22 \end{array}\right] \begin{array}{l} \\ (1\boldsymbol{R_1}+\boldsymbol{R_2}) \\ (-2\boldsymbol{R_1}+\boldsymbol{R_3}) \end{array}$$

$$\left[\begin{array}{ccc|c} 1 & -1 & 0 & 9 \\ 0 & -1 & 2 & 22 \\ 0 & 0 & 5 & 40 \end{array}\right] \begin{array}{l} \text{Interchange} \\ \text{rows 2 and 3} \\ \\ \end{array}$$

$$\left[\begin{array}{ccc|c} 1 & -1 & 0 & 9 \\ 0 & 1 & -2 & -22 \\ 0 & 0 & 1 & 8 \end{array}\right] \begin{array}{l} (-1\boldsymbol{R_2}) \\ \left(\frac{1}{5}\boldsymbol{R_3}\right) \end{array}$$

$$\left[\begin{array}{ccc|c} 1 & 0 & -2 & -13 \\ 0 & 1 & -2 & -22 \\ 0 & 0 & 1 & 8 \end{array}\right] \begin{array}{l} (1\boldsymbol{R_2}+\boldsymbol{R_1}) \\ \\ \\ \end{array}$$

$$\left[\begin{array}{ccc|c} 1 & 0 & 0 & 3 \\ 0 & 1 & 0 & -6 \\ 0 & 0 & 1 & 8 \end{array}\right] \begin{array}{l} (2\boldsymbol{R_3}+\boldsymbol{R_1}) \\ (2\boldsymbol{R_3}+\boldsymbol{R_2}) \\ \\ \end{array}$$

So $x=3,\ y=-6,\ z=8$

38. $3x+2y-z+w=8$

$3x+y-2z-2w=1$

$x+3y+4z+8w=17$

$-x+y+z=-3$

$$\begin{bmatrix} 3 & 2 & -1 & 1 & 8 \\ 3 & 1 & -2 & -2 & 1 \\ 1 & 3 & 4 & 8 & 17 \\ -1 & 1 & 1 & 0 & -3 \end{bmatrix}$$

$$\begin{bmatrix} 1 & 4 & 1 & 1 & 2 \\ 3 & 1 & -2 & -2 & 1 \\ 1 & 3 & 4 & 8 & 17 \\ -1 & 1 & 1 & 0 & -3 \end{bmatrix} \begin{matrix} 2R_4+R_1 \\ \\ \\ \\ \end{matrix}$$

$$\begin{bmatrix} 1 & 4 & 1 & 1 & 2 \\ 0 & -11 & -5 & -5 & -5 \\ 0 & -1 & 3 & 7 & 15 \\ 0 & 5 & 2 & 1 & -1 \end{bmatrix} \begin{matrix} \\ -3R_1+R_2 \\ -R_1+R_3 \\ R_1+R_4 \end{matrix}$$

$$\begin{bmatrix} 1 & 4 & 1 & 1 & 2 \\ 0 & -1 & -1 & -3 & -7 \\ 0 & -1 & 3 & 7 & 15 \\ 0 & 5 & 2 & 1 & -1 \end{bmatrix} \begin{matrix} \\ 2R_4+R_2 \\ \\ \\ \end{matrix}$$

$$\begin{bmatrix} 1 & 4 & 1 & 1 & 2 \\ 0 & 1 & 1 & 3 & 7 \\ 0 & -1 & 3 & 7 & 15 \\ 0 & 5 & 2 & 1 & -1 \end{bmatrix} \begin{matrix} \\ -R_2 \\ \\ \\ \end{matrix}$$

$$\begin{bmatrix} 1 & 0 & -3 & -11 & -26 \\ 0 & 1 & 1 & 3 & 7 \\ 0 & 0 & 4 & 10 & 22 \\ 0 & 0 & -3 & -14 & -36 \end{bmatrix} \begin{matrix} -4R_2+R_1 \\ \\ R_2+R_3 \\ -5R_2+R_4 \end{matrix}$$

$$\begin{bmatrix} 1 & 0 & -3 & -11 & -26 \\ 0 & 1 & 1 & 3 & 7 \\ 0 & 0 & 1 & -4 & -14 \\ 0 & 0 & -3 & -14 & -36 \end{bmatrix} \begin{matrix} \\ \\ R_4+R_3 \\ \\ \end{matrix}$$

$$\begin{bmatrix} 1 & 0 & 0 & -23 & -68 \\ 0 & 1 & 0 & 7 & 21 \\ 0 & 0 & 1 & -4 & -14 \\ 0 & 0 & 0 & -26 & -78 \end{bmatrix} \begin{matrix} 3R_3+R_1 \\ -R_3+R_2 \\ \\ 3R_3+R_4 \end{matrix}$$

$$\begin{bmatrix} 1 & 0 & 0 & -23 & -68 \\ 0 & 1 & 0 & 7 & 21 \\ 0 & 0 & 1 & -4 & -14 \\ 0 & 0 & 0 & 1 & 3 \end{bmatrix} \begin{matrix} \\ \\ \\ R_4 \div (-26) \end{matrix}$$

$$\begin{bmatrix} 1 & 0 & 0 & 0 & 1 \\ 0 & 1 & 0 & 0 & 0 \\ 0 & 0 & 1 & 0 & -2 \\ 0 & 0 & 0 & 1 & 3 \end{bmatrix} \begin{matrix} 23R_4+R_1 \\ -7R_4+R_2 \\ 4R_4+R_3 \\ \\ \end{matrix}$$

Solution is
$x=1$, $y=0$, $z=-2$, $w=3$

Chapter 16, Exercises 16.1

1. $f(x)=3x(x-5)(x-2)$

a) x-intercepts - let $f(x)=0$

$0=3x(x-5)(x-2)$

$x=0,5,2$

b) y-intercepts - let $x=0$

$f(0)=3(0)(0-5)(0-2)$

$f(0)=0$ ie

$y=0$

3. $f(x)=(x-4)(2x-5)(x-1)$

a) x-intercepts - let $f(x)=0$

$0=(x-4)(2x-5)(x-1)$

$x=4,2.5,1$

b) y-intercepts - let $x=0$

$f(0)=(0-4)(2(0)-5)(0-1)$

$f(0)=-20$

$y=-20$

5. $f(x)=x^3-9x$

$f(x)=x(x^2-9)$

$f(x)=x(x+3)(x-3)$

a) x-intercepts - let $f(x)=0$

$0=x(x+3)(x-3)$

$x=0,-3,3$

b) y-intercepts - let $x=0$

$f(0)=(0)^3-9(0)$

$f(0)=0$

$y=0$

7. $f(x)=x^3+5x^2-2x-24$

From the graph, the zeros are: $-4,-3,2$.

$f(x)=(x+4)(x+3)(x-2)$

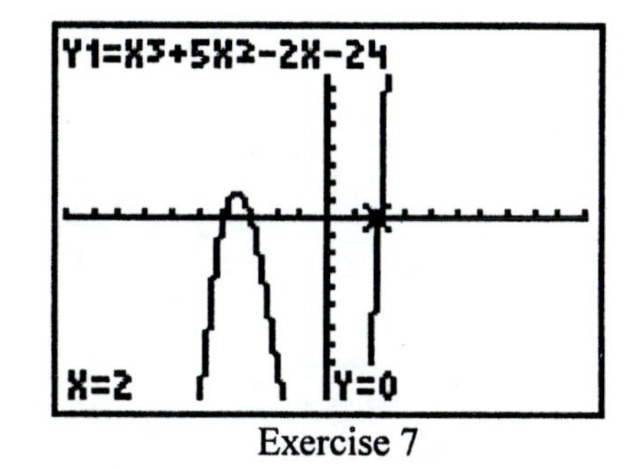

Exercise 7

9. $f(x)=x^3-x^2-9x+9$

The zeros are: $-3,1,3$.

$f(x)=(x+3)(x-1)(x-3)$

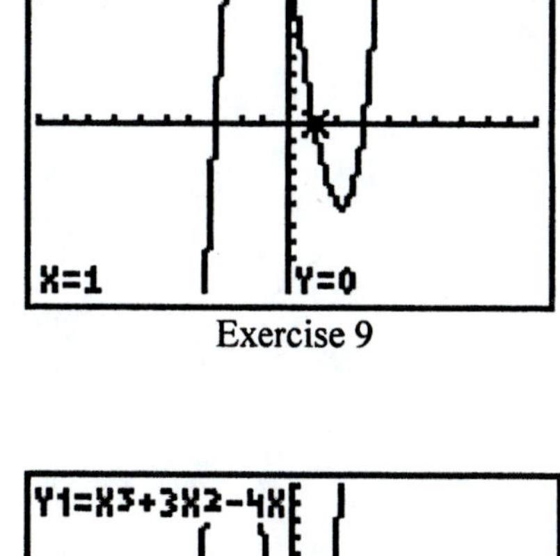

Exercise 9

11. $f(x)=x^3+3x^2-4x$

The zeros are: $-4,0,1$.

$f(x)=(x-0)(x+4)(x-1)$

$f(x)=x(x+4)(x-1)$

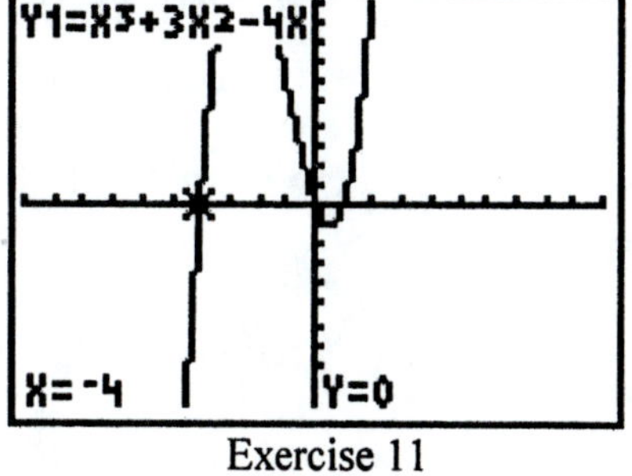

Exercise 11

13. $f(x) = x^3 - 7x^2 + 15x - 9$

The zeros are 1 and 3.
Because the graph is above the x-axis on both sides of $x = 3$, and because $f(x)$ is cubic, 3 is a zero of muliplicity 2.

$f(x) = (x-1)(x-3)^2$

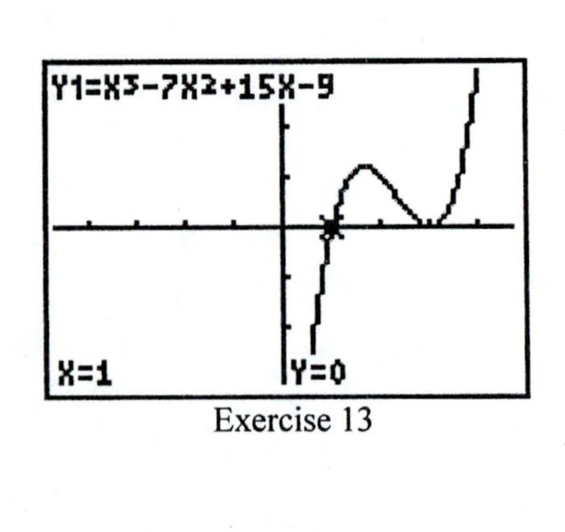

Exercise 13

15. $f(x) = x^3 - 6x^2 + 32$

The zeros are:
$-2, 4$ (of multiplicity 2).

$f(x) = (x+2)(x-4)^2$

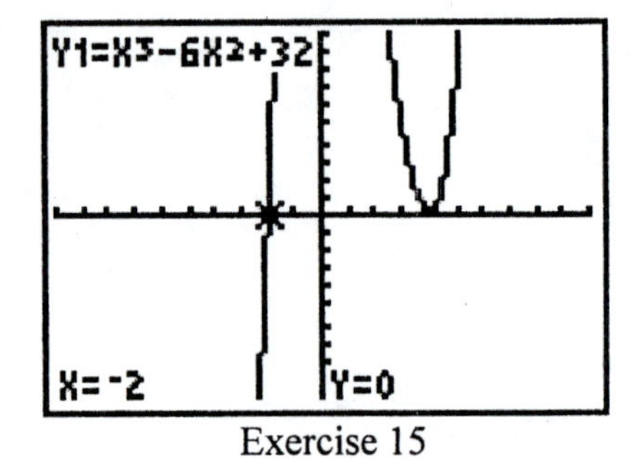

Exercise 15

17. $f(x) = x^4 - 2x^3 - 13x^2 + 14x + 24$

The zeros are: $-3, -1, 2, 4.$

$f(x) = (x+3)(x+1)(x-2)(x-4)$

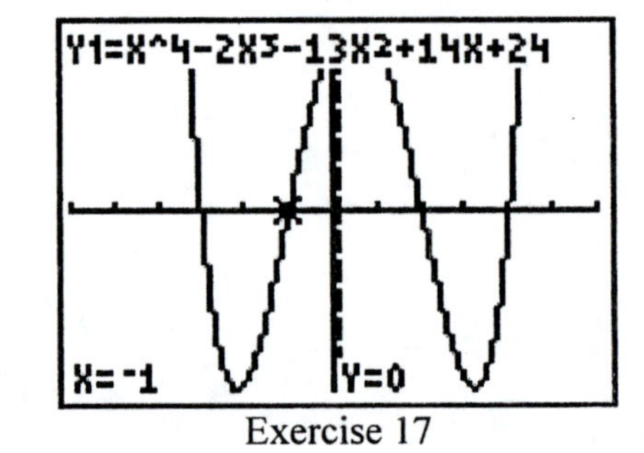

Exercise 17

19. $f(x) = x^4 - 5x^3 + x^2 + 21x - 18$

The zeros are: $-2, 1, 3$ (of multiplicity 2).

$f(x) = (x+2)(x-1)(x-3)^2$

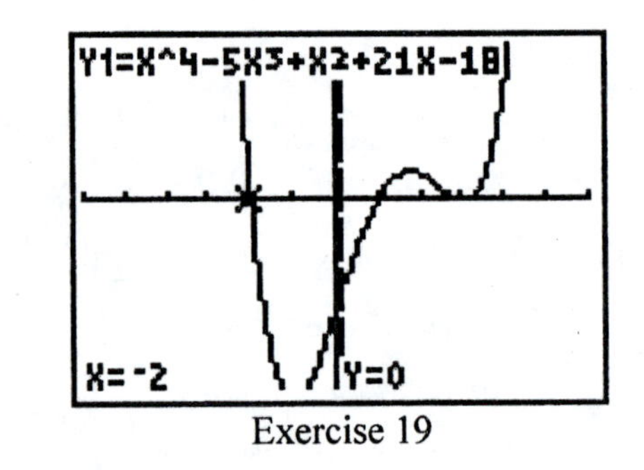

Exercise 19

21. $f(x) = x^4 - 2x^3 - 8x^2 + 18x - 9$

The zeros are: $-3, 1$ (of multiplicity 2) and 3.

$f(x) = (x+3)(x-3)(x-1)^2$

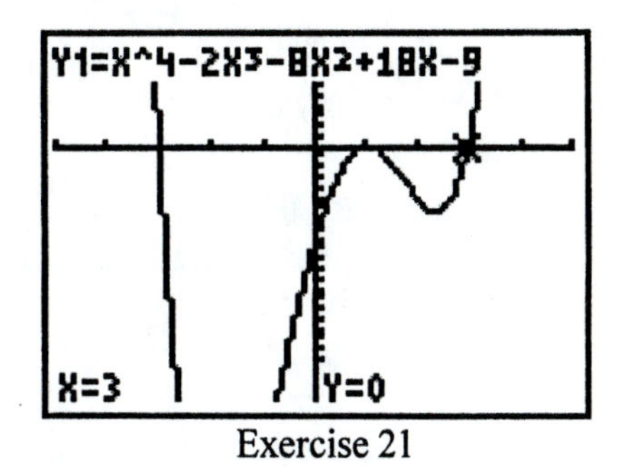

Exercise 21

23. $f(x) = x^4 + 4x^3 - 2x^2 - 12x + 9$

The zeros are: -3 (of multiplicity 2), 1 (of multiplicity 2).

$f(x) = (x+3)^2 (x-1)^2$

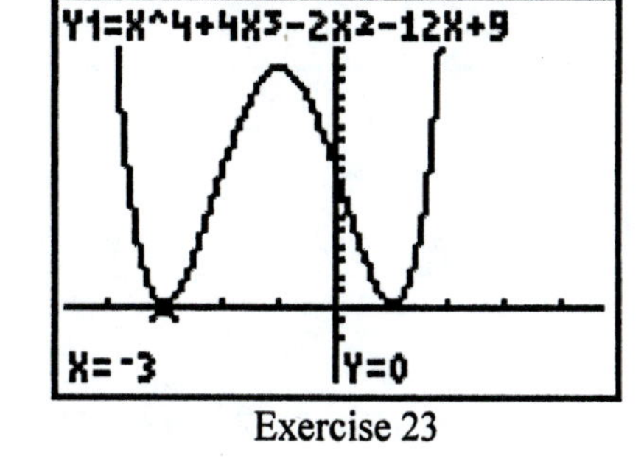

Exercise 23

Exercises 16.2

1. $x^3 + x^2 - 5x + 2 = 0$

Estimates are:

$-3, 0.5, 1.5$

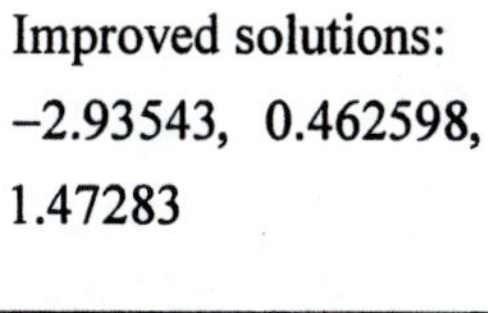

Improved solutions:

-2.93543, 0.462598,

1.47283

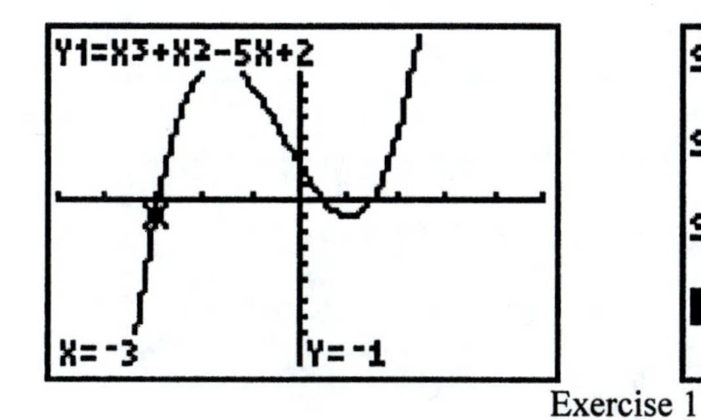

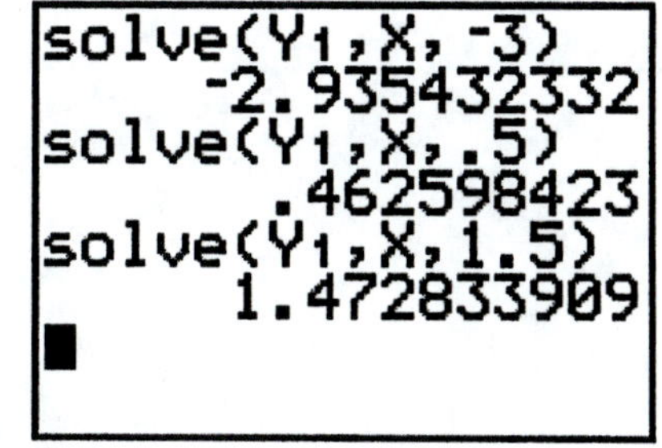

Exercise 1

3. $2x^3 - 5x^2 + x + 1 = 0$

Estimates are:

$-0.3, 0.6, 2$

Improved solutions:

-0.340665, 0.678963,

2.16170

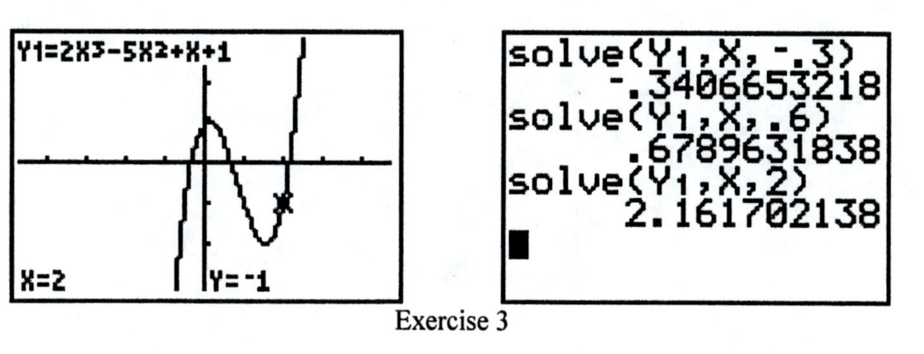

Exercise 3

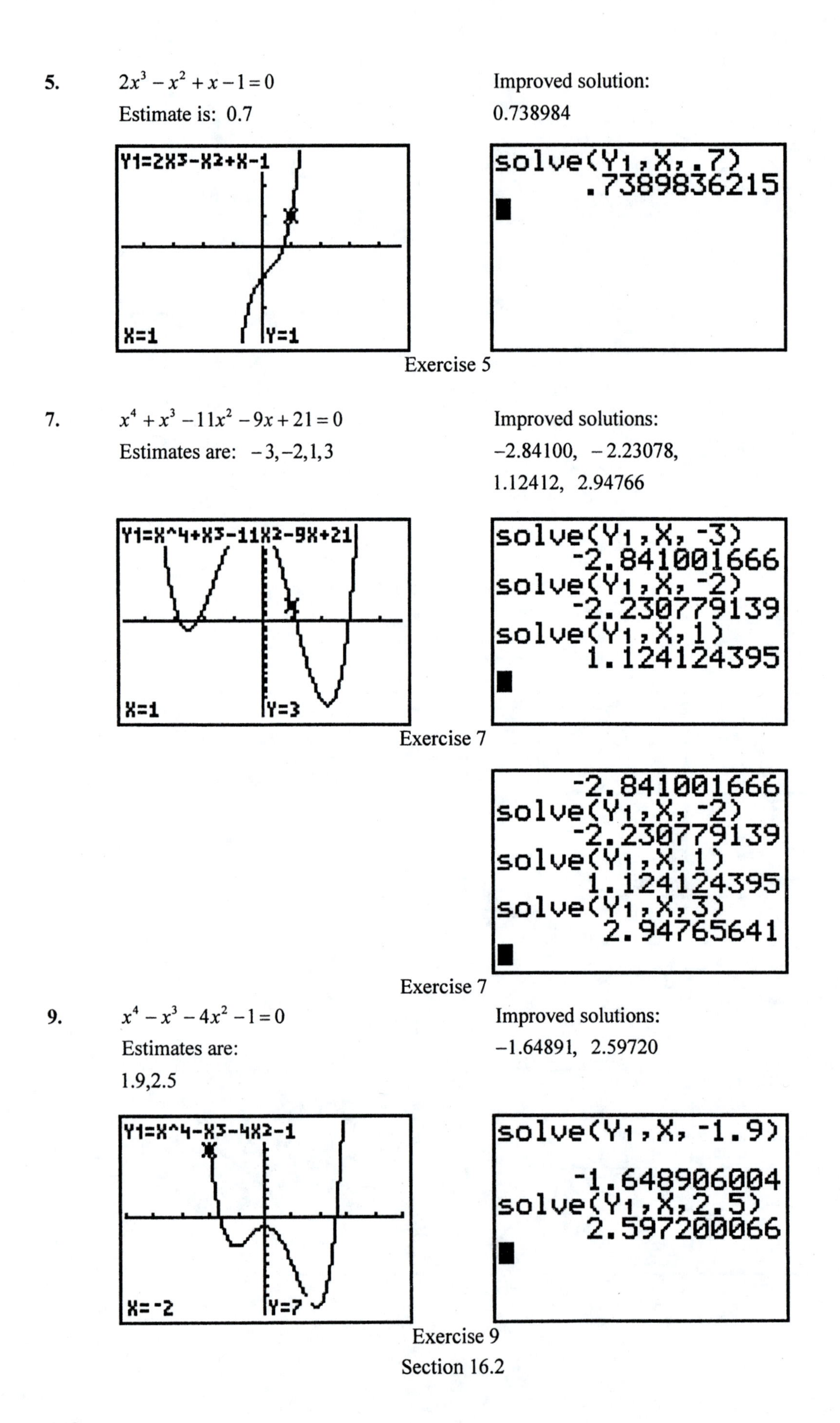

5. $2x^3 - x^2 + x - 1 = 0$

Estimate is: 0.7

Improved solution:

0.738984

Exercise 5

7. $x^4 + x^3 - 11x^2 - 9x + 21 = 0$

Estimates are: $-3, -2, 1, 3$

Improved solutions:

−2.84100, -2.23078,

1.12412, 2.94766

Exercise 7

Exercise 7

9. $x^4 - x^3 - 4x^2 - 1 = 0$

Estimates are:

1.9,2.5

Improved solutions:

−1.64891, 2.59720

Exercise 9

Section 16.2

Exercises 16.3

1. 5 **3.** 4 **5.** 10

7. $x^3 - 6x^2 + x + 34 = 0$

Since $(4 + j)$ is a solution, $(4 - j)$ is also a solution.

$(x-(4+j))(x-(4-j)) =$

$x^2 - (4+j)x - (4-j)x + 16 - j^2 =$

$x^2 - 4x - jx - 4x + jx + 16 + 1 = x^2 - 8x + 17.$

Since $x^2 - 8x + 17$ is a factor of $x^3 - 6x^2 + x + 34$, we divide to find the other factor.

$$
\begin{array}{r}
x+2 \\
x^2 - 8x + 17 \overline{)\, x^3 - 6x^2 + x + 34} \\
\underline{x^3 - 8x^2 + 17x} \\
2x^2 - 16x + 34 \\
\underline{2x^2 - 16x + 34} \\
0
\end{array}
$$

Since $(x+2)$ is a factor, the other zero is -2.

9. $x^4 - 2x^3 + 3x^2 - 2x + 2 = 0$. Since j is a solution, $-j$ is also a solution. Since $(1 - j)$ is a solution, $(1 + j)$ is also a solution.

11. $x^5 - 3x^4 + 2x^3 - 6x^2 + x - 3 = 0$. Since (j) is a solution of multiplicity 2, $(-j)$ is a solution of multiplicity 2. The other solution is 3.

13. $x^4 - 4x^3 + 10x^2 + 68x - 75 = 0$

Since $3 - 4j$ is a solution, $3 + 4j$ is also a solution.

$(x-(3-4j))(x-(3+4j)) = x^2 - (3-4j)x - (3+4j)x + 9 - 16j^2 =$

$x^2 - 3x + 4jx - 3x - 4jx + 9 + 16 = x^2 - 6x + 25.$

Since $x^2 - 6x + 25$ is a factor of $x^4 - 4x^3 + 10x^2 + 68x - 75$, we divide to find the other factor.

$$
\begin{array}{r}
x^2 + 2x - 3 \\
x^2 - 6x + 25 \overline{)\, x^4 - 4x^3 + 10x^2 + 68x - 75} \\
\underline{x^4 - 6x^3 + 25x^2} \\
2x^3 - 15x^2 + 68x \\
\underline{2x^3 - 12x^2 + 50x} \\
-3x^2 + 18x - 75 \\
\underline{-3x^2 + 18x - 75} \\
0
\end{array}
$$

Since $(x^2 + 2x - 3)$ is a factor, solve $x^2 + 2x - 3 = 0$ to find the other solutions. $x^2 + 2x - 3 = (x+3)(x-1)$, so -3 and 1 are the other solutions.

15. Degree 3, solutions include 4 and $2j$.

$$f(x)=(x-4)(x-2j)(x+2j)=0$$
$$f(x)=(x-4)(x^2-4j^2)=0$$
$$f(x)=(x-4)(x^2+4)=0$$
$$f(x)=x^3-4x^2+4x-16=0$$

17. Degree 4, solutions include $2,-1,$ and $1+j$.

$$f(x)=(x-2)(x+1)(x-(1+j))(x-(1-j))=0$$
$$f(x)=(x^2-x-2)(x^2-(1+j)x-(1-j)x+1-j^2)=0$$
$$f(x)=(x^2-x-2)(x^2-x-jx-x+jx+1+1)=0$$
$$f(x)=(x^2-x-2)(x^2-2x+2)=0$$
$$f(x)=x^4-3x^3+2x^2+2x-4=0$$

19. Degree 4, solutions include $(1+j)$ and $(1-2j)$.

$$f(x)=(x-(1+j))(x-(1-j))(x-(1-2j))(x-(1+2j))=0$$
$$f(x)=(x^2-(1+j)x-(1-j)x+1-j^2)(x^2-(1-2j)x-(1+2j)x+1-4j^2)=0$$
$$f(x)=(x^2-x-jx-x+jx+1+1)(x^2-x+2jx-x-2jx+1+4)=0$$
$$f(x)=(x^2-2x+2)(x^2-2x+5)=0$$
$$f(x)=x^4-4x^3+11x^2-14x+10=0$$

21. Degree 5, solutions include $0,(1-3j),$ and $(2j)$.

$$f(x)=(x-0)(x-(1-3j))(x-(1+3j))(x-2j)(x-(-2j))=0$$
$$f(x)=x(x^2-(1-3j)x-(1+3j)x+1-9j^2)(x^2-4j^2)=0$$
$$f(x)=x(x^2-x+3jx-x-3jx+1+9)(x^2+4)=0$$
$$f(x)=x(x^2-2x+10)(x^2+4)=0$$
$$f(x)=x^5-2x^4+14x^3-8x^2+40x=0$$

23. Since complex roots occur in conjugate pairs, there would be five roots for a fourth degree equation.

25. Since complex roots occur in conjugate pairs, there would be four roots for a third degree equation.

Chapter 16 Review

1. seventh

2. fifth

3. twelfth

4. $f(x)=4x(x+7)(2x-1)$

x-intercepts at $f(x)=0$.

$x=0,-7,0.5$

y-intercepts at $x=0$.

$f(0)=4(0)(0+7)(2(0)-1)$

$f(0)=0$

5. $f(x)=(x-2)(x^2+6x+9)$

$f(x)=(x-2)(x+3)^2$

x-intercepts at $f(x)=0$.

$x=2,-3$

y-intercepts at $x=0$.

$f(0)=(0-2)(0+3)^2$

$f(0)=-2(9)=-18$

6. $f(x)=x^3-6x^2$

$f(x)=x^2(x-6)$

x-intercepts at $f(x)=0$.

$x=0,6$

y-intercepts at $x=0$.

$f(0)=0^2(0-6)=0$

7. $f(x)=2x^3+3x^2-44x-105$

y-intercepts at $x=0$.

$f(0)=2(0^3)+3(0^2)-44(0)-105$

$f(0)=-105$

x-intercepts at $f(x)=0$.

Use the solve feature:

$x=-3.5,-3,5$

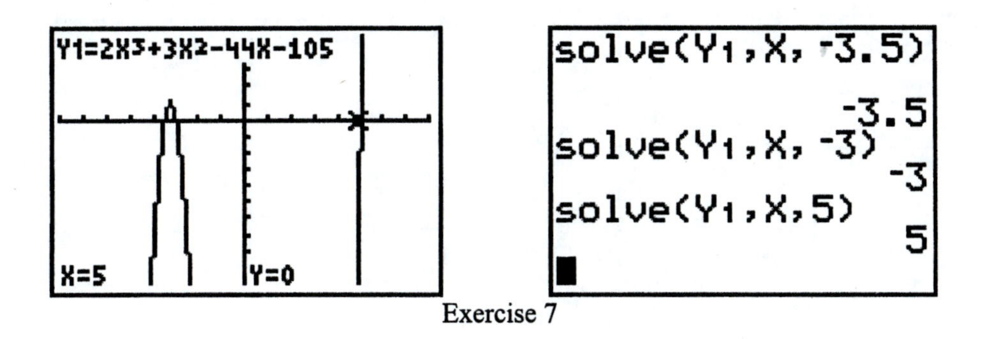

Exercise 7

8. $f(x) = x^3 - 3x^2 - 22x + 24$

y-intercepts at $x = 0$.

$f(0) = (0)^3 - 3(0^2) - 22(0) + 24$

$f(0) = 24$

x-intercepts at $f(x) = 0$.

Use the solve feature:

$x = -4, 1, 6$

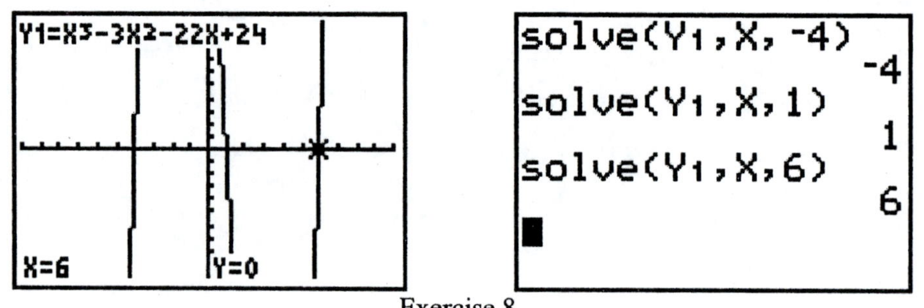

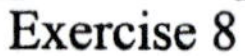
Exercise 8

9. $f(x) = x^3 - 7x^2 + 14x - 8$

zeros are: $x = 1, 2, 4$

$f(x) = (x-1)(x-2)(x-4)$

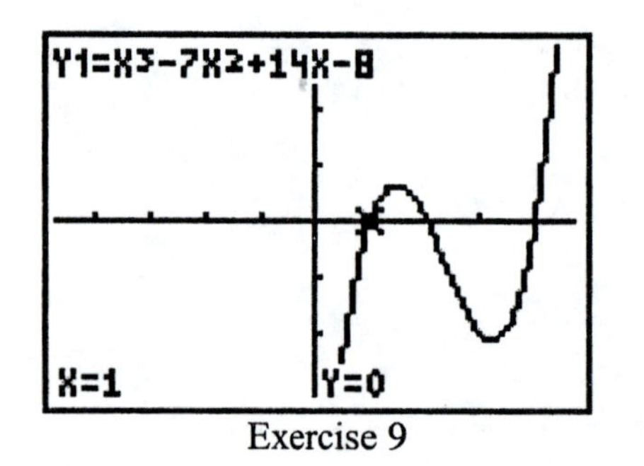

Exercise 9

10. $f(x) = x^3 + 4x^2 + x - 6$

zeros are: $x = -3, -2, 1$

$f(x) = (x+3)(x+2)(x-1)$

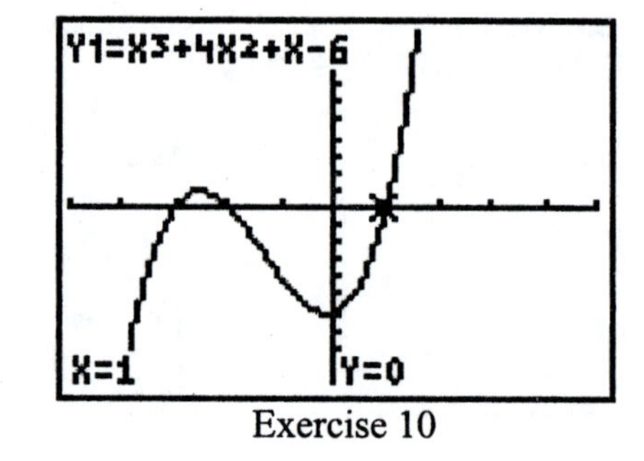

Exercise 10

11. $f(x) = x^4 + 9x^3 + 9x^2 - 49x + 30$

zeros are: $-6, -5, 1$ (multiplicity of 2)

$f(x) = (x+6)(x+5)(x-1)^2$

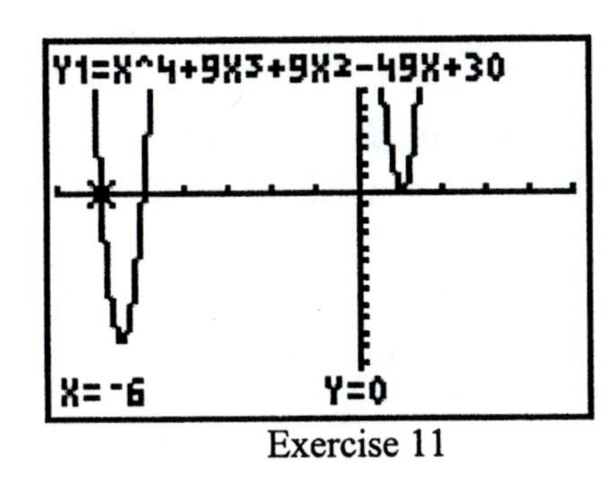

Exercise 11

12. $f(x) = x^4 - 4x^3 - 7x^2 + 34x - 24$

zeros are: $x = -3, 1, 2, 4$

$f(x) = (x+3)(x-1)(x-2)(x-4)$

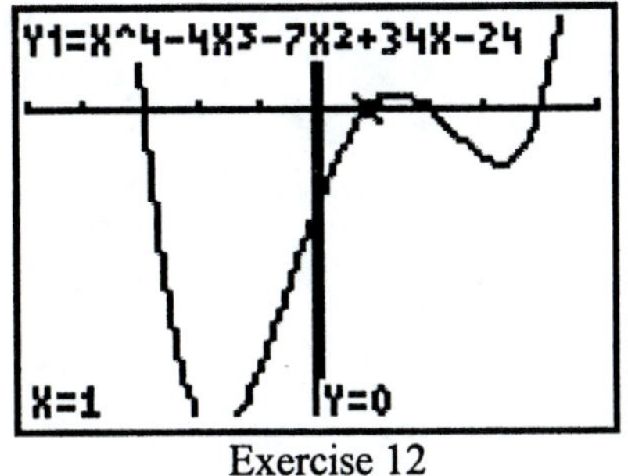

Exercise 12

13. $f(x) = x^4 + 3x^3 - 13x^2 - 51x - 36$

zeros are: $-1, 4$, and -3 (of multiplicity 2).

$f(x) = (x+1)(x-4)(x+3)^2$

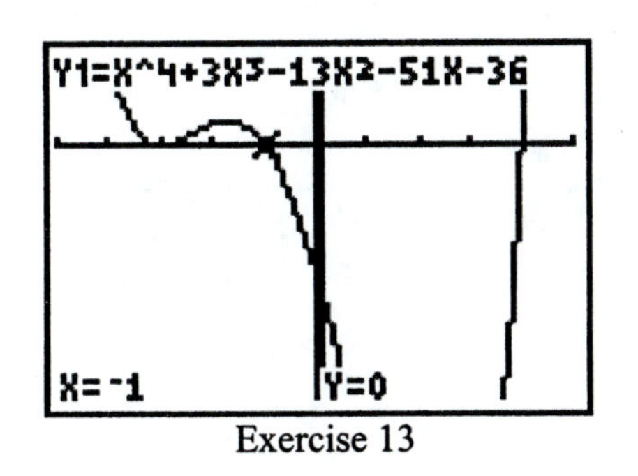

Exercise 13

14. $f(x) = 2x^4 - 35x^3 + 166x^2 + 77x - 1470$

zeros are: $\frac{-5}{2}, 6, 7$ (of multiplicity 2).

Since $x = \frac{-5}{2}$, $2x = -5$, so $2x + 5 = 0$;

ie $2x + 5$ is a factor.

$f(x) = (2x+5)(x-6)(x-7)^2$

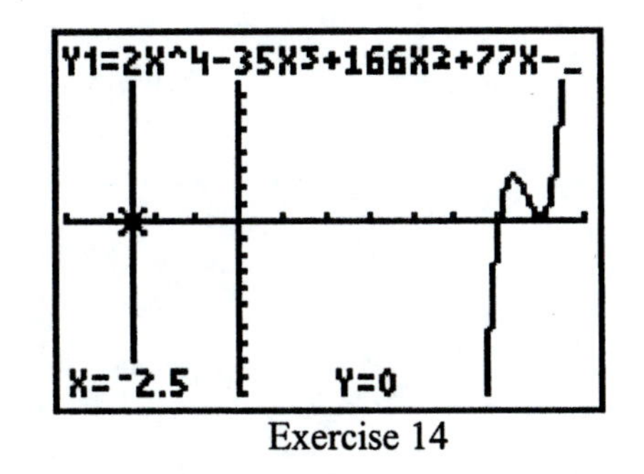

Exercise 14

15. $f(x) = (2x+9)^2(x-4)^3$

$x = \dfrac{-9}{2}$ (multiplicity of 2)

$x = 4$ (multiplicity of 3)

16. $f(x) = (x-6)(x+10)^3$

$x = 6$ (multiplicity of 1)

$x = -10$ (multiplicity of 3)

17. $f(x) = x^4(x-1)^2$

$x = 0$ (multiplicity of 4)

$x = 1$ (multiplicity of 2)

18. $f(x) = (x-3)(x+5)^2$

$x = 3$ (multiplicity of 1)

$x = -5$ (multiplicity of 2)

19. $x^2 + 3x - 1 = 0$

Estimates are: $-3.4, 0.4$

Improved solutions:

-3.30278, 0.302776

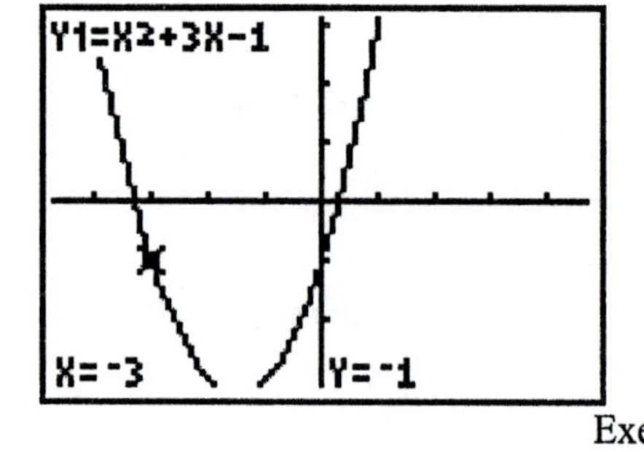

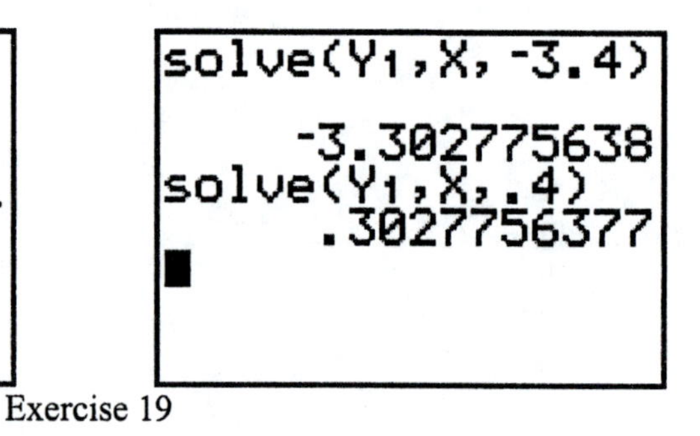

Exercise 19

20. $x^3 - 5x^2 + 6x - 3 = 0$

Estimate is 3.5

Improved solution:

3.54682

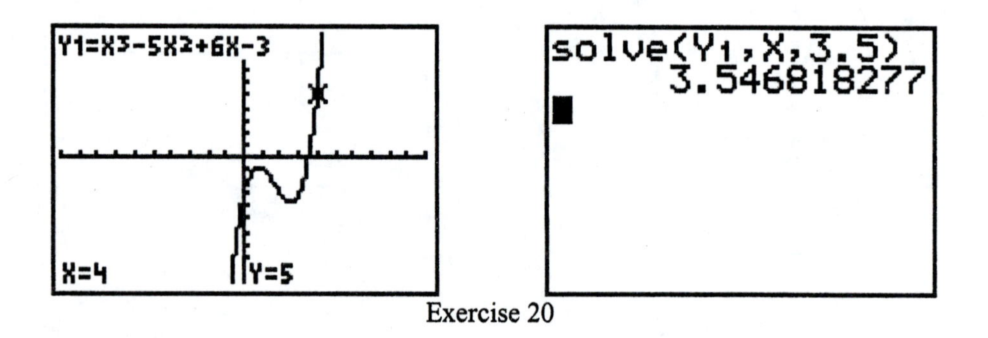

Exercise 20

21. $2x^3 - 8x^2 + 3x + 5 = 0$

Estimates are: $-0.5, 1.2, 3.3$

Improved solutions:
−0.592366, 1.27047,
3.32190

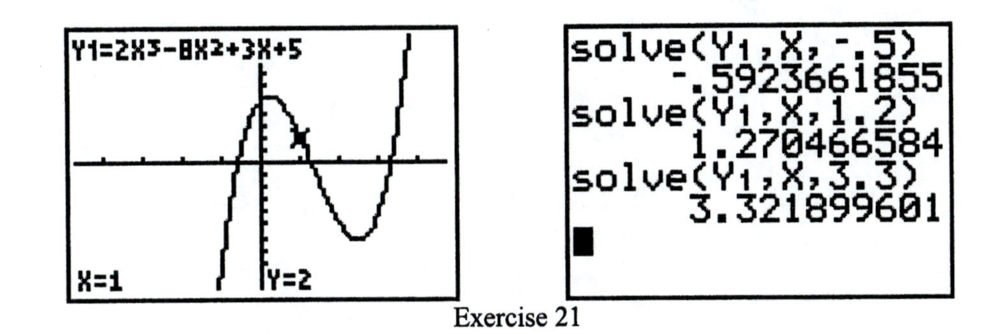

Exercise 21

22. $4x^3 - x^2 + 2x - 8 = 0$

Estimate is 1.2

Improved solution:
1.20754

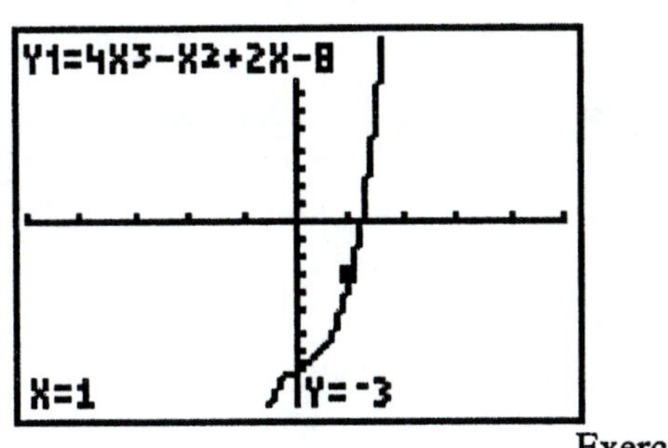

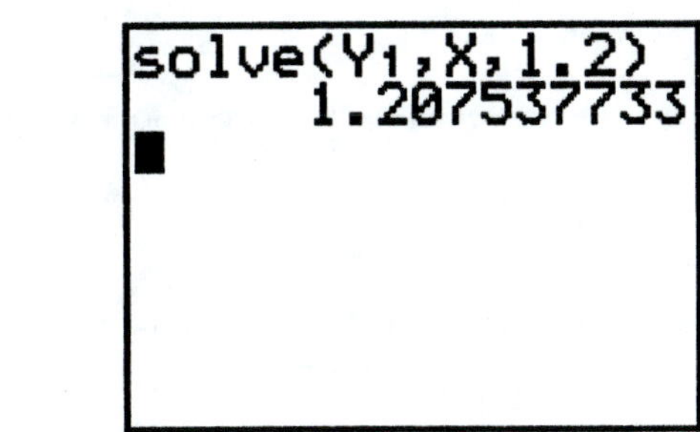

Exercise 22

23. $x^3 + 2x^2 - 5x + 1 = 0$

Estimates are: $-3.5, 0, 1.2$

Improved solutions:
−3.50702, 0.221876,
1.28514

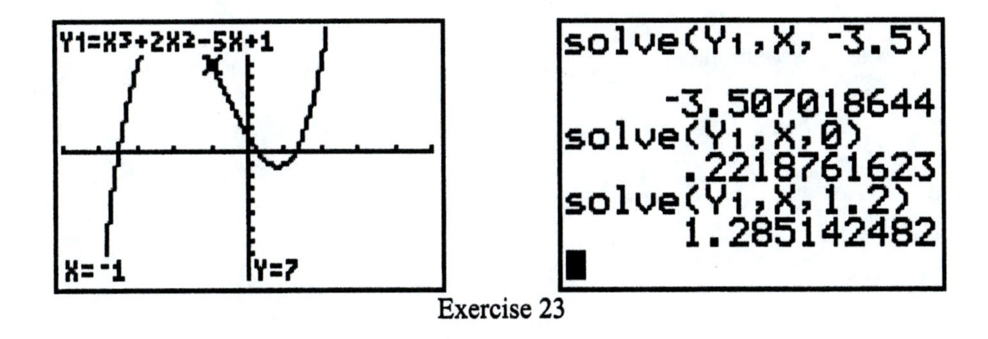

Exercise 23

24. $5x^3 - x^2 - 8x + 3 = 0$

Estimates are: $-1.3, 0.3, 1.1$

Improved solutions:

-1.33507, 0.393780,

1.14129

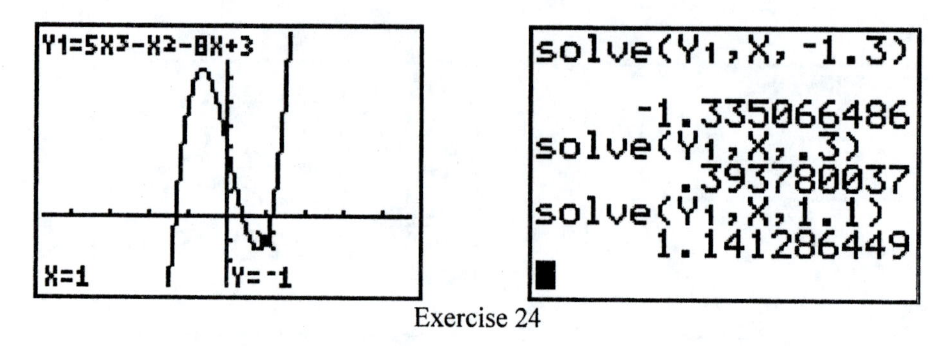

Exercise 24

25. eight

26. five

27. $x^3 - 11x^2 + 40x - 50 = 0$

Since $(3 + j)$ is a solution, $(3 - j)$ is also a solution.

$\left(x - (3 + j)\right)\left(x - (3 - j)\right) =$

$x^2 - (3 + j)x - (3 - j)x + 9 - j^2 =$

$x^2 - 3x - jx - 3x + jx + 9 + 1 = x^2 - 6x + 10.$

Since $x^2 - 6x + 10$ is a factor,

we divide to find the other factor.

$$\begin{array}{r} x - 5 \\ x^2 - 6x + 10 \overline{\smash{\big)} x^3 - 11x^2 + 40x - 50} \\ \underline{x^3 - 6x^2 + 10x} \\ -5x^2 + 30x - 50 \\ \underline{-5x^2 + 30x - 50} \\ 0 \end{array}$$

Thus 5 is the third solution.

28. $x^3 - 7x^2 + 25x - 39 = 0$

Since $(2-3j)$ is a solution, $(2+3j)$ is also a solution.

$\left(x-(2-3j)\right)\left(x-(2+3j)\right) =$

$x^2 - (2-3j)x - (2+3j)x + 4 - 9j^2 =$

$x^2 - 2x + 3jx - 2x - 3jx + 4 + 9 = x^2 - 4x + 13.$

Since $x^2 - 4x + 13$ is a factor,

we divide to find the other factor.

$$
\begin{array}{r}
x-3 \\
x^2-4x+13\overline{)\,x^3-7x^2+25x-39} \\
\underline{x^3-4x^2+13x} \\
-3x^2+12x-39 \\
\underline{-3x^2+12x-39} \\
0
\end{array}
$$

Thus 3 is the third solution.

29. $x^4 + 8x^3 + 25x^2 + 72x + 144 = 0.$ Solutions are: $3j - 3j, -4$ (multiplicity of 2).

30. $x^4 - 2x^3 - 38x^2 + 150x - 175 = 0.$ Since $(2+j)$ is a solution $(2-j)$ is also.

$\left(x-(2+j)\right)\left(x-(2-j)\right) = x^2 - (2+j)x - (2-j)x + 4 - j^2 =$

$x^2 - 2x - jx - 2x + jx + 4 + 1 = x^2 - 4x + 5.$

Divide to find the other factor.

$$
\begin{array}{r}
x^2+2x-35 \\
x^2-4x+5\overline{)\,x^4-2x^3+38x^2+150x-175} \\
\underline{x^4-4x^3+5x^2} \\
2x^3-43x^2+150x \\
\underline{2x^3-8x^2+10x} \\
-35x^2+140x-175 \\
\underline{-35x^2+140x-175} \\
0
\end{array}
$$

Now solve $x^2 + 2x - 35 = 0$

$$(x+7)(x-5) = 0$$

Solutions are: $-7, 5$

31. Degree 3, solutions include 8 and $7j$.

$$f(x)=(x-8)(x-7j)(x+7j)=0$$
$$f(x)=(x-8)(x^2-49j^2)=0$$
$$f(x)=(x-8)(x^2+49)=0$$
$$f(x)=x^3-8x^2+49x-392=0$$

32. Degree 3, solutions include -4 and $(3-2j)$.

$$f(x)=(x+4)(x-(3-2j))(x-(3+2j))=0$$
$$f(x)=(x+4)(x^2-(3-2j)x-(3+2j)x+9-4j^2)=0$$
$$f(x)=(x+4)(x^2-3x+2jx-3x-2jx+9+4)=0$$
$$f(x)=(x+4)(x^2-6x+13)=0$$
$$f(x)=x^3-2x^2-11x+52=0$$

33. Degree 4, solutions include $(2j)$ and $(1+j)$.

$$f(x)=(x-2j)(x+2j)(x-(1+j))(x-(1-j))=0$$
$$f(x)=(x^2-4j^2)(x^2-(1+j)x-(1-j)x+1-j^2)=0$$
$$f(x)=(x^2+4)(x^2-x-jx-x+jx+1+1)=0$$
$$f(x)=(x^2+4)(x^2-2x+2)=0$$
$$f(x)=x^4-2x^3+6x^2-8x+8=0$$

34. Degree 5, solutions include $\frac{3}{5}$ (multiplicity 2), $\frac{1}{2}$ and j. Note: if $x=\frac{3}{5}$, then $5x=3$ and $5x-3=0$, so $5x-3$ is a factor. For $x=\frac{1}{2}$, $2x=1$ and $2x-1=0$, so $2x-1$ is a factor.

$$f(x)=(5x-3)^2(2x-1)(x-j)(x+j)=0$$
$$f(x)=(5x-3)^2(2x-1)(x^2-j^2)=0$$
$$f(x)=(5x-3)^2(2x-1)(x^2+1)=0$$
$$f(x)=50x^5-85x^4+98x^3-94x^2+48x-9=0$$

Chapter 17, Exercises 17.1

The graphs appear in the text answer section.

11. $x \geq 3$

13. $x < -1$

15. $4 < x < 7$

17. $-8 \leq x < -5$

19. $x < -1$ or $x \geq 1$

21.
$$4x < 12$$
$$\frac{1}{4}(4x) < \frac{1}{4}(12)$$
$$x < 3$$

23.
$$-3x \geq 9$$
$$\frac{-1}{3}(-3x) \leq \frac{-1}{3}(9)$$
$$x \leq -3$$

25.
$$3x + 4 < 22$$
$$3x + 4 - 4 < 22 - 4$$
$$3x < 18$$
$$\frac{1}{3}(3x) < \frac{1}{3}(18)$$
$$x < 6$$

27.
$$5 - 2x \geq 17$$
$$5 - 2x - 5 \geq 17 - 5$$
$$-2x \geq 12$$
$$\frac{-1}{2}(-2x) \leq \frac{-1}{2}(12)$$
$$x \leq -6$$

29.
$$6x - 5 \leq 2x + 7$$
$$6x - 5 - 2x \leq 2x + 7 - 2x$$
$$4x - 5 \leq 7$$
$$4x - 5 + 5 \leq 7 + 5$$
$$4x \leq 12$$
$$\frac{1}{4}(4x) \leq \frac{1}{4}(12)$$
$$x \leq 3$$

31.
$$3(x - 2) > 5x - 12$$
$$3x - 6 > 5x - 12$$
$$3x - 6 - 5x > 5x - 12 - 5x$$
$$-2x - 6 > -12$$
$$-2x - 6 + 6 > -12 + 6$$
$$-2x > -6$$
$$\frac{-1}{2}(-2x) < \frac{-1}{2}(-6)$$
$$x < 3$$

33.
$$\frac{x}{4} \geq 2$$
$$4\left(\frac{x}{4}\right) \geq 4(2)$$
$$x \geq 8$$

35.
$$\frac{x}{6} + \frac{3}{8} > \frac{3x}{4} + \frac{7}{12}$$
$$24\left(\frac{x}{6} + \frac{3}{8}\right) > 24\left(\frac{3x}{4} + \frac{7}{12}\right)$$
$$4x + 9 > 18x + 14$$
$$4x + 9 - 18x > 18x + 14 - 18x$$
$$-14x + 9 > 14$$
$$-14x + 9 - 9 > 14 - 9$$
$$-14x > 5$$
$$\frac{-1}{14}(-14x) < \frac{-1}{14}(5)$$
$$x < \frac{-5}{14}$$

37.
$$\frac{2x + 3}{4} < 5x$$
$$4\left(\frac{2x + 3}{4}\right) < 4(5x)$$
$$2x + 3 < 20x$$
$$2x + 3 - 20x < 20x - 20x$$
$$-18x + 3 < 0$$
$$-18x + 3 - 3 < 0 - 3$$
$$-18x < -3$$
$$\frac{-1}{18}(-18x) > \frac{-1}{18}(-3)$$
$$x > \frac{1}{6}$$

39. $\frac{2}{3}(3x+4) \le \frac{4}{5}(x-2)$

$15\left[\frac{2}{3}(3x+4)\right] \le 15\left[\frac{4}{5}(x-2)\right]$

$10(3x+4) \le 12(x-2)$

$30x+40 \le 12x-24$

$30x+40-12x \le 12x-24-12x$

$18x+40 \le -24$

$18x+40-40 \le -24-40$

$18x \le -64$

$\frac{1}{18}(18x) \le \frac{1}{18}(-64)$

$x \le -\frac{32}{9}$

41. $x \ge 5$ and $x \ge 10$

And implies the intersection of the two graphs, which is: $x \ge 10$.

43. $x < 4$ or $x \le -3$

OR implies the union of the two graphs, which is: $x < 4$.

45. $x > 7$ and $x \le 10$

AND implies the intersection of the two graphs which is:

$7 < x \le 10$

47. $x < 4$ or $x > 1$

OR implies the union of the two graphs, which is the entire real number line.

49. $3x+4 < 16$ and $5x+1 \ge -14$

$3x < 12$ and $5x \ge -15$

$x < 4$ and $x \ge -3$

AND implies the intersection, which is:

$-3 \le x < 4$

51. $-9 \le 2x+3 \le 7$

$-9-3 \le 2x+3-3 \le 7-3$

$-12 \le 2x \le 4$

$\frac{1}{2}(-12) \le \frac{1}{2}(2x) \le \frac{1}{2}(4)$

$-6 \le x \le 2$

53. $-2 < 4x-2 < 18$

$-2+2 < 4x-2+2 < 18+2$

$0 < 4x < 20$

$\frac{1}{4}(0) < \frac{1}{4}(4x) < \frac{1}{4}(20)$

$0 < x < 5$

55.

$$-9 \le 1-2x < -1$$
$$-9-1 \le 1-2x-1 < -1-1$$
$$-10 \le -2x < -2$$
$$\frac{-1}{2}(-10) \ge \frac{-1}{2}(-2x) > \frac{-1}{2}(-2)$$
$$5 \ge x > 1$$
$$\text{or } 1 < x \le 5$$

57.

$$15 < 5-3x \le 26$$
$$15-5 < 5-3x-5 \le 26-5$$
$$10 < -3x \le 21$$
$$\frac{-1}{3}(10) > \frac{-1}{3}(-3x) \ge \frac{-1}{3}(21)$$
$$\frac{-10}{3} > x \ge -7$$
$$\text{or } -7 \le x < \frac{-10}{3}$$

Exercises 17.2

1.

$$|x+4| = 10$$
is equivalent to
$$x+4=10 \text{ or } x+4=-10$$
$$x=6 \text{ or } x=-14$$

3.

$$|6-5x| = 4$$
is equivalent to
$$6-5x=4 \text{ or } 6-5x=-4$$
$$-5x=-2 \text{ or } -5x=-10$$
$$x=\frac{2}{5} \text{ or } x=2$$

5.

$$|x-2| = |2x+4|$$
Of the four logical possibilities, we have 2 pairs of equivalent values. ie

A. $x-2=2x+4$

B. $-(x-2)=-(2x+4)$

Note *A* and *B* are equivalent.

C. $(x-2)=-2(2x+4)$

D. $-(x-2)=(2x+4)$

Note *C* and *D* are equivalent.

The *A*, *B* pair:
$$x-2=2x+4$$
$$-x=6$$
$$x=-6$$
OR

The *C*, *D* pair:
$$x-2=-(2x+4)$$
$$x-2=-2x-4$$
$$3x=-2$$
$$x=\frac{-2}{3}$$

7. $|3x-2|=|3x+4|$

There are four logical possibilities:

A. $3x-2=3x+4$

B. $-(3x-2)=-(3x+4)$

Note: A and B are equivalent.

C. $3x-2=-(3x+4)$

D. $-(3x-2)=3x+4$

Note: C and D are equivalent.

The A, B pair:

$3x-2=3x+4$

$-2=4$

No solution.

OR

The C, D pair:

$3x-2=-(3x+4)$

$3x-2=-3x-4$

$6x=-2$

$x=\frac{-1}{3}$

9. $\left|\frac{x+3}{x-4}\right|=8$

$\frac{x+3}{x-4}=8$ or $\left(\frac{x+3}{x-4}\right)=-8$

$x+3=8(x-4)$ or $x+3=-8(x-4)$

$x+3=8x-32$ or $x+3=-8x+32$

$-7x=-35$ or $9x=29$

$x=5$ or $x=\frac{29}{9}$

11. $|x|<3$ is equivalent to

$-3<x<3$

13. $|x-2|<5$ is equivalent to

$-5<x-2<5$

$-5+2<x<5+2$

$-3<x<7$

15. $|2x+3|>1$ is equivalent to

$2x+3>1$ or $2x+3<-1$

$2x>-2$ or $2x<-4$

$x>-1$ or $x<-2$

17. $|3-4x|\le 19$ is equivalent to

$-19\le 3-4x\le 19$

$-19-3\le -4x\le 19-3$

$-22\le -4x\le 16$

$\frac{-22}{-4}\ge\frac{-4x}{-4}\ge\frac{16}{-4}$

$\frac{11}{2}\ge x\ge -4$

or

$-4\le x\le\frac{11}{2}$

19. $|3x+1|\ge 16$ is equivalent to

$3x+1\ge 16$ or $3x+1\le -16$

$3x\ge 15$ or $3x\le -17$

$x\ge 5$ or $x\le\frac{-17}{3}$

21. $|4x-5|<31$ is
equivalent to
$-31<4x-5<31$
$-31+5<4x<31+5$
$-26<4x<36$
$\frac{-26}{4}<x<\frac{36}{4}$
$\frac{-13}{2}<x<9$

23. $|1-6x|\geq 7$ is
equivalent to
$1-6x\geq 7$ or $1-6x\leq -7$
$-6x\geq 6$ or $-6x\leq -8$
$x\leq -1$ or $x\geq \frac{8}{6}$
$x\leq -1$ or $x\geq \frac{4}{3}$

25. $|2x+6|>18$ is
equivalent to
$2x+6>18$ or $2x+6<-18$
$2x>12$ or $2x<-24$
$x>6$ or $x<-12$

27. $|4x-2|<-3$
No solution.
On the left side of the inequality is a positive or zero value. On the right side is a negative value.

Exercises 17.3

1. $(x+3)(x-5)>0$
Points of division are -3 and 5.
For $x<-3$ the signs of the factors are: $(-)(-)$; products is $+$.
For $-3<x<5$ the signs of the factors are $(+)(-)$; product is $-$. For $x>5$ the signs of the factors are: $(+)(+)$; product is $+$.
Because the inequality is >0, we want the intervals where the product is positive, ie $x<-3$ or $x>5$.

3. $(3x-7)(4x+1)\leq 0$
Points of division are $\frac{-1}{4}$ and $\frac{7}{3}$. For $x\leq\frac{-1}{4}$ the signs of the factors are: $(-)(-)$; product is $+$.
For $\frac{-1}{4}\leq x\leq\frac{7}{3}$ the signs of the factors are: $(-)(+)$; product is $-$.
For $x\geq\frac{7}{3}$, the signs of the factors are $(+)(+)$; product is $+$. Because the inequality is ≤ 0, we want the intervals where the product is negative or zero, ie
$\frac{-1}{4}\leq x\leq\frac{7}{3}$.

		Interval	Factor Signs	Product/ Quotient	Satisfies Inequality
5.	$x^2+9x+14>0$	a) $x<-7$	$(-)(-)$	+	Yes
	$(x+7)(x+2)$	b) $-7<x<-2$	$(+)(-)$	−	No
		c) $x>-2$	$(+)(+)$	+	Yes

Solution is $x<-7$ or $x>-2$.

		Interval	Factor Signs	Product/ Quotient	Satisfies Inequality
7.	$2x^2+4<9x$	a) $x<\frac{1}{2}$	$(-)(-)$	+	No
	$2x^2-9x+4<0$	b) $\frac{1}{2}<x<4$	$(-)(+)$	−	Yes
	$(x-4)(2x-1)<0$	c) $x>4$	$(+)(+)$	+	No

Solution is $\frac{1}{2}<x<4$.

		Interval	Factor Signs	Product/ Quotient	Satisfies Inequality
9.	$15x^2\ge 2-x$	a) $x\le\frac{-2}{5}$	$(-)(-)$	+	Yes
	$15x^2+x-2\ge 0$	b) $\frac{-2}{5}\le x\le\frac{1}{3}$	$(-)(+)$	−	No
	$(3x-1)(5x+2)\ge 0$	c) $x\ge\frac{1}{3}$	$(+)(+)$	+	Yes

Solution is $x\le\frac{-2}{5}$ or $x\ge\frac{1}{3}$.

		Interval	Factor Signs	Product/ Quotient	Satisfies Inequality
11.	$4x^2\le 11x+3$	a) $x\le\frac{-1}{4}$	$(-)(-)$	+	No
	$4x^2-11x-3\le 0$	b) $\frac{-1}{4}\le x\le 3$	$(-)(+)$	−	Yes
	$(x-3)(4x+1)\le 0$	c) $x\ge 3$	$(+)(+)$	+	No

Solution is $\frac{-1}{4}\le x\le 3$.

		Interval	Factor Signs	Product/ Quotient	Satisfies Inequality
13.	$4x^2+12x+9>0$	a) $x<\frac{-3}{2}$	$(-)^2$	+	Yes
	$(2x+3)^2>0$	b) $x>\frac{-3}{2}$	$(+)^2$	+	Yes

Solution is all real numbers except $\frac{-3}{2}$.

		Interval	Factor Signs	Product/ Quotient	Satisfies Inequality
15.	$x^2 > 4$	a) $x < -2$	$(-)(-)$	$+$	Yes
	$x^2 - 4 > 0$	b) $-2 < x < 2$	$(-)(+)$	$-$	No
	$(x-2)(x+2)$	c) $x > 2$	$(+)(+)$	$+$	Yes

Solution is $x < -2$ or $x > 2$.

		Interval	Factor Signs	Product/ Quotient	Satisfies Inequality
17.	$x^2 \le 5$	a) $x \le -\sqrt{5}$	$(-)(-)$	$+$	No
	$x^2 - 5 \le 0$	b) $-\sqrt{5} \le x \le \sqrt{5}$	$(-)(+)$	$-$	Yes
	$\left(x-\sqrt{5}\right)\left(x+\sqrt{5}\right) \le 0$	c) $x \ge \sqrt{5}$	$(+)(+)$	$+$	No

Solution is $-\sqrt{5} \le x \le \sqrt{5}$.

		Interval	Factor Signs	Product/ Quotient	Satisfies Inequality
19.	$x^2 > 3x$	a) $x < 0$	$(-)(-)$	$+$	Yes
	$x^2 - 3x > 0$	b) $0 < x < 3$	$(+)(-)$	$-$	No
	$x(x-3) > 0$	c) $x > 3$	$(+)(+)$	$+$	Yes

Solution is $x < 0$ or $x > 3$.

		Interval	Factor Signs	Product/ Quotient	Satisfies Inequality
21.	$x^3 + x^2 \ge 12x$	a) $x \le -4$	$(-)(-)(-)$	$-$	No
	$x^3 + x^2 - 12x \ge 0$	b) $-4 \le x \le 0$	$(-)(-)(+)$	$+$	Yes
	$x(x-3)(x+4) \ge 0$	c) $0 \le x \le 3$	$(+)(-)(+)$	$-$	No
		d) $x \ge 3$	$(+)(+)(+)$	$+$	Yes

Solution is $-4 \le x \le 0$ or $x \ge 3$.

		Interval	Factor Signs	Product/ Quotient	Satisfies Inequality
23.	$x^3 > 6x^2$	a) $x < 0$	$(-)^2(-)$	$-$	No
	$x^3 - 6x^2 > 0$	b) $0 < x < 6$	$(+)^2(-)$	$-$	No
	$x^2(x-6) > 0$	c) $x > 6$	$(+)^2(+)$	$+$	Yes

Solution is $x > 6$.

		Interval	Factor Signs	Product/ Quotient	Satisfies Inequality
25.	$x^4-5x^2+4<0$	a) $x<-2$	$(-)(-)(-)(-)$	+	No
	$(x^2-1)(x^2-4)<0$	b) $-2<x<-1$	$(-)(-)(-)(+)$	−	Yes
	$(x-1)(x+1)(x-2)$	c) $-1<x<1$	$(-)(+)(-)(+)$	+	No
	$(x+2)<0$	d) $1<x<2$	$(+)(+)(-)(+)$	−	Yes
		e) $x>2$	$(+)(+)(+)(+)$	+	No

Solution is $-2<x<-1$ or $1<x<2$.

		Interval	Factor Signs	Product/ Quotient	Satisfies Inequality
27.	$\frac{x+4}{x-2}<0$	a) $x<-4$	$\frac{(-)}{(-)}$	+	No
		b) $-4<x<2$	$\frac{(+)}{(-)}$	−	Yes
		c) $x>2$	$\frac{(+)}{(+)}$	+	No

Solution is $-4<x<2$.

		Interval	Factor Signs	Product/ Quotient	Satisfies Inequality
29.	$\frac{x^2+3x+2}{x+3}\geq 0$	a) $x<-3$	$\frac{(-)(-)}{(-)}$	−	No
	$\frac{(x+1)(x+2)}{x+3}\geq 0$	b) $-3<x\leq -2$	$\frac{(-)(-)}{(+)}$	+	Yes
		c) $-2\leq x\leq -1$	$\frac{(-)(+)}{(+)}$	−	No
		d) $x\geq -1$	$\frac{(+)(+)}{(+)}$	+	Yes

Solution is $-3<x\leq -2$ or $x\geq -1$.

		Interval	Factor Signs	Product/ Quotient	Satisfies Inequality
31.	$\frac{9-x^2}{16-x^2} \le 0$	a) $x < -4$	$\frac{(-)(+)}{(-)(+)}$	+	No
	$\frac{(3+x)(3-x)}{(4+x)(4-x)} \le 0$	b) $-4 < x \le -3$	$\frac{(-)(+)}{(+)(+)}$	−	Yes
		c) $-3 \le x \le 3$	$\frac{(+)(+)}{(+)(+)}$	+	No
		d) $3 \le x < 4$	$\frac{(+)(-)}{(+)(+)}$	−	Yes
		e) $x > 4$	$\frac{(+)(-)}{(+)(-)}$	+	No

Solution is $-4 < x \le -3$ or $3 \le x < 4$.

		Interval	Factor Signs	Product/ Quotient	Satisfies Inequality
33.	$\frac{9-12x+4x^2}{x^2-3x-10} > 0$	a) $x < -2$	$\frac{(-)^2}{(-)(-)}$	+	Yes
	$\frac{(2x-3)^2}{(x-5)(x+2)} > 0$	b) $-2 < x < 1.5$	$\frac{(-)^2}{(-)(+)}$	−	No
		c) $1.5 < x < 5$	$\frac{(+)^2}{(-)(+)}$	−	No
		d) $x > 5$	$\frac{(+)^2}{(+)(+)}$	+	Yes

Solution is $x < -2$ or $x > 5$.

		Interval	Factor Signs	Product/ Quotient	Satisfies Inequality
35.	$\frac{x^3(x+1)(x-2)}{(3-x)^2(x-1)} \le 0$	a) $x \le -1$	$\frac{(-)(-)(-)}{(+)^2(-)}$	+	No
		b) $-1 \le x \le 0$	$\frac{(-)^3(+)(-)}{(+)^2(-)}$	−	Yes
		c) $0 \le x < 1$	$\frac{(+)^3(+)(-)}{(+)^2(-)}$	+	No
		d) $1 < x \le 2$	$\frac{(+)^3(+)(-)}{(+)^2(+)}$	−	Yes
		e) $2 \le x < 3$	$\frac{(+)^3(+)(+)}{(+)^2(+)}$	+	No
		f) $x > 3$	$\frac{(+)^3(+)(+)}{(-)^2(+)}$	+	No

Solution is $-1 \le x \le 0$ or $1 < x \le 2$.

		Interval	Factor Signs	Product/ Quotient	Satisfies Inequality
37.	$\frac{x+1}{x-2} \geq 4$	a) $x < 2$	$\frac{(+)}{(-)}$	$-$	No
	$\frac{x+1}{x-2} - \frac{4(x-2)}{x-2} \geq 0$	b) $2 < x \leq 3$	$\frac{(+)}{(+)}$	$+$	Yes
	$\frac{-3x+9}{x-2} \geq 0$	c) $x \geq 3$	$\frac{(-)}{(+)}$	$-$	No

Solution is $2 < x \leq 3$.

		Interval	Factor Signs	Product/ Quotient	Satisfies Inequality
39.	$\frac{1-x}{x} < 1$	a) $x < 0$	$\frac{(+)}{(-)}$	$-$	Yes
	$\frac{1-x}{x} - 1\frac{(x)}{x} < 0$	b) $0 < x < \frac{1}{2}$	$\frac{(+)}{(+)}$	$+$	No
	$\frac{1-2x}{x} < 0$	c) $x > \frac{1}{2}$	$\frac{(-)}{(+)}$	$-$	Yes

Solution is $x < 0$ or $x > \frac{1}{2}$.

Exercises 17.4

The graphs of the odd numbered problems appear in the text answer section. Below are shown ordered pairs of the equations associated with the inequalities.

1. $y < 2x - 5$

x	y
0	−5
2	−1
4	3

3. $y \geq 8 - 4x$

x	y
1	4
2	0
3	−4

5. $2x + 3y \leq 6$

x	y
0	2
3	0
6	−2

7. $4x - 2y > 8$

x	y
1	−2
3	2
4	4

9. $x \geq 2y$

x	y
0	0
2	1
4	2

11. $x \geq 3$

x	y
3	−2
3	0
3	2

13. $y \le x^2 + 1$

x	y
−2	5
−1	2
0	1
1	2
2	5

15. $y \ge x^3$

x	y
−2	−8
−1	−1
0	0
1	1
2	8

17. $y > 2x + 1$ and $y < x$

x	y
−1	−1
1	3
2	5

x	y
−2	−2
0	0
3	3

19. $x \le 2y$ and $y \le 3x$ and $2x + y \le 5$

x	y
−2	−1
0	0
2	1

x	y
1	3
2	6
3	9

x	y
1	3
2	1
3	−1

21. $y > x^2$ and $y < x + 6$ and $y < 5$

x	y
−2	4
0	0
2	4

x	y
−7	−1
−6	0
−5	1

x	y
2	5
4	5
6	5

Chapter 17 Review

Graphs for the problems in the review exercises are in the text answer section.

5. $x > 2$ **6.** $-1 < x \le 2$ **7.** $x \le 4$ or $x \ge 7$ **8.** $x \le -3$

9.
$$7x \le -28$$
$$\frac{1}{7}(7x) \le \frac{1}{7}(-28)$$
$$x \le -4$$

10.
$$\frac{x}{2} > 7$$
$$2\left(\frac{x}{2}\right) > 2(7)$$
$$x > 14$$

11.
$$-2x \ge 10$$
$$\frac{-1}{2}(-2x) \le \frac{-1}{2}(10)$$
$$x \le -5$$

12.
$$2x + 3 < 9$$
$$2x + 3 - 3 < 9 - 3$$
$$2x < 6$$
$$\frac{1}{2}(2x) < \frac{1}{2}(6)$$
$$x < 3$$

13.
$$4 - 3x < 34$$
$$4 - 3x - 4 < 34 - 4$$
$$-3x < 30$$
$$\frac{-1}{3}(-3x) > \frac{-1}{3}(30)$$
$$x > -10$$

14.
$$6x + 15 \ge 4x - 7$$
$$6x + 15 - 4x \ge 4x - 7 - 4x$$
$$2x + 15 \ge -7$$
$$2x + 15 - 15 \ge -7 - 15$$
$$2x \ge -22$$
$$\frac{1}{2}(2x) \ge \frac{1}{2}(-22)$$
$$x \ge -11$$

15.
$$4(x-3) \ge 3(x+4)$$
$$4x - 12 \ge 3x + 12$$
$$4x - 12 - 3x \ge 3x + 12 - 3x$$
$$x - 12 \ge 12$$
$$x - 12 + 12 \ge 12 + 12$$
$$x \ge 24$$

16.
$$\frac{x}{2} + \frac{3}{4} \le 2x + \frac{1}{6}$$
$$12\left(\frac{x}{2} + \frac{3}{4}\right) \le 12\left(2x + \frac{1}{6}\right)$$
$$6x + 9 \le 24x + 2$$
$$6x + 9 - 24x \le 24x + 2 - 24x$$
$$-18x + 9 \le 2$$
$$-18x + 9 - 9 \le 2 - 9$$
$$-18x \le -7$$
$$\frac{-1}{18}(-18x) \ge \frac{-1}{18}(-7)$$
$$x \ge \frac{7}{18}$$

17. OR means unions of sets. $x \le 6$

18. AND means intersection of sets. No solution

19. OR means unions of sets. $x \le 0$ or $x > 3$

20. AND means intersection of sets. $x \le -1$

21.

$$-7 \le 2x+5 \le 9$$
$$-7-5 \le 2x+5-5 \le 9-5$$
$$-12 \le 2x \le 4$$
$$\frac{1}{2}(-12) \le \frac{1}{2}(2x) \le \frac{1}{2}(4)$$
$$-6 \le x \le 2$$

22.

$$-2 < 1-3x \le 8$$
$$-2-1 < 1-3x-1 \le 8-1$$
$$-3 < -3x \le 7$$
$$\frac{-1}{3}(-3) > \frac{-1}{3}(-3x) \ge \frac{-1}{3}(7)$$
$$1 > x \ge \frac{-7}{3} \text{ or}$$
$$\frac{-7}{3} \le x < 1$$

23.

$$|2x+1| = 9$$
is equivalent to
$$2x+1=9 \text{ or } 2x+1=-9$$
$$2x=8 \text{ or } 2x=-10$$
$$x=4 \text{ or } x=-5$$

24.

$$|2-3x| = 11$$
is equivalent to
$$2-3x=11 \text{ or } 2-3x=-11$$
$$-3x=9 \text{ or } -3x=-13$$
$$x=-3 \text{ or } x=\frac{13}{3}$$

25.

$$|3x+2| = |4x-7|$$
$$\frac{|3x+2|}{|4x-7|}=1$$
$$\left|\frac{3x+2}{4x-7}\right|=1$$
is equivalent to
$$\frac{3x+2}{4x-7}=1 \quad \text{or} \quad \frac{3x+2}{4x-7}=-1$$
$$3x+2=4x-7 \text{ or } 3x+2=-1(4x-7)$$
$$-x=-9 \quad \text{or } 3x+2=-4x+7$$
$$x=9 \quad \text{or } 7x=5$$
$$x=9 \quad \text{or } x=\frac{5}{7}$$

26.

$$\left|\frac{x-2}{x+3}\right|=6$$
is equivalent to
$$\frac{x-2}{x+3}=6 \quad \text{or} \quad \frac{x-2}{x+3}=-6$$
$$x-2=6(x+3) \text{ or } x-2=-6(x+3)$$
$$x-2=6x+18 \text{ or } x-2=-6x-18$$
$$-5x=20 \quad \text{or } 7x=-16$$
$$x=-4 \quad \text{or } x=\frac{-16}{7}$$

27.

$$|x+4| < 8$$
is equivalent to
$$-8 < x+4 < 8$$
$$-12 < x < 4$$

28.

$$|x-3| > 9$$
is equivalent to
$$x-3>9 \text{ or } x-3<-9$$
$$x>12 \quad \text{or } x<-6$$

29.

$$|2x+5| \ge 17$$
is equivalent to
$$2x+5 \ge 17 \text{ or } 2x+5 \le -17$$
$$2x \ge 12 \quad \text{or } 2x \le -22$$
$$x \ge 6 \quad \text{or } x \le -11$$

30. $|3-2x| \le 15$

is equivalent to

$-15 \le 3-2x \le 15$

$-18 \le -2x \le 12$

$9 \ge x \ge -6$ or

$-6 \le x \le 9$

31. $|1-3x| > 16$

is equivalent to

$1-3x > 16$ or $1-3x < -16$

$-3x > 15$ or $-3x < -17$

$x < -5$ or $x > \frac{17}{3}$

32. $|4x-7| \le 17$

$-17 \le 4x-7 \le 17$

$-10 \le 4x \le 24$

$\frac{-10}{4} \le x \le 6$

$\frac{-5}{2} \le x \le 6$

		Interval	Factor Signs	Product/ Quotient	Satisfies Inequality
33.	$(2x+1)(x-4) \le 0$	a) $x \le \frac{-1}{2}$	$(-)(-)$	+	No
		b) $\frac{-1}{2} \le x \le 4$	$(+)(-)$	−	Yes
		c) $x \ge 4$	$(+)(+)$	+	No

Solution is $\frac{-1}{2} \le x \le 4$.

		Interval	Factor Signs	Product/ Quotient	Satisfies Inequality
34.	$3x^2+12 > 20x$	a) $x < \frac{2}{3}$	$(-)(-)$	+	Yes
	$3x^2-20x+12 > 0$	b) $\frac{2}{3} < x < 6$	$(-)(+)$	−	No
	$(x-6)(3x-2) > 0$	c) $x > 6$	$(+)(+)$	+	Yes

Solution is $x < \frac{2}{3}$ or $x > 6$.

		Interval	Factor Signs	Product/ Quotient	Satisfies Inequality
35.	$x^2 \le 49$	a) $x \le -7$	$(-)(-)$	+	No
	$x^2 - 49 \le 0$	b) $-7 \le x \le 7$	$(-)(+)$	−	Yes
	$(x-7)(x+7) \le 0$	c) $x \ge 7$	$(+)(+)$	+	No

Solution is $-7 \le x \le 7$.

		Interval	Factor Signs	Product/ Quotient	Satisfies Inequality
36.	$x^3 \ge 36x$	a) $x \le -6$	$(-)(-)(-)$	−	No
	$x^3 - 36x \ge 0$	b) $-6 \le x \le 0$	$(-)(-)(+)$	+	Yes
	$x(x-6)(x+6) \ge 0$	c) $0 \le x \le 6$	$(+)(-)(+)$	−	No
		d) $x \ge 6$	$(+)(+)(+)$	+	Yes

Solution is $-6 \le x \le 0$ or $x \ge 6$.

		Interval	Factor Signs	Product/ Quotient	Satisfies Inequality
37.	$x^3 + 8x^2 + 16x > 0$	a) $x < -4$	$(-)(-)^2$	−	No
	$x(x+4)^2 > 0$	b) $-4 < x < 0$	$(-)(+)^2$	−	No
		c) $x > 0$	$(+)(+)^2$	+	Yes

Solution is $x > 0$.

		Interval	Factor Signs	Product/ Quotient	Satisfies Inequality
38.	$\dfrac{x+4}{x-9} < 0$	a) $x < -4$	$\dfrac{(-)}{(-)}$	+	No
		b) $-4 < x < 9$	$\dfrac{(+)}{(-)}$	−	Yes
		c) $x > 9$	$\dfrac{(+)}{(+)}$	+	No

Solution is $-4 < x < 9$.

		Interval	Factor Signs	Product/ Quotient	Satisfies Inequality
39.	$\frac{3x^2+8x-3}{x-3}\le 0$	a) $x\le -3$	$\frac{(-)(-)}{(-)}$	$-$	Yes
	$\frac{(3x-1)(x+3)}{x-3}\le 0$	b) $-3\le x\le \frac{1}{3}$	$\frac{(-)(+)}{(-)}$	$+$	No
		c) $\frac{1}{3}\le x<3$	$\frac{(+)(+)}{(-)}$	$-$	Yes
		d) $x>3$	$\frac{(+)(+)}{(+)}$	$+$	No

Solution is $x\le -3$ or $\frac{1}{3}\le x<3$.

		Interval	Factor Signs	Product/ Quotient	Satisfies Inequality
40.	$\frac{(2x-3)^2}{16-8x-3x^2}\ge 0$	a) $x<-4$	$\frac{(-)^2}{(-)(+)}$	$-$	No
	$\frac{(2x-3)^2}{(4+x)(4-3x)}\ge 0$	b) $-4<x<\frac{4}{3}$	$\frac{(-)^2}{(+)(+)}$	$+$	Yes
		c) $\frac{4}{3}<x\le \frac{3}{2}$	$\frac{(-)^2}{(+)(-)}$	$-$	No
		d) $x\ge \frac{3}{2}$	$\frac{(+)^2}{(+)(-)}$	$-$	No

Solution is $-4<x<\frac{4}{3}$.

		Interval	Factor Signs	Product/ Quotient	Satisfies Inequality
41.	$\frac{1}{x+5}\le 10$	a) $x<-5$	$\frac{(+)}{(-)}$	$-$	Yes
	$\frac{1}{x+5}-\frac{10(x+5)}{x+5}\le 0$	b) $-5<x\le -4.9$	$\frac{(+)}{(+)}$	$+$	No
	$\frac{-10x-49}{x+5}\le 0$	c) $x\ge -4.9$	$\frac{(-)}{(+)}$	$-$	Yes

Solution is $x<-5$ or $x\ge -4.9$.

		Interval	Factor Signs	Product/ Quotient	Satisfies Inequality
43.	$\frac{2x+1}{x-4}\le 3$	a) $x<4$	$\frac{(+)}{(-)}$	$-$	Yes
	$\frac{2x+1}{x-4}-\frac{3(x-4)}{x-4}\le 0$	b) $4<x\le 13$	$\frac{(+)}{(+)}$	$+$	No
	$\frac{-x+13}{x-4}\le 0$	c) $x\ge 13$	$\frac{(-)}{(+)}$	$-$	Yes

Solution is $x<4$ or $x\ge 13$.

Chapter 18, Exercises 18.1

1.
$2, 5, 8, ..., n = 6$
$d = 5 - 2 = 3$
$\ell = a + (n-1)d$
$\ell = 2 + (6-1)(3)$
$\ell = 2 + 5(3) = 17$

3.
$3, 4.5, 6, ..., n = 15$
$d = 4.5 - 3 = 1.5$
$\ell = a + (n-1)d$
$\ell = 3 + (15-1)(1.5)$
$\ell = 3 + (14)(1.5) = 24$

5.
$4, -5, -14, ..., n = 12$
$d = -5 - 4 = -9$
$\ell = a + (n-1)d$
$\ell = 4 + (12-1)(-9)$
$\ell = 4 + (11)(-9) = -95$

7.
Corresponds to problem 1.
$n = 6,\ a = 2,\ \ell = 17$
$S_n = \frac{n}{2}(a + \ell)$
$S_6 = \frac{6}{2}(2 + 17)$
$S_6 = 3(19) = 57$

9.
Corresponds to problem 3.
$n = 15,\ a = 3,\ \ell = 24$
$S_n = \frac{n}{2}(a + \ell)$
$S_{15} = \frac{15}{2}(3 + 24)$
$S_{15} = 7.5(27) = 202.5$

11.
Corresponds to problem 5.
$n = 12,\ a = 4,\ \ell = -95$
$S_n = \frac{n}{2}(a + \ell)$
$S_{12} = \frac{12}{2}(4 + (-95))$
$S_{12} = 6(-91) = -546$

13.
$a = 2,\ d = -3$
$2,\ 2 + (-3),\ 2 + 2(-3),$
$2 + 3(-3),\ 2 + 4(-3)$
ie $2, -1, -4, -7, -10$

15.
$a = 5,\ d = \frac{2}{3}$
$5,\ 5 + \frac{2}{3},\ 5 + 2\left(\frac{2}{3}\right),$
$5 + 3\left(\frac{2}{3}\right),\ 5 + 4\left(\frac{2}{3}\right)$
ie $5,\ 5\frac{2}{3},\ 6\frac{1}{3},\ 7,\ 7\frac{2}{3}$

17.
$n = 10,\ \ell = 12,\ S_{10} = 80$
$S_n = \frac{n}{2}(a + \ell)$
$S_{10} = \frac{10}{2}(a + 12)$
$80 = 5a + 60$
$20 = 5a$
$4 = a$

19.
$a = 1,\ d = 2,\ n = 1000$
$\ell = a + (n-1)d$
$\ell = 1 + (1000-1)(2)$
$\ell = 1999$
$S_n = \frac{n}{2}(a + \ell)$
$S_{1000} = \frac{1000}{2}(1 + 1999)$
$S_{1000} = 1{,}000{,}000$

21.
$a = \$24{,}000,\ d = \$800,\ n = 10$
$\ell = a + (n-1)d$
$\ell = 24{,}000 + (10-1)(800)$
$\ell = 24{,}000 + 9(800)$
$\ell = \$31{,}200$

Exercises 18.2

1. $20, \frac{20}{3}, \frac{20}{9}, \ldots, n = 8$

$\ell = ar^{n-1},\ a = 20,\ r = \frac{1}{3}$

$\ell = 20\left(\frac{1}{3}\right)^{8-1} = 20\left(\frac{1}{3}\right)^{7}$

$\ell = \frac{20}{2187}$

3. $\sqrt{2}, 2, 2\sqrt{2}, \ldots, n = 6$

$a = \sqrt{2},\ r = \sqrt{2}$

$\ell = ar^{n-1}$

$\ell = \sqrt{2}\left(\sqrt{2}\right)^{6-1}$

$\ell = \sqrt{2}\left(\sqrt{2}\right)^{5} = 8$

5. $8, -4, 2, \ldots, n = 10$

$a = 8,\ r = \frac{-1}{2}$

$\ell = ar^{n-1}$

$\ell = 8\left(\frac{-1}{2}\right)^{10-1} = 8\left(\frac{-1}{2}\right)^{9}$

$\ell = \frac{-1}{64}$

7. Corresponds to problem 1.

$a = 20,\ r = \frac{1}{3},\ n = 8$

$S_n = \frac{a\left(1-r^n\right)}{1-r}$

$S_8 = \frac{20\left(1-\left(\frac{1}{3}\right)^8\right)}{1-\frac{1}{3}}$

$S_8 = \frac{20\left(1-\frac{1}{6561}\right)}{\frac{2}{3}}$

$S_8 = 20\left(\frac{3}{2}\right)\left(\frac{6560}{6561}\right)$

$S_8 = 30\left(\frac{6560}{6561}\right)$

$S_8 = \frac{65,600}{2187}$

9. Corresponds to problem 3.

$a = \sqrt{2},\ r = \sqrt{2},\ n = 6$

$S_n = \frac{a\left(1-r^n\right)}{1-r}$

$S_6 = \frac{\sqrt{2}\left(1-\left(\sqrt{2}\right)^6\right)}{1-\sqrt{2}}$

$S_6 = \frac{\sqrt{2}(1-8)}{1-\sqrt{2}}$

$S_6 = \frac{-7\sqrt{2}}{\left(1-\sqrt{2}\right)} \cdot \frac{\left(1+\sqrt{2}\right)}{\left(1+\sqrt{2}\right)}$

$S_6 = \frac{-7\sqrt{2}\left(1+\sqrt{2}\right)}{1-2}$

$S_6 = 7\sqrt{2} + 7(2)$

$S_6 = 7\sqrt{2} + 14$

11. Corresponds to problem 5.

$a = 8,\ r = \frac{-1}{2},\ n = 10$

$S_n = \frac{a\left(1-r^n\right)}{1-r}$

$S_{10} = \frac{8\left(1-\left(\frac{-1}{2}\right)^{10}\right)}{1-\left(\frac{-1}{2}\right)}$

$S_{10} = \frac{8\left(1-\frac{1}{1024}\right)}{\frac{3}{2}}$

$S_{10} = 8\left(\frac{2}{3}\right)\left(\frac{1023}{1024}\right)$

$S_{10} = \frac{341}{64}$

13. $a = 3,\ r = \frac{1}{2}$

$3, 3\left(\frac{1}{2}\right), 3\left(\frac{1}{2}\right)^2,$

$3\left(\frac{1}{2}\right)^3, 3\left(\frac{1}{2}\right)^4$

ie: $3, \frac{3}{2}, \frac{3}{4}, \frac{3}{8}, \frac{3}{16}$

15. $a = 5,\ r = \frac{-1}{4}$

$5, 5\left(\frac{-1}{4}\right), 5\left(\frac{-1}{4}\right)^2,$

$5\left(\frac{-1}{4}\right)^3, 5\left(\frac{-1}{4}\right)^4$

ie: $5, \frac{-5}{4}, \frac{5}{16}, \frac{-5}{64}, \frac{5}{256}$

17. $a = -4,\ r = 3$

$-4, -4(3), -4(3)^2,$

$-4(3)^3, -4(3)^4$

ie: $-4, -12, -36, -108, -324$

19. $a = 6,\ \ell = \frac{3}{4},\ n = 4$

$\ell = ar^{n-1}$

$\frac{3}{4} = 6r^3$

$\frac{1}{8} = r^3$

$r = \frac{1}{2}$

21. The amount at the end of the first year is $\$1000(1.0575) = \$1057.50$, so $a = \$1057.50$ $r = 1.0575$ and $n = 10$.

$S_n = \frac{a(1-r^n)}{1-r}$

$S_{10} = \frac{1057.50(1-1.0575^{10})}{1-1.0575}$

$S_{10} = \$13{,}776$

23. Bounce one results in the second height, thus, bounce five results in the sixth height.

$r = \frac{1}{2},\ n = 6,\ a = 12\,\text{ft}$

$\ell = ar^{n-1} = 12\left(\frac{1}{2}\right)^{6-1} = 12\left(\frac{1}{2}\right)^5 = \frac{12}{32} = \frac{3}{8}\,\text{ft}$

25. After minute 8 is the beginning of minute 9.

$a = 90°C,\ r = \frac{4}{5},\ n = 9$

$\ell = ar^{n-1} = 90\left(\frac{4}{5}\right)^{9-1} = 90\left(\frac{4}{5}\right)^8 = 15.1°C$

27. $4 + \frac{4}{7} + \frac{4}{49} + \ldots + 4\left(\frac{1}{7}\right)^{n-1} + \ldots$

$a = 4,\ r = \frac{1}{7},\ S = \frac{a}{1-r}.$

Thus $S = \frac{4}{1-\frac{1}{7}} = \frac{4}{\frac{6}{7}} = \frac{14}{3}$

29. $3 - \frac{3}{8} + \frac{3}{64} - \ldots + 3\left(\frac{-1}{8}\right)^{n-1} + \ldots$

$a = 3,\ r = \frac{-1}{8},\ S = \frac{a}{1-r}.$

Thus $S = \frac{3}{1-\left(\frac{-1}{8}\right)} = \frac{3}{\frac{9}{8}} = \frac{8}{3}$

31. $4+12+36+\ldots+4(3)^{n-1}+\ldots$

Since $r=3$, there is no sum.

33. $5+1+0.2+0.04+\ldots+5(0.2)^{n-1}$

$a=5,\ r=0.2$

$$S=\frac{a}{1-r}=\frac{5}{1-0.2}=6.25$$

35. $0.333\ldots$

$0.33\overline{3}=0.3+0.03+0.003\ldots$

Then $a=0.3$ and $r=0.1$

$$S=\frac{a}{1-r}=\frac{0.3}{1-0.1}=\frac{0.3}{0.9}=\frac{1}{3}$$

37. $0.012\overline{12}$

$0.012\overline{12}=0.012+0.00012+0.0000012\ldots$

$a=0.012,\ r=0.01$

$$S=\frac{a}{1-r}=\frac{0.012}{(1-0.01)}=\frac{0.012}{0.99}=\frac{12}{990}=\frac{2}{165}$$

39. $0.866\overline{6}$

$0.866\overline{6}=0.8+0.06+0.006+\ldots$

For $0.06+0.006+\ldots,\ a=0.06,\ r=0.1,$

$$S=\frac{a}{1-r}=\frac{0.06}{1-0.1}=\frac{0.06}{0.9}=\frac{6}{90}=\frac{1}{15}$$

Thus $0.866\overline{6}=0.8+\frac{1}{15}=\frac{8}{10}+\frac{1}{15}=\frac{12}{15}+\frac{1}{15}=\frac{13}{15}$

Section 18.3

Each problem in this section will be done two ways – the first, using the formulas illustrated in the text; the second method will use the notation ${}_nC_r$, which is found on both the TI-83 and TI-89 \ calculators. For the TI-83, it is under MATH, PRB. Enter n before the symbol ${}_nC_r$. After ${}_nC_r$, enter r. For the TI-89,it is under CATALOG alpha n. After getting the symbol ${}_nC_r$, enter n,r). For $(a+b)^n$, the kth term is ${}_nC_{k-1}(a)^{n-(k-1)}(b)^{k-1}$.

The formula we are using is in section 18.3.

1. $(3x+y)^3=(3x)^3+3(3x)^2(y)^1+\frac{3(2)}{2!}(3x)^1y^2+\frac{3(2)(1)}{3!}(3x)^0y^3$

$(3x+y)^3=27x^3+27x^2y+9xy^2+y^3$

Method 2

$(3x+y)^3={}_3C_0(3x)^3y^0+{}_3C_1(3x)^2y^1+{}_3C_2(3x)^1y^2+{}_3C_3(3x)^0y^3$

$(3x+y)^3=1(27x^3)+3(9x^2)y+3(3x)y^2+1(1)y^3$

$(3x+y)^3=27x^3+27x^2y+9xy^2+y^3$

3. $$(a-2)^5 = a^5 + 5a^4(-2)^1 + \frac{5(4)}{2!}a^3(-2)^2 + \frac{5(4)(3)}{3!}a^2(-2)^3 + \frac{5(4)(3)(2)}{4!}a^1(-2)^4 + \frac{5(4)(3)(2)(1)}{5!}a^0(-2)^5$$

$$(a-2)^5 = a^5 - 10a^4 + 40a^3 - 80a^2 + 80a - 32$$

Method 2

$$(a-2)^5 = {}_5C_0(a)^5(-2)^0 + {}_5C_1(a^4)(-2)^1 + {}_5C_2a^3(-2)^2 + {}_5C_3a^2(-2)^3 + {}_5C_4a^1(-2)^4 + {}_5C_5a^0(-2)^5$$

$$(a-2)^5 = a^5 + 5a^4(-2) + 10a^3(-2)^2 + 10a^2(-2)^3 + 5a(-2)^4 + 1(1)(-2)^5$$

$$(a-2)^5 = a^5 - 10a^4 + 40a^3 - 80a^2 + 80a - 32$$

5. $$(2x-1)^4 = (2x)^4 + 4(2x)^3(-1)^1 + \frac{4(3)}{2!}(2x)^2(-1)^2 + \frac{4(3)(2)}{3!}(2x)^1(-1)^3 + \frac{4(3)(2)(1)}{4!}(2x)^0(-1)^4$$

$$(2x-1)^4 = 16x^4 - 32x^3 + 24x^2 - 8x + 1$$

Method 2

$$(2x-1)^4 = {}_4C_0(2x)^4 + {}_4C_1(2x)^3(-1)^1 + {}_4C_2(2x)^2 + (-1)^2 + {}_4C_3(2x)^1(-1)^3 + {}_4C_4(2x)^0(-1)^4$$

$$(2x-1)^4 = 1(2x)^4 + 4(2x)^3(-1)^1 + 6(2x)^2(-1)^2 + 4(2x)(-1)^3 + 1(2x)^0(-1)^4$$

$$(2x-1)^4 = 16x^4 - 32x^3 + 24x^2 - 8x + 1$$

Section 18.3

7. $$(2a+3b)^6 = (2a)^6 + 6(2a)^5(3b)^1 + \frac{6\cdot 5}{2!}(2a)^4(3b)^2 + \frac{6\cdot 5\cdot 4}{3!}(2a)^3(3b)^3 + \frac{6\cdot 5\cdot 4\cdot 3}{4!}(2a)^2(3b)^4 + \frac{6\cdot 5\cdot 4\cdot 3\cdot 2}{5!}(2a)^1(3b)^5 + \frac{6\cdot 5\cdot 4\cdot 3\cdot 2\cdot 1}{6!}(2a)^0(3b)^6$$

$$(2a+3b)^6 = 64a^6 + 576a^5b + 2160a^4b^2 + 4320a^3b^3 + 4860a^2b^4 + 2916ab^5 + 729b^6$$

Method 2

$$(2a+3b)^6 = {}_6C_0(2a)^6(3b)^0 + {}_6C_1(2a)^5(3b)^1 + {}_6C_2(2a)^4(3b)^2 + {}_6C_3(2a)^3(3b)^3 + {}_6C_4(2a)^2(3b)^4 + {}_6C_5(2a)^1(3b)^5 + {}_6C_6(3b)^6$$

$$(2a+3b)^6 = 1(64a^6) + 6(32)(a^5)(3)b + 15(16a^4)(9b^2) + 20(2^3)a^3(3^3)b^3 + 15(2^2)a^2(3^4)b^4 + 6(2a)(3^5b^5) + 1(3^6)b^6$$

$$(2a+3b)^6 = 64a^6 + 576a^5b + 2160a^4b^2 + 4320a^3b^3 + 4860a^2b^4 + 2916ab^5 + 729b^6$$

9. $$\left(\frac{2}{3}x-2\right)^5 = \left(\frac{2}{3}x\right)^5 + 5\left(\frac{2}{3}x\right)^4(-2)^1 + \frac{5\cdot 4}{2!}\left(\frac{2}{3}x\right)^3(-2)^2 + \frac{5\cdot 4\cdot 3}{3!}\left(\frac{2}{3}x\right)^2(-2)^3 + \frac{5\cdot 4\cdot 3\cdot 2}{4!}\left(\frac{2}{3}x\right)(-2)^4 + \frac{5\cdot 4\cdot 3\cdot 2\cdot 1}{5!}\left(\frac{2}{3}x\right)^0(-2)^5$$

$$\left(\frac{2}{3}x-2\right)^5 = \frac{32}{243}x^5 - \frac{160}{81}x^4 + \frac{320}{27}x^3 - \frac{320}{9}x^2 + \frac{160}{3}x - 32$$

Method 2

$$\left(\frac{2}{3}x-2\right)^5 = {}_5C_0\left(\frac{2}{3}x\right)^5 + {}_5C_1\left(\frac{2}{3}x\right)^4(-2)^1 + {}_5C_2\left(\frac{2}{3}x\right)^3(-2)^2 + {}_5C_3\left(\frac{2}{3}x\right)^2(-2)^3 + {}_5C_4\left(\frac{2}{3}x\right)(-2)^4 + {}_5C_5\left(\frac{2}{3}x\right)^0(-2)^5$$

$$\left(\frac{2}{3}x-2\right)^5 = \left(\frac{2}{3}\right)^5x^5 + 5\left(\frac{2}{3}\right)^4x^4(-2) + 10\left(\frac{2}{3}\right)^3x^3(-2)^2 + 10\left(\frac{2}{3}\right)^2x^2(-2)^3 + 5\left(\frac{2}{3}\right)x(-2)^4 + 1(-2)^5$$

$$\left(\frac{2}{3}x-2\right)^5 = \frac{32}{243}x^5 - \frac{160}{81}x^4 + \frac{320}{27}x^3 - \frac{320}{9}x^2 + \frac{160}{3}x - 32$$

11. $$\left(a^{1/2}+3b^2\right)^4=\left(a^{1/2}\right)^4+4\left(a^{1/2}\right)^3\left(3b^2\right)^1+\frac{4\cdot3}{2!}\left(a^{1/2}\right)^2\left(3b^2\right)^2$$
$$+\frac{4\cdot3\cdot2}{3!}\left(a^{1/2}\right)\left(3b^2\right)^3+\frac{4\cdot3\cdot2\cdot1}{4!}\left(a^{1/2}\right)^0\left(3b^2\right)^4$$
$$\left(a^{1/2}+3b^2\right)^4=a^2+12a^{3/2}b^2+54ab^4+108a^{1/2}b^6+81b^8$$

Method 2

$$\left(a^{1/2}+3b^2\right)^4={}_4C_0\left(a^{1/2}\right)^4\left(3b^2\right)^0+{}_4C_1\left(a^{1/2}\right)^3\left(3b^2\right)^1$$
$$+{}_4C_2\left(a^{1/2}\right)^2\left(3b^2\right)^2+{}_4C_3\left(a^{1/2}\right)\left(3b^2\right)^3+{}_4C_4\left(a^{1/2}\right)^0\left(3b^2\right)^4$$
$$\left(a^{1/2}+3b^2\right)^4=\left(a^{1/2}\right)^4\left(3b^2\right)^0+4\left(a^{1/2}\right)^3\left(3b^2\right)+6\left(a^{1/2}\right)^2\left(3^2\right)\left(b^4\right)$$
$$+4\left(a^{1/2}\right)\left(3^3b^6\right)+1\left(3^4\right)b^8$$
$$\left(a^{1/2}+3b^2\right)^4=a^2+12a^{3/2}b^2+54ab^4+108a^{1/2}b^6+81b^8$$

13. $$\left(\frac{x}{y}-\frac{2}{z}\right)^4=\left(\frac{x}{y}\right)^4+4\left(\frac{x}{y}\right)^3\left(\frac{-2}{z}\right)^1+\frac{4\cdot3}{2!}\left(\frac{x}{y}\right)^2\left(\frac{-2}{z}\right)^2$$
$$+\frac{4\cdot3\cdot2}{3!}\left(\frac{x}{y}\right)\left(\frac{-2}{z}\right)^3+\frac{4\cdot3\cdot2\cdot1}{4!}\left(\frac{x}{y}\right)^0\left(\frac{-2}{z}\right)^4$$
$$\left(\frac{x}{y}-\frac{2}{z}\right)^4=\frac{x^4}{y^4}-\frac{8x^3}{y^3z}+\frac{24x^2}{y^2z^2}-\frac{32x}{yz^3}+\frac{16}{z^4}$$

Method 2

$$\left(\frac{x}{y}-\frac{2}{z}\right)^4={}_4C_0\left(\frac{x}{y}\right)^4\left(\frac{-2}{z}\right)^0+{}_4C_1\left(\frac{x}{y}\right)^3\left(\frac{-2}{z}\right)+{}_4C_2\left(\frac{x}{y}\right)^2\left(\frac{-2}{z}\right)^2$$
$$+{}_4C_3\left(\frac{x}{y}\right)\left(\frac{-2}{z}\right)^3+{}_4C_4\left(\frac{x}{y}\right)^0\left(\frac{-2}{z}\right)^4$$
$$\left(\frac{x}{y}-\frac{2}{z}\right)^4=\frac{x^4}{y^4}+4\left(\frac{-2x^3}{y^3z}\right)+6\frac{x^2}{y^2}\left(\frac{4}{z^2}\right)$$
$$+4\left(\frac{x}{y}\right)\left(\frac{-8}{z^3}\right)+1\left(\frac{16}{z^4}\right)$$
$$\left(\frac{x}{y}-\frac{2}{z}\right)^4=\frac{x^4}{y^4}-\frac{8x^3}{y^3z}+\frac{24x^2}{y^2z^2}-\frac{32x}{yz^3}+\frac{16}{z^4}$$

15. $(x-y)^9$; 6th term. $k=6, n=9, a=x, 6=-y$

$$k\text{th term} = \frac{n(n-1)(n-2)...(n-[k-2])}{(k-1)!}a^{n-(k-1)}b^{k-1}$$

$$6\text{th term} = \frac{9(8)(7)(6)(5)}{(6-1)!}x^{9-5}(-y)^5$$

$$6\text{th term} = -126x^4y^5$$

Method 2

$$k\text{th term} = {}_nC_{k-1}a^{n-(k-1)}b^{k-1}$$

$$6\text{th term} = {}_9C_5(x)^4(-y)^5$$

$$6\text{th term} = -126x^4y^5$$

17. $(2x-y)^{13}$; 9th term. $k=9, n=13, a=2x, b=-y$

$$k\text{th term} = \frac{n(n-1)(n-2)...(n-[k-2])}{(k-1)!}a^{n-(k-1)}b^{k-1}$$

$$9\text{th term} = \frac{13\cdot 12\cdot 11\cdot 10\cdot 9\cdot 8\cdot 7\cdot 6}{(9-1)!}(2x)^5(-y)^8$$

$$9\text{th term} = 1287(2^5)x^5(-y)^8 = 41{,}184x^5y^8$$

Method 2

$$k\text{th term} = {}_nC_{k-1}a^{n-(k-1)}b^{k-1}$$

$$9\text{th term} = {}_{13}C_8(2x)^5(-y)^8 = 1287(2^5)x^5(-y)^8 = 41{,}184x^5y^8$$

19. $(2x+y^2)^7$; 5th term. $n=7, k=5, a=2x, b=y^2$

$$k\text{th term} = \frac{n(n-1)(n-2)...(n-[k-2])}{(k-1)!}a^{n-(k-1)}b^{k-1}$$

$$5\text{th term} = \frac{7\cdot 6\cdot 5\cdot 4}{4!}(2x)^3(y^2)^4$$

$$5\text{th term} = 35(2^3)x^3(y^8) = 280x^3y^8$$

Method 2

$$k\text{th term} = {}_nC_{k-1}a^{n-(k-1)}b^{k-1}$$

$$5\text{th term} = {}_7C_4(2x)^3(y^2)^4 = 35(2^3)x^3y^8 = 280x^3y^8$$

21. $(3x+2y)^6$; middle term. There are 7 terms. The middle is the 4th. $n=6,\ k=4,\ a=3x,\ b=2y$

$$k\text{th term} = \frac{n(n-1)(n-2)...(n-[k-2])}{(k-1)!}a^{n-(k-1)}b^{k-1}$$

$$4\text{th term} = \frac{6\cdot5\cdot4}{(4-1)!}(3x)^3(2y)^3 = 20(3)^3x^3(2)^3y^3$$

$$4\text{th term} = 4320x^3y^3$$

Method 2

$$k\text{th term} = {}_nC_{k-1}a^{n-(k-1)}b^{k-1}$$

$$4\text{th term} = {}_6C_3(3x)^3(2y)^3 = 20(3)^3x^3(2)^3y^3$$

$$4\text{th term} = 4320x^3y^3$$

23. $(2x-1)^{10}$; term containing x^5. If x is to the 5th power, then (-1) is also, since the sum of the exponents is n, ie 10. In the kth term, b's exponent is $k-1$. So, k must be 6. We are seeking the 6th term.

$n=10,\ k=6,\ a=2x,\ b=-1$

$$k\text{th term} = \frac{n(n-1)(n-2)...(n-[k-2])}{(k-1)!}a^{n-(k-1)}b^{k-1}$$

$$6\text{th term} = \frac{10\cdot9\cdot8\cdot7\cdot6}{(6-1)!}(2x)^5(-1)^5 = 252(2^5)x^5(-1)$$

$$6\text{th term} = -8064x^5$$

Method 2

$$k\text{th term} = {}_nC_{k-1}a^{n-(k-1)}b^{k-1}$$

$$6\text{th term} = {}_{10}C_5(2x)^5(-1)^5 = 252(2^5)x^5(-1)$$

$$6\text{th term} = -8064x^5$$

Chapter 18 Review

1. $3, 7, 11, 15, \ldots, n = 12$. Arithmetic: $d = 7 - 3 = 4$.

$\ell = a + (n-1)d$; $\ell = 3 + (12-1)(4) = 47$

2. $4, 2, 1, \frac{1}{2}, \ldots, n = 7$. Geometric: $r = \frac{1}{2}$

$\ell = ar^{n-1}$, $\ell = 4\left(\frac{1}{2}\right)^6 = \frac{1}{16}$

3. $\sqrt{3}, -3, 3\sqrt{3}, -9, \ldots, n = 8$. Geometric: $r = -\sqrt{3}$

$l = ar^{n-1}$; $\ell = \sqrt{3}\left(-\sqrt{3}\right)^7 = -81$

4. $4, -2, -8, -14, \ldots, n = 12$. Arithmetic : $d = -6$

$\ell = a + (n-1)d$; $\ell = 4 + (12-1)(-6) = -62$

5. $6, 2, \frac{2}{3}, \frac{2}{9}, \ldots, n = 6$. Geometric: $r = \frac{1}{3}$

$\ell = ar^{n-1}$; $\ell = 6\left(\frac{1}{3}\right)^5 = \frac{2}{81}$

6. $5, 15, 25, 35, \ldots, n = 10$. Arithmetic: $d = 10$

$\ell = a + (n-1)d$; $\ell = 5 + (10-1)(10) = 95$

7. Corresponds to problem 1.

$$S_n = \frac{n}{2}(a + \ell)$$

$$S_{12} = \frac{12}{2}(3 + 47) = 300$$

8. Corresponds to problem 2.

$$S_n = \frac{a(1 - r^n)}{1 - r}$$

$$S_{12} = \frac{4\left(1 - \left(\frac{1}{2}\right)^7\right)}{1 - \frac{1}{2}} = \frac{127}{16}$$

9. Corresponds to problem 3.

$$S_n = \frac{a(1-r^n)}{1-r}$$

$$S_8 = \frac{\sqrt{3}\left(1-\left(-\sqrt{3}\right)^8\right)}{1-\left(-\sqrt{3}\right)}$$

$$S_8 = \frac{\sqrt{3}(1-81)}{1+\sqrt{3}} = \frac{-80\sqrt{3}}{1+\sqrt{3}}$$

10. Corresponds to problem 4.

$$S_n = \frac{n}{2}(a+\ell)$$

$$S_{12} = \frac{12}{2}(4+(-62)) = -348$$

11. Corresponds to problem 5.

$$S_n = \frac{a(1-r^n)}{1-r}$$

$$S_6 = \frac{6\left(1-\left(\frac{1}{3}\right)^6\right)}{1-\frac{1}{3}} = \frac{728}{81}$$

12. Corresponds to problem 6.

$$S_n = \frac{n}{2}(a+\ell)$$

$$S_{10} = \frac{10}{2}(5+95) = 500$$

13. $2, 4, 6, \ldots, n = 1000$. Arithmetic: $d = 2$.

$$\ell = a + (n-1)d$$

$$\ell = 2 + (1000-1)(2) = 2000$$

$$S_n = \frac{n}{2}(a+\ell)$$

$$S_{1000} = \frac{1000}{2}(2+2000) = 1{,}001{,}000$$

14. The amount at the end of the first year is $\$500(1.06) = \$530$, so $a = \$530$. $r = 1.06$, and $n = 5$.

$$S_n = \frac{a(1-r^n)}{1-r}$$

$$S_5 = \frac{\$530(1-1.06^5)}{1-1.06} = \$2988$$

15. $3 + 6 + 12 + \ldots; r = 2.$

Since $r \geq 1$, no sum.

16. $5 + \frac{5}{7} + \frac{5}{49} + \ldots;\ r = \frac{1}{7}.$

$$S = \frac{a}{1-r} = \frac{5}{1-\frac{1}{7}} = \frac{35}{6}$$

17. $2 - \frac{2}{3} + \frac{2}{9} - \ldots;\ r = \frac{-1}{3}.$

$$S = \frac{a}{1-r} = \frac{2}{1+\frac{1}{3}} = \frac{3}{2}$$

18. $3 + \frac{9}{2} + \frac{27}{4} + \ldots;\ r = \frac{3}{2}.$

Since $r \geq 1$, no sum.

19. $0.454545... = 0.45 + 0.0045 + 0.000045 + ...;$

$r = 0.01,\ S = \frac{a}{1-r} = \frac{0.45}{1-0.01} = \frac{5}{11}$

20. $0.9212121... = 0.9 + 0.021 + 0.00021 + ...$

For: $0.021 + 0.00021 + ...,\ r = 0.01;$

$S = \frac{a}{1-r} = \frac{0.021}{1-0.01} = \frac{7}{330}$

So the whole sum is $0.9 + \frac{7}{330} = \frac{152}{165}$

21.

$$(a-b)^6 = a^6 + 6(a)^5(-b)^1 + \frac{6.5}{2!}(a)^4(-b)^2$$
$$+\frac{6\cdot5\cdot4}{3!}(a)^3(-b)^3 + \frac{6\cdot5\cdot4\cdot3}{4!}(a)^2(-b)^4$$
$$+\frac{6\cdot5\cdot4\cdot3\cdot2}{5!}(a)^1(-b)^5 + \frac{6\cdot5\cdot4\cdot3\cdot2\cdot1}{6!}(-b)^6$$
$$(a-b)^6 = a^6 - 6a^5b + 15a^4b^2 - 20a^3b^3 + 15a^2b^4$$
$$-6ab^5 + b^6$$

Method 2

$$(a-b)^6 = {}_6C_0a^6(-b)^0 + {}_6C_1a^5(-b) + {}_6C_2a^4(-b)^2 + {}_6C_3a^3(-b)^3$$
$$+ {}_6C_4a^2(-b)^4 + {}_6C_5a(-b)^5 + {}_6C_6a^0(-b)^6$$
$$(a-b)^6 = a^6 - 6a^5b + 15a^4b^2 - 20a^3b^3 + 15a^2b^4 - 6ab^5 + b^6$$

22.

$$(2x^2-1)^5 = (2x^2)^5 + 5(2x^2)^4(-1)^1 + \frac{5\cdot4}{2!}(2x^2)^3(-1)^2$$
$$+\frac{5\cdot4\cdot3}{3!}(2x^2)^2(-1)^3 + \frac{5\cdot4\cdot3\cdot2}{4!}(2x^2)^1(-1)^4 + \frac{5\cdot4\cdot3\cdot2\cdot1}{5!}(-1)^5$$
$$(2x^2-1)^5 = 32x^{10} - 80x^8 + 80x^6 - 40x^4 + 10x^2 - 1$$

Method 2

$$(2x^2-1)^5 = {}_5C_0(2x^2)^5(-1)^0 + {}_5C_1(2x^2)^4(-1)^1 + {}_5C_2(2x^2)^3(-1)^2$$
$$+ {}_5C_3(2x^2)^2(-1)^3 + {}_5C_4(2x^2)^1(-1)^4 + {}_5C_5(2x^2)^0(-1)^5$$
$$(2x^2-1)^5 = 2^5x^{10} - 5(2^4)x^8 + 10(2^3)x^6 - 10(2^2)x^4 + 5(2x^2) - 1$$
$$(2x^2-1)^5 = 32x^{10} - 80x^8 + 80x^6 - 40x^4 + 10x^2 - 1$$

23. $(2x+3y)^4 = (2x)^4 + 4(2x)^3(3y) + \frac{4\cdot 3}{2!}(2x)^2(3y)^2$

$$+\frac{4\cdot 3\cdot 2}{3!}(2x)(3y)^3 + \frac{4\cdot 3\cdot 2\cdot 1}{4!}(2x)^0(3y)^4$$

$(2x+3y)^4 = 16x^4 + 96x^3y + 216x^2y^2 + 216xy^3 + 81y^4$

Method 2

$(2x+3y)^4 = {}_4C_0(2x)^4(3y)^0 + {}_4C_1(2x)^3(3y)^1 + {}_4C_2(2x)^2(3y)^2$

$$+{}_4C_3(2x)^1(3y)^3 + {}_4C_4(2x)^0(3y)^4$$

$(2x+3y)^4 = 2^4x^4 + 4(2^3)x^3(3y) + 6(2^2)x^2(3^2)y^2$

$$+4(2x)(3^3)y^3 + 1(3^4)y^4$$

$(2x+3y)^4 = 16x^4 + 96x^3y + 216x^2y^2 + 216xy^3 + 81y^4$

24. $(1+x)^8 = 1^8 + 8(1)^7(x) + \frac{8\cdot 7}{2!}(1)^6(x)^2 + \frac{8\cdot 7\cdot 6}{3!}(1)^5(x)^3$

$$+\frac{8\cdot 7\cdot 6\cdot 5}{4!}(1)^4x^4 + \frac{8\cdot 7\cdot 6\cdot 5\cdot 4}{5!}(1)^3(x)^5$$

$$+\frac{8\cdot 7\cdot 6\cdot 5\cdot 4\cdot 3}{6!}(1)^2(x^6) + \frac{8\cdot 7\cdot 6\cdot 5\cdot 4\cdot 3\cdot 2}{7!}(1)(x^7)$$

$$+\frac{8\cdot 7\cdot 6\cdot 5\cdot 4\cdot 3\cdot 2\cdot 1}{8!}(1)^0(x^8)$$

$(1+x)^8 = 1 + 8x + 28x^2 + 56x^3 + 70x^4 + 56x^5 + 28x^6 + 8x^7 + x^8$

Method 2

$(1+x)^8 = {}_8C_0(1)^8(x)^0 + {}_8C_1(1)^7(x)^1 + {}_8C_2(1)^6(x)^2$

$$+{}_8C_3(1)^5(x)^3 + {}_8C_4(1)^4(x)^4 + {}_8C_5(1)^3(x)^5$$

$$+{}_8C_6(1)^2(x)^6 + {}_8C_7(1)^2(x)^7 + {}_8C_8(1)^0(x)^8$$

$(1+x)^8 = 1 + 8x + 28x^2 + 56x^3 + 70x^4 + 56x^5 + 28x^6 + 8x^7 + x^8$

25. $(1-3x)^5$; 3rd term

kth term $= \frac{n(n-1)(n-2)\ldots(n-[k-2])}{(k-1)!}a^{n-(k-1)}b^{k-1}$

3rd term $= \frac{5\cdot 4}{(3-1)!}(1)^3(-3x)^2 = 90x^2$

Method 2

kth term $= {}_nC_{k-1}a^{n-(k-1)}b^{k-1}$

3rd term $= {}_5C_2(1)^3(-3x)^2 = 10(-3)^2x^2 = 90x^2$

26. $(a+4b)^6$; 4th term

$$k\text{th term} = \frac{n(n-1)(n-2)...(n-[k-2])}{(k-1)!}a^{n-(k-1)}b^{k-1}$$

$$4\text{th term} = \frac{6\cdot5\cdot4}{3!}a^3(4b)^3 = 1280a^3b^3$$

Method 2

$$k\text{th term} = {}_nC_{k-1}a^{n-(k-1)}b^{k-1}$$

$$4\text{th term} = {}_6C_3a^3(4b)^3 = 20(4^3)a^3b^3 = 1280a^3b^3$$

27. $(x+2b^2)^{10}$; middle term. There are 11 terms; the middle one is the 6th term.

$$k\text{th term} = \frac{n(n-1)(n-2)...(n-[k-2])}{(k-1)!}a^{n-(k-1)}b^{k-1}$$

$$6\text{th term} = \frac{10\cdot9\cdot8\cdot7\cdot6}{(6-1)!}(x)^5(2b^2)^5 = 8064x^5b^{10}$$

Method 2

$$k\text{th term} = {}_nC_{k-1}a^{n-(k-1)}b^{k-1}$$

$$6\text{th term} = {}_{10}C_5x^5(2b^2)^5 = 252(2^5)x^5b^{10} = 8064x^5b^{10}$$

28. $(3x^2-1)^{12}$; term containing x^{16}. This term will contain $(3x^2)^8$; ie $n-(k-1)=8$. Since $n=12$, $k=5$. We are seeking the 5th term.

$$k\text{th term} = \frac{n(n-1)(n-2)...(n-[k-2])}{(k-1)!}a^{n-(k-1)}b^{k-1}$$

$$5\text{th term} = \frac{12\cdot11\cdot10\cdot9}{(5-1)!}(3x^2)^8(-1)^4$$

$$5\text{th term} = 495(3^8)x^{16} = 3,247,695x^{16}$$

Method 2

$$k\text{th term} = {}_nC_{k-1}a^{n-(k-1)}b^{k-1}$$

$$5\text{th term} = {}_{12}C_4(3x^2)^8(-1)^4 = 495(3^8)x^{16} = 3,247,695x^{16}$$

Chapter 19, Exercises 19.1

The tables and drawings for problems 1-13 appear in the text answer section.

Exercises 19.2

1. Arranged in ascending order, the numbers are 17, 17, 17, 19, 21, 21, 23, 25, 26, 28.
$\bar{x} = \frac{\text{sum of numbers}}{10}$; $\bar{x} = \frac{214}{10} = 21.4$. The median is the mean of the numbers in the fifth and sixth positions. The median is $\frac{21+21}{2} = 21$. The mode is 17.

3. Arranged in ascending order, the numbers are 2432, 2468, 2497, 2594, 2639.
$\bar{x} = \frac{\text{sum of numbers}}{5}$; $\bar{x} = \frac{12,630}{5} = 2526$. The median is 2497. There is no mode.

5. Arranged in ascending order, the numbers are 1.5, 1.8, 1.9, 2.0, 2.1, 2.5, 2.5, 3.1, 3.2.
$\bar{x} = \frac{\text{sum of numbers}}{9}$; $\bar{x} = \frac{20.6}{9} = 2.3$. The median is 2.1. The mode is 2.5.

7. Arranged in ascending order, the numbers are 9, 14, 19, 23, 23, 27, 27, 31, 34.
$\bar{x} = \frac{\text{sum of numbers}}{9}$; $\bar{x} = \frac{207}{9} = 23$. The median is 23. There are two modes, 23 and 27.

9. Arranged in ascending order, the numbers are 0, 2, 3, 3, 3, 3, 3, 5, 7, 8, 9, 9, 9, 12, 13, 14, 16, 16, 17, 17, 17, 18, 18, 20, 21, 21, 23, 25, 26, 27, 27, 27, 27, 28, 28, 29. The median is 17. There are two modes, 3 and 27.

11. Arranged in ascending order, the numbers are 1, 6, 7, 9, 11, 11, 11, 12, 12, 18, 19, 19, 19, 20, 21, 22, 24, 26, 28, 29, 31, 34, 35, 44, 48. The median is 19. There are two modes, 11 and 19.

13. Arranged in ascending order, the numbers are 220, 222, 225, 225, 228, 230, 231, 234, 240, 243, 245, 246, 250, 252, 253, 254, 260, 261, 268, 268, 268, 272, 273, 279, 279. The median is 250. The mode is 268.

15. $\bar{x} = \frac{\text{sum of numbers}}{35} = \frac{548}{35} = 15.7$

17. $\bar{x} = \frac{\text{sum of numbers}}{20} = \frac{194}{20} = 9.7$

19. The sum of the temperatures is $1(28)+2(35)+8(42)+12(56)+5(67)+3(75)=1666$.

The sum of the number of temperatures is $1+2+8+12+5+3=31$. $\bar{x}=\frac{1666}{31}=54$.

For 31 temperatures, the median is in the 16h position. The median is 56. The mode is 56 (which occurred 12 times).

Exercises 19.3

Use a calculator as demonstrated on pages 542-543.

1. Range: $12-3=9$; $s_x=2.83$

3. Range: $35-19=16$; $s_x=5.29$

5. Range: $77-58=19$ $s_x=6.14$

7. Range: $94-78=16$; $s_x=6.40$

9. Range: $875-858=17$; $s_x=6.78$

11. Range: $91-47=44.0$; $s_x=16.1$

13. Range: $472-389=83.0$; $s_x=31.6$

15. Range: $96.5-94.0=2.50$; $s_x=0.969$

17. Range: $203-194=9.00$; $s_x=3.32$

19. Range: $508-483=25.0$; $s_x=7.64$

Exercises 19.4

1. $\bar{x}=49.3$; $s_x=20.8$

$\bar{x}-1s_x=28.5$; $\bar{x}+1s_x=70.1$

Data between 28.5 and 70.1 are 38, 43, 44, 57. This is 4 of 8 data points or 50.0%. No, 50% is not within 2% of 68.2%.

3. $\bar{x}=46.3$; $s_x=10.1$

$\bar{x}=1s_x=36.2$; $\bar{x}+1s_x=56.4$

Data between 36.2 and 56.4 are 38, 43, 47, 48, 48, 55. This is 6 of 9 data points or 66.7%. Yes, 66.7% is within 2% of 68.2%.

5. $\bar{x}=34.7$; $s_x=11.0$

$\bar{x}-1s_x=23.7$; $\bar{x}+1s_x=45.7$

Data between 23.7 and 45.7 are 24, 32, 37, 38, 45. This is 4 of 7 data points or 57.1%. No, 57.1% is not within 2% of 68.2%.

7. $\bar{x} = 12.0;\ s_x = 1.42$

$\bar{x} - 1s_x = 10.58;\ \bar{x} + 1s_x = 13.42$

Data between 10.58 and 13.42 are 11 (31 times), 12 (38 times), 13 (27 times). This is 96 data points out of 140 or 68.6%.

$\bar{x} - 2s_x = 9.16;\ \bar{x} + 2s_x = 14.84$

Data between 9.16 and 14.84 are 10 (16 times), 11 (31 times), 12 (38 times), 13 (27 times), 14 (19 times). This is 131 data points out of 140 or 93.6%. Yes, for each.

9. $\bar{x} = 67.9;\ s_x = 1.39$

$\bar{x} - 1s_x = 66.51;\ \bar{x} + 1s_x = 69.29$

Data between 66.51 and 69.29 are 67 (27 times), 68 (47 times), 69 (29 times). This is 103 data points out of 151 or 68.2%.

$\bar{x} - 2s_x = 65.12;\ \bar{x} + 2s_x = 70.68$

Data between 65.12 and 70.68 are 66 (21 times), 67 (27 times), 68 (47 times), 69 (29 times), 70 (18 times). This is 142 data points out of 151 or 94.0%. Yes, for each.

11. $\bar{x} = 26.0;\ s_x = 1.41$

$\bar{x} - 1s_x = 24.59;\ \bar{x} + 1s_x = 27.41$

Data between 24.59 and 27.41 are 25 (45 times), 26 (79 times), 27 (48 times). This is 172 data points out of 259, or 66.4%.

$\bar{x} - 2s_x = 23.18;\ \bar{x} + 2s_x = 28.82$

Data between 23.18 and 28.82 are 24 (35 times), 25 (45 times), 26 (79 times), 27 (48 times), 28 (37 times). This is 244 data points out of 259 or 94.2%. Yes, for each.

13. $\bar{x} = 72.5\text{ in.};\ s_x = 3.95$

$\bar{x} - 1s_x = 68.55;\ \bar{x} + 1s_x = 76.45$

For a normally distributed group, 50% of the 3500 students would have a height at or below $\bar{x} = 72.5$ inches. Thus, it is not possible for 68% of them to be at least 76 inches tall.

15. $\bar{x} = 20.0\text{ in.};\ s_x = 0.125$

95% should be between $\bar{x} - 2s_x$ and $\bar{x} + 2s_x$, ie between 19.75 inches and 20.25 inches.

Section 19.4

Exercises 19.5

1. Use $ax+b=y$

$a(2)+b=1$

$a(4)+b=6$

$a(6)+b=8$

$a(7)+b=12$

$a(9)+b=14$

which in matrix form is $\begin{bmatrix}2 & 1\\4 & 1\\6 & 1\\7 & 1\\9 & 1\end{bmatrix}\begin{bmatrix}a\\b\end{bmatrix}=\begin{bmatrix}1\\6\\8\\12\\14\end{bmatrix}$.

Multiply both sides of the matrix equation by the transpose of the coefficient matrix.

$$\begin{bmatrix}2 & 4 & 6 & 7 & 9\\1 & 1 & 1 & 1 & 1\end{bmatrix}\begin{bmatrix}2 & 1\\4 & 1\\6 & 1\\7 & 1\\9 & 1\end{bmatrix}\begin{bmatrix}a\\b\end{bmatrix}=\begin{bmatrix}2 & 4 & 6 & 7 & 9\\1 & 1 & 1 & 1 & 1\end{bmatrix}\begin{bmatrix}1\\6\\8\\12\\14\end{bmatrix}.$$

$$\begin{bmatrix}186 & 28\\28 & 5\end{bmatrix}\begin{bmatrix}a\\b\end{bmatrix}=\begin{bmatrix}284\\41\end{bmatrix}$$

Find $\begin{bmatrix}a\\b\end{bmatrix}$ by multiplying the inverse of $\begin{bmatrix}186 & 28\\28 & 5\end{bmatrix}$ by $\begin{bmatrix}284\\41\end{bmatrix}$

$$\begin{bmatrix}a\\b\end{bmatrix}=\begin{bmatrix}1.9\\-2.2\end{bmatrix}$$

So, $y=1.9x-2.2$

3. Use $ax + b = y$

$$1a + b = 8$$
$$4a + b = 17$$
$$5a + b = 19$$
$$8a + b = 26$$
$$10a + b = 33$$
$$13a + b = 41$$

which in matrix form is $\begin{bmatrix} 1 & 1 \\ 4 & 1 \\ 5 & 1 \\ 8 & 1 \\ 10 & 1 \\ 13 & 1 \end{bmatrix} \begin{bmatrix} a \\ b \end{bmatrix} = \begin{bmatrix} 8 \\ 17 \\ 19 \\ 26 \\ 33 \\ 41 \end{bmatrix}$.

Multiply both sides of the matrix equation by the transpose of the coefficient matrix.

$$\begin{bmatrix} 1 & 4 & 5 & 8 & 10 & 13 \\ 1 & 1 & 1 & 1 & 1 & 1 \end{bmatrix} \begin{bmatrix} 1 & 1 \\ 4 & 1 \\ 5 & 1 \\ 8 & 1 \\ 10 & 1 \\ 13 & 1 \end{bmatrix} \begin{bmatrix} a \\ b \end{bmatrix} = \begin{bmatrix} 1 & 4 & 5 & 8 & 10 & 13 \\ 1 & 1 & 1 & 1 & 1 & 1 \end{bmatrix} \begin{bmatrix} 8 \\ 17 \\ 19 \\ 26 \\ 33 \\ 41 \end{bmatrix}$$

$$\begin{bmatrix} 375 & 41 \\ 41 & 6 \end{bmatrix} \begin{bmatrix} a \\ b \end{bmatrix} = \begin{bmatrix} 1242 \\ 144 \end{bmatrix}$$

$$\begin{bmatrix} a \\ b \end{bmatrix} = \begin{bmatrix} 2.7 \\ 5.4 \end{bmatrix}$$

So, $y = 2.7x + 5.4$

5. Use $ax^2 + bx + c = y$

$$a(2)^2 + 2b + c = 3$$
$$a(3)^2 + 3b + c = 9$$
$$a(4)^2 + 4b + c = 20$$
$$a(6)^2 + 6b + c = 41$$
$$a(10)^2 + 10b + c = 117$$

which in matrix form is $\begin{bmatrix} 4 & 2 & 1 \\ 9 & 3 & 1 \\ 16 & 4 & 1 \\ 36 & 6 & 1 \\ 100 & 10 & 1 \end{bmatrix} \begin{bmatrix} a \\ b \\ c \end{bmatrix} = \begin{bmatrix} 3 \\ 9 \\ 20 \\ 41 \\ 117 \end{bmatrix}$.

Multiply both sides of the matrix equation by the transpose of the coefficient matrix.

$$\begin{bmatrix} 4 & 9 & 16 & 36 & 100 \\ 2 & 3 & 4 & 6 & 10 \\ 1 & 1 & 1 & 1 & 1 \end{bmatrix} \begin{bmatrix} 4 & 2 & 1 \\ 9 & 3 & 1 \\ 16 & 4 & 1 \\ 36 & 6 & 1 \\ 100 & 10 & 1 \end{bmatrix} \begin{bmatrix} a \\ b \\ c \end{bmatrix} = \begin{bmatrix} 4 & 9 & 16 & 36 & 100 \\ 2 & 3 & 4 & 6 & 10 \\ 1 & 1 & 1 & 1 & 1 \end{bmatrix} \begin{bmatrix} 3 \\ 9 \\ 20 \\ 41 \\ 117 \end{bmatrix}$$

$$\begin{bmatrix} 11649 & 1315 & 165 \\ 1315 & 165 & 25 \\ 165 & 25 & 5 \end{bmatrix} \begin{bmatrix} a \\ b \\ c \end{bmatrix} = \begin{bmatrix} 13589 \\ 1529 \\ 190 \end{bmatrix}$$

$$\begin{bmatrix} a \\ b \\ c \end{bmatrix} = \begin{bmatrix} 1.1 \\ 0.72 \\ -2.7 \end{bmatrix}$$

So, $y = 1.1x^2 - 0.72x - 2.7$

7. Use $ax^2+bx+c=y$

$$a(1)^2+b(1)+c=3$$
$$a(2)^2+b(2)+c=5$$
$$a(4)^2+b(4)+c=20$$
$$a(6)^2+b(6)+c=54$$
$$a(7)^2+b(7)+c=76$$
$$a(9)^2+b(9)+c=132$$

which in matrix form is $\begin{bmatrix}1&1&1\\4&2&1\\16&4&1\\36&6&1\\49&7&1\\81&9&1\end{bmatrix}\begin{bmatrix}a\\b\\c\end{bmatrix}=\begin{bmatrix}3\\5\\20\\54\\76\\132\end{bmatrix}$

Multiply both sides of the matrix equation by the transpose of the coefficient matrix.

$$\begin{bmatrix}1&4&16&36&49&81\\1&2&4&6&7&9\\1&1&1&1&1&1\end{bmatrix}\begin{bmatrix}1&1&1\\4&2&1\\16&4&1\\36&6&1\\49&7&1\\81&9&1\end{bmatrix}\begin{bmatrix}a\\b\\c\end{bmatrix}=\begin{bmatrix}1&4&16&36&49&81\\1&2&4&6&7&9\\1&1&1&1&1&1\end{bmatrix}\begin{bmatrix}3\\5\\20\\54\\76\\132\end{bmatrix}$$

$$\begin{bmatrix}10531&1361&187\\1361&187&29\\187&29&6\end{bmatrix}\begin{bmatrix}a\\b\\c\end{bmatrix}=\begin{bmatrix}16703\\2137\\290\end{bmatrix}$$

Multiply both sides by the inverse of $\begin{bmatrix}10531&1361&187\\1361&187&29\\187&29&6\end{bmatrix}$

$\begin{bmatrix}a\\b\\c\end{bmatrix}=\begin{bmatrix}2.0\\-4.1\\4.9\end{bmatrix}$. Thus $y=2.0x^2-4.1x+4.9$.

For problems 9 through 15, a calculator routine is used.

9. Use QuadReg. $y=-1.0x^2+6.5x+11$

11. Use PowerReg (PwrReg). $y=50,\bar{0}00x^{-0.982}$

13. Use ExpReg. $y=325(5.00)^x$

15. Use LinReg $(ax+b)$. $y=0.641x+0.308$

Exercises 19.6

1.
$UTL = 38.0004$
$UCL = 38.002$
$LCL = 37.998$
$LTL = 37.9996$
No, the control limits are not within the tolerance limits.

3.
$UTL = 17.002$
$UCL = 17.003$
$LCL = 16.997$
$LTL = 16.998$
No, the control limits are not with the tolerance limits.

5.
$UTL = 28.002$
$UCL = 28.0006$
$LCL = 27.9994$
$LTL = 27.998$
Yes, the control limits are within the tolerance limits.

7.
$UTL = 52.005$
$UCL = 52.006$
$LCL = 51.994$
$LTL = 51.995$
No, the control limits are not within the tolerance limits.

9. common cause

11. capable

Chapter 19 Review

1-3. The solutions are in the textbook.

4. In ascending order, the numbers are 16, 18, 18, 21, 23, 24, 25, 27, 31.
$\bar{x} = \frac{\text{sum of numbers}}{9} = \frac{203}{9} = 22.6$. The median is 23.0. The mode is 18.0

5. In ascending order, the numbers are 8, 9, 10, 13, 14, 15, 15, 15, 19.
$\bar{x} = \frac{118}{9} = 13.1$. The median is 14.0. The mode is 15.0.

6. In ascending order, the numbers are 33, 34, 35, 36, 37, 38, 38, 40.
$\bar{x} = \frac{291}{8} = 36.4$. The median is $\frac{36+37}{2} = 36.5$. The mode is 38.0.

7. In ascending order, the numbers are 18, 19, 20, 21, 21, 23, 25, 29.
$\bar{x} = \frac{176}{8} = 22.0$. The median is $\frac{21+21}{2} = 21.0$. The mode is 21.0.

8. $\bar{x} = \frac{\text{sum of measurements}}{57} = \frac{1127}{57} = 19.8$. The median is the 29th measurement, ie 20.0. The mode is 20.0.

9. Range: $14 - 8 = 6.00$; $s_x = 2.28$

10. Range: $28 - 18 = 10.0$; $s_x = 4.03$

11. Range: $39 - 25 = 14.0$; $s_x = 5.34$

12. Range: $17 - 6 = 11.0$; $s_x = 3.92$

13. $\bar{x} = 56.9$; $s_x = 1.38$
$\bar{x} - 1s_x = 55.52$; $\bar{x} + 1s_x = 58.28$
Data between 55.52 and 58.28 are 56 (28 times), 57 (47 times), 58 (27 times). This is 102 data points out of 149, ie 68.5%.

$\bar{x} - 2s_x = 54.14$; $\bar{x} + 2s_x = 59.66$
Data between 54.14 and 59.66 are 55 (22 times), 56 (28 times), 57 (47 times), 58 (27 times), 59 (17 times). This is 141 data points out of 149, ie 94.6%.
Yes, for both.

14. $\bar{x} = 29.9$; $s_x = 1.40$
$\bar{x} - 1s_x = 28.5$; $\bar{x} + 1s_x = 31.3$
Data between 28.5 and 31.3 are 29 (35 times), 30 (62 times), 31 (37 times). This is 134 data points out of 198, ie 67.7%.

$\bar{x} - 2s_x = 27.1$; $\bar{x} + 2s_x = 32.7$
Data between 27.1 and 32.7 are 28 (29 times), 29 (35 times), 30 (62 times), 31 (37 times), 32 (23 times). This is 186 out of 198, ie 93.9%. Yes, for both.

15. Use $ax + b = y$

$a(1) + b = 2$
$a(2) + b = 8$
$a(3) + b = 13$
$a(5) + b = 20$
$a(6) + b = 26$
$a(7) + b = 38$

which in matrix form is $\begin{bmatrix} 1 & 1 \\ 2 & 1 \\ 3 & 1 \\ 5 & 1 \\ 6 & 1 \\ 7 & 1 \end{bmatrix} \begin{bmatrix} a \\ b \end{bmatrix} = \begin{bmatrix} 2 \\ 8 \\ 13 \\ 20 \\ 26 \\ 38 \end{bmatrix}$

Multiply both sides of the equation by the transpose of the coefficient matrix.

$$\begin{bmatrix} 1 & 2 & 3 & 5 & 6 & 7 \\ 1 & 1 & 1 & 1 & 1 & 1 \end{bmatrix} \begin{bmatrix} 1 & 1 \\ 2 & 1 \\ 3 & 1 \\ 5 & 1 \\ 6 & 1 \\ 7 & 1 \end{bmatrix} \begin{bmatrix} a \\ b \end{bmatrix} = \begin{bmatrix} 1 & 2 & 3 & 5 & 6 & 7 \\ 1 & 1 & 1 & 1 & 1 & 1 \end{bmatrix} \begin{bmatrix} 2 \\ 8 \\ 13 \\ 20 \\ 26 \\ 38 \end{bmatrix}$$

$$\begin{bmatrix} 124 & 24 \\ 24 & 6 \end{bmatrix} \begin{bmatrix} a \\ b \end{bmatrix} = \begin{bmatrix} 579 \\ 107 \end{bmatrix}.$$

Multiply both sides by the inverse of the 2×2 matrix.

$\begin{bmatrix} a \\ b \end{bmatrix} = \begin{bmatrix} 5.4 \\ -3.7 \end{bmatrix}$. Thus $y = 5.4x - 3.7$

16. Use $ax + b = y$

$a(-2) + b = -5$

$a(1) + b = 7$

$a(2) + b = 10$

$a(3) + b = 15$

$a(4) + b = 19$

$a(5) + b = 21$

which in matrix form is $\begin{bmatrix} -2 & 1 \\ 1 & 1 \\ 2 & 1 \\ 3 & 1 \\ 4 & 1 \\ 5 & 1 \end{bmatrix}\begin{bmatrix} a \\ b \end{bmatrix} = \begin{bmatrix} -5 \\ 7 \\ 10 \\ 15 \\ 19 \\ 21 \end{bmatrix}$

Multiply both sides of the equation by the transpose of the coefficient matrix.

$$\begin{bmatrix} -2 & 1 & 2 & 3 & 4 & 5 \\ 1 & 1 & 1 & 1 & 1 & 1 \end{bmatrix}\begin{bmatrix} -2 & 1 \\ 1 & 1 \\ 2 & 1 \\ 3 & 1 \\ 4 & 1 \\ 5 & 1 \end{bmatrix}\begin{bmatrix} a \\ b \end{bmatrix} = \begin{bmatrix} -2 & 1 & 2 & 3 & 4 & 5 \\ 1 & 1 & 1 & 1 & 1 & 1 \end{bmatrix}\begin{bmatrix} -5 \\ 7 \\ 10 \\ 15 \\ 19 \\ 21 \end{bmatrix}$$

$$\begin{bmatrix} 59 & 13 \\ 13 & 6 \end{bmatrix}\begin{bmatrix} a \\ b \end{bmatrix} = \begin{bmatrix} 263 \\ 67 \end{bmatrix}.$$

Multiply both sides by the inverse of the 2×2 matrix.

$\begin{bmatrix} a \\ b \end{bmatrix} = \begin{bmatrix} 3.8 \\ 2.9 \end{bmatrix}$. Thus $y = 3.8x + 2.9$

17. Use $ax + b = y$

$a(1) + b = 3$

$a(2) + b = 5$

$a(3) + b = 7$

$a(4) + b = 12$

$a(5) + b = 15$

$a(7) + b = 21$

which in matrix form is $\begin{bmatrix} 1 & 1 \\ 2 & 1 \\ 3 & 1 \\ 4 & 1 \\ 5 & 1 \\ 7 & 1 \end{bmatrix} \begin{bmatrix} a \\ b \end{bmatrix} = \begin{bmatrix} 3 \\ 5 \\ 7 \\ 12 \\ 15 \\ 21 \end{bmatrix}$

Multiply both sides of the equation by the transpose of the coefficient matrix.

$$\begin{bmatrix} 1 & 2 & 3 & 4 & 5 & 7 \\ 1 & 1 & 1 & 1 & 1 & 1 \end{bmatrix} \begin{bmatrix} 1 & 1 \\ 2 & 1 \\ 3 & 1 \\ 4 & 1 \\ 5 & 1 \\ 7 & 1 \end{bmatrix} \begin{bmatrix} a \\ b \end{bmatrix} = \begin{bmatrix} 1 & 2 & 3 & 4 & 5 & 7 \\ 1 & 1 & 1 & 1 & 1 & 1 \end{bmatrix} \begin{bmatrix} 3 \\ 5 \\ 7 \\ 12 \\ 15 \\ 21 \end{bmatrix}$$

$$\begin{bmatrix} 104 & 22 \\ 22 & 6 \end{bmatrix} \begin{bmatrix} a \\ b \end{bmatrix} = \begin{bmatrix} 304 \\ 63 \end{bmatrix}.$$

Multiply both sides by the inverse of the 2×2 matrix.

$\begin{bmatrix} a \\ b \end{bmatrix} = \begin{bmatrix} 3.1 \\ -9.7 \end{bmatrix}$. Thus $y = 3.1x - 0.97$

18. Use $ax + b = y$

$a(1) + b = 2$
$a(2) + b = 3$
$a(3) + b = 6$
$a(4) + b = 8$
$a(5) + b = 10$
$a(7) + b = 14$

which in matrix form is $\begin{bmatrix} 1 & 1 \\ 2 & 1 \\ 3 & 1 \\ 4 & 1 \\ 5 & 1 \\ 7 & 1 \end{bmatrix} \begin{bmatrix} a \\ b \end{bmatrix} = \begin{bmatrix} 2 \\ 3 \\ 6 \\ 8 \\ 10 \\ 14 \end{bmatrix}$

Multiply both sides of the equation by the transpose of the coefficient matrix.

$$\begin{bmatrix} 1 & 2 & 3 & 4 & 5 & 7 \\ 1 & 1 & 1 & 1 & 1 & 1 \end{bmatrix} \begin{bmatrix} 1 & 1 \\ 2 & 1 \\ 3 & 1 \\ 4 & 1 \\ 5 & 1 \\ 7 & 1 \end{bmatrix} \begin{bmatrix} a \\ b \end{bmatrix} = \begin{bmatrix} 1 & 2 & 3 & 4 & 5 & 7 \\ 1 & 1 & 1 & 1 & 1 & 1 \end{bmatrix} \begin{bmatrix} 2 \\ 3 \\ 6 \\ 8 \\ 10 \\ 14 \end{bmatrix}$$

$$\begin{bmatrix} 104 & 22 \\ 22 & 6 \end{bmatrix} \begin{bmatrix} a \\ b \end{bmatrix} = \begin{bmatrix} 206 \\ 43 \end{bmatrix}.$$

Multiply both sides by the inverse of the 2×2 matrix.

$\begin{bmatrix} a \\ b \end{bmatrix} = \begin{bmatrix} 2.1 \\ -0.43 \end{bmatrix}$. Thus $y = 2.1x - 0.43$

19. Use $ax^2 + bx + c = y$

$$a(1)^2 + b(1) + c = 7$$
$$a(2)^2 + b(2) + c = 13$$
$$a(3)^2 + b(3) + c = 21$$
$$a(4)^2 + b(4) + c = 34$$
$$a(5)^2 + b(5) + c = 45$$
$$a(6)^2 + b(6) + c = 58$$

which in matrix form is $\begin{bmatrix} 1 & 1 & 1 \\ 4 & 2 & 1 \\ 9 & 3 & 1 \\ 16 & 4 & 1 \\ 25 & 5 & 1 \\ 36 & 6 & 1 \end{bmatrix} \begin{bmatrix} a \\ b \\ c \end{bmatrix} = \begin{bmatrix} 7 \\ 13 \\ 21 \\ 34 \\ 45 \\ 58 \end{bmatrix}$

Multiply both sides of the equation by the transpose of the coefficient matrix.

$$\begin{bmatrix} 1 & 4 & 9 & 16 & 25 & 36 \\ 1 & 2 & 3 & 4 & 5 & 6 \\ 1 & 1 & 1 & 1 & 1 & 1 \end{bmatrix} \begin{bmatrix} 1 & 1 & 1 \\ 4 & 2 & 1 \\ 9 & 3 & 1 \\ 16 & 4 & 1 \\ 25 & 5 & 1 \\ 36 & 6 & 1 \end{bmatrix} \begin{bmatrix} a \\ b \\ c \end{bmatrix} = \begin{bmatrix} 1 & 4 & 9 & 16 & 25 & 36 \\ 1 & 2 & 3 & 4 & 5 & 6 \\ 1 & 1 & 1 & 1 & 1 & 1 \end{bmatrix} \begin{bmatrix} 7 \\ 13 \\ 21 \\ 34 \\ 45 \\ 58 \end{bmatrix}$$

$$\begin{bmatrix} 2275 & 441 & 91 \\ 441 & 91 & 21 \\ 91 & 21 & 6 \end{bmatrix} \begin{bmatrix} a \\ b \\ c \end{bmatrix} = \begin{bmatrix} 4005 \\ 805 \\ 178 \end{bmatrix}.$$

$\begin{bmatrix} a \\ b \\ c \end{bmatrix} = \begin{bmatrix} 0.84 \\ 4.5 \\ 1.1 \end{bmatrix}$. Thus $y = 0.84x^2 + 4.5x + 1.1$

20. Use $ax^2 + bx + c = y$

$$a(2)^2 + b(2) + c = 4$$
$$a(3)^2 + b(3) + c = 11$$
$$a(4)^2 + b(4) + c = 22$$
$$a(5)^2 + b(5) + c = 34$$
$$a(6)^2 + b(6) + c = 43$$
$$a(7)^2 + b(7) + c = 58$$

which in matrix form is $\begin{bmatrix} 4 & 2 & 1 \\ 9 & 3 & 1 \\ 16 & 4 & 1 \\ 25 & 5 & 1 \\ 36 & 6 & 1 \\ 49 & 7 & 1 \end{bmatrix} \begin{bmatrix} a \\ b \\ c \end{bmatrix} = \begin{bmatrix} 4 \\ 11 \\ 22 \\ 34 \\ 43 \\ 58 \end{bmatrix}$

Multiply both sides of the equation by the transpose of the coefficient matrix.

$$\begin{bmatrix} 4 & 9 & 16 & 25 & 36 & 49 \\ 2 & 3 & 4 & 5 & 6 & 7 \\ 1 & 1 & 1 & 1 & 1 & 1 \end{bmatrix} \begin{bmatrix} 4 & 2 & 1 \\ 9 & 3 & 1 \\ 16 & 4 & 1 \\ 25 & 5 & 1 \\ 36 & 6 & 1 \\ 49 & 7 & 1 \end{bmatrix} \begin{bmatrix} a \\ b \\ c \end{bmatrix} = \begin{bmatrix} 4 & 9 & 16 & 25 & 36 & 49 \\ 2 & 3 & 4 & 5 & 6 & 7 \\ 1 & 1 & 1 & 1 & 1 & 1 \end{bmatrix} \begin{bmatrix} 4 \\ 11 \\ 22 \\ 34 \\ 43 \\ 58 \end{bmatrix}$$

$$\begin{bmatrix} 3248 & 532 & 91 \\ 644 & 112 & 21 \\ 139 & 27 & 6 \end{bmatrix} \begin{bmatrix} a \\ b \\ c \end{bmatrix} = \begin{bmatrix} 3953 \\ 791 \\ 172 \end{bmatrix}.$$

$\begin{bmatrix} a \\ b \\ c \end{bmatrix} = \begin{bmatrix} 0.57 \\ 5.7 \\ -1\overline{0} \end{bmatrix}$. Thus $y = 0.57x^2 + 5.7 - 1\overline{0}$

21. Use $y = ab^x$, ExpReg.
$y = 2.1(3.0)^x$

22. $y = ab^x$, ExpReg.
$y = 2.0(0.63)^x$

23. $y = ax^b$, PowerReg.
$y = 51{,}500x^{-1.02}$

24.
$UTL = 14.005$
$UCL = 14.003$
$LCL = 13.997$
$LTL = 13.995$
Yes, the control limits are within the tolerance limits.

25.
$UTL = 21.001$
$UCL = 21.005$
$LCL = 20.995$
$LTL = 20.999$
No, the control limits are not within the tolerance limits.

Chapter 20, Exercises 20.1

1. **3.**

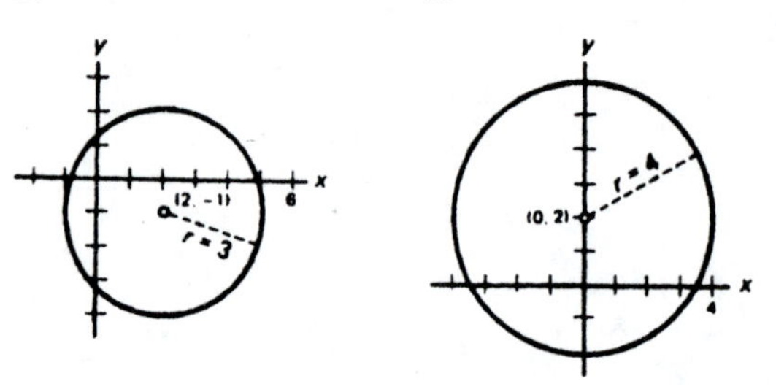

5. $(x-1)^2 + (y+1)^2 = 16$

7. $(x+2)^2 + (y+4)^2 = 34$

9. $x^2 + y^2 = 36$

11. $C(0, 0); r = 4$

13. $C(-3, 4); r = 4$

15. $x^2 - 8x + y^2 + 12y = 8$
$x^2 - 8x + 16 + y^2 + 12y + 36 = 8 + 16 + 36$
$(x-4)^2 + (y+6)^2 = 60$
$C(4, -6); r = 2\sqrt{15}$

17. $x^2 - 12x + y^2 - 2y = 12$
$x^2 - 12x + 36 + y - 2y + 1 = 12 + 36 + 1$
$(x-6)^2 + (y-1)^2 = 49; C(6, 1); r = 7$

19. $x^2 + 7x + y^2 + 3y = 9$
$x^2 + 7x + 49/4 + y^2 + 3y + 9/4 = 9 + 49/4 + 9/4$
$(x + 7/2)^2 + (y + 3/2)^2 = 47/2;$ $C(-7/2, -3/2); r = \sqrt{94}/2$

21. $\sqrt{(0-1)^2 + (y-4)^2} = \sqrt{(0+3)^2 + (y-2)^2}$
$1 + y^2 - 8y + 16 = 9 + y^2 - 4y + 4$
$4 = 4y \text{ or } y = 1$
$C(0, 1); r = \sqrt{(1-0)^2 + (4-1)^2} = \sqrt{10}$
$x^2 + (y-1)^2 = 10 \text{ or } x^2 + y^2 - 2y - 9 = 0$

23. $x^2 + y^2 + Dx + Ey + F = 0$; substitute (3, 1), (0, 0) and (8, 4)
$3D + E + F = -10$
$F = 0$
$8D + 4E + F = -80$

Solve simultaneously: $D = 10$,
$E = -40, F = 0$
$x^2 + y^2 + 10x - 40y = 0$
$(x+5)^2 + (y-20)^2 = 425$
$C(-5, 20); r = 5\sqrt{17}$

Exercises 20.2

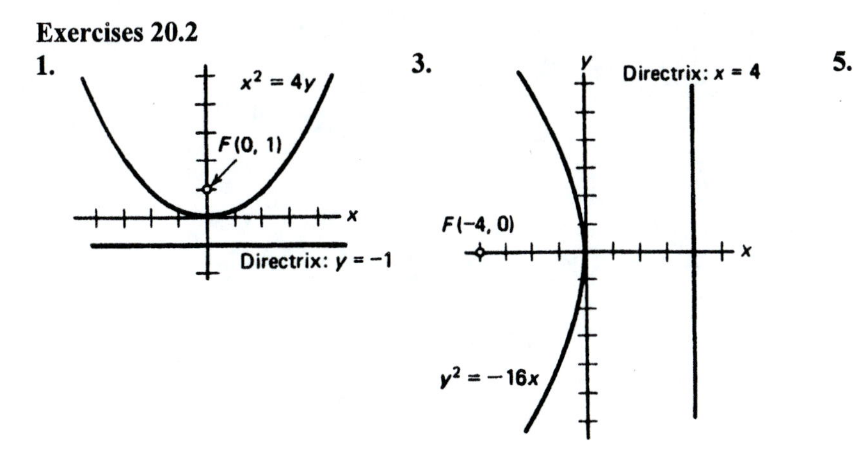

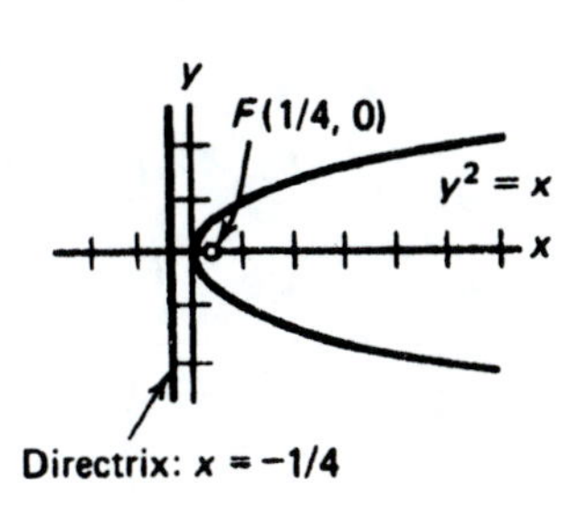

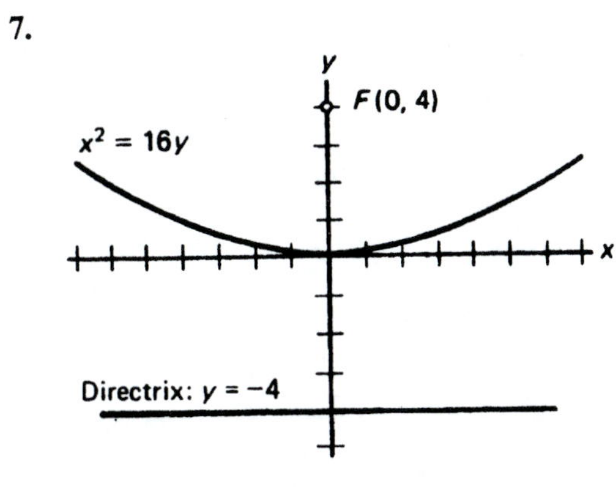

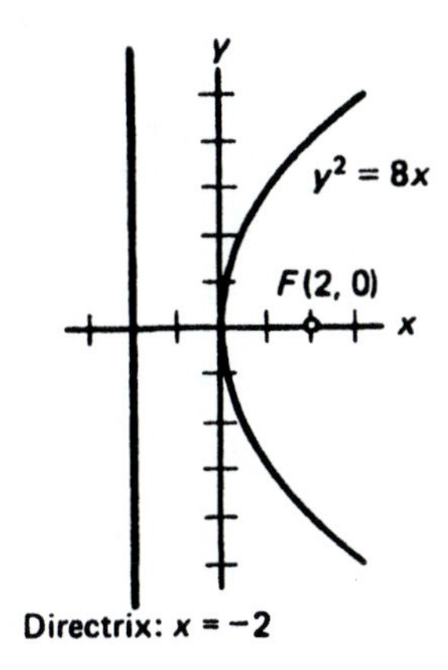

11. $y^2 = 8x$ **13.** $y^2 = -32x$ **15.** $x^2 = 24y$ **17.** $y^2 = -16x$

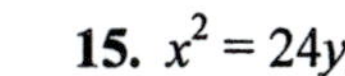

19. $\sqrt{(x+1)^2 + (y-3)^2} = \sqrt{(x-3)^2 + (y-y)^2}$;
$y^2 - 6y + 8x + 1 = 0$

21. $x^2 = 4py$
$200^2 = 4p(-16)$; $p = -625$; $x^2 = -2500y$
at 50 m, $50^2 = -625y_1$; $y_1 = -1$ so 15 m high
at 150 m, $150^2 = -625y_2$; $y_2 = -9$ m so 7 m high

23. $x^2 = 32y$

25. $y = 2x^2 + 7x - 15$

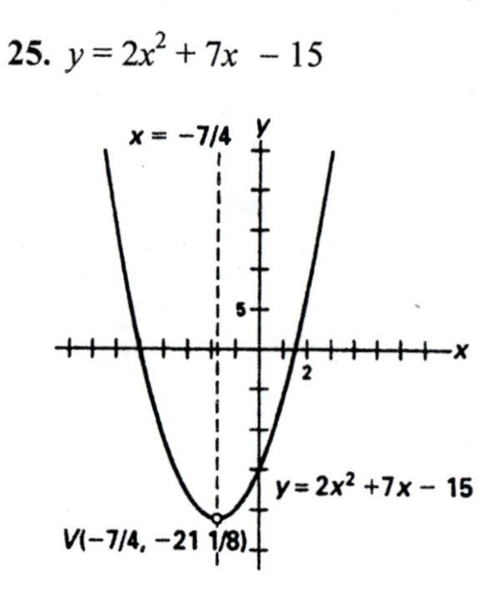

27. $y = -2x^2 + 4x + 16$

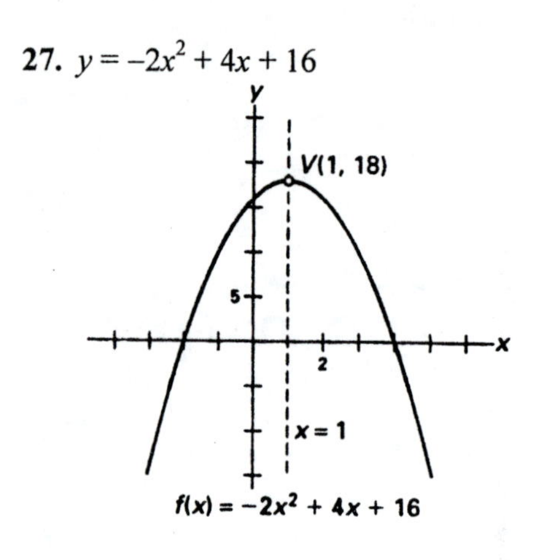

29. a) Maximum occurs at $x = -\dfrac{b}{2a} = 512$; $h(512) = 1024$ meters.

b) horizontal and vertical directions from 0 to 1024 metres.

31. $2\ell + 2w = 240; A = \ell w = \ell(120 - \ell), A = -\ell^2 + 120\ell$

maximum at $\ell = \dfrac{-b}{2a} = 60, A = 60(60) = 3600$ sq. m.

Exercises 20.3

1.

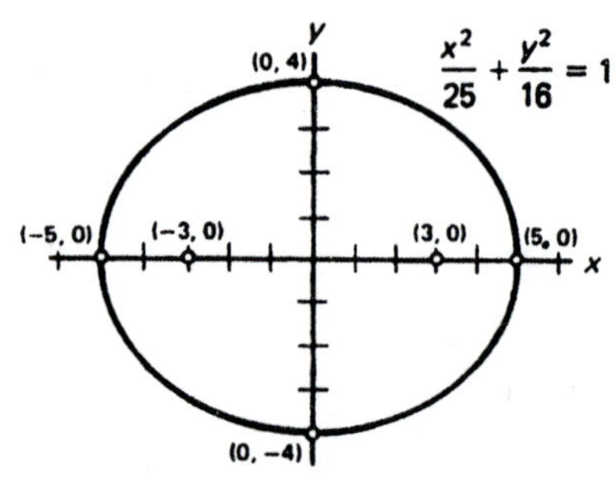

3.

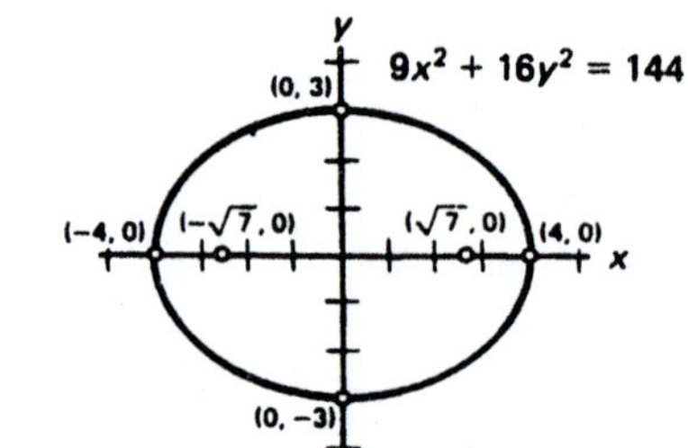

5.

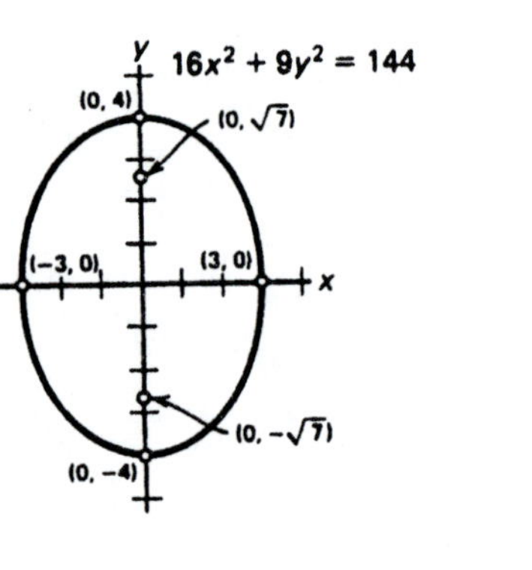

7.

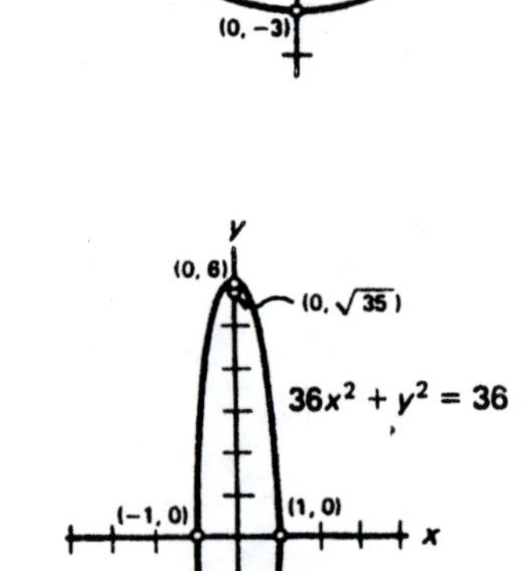
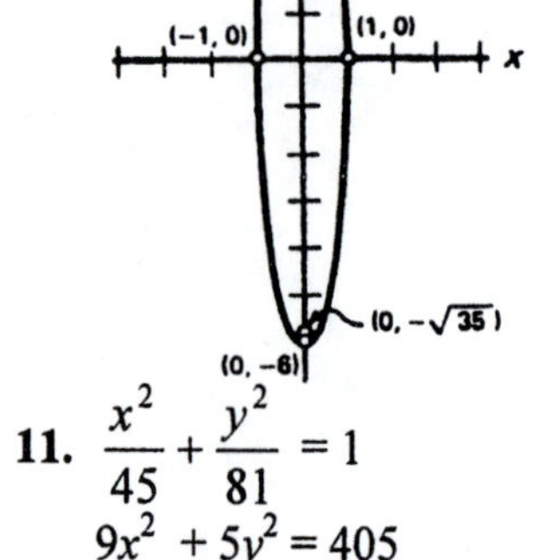

9. $\dfrac{x^2}{16} + \dfrac{y^2}{12} = 1$

$3x^2 + 4y^2 = 48$

11. $\dfrac{x^2}{45} + \dfrac{y^2}{81} = 1$

$9x^2 + 5y^2 = 405$

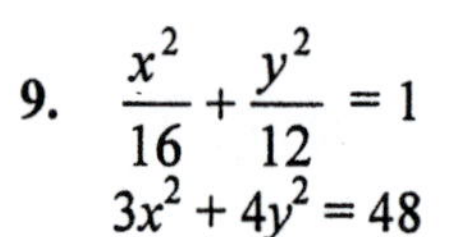

13. $\dfrac{x^2}{36}+\dfrac{y^2}{25}=1$
$25x^2+36y^2=900$

15. $\dfrac{x^2}{39}+\dfrac{y^2}{64}=1$
$64x^2+39y^2=2496$

17. $\dfrac{x^2}{5600^2}+\dfrac{y^2}{5000^2}=1;\ 625x^2+784y^2=1.96\times 10^{10}$

Exercises 20.4

1.

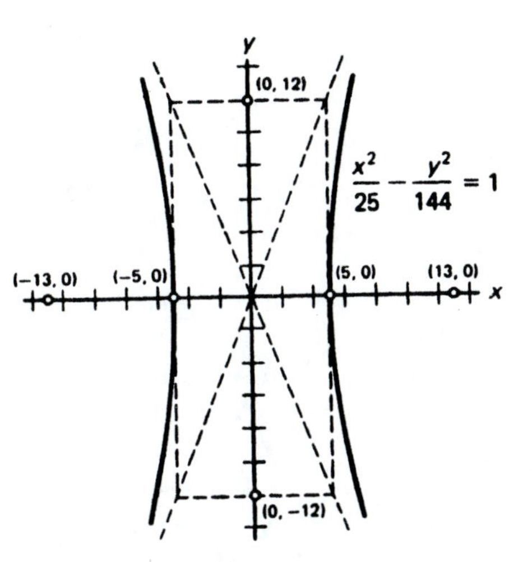

3.

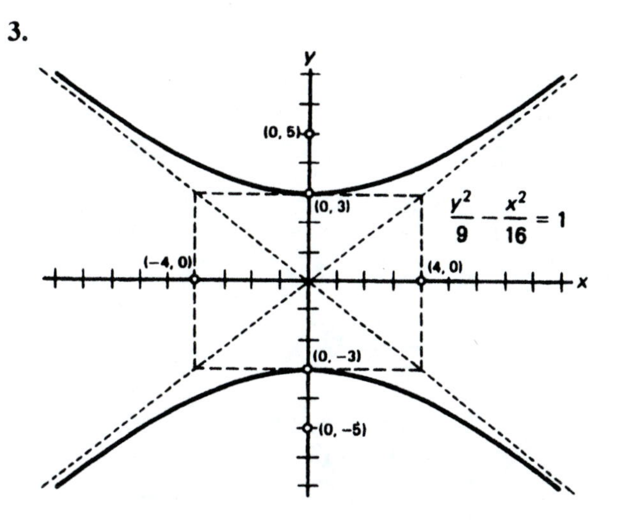

5.

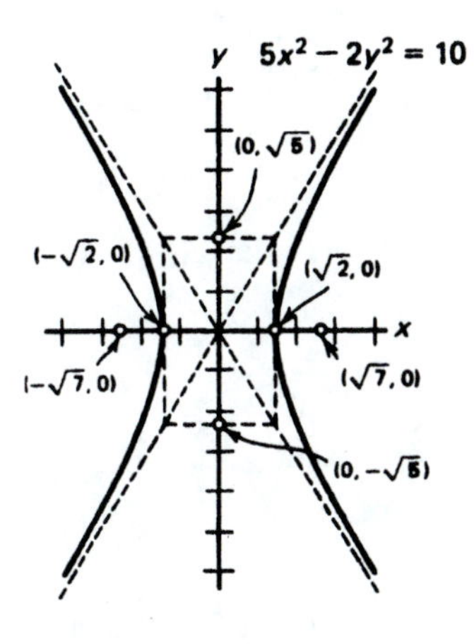

7.

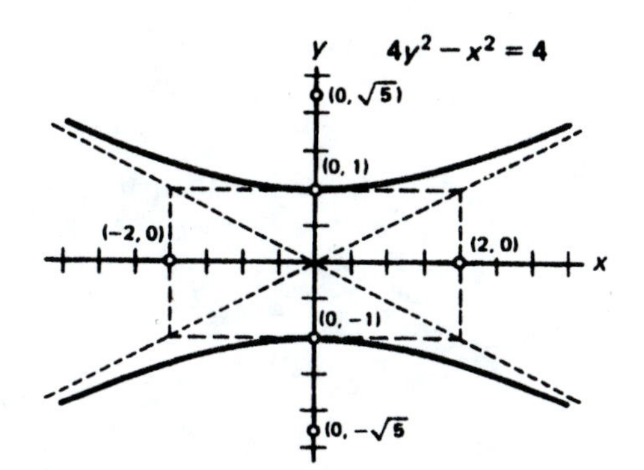

9. $\dfrac{x^2}{16}-\dfrac{y^2}{20}=1$
$5x^2-4y^2=80$

11. $\dfrac{y^2}{36}-\dfrac{x^2}{28}=1$
$7y^2-9x^2=252$

13. $\dfrac{x^2}{9}-\dfrac{y^2}{25}=1$
$25x^2-9y^2=225$

15. $\dfrac{x^2}{25} - \dfrac{y^2}{11} = 1$

$11x^2 - 25y^2 = 275$

17.

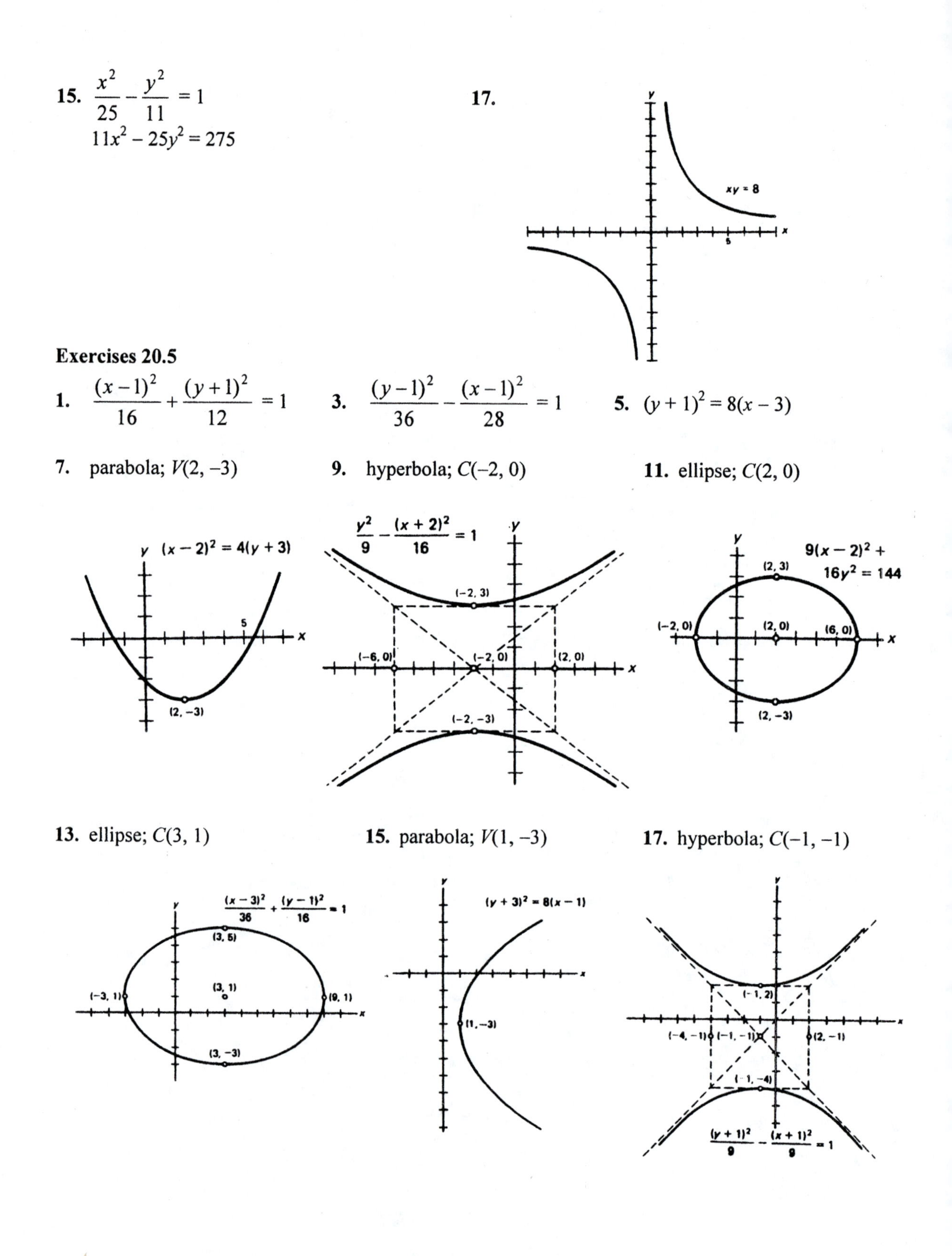

Exercises 20.5

1. $\dfrac{(x-1)^2}{16} + \dfrac{(y+1)^2}{12} = 1$

3. $\dfrac{(y-1)^2}{36} - \dfrac{(x-1)^2}{28} = 1$

5. $(y+1)^2 = 8(x-3)$

7. parabola; $V(2, -3)$

9. hyperbola; $C(-2, 0)$

11. ellipse; $C(2, 0)$

13. ellipse; $C(3, 1)$

15. parabola; $V(1, -3)$

17. hyperbola; $C(-1, -1)$

19. parabola; $V(2, -1)$

$(x-2)^2 = -2(y+1)$

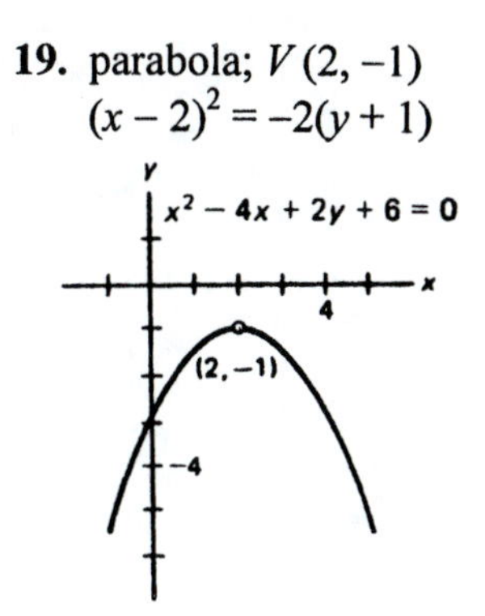

21. ellipse; $C(-2, 1)$

$$\frac{(x+2)^2}{16} + \frac{(y-1)^2}{4} = 1$$

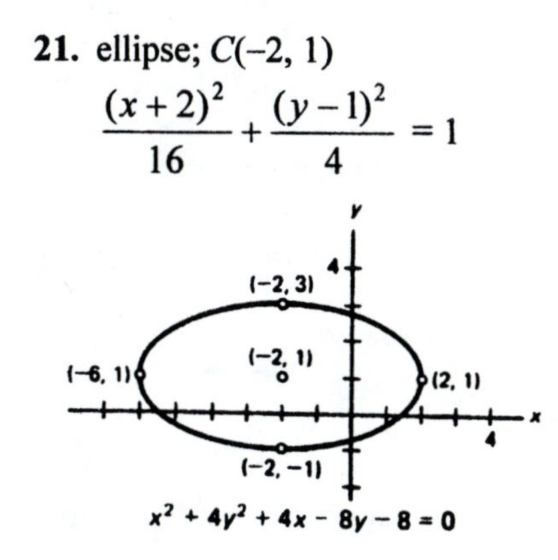

23. hyperbola; $C(1, 1)$

$$\frac{(x-1)^2}{1} - \frac{(y-1)^2}{4} = 0$$

$4x^2 - y^2 - 8x + 2y + 3 = 0$

25. hyperbola; $C(-3, 3)$

$$\frac{(y-3)^2}{4} - \frac{(x+3)^2}{25} = 1$$

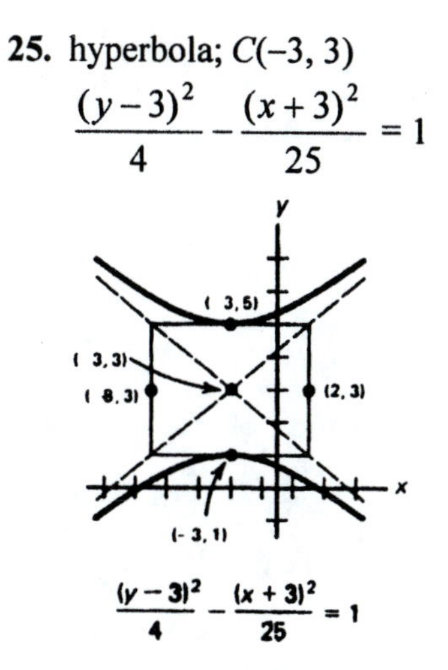

27. parabola; $V(-8, -2)$

$(x+8)^2 = -12(y+2)$

$(x+8)^2 = 12(y+2)$

29. ellipse; $C(-6, -2)$

$$\frac{(x+6)^2}{16} + \frac{(y+2)^2}{64} = 1$$

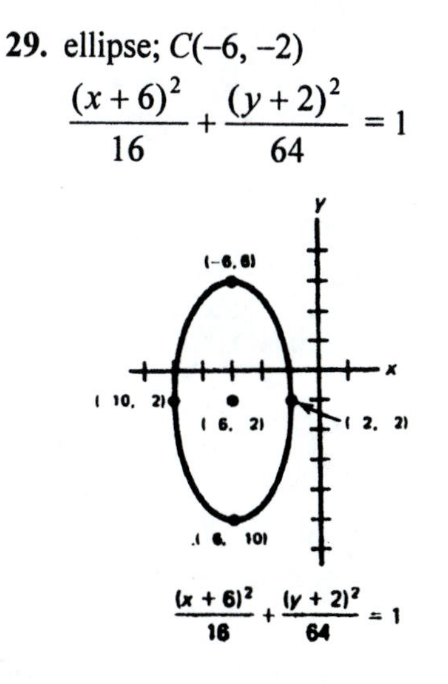

Exercises 20.6

1. A and C both positive and not equal, thus ellipse

3. $A = 0$ and C not zero, thus parabola

5. $A > 0, B < 0$, thus hyperbola

7. $A = C$, thus circle

9. $A = C$, thus circle

11. A and C both positive and not equal, thus ellipse

13. $A > 0, B < 0$, thus hyperbola

15. $A = 0$ and C not zero, thus parabola

Exercises 20.7

1. $x^2 = 3y$ and $y = 2x - 3$; substitute, $x^2 = 3(2x - 3)$,
   $x^2 - 6x + 9 = 0, x = 3, (3, 3)$.

3. $x^2 + 4x + y^2 - 8 = 0$ and $x^2 + y^2 = 4$; subtract, $4x - 4 = 0$,
   $x = 1, (1, \sqrt{3}), (1, -\sqrt{3})$.

5. $y^2 - x^2 = 12$ and $x^2 = 4y$; substitute, $y^2 - 4y - 12 = 0$,
   $(y - 6)(y + 2) = 0, y = 6$, y cannot be negative, $(2\sqrt{6}, 6), (-2\sqrt{6}, 6)$.

7. $x^2 + y^2 = 4$ and $x^2 - y^2 = 4$; add, $2x^2 = 8, x \pm 2, (2, 0), (-2, 0)$.

9. $x^2 = 6y$ and $y = 6$; substitute, $x^2 = 6(6), x = \pm 6, (6, 6), (-6, 6)$.

11. $y^2 = 4x + 12$ and $y^2 = -4x - 4$; substitute, $4x + 12 = -4x - 4, 8x = -16, x = -2, (-2, 2), (-2, -2)$.

13. $x^2 + y^2 = 36$ and $y = x^2$; substitute, $y + y^2 = 36, y = \dfrac{-1 \pm \sqrt{145}}{2}$,
   $y = \dfrac{-1 + \sqrt{145}}{2} \approx 5.5$, y cannot be negative. Approximate points $(2.3, 5.5), (-2.3, 5.5)$.

15. $x^2 - y^2 = 9$ and $x^2 + 9y^2 = 169$; subtract, $10y^2 = 160, y^2 = 16, y = \pm 4, (5, 4), (5, -4) (-5, 4)$,
   $(-5, -4)$.

17. $x^2 + y^2 = 17$ and $xy = 4$; substitute, $\dfrac{16}{2} + y^2 = 17$,
   $y^4 - 17y^2 + 16 = 0, (y^2 - 16)(y^2 - 1) = 0, y = \pm 4$ or ± 1.
   $(1, 4), (-1, -4), (4, 1), (-4, -1)$.

Exercises 20.8

1.

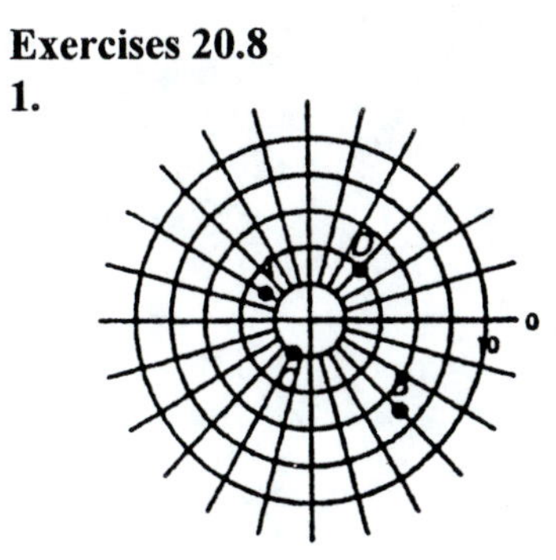

3.

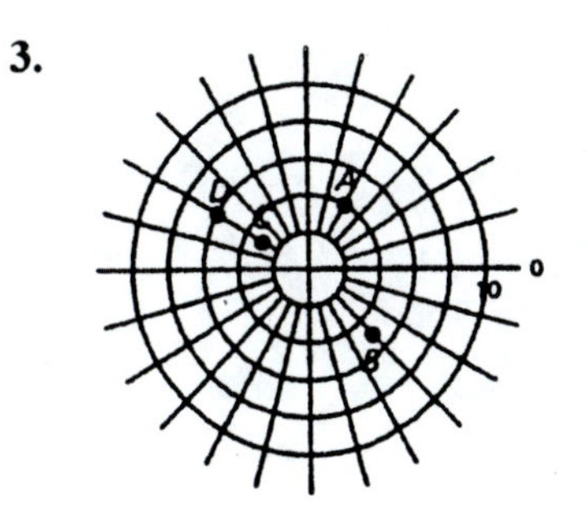

5. $(-3, 240°), (-3, -120°), (3, -300°)$

7. $(5, 135°), (-5, -45°), (5, -225°)$

9. $(-4, -315°), (-4, 45°), (4, 225°)$

11. $(-3, 7\pi/6), \left(-3, \dfrac{-5\pi}{6}\right), \left(3, \dfrac{-11\pi}{6}\right)$

13. $(9, 5\pi/3), (9, -\pi/3), (-9, -4\pi/3)$

15. $(4, -3\pi/4), (-4, \pi/4), (4, 5\pi/4)$

17. $r = 10 \sin \theta$

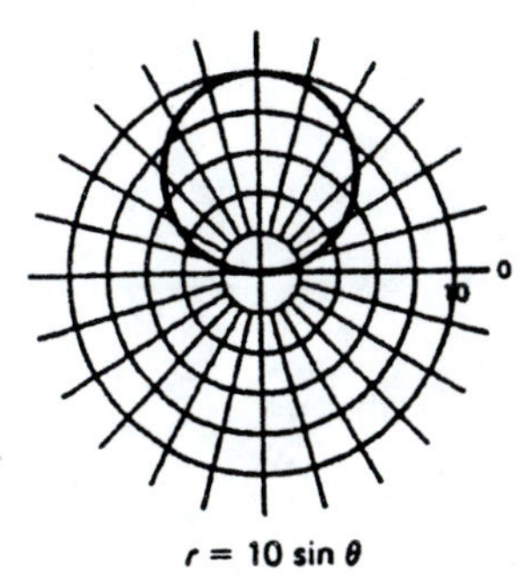

$r = 10 \sin \theta$

19. $r = 4 + 4 \cos \theta$

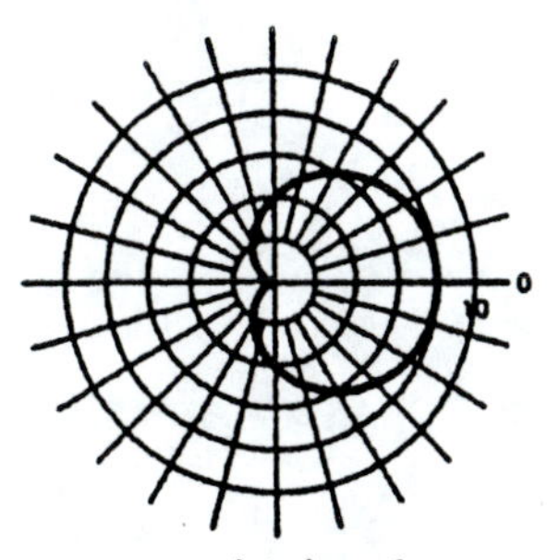

$r = 4 + 4 \cos \theta$

21. $r \cos \theta = 4$

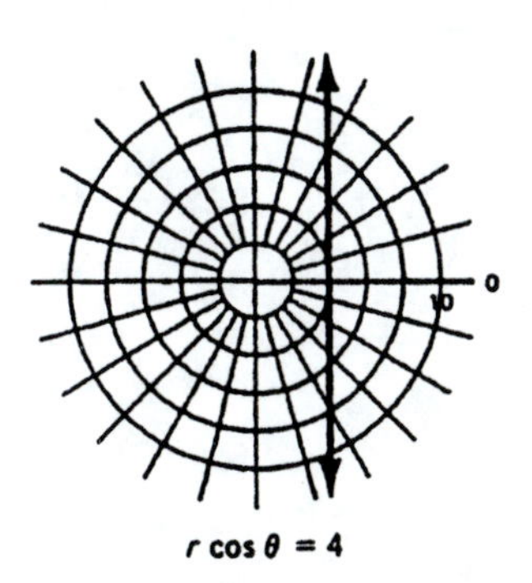

$r \cos \theta = 4$

23. $r = -10 \cos \theta$

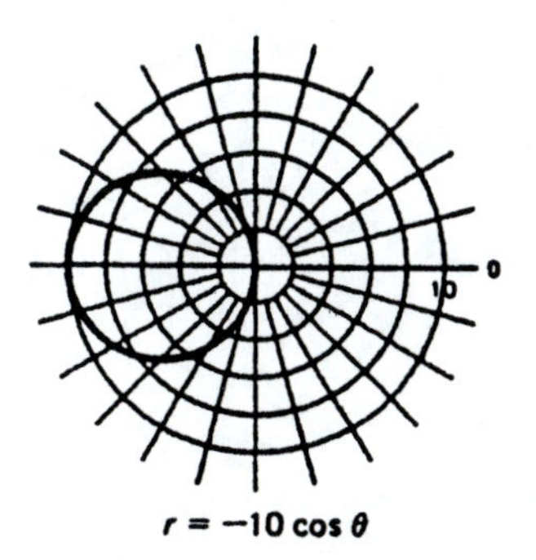

$r = -10 \cos \theta$

25. $r = 4 - 4\sin\theta$ **27.** $r = \theta$

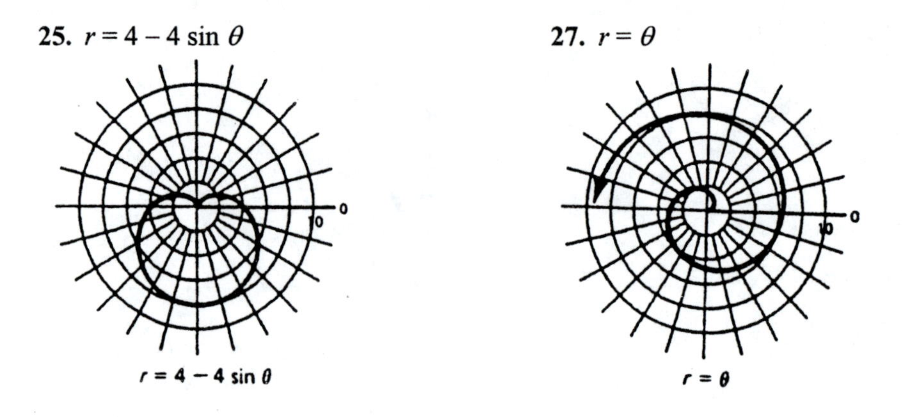

Note: A sample of the different kinds of exercises are given here. The other exercises are done in a similar manner.

29. $x = r\cos\theta$, $x = 3\cos 30°$, $x = \dfrac{3\sqrt{3}}{2}$; $y = r\sin\theta$, $y = 3\sin 30°$, $y = 3\left(\dfrac{1}{2}\right) = \dfrac{3}{2}$. $\left(\dfrac{3\sqrt{3}}{2}, \dfrac{3}{2}\right)$

31. $x = r\cos\theta$, $x = 2\cos\dfrac{\pi}{3}$, $x = 2\left(\dfrac{1}{2}\right) = 1$; $y = r\sin\theta$, $y = 2\sin\dfrac{\pi}{3}$, $y = 2\left(\dfrac{\sqrt{3}}{2}\right) = \sqrt{3}$. $(1, \sqrt{3})$

33. $(2\sqrt{3}, -2)$ **35.** $(0, 6)$ **37.** $(2.5, -4.33)$ **39.** $(1.4, 1.4)$

41. $x^2 + y^2 = r^2$, $5^2 + 5^2 = r^2$, $50 = r^2$, $r = 5\sqrt{2} \approx 7.1$;
$\theta = \arctan\dfrac{y}{x}$, $\theta = \arctan\dfrac{5}{5}$, $\theta = 45°$; $(7.1, 45°)$.

43. $(4, 90°)$ **45.** $(4, 240°)$ **47.** $\left(4\sqrt{2}, \dfrac{3\pi}{4}\right)$

49. $x^2 + y^2 = r^2$, $\left(\sqrt{-6}^{\,2}\right) + (\sqrt{2})^2 = r^2$, $6 + 2 = r^2$, $r = 2\sqrt{2}$;

$\theta = \arctan\dfrac{y}{x}$, $\theta = \arctan\dfrac{\sqrt{2}}{-\sqrt{6}}$, $\theta = \arctan\left(-\dfrac{1}{\sqrt{3}}\right)$, $\theta = \dfrac{5\pi}{6}$, $\left(2\sqrt{2}, \dfrac{5\pi}{6}\right)$

51. $\left(4, \dfrac{3\pi}{2}\right)$ **53.** $r\cos\theta = 3$ **55.** $r = 6$

57. $x^2 + y^2 + 2x + 5y = 0$, $r^2 + 2r\cos\theta + 5r\sin\theta = 0$, $r + 2\cos\theta + 5\sin\theta = 0$

59. $4r\cos\theta - 3r\sin\theta = 12$, $r = \dfrac{12}{(4\cos\theta - 3\sin\theta)}$.

Section 20.8

61. $9(r\cos\theta)^2 + 4(r\sin\theta)^2 = 36$, $9r^2\cos^2\theta + 4r^2\sin^2\theta = 36$,
$r^2(9(1-\sin^2\theta) + 4\sin^2\theta) = 36$, $r^2(9 - 5\sin^2\theta) = 36$, $r^2 = \dfrac{36}{(9-5\sin^2\theta)}$.

63. $(r\cos\theta)^3 = 4(r\sin\theta)^2$, $r^3\cos^3\theta = 4r^2\sin^2\theta$, $r = \dfrac{4\sin^2\theta}{\cos^3\theta}$, $r = 4\sec\theta\tan^2\theta$.

65. $y = -3$

67. $x^2 + y^2 = 25$

69. $\tan\theta = \tan\dfrac{\pi}{4}$, $\dfrac{y}{x} = 1$, $y = x$

71. $r^2 = 5r\cos\theta$. $x^2 + y^2 = 5x$

73. $r^2 = 6r\cos\left(\theta + \dfrac{\pi}{3}\right)$, $r^2 = 6r\left[\cos\theta\cos\dfrac{\pi}{3} - \sin\theta\sin\dfrac{\pi}{3}\right]$,
$r^2 = 6r\left[\dfrac{1}{2}\cos\theta - \dfrac{\sqrt{3}}{2}\sin\theta\right]$, $x^2 + y^2 = 3x - 3\sqrt{3}y$

75. $r\sin\theta\sin\theta = 3\cos\theta$, $y = \dfrac{3\cos\theta}{\sin\theta}$, $y = 3\cot\theta$, $y = 3\left(\dfrac{x}{y}\right)$ $y^2 = 3x$

77. $r^2(2\sin\theta\cos\theta) = 2$, $2r\sin\theta r\cos\theta = 2$, $xy = 1$

79. $r^4 = 2r\sin\theta r\cos\theta$, $\left(x_2^2 + y^2\right)^2 = 2xy$, $x^4 + 2x^2y^2 + y^4 - 2xy = 0$

81. $r^2 = \tan^2\theta$, $x^2 + y^2 = \left(\dfrac{y}{x}\right)^2$, $x^2(x^2 + y^2) = y^2$

83. $r + r\sin\theta = 3$, $\sqrt{x^2 + y^2} = 3 - y$, $x^2 + y^2 = (3-y)^2$,
$x^2 + y^2 = 9 - 6y + y^2$, $x^2 + 6y - 9 = 0$

85. $r = 4\sin(2\theta + \theta)$, $r = 4[\sin 2\theta\cos\theta + \cos 2\theta\sin\theta]$
$r = 4[2\sin\theta\cos\theta\cos\theta + \sin\theta(\cos^2\theta - \sin^2\theta)]$,
$r = 4[2\sin\theta\cos^2\theta + \sin\theta\cos^2\theta - \sin^3\theta]$,
$r = 4[3\sin\theta\cos^2\theta - \sin^3\theta]$
$r^4 = 12r\sin\theta r^2\cos^2\theta - 4r^3\sin^3\theta$
$(x^2 + y^2)^2 = 12yx^2 - 4y^3$

87. $r^2 = 2r + 4r\sin\theta$, $x^2 + y^2 = 2\left(\pm\sqrt{x^2 + y^2} + 4y\right)$

$(x^2 + y^2 - 4y)^2 = 2\left(\pm\sqrt{x^2 + y^2}\right)^2$
$x^4 + 2x^2y^2 - 8x^2y - 8y^3 + 16y^2 + y^4 = 4(x^2 + y^2)$
$x^4 + 2x^2y^2 - 8x^2y - 8y^3 + 12y^2 + y^4 - 4x^2 = 0$

Section 20.8

89. $\sqrt{13}$

91. $P_1(r_1 \cos\,\theta_1, r_1\sin\theta_1), P_2(r_2\cos\theta_2, r_2\sin\theta_2);$

$d^2 = (r_1\cos\theta_1 - r_2\cos\theta_2)^2 + (r_1\sin\theta_1 - r_2\sin\theta)^2$

$d^2 = r_1^2\cos^2\theta_1 - 2r_1\cos\theta_1\; r_2\cos_2 + r_2^2\cos^2\theta_2 + r_1^2\sin^2\theta_1 - 2r_1\sin\theta_1 r_2\sin\theta_2 + r_2^2\sin^2\theta_2$

$d^2 = r_1^2\left(\sin^2\theta_1 + \cos^2\theta_1\right) + r_2^2\left(\sin^2\theta_2 + \cos^2\theta_2\right) - 2r_1r_2(\cos\theta_1\cos\theta_2 + \sin\theta_1\sin\theta_2)$

$d^2 = r_1^2 + r_2^2 - 2r_1r_2\cos(\theta_1 - \theta_2)$

$d = \sqrt{r_1^2 + r_2^2 - 2r_1r_2\cos(\theta_1 - \theta_2)}$

Exercises 20.9

1.

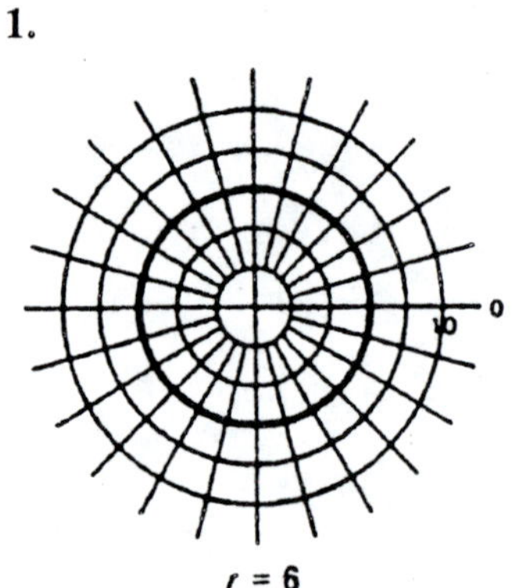

$r = 6$

3.

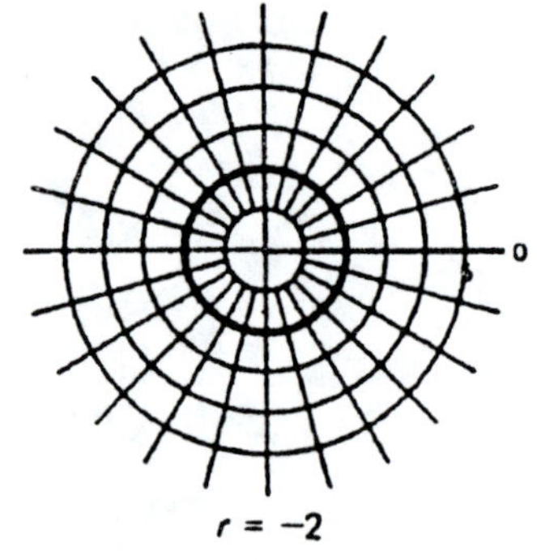

$r = -2$

5.

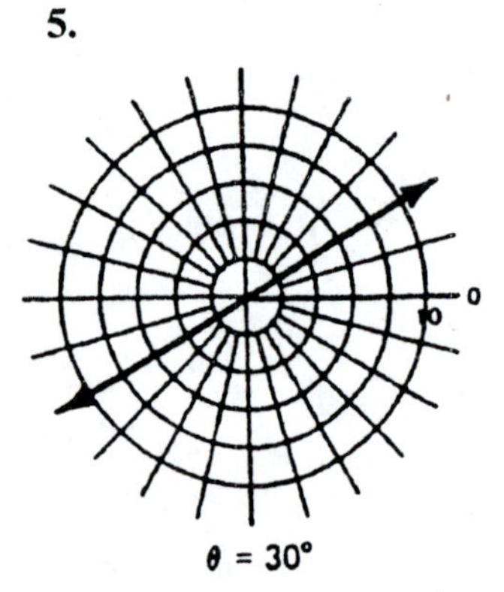

$\theta = 30°$

7.

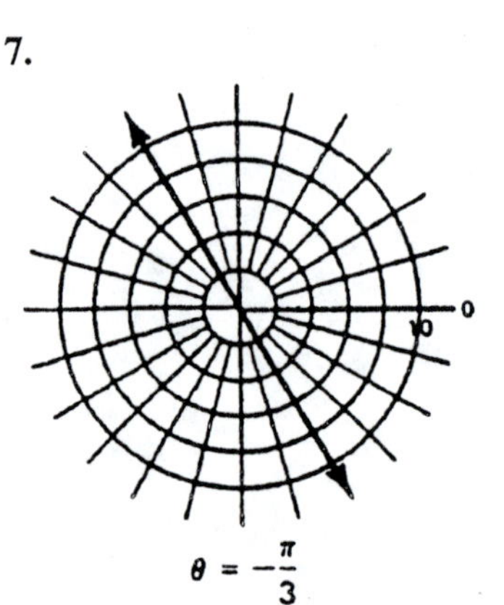

$\theta = -\frac{\pi}{3}$

9.

$r = 5\sin\theta$

11.

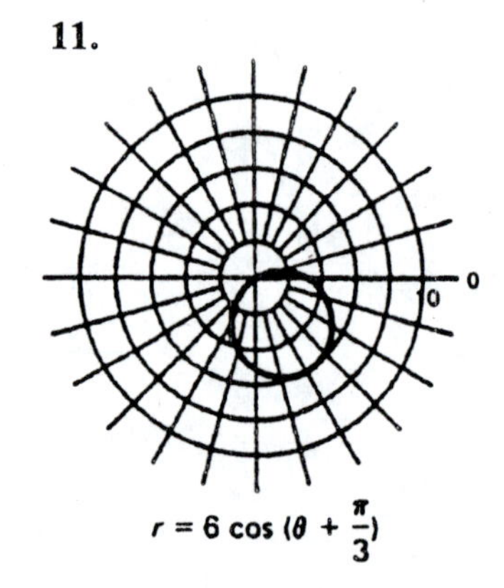

$r = 6\cos(\theta + \frac{\pi}{3})$

13.

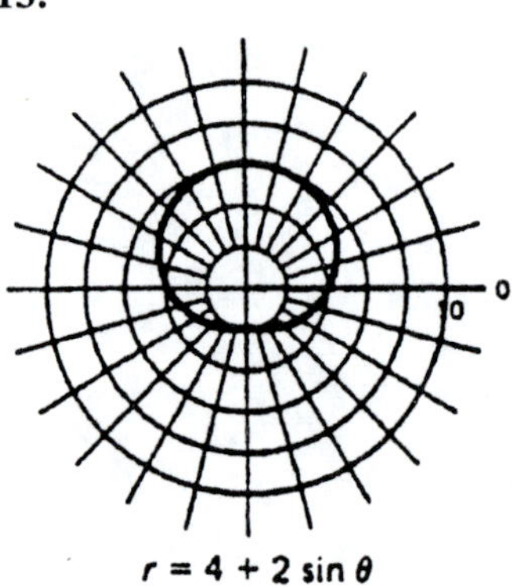

$r = 4 + 2\sin\theta$

15.

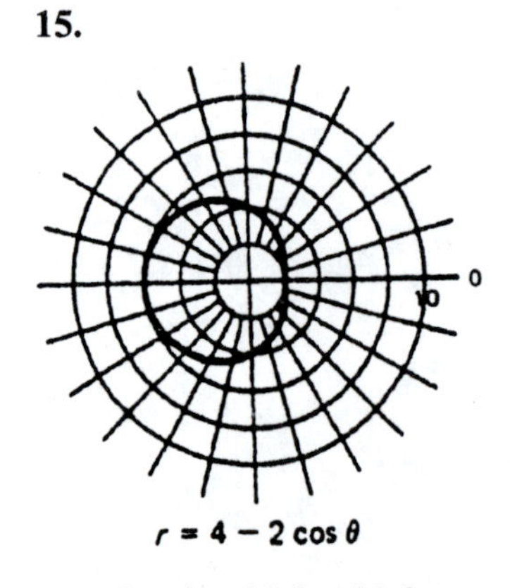

$r = 4 - 2\cos\theta$

17.

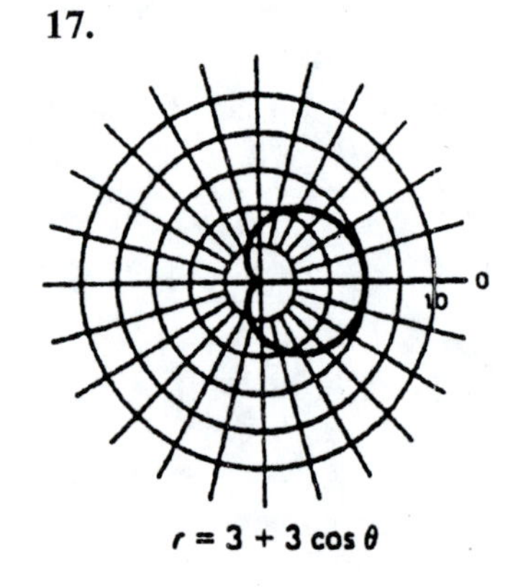

$r = 3 + 3\cos\theta$

19.

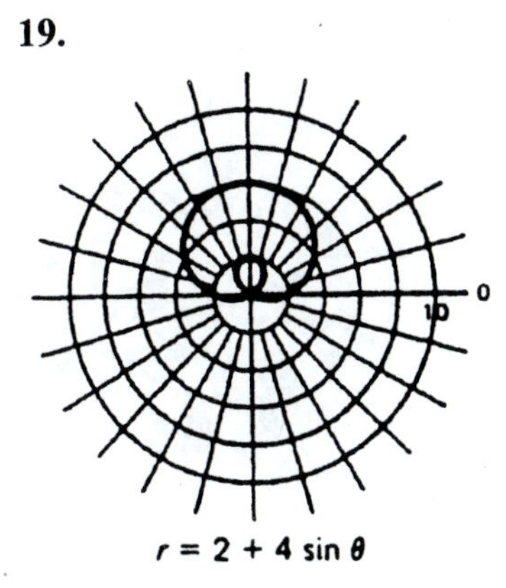

$r = 2 + 4 \sin \theta$

21.

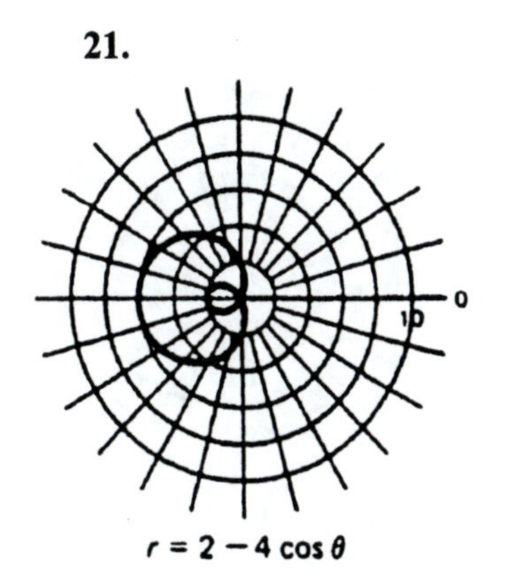

$r = 2 - 4 \cos \theta$

23.

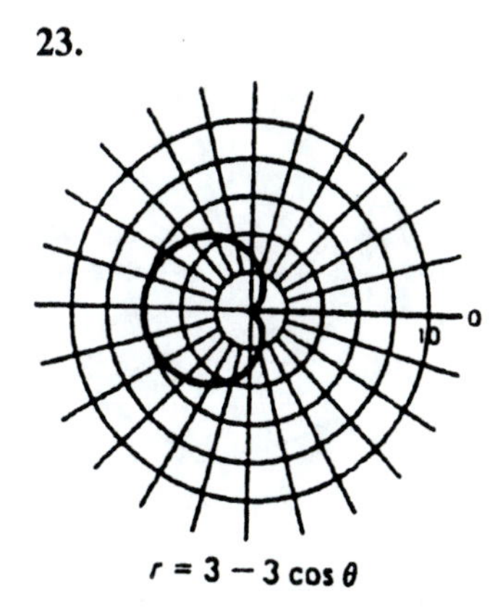

$r = 3 - 3 \cos \theta$

25.

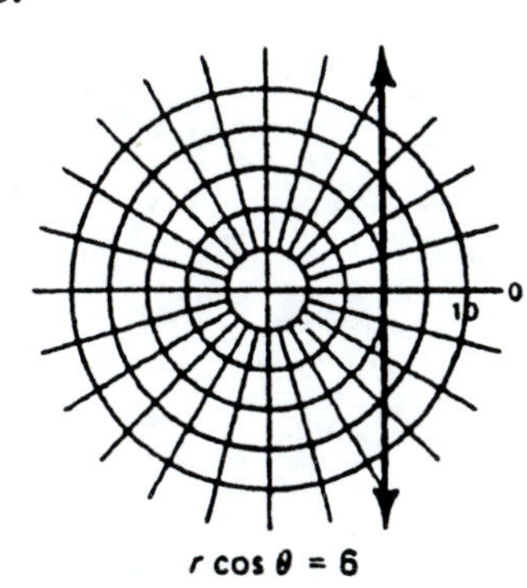

$r \cos \theta = 6$

27.

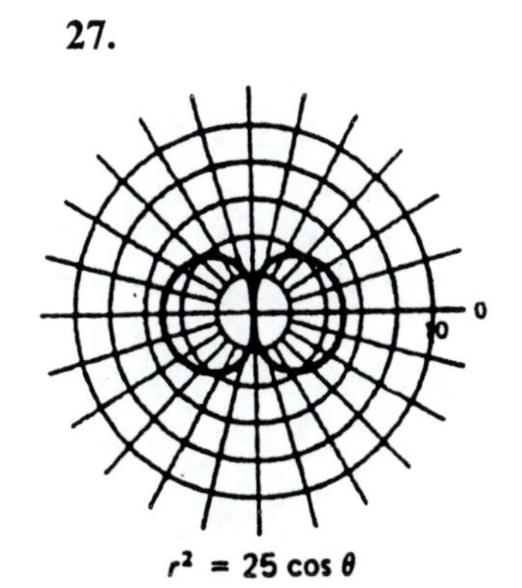

$r^2 = 25 \cos \theta$

29

$r^2 = 9 \sin 2\theta$

31.

$r^2 = -36 \cos 2\theta$

33.

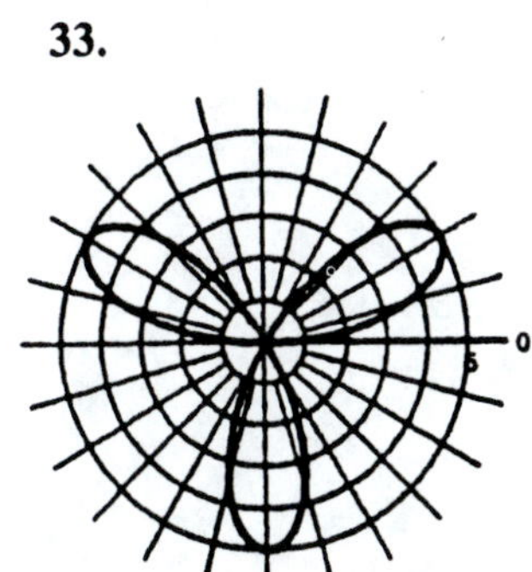

$r = 5 \sin 3\theta$

35.

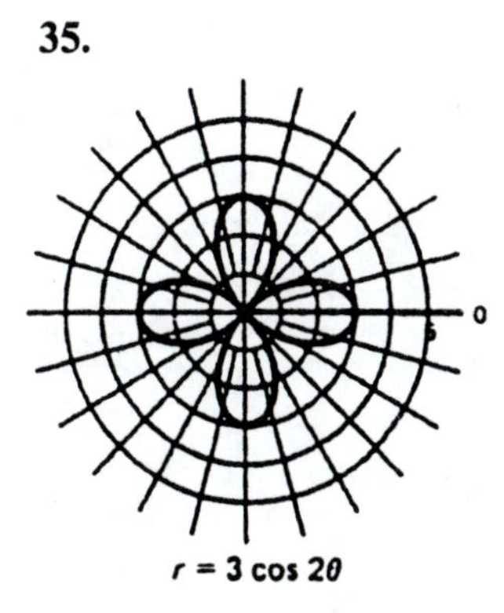

$r = 3 \cos 2\theta$

37.

$r = 9 \sin^2 \theta$

39.

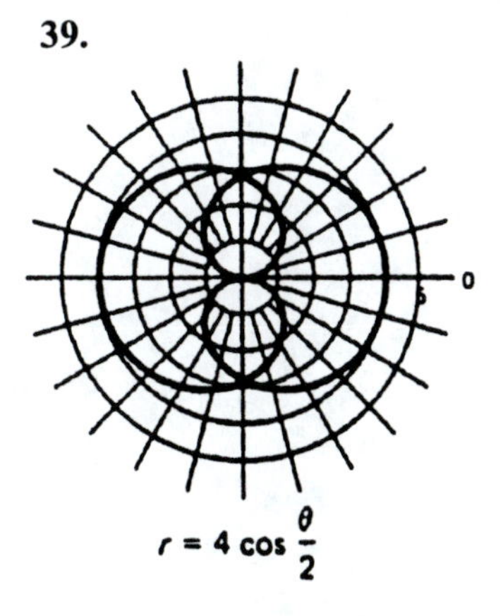

$r = 4 \cos \dfrac{\theta}{2}$

41.

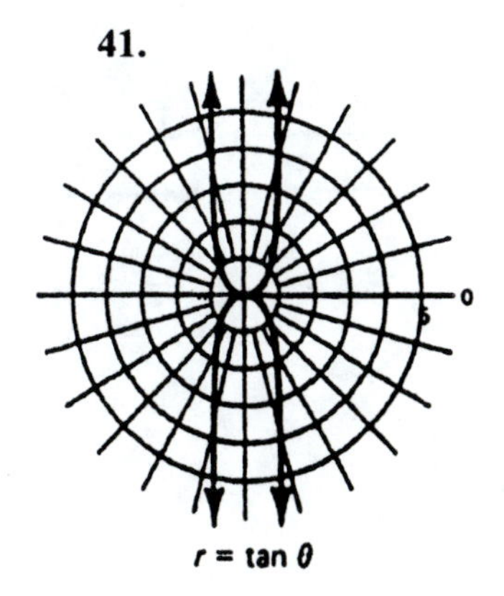

$r = \tan \theta$

Section 20.9

43.

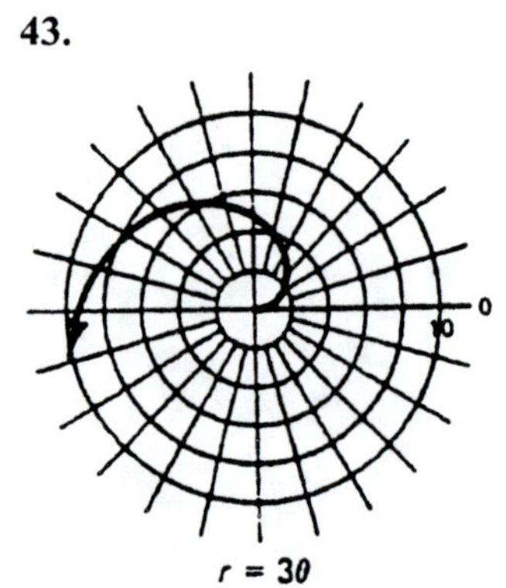

45.

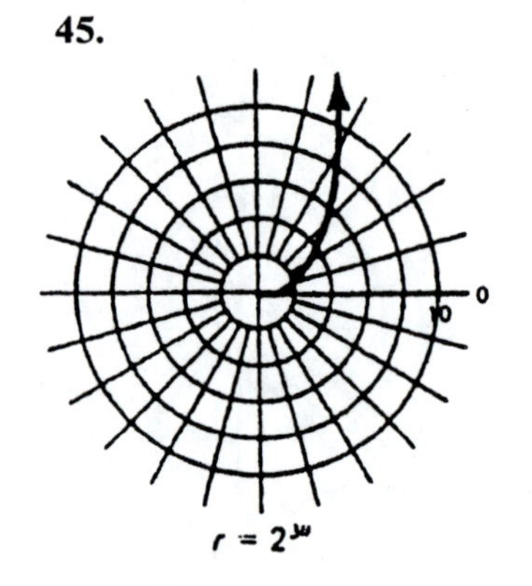

47.

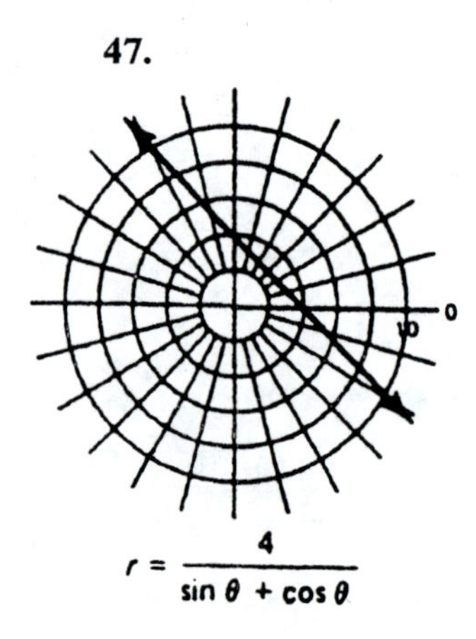

49.

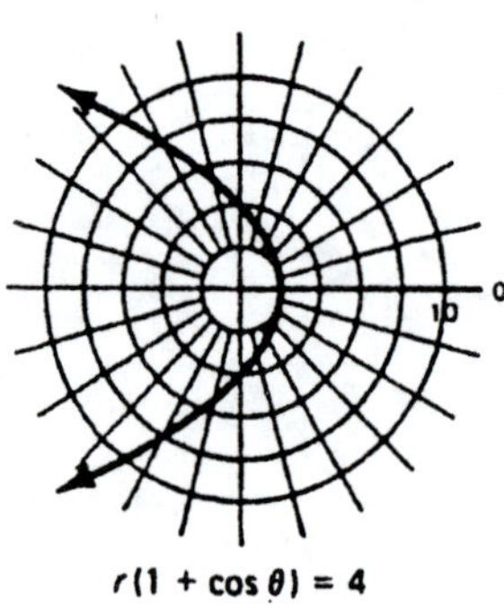

Chapter 20 Review

1. $(x-5)^2+(y+7)^2=36$
$x^2+y^2-10x+14y+38=0$

2. $x^2-8x+y^2+6y=24$
$x^2-8x+16+y^2+6y+9=24+16+9$
$(x-4)^2+(y+3)^2=49$, thus $C(4,-3)$; $r=7$

3. $F(0, 3/2)$; directrix $y=-3/2$

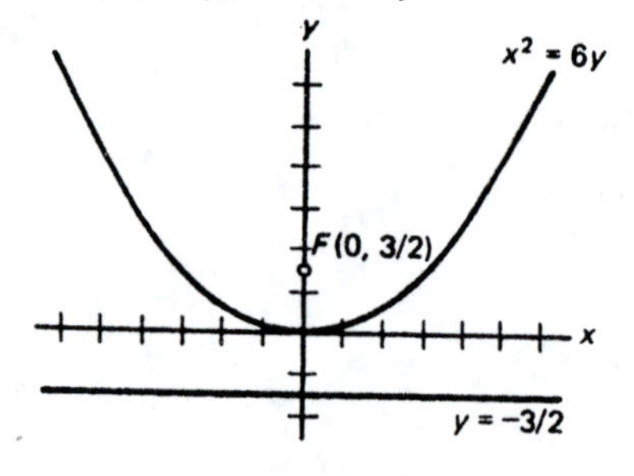

4. $y^2=-16x$

5. $(y-3)^2=8(x-2)$
$y^2-6y-8x+25=0$

6. $\dfrac{x^2}{49}+\dfrac{y^2}{4}=1$; $V(\pm 7, 0)$; $F\left(\pm 3\sqrt{5},0\right)$

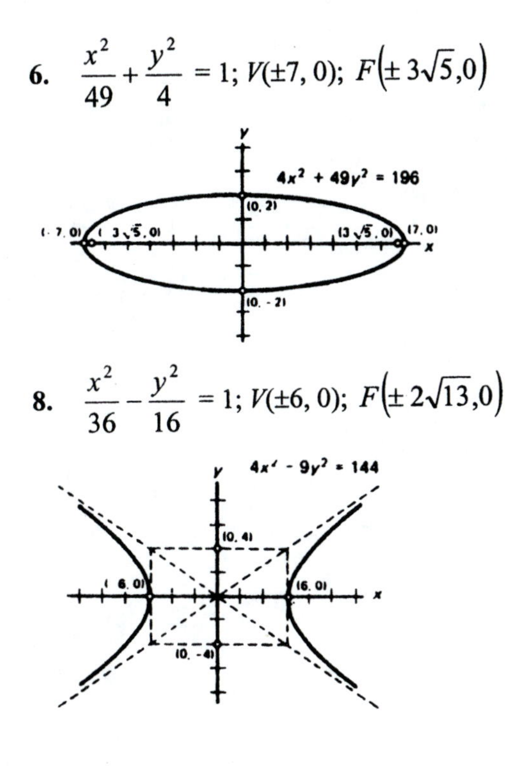

7. $\dfrac{x^2}{4}+\dfrac{y^2}{16}=1$

$4x^2+16y^2=16$

8. $\dfrac{x^2}{36}-\dfrac{y^2}{16}=1$; $V(\pm 6, 0)$; $F\left(\pm 2\sqrt{13},0\right)$

9. $\dfrac{y^2}{25}-\dfrac{x^2}{16}=1$

$16y^2-25x^2=400$

10. $\dfrac{(x-3)^2}{9}+\dfrac{(y+4)^2}{25}=1$

11. $\dfrac{(x+7)^2}{81}-\dfrac{(y-4)^2}{9}=1$

12. $16x^2-64x-4y^2-24y=-12$

$16(x^2-4x)-4(y^2+6y)=-12$

$16(x^2-4x+4)-4(y^2+6y+9)$

$=-12-36+64$

$16(x-2)^2-4(y+3)^2=16$

$\dfrac{(x-2)^2}{1}-\dfrac{(y+3)^2}{4}=1$

hyperbola; $C(2,-3)$

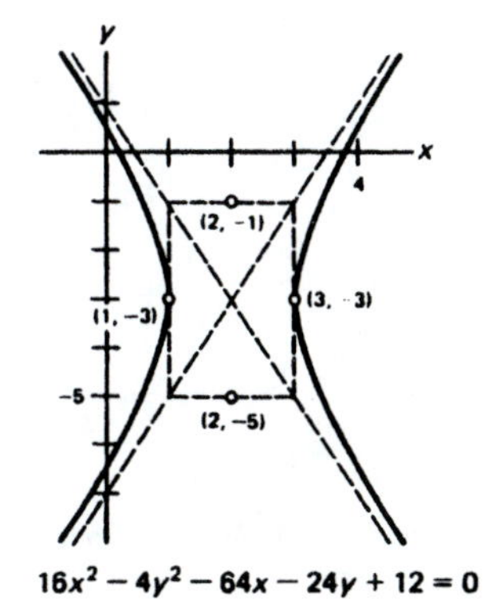

13. $y^2+4y+x=0$

$x=2y$

$y^2+4y+2y=0$

$y^2+6y=0$

$y(y+6)=0$

$y=0$ or -6

$(0, 0), (-12, -6)$

14. $3x^2-4y^2=36$

$5x^2-8y^2=56$

$-6x^2+8y^2=-72$

$5x^2-8y^2=56$

$-x^2=-16$

$x=4$ or -4

$\left(4\sqrt{3}\right), \left(4,-\sqrt{3}\right), \left(-4,\sqrt{3}\right), \left(-4,-\sqrt{3}\right)$

15.

16.

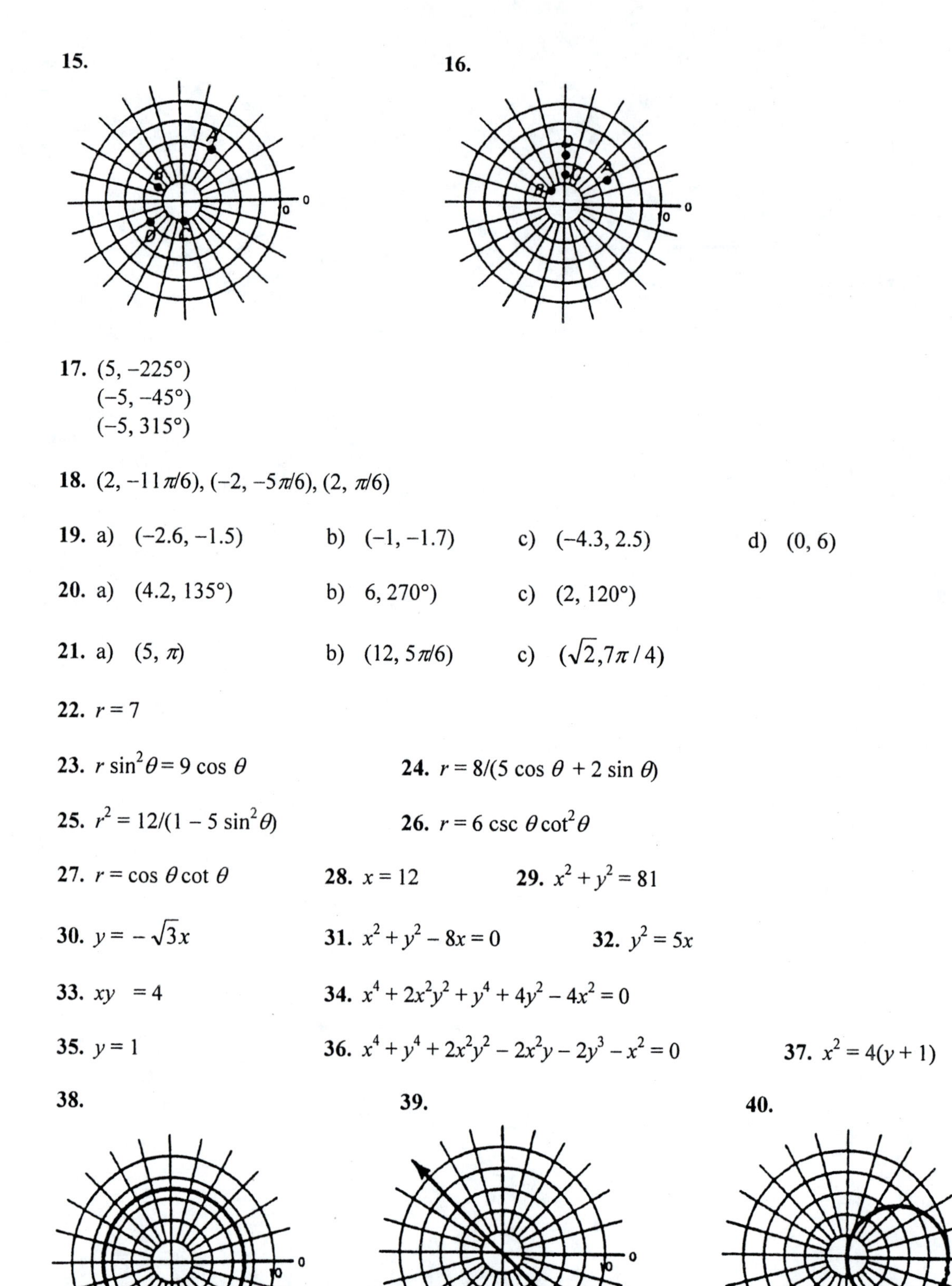

17. $(5, -225°)$
$(-5, -45°)$
$(-5, 315°)$

18. $(2, -11\pi/6)$, $(-2, -5\pi/6)$, $(2, \pi/6)$

19. a) $(-2.6, -1.5)$ b) $(-1, -1.7)$ c) $(-4.3, 2.5)$ d) $(0, 6)$

20. a) $(4.2, 135°)$ b) $6, 270°)$ c) $(2, 120°)$

21. a) $(5, \pi)$ b) $(12, 5\pi/6)$ c) $(\sqrt{2}, 7\pi/4)$

22. $r = 7$

23. $r \sin^2\theta = 9 \cos \theta$ **24.** $r = 8/(5 \cos \theta + 2 \sin \theta)$

25. $r^2 = 12/(1 - 5 \sin^2\theta)$ **26.** $r = 6 \csc \theta \cot^2\theta$

27. $r = \cos \theta \cot \theta$ **28.** $x = 12$ **29.** $x^2 + y^2 = 81$

30. $y = -\sqrt{3}x$ **31.** $x^2 + y^2 - 8x = 0$ **32.** $y^2 = 5x$

33. $xy = 4$ **34.** $x^4 + 2x^2y^2 + y^4 + 4y^2 - 4x^2 = 0$

35. $y = 1$ **36.** $x^4 + y^4 + 2x^2y^2 - 2x^2y - 2y^3 - x^2 = 0$ **37.** $x^2 = 4(y + 1)$

38. **39.** **40.**

$r = 7$ $\theta = -\pi/4$ $r = 5 \cos \theta$

Chapter 20 Review

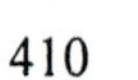

41.

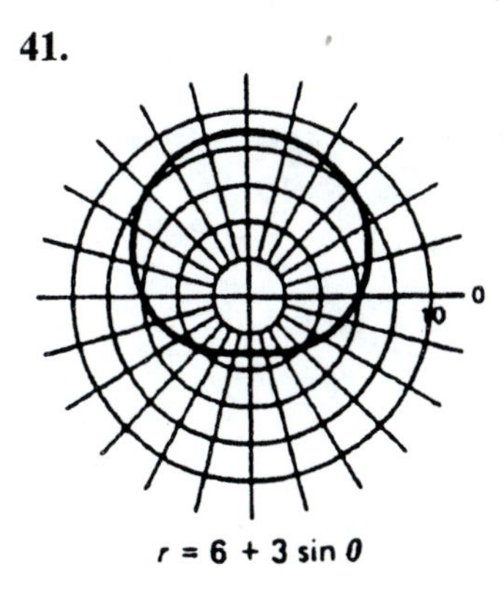

$r = 6 + 3 \sin \theta$

42.

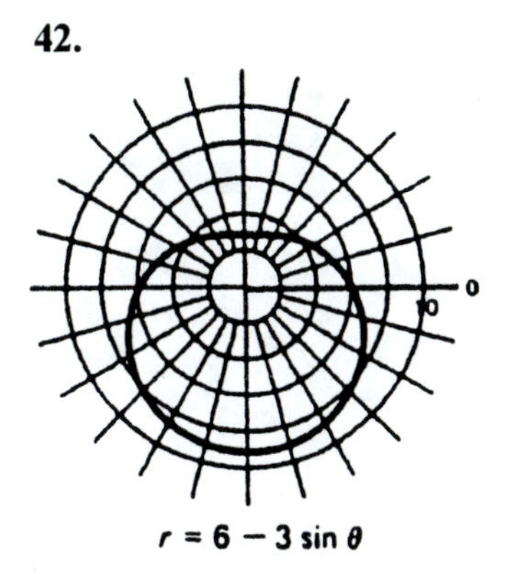

$r = 6 - 3 \sin \theta$

43.

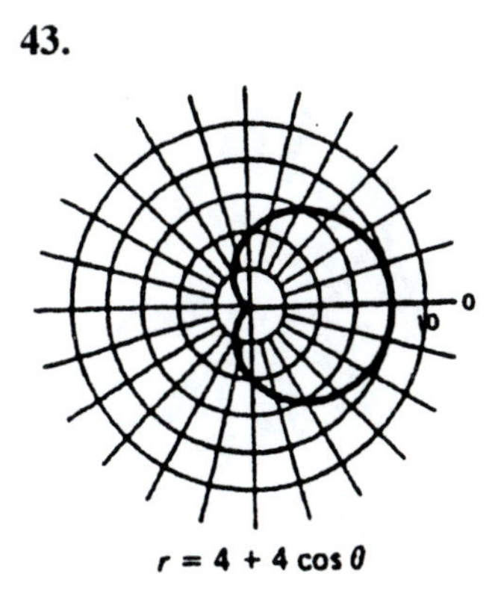

$r = 4 + 4 \cos \theta$

44.

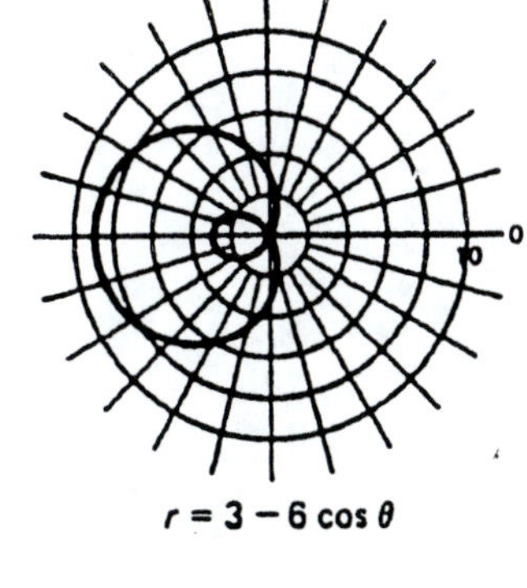

$r = 3 - 6 \cos \theta$

45.

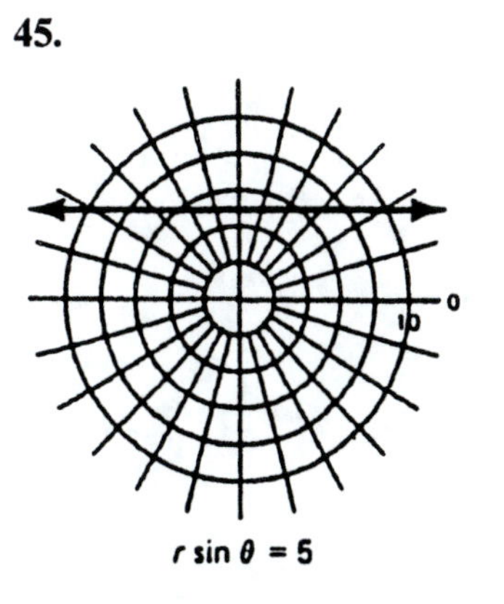

$r \sin \theta = 5$

46.

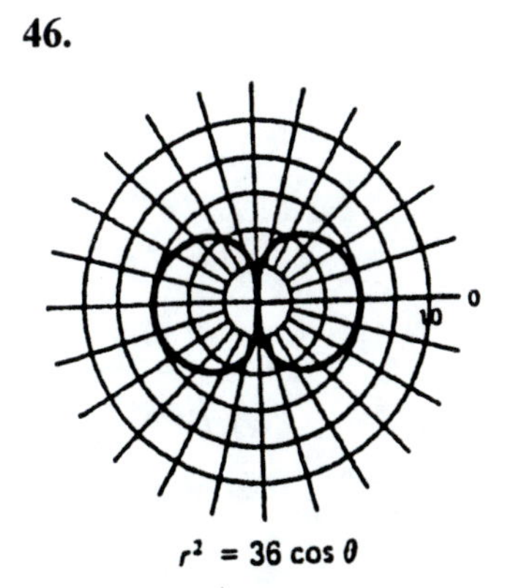

$r^2 = 36 \cos \theta$

47.

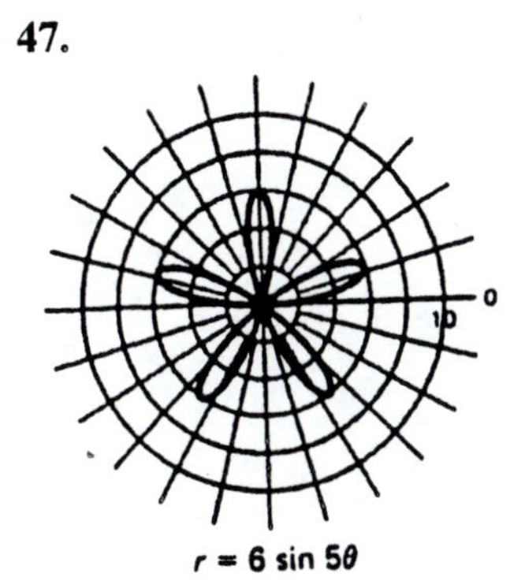

$r = 6 \sin 5\theta$

48.

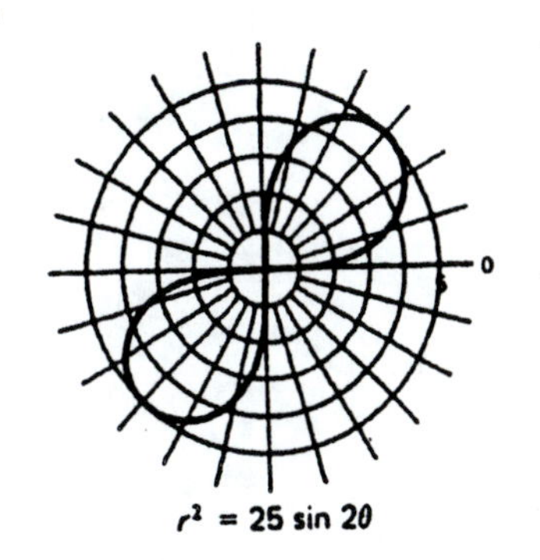

$r^2 = 25 \sin 2\theta$

49.

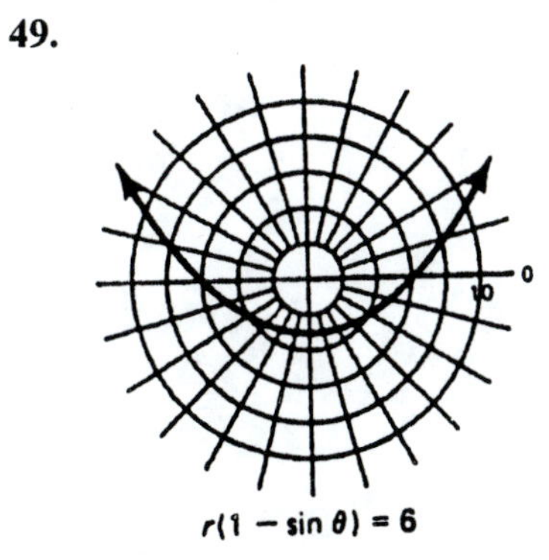

$r(1 - \sin \theta) = 6$

Chapter 21, Exercises 21.1

1. $f(1) = 2(1)^2 + 7 = 9$

3. $h(-2) = 3(-2)^3 - 2(-2) + 4 = -16$

5. $f(-2) = \dfrac{2^2 - 3}{2+5} = \dfrac{1}{7}$

7. $f(-5) = \sqrt{(-5)^2 + 3} = 2\sqrt{7}$

9. $f(h + 3) = 3(h + 3) - 2 = 3h + 7$

11. $f(2 + \Delta t) = 3(2 + \Delta t)^2 + 2(2 + \Delta t) - 5 = 11 + 14\Delta t + 3(\Delta t)^2$

13. a) $\Delta s = f(t + \Delta t) - f(t) = [3(t + \Delta t) - 4] - [3t - 4] = 3\Delta t$

b) $\bar{v} = \dfrac{\Delta s}{\Delta t} = \dfrac{3\Delta t}{\Delta t} = 3$

15. a) $\Delta s = [2(t + \Delta t)^2 + 5] - [2t^2 + 5] = 4t\,\Delta t + 2(\Delta t)^2$

b) $\bar{v} = \dfrac{4t\Delta t + 2(\Delta t)^2}{\Delta t} = 4t + 2\Delta t$

17. a) $\Delta s = [(t + \Delta t)^2 - 2(t + \Delta t) + 8] - [t^2 - 2t + 8] = 2t\,\Delta t + (\Delta t)^2 - 2\Delta t$

b) $\bar{v} = \dfrac{2t\Delta t + (\Delta t)^2 - 2\Delta t}{\Delta t} = 2t + \Delta t - 2$

19. $\bar{v} = \dfrac{\Delta s}{\Delta t} = \dfrac{f(7) - f(3)}{4} = \dfrac{[5(7)^2 + 6] - [5(3)^2 + 2 + 4]}{4} = \dfrac{251 - 51}{4} = 50 \text{ m/s}$

$t = 3, \Delta t = 4$

21. $\bar{v} = \dfrac{\Delta s}{\Delta t} = \dfrac{f(4) - f(2)}{2} = \dfrac{[3(4)^2 - 4 + 4] - [3(2)^2 - 2 + 4]}{2} = \dfrac{48 - 14}{2} = 17 \text{ m/s}$

$t = 2, \Delta t = 2$

23. $\bar{v} = \dfrac{\Delta s}{\Delta t} = \dfrac{0.5 \text{ m} - 0.2 \text{ m}}{0.5 \times 10^{-6} \text{ s}} = 6 \times 10^5 \text{ m/s}$

25. $i_{av} = C\dfrac{\Delta V}{\Delta t} = 10\ \mu\text{F}\left[\dfrac{(5^2 + 3(5) + 160) - (2^2 + 3(2) + 160)}{3s}\right] = 100\ \mu\text{A}$

27. $\bar{v} = \dfrac{f(2 + \Delta t) - f(t)}{\Delta t} = \dfrac{[3(2 + \Delta t)^2 - 6(2 + \Delta t) + 1] - [3(2)^2 - 6(2) + 1]}{\Delta t}$

$= \dfrac{12\Delta t + 3(\Delta t)^2 - 6\Delta t}{\Delta t} = 6 + 3\Delta t$, when $\Delta t \to 0$, $v = 6$ m/s

29. $\bar{v} = \dfrac{f(1+\Delta t) - f(1)}{\Delta t} = \dfrac{[5(1+\Delta t)^2 - 7] - [5(1)^2 - 7]}{\Delta t} = \dfrac{10\Delta t + 5(\Delta t)^2}{\Delta t}$
$= 10 + 5\Delta t$, when $\Delta t \to 0$, $v = 10$ m/s

31. $\bar{v} = \dfrac{f(3+\Delta t) - f(3)}{\Delta t} = \dfrac{\dfrac{1}{2(3+\Delta t)} - \dfrac{1}{2(3)}}{\Delta t} = \dfrac{1}{\Delta t} \cdot \dfrac{-2\,\Delta t}{6(6+\Delta t)} = \dfrac{-2}{6(6+\Delta t)}$,

when $\Delta t \to 0$, $v = -\dfrac{1}{18}$ m/s

33. $\bar{v} = \dfrac{f(4+\Delta t) - f(4)}{\Delta t} = \dfrac{\dfrac{1}{(4+\Delta t) - 2} - \dfrac{1}{4-2}}{\Delta t} = \dfrac{1}{\Delta t} \cdot \dfrac{-\Delta t}{2(2+\Delta t)} = \dfrac{-1}{2(2+\Delta t)}$,

when $\Delta t \to 0$, $v = -\dfrac{1}{4}$ m/s

35. $\bar{v} = \dfrac{\Delta s}{\Delta t} = \dfrac{16(2+\Delta t)^2 - 16(2)}{\Delta t} = \dfrac{64\Delta t + 16(\Delta t)^2}{\Delta t} = 64 + 16\Delta t$, when $\Delta t \to 0$,
$v = 64$ ft/s

Section 21.2

1. $\lim\limits_{x\to 2} \dfrac{x^2 - 4}{x - 2} = \lim\limits_{x\to 2}(x+2) = 4$

3. $\lim\limits_{x\to\infty} \dfrac{3x+2}{x} = \lim\limits_{x\to\infty}(3 + 2/x) = 3$

5. $\lim\limits_{x\to 0} \dfrac{\sin x}{x} = 1$

7. $\lim\limits_{x\to 2} (x^2 - 5x) = 2^2 - 5(2) = -6$

9. $\lim\limits_{x\to -1} (2x^3 + 5x^2 - 2) = 2(-1)^3 + 5(-1)^2 - 2 = 1$

11. $\lim\limits_{x\to 1} \dfrac{x^2 - 1}{x - 1} = \lim\limits_{x\to 1}(x+1) = 2$

13. $\lim\limits_{x\to -3/2} \dfrac{4x^2 - 9}{2x + 3} = \lim\limits_{x\to -3/2}(2x - 3) = -6$

15. $\lim\limits_{x\to -1} \sqrt{2x+3} = 1$

17. $\lim\limits_{x\to 6} \sqrt{4 - x}$ = no limit (does not exist)

19. $\lim\limits_{x\to\infty} \dfrac{1}{2x} = 0$

21. $\lim\limits_{x\to\infty} \dfrac{3x^2 - 5x + 2}{4x^2 + 8x - 11} = \lim\limits_{x\to\infty} \dfrac{3 - \dfrac{5}{x} + \dfrac{2}{x^2}}{4 + \dfrac{8}{x} - \dfrac{11}{x^2}} = \dfrac{3}{4}$

23. $v = \lim_{\Delta t \to \infty} \dfrac{f(2+\Delta t) - f(2)}{\Delta t} = \lim_{\Delta t \to \infty} \dfrac{[4(2+\Delta t)^2 - 3(2+\Delta t)] - [4(2)^2 - 3(2)]}{\Delta t}$

$= \lim_{\Delta t \to 0} \dfrac{13\Delta t + 4(\Delta t)^2}{\Delta t} = \lim_{\Delta t \to 0}(13 + 4\Delta t) = 13$

25. $v = \lim_{\Delta t \to 0} \dfrac{f(4+\Delta t) - f(4)}{\Delta t} = \lim_{\Delta t \to 0} \dfrac{[(4+\Delta t)^2 + 3(4+\Delta t) - 10] - [4^2 + 2]}{\Delta t}$

$= \lim_{\Delta t \to 0} \dfrac{11\Delta t + (\Delta t)^2}{\Delta t} = \lim_{\Delta t \to 0}(11 + \Delta t) = 11$

27. $\lim_{x \to 2} x^2 + \lim_{x \to 2} x = 4 + 2 = 6$

29. $\lim_{x \to 1} 4x^2 + \lim_{x \to 1} 100x + \lim_{x \to 1}(-2) = 102$

31. $\lim_{x \to 1}(x+3) \cdot \lim_{x \to 1}(x-4) = 4(-3) = -12$

33. $\lim_{x \to -2}(x^2 + 3x + 1) \cdot \lim_{x \to -2}(x^4 - 2x^2 + 3) = (-1)(11) = -11$

35. $\dfrac{\lim_{x \to 2}(x^2 + 3x + 2)}{\lim_{x \to 2}(x^2 + 1)} = \dfrac{12}{5}$

37. $\lim_{x \to -7} \dfrac{x^2 - 49}{x + 7} = \lim_{x \to -7}(x - 7) = -14$

39. $\lim_{x \to 5/2} \dfrac{4x^2 - 25}{2x - 5} = \lim_{x \to 5/2}(2x + 5) = 10$

41. $\dfrac{\lim_{x \to 3}(x^2 + 3x + 1)\lim_{x \to 3}(x + 5)}{\lim_{x \to 3}(x - 2)} = \dfrac{(19)(8)}{1} = 152$

43. $\lim_{h \to 0} \dfrac{(x+h)^2 - x^2}{h} = \lim_{h \to 0} \dfrac{2xh + h^2}{h} = \lim_{h \to 0}(2x + h) = 2x$

45. $\lim_{h \to 0} \dfrac{\dfrac{1}{x+h} - \dfrac{1}{x}}{h} = \lim_{h \to 0} \dfrac{\dfrac{x - (x+h)}{x(x+h)}}{h} = \lim_{h \to 0} \dfrac{-1}{x(x+h)} = -\dfrac{1}{x^2}$

47. $\lim_{x \to a} \dfrac{(\sqrt{x} - \sqrt{a})(\sqrt{x} + \sqrt{a})}{(x-a)(\sqrt{x} + \sqrt{a})} = \lim_{x \to a} \dfrac{x - a}{(x-a)(\sqrt{x} + \sqrt{a})} = \lim_{x \to a} \dfrac{1}{\sqrt{x} + \sqrt{a}} = \dfrac{1}{2\sqrt{a}}$

49. Does not exist

51. b

53. Does not exist

55. b

57. no

59. no

61. no

63. no

65. Does not exist

67. 0

69. b

71. Does not exist

Section 21.2

Section 21.3

1. $m_{\tan} = \lim_{\Delta x \to 0} \frac{(3+\Delta x)^2 - 3^2}{\Delta x} = \lim_{\Delta x \to 0} \frac{6\Delta x + (\Delta x)^2}{\Delta x} = \lim_{\Delta x \to 0} (6 + \Delta x) = 6$

3. $m_{\tan} = \lim_{\Delta x \to 0} \frac{[3(-1+\Delta x)^2 - 4] - [3(-1)^2 - 4]}{\Delta x} = \lim_{\Delta x \to 0} \frac{-6(\Delta x) + 3(\Delta x)^2}{\Delta x} = \lim_{\Delta x \to 0} (-6 + 3\Delta x) = -6$

5. $m_{\tan} = \lim_{\Delta x \to 0} \frac{[2(2+\Delta x)^2 + (2+\Delta x) - 3] - [2(2)^2 + 2 - 3)]}{\Delta x} = \lim_{\Delta x \to 0} \frac{9\Delta x + 2(\Delta x)^2}{\Delta x} = \lim_{\Delta x \to 0} (9 + 2\Delta x) = 9$

7. $m_{\tan} = \lim_{\Delta x \to 0} \frac{[-4(1+\Delta x)^2 + 3(1+\Delta x) - 2] - [-4(1)^2 + 3(1) - 2]}{\Delta x}$

$= \lim_{\Delta x \to 0} \frac{-5\Delta x - 4(\Delta x)^2}{\Delta x} = \lim_{\Delta x \to 0} (-5 - 4\Delta x) = -5$

9. $m_{\tan} = \lim_{\Delta x \to 0} \frac{(2+\Delta x)^3 - 2^3}{\Delta x} = \lim_{\Delta x \to 0} \frac{12\Delta x + 6(\Delta x)^2 + (\Delta x)^3}{\Delta x} = \lim_{\Delta x \to 0} [12 + 6\Delta x + (\Delta x)^2] = 12$

11. $m_{\tan} = -4, (-2, 4)$
$y - 4 = -4(x + 2)$
$4x + y + 4 = 0$

13. $m_{\tan} = -8, (-2, 5)$
$y - 5 = -8(x + 2)$
$8x + y + 11 = 0$

15. $m_{\tan} = -13, (-1, 10)$
$y - 10 = -13(x + 1)$
$13x + y + 3 = 0$

17. $m_{\tan} = -13$; $x = 3$ then $y = -10$; $y + 10 = -13(x - 3)$ or $13x + y - 29 = 0$

19. $m_{\tan} = 4$, $x = 1$ then $y = 1$; $y - 1 = 4(x - 1)$; $4x - y - 3 = 0$

21. $m_{\tan} = \lim_{\Delta x \to 0} \frac{(x+\Delta x)^2 - x^2}{\Delta x} = \lim_{\Delta x \to 0} \frac{2x(\Delta x) + (\Delta x)^2}{\Delta x} = \lim_{\Delta x \to 0} (2x + \Delta x) = 2x$

$m_{\tan} = -1/3$; thus $2x = -1/3$ so $x = -1/6$ and $(-1/6, 1/36)$

23. $m_{\tan} = \lim_{\Delta x \to 0} \frac{[(x+\Delta x)^3 + (x+\Delta x)] - [x^3 + x]}{\Delta x} = \lim_{\Delta x \to 0} \frac{3x^2(\Delta x) + 3x(\Delta x)^2 + (\Delta x)^3 + \Delta x}{\Delta x}$

$= \lim_{\Delta x \to 0} [3x^2 + 3x(\Delta x) + (\Delta x)^2 + 1] = 3x^2 + 1$

$m_{\tan} = 4$ thus $3x^2 + 1 = 4$ so $x = \pm 1$; $(1, 2)$ and $(-1, -2)$

Section 21.4

Note: Samples of the different kinds of exercises are given here. The other exercises are done in a similar manner.

1. $\dfrac{dy}{dx} = \lim_{\Delta x \to 0} \dfrac{f(x+\Delta x) - f(x)}{\Delta x} = \lim_{\Delta x \to 0} \dfrac{[3(x+\Delta x)+4] - [3x+4]}{\Delta x} = \lim_{\Delta x \to 0} \dfrac{3\Delta x}{\Delta x} = \lim_{\Delta x \to 0} 3 = 3$

3. -2 **5.** $6x$ **7.** $2x - 2$

9. $\dfrac{dy}{dx} = \lim_{\Delta x \to 0} \dfrac{f(x+\Delta x) - f(x)}{\Delta x} = \lim_{\Delta x \to 0} \dfrac{[3(x+\Delta x)^2 - 4(x+\Delta x) + 1] - [3x^2 - 4x + 1]}{\Delta x}$

$$= \lim_{\Delta x \to 0} \frac{3x^2 + 6x\Delta x + 3(\Delta x)^2 - 4x - 4\Delta x + 1 - 3x^2 + 4x - 1}{\Delta x}$$

$$= \lim_{\Delta x \to 0} \frac{6x(\Delta x) + 3(\Delta x)^2 - 4\Delta x}{\Delta x} = \lim_{\Delta x \to 0} [6x + 3\Delta x - 4] = 6x - 4$$

11. $-12x$ **13.** $3x^2 + 4$ **15.** $-1/x^2$

17. $\dfrac{dy}{dx} = \lim_{\Delta x \to 0} \dfrac{f(x+\Delta x) - f(x)}{\Delta x} = \lim_{\Delta x \to 0} \dfrac{\dfrac{2}{(x+\Delta x)-3} - \dfrac{2}{x-3}}{\Delta x} = \lim_{\Delta x \to 0} \dfrac{2(x-3) - 2(x+\Delta x-3)}{(\Delta x)(x-3)(x+\Delta x-3)}$

$$= \lim_{\Delta x \to 0} \frac{-2\Delta x}{(\Delta x)(x-3)(x+\Delta x-3)} = \lim_{\Delta x \to 0} \frac{-2}{(x-3)(x+\Delta x-3)} = \frac{-2}{(x-3)^2}$$

19. $-2/x^3$ **21.** $2x/(4 - x^2)^2$

23. $\dfrac{dy}{dx} = \lim_{\Delta x \to 0} \dfrac{f(x+\Delta x) - f(x)}{\Delta x} = \lim_{\Delta x \to 0} \dfrac{\sqrt{(x+\Delta x)+1} - \sqrt{x+1}}{\Delta x}$

$$= \lim_{\Delta x \to 0} \left[\frac{\sqrt{(x+\Delta x)+1} - \sqrt{x+1}}{\Delta x} \cdot \frac{\sqrt{(x+\Delta x)+1} + \sqrt{x+1}}{\sqrt{(x+\Delta x+1} + \sqrt{x+1}} \right]$$

$$= \lim_{\Delta x \to 0} \frac{[(x+\Delta x)+1] - [x+1]}{\Delta x\left[\sqrt{(x+\Delta x)+1} + \sqrt{x+1}\right]} = \lim_{\Delta x \to 0} \frac{\Delta x}{\Delta x\sqrt{x+\Delta x+1} + \sqrt{x+1}}$$

$$= \lim_{\Delta x \to 0} \frac{1}{\sqrt{x+\Delta x+1} + \sqrt{x+1}} = \frac{1}{2\sqrt{x+1}}$$

25. $-1/\sqrt{1-2x}$

27. $\frac{dy}{dx} = \lim_{\Delta x \to 0} \frac{f(x+\Delta x) - f(x)}{\Delta x} = \lim_{\Delta x \to 0} \frac{\frac{1}{\sqrt{x+\Delta x - 1}} - \frac{1}{\sqrt{x-1}}}{\Delta x}$

$= \lim_{\Delta x \to 0} \frac{\sqrt{x-1} - \sqrt{x+\Delta x - 1}}{\Delta x \sqrt{x+\Delta x - 1}\sqrt{x-1}} \cdot \frac{\sqrt{x-1} + \sqrt{x+\Delta x - 1}}{\sqrt{x-1} + \sqrt{x+\Delta x - 1}}$

$= \lim_{\Delta x \to 0} \frac{(x-1) - (x+\Delta x - 1)}{\Delta x \sqrt{x+\Delta x - 1}\sqrt{x-1}\,(\sqrt{x-1} + \sqrt{x+\Delta x - 1})}$

$= \lim_{\Delta x \to 0} \frac{-\Delta x}{\Delta x \sqrt{x+\Delta x - 1}\sqrt{x-1}\,(\sqrt{x-1} + \sqrt{x+\Delta x - 1})}$

$= \lim_{\Delta x \to 0} \frac{-1}{\sqrt{x+\Delta x - 1}\sqrt{x-1}\,(\sqrt{x-1} + \sqrt{x+\Delta x - 1})}$

$= \frac{-1}{2(x-1)\sqrt{x-1}} = \frac{-1}{2(x-1)^{3/2}}$

29. 6 **31.** 3/2 **33.** $y = x - 4$ **35.** $x - 4y = 3$

37. (4, 1), (2, −1) **39.** (−3, 1)

Section 21.5

1. 0 **3.** $5x^4$ **5.** 4 **7.** −3 **9.** $10x$ **11.** $2x - 3$

13. $8x - 3$ **15.** $-16x$ **17.** $9x^2 + 4x - 6$ **19.** $20x^4 - 6x^2 + 1$

21. $20x^7 - 6x^4 + 30x^3 - 3x^2$ **23.** $4\sqrt{7}x^3 - 3\sqrt{5}x^2 - \sqrt{3}$

25. $f'(x) = 6x + 2$
$f'(-1) = 6(-1) + 2 = -4$

27. $f'(x) = 6x^2 - 12x + 2$
$f'(-3) = 6(-3)^2 - 12(-3) + 2 = 92$

29. $f'(x) = 20x^4 + 6x$
$f'(1) = 20(1)^4 + 6(1) = 26$

31. $f'(x) = 20x^3 + 24x^2 + 2$
$f'(0) = 2$

33. $f'(x) = -16x - 15x^2 + 30x^5$
$f'(3) = -16(3) - 15(3)^2 + 30(3)^5 = 7107$

35. $m_{\tan} = 3x^2 + 8x - 1 \Big|_{x=-2} = -5$; $(-2, 12)$ $y - 12 = -5(x + 2)$
$5x + y - 2 = 0$

37. $p = 30i^2$; $\frac{dp}{dt} = 60i \Big|_{i=2} = 120$ W/A

39. $\frac{dV}{dr} = i \Big|_{i=0.4} = 0.4$ V/Ω

41. $\frac{3}{2}x^{1/2}$

43. $-4x^{-5}$

45. $120x^{19}$

47. $-112x^{-9}$

49. $-\frac{5}{3}x^{-4/3}$

51. $\frac{dV}{dr} = i \Big|_{i=0.5} = 0.5$ V/Ω

Section 21.6

1. $y' = x^2(2) + (2x + 1)(2x) = 6x^2 + 2x$
or $y = 2x^3 + x^2$, thus $y' = 6x^2 + 2x$

3. $y' = 2x(8x + 3) + 2(4x^2 + 3x - 5)$ or $y = 8x^3 + 6x^2 - 10x$
$= 24x^2 + 12x - 10$ $\quad y' = 24x^2 + 12x - 10$

5. $y' = 5(2x + 3) + 2(5x - 4)$ or $y = 10x^2 + 7x - 12$
$= 20x + 7$ $\quad y' = 20x + 7$

7. $y' = (4x + 7)(2x) + (x^2 - 1)(4)$ or $y = 4x^3 + 7x^2 - 4x - 7$
$= 12x^2 + 14x - 4$ $\quad y' = 12x^2 + 14x - 4$

9. $y' = (x^2 + 3x + 4)(3x^2 - 4) + (x^3 - 4x)(2x + 3)$
$= 5x^4 + 12x^3 - 24x - 16$

11. $y' = (x^4 - 3x^2 - x)(6x^2 - 4) + (2x^3 - 4x)(4x^3 - 6x - 1)$
$= 14x^6 - 50x^4 - 8x^3 + 36x^2 + 8x$

13. $y' = \frac{(2x+5)((1) - x(2)}{(2x+5)^2} = \frac{5}{(2x+5)^2}$

15. $y' = \frac{(x^2+x)(0) - (1)(2x+1)}{(x^2+x)^2} = \frac{-2x-1}{(x^2+x)^2}$

17. $y' = \frac{3(2x+4) - 2(3x-1)}{(2x+4)^2} = \frac{14}{(2x+4)^2} = \frac{7}{2(x+2)^2}$

19. $y' = \frac{(2x+1)(2x) - x^2(2)}{(2x+1)^2} = \frac{2x^2+2x}{(2x+1)^2}$

21. $y' = \frac{(x^2+x+1)(1) - (x-1)(2x+1)}{(x^2+x+1)^2} = \frac{-x^2+2x+2}{(x^2+x+1)^2}$

23. $$y' = \frac{(3x^3-4x^2)(8x)-(4x^2+9)(9x^2-8x)}{(3x^3-4x^2)^2} = \frac{-12x^4-81x^2+72x}{(3x^3-4x^2)^2} = \frac{-12x^3-81x+72}{x^3(3x-4)^2}$$

25. $f'(x) = (x^2 - 4x + 3)(3x^2 - 5) + (x^3 - 5x)(2x - 4) = 5x^4 - 16x^3 - 6x^2 + 40x - 15$
$f'(2) = 5(2)^4 - 16(2)^3 - 6(2)^2 + 40(2) - 15 = -7$

27. $$f'(x) = \frac{(x+2)(3)-(3x-4)(1)}{(x+2)^2} = \frac{10}{(x+2)^2}; f'(-1) = 10$$

29. $$y' = \frac{(2-5x)(1)-(x-3)(-5)}{(2-5x)^2} = \frac{-13}{(2-5x)^2}; m_{\tan} = f'(2) = -\frac{13}{64}$$

31. $$y' = \frac{(x-2)-(x+3)}{(x-2)^2} = \frac{-5}{(x-2)^2}; m_{\tan} = f'(3) = -5 \text{ at } (3, 6)$$
then $y - 6 = -5(x - 3)$ or $5x + y - 21 = 0$

33. $$\left.\frac{dV}{dt}\right|_{t=3} = i\frac{dr}{dt} + r\frac{di}{dt} = \left.(6 + 0.02t^3)(-0.05) + (20 - 0.05t)(0.6t^2)\right|_{t=3}$$
$$= 10.4 \text{ V/s}$$

Exercises 21.7

1. $y' = 40(4x + 3)^{39}(4) = 160(4x + 3)^{39}$

3. $y' = 5(3x^2 - 7x + 4)^4(6x - 7) = 5(6x - 7)(3x^2 - 7x + 4)^4$

5. $$y' = -4(x^3 + 3)^{-5}(3x^2) = \frac{-12x^2}{(x^3+3)^5}$$

7. $$y' = \frac{1}{2}(5x^2 - 7x + 2)^{-1/2}(10x - 7) = \frac{10x-7}{2\sqrt{5x^2-7x+2}}$$

9. $$y' = \frac{2}{3}(8x^3 + 3x)^{-1/3}(24x^2 + 3) = \frac{16x^2+2}{\sqrt[3]{8x^3+3x}}$$

11. $$y' = -\frac{3}{4}(2x+3)^{-7/4}(2) = \frac{-3}{2(2x+3)^{7/4}}$$

13. $$\begin{aligned} y' &= 3x\left[4(4x+5)^3(4)\right] + (4x+5)^4(3) \\ &= 48x(4x+5)^3 + 3(4x+5)^4 \\ &= 3(4x+5)^3[16x + (4x+5)] \\ &= 3(4x+5)^3[20x+5] \\ &= 15(4x+5)^3(4x+1) \end{aligned}$$

15. $y' = x^3\left[3(x^3 - x)^2(3x^2 - 1)\right] + (x^3 - x)^3(3x^2)$
$= 3x^3(3x^2 - 1)(x^3 - x)^2 + 3x^2(x^3 - x)^3$
$= 3x^2(x^3 - x)^2\left[x(3x^2 - 1) + (x^3 - x)\right]$
$= 3x^2(x^3 - x)^2\left[3x^2 - x + x^3 - x\right]$
$= 3x^2(x^3 - x)^2(4x^3 - 2x)$
$= 6x^2(x[x^2 - 1])^2(x[2x^2 - 1])$
$= 6x^5(x^2 - 1)^2(2x^2 - 1)$

17. $y' = (2x + 1)^2\left[2(x^2 + 1)(2x)\right] + (x^2 + 1)^2[2(2x + 1)(2)]$
$= 4x(2x + 1)^2(x^2 + 1) + 4(2x + 1)(x^2 + 1)^2$
$= 4(2x + 1)(x^2 + 1)\left[x(2x + 1) + (x^2 + 1)\right]$
$= 4(2x + 1)(x^2 + 1)\left[2x^2 + x + x^2 + 1\right]$
$= 4(2x + 1)(x^2 + 1)(3x^2 + x + 1)$

19. $y' = (x^2 + 1)\left(\frac{1}{2}\right)(9x^2 - 2x)^{-1/2}(18x - 2) + (9x^2 - 2x)^{1/2}(2x)$

$= (9x^2 - 2x)^{-1/2}[(9x - 1)(x^2 + 1) + 2x(9x^2 - 2x)] = \frac{27x^3 - 5x^2 + 9x - 1}{\sqrt{9x^2 - 2x}}$

21. $y' = (3x + 4)^{3/4}(8x) + (4x^2 + 8)\left(\frac{3}{4}\right)(3x + 4)^{-1/4}(3)$

$= (3x + 4)^{-1/4}[(3x + 4)(8x) + 9(x^2 + 2)] = \frac{33x^2 + 32x + 18}{(3x + 4)^{1/4}}$

23. $y' = -4x^{-5} - 4(2x + 1)^3(2)$
$= \frac{-4}{x^5} - 8(2x + 1)^3$

25. $y' = \frac{(3x - 1)^2(10x) - 5x^2(2)(3x - 1)(3)}{(3x - 1)^4}$

$= \frac{10x(3x - 1)[(3x - 1) - 3x]}{(3x - 1)^4} = \frac{10x[-1]}{(3x - 1)^3} = \frac{-10x}{(3x - 1)^3}$

27. $y' = \frac{(4x^2 - 3x)(4)(x^3 + 2)^3(3x^2) - (x^3 + 2)^4(8x - 3)}{(4x^2 - 3x)^2}$

$= \frac{(x^3 + 2)^3[12x^2(4x^2 - 3x) - (x^3 + 2)(8x - 3)]}{(4x^2 - 3x)^2} = \frac{(x^3 + 2)^3(40x^4 - 33x^3 - 16x + 6)}{(4x^2 - 3x)^2}$

29. $y' = \dfrac{(2x-1)^3(5)(3x+2)^4(3)-(3x+2)^5(3)(2x-1)^2(2)}{(2x-1)^6}$

$= \dfrac{3(2x-1)^2(3x+2)^4[5(2x-1)-2(3x+2)]}{(2x-1)^6} = \dfrac{3(3x+2)^4(4x-9)}{(2x-1)^4}$

31. $y' = \dfrac{(4x+3)^{1/2}\left(\dfrac{2}{3}\right)(3x-1)^{-1/3}(3)-(3x-1)^{2/3}\left(\dfrac{1}{2}\right)(4x+3)^{-1/2}(4)}{\left(\sqrt{4x+3}\right)^2}$

$= \dfrac{2(4x+3)^{-1/2}(3x-1)^{-1/3}[(4x+3)-(3x-1)]}{4x+3} = \dfrac{2(x+4)}{(3x-1)^{1/3}(4x+3)^{3/2}}$

33. $y' = 4\left(\dfrac{1+x}{1-x}\right)^3\left[\dfrac{(1-x)(1)-(1+x)(-1)}{(1-x)^2}\right] = \dfrac{4(1+x)^3}{(1-x)^3}\left[\dfrac{1-x+1+x}{(1-x)^2}\right]$

$= \dfrac{4(1+x)^3}{(1-x)^3}\left[\dfrac{2}{(1-x)^2}\right] = \dfrac{8(1+x)^3}{(1-x)^5}$

35. $s' = \dfrac{(t^2-1)^{1/2}(1)-(t+1)\left(\dfrac{1}{2}\right)(t^2-1)^{-1/2}(2t)}{\left(\sqrt{t^2-1}\right)^2} = \dfrac{(t^2-1)^{-1/2}[(t^2-1)-t(t+1)]}{t^2-1}$

$= \dfrac{-1-t}{(t^2-1)^{3/2}}$ thus $s'(3) = \dfrac{-1-3}{8^{3/2}} = 0.177$ m/s

Section 21.8

Note: To save space, we will let $y' = dy/dx$.

1. $4 + 3y' = 0$
$y' = -4/3$

3. $2x - 2yy' = 0$
$y' = x/y$

5. $2x + 2yy' + 4y' = 0$
$(2y+4)y' = -2x$
$y' = \dfrac{-2x}{2y+4} = \dfrac{-x}{y+2}$

7. $6x - 3y^2y' - 3xy' - 3y = 0$
$(y^2+x)y' = 2x - y$
$y' = \dfrac{2x-y}{y^2+x}$

9. $4y^3y' - 2xyy' - y^2 + 2x = 0$
$(4y^3 - 2xy)y' = y^2 - 2x$
$y' = \dfrac{y^2-2x}{4y^3-2xy}$

11. $4y^3 - 4yx^2y' - 4y^2x + 6x = 0$
$(2y^3 - 2yx^2)y' = 2xy^2 - 3x$
$y' = \dfrac{2xy^2-3x}{2y^3-2x^2y}$

13. $2(y^2+2)(2yy') = 3(x^3-4x)^2(3x^2-4)$
$y' = \dfrac{3(x^3-4x)^2(3x^2-4)}{4y(y^2+2)}$

15. $3(x+y)^2(1+y') = 2(x-y+4)(1-y')$

$(3x^2+6xy+3y^2)(1+y') = 2(x-y+4)(1-y')$

$(3x^2+6xy+3y^2)y' + 2(x-y+4)y' = 2(x-y+4) - (3x^3+6xy+3y^2)$

$$y' = \frac{2x-2y+8-3x^2-6xy-3y^2}{3x^2+6xy+3y^2+2x-2y+8}$$

17. $$\frac{(x-y)(1+y')-(x+y)(1-y')}{(x-y)^2} = 2yy'$$

$$\frac{(x-y)+(x-y)y'-(x+y)+(x+y)y'}{(x-y)^2} = 2yy'$$

$$\frac{-2y+2xy'}{(x-y)^2} = 2yy'$$

$-2y+2xy' = (x-y)^2 2yy'$

$xy' - y(x-y)^2y' = y$

$[x-y(x-y)^2]y' = y$

$$y' = \frac{y}{x-y(x-y)^2}$$

19. $8x + 10yy' = 0$

$$m_{\tan} = y' = -\frac{4x}{5y}\Bigg|_{\substack{x=2\\y=-2}} = \frac{4}{5}$$

21. $2x + 2yy' - 6 - 2y' = 0$

$(y-1)y' = 3-x$

$$m_{\tan} = y' = \frac{3-x}{y-1}\Bigg|_{\substack{x=2\\y=4}} = \frac{1}{3}$$

23. $m_{\tan} = y' = 6x+4\Big|_{x=0} = 4;$

$(0, 9);\ y-9 = 4(x-0)$

$4x - y + 9 = 0$

25. $2yy' + 3xy' + 3y = 0$

$(2y+3x)y' = -3y$

$$m_{\tan} = y' = \frac{-3y}{2y+3x}\Bigg|_{\substack{x=1\\y=-4}} = -\frac{12}{5}$$

$y+4 = (-12/5)(x-1)$

$12x + 5y + 8 = 0$

27. **a)** $y = \dfrac{-x^2}{2} + \dfrac{7}{2}, y' = -x, y' = -(1), y' = -1$

b) $2x + 2y' = 0, y' = -x, y' = -1$

29. **a)** $y = \sqrt{x-2}, y' = \dfrac{1}{2}(x-2)^{-1/2}, y' = \dfrac{1}{2\sqrt{11-2}},$

$y' = \dfrac{1}{2(3)}, y' = \dfrac{1}{6}$

b) $2yy' = 1, y' = \dfrac{1}{2y}, y' = \dfrac{1}{2(3)}, y' = \dfrac{1}{6}$

Section 21.10

1. $y' = 5x^4 + 6x; y'' = 20x^3 + 6; y''' = 60x^2; y^{(4)} = 120x$

3. $y' = 25x^4 + 6x^2 - 8; y' = 100x^3 + 12x; y''' = 300x^2 + 12; y^{(4)} = 600x$

5. $y' = -\dfrac{1}{x^2}; y'' = \dfrac{2}{x^3}; y''' = -\dfrac{6}{x^4}$

7. $y' = 9(3x-5)^2; y'' = 54(3x-5); y''' = 162$

9. $y' = \dfrac{1}{2}(3x+2)^{-1/2}(3); y'' = (3/2)(-1/2)(3x+2)^{-3/2}(3) = (-9/4)(3x+2)^{-3/2}$

$y''' = (-9/4)(-3/2)(3x+2)^{-5/2}(3) = (81/8)(3x+2)^{-5/2}$

$y^{(4)} = (81/8)(-5/2)(3x+2)^{7/2}(3) = (-1215/16)(3x+2)^{-7/2} = \dfrac{-1215}{16(3x+2)^{7/2}}$

11. $y = (x^2+1)^{-1};$

$$\begin{aligned} y' &= -1(x^2+1)^{-2}(2x) = -2x(x^2+1)^{-2} \\ y'' &= -2x[-2(x^2+1)^{-3}(2x)] - 2(x^2+1)^{-2} \\ &= -2(x^2+1)^{-3}[-4x^2 + (x^2+1)] \\ &= \frac{6x^2-2}{(x^2+1)^3} \end{aligned}$$

13. $y' = \dfrac{(x-1)-(x+1)}{(x-1)^2} = \dfrac{-2}{(x-1)^2} = -2(x-1)^{-2}; \; y'' = 4(x-1)^{-3} = \dfrac{4}{(x-1)^3}$

15. $2x + 2yy' = 0; \; y' = -x/y; \; y'' = \dfrac{-y + xy'}{y^2} = \dfrac{-y + x(-x/y)}{y^2} = \dfrac{y^2 + x^2}{y^3} = \dfrac{-1}{y^3}$

17. $2x - xy' - y + 2yy' = 0$

$$y' = \frac{y-2x}{2y-x} = \frac{2x-y}{x-2y}$$

$$y'' = \frac{(x-2y)(2-y') - (2x-y)(1-2y')}{(x-2y)^2}$$

$$y'' = \frac{(x-2y)\left\{2 - \dfrac{2x-y}{x-2y}\right\} - (2x-y)\left\{1 - 2\left\{\dfrac{2x-y}{x-2y}\right\}\right\}}{(x-2y)^2} \cdot \frac{(x-2y)}{(x-2y)}$$

$$= \frac{2(x-2y)^2 - (2x-y)(x-2y) - (2x-y)[(x-2y) - 2(2x-y)]}{(x-2y)^3}$$

$$= \frac{2(x-2y)^2 - (2x-y)(x-2y) - (2x-y)(x-2y) + 2(2x-y)^2}{(x-2y)^3}$$

$$= \frac{6x^2 - 6xy + 6y^2}{(x-2y)^3} = \frac{6(x^2 - xy + y)^2}{(x-2y)^3} = \frac{6}{(x-2y)^3} \quad \text{(since } x^2 - xy + y^2 = 1\text{)}$$

19. $\frac{1}{2}x^{-1/2} + \frac{1}{2}y^{-1/2}y' = 0; \quad y' = -\frac{y^{1/2}}{x^{1/2}}$

$$y'' = \frac{-\frac{1}{2}x^{1/2}y^{-1/2}y' + \frac{1}{2}y^{1/2}x^{-1/2}}{(x^{1/2})^2} = \frac{-x^{1/2}y^{-1/2}\left\{\dfrac{-y^{1/2}}{x^{1/2}}\right\} + \dfrac{y^{1/2}}{x^{1/2}}}{2x}$$

$$= \frac{1 + \dfrac{y^{1/2}}{x^{1/2}}}{2x} = \frac{x^{1/2} + y^{1/2}}{2x^{3/2}} = \frac{1}{2x^{3/2}}$$

21. $-\frac{1}{x^2} + \frac{1}{y^2} \cdot y' = 0$

$$y' = \frac{1/x^2}{1/y^2} = \frac{y^2}{x^2}$$

$$y'' = \frac{x^2(2yy') - y^2(2x)}{(x^2)^2} = \frac{2x^2y[y^2/x^2] - 2xy^2}{x^4} = \frac{2y^3 - 2xy^2}{x^4}$$

23. $2(1+y)y' = 1$

$$y' = \frac{1}{2(1+y)}$$

$$y'' = -\frac{1}{2}(1+y)^{-2}y' = -\frac{1}{2} \cdot \frac{1}{(1+y)^2} \cdot \frac{1}{2(1+y)} = -\frac{1}{4(1+y)^3}$$

25. $v = \dfrac{ds}{dt} = 2t^3 - 18t^2 - 8t$

$a = \dfrac{dv}{dt} = 6t^2 - 36t - 8$

27. $v = \dfrac{1}{2}(6t-4)^{-1/2}(6) = 3(6t-2)^{-1/2}$

$a = (-3/2)(6t-4)^{-3/2}(6) = \dfrac{-9}{(6t-4)^{3/2}}$

29. $2x - xy' - y + 2yy' = 0$

$y' = \dfrac{y-2x}{2y-x}$

For (1, 3), $m_{\tan} = y' = 1/5$

$y - 3 = \dfrac{1}{5}(x-1)$

$x - 5y + 14 = 0$

If $x = 1$, then $1 - y + y^2 = 7$

$y^2 - y - 6 = 0$ or $y = 3, -2$

For (1, –2) $m_{\tan} = y' = 4/5$

$y + 2 = \dfrac{4}{5}(x-1)$

$4x - 5y - 14 = 0$

31. $m_{\tan} = y' = \dfrac{-b^2x}{a^2y}$

Equation of tangent to ellipse at (x_o, y_o) is given by

$y - y_o = \dfrac{-b^2x}{a^2y}(x - x_o)$

$a^2y^2 - a^2yy_o = -b^2x^2 + b^2xx_o$

$b^2xx_o + a^2yy_o = a^2y^2 + b^2x^2$

but $a^2y^2 + b^2x^2 = a^2b^2$ so

$b^2xx_o + a^2yy_o = a^2b^2$

$\dfrac{x_ox}{a^2} + \dfrac{y_oy}{b^2} = 1$

Chapter 21 Review

1. a) $\Delta s = f(t+\Delta t) - f(t) = [3(t+\Delta t)^2 + 4] - [3t^2 + 4]$
$= 3t^2 + 6t\Delta t + 3(\Delta t)^2 + 4 - 3t^2 - 4 = 6t\Delta t + 3(\Delta t)^2$

b) $\bar{v} = \dfrac{\Delta s}{\Delta t} = \dfrac{6t\Delta t + 3(\Delta t)^2}{\Delta t} = 6t + 3\Delta t$

2. a) $\Delta s = f(t+\Delta t) - f(t) = [5(t+\Delta t)^2 - 6] - [5t^2 - 6]$
$= 5t^2 + 10t\Delta t + 5(\Delta t)^2 - 6 - 5t^2 + 6 = 10t\Delta t + 5(\Delta t)^2$

b) $\bar{v} = \dfrac{\Delta s}{\Delta t} = \dfrac{10t\Delta t + 5(\Delta t)^2}{\Delta t} = 10t + 5\Delta t$

3. a) $\Delta s = f(t+\Delta t) - f(t) = [(t+\Delta t)^2 - 3(t+\Delta t) + 5] - [t^2 - 3t + 5]$
$= t^2 + 2t\Delta t + (\Delta t)^2 - 3t - 3\Delta t + 5 - t^2 + 3t - 5 = 2t\Delta t + (\Delta t)^2 - 3\Delta t$

b) $\bar{v} = \dfrac{\Delta s}{\Delta t} = \dfrac{2t\Delta t + (\Delta t)^2 - 3\Delta t}{\Delta t} = 2t + \Delta t - 3$

4. a) $\Delta s = f(t+\Delta t) - f(t) = [3(t+\Delta t)^2 - 6(t+\Delta t) + 8] - [3t^2 - 6t + 8]$
$= 3t^2 + 6t\Delta t + 3(\Delta t)^2 - 6t - 6\Delta t + 8 - 3t^2 + 6t - 8$
$= 6t\Delta t + 3(\Delta t)^2 - 6\Delta t$

b) $\bar{v} = \dfrac{\Delta s}{\Delta t} = \dfrac{6t\Delta t + 3(\Delta t)^2 - 6\Delta t}{\Delta t} = 6t + 3\Delta t - 6$

5. $\bar{v} = \frac{\Delta s}{\Delta t} = \frac{f(t+\Delta t) - f(t)}{\Delta t} = \frac{f(2+5) - f(2)}{5} = \frac{140-5}{5} = 27$ m/s

6. $\bar{v} = \frac{f(3) - f(1)}{2} = 20$ m/s

7. $\bar{v} = \frac{f(5) - f(2)}{3} = 10$ m/s

8. $\bar{v} = \frac{f(7) - f(3)}{4} = 33$ m/s

9. $$v = \lim_{\Delta t \to 0} \frac{f(t+\Delta t) - f(t)}{\Delta t} = \lim_{\Delta t \to 0} \frac{f(2+\Delta t) - f(2)}{\Delta t}$$
$$= \lim_{\Delta t \to 0} \frac{[3(2+\Delta t)^2 - 7] - [3(2)^2 - 7]}{\Delta t} = \lim_{\Delta t \to 0} \frac{12\Delta t + 3(\Delta t)^2}{\Delta t}$$
$$= \lim_{\Delta t \to 0} (12 + 3\Delta t) = 12 \text{ m/s}$$

10. $$v = \lim_{\Delta t \to 0} \frac{f(1+\Delta t) - f(1)}{\Delta t} = \lim_{\Delta t \to 0} \frac{[5(1+\Delta t)^2 - 3] - [5(1)^2 - 3]}{\Delta t}$$
$$= \lim_{\Delta t \to 0} \frac{10\Delta t + 5(\Delta t)^2}{\Delta t} = \lim_{\Delta t \to 0} (10 + 5\Delta t) = 10 \text{ m/s}$$

11. $$v = \lim_{\Delta t \to 0} \frac{f(2+\Delta t) - f(2)}{\Delta t}$$
$$= \lim_{\Delta t \to 0} \frac{[2(2+\Delta t)^2 - 4(2+\Delta t) + 7] - [2(2)^2 - 4(2) + 7]}{\Delta t}$$
$$= \lim_{\Delta t \to 0} \frac{4\Delta t + 2(\Delta t)^2}{\Delta t} = \lim_{\Delta t \to 0} (4 + 2\Delta t) = 4 \text{ m/s}$$

12. $$v = \lim_{\Delta t \to 0} \frac{f(3+\Delta t) - f(3)}{\Delta t}$$
$$= \lim_{\Delta t \to 0} \frac{[4(3+\Delta t)^2 - 7(3+\Delta t) + 2] - [4(3)^2 - 7(3) + 2]}{\Delta t}$$
$$= \lim_{\Delta t \to 0} \frac{17\Delta t + 4(\Delta t)^2}{\Delta t} = \lim_{\Delta t \to 0} (17 + 4\Delta t) = 17 \text{ m/s}$$

13. $\lim_{x \to 3} (2x^2 - 5x + 1) = 4$

14. $\lim_{x \to -2} (x^2 + 4x - 7) = -11$

15. $\lim_{x \to -2} (x - 2) = -4$

16. $\lim_{x \to 5} (5 + x) = 10$

17. No limit ($\sqrt{-8}$ is not real.)

18. $\lim_{x \to -3} \sqrt{6 + x^2} = \sqrt{15}$

19. $\lim_{x \to 2} \frac{5x^2 + 2}{3x^2 - 2x + 1} = \frac{22}{9}$

20. $\lim_{x \to -3} \frac{2x^2 - 4x + 7}{x^3 - x} = -\frac{37}{24}$

21. $\lim_{x\to -3} (x^2 - 4x + 3)(2x^2 + 5x + 4) = 24(7) = 168$

22. $\lim_{x\to 2} (x^3 + x - 2)(x^3 + x^2 + x) = (8)(14) = 112$

23. $\lim_{h\to 0} \frac{2-(2+h)}{2(2+h)h} = \lim_{h\to 0} \frac{-h}{2(2+h)h} = \lim_{h\to 0} \frac{-1}{2(2+h)} = -\frac{1}{4}$

24. $\lim_{h\to 0} \frac{\sqrt{1+h}-1}{h} \cdot \frac{\sqrt{1+h}+1}{\sqrt{1+h}+1} = \lim_{h\to 0} \frac{h}{h(\sqrt{1+h}+1)} = \lim_{h\to 0} \frac{1}{\sqrt{1+h}+1} = \frac{1}{2}$

25. $\lim_{x\to\infty} \frac{(5x^2 - 2x + 3)/(x^2)}{(2x^2 - 4)/(x^2)} = \lim_{x\to\infty} \frac{5 - \frac{2}{x} + \frac{3}{x^2}}{2 - \frac{4}{x^2}} = \frac{5}{2}$

26. $\lim_{x\to\infty} \frac{(7x^3 - 4x + 2)/(x^3)}{(10x^3 - x^2 + 5)/(x^3)} = \lim_{x\to\infty} \frac{7 - \frac{4}{x^2} + \frac{2}{x^3}}{10 - \frac{1}{x} + \frac{5}{x^3}} = \frac{7}{10}$

27. Does not exist **28.** c **29.** c **30.** Does not exist **31.** No **32.** Yes

33. $m_{\tan} = y' = 6x - 4 \Big|_{x=1} = -10$

tangent line at $(-1, 12)$
$y - 12 = -10(x + 1)$
$10x + y - 2 = 0$

34. $m_{\tan} = y' = 2x - 5 \Big|_{x=2} = -1$

tangent line at $(2, -18)$
$y + 18 = -1(x - 2)$
$x + y + 16 = 0$

35. $m_{\tan} = y' = 4x + 2 \Big|_{x=3} = 14$

tangent line at $(3, 31)$
$y - 31 = 14(x - 3)$
$14x - y - 11 = 0$

36. $m_{\tan} = y' = 8x - 8 \Big|_{x=2} = -24$

tangent line at (−2, 35)
$y - 35 = -24(x + 2)$
$24x + y + 13 = 0$

37. $v = \dfrac{ds}{dt} = -\dfrac{6}{t^3}\Big|_{t=5} = -\dfrac{6}{125}$ cm/s or −0.048 cm/s

38. $v = \dfrac{ds}{dt} = -32t\Big|_{t=2} = -64$ ft/s

39. $y' = 20x^3 - 9x^2 + 4x + 5$

40. $y' = 100x^{99} + 400x^4$

41. $y' = (x^3 + 4)(3x^2 - 1) + (x^3 - x + 1)(3x^2)$
$= 6x^5 - 4x^3 + 15x^2 - 4$

42. $y' = (3x^2 - 5)(5x^4 + 2x - 4) + (x^5 + x^2 - 4x)(6x)$
$= 21x^6 - 25x^4 + 12x^3 - 36x^2 - 10x + 20$

43. $y' = \dfrac{(3x-4)(2x) - (x^2+1)(3)}{(3x-4)^2} = \dfrac{3x^2 - 8x - 3}{(3x-4)^2}$

44. $y' = \dfrac{(3x^4+2)(2-2x) - (2x - x^2)(12x^3)}{(3x^4+2)^2} = \dfrac{6x^5 - 18x^4 - 4x + 4}{(3x^4+2)^2}$

45. $y' = 5(3x^2 - 8)^4(6x) = 30x(3x^2 - 8)^4$

46. $y' = \dfrac{3}{4}(x^4 + 2x^3 + 7)^{-1/4}(4x^3 + 6x^2)$

47. $y' = -4(3x + 5)^{-5}(3) = \dfrac{-12}{(3x+5)^5}$

48. $y' = \dfrac{(x+3)^2(1/2)(7x^2-5)^{-1/2}(14x) - (7x^2-5)^{1/2}(2)(x+3)}{[(x+3)^2]^2}$

$= \dfrac{(x+3)(7x^2-5)^{-1/2}[7x(x+3) - 2(7x^2-5)]}{(x+3)^4} = \dfrac{-7x^2 + 21x + 10}{(7x^2-5)^{1/2}(x+3)^3}$

49. $y' = \dfrac{(x+5)[(x\left(\frac{1}{2}\right)(2-3x)^{-1/2}(-3) + (2-3x)^{1/2} - x(2-3x)^{1/2}}{(x+5)^2}$

$= \dfrac{\frac{1}{2}(2-3x)^{-1/2}\{(x+5)[-3x+2(2-3x)] - 2x(2-3x)\}}{(x+5)^2} = \dfrac{-3x^2-45x+20}{2(x+5)^2\sqrt{2-3x}}$

50. $2x - 4y^3 - 12xy^2y' + 2yy' = 0$

$(2y - 12xy^2)y' = 4y^3 - 2x$

$y' = \dfrac{2y^3 - x}{y - 6xy^2}$

51. $4y^3y' - 2yy' = 2xy' + 2y$

$(4y^3 - 2y - 2x)y' = 2y$

$y' = \dfrac{y}{2y^3 - y - x}$

52. $3(y^2+1)^2(2yy') = 8x$

$y' = \dfrac{4x}{3y(y^2+1)^2}$

53. $4(y+2)^3y' = 3(2x^3-3)^2(6x)^2$

$y' = \dfrac{9x^2(2x^3-3)^2}{2(y+2)^3}$

54. $f'(x) = (x^2+2)^3 + (x+1)(3)(x^2+2)^2(2x)$

$= (x^2+2)^3 + 6x(x+1)(x^2+2)^2$

$= (x^2+2)^2(7x^2+6x+2)$

$f'(-2) = (36)(18) = 648$

55. $f'(x) = \dfrac{(x+3)(2x) - (x^2-8)(1)}{(x+3)^2}$

$= \dfrac{x^2+6x+8}{(x+3)^2} \qquad f'(-4) = 0$

56. $f'(x) = \dfrac{(x-2)(2x+3) - (x^2+3x-2)(1)}{(x-2)^2}$

$= \dfrac{x^2-4x-4}{(x-2)^2} \qquad f'(3) = -7$

57. $f'(x) = \dfrac{(x+5)(1/2)(3x^2-1)^{-1/2}(6x) - (3x^2-1)^{1/2}(1)}{(x+5)^2}$

$= \dfrac{(3x^2-1)^{-1/2}[3x(x+5) - (3x^2-1)]}{(x+5)^2} = \dfrac{15x+1}{(x+5)^2\sqrt{3x-1}}$

$f'(1) = \dfrac{4}{9\sqrt{2}}$ or $\dfrac{2\sqrt{2}}{9}$

58. $m_{\tan} = 6x + 1 \Big|_{x=-2} = -11$

tangent line at $(-2, 8)$ is

$y - 8 = -11(x + 2)$

$11x + y + 14 = 0$

59. $m_{\tan} = y' = 3x^2 - 1 \Big|_{x=-3} = 26$

tangent line at $(-3, -16)$ is

$y + 16 = 26(x + 3)$

$26x - y + 62 = 0$

60. $y' = \dfrac{(x-3)^{1/2}(2x) - (x^2-2)(1/2)(x-3)^{-1/2}}{[(x-3)^{1/2}]^2}$

$= \dfrac{(1/2)(x-3)^{-1/2}[4x(x-3) - (x^2-2)]}{x-3} = \dfrac{3x^2-12x+2}{2(x-3)^{3/2}}$

$m_{\tan} = y' \Big|_{x=4} = 1$; tangent line at $(4, 14)$ is $y - 14 = 1(x - 4)$ or $x - y + 10 = 0$

61. $y' = \dfrac{(x-2)(1/2)(x^2+7)^{-1/2}(2x) - (x^2+7)^{1/2}(1)}{(x-2)^2}$

$= \dfrac{(x^2+7)^{-1/2}[x(x-2) - (x^2+7)]}{(x-2)^2} = \dfrac{-2x-7}{\sqrt{x^2+7}(x-2)^2}$

$m_{\tan} = y' \Big|_{x=3} = -\dfrac{13}{4}$; tangent line at $(3, 4)$ is $y - 4 = -\dfrac{13}{4}(x-3)$

$13x + 4y - 55 = 0$

62. $v = s'(t) = 6t - 8 \Big|_{t=2} = 4$ m/s

63. $v = s'(t) = 3t^2 - 18t \Big|_{t=-2} = 48$ m/s

64. $v = s'(t) = \dfrac{(t+1)^{1/2}(2t) - (t^2-5)(1/2)(t+1)^{-1/2}}{[(t+1)^{1/2}]^2}$

$= \dfrac{(1/2)(t+1)^{-1/2}[(t+1)(4t) - (t^2-5)]}{t+1} = \dfrac{3t^2+4t+5}{2(t+1)^{3/2}}\Bigg|_{t=3} = 2.75$ m/s

65. $v = s'(t) = \dfrac{(t+5)(1/2)(t^2-3)^{-1/2}(2t) - (t^2-3)^{1/2}(1)}{(t+5)^2}$

$= \dfrac{(t^2-3)^{-1/2}[t(t+5)-(t^2-3)]}{(t+5)^2} = \dfrac{5t+3}{(t^2-3)^{1/2}(t+5)^2}\Bigg|_{t=4}$

$= \dfrac{23}{81\sqrt{13}} \approx 0.0788 \text{ m/s}$

66. $m_{\tan} = \dfrac{(x+3)(2x)-(x^2-6)(1)}{(x+3)^2} = \dfrac{x^2+6x+6}{(x+3)^2}\Bigg|_{x=-4} = -2$

tangent line at $(-4, -10)$ is $y + 10 = -2(x+4)$ or $2x + y + 18 = 0$

67. $i = \dfrac{dq}{dt} = 3000t^2 + 50 \Big|_{t=0.01} = 50.3 \text{ A}$

68. $\dfrac{dc}{dT} = 0.5 + 0.000012T$

69. $\dfrac{df}{dC} = -\dfrac{1}{2}\cdot\dfrac{1}{2\pi}(LC)^{-3/2}(L) = \dfrac{-L}{4\pi\sqrt{(LC)}^3} = \dfrac{-L}{4\pi LC\sqrt{LC}} = \dfrac{-1}{4\pi C\sqrt{LC}}$

70. $y' = 24x^5 - 32x^3 + 27x^2 - 6;\ y'' = 120x^4 - 96x^2 + 54x;$
$y''' = 480x^3 - 192x + 54;\ y^{(4)} = 1440x^2 - 192$

71. $y' = (1/2)(2x-3)^{-1/2}(2) = (2x-3)^{-1/2}$

$y'' = (-1/2)(2x-3)^{-3/2}(2) = \dfrac{-1}{(2x-3)^{3/2}}$

72. $y' = -3(2x^2+1)^{-2}(4x) = \dfrac{-12x}{(2x^2+1)^2}$

$y'' = \dfrac{(2x^2+1)^2(-12)-(-12x)(2)(2x^2+1)(4x)}{(2x^2+1)^4} = \dfrac{-12(2x^2+1)[(2x^2+1)-8x^2]}{(2x^2+1)^4} = \dfrac{72x^2-12}{(2x^2+1)^3}$

73. $2yy' + 2xy' + 2y = 0$

$y' = -\dfrac{y}{y+x}$

$y'' = \dfrac{(y+x)(-y')-(-y)(y'+1)}{(y+x)^2} = \dfrac{(y+x)\left\{\dfrac{y}{y+x}\right\} + y\left\{\dfrac{-y}{y+x}+1\right\}}{(y+x)^2}\cdot\dfrac{y+x}{y+x}$

$= \dfrac{y(y+x)+y[-y+(y+x)]}{(y+x)^3} = \dfrac{y^2+yx-y^2+y^2+xy}{(y+x)^3} = \dfrac{y^2+2xy}{(x+y)^3} = \dfrac{4}{(x+y)^3}$

74. $x^{-1/2} - y^{-1/2} = 1$

$(-1/2)x^{-3/2} + (1/2)y^{-3/2}y' = 0$

$$y' = \frac{x^{-3/2}}{y^{-3/2}} = \frac{y^{3/2}}{x^{3/2}}$$

$$y'' = \frac{x^{3/2}(3/2)y^{1/2}y' - y^{3/2}(3/2)x^{1/2}}{(x^{3/2})^2} = \frac{3\left[x^{3/2}y^{1/2}(y^{3/2}/x^{3/2}) - y^{3/2}x^{1/2}\right]}{2x^3} = \frac{3(y^2 - y^{3/2}x^{1/2})}{2x^3}$$

75. $v = \dfrac{ds}{dt} = (1/4)(2t+3)^{-3/4}(2) = (1/2)(2t+3)^{-3/4}$

$a = \dfrac{dv}{dt} = (-3/8)(2t+3)^{-7/4}(2) = (-3/4)(2t+3)^{-7/4}$

Chapter 22, Exercises 22.1

KEY: I = intercepts; A = asymptotes; S = symmetry; D = domain;
a = curve above the x-axis; b = curve below the x-axis

In Exercises 1-12, the domain is the set of real numbers; there are no asymptotes and no symmetry, except number 9, which has symmetry with respect to the y-axis.

1. I: $x = 0, x = -1, x = 4$
$y = 0$

3. I: $x = -5, 1, 3$; $y = 15$

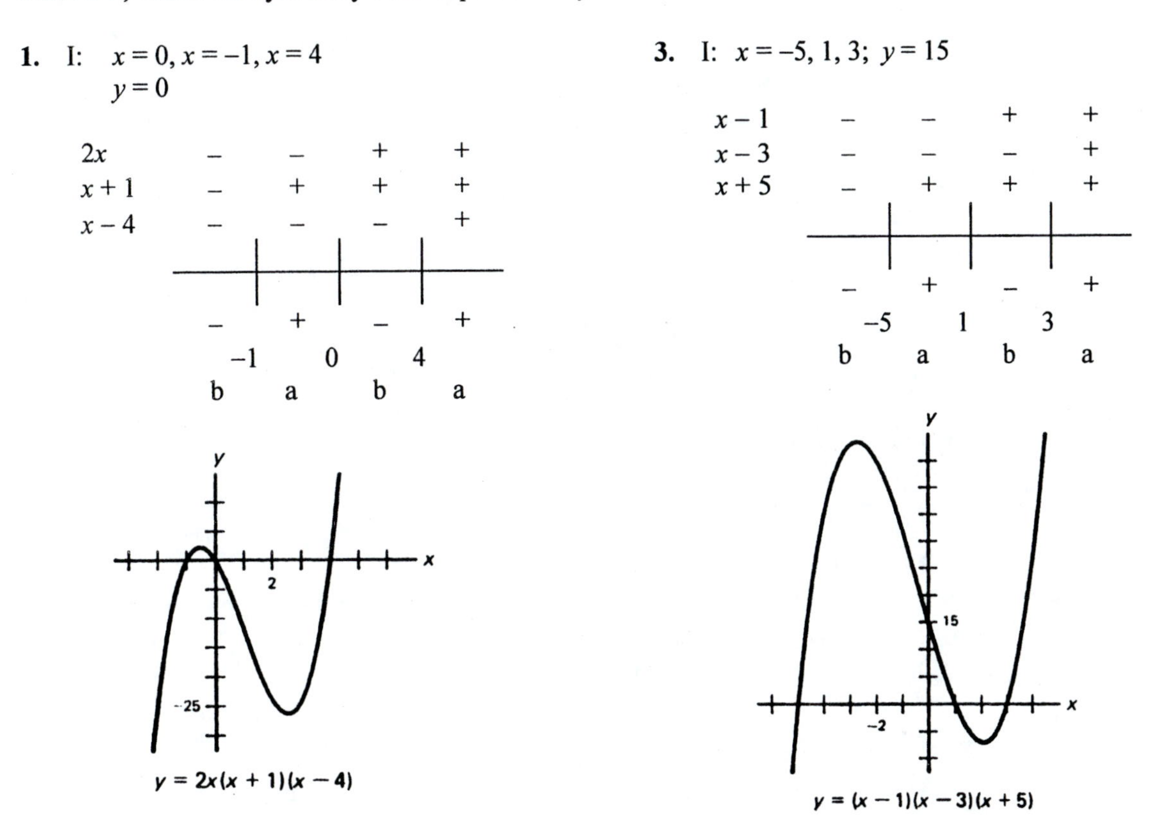

$y = 2x(x+1)(x-4)$

$y = (x-1)(x-3)(x+5)$

5. $y = x(x + 5)(x - 3)$
I: $x = -5, 0, 3$: $y = 0$

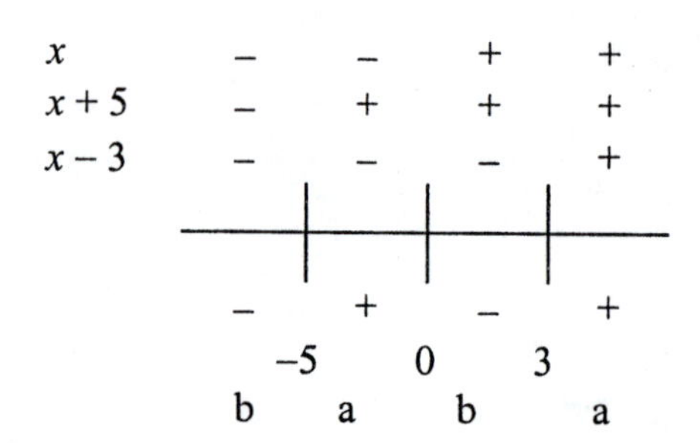

x	−	−	+	+
$x + 5$	−	+	+	+
$x - 3$	−	−	−	+
	−	+	−	+

	−5	0	3	
b	a	b	a	

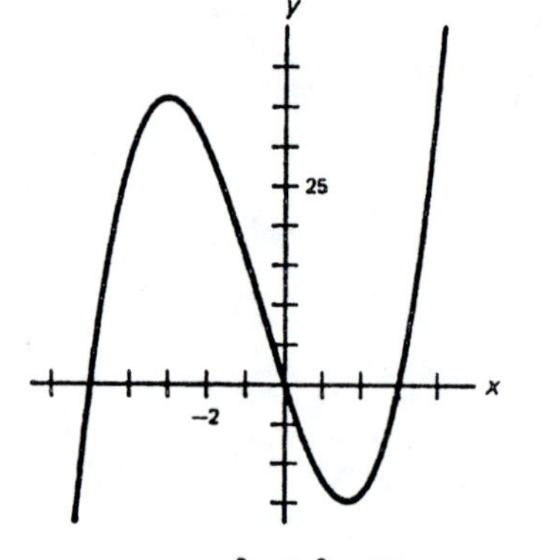

$y = x^3 + 2x^2 - 15x$

7. I: $x = 0, -1, 3/2$; $y = 0$

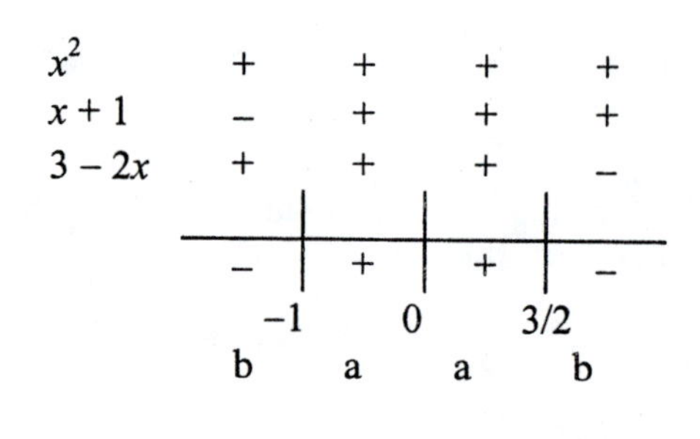

x^2	+	+	+	+
$x + 1$	−	+	+	+
$3 - 2x$	+	+	+	−
	−	+	+	−

−1	0	3/2	
b	a	a	b

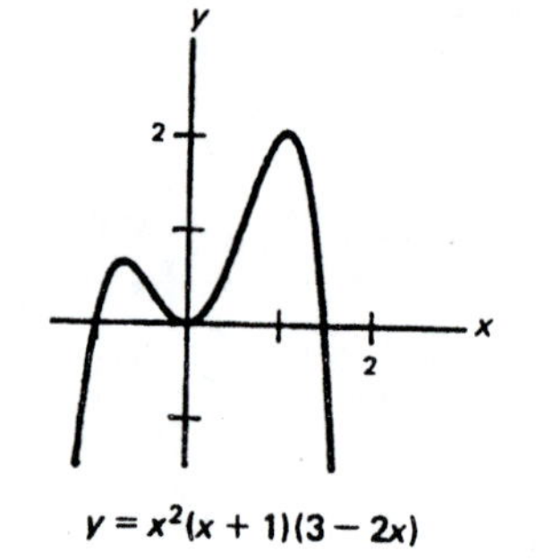

$y = x^2(x + 1)(3 - 2x)$

9. $y = (x + 2)(x + 3)(x - 2)(x - 3)$
I: $x = -2, -3, 2, 3$; $y = 36$

$x + 2$	−	−	+	+	+
$x + 3$	−	+	+	+	+
$x - 2$	−	−	−	+	+
$x - 3$	−	−	−	−	+
	+	−	+	−	+

−3	−2	2	3	
a	b	a	b	a

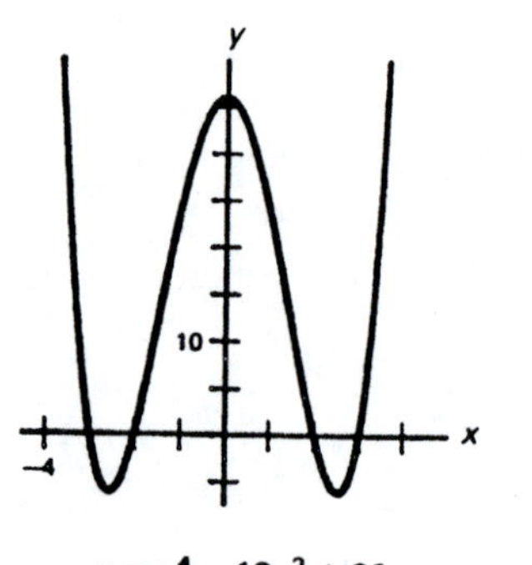

$y = x^4 - 13x^2 + 36$

11. I: $x = -4, 0, 2$; $y = 0$

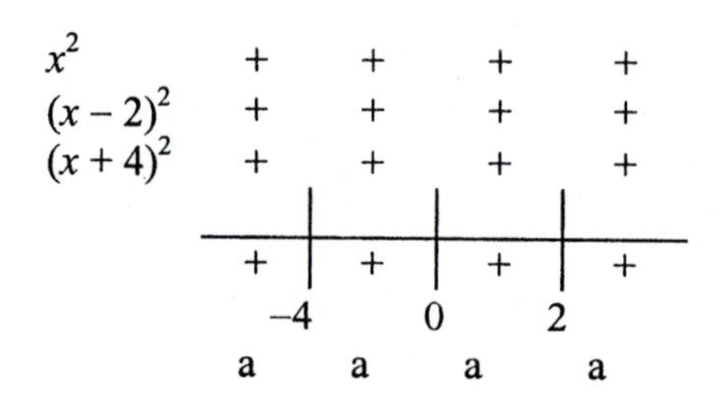

x^2	+	+	+	+
$(x - 2)^2$	+	+	+	+
$(x + 4)^2$	+	+	+	+
	+	+	+	+

−4	0	2	
a	a	a	a

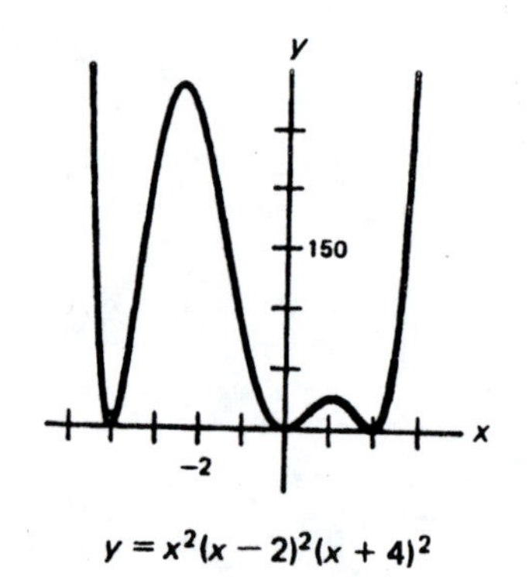

$y = x^2(x - 2)^2(x + 4)^2$

13. I: $y = 3$

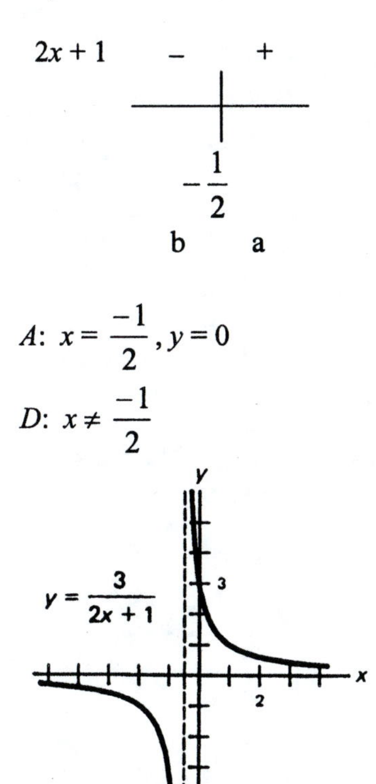

A: $x = \frac{-1}{2}, y = 0$

D: $x \neq \frac{-1}{2}$

15. I: $x = 0, y = 0$

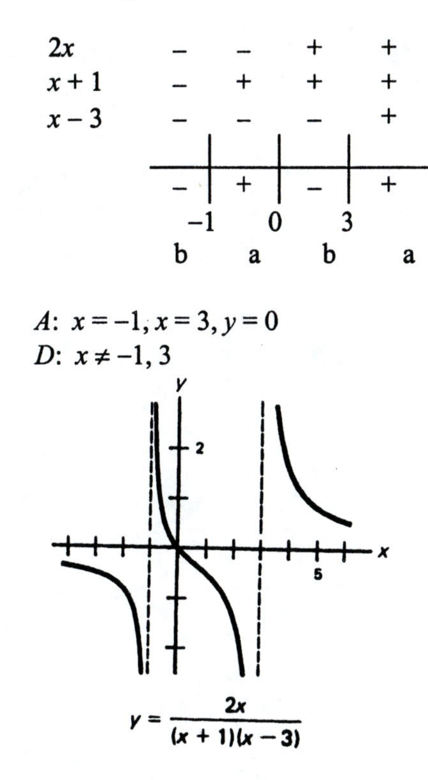

A: $x = -1, x = 3, y = 0$

D: $x \neq -1, 3$

17. I: $y = 3/4$; graph always above the x-axis;

A: $y = 0$;

S: y-axis; $D = \{\text{reals}\}$

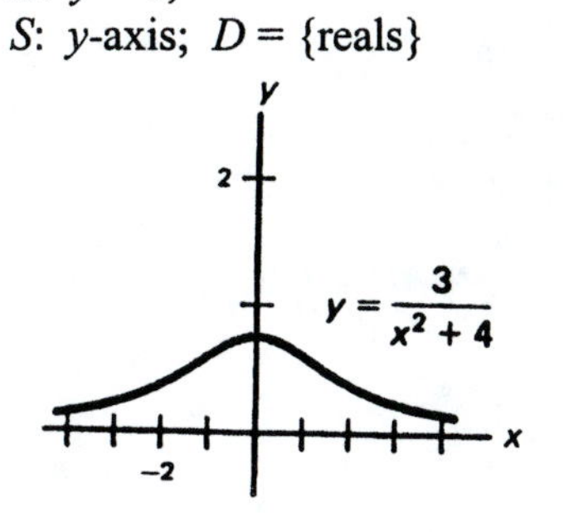

19. I: $x = 0, y = 0$

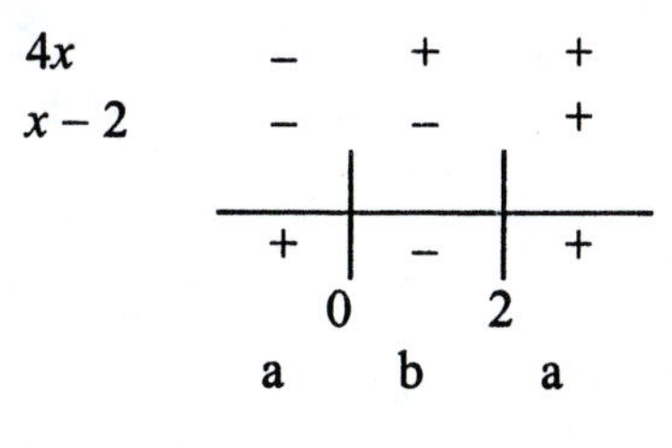

A: $x = 2, y = 4$; D: $x \neq 2$

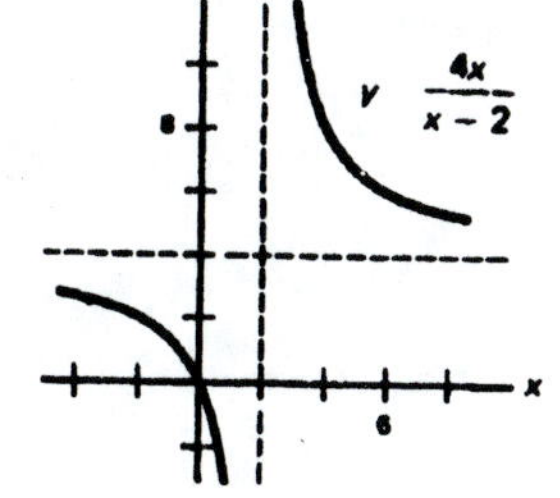

21. I: $x = 0, y = 0$

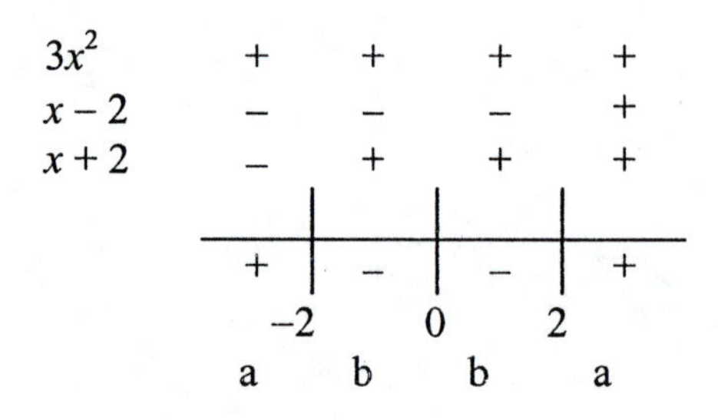

$3x^2$	+	+	+	+
$x-2$	−	−	−	+
$x+2$	−	+	+	+
	+	−	−	+
	-2	0	2	
	a	b	b	a

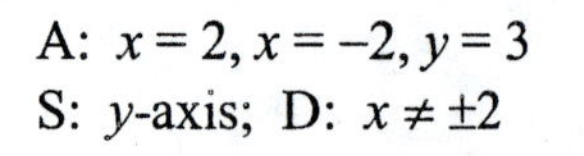

A: $x = 2, x = -2, y = 3$
S: y-axis; D: $x \neq \pm 2$

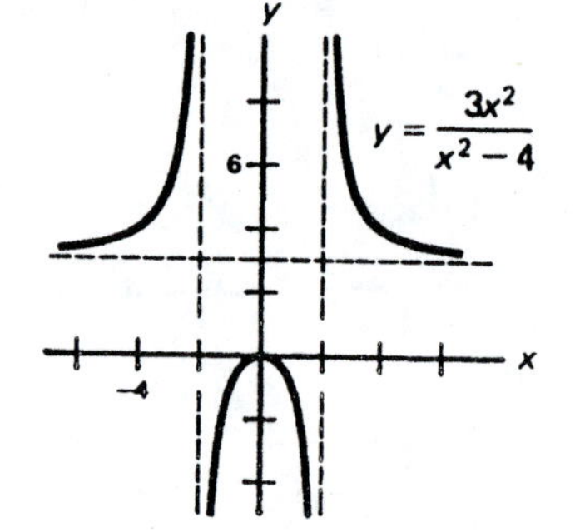

23. I: $x = \pm\sqrt{6}, y = -3$

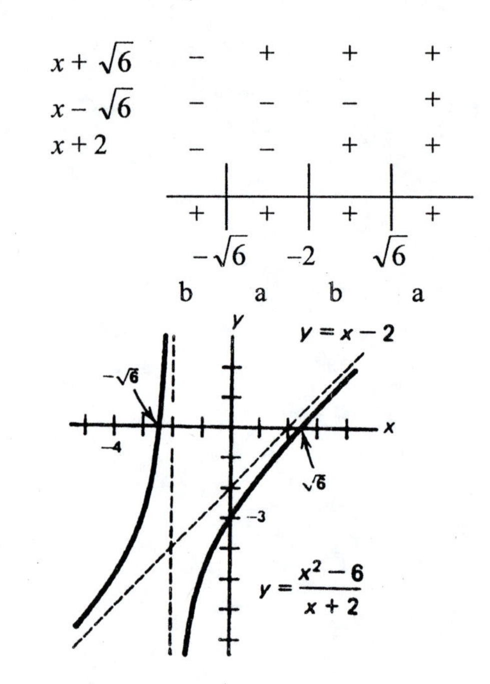

$x+\sqrt{6}$	−	+	+	+
$x-\sqrt{6}$	−	−	−	+
$x+2$	−	−	+	+
	+	+	+	+
	$-\sqrt{6}$	-2	$\sqrt{6}$	
	b	a	b	a

25. I: $x = 3/2, -1; y = 3/4$

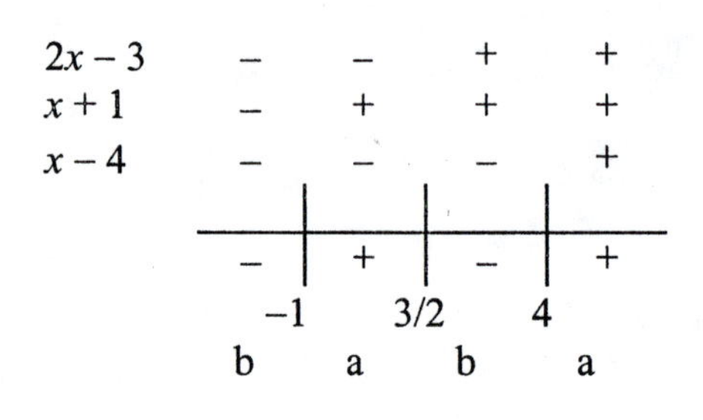

$2x-3$	−	−	+	+
$x+1$	−	+	+	+
$x-4$	−	−	−	+
	−	+	−	+
	-1	$3/2$	4	
	b	a	b	a

A: $x = 4, y = 2x + 7$;
D: $x \neq 4$

27. I: $x = -4, y = 2$
Graph always above x-axis.
D: $x \geq -4$

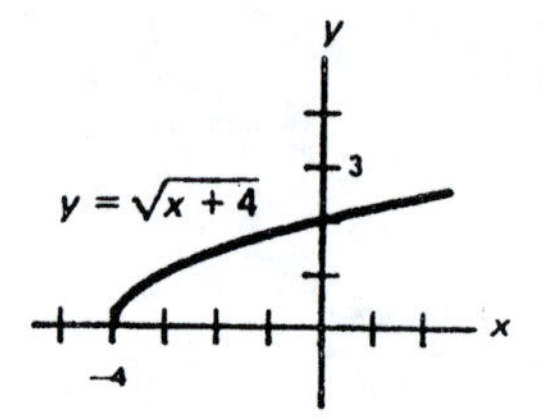

Section 22.1

29. I: $x = 0, y = 0$

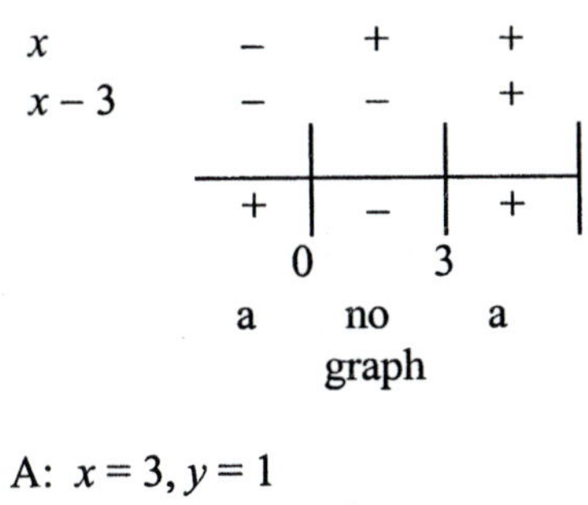

A: $x = 3, y = 1$
D: $x \le 0, x > 3$

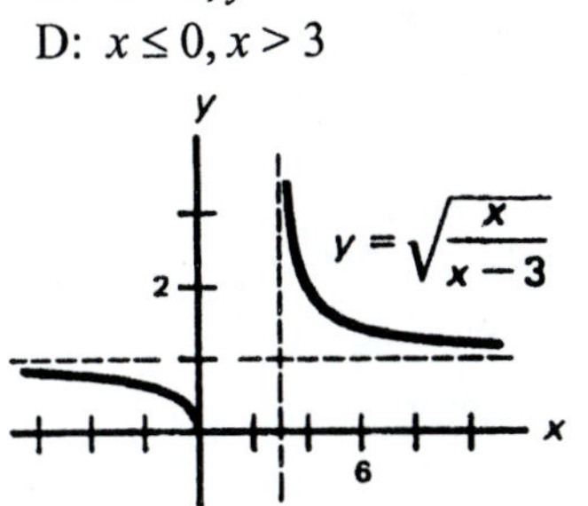

31. I: $x = -9, y = \pm 3$
S: x-axis
D: $x \ge 9$

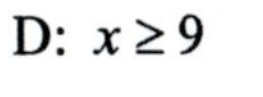

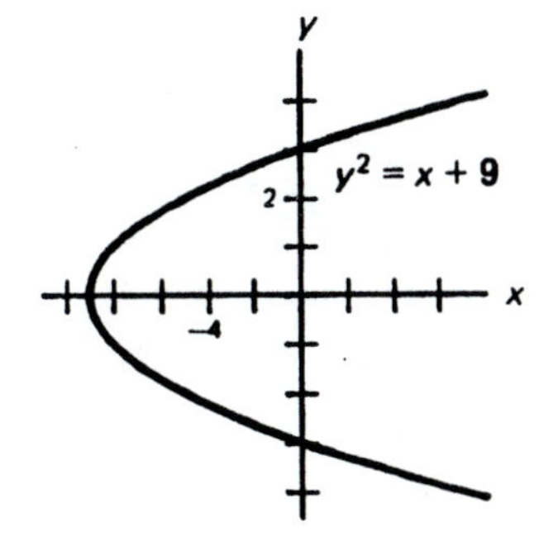

33. $y = \pm\sqrt{x/(x+4)}$ I: $x = 0, y = 0$

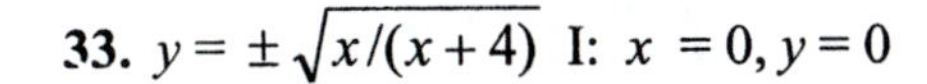

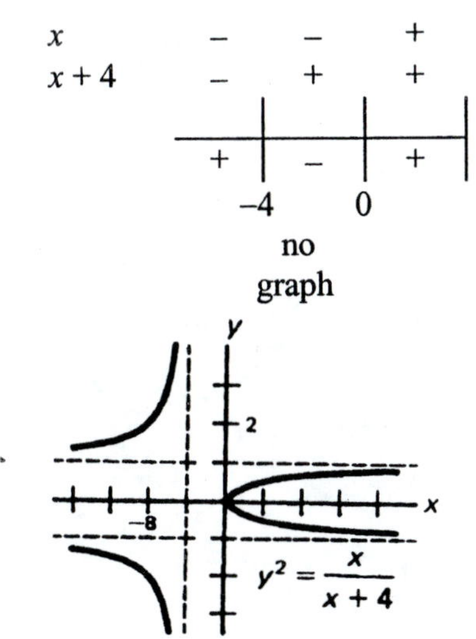

35. $y = \pm\sqrt{x^2/(x^2+4)}$

I: $x = 0, y = 0$
S: x-axis, y-axis, origin
A: $y = \pm 1$; D: {reals)

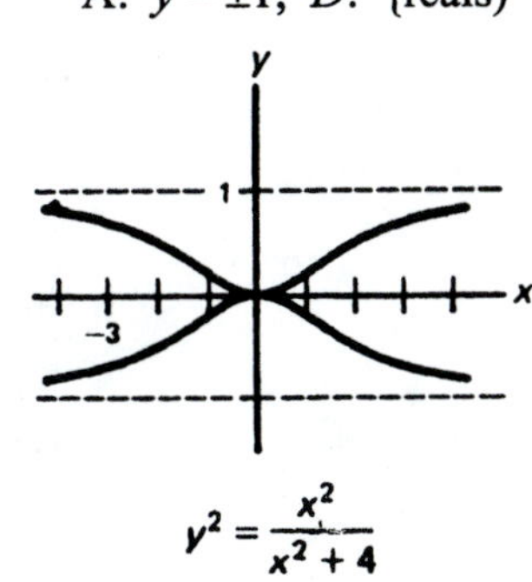

$y^2 = \dfrac{x^2}{x^2 + 4}$

Exercises 22.2

KEY: d = decreasing, i = increasing, and same as in Section 3.1.

1. $f'(x) = 2x + 6 = 0;\ x = -3$

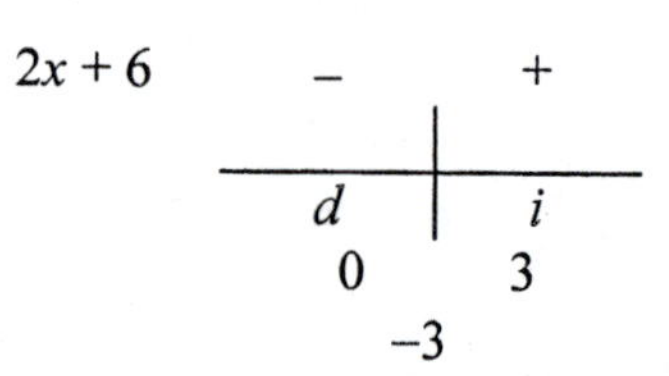

$f(-3) = -25$; $(-3, -25)$ min

I: $x = -8, 2$; $y = -16$

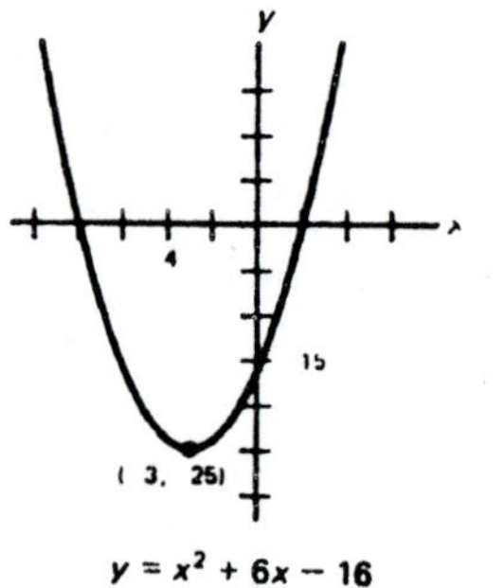

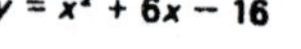

3. $f'(x) = -4 - 6x = 0;\ x = -2/3$

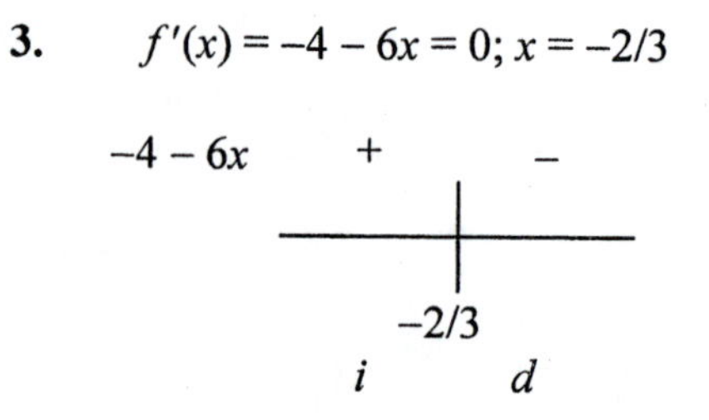

$f(-2/3) = 5\ 1/3$; $(-2/3, 16/3)$ max

I: $y = 4$; $x = -2, 2/3$

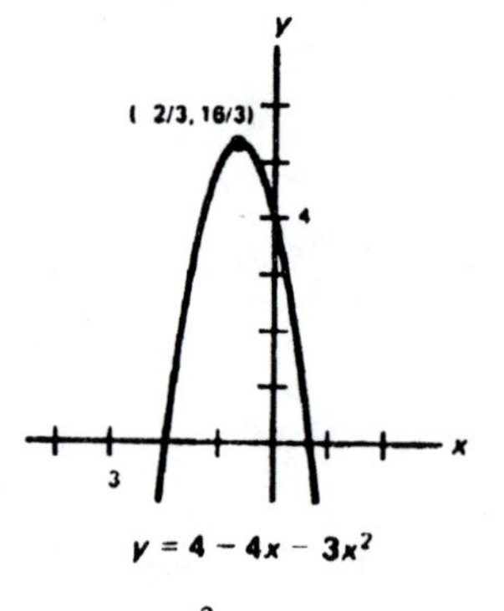

$y = 4 - 4x - 3x^2$

5. $f'(x) = 3x^2 + 6x = 3x(x + 2) = 0$

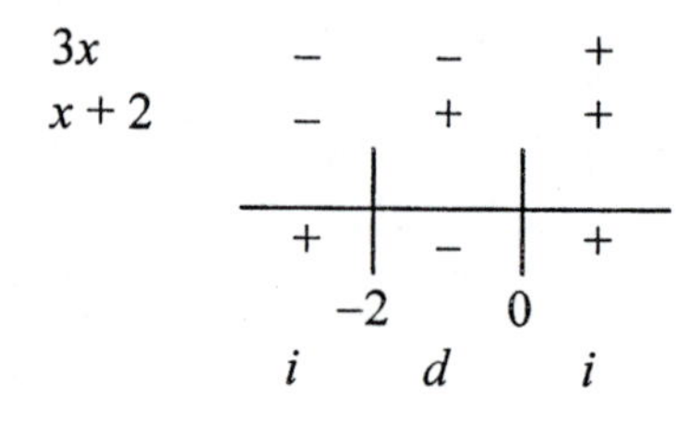

max $(-2, 8)$; min: $(0, 4)$

I: $y = 4$

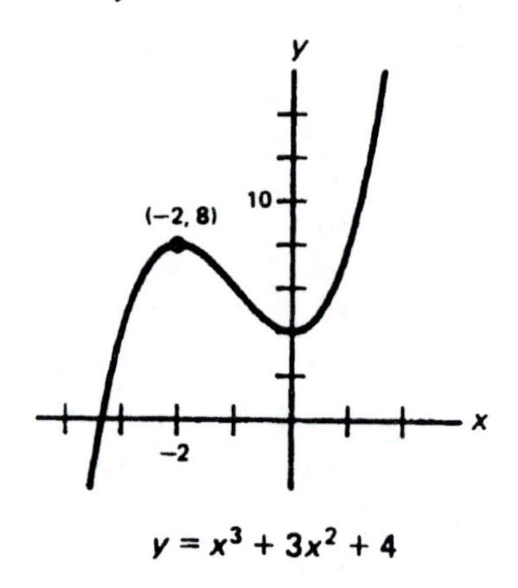

$y = x^3 + 3x^2 + 4$

7. $f'(x) = x^2 - 9 = (x - 3)(x + 3) = 0$

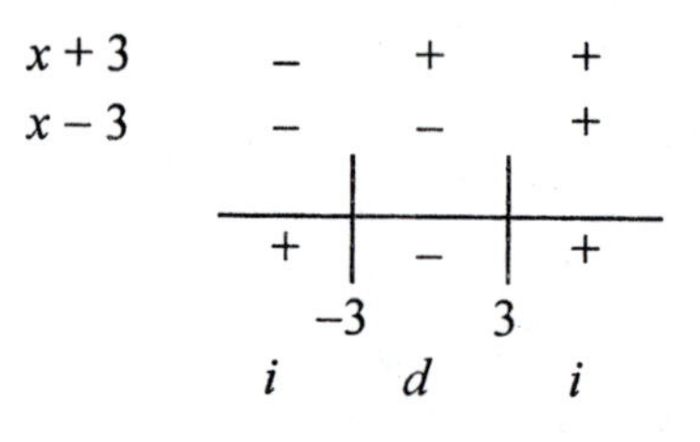

max $(-3, 14)$; min $(3, -22)$

I: $y = -4$

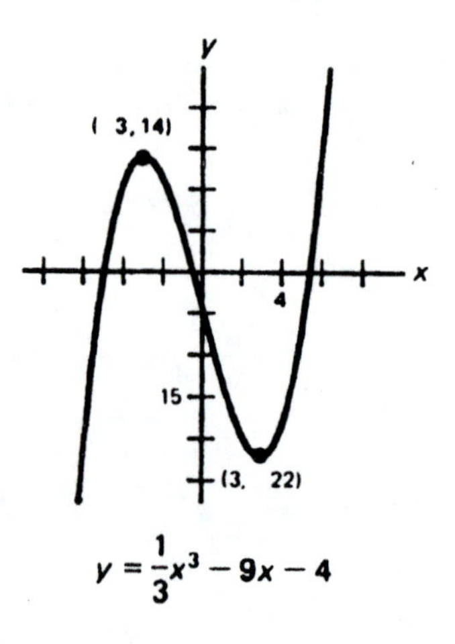

$y = \frac{1}{3}x^3 - 9x - 4$

9. $f'(x) = 15x^4 - 15x^2 = 15x^2(x-1)(x+1)$

$15x^2$	+	+	+	+
$x-1$	−	−	−	+
$x+1$	−	+	+	+
	+	−	−	+
	-1	0	1	
	i	d	d	i

max $(-1, 2)$; min $(1, -2)$;
I: $x = 0, y = 0$

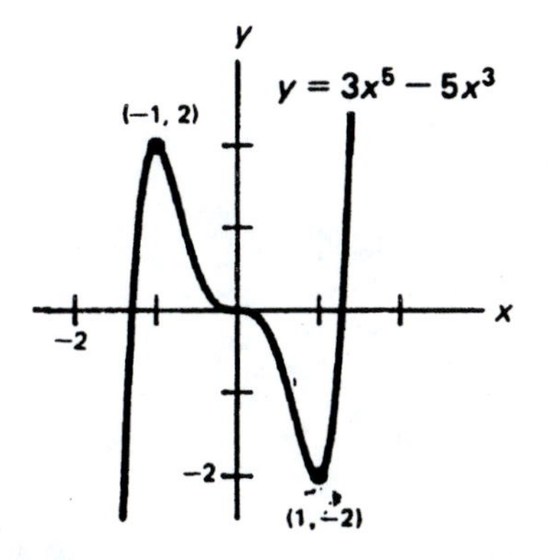

11. $f'(x) = 5(x-2)^4$

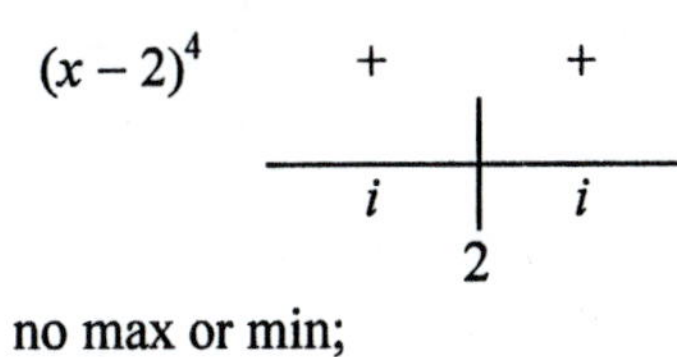

$(x-2)^4$	+	+
	i	i
	2	

no max or min;
I: $x = 2, y = -32$

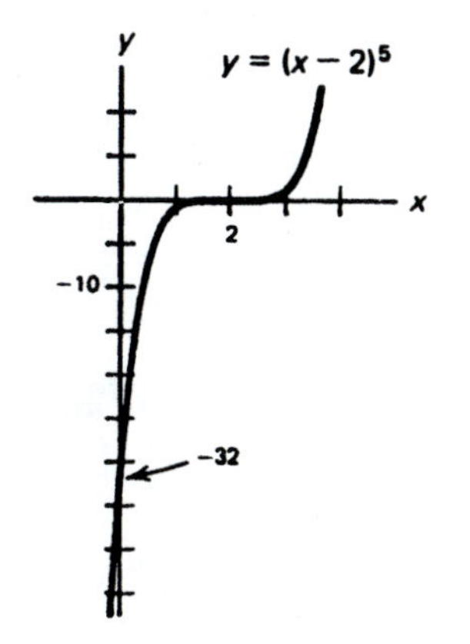

13. $y' = 8x(x^2-1)^3$
$= 8x'(x+1)^3(x-1)^3$

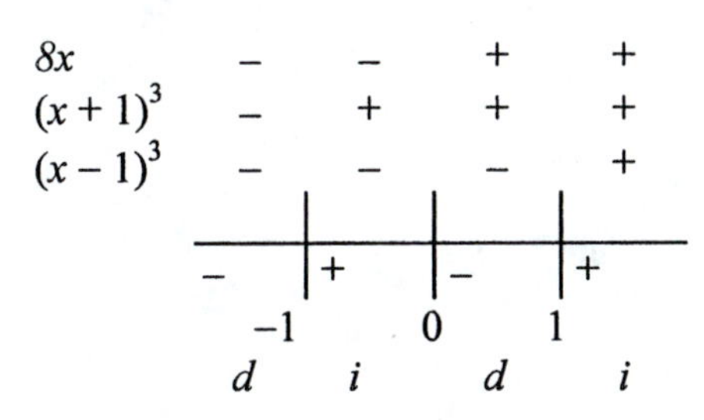

$8x$	−	−	+	+
$(x+1)^3$	−	+	+	+
$(x-1)^3$	−	−	−	+
	−	+	−	+
	-1	0	1	
	d	i	d	i

min $(-1, 0)$, $(1, 0)$; max $(0, 1)$
I: $x = \pm 1, y = 1$; S: y-axis

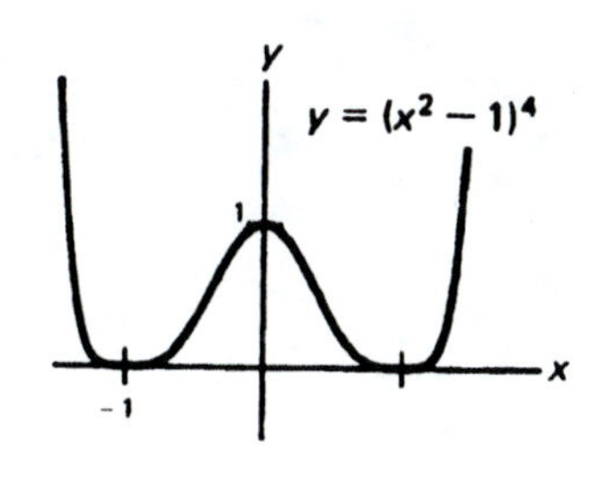

15. 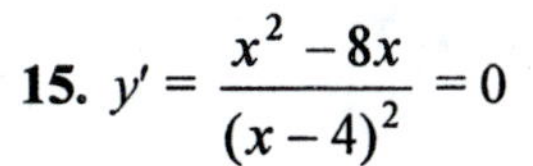

$$y' = \frac{x^2 - 8x}{(x-4)^2} = 0$$

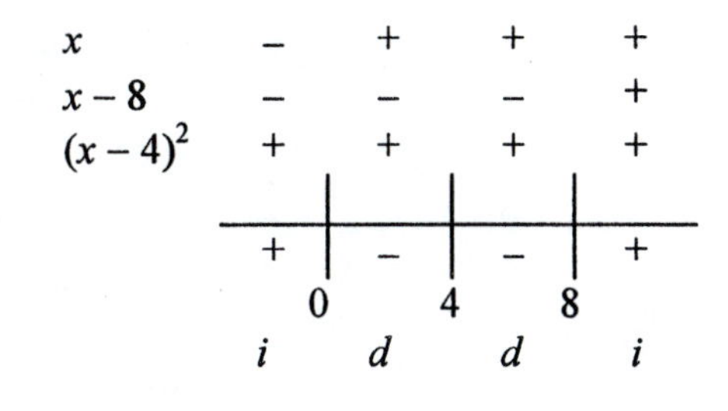

x	−	+	+	+
$x-8$	−	−	−	+
$(x-4)^2$	+	+	+	+
	+	−	−	+
	0	4	8	
	i	d	d	i

max $(0, 0)$ min $(8, 16)$
I: $(0, 0)$ A: $x = 4, y = x + 4$, D: $x \neq 4$

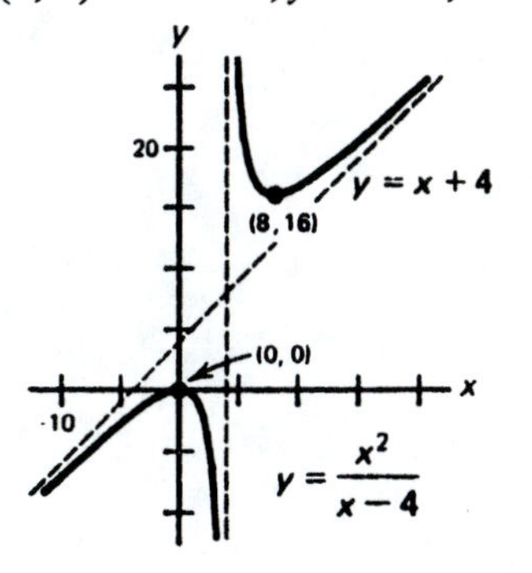

Section 22.2

17. $y' = \dfrac{1}{(x+1)^2}$

$(x+1)^2$ + +

i | i

−1

no max or min; I: (0, 0)
A: $x = -1, y = 1$
D: $x \neq -1$

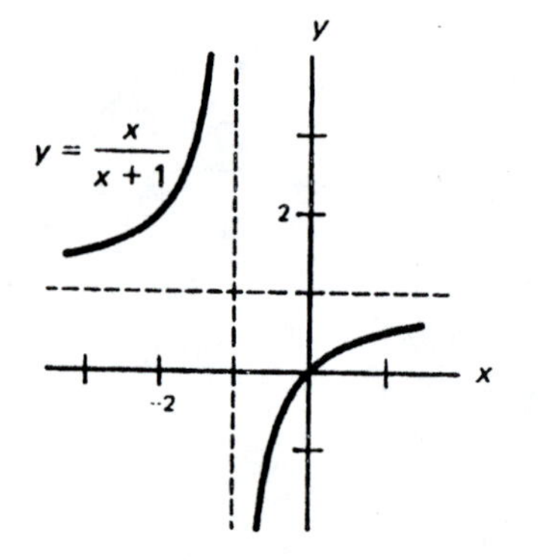

19. $y' = 1 - \dfrac{1}{x^2} = \dfrac{x^2 - 1}{x^2} = \dfrac{(x-1)(x+1)}{x^2}$

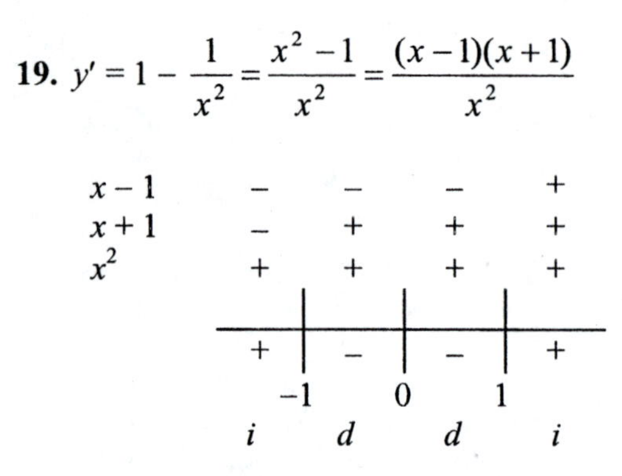

max (−1, −2) min (1, 2)
A: $x = 0, y = x$ D: $x \neq 0$

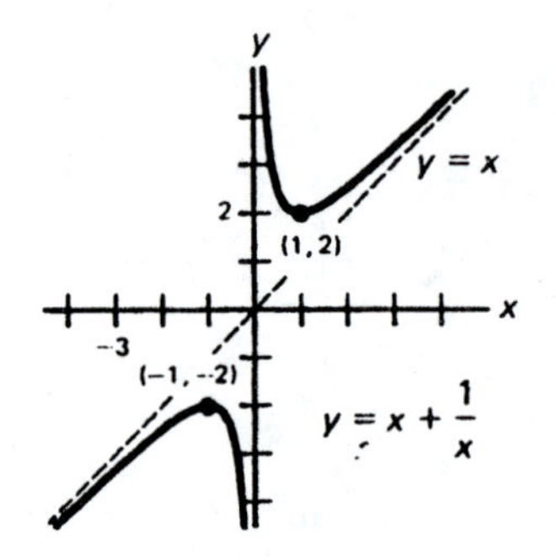

21. $y' = (1/2)x^{-1/2} = 1/(2\sqrt{x})$

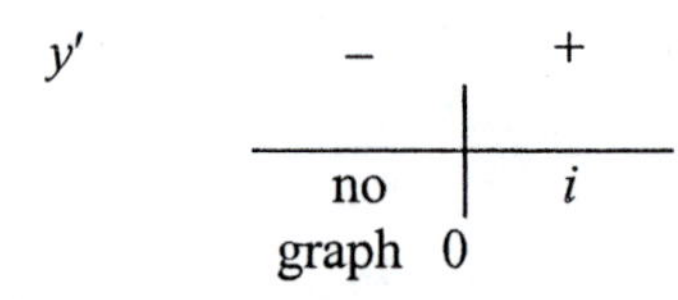

min (0, 0); D: $x \geq 0$

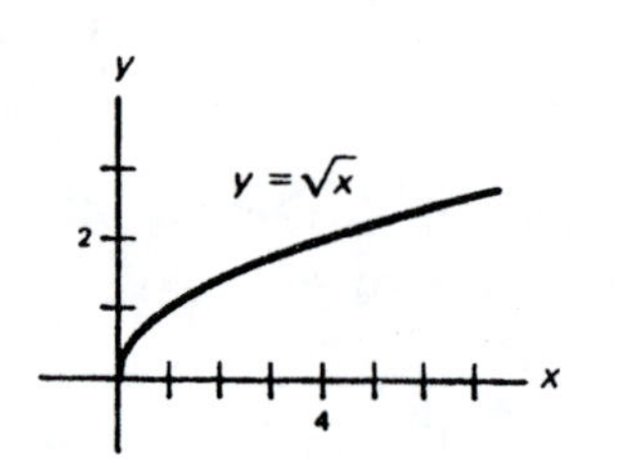

23. $y' = (2/3)x^{-1/3}$

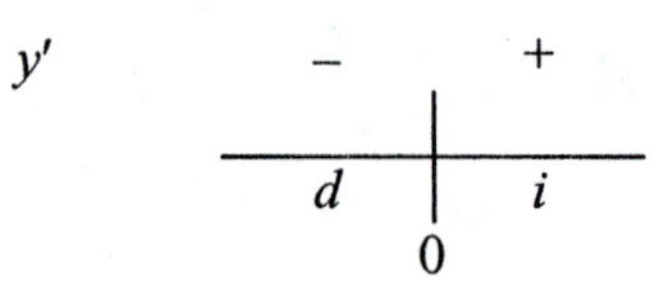

min (0, 0) I: (0, 0); S: y-axis
D: reals

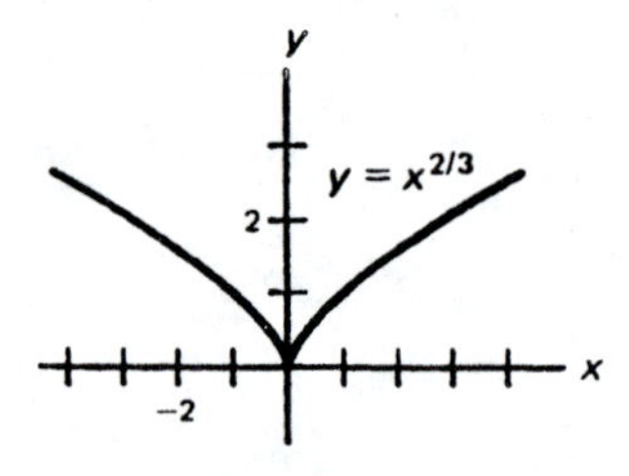

Exercises 22.3

KEY: up = concave upward; dn = concave downward; and same as in previous sections. Use steps as needed. Not all steps shown.

1. $y' = 2x - 4 = 0;\ x = 2$

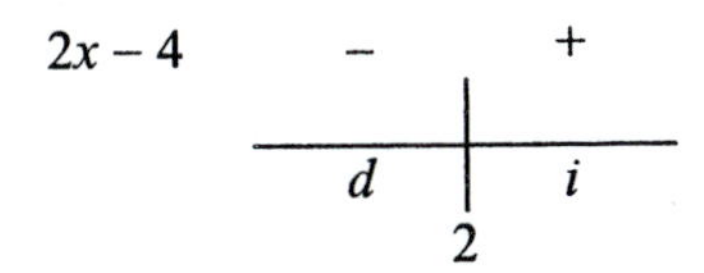

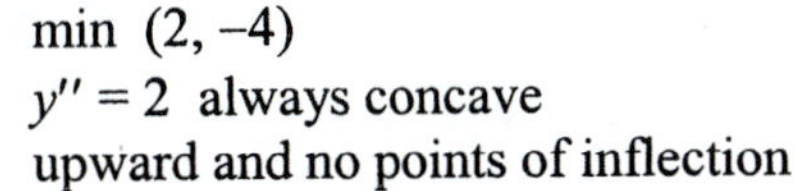

min $(2, -4)$
$y'' = 2$ always concave upward and no points of inflection

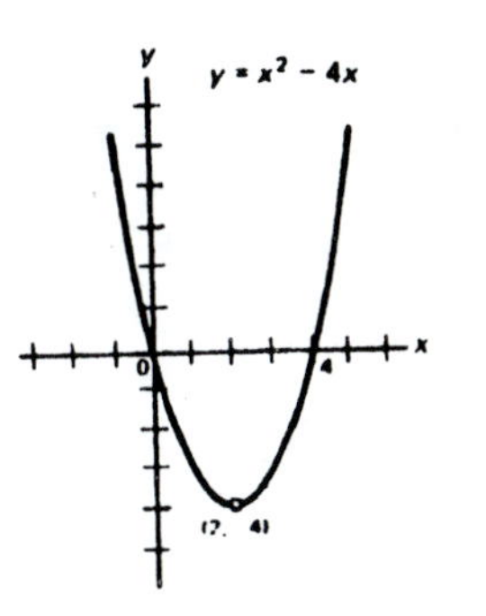

3. $y' = 4x^3 = 0$

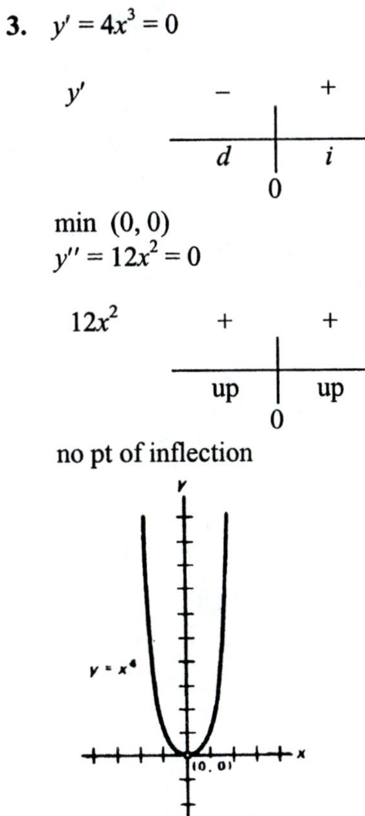

y'	$-$	$+$
	d	i
	0	

min $(0, 0)$
$y'' = 12x^2 = 0$

$12x^2$	$+$	$+$
	up	up
	0	

no pt of inflection

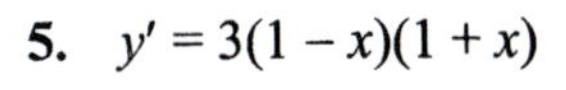

5. $y' = 3(1 - x)(1 + x)$

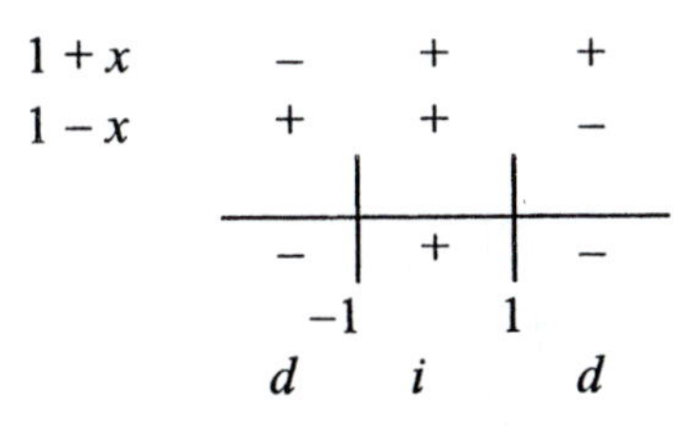

$1 + x$	$-$	$+$	$+$
$1 - x$	$+$	$+$	$-$
	$-$	$+$	$-$
	-1		1
	d	i	d

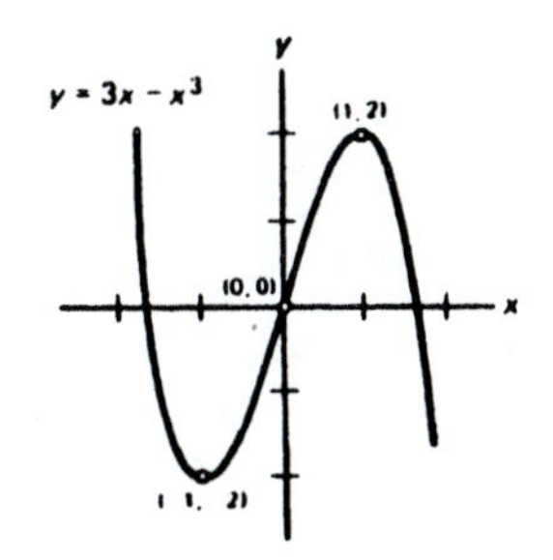

max $(1, 2)$ min $(-1, -2)$

$y'' = -6x$	$+$	$-$
	up	down
	0	

pt. of inflection $(0, 0)$

7. $y' = -2x + 4$

$2x + 4$ +   –

i | d

2

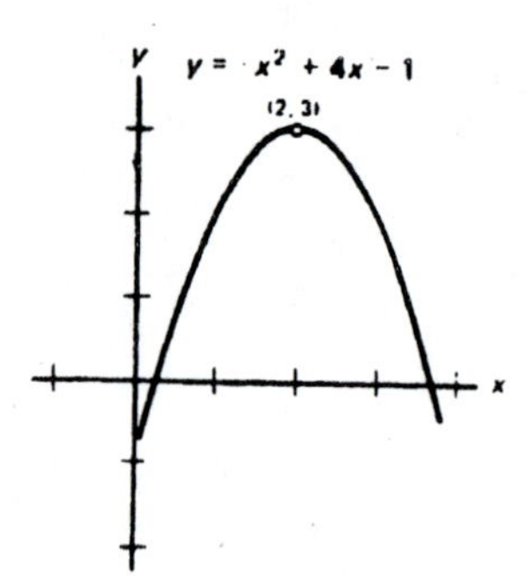

max (2, 3)
$y'' = -2$ Thus graph concave downward for all x; no points of inflection.

9. $y' = 4x(x - 2)(x + 2)$

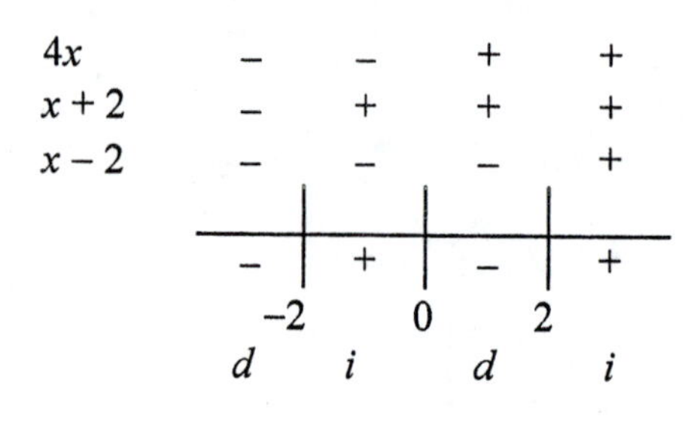

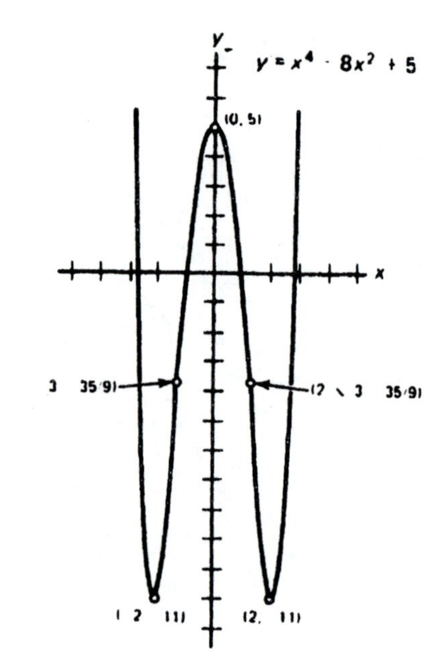

min (–2, –11), (2, 11); max (0, 5)
$y'' = 12x^2 - 16 = 4(\sqrt{3}x - 2)(\sqrt{3}x + 2)$

$\sqrt{3}x - 2$ –   –   +
$\sqrt{3}x + 2$ –   +   +

+ | – | +

$-2/\sqrt{3}$ $2/\sqrt{3}$

up   dn   up

inf. pts. $(-2/\sqrt{3}, -35/9)$, $(2/\sqrt{3}, -35/9)$

11. $y' = -2/x^3$

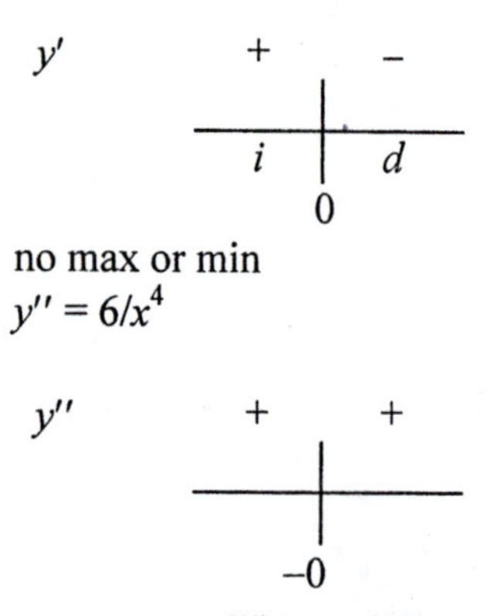

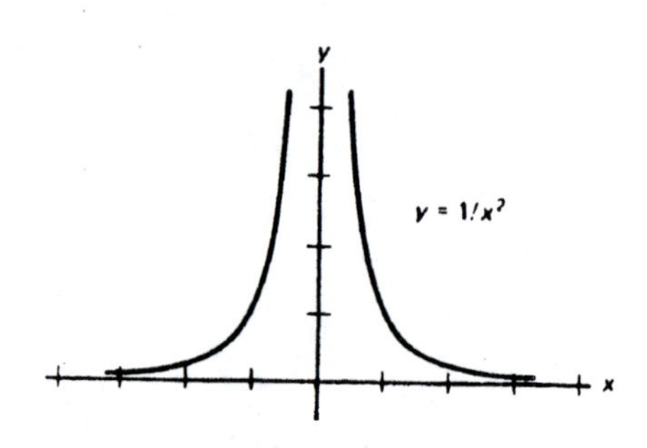

no pts. of inflection

Section 22.3

13. $y' = 2/(x+2)^2$

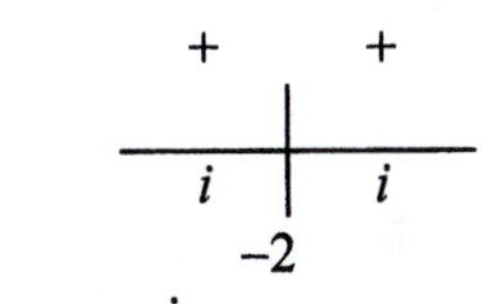

no max or min

$y'' = -4/(x+2)^3$

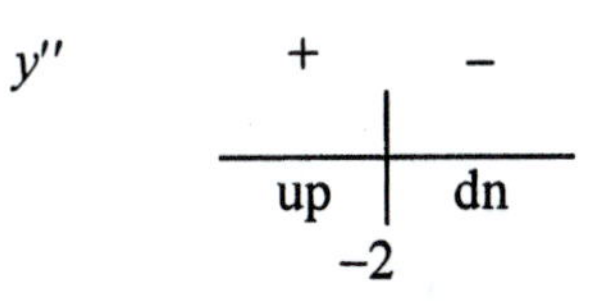

no points of inflection

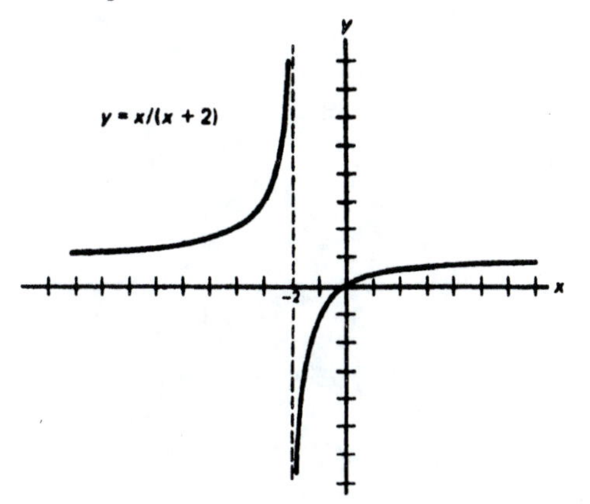

15. $y' = 3/(x+1)^4$

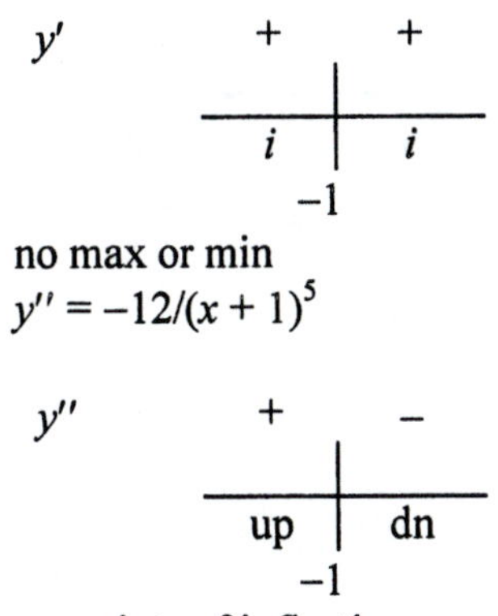

no points of inflection

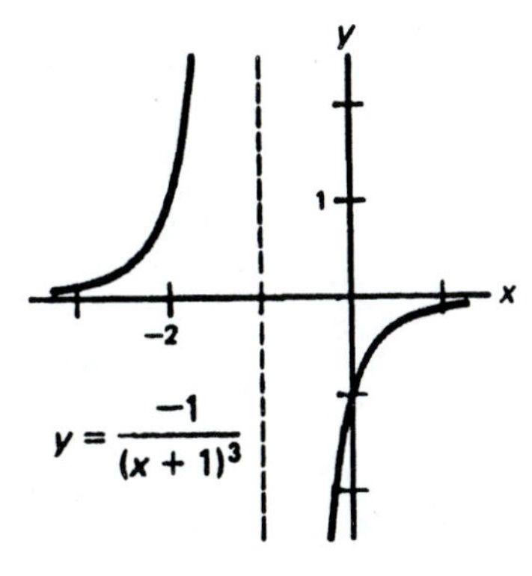

17. $y' = \dfrac{1-x^2}{(x^2+1)^2} = \dfrac{(1+x)(1-x)}{(x^2+1)^2}$

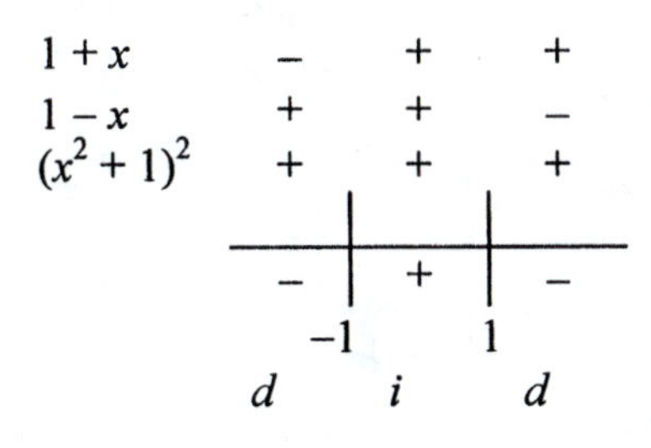

max (1, 1/2); min (−1, −1/2)

$$y'' = \frac{-2x(3-x^2)}{(x^2+1)^3} = \frac{-2x(\sqrt{3}-x)(\sqrt{3}+x)}{(x^2+1)^3}$$

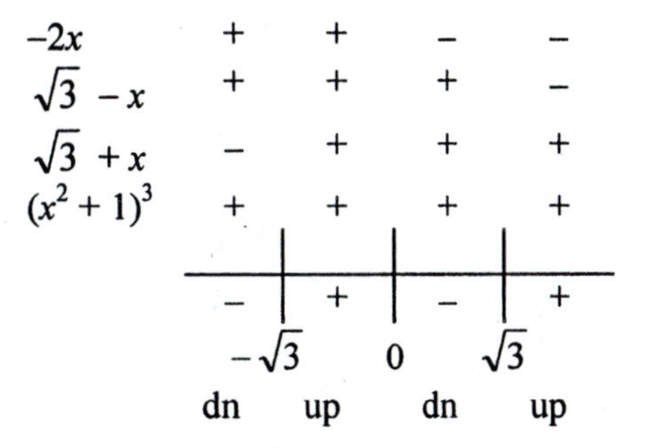

points of inflection: $(-\sqrt{3}, -\sqrt{3}/4)$, $(0, 0)$ and $(\sqrt{3}, \sqrt{3}/4)$

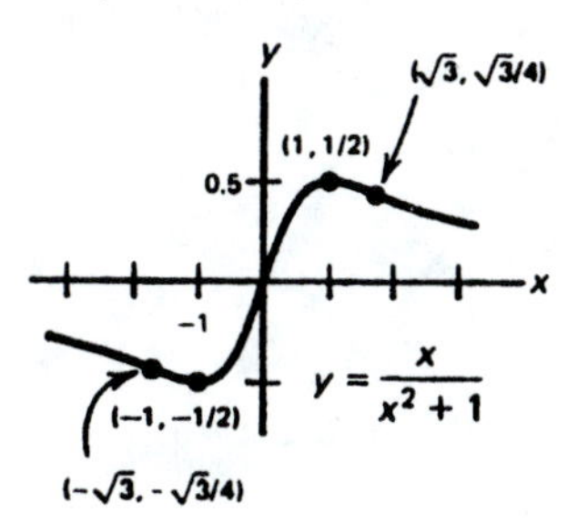

19. $y' = 5/(x+3)^2$

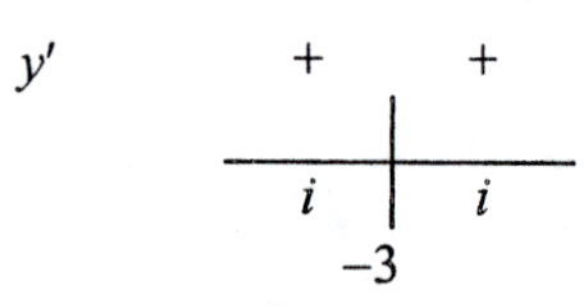

no max or min

$y'' = -10/(x+3)^3$

y''	+	−
	up	dn
	−3	

no points of inflection

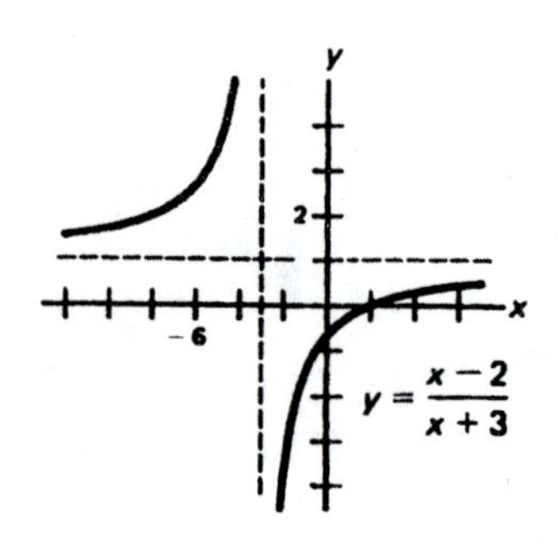

21. $y' = -8x/(x^2+4)^2$

$-8x$	+	−
$(x^2+4)^2$	+	+
	+	−
	i 0	d

max (0, 1)

$$y'' = \frac{-8(4-3x^2)}{(x^2+4)^3}$$

$$= \frac{-8(2+\sqrt{3}x)(2-\sqrt{3}x)}{(x^2+4)^3}$$

-8	−	−	−
$2+\sqrt{3}\,x$	−	+	+
$2-\sqrt{3}\,x$	+	+	−
$(x^2+4)^3$	+	+	+
	+	−	+
	$-2/\sqrt{3}$	$2/\sqrt{3}$	
	up	dn	up

pts. of inflection $(-2/\sqrt{3}, 3/4)$; $(2/\sqrt{3}, 3/4)$

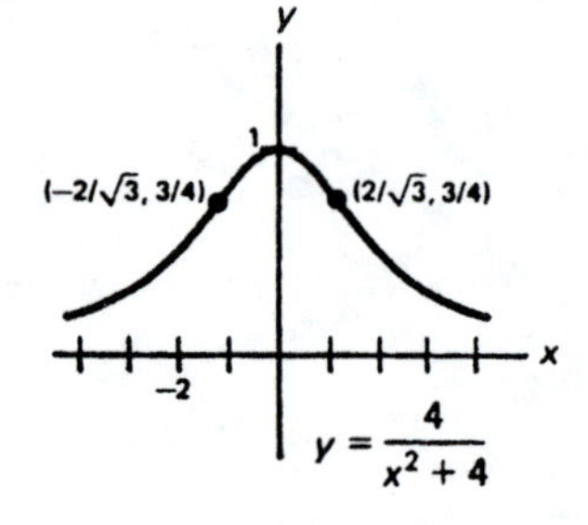

23. $y' = \dfrac{2-x}{x^3}$

$2-x$	+	+	−
x^3	−	+	+
	−	+	−
	0	2	
	d	i	d

max (2, 1/4)

$y'' = \dfrac{2(x-3)}{x^4}$

$2(x-3)$	−	−	+
x^4	+	+	+
	dn	dn	up
	0	3	

pt of inflection (3, 2/9)

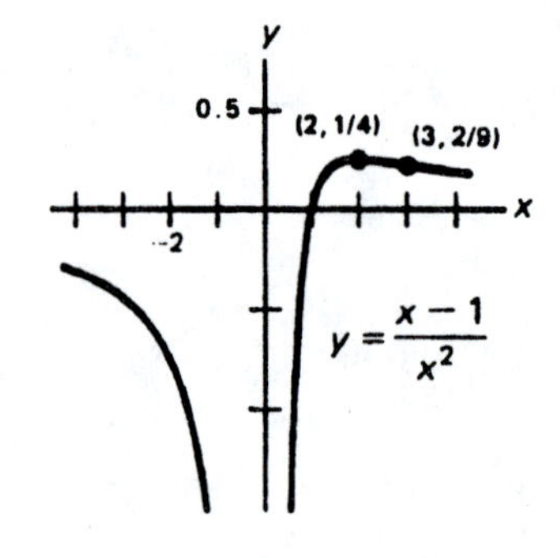

Exercises 22.4

1.

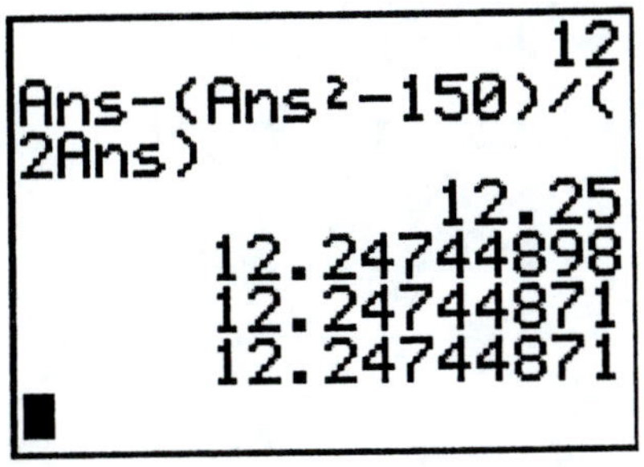

```
                12
Ans-(Ans²-150)/(
2Ans)
             12.25
       12.24744898
       12.24744871
       12.24744871
```

3.

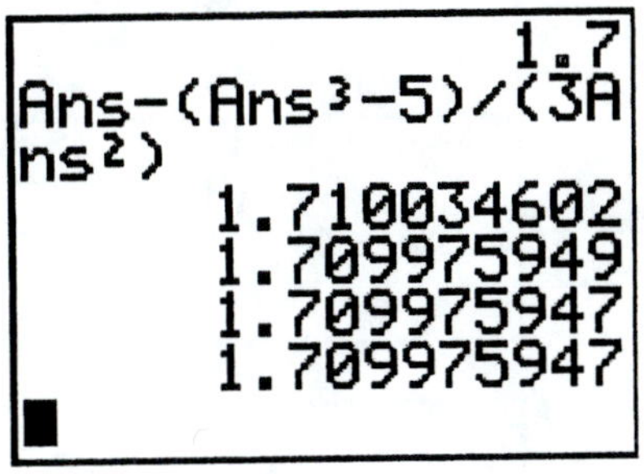

```
               1.7
Ans-(Ans³-5)/(3A
ns²)
       1.710034602
       1.709975949
       1.709975947
       1.709975947
```

5.

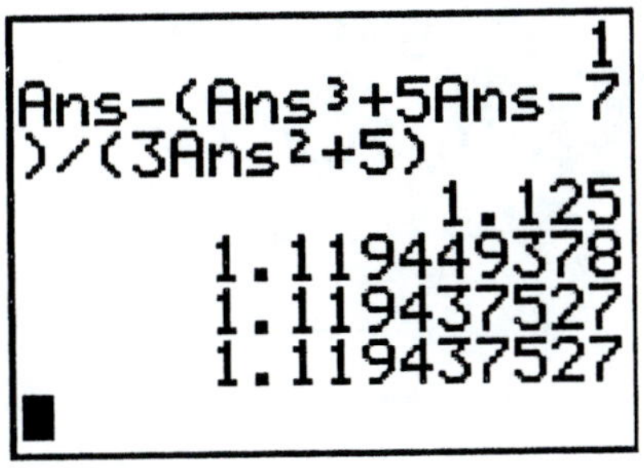

```
                 1
Ans-(Ans³+5Ans-7
)/(3Ans²+5)
             1.125
       1.119449378
       1.119437527
       1.119437527
```

7.

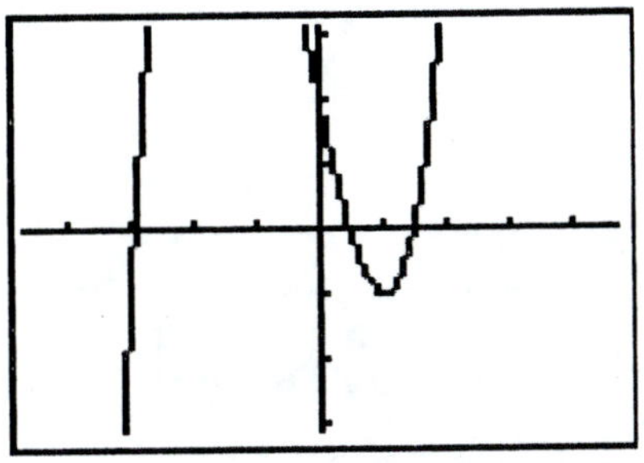

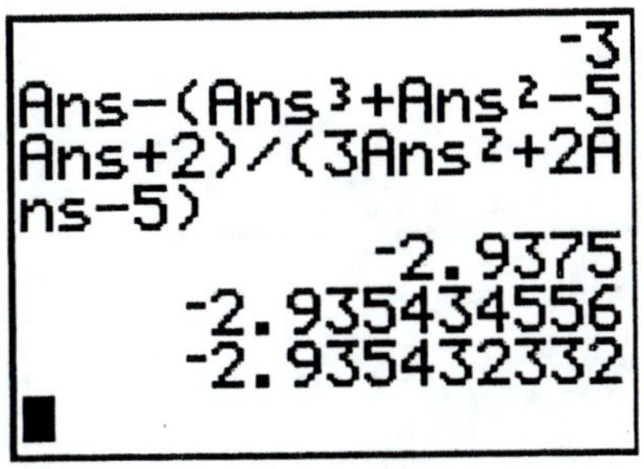

```
                -3
Ans-(Ans³+Ans²-5
Ans+2)/(3Ans²+2A
ns-5)
           -2.9375
      -2.935434556
      -2.935432332
```

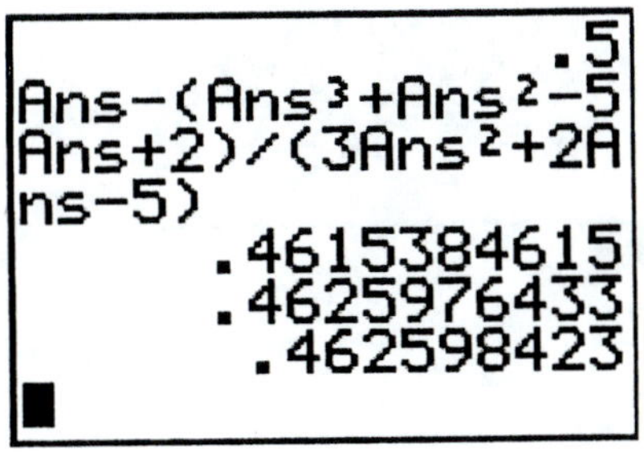

```
                .5
Ans-(Ans³+Ans²-5
Ans+2)/(3Ans²+2A
ns-5)
      .4615384615
      .4625976433
       .462598423
```

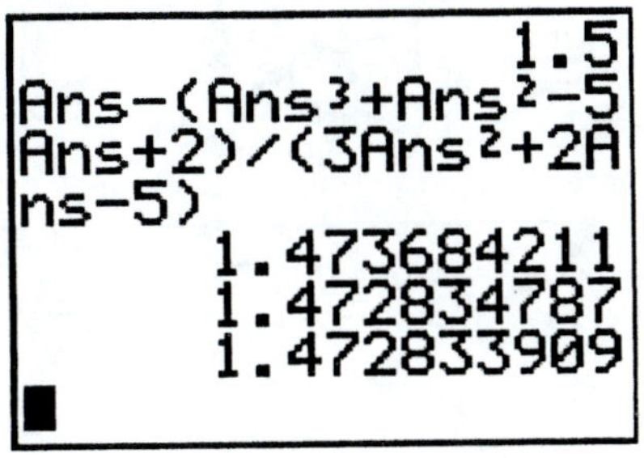

```
               1.5
Ans-(Ans³+Ans²-5
Ans+2)/(3Ans²+2A
ns-5)
       1.473684211
       1.472834787
       1.472833909
```

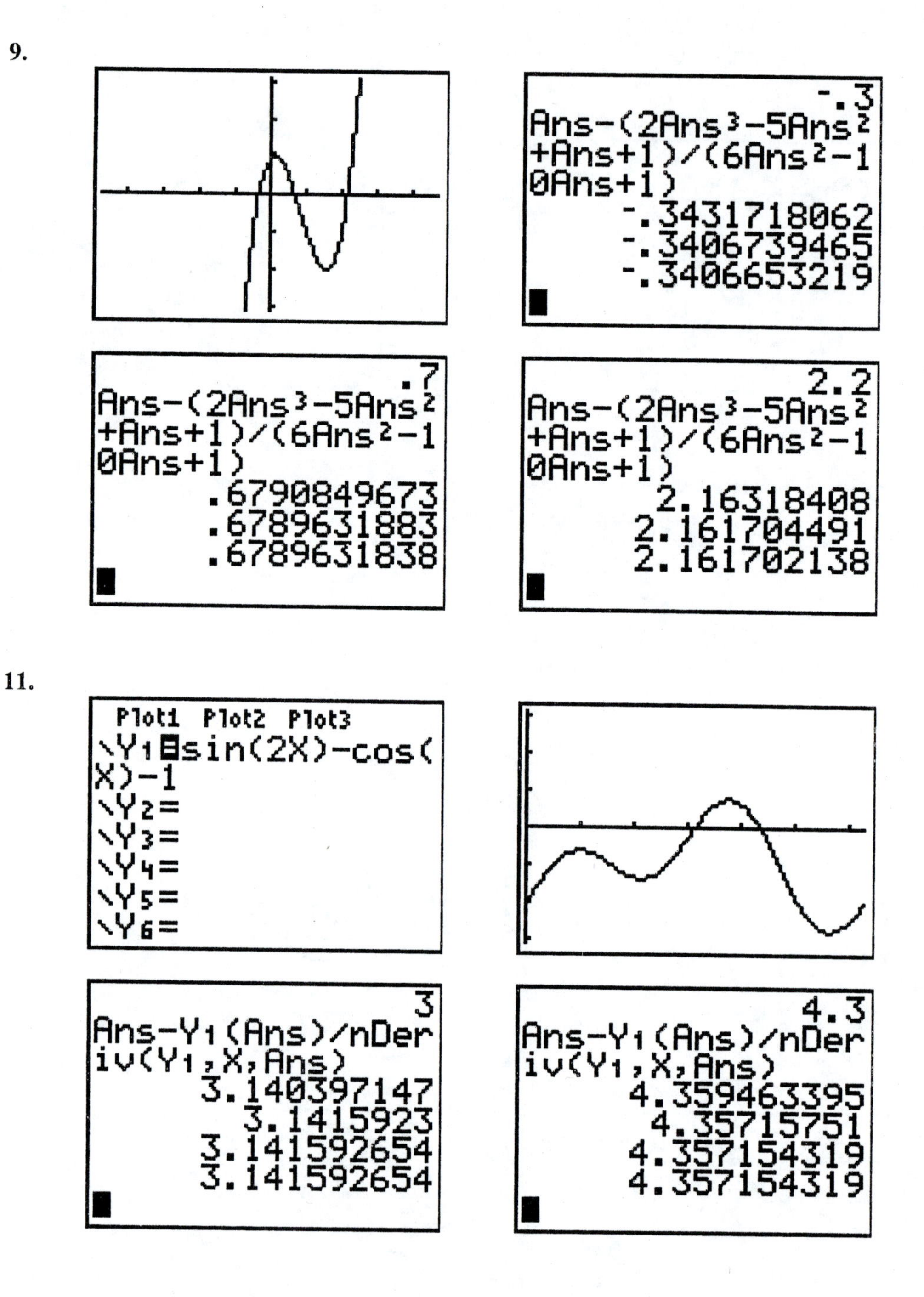

-.3
Ans-(2Ans³-5Ans²
+Ans+1)/(6Ans²-1
0Ans+1)
-.3431718062
-.3406739465
-.3406653219
.7
Ans-(2Ans³-5Ans²
+Ans+1)/(6Ans²-1
0Ans+1)
.6790849673
.6789631883
.6789631838
2.2
Ans-(2Ans³-5Ans²
+Ans+1)/(6Ans²-1
0Ans+1)
2.16318408
2.161704491
2.161702138
11.
Plot1 Plot2 Plot3
\Y1=sin(2X)-cos(
X)-1
\Y2=
\Y3=
\Y4=
\Y5=
\Y6=
3
Ans-Y1(Ans)/nDer
iv(Y1,X,Ans)
3.140397147
3.1415923
3.141592654
3.141592654
4.3
Ans-Y1(Ans)/nDer
iv(Y1,X,Ans)
4.359463395
4.35715751
4.357154319
4.357154319

13.

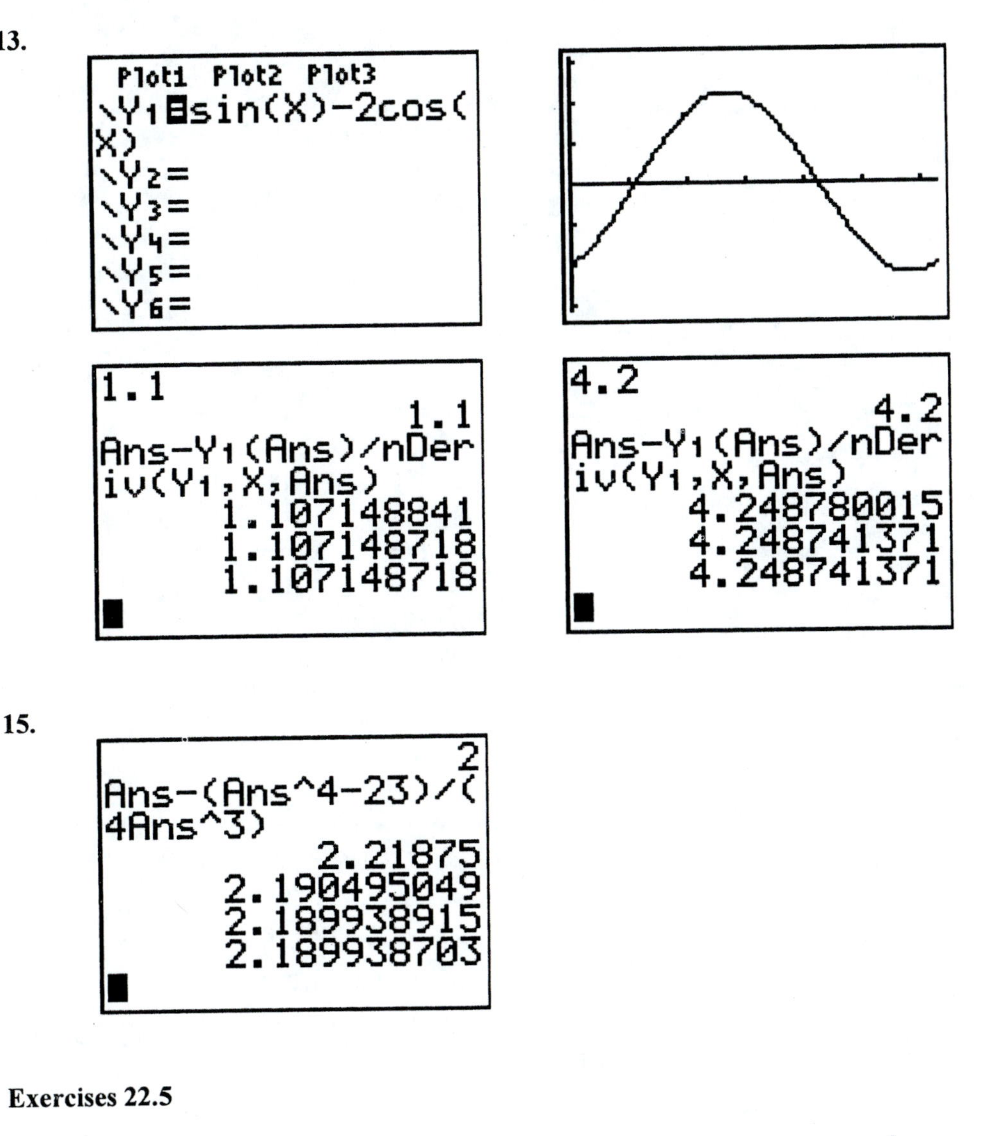

Exercises 22.5

1. $P = xy$ and $x + y = 56$
   $P = x(56 - x) = 56x - x^2$
   $dP/dx = 56 - 2x = 0$
   $x = 28, y = 28$

3. $V = lwh = (3 - 2x)^2 x$
   $dV/dx = x \cdot 2(3 - 2x)(-2) + (3 - 2x)^2$
   $= (3 - 2x)[-4x + (3 - 2x)]$
   $= (3 - 2x)(3 - 6x) = 0$
   max $x = 1/2$;
   dimensions $2 \times 2 \times 0.5$

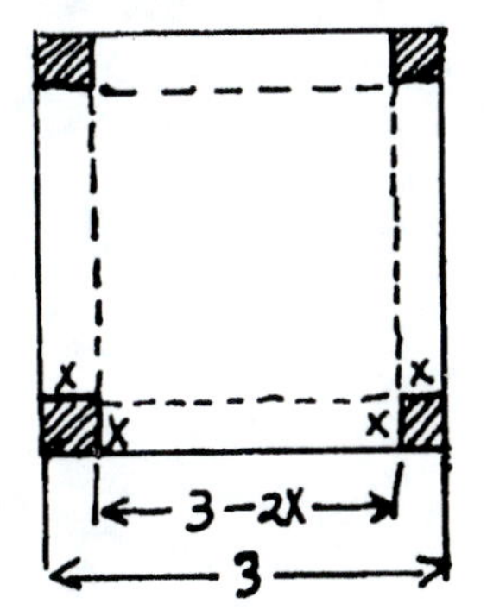

5. $P = 2x + y; A = xy$

$A = x(800 - 2x) = 800x - 2x^2$

$dA/dx = 800 - 4x = 0$

$x = 200, y = 400$

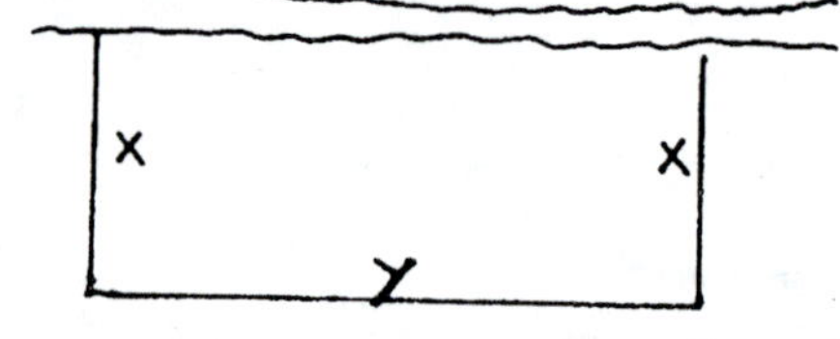

7. $A = \ell w$; $P = 2\ell + 2w = 36$

$\ell + w = 18$

$A = \ell(18 - \ell) = 18\ell - \ell^2$

$dA/D\ell = 18 - 2\ell = 0$

$\ell = 9, w = 9$

Since $A'' = -2 < 0$, $\ell = 9$ is a max.

9.

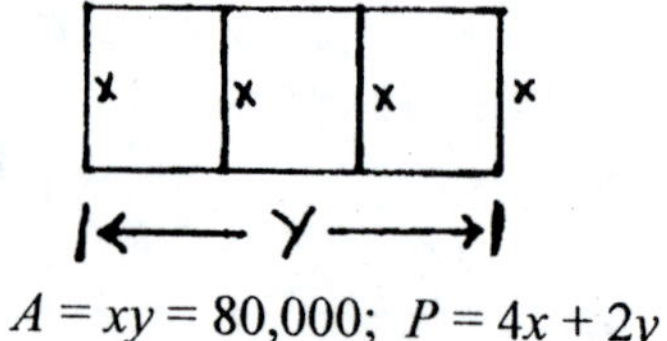

$A = xy = 80{,}000$; $P = 4x + 2y$

$P = 4x + 2(80{,}000/x)$

$dP/dx = 4 - 160{,}000/x^2 = 0$

$x = 200, y = 400$

$P = 4(200) + 2(400) = 1600$

11.

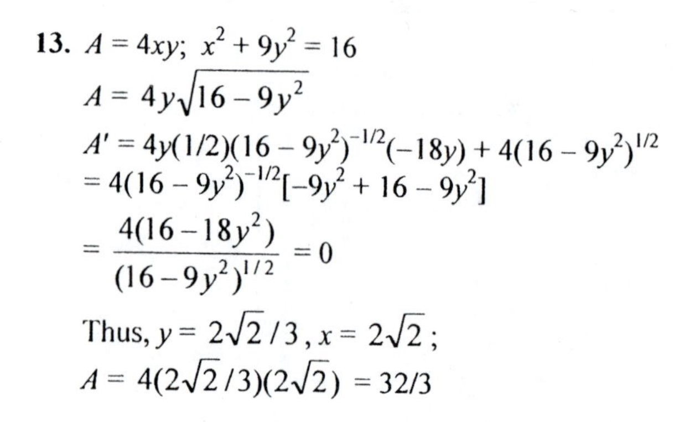

$A = \ell w = 2x(36 - x^2) = 72x - 2x^3$

$A' = 72 - 6x^2 = 0$

$x = \pm 2\sqrt{3}, y = 24$

$A = (2)(2\sqrt{3})(24) = 96\sqrt{3}$

Since $A'' = -12x\Big|_{2\sqrt{3}} = -24\sqrt{3} < 0$

A is a max.

13. $A = 4xy;\ x^2 + 9y^2 = 16$

$A = 4y\sqrt{16 - 9y^2}$

$A' = 4y(1/2)(16 - 9y^2)^{-1/2}(-18y) + 4(16 - 9y^2)^{1/2}$

$= 4(16 - 9y^2)^{-1/2}[-9y^2 + 16 - 9y^2]$

$= \dfrac{4(16 - 18y^2)}{(16 - 9y^2)^{1/2}} = 0$

Thus, $y = 2\sqrt{2}/3$, $x = 2\sqrt{2}$;

$A = 4(2\sqrt{2}/3)(2\sqrt{2}) = 32/3$

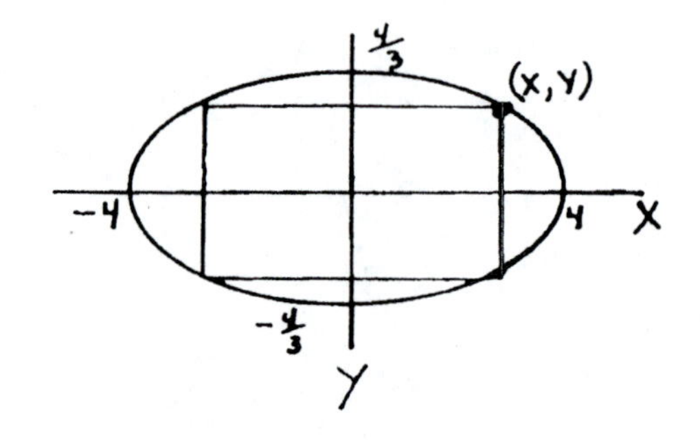

15. $\dfrac{dP}{dR} = \dfrac{(R+4)^2(36) - 72R(R+4)}{(R+4)^4} = \dfrac{(R+4)[36(R+4) - 72R]}{(R+4)^4} = \dfrac{-36R+144}{(R+4)^3} = 0$ thus $R = 4\ \Omega$

17. $A = 4xy;\ x^2 + y^2 = r^2$

$A = 4x(r^2 - x^2)^{1/2}$

$A' = 4(r^2 - x^2)^{1/2} + 4x(1/2)(r^2 - x^2)^{-1/2}(-2x)$

$= 4(r^2 - x^2)^{-1/2}[(r^2 - x^2) - x^2]$

$= \dfrac{4(r^2 - 2x^2)}{\sqrt{r^2 - x^2}} = 0$ thus $x = r\sqrt{2}/2,\ y = r\sqrt{2}/2$

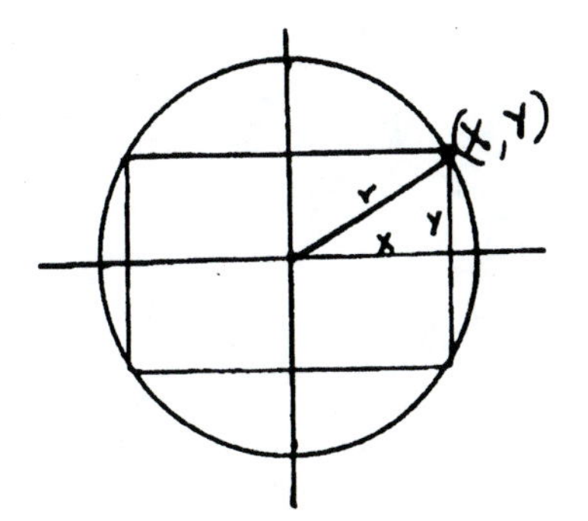

Since $A''(r\sqrt{2}/2) < 0$ and $x = y$, the max area is a square.

19. $m = y' = 6x - 6x^2$

$dm/dx = 6 - 12x = 0$

$x = 1/2$

Since $m'' = -12 < 0$, $x = 1/2$ gives a max slope and

$m = 6(1/2) - 6(1/2)^2 = 3/2$

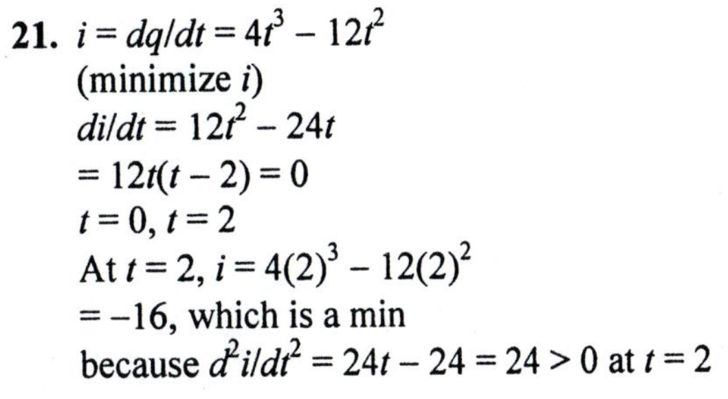

21. $i = dq/dt = 4t^3 - 12t^2$

(minimize i)

$di/dt = 12t^2 - 24t$

$= 12t(t-2) = 0$

$t = 0, t = 2$

At $t = 2$, $i = 4(2)^3 - 12(2)^2$

$= -16$, which is a min

because $d^2i/dt^2 = 24t - 24 = 24 > 0$ at $t = 2$

23. $V = x^2y = 4000;\ A = 4xy + x^2$

$A = 4x(4000/x^2) + x^2 = 16000/x + x^2$

$A' = -16000/x^2 + 2x = 0$

$2x = 16000/x^2$

$x^3 = 8000$

$x = 20;\ y = 10$

dimensions: $20 \times 20 \times 10$

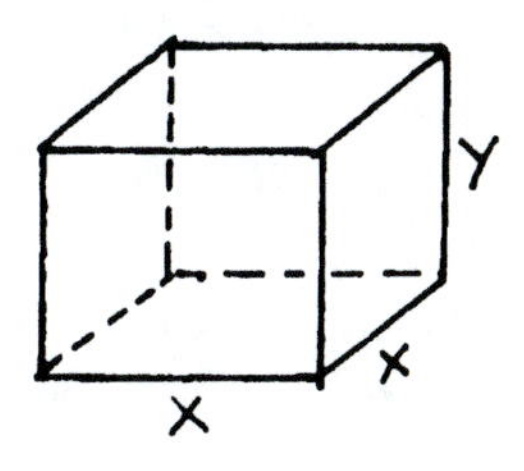

25. $P = 36.75x - (0.005x^3 + 0.45x^2 + 12.75x) = -0.005x^3 - 0.45x^2 + 24x$

$P' = -0.015x^2 - 0.9x + 24 = 0$

$x = \dfrac{0.9 \pm \sqrt{(-0.9)^2 - 4(-0.015)(24)}}{2(-0.015)} = \dfrac{0.9 \pm \sqrt{2.25}}{-0.03} = 20$

27. $V = \pi r^2 h;\ A = \pi r^2 + 2\pi rh = 24\pi$

$h = \dfrac{24 - r^2}{2r}$

$V = \pi r^2(24 - r^2)/2r = \pi(12r - r^3/2)$

$V' = \pi(12 - 3r^2/2) = 0;\ 12 = 3r^2/2$

$r^2 = 8;\ r = 2\sqrt{2}$,

$h = (24 - 8)/(2\sqrt{2}) = 4\sqrt{2}$

29. $S = kwd^2$; $r^2 = d^2/4 + w^2/4$

$$d^2 = 4r^2 - w^2$$

$$S = kw(4r^2 - w^2) = 4kwr^2 - kw^3$$

$$dS/dw = 4kr^2 - 3kw^2 = 0$$

$$4r^2 - 3w^2 = 0$$

$$w^2 = 4r^2/3;\ w = 2r/\sqrt{3}$$

$$d^2 = 4r^2 - 4r^2/3 = 8r^2/3$$

$$d = 2r\sqrt{2}/\sqrt{3} = 2r\sqrt{6}/3$$

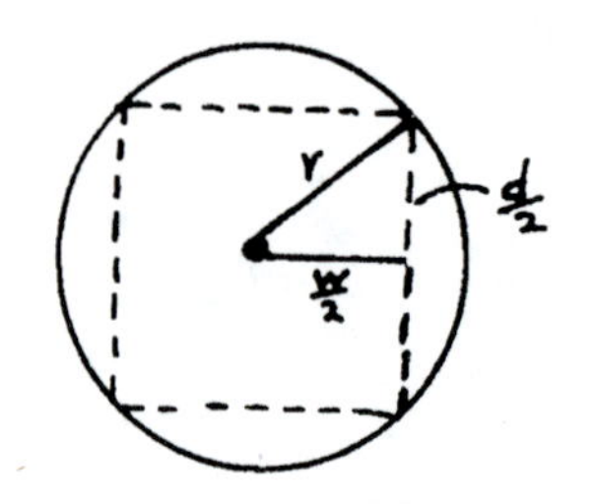

31. Use similar triangles:

$$\frac{8}{12} = \frac{x}{10 - x}$$

$$80 - 8x = 12x$$

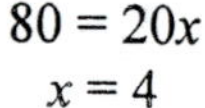

$$80 = 20x$$

$$x = 4$$

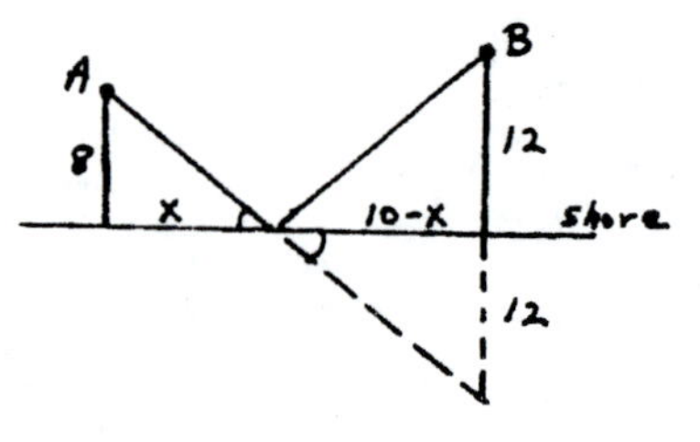

Exercises 22.6

1. $\frac{dy}{dt} = 6x\frac{dx}{dt},\quad \frac{dy}{dt} = 6(3)(2),\quad \frac{dy}{dt} = 36.$

3. $x^2 + 4^2 = 25$ and $x > 0$ implies $x = 3$,

$$2x\frac{dx}{dt} + 2y\frac{dy}{dt} = 0,\quad \frac{dx}{dt} = \frac{-y}{x}\frac{dy}{dt},\quad \frac{dx}{dt} = \frac{-4}{3}\left(\frac{9}{2}\right) = -6$$

5. $dT/dt = 0.3°\text{C/s}$, find dR/dt when $T = 100°\text{C}$
$dR/dt = 0.002T\ dT/dt = 0.002(100)(0.3) = 0.06\Omega/\text{s}$

7. $A = s^2$; find dA/dt when $ds/dt = 0.08$ cm/min and $s = 12$ cm
$dA/dt = 2s\ ds/dt = 2(0.08)(12) = 1.92\ \text{cm}^2/\text{min}$

9. $V = e^3$; find dV/dt when $ds/dt = 0.1$ cm/min and $e = 12$ cm
$dV/dt = 3e^2\ de/dt = 3(12^2)(0.1) = 43.2\ \text{cm}^3/\text{min}$

11. $A = \sqrt{3}s^2/4$; find dA/dt when $ds/dt = -0.04$ cm/min and $s = 8$ cm
$dA/dt = (\sqrt{3}/2)s\ ds/dt = (\sqrt{3}/2)(8)(-0.04) = -0.277\ \text{cm}^2/\text{min}$

13. $h = 3r/4$; $V = (1/3)\pi r^2 h = (1/3)\pi r^2(3r/4) = \pi r^3/4$
Find dr/dt when $dV/dt = 5\ \text{m}^3/\text{min}$, $h = 10$ m,
and $r = 40/3$ m; $dV/dt = (3/4)\pi r^2 dr/dt$;
$5 = (3/4)\pi(40/3)^2 dr/dt$; $dr/dt = 3/(80\pi)$ m/s

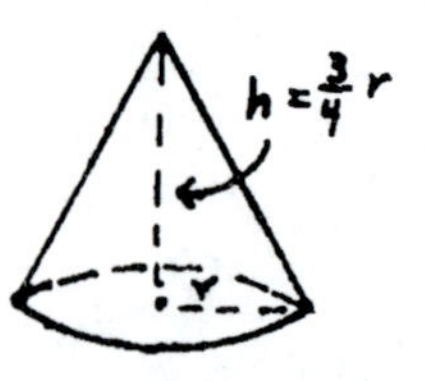

15. $P = I^2R$; find dP/dt when $I = 8A$, $dI/dt = -0.4\ A/s$, $R = 75\ \Omega$, and $dR/dt = 5\ \Omega/\text{s}$.
$dP/dt = I^2\ dR/dt + 2IR\ dI/dt = (8^2)(5) + 2(8)(75)(-0.4) = -160$ W/s

17. $r = 75$ mm (fixed) = 7.5 cm; $V = \pi r^2 h = \pi(7.5)^2 h$
Find dh/dt when $dV/dt = 90$ cm^3/s
$dV/dt = \pi(7.5)^2 dh/dt$; $dh/dt = 90/[(7.5)^2\pi] = 0.509$ cm/s

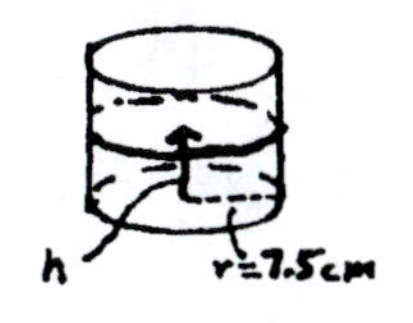

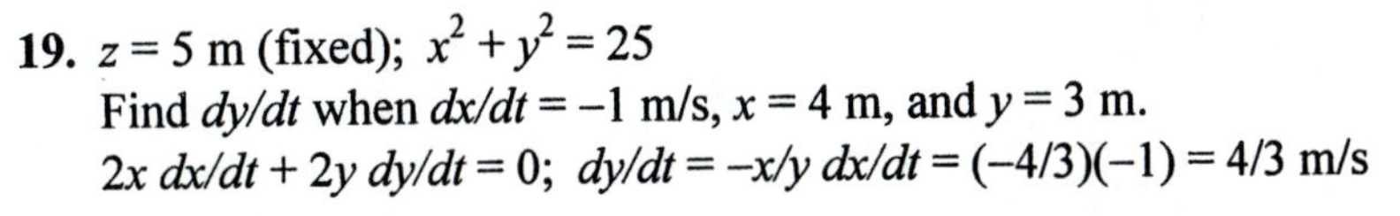

19. $z = 5$ m (fixed); $x^2 + y^2 = 25$
Find dy/dt when $dx/dt = -1$ m/s, $x = 4$ m, and $y = 3$ m.
$2x\ dx/dt + 2y\ dy/dt = 0$; $dy/dt = -x/y\ dx/dt = (-4/3)(-1) = 4/3$ m/s

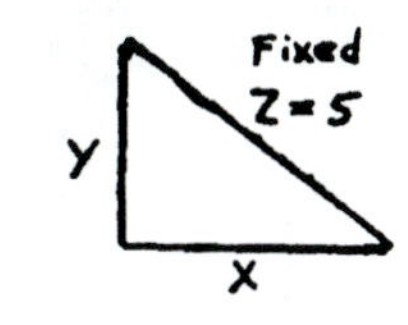

21. $PV^{1.4} = k$; $P = kV^{-1.4}$; find dP/dt when $P = 60$ lb/in^2 · $V = 56$ in^3,
$dV/dt = -8$ in^3/min; $dP/dt = -1.4kV^{-2.4}\ dV/dt$
$= -1.4(PV^{1.4})V^{-2.4}\ dV/dt$
$= [-1.4P/V]\ dV/dt = [-1.4(60)/56](-8) = 12$ lb/in^2/min

23. $R = \dfrac{120R_2}{120 + R_2}$; find dR/dt when $dR_2/dt = -15\ \Omega/s$, $R_2 = 180\ \Omega$

$$\frac{dR}{dt} = \frac{(120 + R_2)120dR_2/dt - 120R_2dR_2/dt}{(120 + R_2)^2_{52}}$$

$$= \frac{(300)(120)(-15) - (120)(180)(-15)}{(120 + 180)^2} = -2.4\ \Omega/\text{s}$$

25. $P = 30i^2$; $dP/di = 60i = 60(4) = 240$ AΩ

27. $V = \dfrac{0.02i}{(0.01)^2} = 200i$; find $di/dt = 0.04$ A/s; $dV/dt = 200\ di/dt = 200(0.04) = 8$ V/s

29. $\dfrac{h}{r} = \dfrac{8}{2} = 4 \qquad r = h/r$
$V = (1/3)\pi r^2 h = (1/3)\pi(h/4)^2 h = (1/48)\pi h^3$
$dV/dt = (\pi/16)h^2 dh/dt$

a) $2\pi = (\pi/16)(2^2)dh/dt$; $dh/dt = 8$ m/min
b) $2\pi = (\pi/16)(6^2)\ dh/dt$; $dh/dt = 8/9$ m/min

Exercises 22.7

1. $dy/dx = 10x - 24x^2$; $dy = (10x - 24x^2)dx$

3. $dy/dx = \dfrac{(2x-1)(1) - (x+3)(2)}{(2x-1)^2} = \dfrac{-7}{(2x-1)^2}$;

$$dy = \frac{-7}{(2x-1)^2}dx$$

5. $dy/dt = 4(2t^2 + 1)^3(4t) = 16t(2t^2 + 1)^3$;
$dy = 16t(2t^2 + 1)^3 dt$

7. $ds/dt = -2(t^4 - t^{-2})^{-3}(4t^3 + 2t^{-3})$;

$ds = -4t^{-3}(t^4 - t^{-2})^{-3}(2t^6 + 1)dt$ or $ds = \dfrac{-4t^3(2t^6+1)}{(t^6-1)^3}dt$

9. $2x + 8y\,dy/dx = 0$; $dy/dx = -x/(4y)$; $dy = (-x/(4y))\,dx$

11. $3(x+y)^2(1+y') = (1/2)x^{-1/2} + (1/2)y^{-1/2}y'$

$3(x+y)^2 + [3(x+y)^2]y' = (1/2)x^{-1/2} + (1/2)y^{-1/2}y'$

$6(x+y)^2 - x^{-1/2} = [y^{-1/2} - 6(x+y)^2]y'$

$y' = \dfrac{6(x+y)^2 - x^{-1/2}}{y^{-1/2} - 6(x+y)^2}$; $dy = \dfrac{6(x+y)^2 - x^{-1/2}}{y^{-1/2} - 6(x+y)^2}dx$

13. $dy = 32x^3\,dx = 32(3^3)(0.05) = 43.2$

15. $dV = (3r^2 - 6r)dr = [3 \cdot 4^2 - 6 \cdot 4](0.05) = 1.2$

17. $dV = 4\pi r^2\,dr = 4\pi(15)^2(0.1) = 282.7$

19. **a)** $dA = 2s\,ds = 2(12)(0.05) = 1.20\text{ cm}^2$

b) $\Delta A = (12.05)^2 - 12^2 = 1.2025\text{ cm}^2$

c) $\dfrac{dA}{A} \cdot 100\% = \dfrac{1.20}{144} \cdot 100\% = 0.833\%$

21. **a)** $V = (4/3)\pi r^3$; $dV = 4\pi r^2 dr = 4\pi(13)^2(0.04) = 84.9\text{ m}^3$

b) steel $= (7800\text{ kg/m}^3)(84.9\text{ m}^3) = 662{,}000\text{ kg}$

23. $dP = 16d\,dd = 16(3.750)(0.005) = 0.3\text{ hp}$

25. $dv = (20/3)p^{-1/3}dP = (20/3)(125)^{-1/3}(3) = 4\text{ V}$

Chapter 22 Review

1. $y = x(x-4)(x+4)$
I: $x = 0, 4, -4; y = 0$

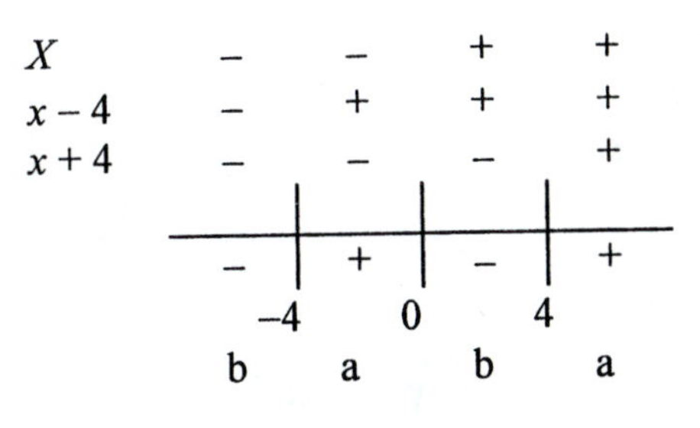

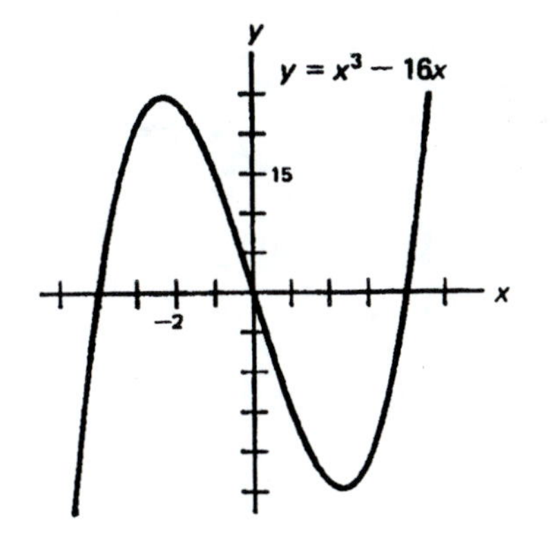

2. $y = \sqrt{-2-4x}$
I: $x = -1/2$; D: $x \le -1/2$

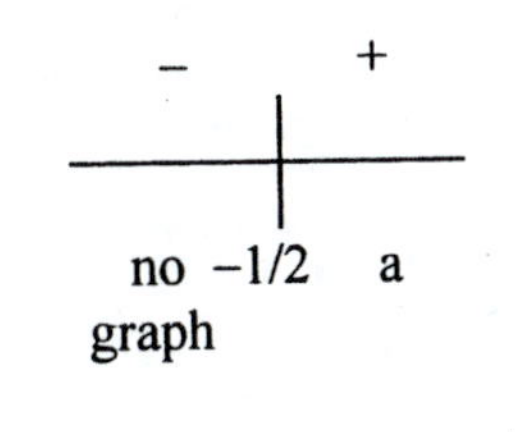

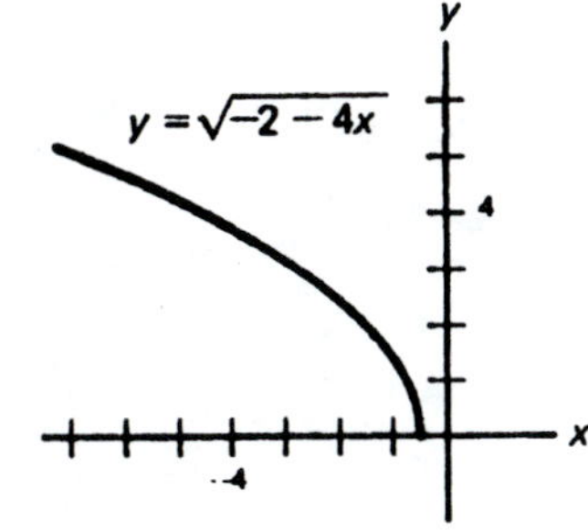

3. $y = (x+2)(5-x)(5+x)$
I: $x = -2, 5, -5; y = 50$

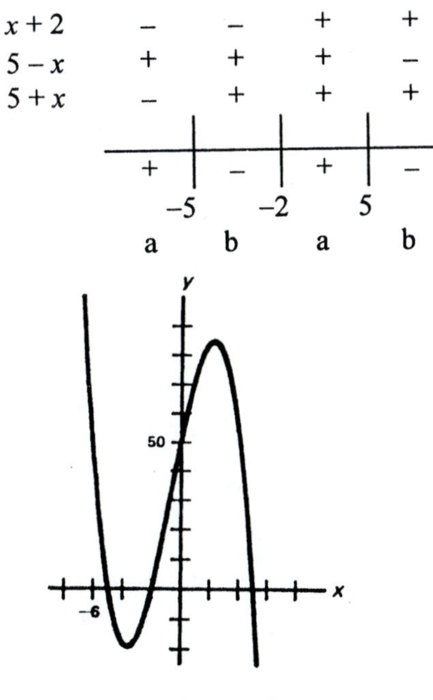

$y = (x + 2)(25 - x^2)$

4. $y = x(x+4)(x-1)^2$
I: $x = 0, 1, -4; y = 0$

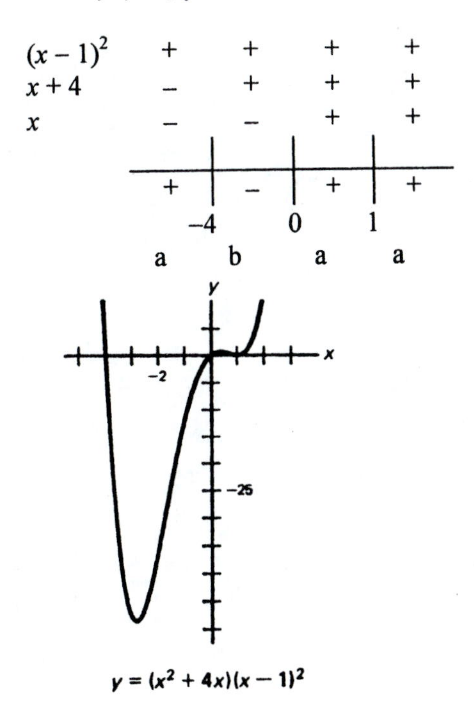

$y = (x^2 + 4x)(x - 1)^2$

5. $y = \pm\sqrt{x-4}$

I: $x = 4$; D: $x \geq 4$

S: x-axis

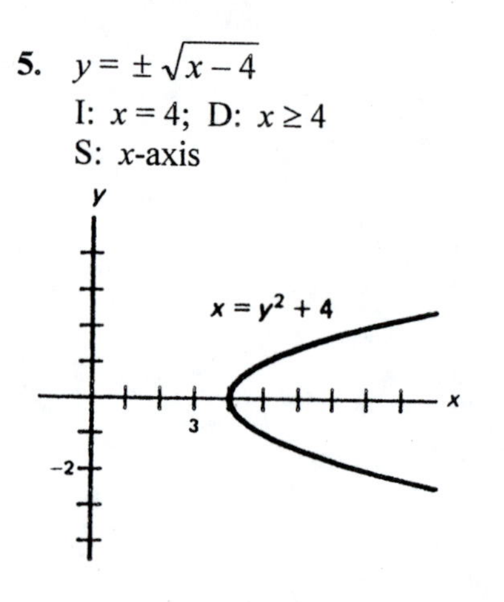

6. I: $x = 2$; $y = 1/2$

D: $x \neq -4, 1$ A: $x = -4, x = 1, y = 0$

$x-2$	−	−	+	+
$x+4$	−	+	+	+
$x-1$	−	−	−	+
	−	+	−	+
		−4	1	2
	b	a	b	a

$$y = \frac{x-2}{(x+4)(x-1)}$$

7. I: $x = 0, y = 0$;

D: $x \neq 1, -1$

A: $x = \pm 1, y = 2$;

S: y-axis

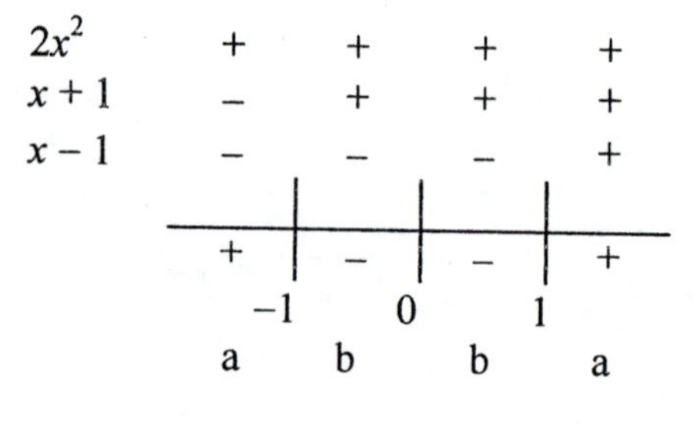

$2x^2$	+	+	+	+
$x+1$	−	+	+	+
$x-1$	−	−	−	+
	+	−	−	+
		−1	0	1
	a	b	b	a

$$y = \frac{2x^2}{x^2-1}$$

8. $y = \dfrac{(x+4)(x-3)}{x+3}$; I: $x = -4, 3$; $y = -4$

D: $x \neq -3$; A: $x = -3, y = x - 2$

$x+4$	−	+	+	+
$x-3$	−	−	−	+
$x+3$	−	−	+	+
	−	+	−	+
		−4	−3	3
	b	a	b	a

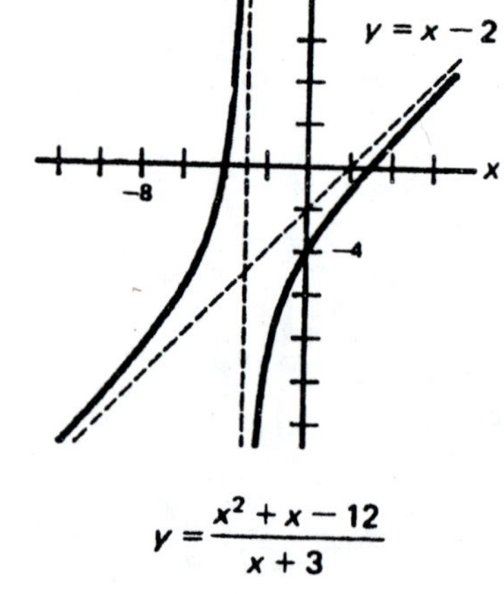

$$y = \frac{x^2+x-12}{x+3}$$

9. I: $x = 0, y = 0$; D: {reals}
S: origin; A: $y = 0$

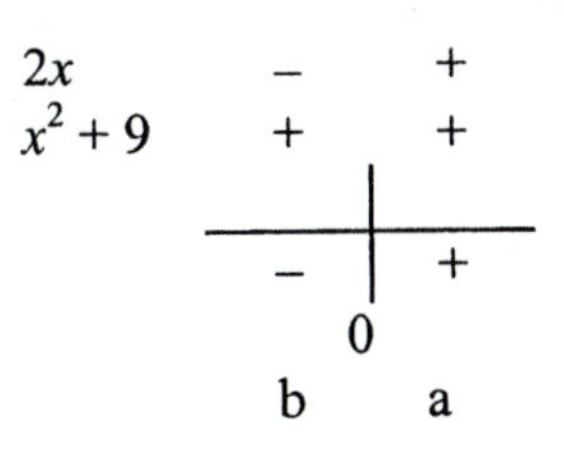

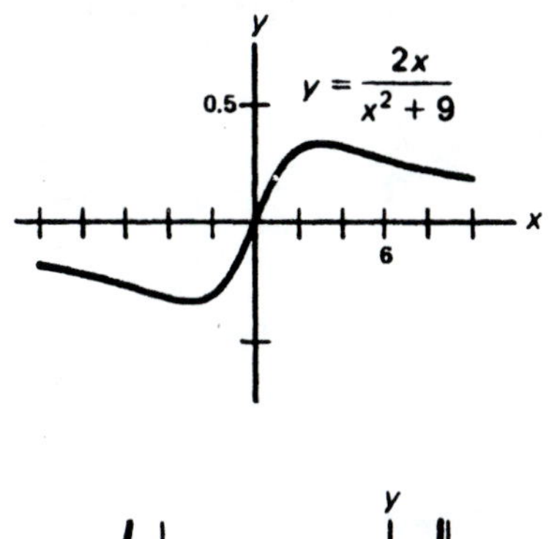

10. $y = \pm\sqrt{x/[(1-x)(x+4)]}$
I: $x = 0, y = 0$
D: $x < -4, 0 \le x < 1$
S: x-axis A: $x = 1, x = -4, y = 0$
$x = -4 \quad y = 0$

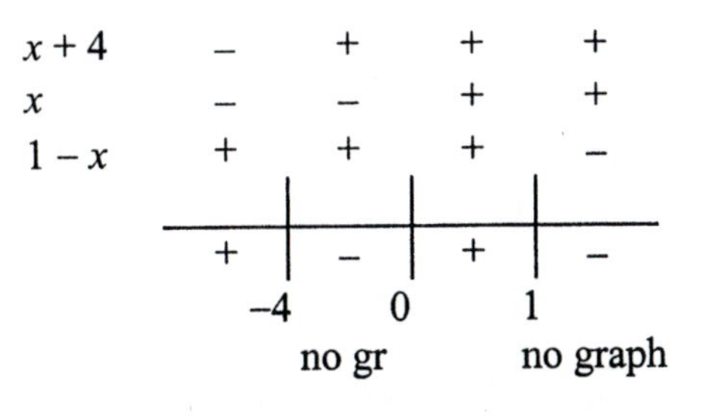

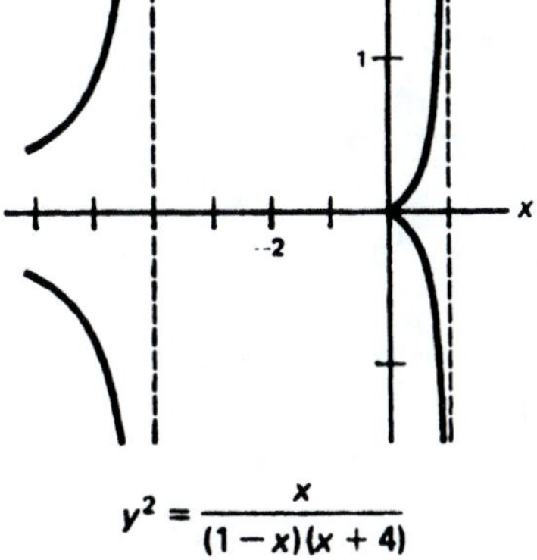

$$y^2 = \frac{x}{(1-x)(x+4)}$$

11. $y' = 6 - 3x^2$
$= 3(\sqrt{2} - x)(\sqrt{2} + x)$

$\sqrt{2} - x$	+	+	–
$\sqrt{2} + x$	–	+	+
	–	+	–

$-\sqrt{2}$ $\sqrt{2}$

d i d

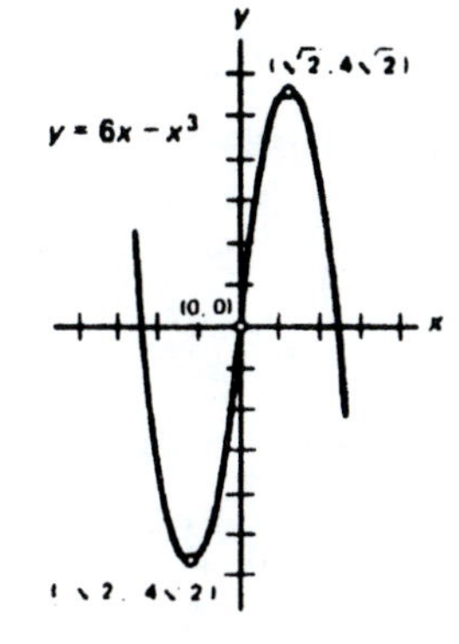

max $(\sqrt{2}, 4\sqrt{2})$, min $(-\sqrt{2}, -4\sqrt{2})$
$y'' = -6x$

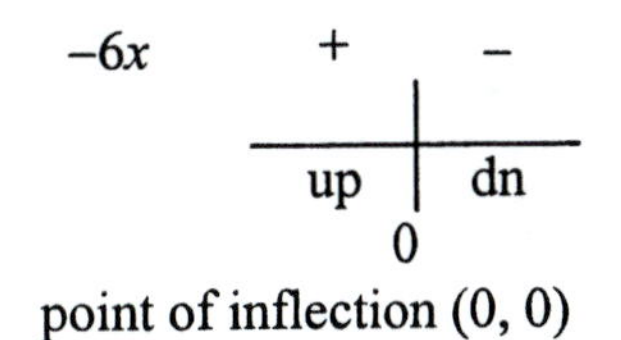

point of inflection (0, 0)

12. $y' = 2x - 3$

y'	$-$	$+$
	d	i
	3/2	

min (3/2, −25/4)
$y'' = 2$; concave up for all x;
no points of inflection

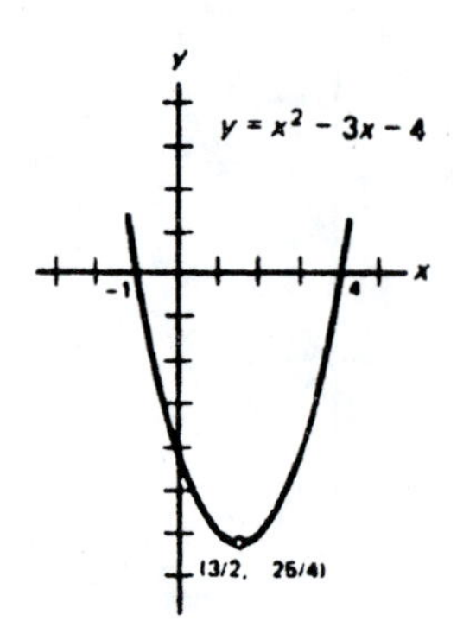

13. $y' = 3x^2$

y'	$+$	$+$
	i	i
	0	

no max or min
$y'' = 6x$

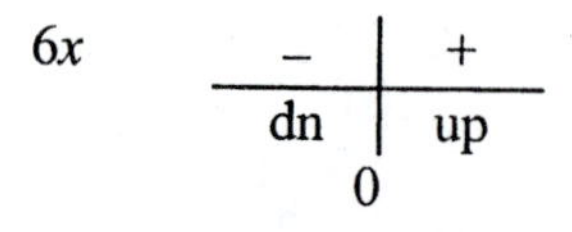

$6x$	$-$	$+$
	dn	up
	0	

point of inflection (0, −7)

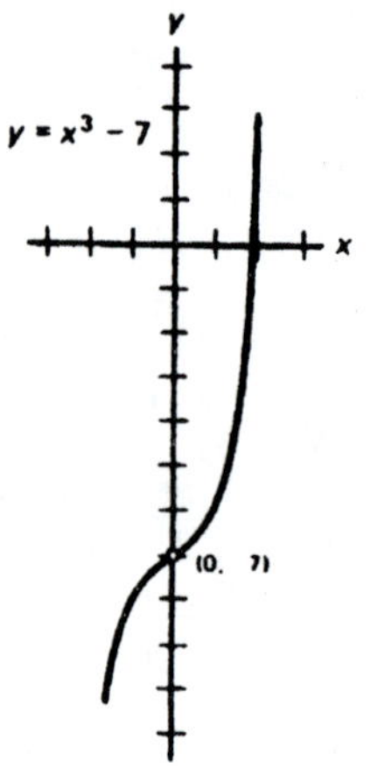

14. $y' = 6x^2 - 18x - 24 = 6(x^2 - 3x - 4)$
$= 6(x - 4)(x + 1)$

$x + 1$	$-$	$+$	$+$
$x - 4$	$-$	$-$	$+$
	$+$	$-$	$+$
	−1	4	
	i	d	i

max(−1, 11); min (4, −114)
$y'' = 12x - 18$

y''	$-$	$+$
	dn	up
	3/2	

point of inf. (3/2, −51.5)

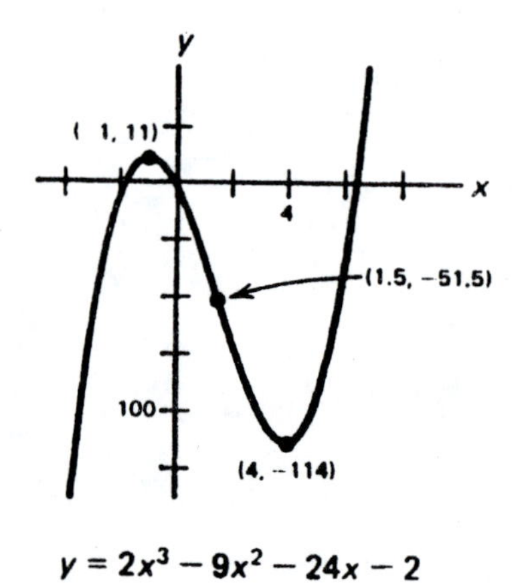

$y = 2x^3 - 9x^2 - 24x - 2$

15. $y' = -2/(x+1)^3$

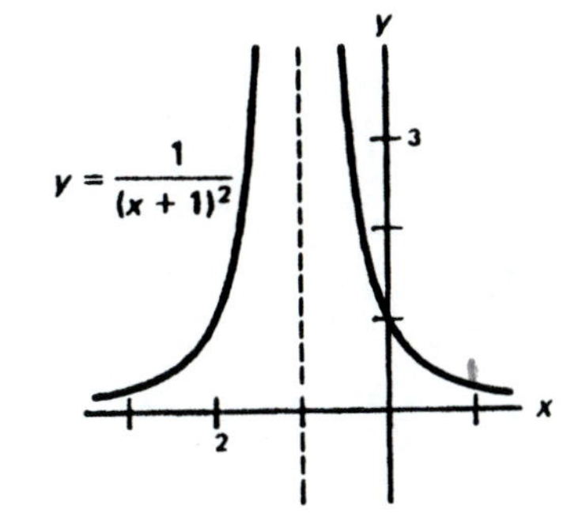

y'	$+$	$-$
	i	d
	-1	

no max or min

$y'' = 6/(x+1)^4$

y''	$+$	$+$
	up	dn
	-1	

no points of inflection

A: $y = 0$

16. $y' = 10x/(x^2+4)^2$

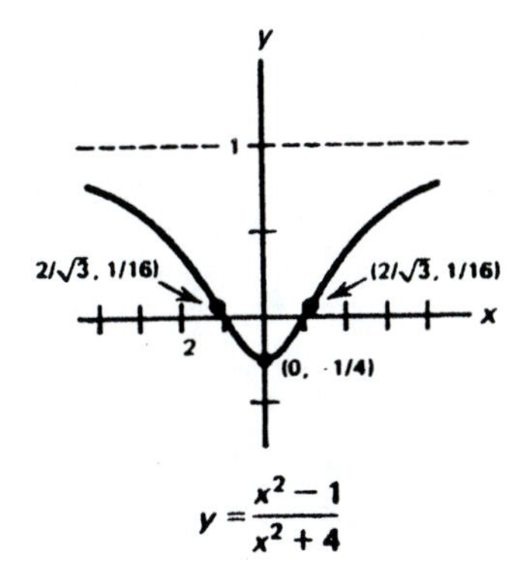

$10x$	$-$	$+$
$(x^2+4)^2$	$+$	$+$
	$-$	$+$
	0	
	d	i

min $(0, -1/4)$

$$y'' = \frac{10(4-3x^2)}{(x^2+4)^3}$$

$$= \frac{10(2+\sqrt{3}x)(2-\sqrt{3}x)}{(x^2+4)^3}$$

$2+\sqrt{3}x$	$-$	$+$	$+$
$2-\sqrt{3}x$	$+$	$+$	$+$
$(x^2+4)^3$	$+$	$+$	$+$
	$-$	$+$	$-$
	$-2/\sqrt{3}$	$2/\sqrt{3}$	
	dn	up	dn

points of inflection $(2/\sqrt{3}, 1/16)$.
$(-2/\sqrt{3}, 1/16)$

17. $y' = -10(x^2+1)^{-2}(2x)$

$$= \frac{-20x}{(x^2+1)^2}$$

$-20x$	+	−
$(x^2+1)^2$	+	+
	+	−
	0	
	i	d

max (0, 10)

$$y'' = \frac{20(3x^2-1)}{(x^2+1)^3}$$

$$= \frac{20(\sqrt{3}x-1)(\sqrt{3}x+1)}{(x^2+1)^3}$$

$\sqrt{3}x - 1$	−	−	+
$\sqrt{3}x + 1$	−	+	+
$(x^2+1)^3$	+	+	+
	+	−	+
	$-1/\sqrt{3}$	$1/\sqrt{3}$	
	up	dn	up

points of inflection $(-1/\sqrt{3}, 15/2)$

$(1/\sqrt{3}, 15/2)$

A: $y = 0$

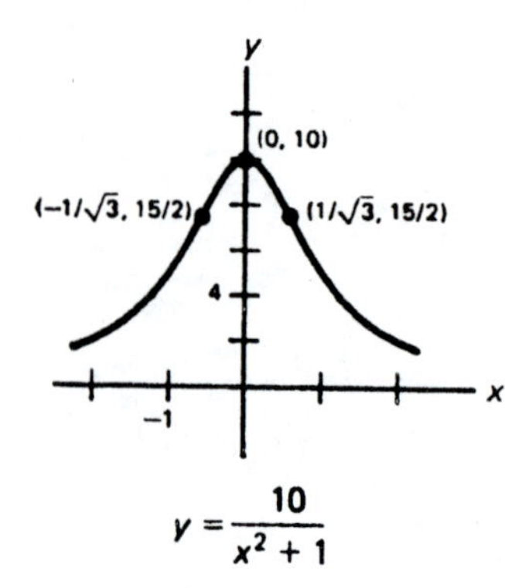

$y = \dfrac{10}{x^2+1}$

18. $y' = \dfrac{-x-2}{x^3}$

$-x-2$	+	−	−
x^3	−	−	+
	−	+	−
	−2	0	
	d	i	d

min(−2, −1/4)

$$y'' = \frac{2x+6}{x^4}$$

$2x+6$	−	+	+
x^4	+	+	+
	−	+	+
	−3	0	
	dn	up	up

point of inflection (−3, −2/9)

A: $y = 0$

$y = \dfrac{x+1}{x^2}$

(−2, −1/4)

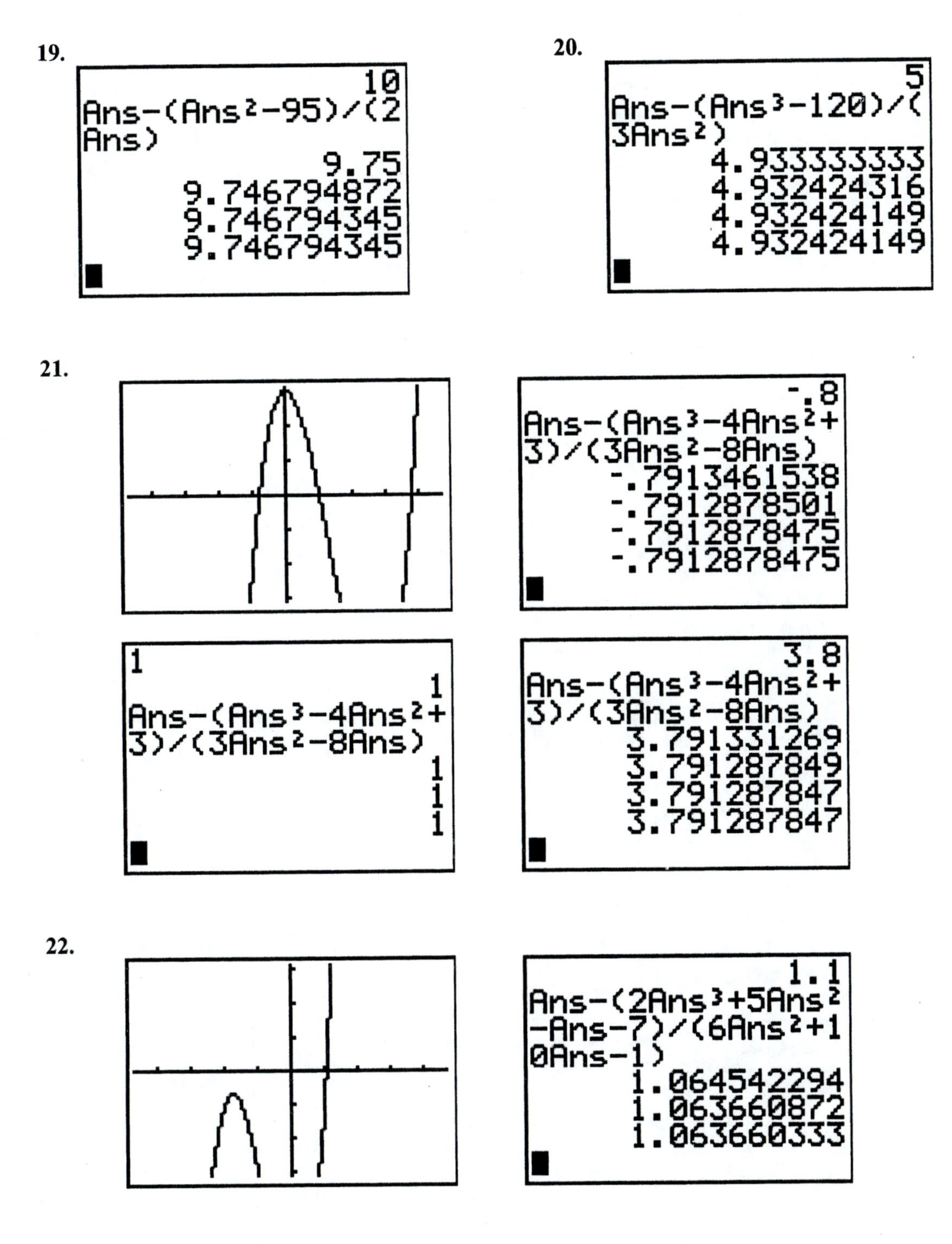

23. $y' = 240 - 32t = 0$; $t = 7.5$ thus max ht y
$= 240(7.5) - 16(7.5)^2 = 900$ ft

24. $a^2 + b^2 = 400$; $A = (1/2)ab$

$A = (1/2)a\sqrt{400 - a^2} = (1/2)a(400 - a^2)^{1/2}$

$A' = (1/2)a[(1/2)(400 - a^2)^{-1/2}(-2a)] + (1/2)(400 - a^2)^{1/2}$

$= (1/2)(400 - a^2)^{-1/2}[-a^2 + (400 - a^2)]$

$= \dfrac{400 - 2a^2}{2\sqrt{400 - a^2}} = 0$; thus $2a^2 = 400$ or $a = 10\sqrt{2}$

$b = \sqrt{400 - a^2} = 10\sqrt{2}$

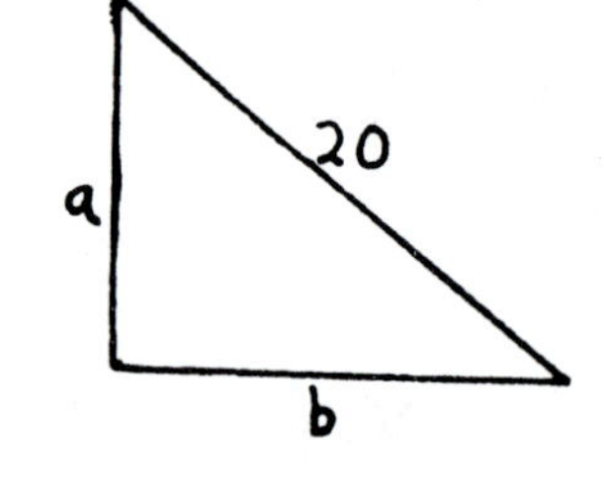

25. $dp/dr = 3 - 3r^2 = 0$

$r = 1\ \Omega$; $d^2p/dr^2 = -6r$

$= -6(1) = -6 < 0$

max power

26. $\dfrac{6 - y}{y} = \dfrac{x}{8 - x}$

$0 = 48 - 8y - 6x$ so $x = (24 - 4y)/3$

$A = xy = y(24 - 4y)/3 = 8y - 4y^2/3$

$dA/dy = 8 - 8y/3 = 0$

Thus, $y = 3$, $x = 4$; $A = (3)(4) = 12\ \text{m}^2$

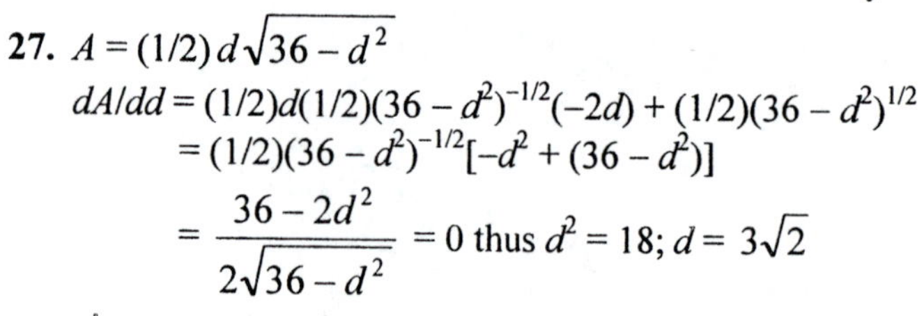

27. $A = (1/2)d\sqrt{36 - d^2}$

$dA/dd = (1/2)d(1/2)(36 - d^2)^{-1/2}(-2d) + (1/2)(36 - d^2)^{1/2}$

$= (1/2)(36 - d^2)^{-1/2}[-d^2 + (36 - d^2)]$

$= \dfrac{36 - 2d^2}{2\sqrt{36 - d^2}} = 0$ thus $d^2 = 18$; $d = 3\sqrt{2}$

gives max A and max V.

28. $d = \sqrt{(x - 1)^2 + (\sqrt{x} - 0)^2} = \sqrt{x^2 - x + 1}$

$dd/dx = (1/2)(x^2 - x + 1)^{-1/2}(2x - 1)$

$= \dfrac{2x - 1}{2\sqrt{x^2 - x + 1}} = 0$

$2x - 1 = 0$; $x = 1/2$, $y = \sqrt{2}/2$

point $(1/2, \sqrt{2}/2)$

29. $V = (4/3)\pi r^3$; $dV/dt = 4\pi r^2\, dr$; $2 = 4\pi(12)^2\, dr$; $dr = 1/(288\pi)$ ft/s

30. $V = 0.06i/(0.01)^2$; $dV/dt = 600\, di/dt = 600(0.03) = 18$ V/s

31. $i = \dfrac{1.5}{R + 0.3}$

$\dfrac{di}{dt} = \dfrac{-1.5}{(R+0.3)^2}\dfrac{dR}{dt}$

$= \dfrac{-1.5}{(8.3)^2}(0.4)$

$= -0.0087$ A/s

32. $A = \pi r^2$

$dA/dt = 2\pi r\, dr/dt$

$= 2\pi(8)(-0.05)$

$= -2.51 \text{ cm}^2/\text{min}$

33. $A = \pi r^2$

$dA/dt = 2\pi r\, dr/dt$

$4 = 2\pi(2.5)\, dr/dt$

$dr/dt = 0.255$ km/d

34. $x = 12$ km (fixed)

$12^2 + y^2 = z^2$

$2y\, dy/dt = 2z\, dz/dt$

$\dfrac{dz}{dt} = \dfrac{y\, dy/dt}{z}$

$= \dfrac{10(3)}{2\sqrt{61}} = 1.92$ km/s

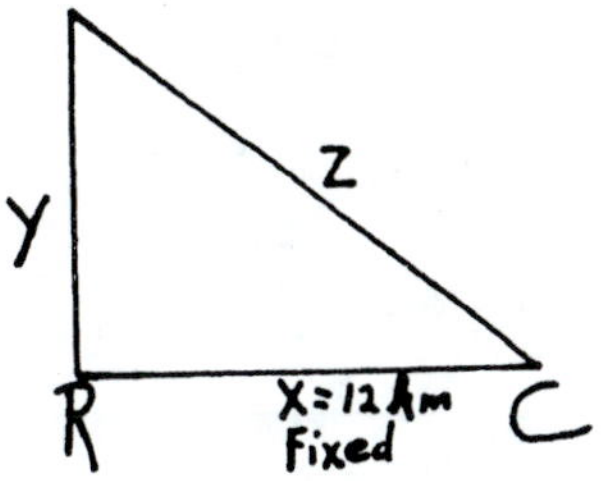

Y

Z

R

X=12 km
Fixed

C

35. $dy/dx = 20x^4 - 18x^2 + 2$

$dy = (20x^4 - 18x^2 + 2)dx$

36. $y' = (-2/3)(3x - 5)^{-5/3}(3)$

$dy = -2(3x - 5)^{-5/3}dx$

37. $\dfrac{ds}{dt} = \dfrac{(5t+1)(6t) - (3t^2 - 4)(5)}{(5t+1)^2}$

$ds = \dfrac{15t^2 + 6t + 20}{(5t+1)^2}dt$

38. $2(x^2 + y^2)(2x + 2y\,y') = y' + 2$

$4x(x^2 + y^2) + 4y(x^2 + y^2)y' = y' + 2$

$4x(x^2 + y^2) - 2 = [1 - 4y(x^2 + y^2)]y'$

$dy = \dfrac{4x(x^2 + y^2) - 2}{1 - 4y(x^2 + y^2)}dx$

39. $ds = (6t - 5)dt$

$= [6(9.5) - 5](0.05) = 2.6$

40. $dy/dx = (-3/4)(8x + 3)^{-7/4}(8)$

$dy = -6(8x + 3)^{-7/4}\, dx$

$= -6[8(10) + 3]^{-7/4}(0.06)$

$= -0.000158$

41. $V = (4/3)\pi r^3$

$dV = 4\pi r^2\, dr$

$= 4\pi(6)^2(0.1)$

$= 45.2 \text{ in}^3$

42. $ds = (t^2 - 3)dt = (3^2 - 3)(0.05) = 0.3$ m

43. $V = e^3$; $dV = 3e^2de = 3(192)^2(0.02) = 2211.84 \text{ in}^3 = 9.58$ gal

44. $V = (4/3)\pi r^3$; $dV = 4\pi r^2\, dr = 4\pi(4.00)^2(0.04) = 8.04 \text{ cm}^3$

45. $F = k/x^2$; $dF = (-2k/x^3)dx = (-2k/(0.030)^3)(0.001) = -74.1\, k$ N

Chapter 23, Exercises 23.1

1. $y' = 7\cos 7x$ **3.** $y' = -10\sin 5x$ **5.** $y' = 6x^2\cos x^3$

7. $y' = -24x\sin 4x^2$ **9.** $y' = -4\cos(1-x)$ **11.** $y' = 6x\cos(x^2+4)$

13. $y' = -4(10x+1)\sin(5x^2+x)$ **15.** $y' = -4(x^3-x)\sin(x^4-2x^2+3)$

17. $y' = -6\cos(3x-1)\sin(3x-1)$ **19.** $y' = 6\sin^2(2x+3)\cos(2x+3)$

21. $y' = 4(2x-5)\cos(2x-5)^2$ **23.** $y' = -12x^2(x^3-4)^3\sin(x^3-4)^4$

25. $y' = \cos 3x\cos x - 3\sin x\sin 3x$

27. $y' = 5\cos 5x\cos 6x - 6\sin 5x\sin 6x$

29. $y' = -7\sin 7x\cos 4x - 4\cos 7x\sin 4x$

31. $y' = -3x^2\sin(x^2+2x)\sin x^3 + 2(x+1)\cos x^3\cos(x^2+2x)$

33. $y' = 5(x^2+3x)\cos(5x-2) + (2x+3)\sin(5x-2)$

35. $y' = \dfrac{5x\cos 5x - \sin 5x}{x^2}$ **37.** $y' = \dfrac{2x\cos 3x + 3(x^2-1)\sin 3x}{\cos^2 3x}$

39. $y' = 5\cos 5x - 6\sin 6x$ **41.** $y' = (2x-3)\cos(x^2-3x) - 4\sin 4x$

43. $y' = \dfrac{\cos^2 x + \sin^2 x}{\cos^2 x} = \dfrac{1}{\cos^2 x} = \sec^2 x$ **45.** $y' = -\sin x;\ y'' = -\cos x$

47. $y' = \cos x;\ y'' = -\sin x;\ y''' = -\cos x$

49. $y' = 5\cos 5x - 6\sin 6x;\ y'' = -25\sin 5x - 36\cos 6x$

51. $y' = 12\cos 3x\Big|_{x=\pi/18} = 12\cos \pi/6 = 12(\sqrt{3}/2) = 6\sqrt{3}$

53. $y = -10\sin 5x\Big|_{x=\pi/10} = -10\sin 5(\pi/10) = -10\sin \pi/2 = -10$

Thus, $(\pi/10, 0)$; $y - 0 = -10(x - \pi/10$; $10x + y = \pi$

Exercises 23.2

1. $y' = 3\sec^2 3x$ **3.** $y' = 7\sec 7x\tan 7x$ **5.** $y' = -6x\csc^2(3x^2-7)$

7. $y' = -9\csc(3x-4)\cot(3x-4)$ **9.** $y' = 10\tan(5x-2)\sec^2(5x-2)$

11. $y' = 12\cot^2 2x(-2\csc^2 2x) = -24\cot^2 2x\csc^2 2x$

13. $y' = (1/2)(x^2 + x)^{-1/2}(2x + 1) \sec\sqrt{x^2 + x}\ \tan\sqrt{x^2 + x}$

$= \dfrac{2x+1}{2\sqrt{x^2+x}} \sec\sqrt{x^2 + x} \tan\sqrt{x^2 + x}$

15. $y' = \dfrac{-3x\csc x\cot x - 3\csc x}{9x^2} = \dfrac{-3\csc x(\cot x + 1)}{9x^2}$

$= \dfrac{-\csc x(\cot x + 1)}{3x^2}$

17. $y' = 3\sec^2 3x - 2x\sec(x^2 + 1)\tan(x^2 + 1)$

19. $y' = \sec^3 x + \sec x\tan^2 x = \sec x(\sec^2 x + \tan^2 x)$

$= \sec x[\sec^2 x + (\sec^2 x - 1)] = \sec x(2\sec^2 x - 1)$

21. $y = \sin^2 x\dfrac{\cos x}{\sin x} = \sin x\cos x$; $y' = -\sin^2 x + \cos^2 x = \cos 2x$

23. $y' = x\sec x\tan x + \sec x = \sec x(x\tan x + 1)$

25. $y' = 2x + 2x^2\tan x\sec^2 x + 2x\tan^2 x = 2x[1 + x\tan x\sec^2 x + \tan^2 x]$

$2x[\sec^2 x + x\tan x\sec^2 x] = 2x\sec^2 x(x\tan x + 1)$

27. $y' = -3\csc^2 3x - 3\cot 3x\csc 3x\cot 3x = -3\csc 3x(\csc^2 3x + \cot^2 3x)$

$= -3\csc 3x[\csc^2 3x + (\csc^2 3x - 1)] = -3\csc 3x(2\csc^2 3x - 1)$

29. $y' = 3\csc^2 3x\cos 3x + \sin 3x(-2\csc 3x \cdot 3\csc 3x\cot 3x)$

$= 3\csc^2 3x\cos 3x - 6\csc 3x\cot 3x$

$= 3\csc 3x(\csc 3x\cos 3x - 2\cot 3x)$

$= 3\csc 3x\left\{\dfrac{\cos 3x}{\sin 3x} - 2\cot 3x\right\} = 3\csc 3x(\cot 3x - 2\cot 3x)$

$= -3\csc 3x\cot 3x$

31. $y' = 2(\sin x - \cos x)(\cos x + \sin x) = 2(\sin^2 x - \cos^2 x) = -2\cos 2x$

33. $y' = 4(x + \sec^2 3x)^3(1 + 2\sec 3x \cdot \sec 3x\tan 3x \cdot 3)$

$= 4(x + \sec^2 3x)^3(1 + 6\sec^2 3x\tan 3x)$

35. $y' = 3(\sec x + \tan x)^2(\sec x\tan x + \sec^2 x)$

$= 3\sec x(\sec x + \tan x)^2(\sec x + \tan x) = 3\sec x(\sec x + \tan x)^3$

37. $y' = \cos(\tan x) \cdot \sec^2 x$

39. $y' = -\sin x\sec^2(\cos x)$

41. $y' = 2\sin(\cos x)\cos(\cos x)(-\sin x) = -2\sin x\sin(\cos x)\cos(\cos x)$

43. $y' = \dfrac{\cos x \sec^2 x + \tan x \sin x}{\cos^2 x} = \dfrac{\cos x \dfrac{1}{\cos^2 x} + \dfrac{\sin x}{\cos x}\sin x}{\cos^2 x}$

$$= \frac{\dfrac{1+\sin^2 x}{\cos x}}{\cos^2 x} = \frac{1+\sin^2 x}{\cos^3 x}$$

45. $y = \dfrac{\sin^2 x}{\dfrac{\sin^2 x}{\cos^2 x}} = \cos^2 x$; $y' = 2\cos x(-\sin x) = -\sin 2x$

47. $y' = \dfrac{(1+\tan x)\cos x - \sin x \sec^2 x}{(1+\tan x)^2}$

$$= \frac{\cos x + \cos x \tan x - \sin x \sec^2 x}{(1+\tan x)^2}$$

$$= \frac{\cos x + \sin x - \sin x \sec^2 x}{(1+\tan x)^2}$$

49. $y' = 3\sec^2 3x$; $y'' = 6\sec 3x \sec 3x \tan 3x\,(3) = 18\sec^2 3x \tan 3x$

51. $y' = -x\csc^2 x + \cot x$; $y'' = -x \cdot 2\csc x\,(-\csc x \cot x) - \csc^2 x - \csc^2 x$
$= 2x\csc^2 x \cot x - 2\csc^2 x = 2\csc^2 x\,(x\cot x - 1)$

53. $y' = \sec^2 x \Big|_{x=\pi/4} = \sec^2(\pi/4) = (\sqrt{2})^2 = 2$

55. $\dfrac{d}{du}(\cot u) = \dfrac{d}{du}\left\{\dfrac{\cos u}{\sin u}\right\} = \dfrac{-\sin^2 u - \cos^2 u}{\sin^2 u} = \dfrac{-1}{\sin^2 u} = -\csc^2 u$

Exercises 23.3

1. $y' = \dfrac{5}{\sqrt{1-25x^2}}$

3. $y' = \dfrac{3}{1+9x^2}$

5. $y' = -\dfrac{-1}{|1-x|\sqrt{(1-x)^2-1}} = \dfrac{1}{|1-x|\sqrt{x^2-2x}}$

7. $y' = 3 \cdot \dfrac{-1}{\sqrt{1-(x-1)^2}} = \dfrac{-3}{\sqrt{2x-x^2}}$

9. $y' = 2 \cdot \dfrac{-1(6x)}{1+9x^4} = \dfrac{-12x}{1+9x^4}$

11. $y' = 5 \cdot \dfrac{3x^2}{|x^3|\sqrt{x^6-1}} = \dfrac{15}{|x|\sqrt{x^6-1}}$

13. $y' = 3\arcsin^2 x \cdot \dfrac{1}{\sqrt{1-x^2}} = \dfrac{3\arcsin^2 x}{\sqrt{1-x^2}}$

15. $y' = 2(2\arccos 3x)\dfrac{-1(3x)}{\sqrt{1-9x^2}} = \dfrac{-12\arccos 3x}{\sqrt{1-9x^2}}$

17. $y' = 12\arctan^3\sqrt{x}\,\dfrac{1}{1+x}(1/2)x^{-1/2} = \dfrac{6\arctan^3\sqrt{x}}{\sqrt{x}(1+x)}$

19. $y' = \dfrac{1}{\sqrt{1-x^2}} - \dfrac{1}{\sqrt{1-x^2}} = 0$

21. $y' = (1/2)(1-x^2)^{-1/2}(-2x) + \dfrac{1}{\sqrt{1-x^2}} = \dfrac{-x}{\sqrt{1-x^2}} + \dfrac{1}{\sqrt{1-x^2}} = \dfrac{1-x}{\sqrt{1-x^2}}$

23. $y' = \dfrac{3x}{\sqrt{1-9x^2}} + \arcsin 3x$

25. $y' = \dfrac{x}{1+x^2} + \arctan x$

27. $y' = x \cdot \dfrac{1}{\sqrt{1-x^2}} + \arcsin x + (1/2)(1-x^2)^{-1/2}(-2x)$

$= \dfrac{x}{\sqrt{1-x^2}} + \arcsin x - \dfrac{x}{\sqrt{1-x^2}} = \arcsin x$

29. $y' = \dfrac{\arcsin x - x \cdot \dfrac{1}{\sqrt{1-x^2}}}{\arcsin^2 x} \cdot \dfrac{\sqrt{1-x^2}}{\sqrt{1-x^2}} = \dfrac{\sqrt{1-x^2}\arcsin x - x}{\sqrt{1-x^2}\arcsin^2 x}$

31. $y' = \left.\dfrac{1}{\sqrt{1-x^2}}\right|_{x=1/2} = 2/\sqrt{3}$

33. $y' = \left.\dfrac{x}{x^2+1} + \arctan\right|_{x=-1} = -\dfrac{1}{2} - \dfrac{\pi}{4} = \dfrac{-\pi-2}{4}$

35. If $y = \arccos u$, then $u = \cos y$ where $0 \le y \le \pi$.
So $du/dx = -\sin y\, dy/dx$
Then $\dfrac{dy}{dx} = \dfrac{-1}{\sin y}\dfrac{du}{dx}$ but $\sin y = \sqrt{1-\cos y}$ since $0 \le y \le \pi$.

$$\frac{dy}{dx} = -\frac{1}{\sqrt{1-u^2}}\frac{du}{dx}$$

Exercises 23.4

1. $y' = \frac{1}{4x-3}\log e(4)\frac{4\log e}{4x-3}$

3. $y' = \frac{1}{3x}\log_2 e(3) = \frac{\log_2 e}{x}$

5. $y' = \frac{1}{2x^3-3}(6x^2) = \frac{6x^2}{2x^3-3}$

7. $y' = \frac{1}{\tan 3x}3\sec^2 3x = \frac{3\sec^2 3x}{\tan 3x}$
or 3 sec $3x$ csc $3x$

9. $y' = \frac{x\cos x + \sin x}{x\sin x}$

11. $y' = \frac{1}{\sqrt{3x-2}}(1/2)(3x-2)^{-1/2}(3) = \frac{3}{2(3x-2)}$

13. $y' = \frac{x^2+1}{x^3}\cdot\frac{(x^2+1)(3x^2)-x^3(2x)}{(x^2+1)^2} = \frac{x^2+1}{x^3}\cdot\frac{x^2(x^2+3)}{(x^2+1)^2} = \frac{x^2+3}{x(x^2-1)}$

15. $y' = \sec^2(\ln x)\cdot\frac{1}{x} = \frac{\sec^2(\ln x)}{x}$

17. $y' = \frac{1}{\ln x}\cdot\frac{1}{x} = \frac{1}{x\ln x}$

19. $y' = \frac{1}{1+\ln^2 x^2}\cdot\frac{1}{x^2}(2x) = \frac{2}{x(1+\ln^2 x^2)}$

21. $y' = \frac{1}{\arccos^2 x}(2\arccos x)\frac{-1}{\sqrt{1-x^2}} = \frac{-2}{(\arccos x)\sqrt{1-x^2}}$

23. $\ln y = \ln(3x+2) + 2\ln(6x-1) + \ln(x-4)$

$\frac{1}{y}\cdot y' = \frac{3}{3x+2} + \frac{12}{6x-1} + \frac{1}{x-4}$

$y' = (3x+2)(6x-1)^2(x-4)\left\{\frac{3}{3x+2} + \frac{12}{6x-1} + \frac{1}{x+4}\right\}$

25. $\ln y = \ln(x+1) + \ln(2x+1) - \ln(3x-4) - \ln(1-8x)$

$\frac{1}{y}\cdot y' = \frac{1}{x+1} + \frac{2}{2x+1} - \frac{3}{3x-4} + \frac{1}{1-8x}$

$y' = \frac{(x+1)(2x+1)}{(3x-4)(1-8x)}\left\{\frac{1}{x+1} - \frac{3}{3x-4} + \frac{2}{2x+1} + \frac{1}{1-8x}\right\}$

27. $\ln y = \ln x^x = x\ln x$; $y'/y = x(1/x) + \ln x$; $y' = x^x(1+\ln x)$

29. $\ln y = \ln x^{2/x} = (2/x)\ln x$; $y'/y = \frac{2}{x}\cdot\frac{1}{x} - \frac{2\ln x}{x^2}$; $y' = 2x^{2/x}\frac{(1-\ln x)}{x^2}$

31. $\ln y = x\ln(\sin x)$; $y'/y = x\dfrac{1}{\sin x}(\cos x) + \ln(\sin x)$; $y' = (\sin x)^x[x\cot x + \ln(\sin x)]$

33. $\ln y = x^2 \ln(1+x)$; $y'/y = x^2\dfrac{1}{1+x} + 2x\ln(1+x)$

$$y' = (1+x)^{x^2}\left\{\frac{x^2}{1+x} + 2x\ln(1+x)\right\}$$

35. $m = y'\Big|_{x=1} = \dfrac{1}{x}\Big|_{x=1} = 1$ Thus, $y = x - 1$

37. $m = y'\Big|_{x=\pi/6} = \cot x\Big|_{x=\pi/6} = \sqrt{3}$; at $x = \pi/6$, $y = -\ln 2$

$y + \ln 2 = \sqrt{3}(x - \pi/6)$ or $y = \sqrt{3}(x - \pi/6) - \ln 2$

Exercises 23.5

1. $y' = 5e^{5x}$

3. $y' = 12x^2 e^{x^3}$

5. $y' = \dfrac{(3)10^{3x}}{\log e}$

7. $y' = \dfrac{-6}{e^{6x}}$

9. $y' = \dfrac{e^{\sqrt{x}}}{2\sqrt{x}}$

11. $y' = (\cos x)e^{\sin x}$

13. $y' = 6x(e^{x^2-1})(2x) + 6e^{x^2-1} = 6e^{x^2-1}(2x^2+1)$

15. $y' = (\cos x)(e^{3x^2})(6x) + (e^{3x^2})(-\sin x) = e^{3x^2}(6x\cos x - \sin x)$

17. $y' = \dfrac{1}{\cos e^{5x}}(-\sin e^{5x})(e^{5x})(5) = \dfrac{-5e^{5x}\sin e^{5x}}{\cos e^{5x}} = -5e^{5x}\tan e^{5x}$

19. $y' = e^x + e^{-x}$

21. $y' = \dfrac{(3e^x - x)(4x) - (2x^2)(3e^x - 1)}{(3e^x - x)^2} = \dfrac{-6e^x x^2 + 2x^2 - 4x^2 + 12xe^x}{(3e^x - x)^2} = \dfrac{12xe^x - 6e^x x^2 - 2x^2}{(3e^x - x)^2}$

$$y' = \frac{6xe^x(2-x) - 2x^2}{(3e^x - x)^2}$$

23. $y' = \dfrac{6e^{3x}}{1+e^{6x}}$

25. $y' = 3(\arccos^2 e^{-2x})\dfrac{-1}{\sqrt{1-e^{-4x}}}(-2) = \dfrac{6e^{-2x}\arccos^2 e^{-2x}}{\sqrt{1-e^{-4x}}}$

27. $y' = xe^x + e^x - e^x = xe^x$

29. $y = x^2$; $y' = 2x$

Exercises 23.6

1. $\lim_{x\to 0}\frac{\sin x}{5x}$

$=\lim_{x\to 0}\frac{\cos x}{5}=\frac{1}{5}$

3. $\lim_{x\to 1}\frac{\ln x}{x^2-5x+4}$

$=\lim_{x\to 1}\frac{\frac{1}{x}}{2x-5}$

$=\lim_{x\to 1}\frac{1}{2x^2-5x}=\frac{-1}{3}$

5. $\lim_{x\to 0}\frac{\ln(\cos x)}{x^2}$

$=\lim_{x\to 0}\frac{\frac{1}{\cos x}\cdot -\sin x}{2x}$

$=\lim_{x\to 0}\frac{-\tan x}{2x}$

$=\lim_{x\to 0}\frac{-\sec^2 x}{2}=\frac{-1}{2}$

7. $\lim_{x\to 0}\frac{x-\sin x}{x^3}$

$=\lim_{x\to 0}\frac{1-\cos x}{3x^2}$

$=\lim_{x\to 0}\frac{\sin x}{6x}$

$=\lim_{x\to 0}\frac{\cos x}{6}=\frac{1}{6}$

9. $\lim_{x\to 0}x\cot x$

$=\lim_{x\to 0}\frac{x}{\tan x}$

$=\lim_{x\to 0}\frac{x}{\sec^2 x}=1$

11. $\lim_{x\to\infty}\frac{5x^2+7x-1}{2x^2-x+9}$

$=\lim_{x\to\infty}\frac{10x+7}{4x-1}$

$=\lim_{x\to\infty}\frac{10}{4}=\frac{5}{2}$

13. $\lim_{x\to\infty}\frac{\ln x}{\sqrt{x}}$

$=\lim_{x\to\infty}\frac{\frac{1}{x}}{\frac{1}{2}x^{\frac{-1}{2}}}$

$=\lim_{x\to\infty}\frac{2}{\sqrt{x}}=0$

15. $\lim_{t\to\infty}t^2e^{-5t}$

$=\lim_{t\to\infty}\frac{t^2}{e^{5t}}$

$=\lim_{t\to\infty}\frac{2t}{5e^{5t}}$

$=\lim_{t\to\infty}\frac{2}{25e^{5t}}=0$

17. $\lim_{x\to\infty}5x^{-2}e^{x^2}$

$=\lim_{x\to\infty}\frac{5e^{x^2}}{x^2}$

$=\lim_{x\to\infty}\frac{5(2x)e^{x^2}}{2x}$

$=\lim_{x\to\infty}5e^{x^2}$

does not exist

19. $\lim_{x\to 0}(\csc x-\cot x)$

$=\lim_{x\to 0}\left(\frac{1}{\sin x}-\frac{\cos x}{\sin x}\right)$

$=\lim_{x\to 0}\frac{1-\cos x}{\sin x}$

$=\lim_{x\to 0}\frac{\sin x}{\cos x}=0$

21. $\lim_{x\to\frac{\pi}{2}} \frac{\sec x}{\tan x}$

$= \lim_{x\to\frac{\pi}{2}} \frac{\sec x \tan x}{\sec^2 x}$

$= \lim_{x\to\frac{\pi}{2}} \frac{\tan x}{\sec x}$

$= \lim_{x\to\frac{\pi}{2}} \frac{\sec^2 x}{\sec x \tan x}$

$= \lim_{x\to\frac{\pi}{2}} \frac{\sec x}{\tan x}$

This is the original problem.

First use algebra to simplify:

$\lim_{x\to\frac{\pi}{2}} \frac{\sec x}{\tan x}$

$= \lim_{x\to\frac{\pi}{2}} \frac{1}{\cos x} \cdot \frac{\cos x}{\sin x}$

$= \lim_{x\to\frac{\pi}{2}} \frac{1}{\sin x} = 1$

Exercises 23.7

1. I: $x = 3\pi/4 + 2n\pi$; $x = 7\pi/4 + 2n\pi$

$y = 0$

$y' = \cos x - \sin x = 0$; $\tan x = 1$

$x = \pi/4 = n\pi$; when n is even,

$y = 1/\sqrt{2} + 1/\sqrt{2} = \sqrt{2}$

$y'' = -\sin x - \cos x = -1/\sqrt{2} - 1/\sqrt{2} = -\sqrt{2} < 0$

Thus $(\pi/4 + 2n\pi, \sqrt{2})$ are relative maxima.

When n is odd, $y = -1/\sqrt{2} - 1/\sqrt{2} = -\sqrt{2}$;

$y'' = 1/\sqrt{2} + 1/\sqrt{2} = \sqrt{2} > 0$

Thus $(\pi/4 + (2n+1)\pi, -\sqrt{2})$ are relative minima.

Points of inflection: $y'' = 0$ or $\tan x = -1$; $x = 3\pi/4 + n\pi$;

for n even, $y = 1/\sqrt{2} - 1/\sqrt{2} = 0$. For n odd, $y = -1/\sqrt{2} + 1/\sqrt{2} = 0$.

Thus $(3\pi/4 + n\pi, 0)$ are points of inflection.

3. I: $x = 1$; no y-intercept

$y' = \frac{x(1/x) - \ln x}{x^2} = \frac{1 - \ln x}{x^2} = 0$;

$\ln x = 1$;

$x = e, y = 1/e$

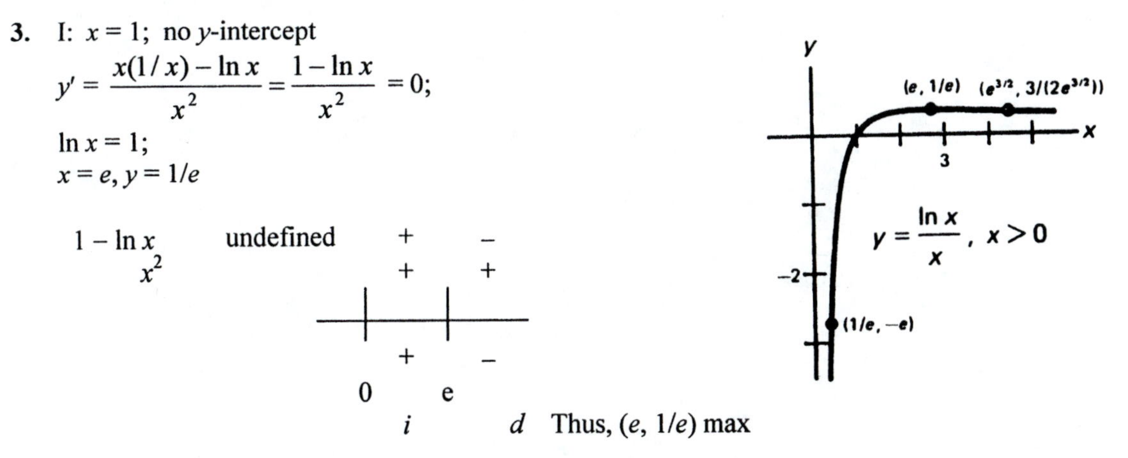

	undefined		
$1 - \ln x$		+	−
x^2		+	+
		+	−
	0		e
		i	d

Thus, $(e, 1/e)$ max

$$y'' = \frac{x^2(-1/x) - (1 - \ln x)(2x)}{x^4} = \frac{x(-3 + 2\ln x)}{x^4} = \frac{-3 + 2\ln x}{x^3} = 0$$

$-3 + 2\ln x = 0;\ \ln x = 3/2;\ x = e^{3/2}$

$f''(x)$		$-$		$+$
	0	dn	$e^{3/2}$	up

Thus $(e^{3/2}, 3e^{-3/2}/2)$ is point of inflection.

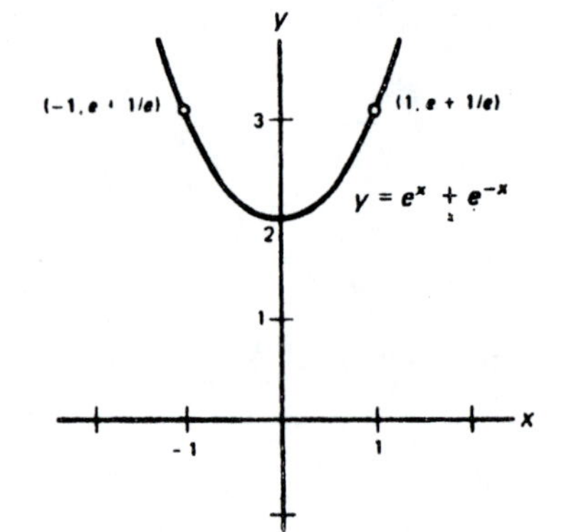

5. I: $y = 2$; no x-intercept
$y' = e^x - e^{-x} = 0; x = -x;$
$x = 0, y = 2;$
$y'' = e^x + e^{-x} > 0$ for all x;
Thus (0, 2) min, graph is concave upward and no points of inflection.

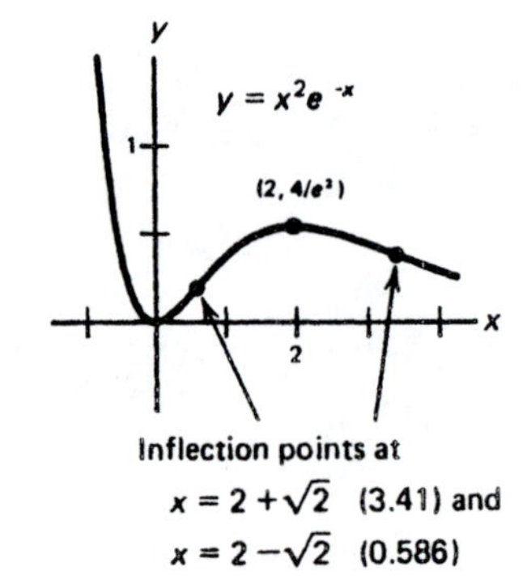

7. I: $x = 0;\ y = 0$
$y' = -x^2e^{-x} + 2xe^{-x}$
$= xe^{-x}(2 - x) = 0$
Note: $e^{-x} > 0$

x	$-$		$+$		$+$
$2 - x$	$+$		$+$		$-$
$f'(x)$	$-$	0	$+$	2	$-$
	d		i		d

(0, 0) min; $(2, 4/e^2)$ max

$y'' = e^{-x}(2 - 2x) - e^{-x}(2x - x^2) = e^{-x}(2 - 4x + x^2) = 0;$
$x = 2 \pm \sqrt{2}$

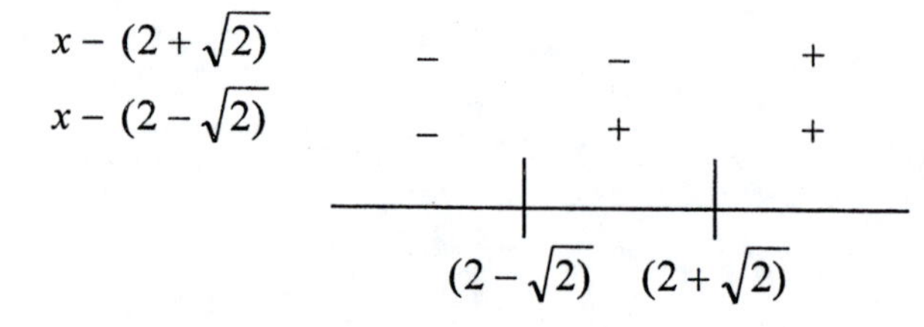

(3.41, 17.8) and (0.586, −1.8) are points of inflection.

9. no intercepts

$x = 0$ is a vertical asymptote

$$y' = \frac{e^x}{(1-e^x)^2} \quad y' > 0 \text{ for all } x \text{ except } 0.$$

no max or min.

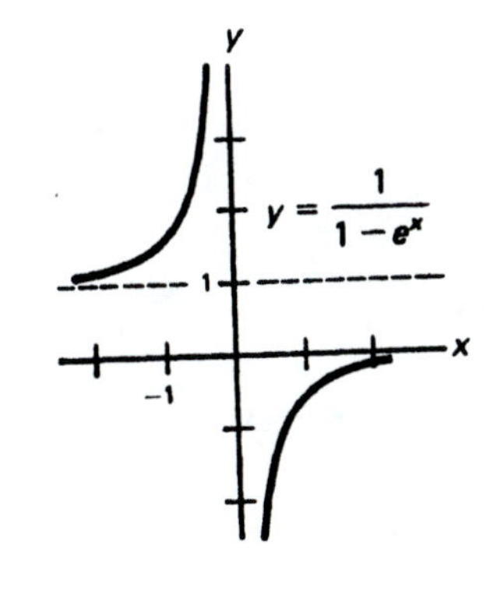

$$y'' = \frac{(1-e^x)^2 e^x - e^x[2(1-e^x)(-e^x)]}{(1-e^x)^4}$$

$$= \frac{e^x(1-e^x)[(1-e^x)+2e^x]}{(1-e^x)^4} = \frac{e^x(1+e^x)}{(1-e^x)^3};$$

$y'' > 0$ for all x except 0.

no points of inflection.

11. $y' = x(-2e^{-2x}) + e^{-2x} = e^{-2x}(1-2x) = 0;$

Note: $e^{nx} > 0$ for all n and x and will not be in sign charts.

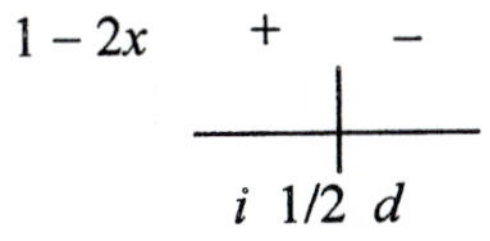

Thus, (1/2, 1/(2*e*)) max

13. $y' = \frac{2x^{2x}x - 2x^2e^{2x}}{e^{4x}} = \frac{2xe^{2x}(1-x)}{e^{4x}} = \frac{2x(1-x)}{e^{2x}} = 0$

$2x$	−	+	+
$1-x$	+	+	−
	d	*i*	*d*

Sign chart divisions: 0, 1

Thus, (0, 0) is a min; $(1, 1/e^2)$ is a max.

15. $y' = \frac{x^2(1/x) - (\ln x)(2x)}{x^4} = \frac{x(1-2\ln x)}{x^4} = \frac{1-2\ln x}{x^3} = 0$

$1-2\ln x$		+	−
x^3		+	+
	not defined	0 *i* $e^{1/2}$	*d*

$1 - 2\ln x = 0$

$\ln x = 1/2$

$x = e^{1/2}$

Thus $(e^{1/2}, 1/(2e))$ is a max.

Chapter 23 Review

1. $y' = 2x \cos(x^2 + 3)$

2. $y' = -8 \sin 8x$

3. $y' = 3\cos^2(5x-1)[-\sin(5x-1)](5) = -15\cos^2(5x-1)\sin(5x-1)$

4. $y' = \sin 3x(-2\sin 2x) + \cos 2x(3\cos 3x) = -2\sin 2x \sin 3x + 3\cos 2x \cos 3x$

5. $y' = 3\sec^2(3x-2)$

6. $y' = 4\sec(4x+3)\tan(4x+3)$

7. $y' = -12x\csc^2 6x^2$

8. $y' = -2(16x+1)\csc(8x^2+x)\csc(8x^2+x)\cot(8x^2+x) = -2(16x+1)\csc^2(8x^2+x)\cot(8x^2+x)$

9. $y' = \sec^2 x \cos x + \sin x(2\sec x \cdot \sec x \tan x)$

$$= \frac{1}{\cos^2 x}\cos x + 2\sin x \frac{1}{\cos^2 x}\frac{\sin x}{\cos x} = \frac{1}{\cos x} + \frac{2\sin^2 x}{\cos^3 x}$$

$$= \frac{\cos^2 x + 2\sin^2 x}{\cos^3 x} = \frac{\cos^2 x + 2(1-\cos^2 x)}{\cos^3 x}$$

$$= \frac{2-\cos^2 x}{\cos^3 x} = \frac{2}{\cos^3 x} - \frac{1}{\cos x} = 2\sec^3 x - \sec x$$

10. $y' = 2x - 2\csc x(-\csc x \cot x) = 2x + 2\csc^2 x \cot x$

11. $y' = \sec^2(\sec x)\sec x \tan x$

12. $y' = \dfrac{(1+\sin x)(-\sin x) - \cos x(\cos x)}{(1+\sin x)^2} = \dfrac{-\sin x - \sin^2 x - \cos^2 x}{(1+\sin x)^2}$

$$= \frac{-(\sin x + 1)}{(1+\sin x)^2} = \frac{-1}{1+\sin x}$$

13. $y' = 3(1-\sin x)^2(-\cos x) = -3\cos x(1-\sin x)^2$

14. $y' = 2(1+\sec 4x)\sec 4x \tan 4x(4) = 8(1+\sec 4x)\sec 4x \tan 4x$

15. $y' = \dfrac{3x^2}{\sqrt{1-x^6}}$

16. $y' = \dfrac{3}{1+9x^2}$

17. $y' = -3\dfrac{1}{\sqrt{1-(1/(2x))^2}}\dfrac{-1}{2x^2} = \dfrac{(-3/2)(-1/x^2)}{\dfrac{\sqrt{4x^2-1}}{2|x|}} = \dfrac{3}{2x^2}\dfrac{2|x|}{\sqrt{4x^2-1}} = \dfrac{3}{|x|\sqrt{4x^2-1}}$

18. $y' = \dfrac{2}{|4x|\sqrt{16x^2-1}}(4) = \dfrac{2}{|x|\sqrt{16x^2-1}}$

19. $y' = 2\text{ arcsin } 3\sqrt{x}\dfrac{(3/2)x^{-1/2}}{\sqrt{1-9x}} = \dfrac{3\arcsin 3\sqrt{x}}{\sqrt{x}\sqrt{1-9x}} = \dfrac{3\arcsin 3\sqrt{x}}{\sqrt{x-9x^2}}$

20. $y' = \dfrac{x}{\sqrt{1-x^2}} + \arcsin x$

21. $y' = \dfrac{6x^2}{2x^3-4} = \dfrac{3x^2}{x^3-2}$

22. $y' = \dfrac{4\log_3 e}{4x+1}$

23. $y = 2\ln x - \ln(x^2+3)$; $y' = \dfrac{2}{x} - \dfrac{2x}{x^2+3} = \dfrac{6}{x(x^2+3)}$

24. $y' = -\sin(\ln x)\cdot(1/x) = (-1/x)\sin(\ln x)$

25. $\ln y = (1/2)\ln(x+1) + \ln(3x-4) - 2\ln x - \ln(x+2)$

$$y'/y = \frac{1}{2}\cdot\frac{1}{x+1} + \frac{3}{3x-4} - \frac{2}{x} - \frac{1}{x+2}$$

$$y' = \frac{\sqrt{x+1}(3x-4)}{x^2(x+2)}\left\{\frac{1}{2(x+1)} + \frac{3}{3x-4} - \frac{2}{x} - \frac{1}{x+2}\right\}$$

26. $\ln y = (1-x)\ln x$; $y'/y = \dfrac{1-x}{x} - \ln x$;

$$y' = x^{1-x}[(1-x)/x - \ln x]$$

27. $y' = 2xe^{x^2} + 5$

28. $y' = \dfrac{3(8^{3x})}{\log_8 e}$

29. $y' = -\csc^2 e^{2x}(e^{2x})(2) = -2e^{2x}\csc^2 e^{2x}$

30. $y' = e^{\sin x^2}(\cos x^2)(2x) = 2x\cos x^2 e^{\sin x^2}$

31. $y' = \dfrac{1}{\sqrt{1-e^{-8x}}}(e^{-4x})(-4) = \dfrac{-4e^{-4x}}{\sqrt{1-e^{-8x}}}$

32. $y' = x^3e^{-4x}(-4) + e^{-4x}(3x^2) = -4e^{-4x}x^3 + 3x^2e^{-4x} = e^{-4x}x^2(3-4x)$

33. I: $x = 0, y = 0$

$y' = -xe^{-x} + e^{-x} = e^{-x}(1 - x) = 0; x = 1$

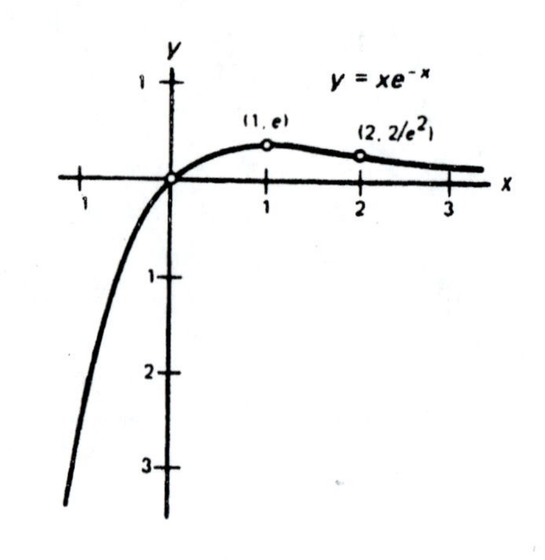

$1 - x$	$+$		$-$
	i	1	d

Thus, $(1, 1/e)$ max

$y'' = xe^{-x} - 2e^{-x} = e^{-x}(x - 2) = 0; x = 2$

$x - 2$	$-$		$+$
	dn	2	up

Thus $(2, 2/e^2)$ is point of inflection.

34. I: $y = 1$; $y' = -2xe^{-x^2}$

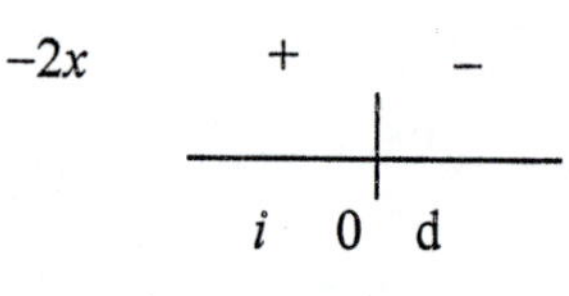

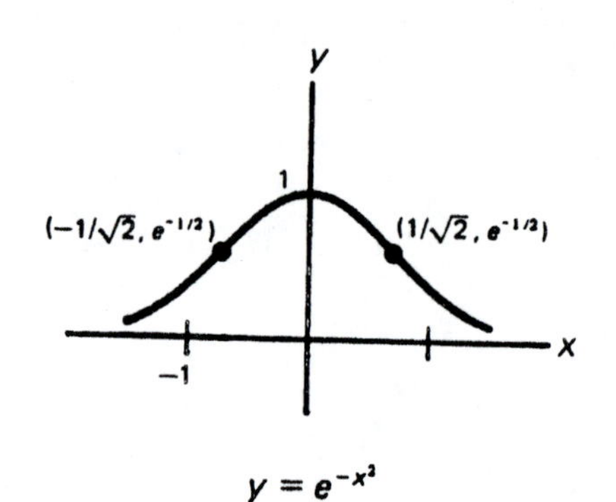

$y = e^{-x^2}$

Thus, $(0, 1)$ is max

$$y'' = -4x^2e^{-x^2} - 2e^{-x^2}$$
$$= -2e^{-x^2}(1 - 2x^2) = 0$$
$$x^2 = 1/2;\ x = \pm 1/\sqrt{2}$$

$2e^{-x^2}$	$-$		$-$		$-$
$1 - \sqrt{2}x$	$+$		$+$		$-$
$1 + \sqrt{2}x$	$-$		$+$		$+$
		$-1/\sqrt{2}$		$1/\sqrt{2}$	
	up		dn		up

Thus $(-1/\sqrt{2}, e^{-1/2})$ and $(1/\sqrt{2}, e^{-1/2})$ are points of inflection.

35. $V_L = L\,di/dt = 2(-100 \sin 5t)(5) = -1000 \sin 5t$

36. $p = dW/dt = 120 \cos 3t\,(-\sin 3t)(3) = -360 \sin 3t \cos 3t = -180 \sin 6t$

37. $y = 2 \ln x;\ m = y' = 2/x \Big|_{x=1} = 2;\ y - 0 = 2(x - 1);$
$y = 2x - 2$

38. $v = ds/dt = e^{\sin t}(\cos t) = \cos t\, e^{\sin t}$

39. $v = 4e^{-t};\ a = -4e^{-t}\Big|_{t=1/2} = -4e^{-1/2} = -2.43$

40.
$$\lim_{x\to 2}\frac{\ln(x-1)}{\sqrt{x-2}}$$
$$= \lim_{x\to 2}\frac{\dfrac{1}{(x-1)}}{0.5(x-2)}$$
$$= \lim_{x\to 2}\frac{2\sqrt{x-2}}{x-1}$$
$$= 0$$

41.
$$\lim_{x\to\infty} xe^{-3x}$$
$$= \lim_{x\to\infty}\frac{x}{e^{3x}}$$
$$= \lim_{x\to\infty}\frac{1}{3e^{3x}}$$
$$= 0$$

42.
$$\lim_{x\to 0}\frac{x\sin x}{1-\cos x}$$
$$= \lim_{x\to 0}\frac{x\cos x+\sin x}{\sin x}$$
$$= \lim_{x\to 0}\frac{-x\sin x+\cos x+\cos x}{\cos x}$$
$$= \lim_{x\to 0}\frac{-x\sin x+2\cos x}{\cos x}$$
$$= 2$$

43.
$$\lim_{x\to\pi/2}\frac{\cos^2 x}{\sin x-1}$$
$$= \lim_{x\to\pi/2}\frac{-\cos^2 x(1+\sin x)}{(1-\sin x)(1+\sin x)}$$
$$= \lim_{x\to\pi/2}\frac{-\cos^2 x(1+\sin x)}{1-\sin x^2}$$
$$= \lim_{x\to\pi/2}\frac{-\cos^2 x(1+\sin x)}{\cos x^2}$$
$$= \lim_{x\to\pi/2} -(1+\sin x)$$
$$= -2$$

Chapter 24, Exercises 24.1

1. $\dfrac{x^8}{8}+C$ **3.** $\dfrac{x^9}{3}+C$ **5.** $4x+C$ **7.** $\int 9x^{5/6}dx = (54/11)x^{11/6}+C$

9. $\int 6x^{-3}dx = -3x^{-2}+C = -3/x^2+C$ **11.** $(5/3)x^3 - 6x^2 + 8x + C$

13. $\int(3x^2 - x + 5x^{-3})dx = x^3 - (1/2)x^2 - 5/(2x^2) + C$

15. $\int(4x^4 - 12x^2 + 9)dx = (4/5)x^5 - 4x^3 + 9x + C$

17. $\int(6x+2)^{1/2}dx = (1/6)\int(6x+2)^{1/2}\cdot 6\,dx = (1/6)(2/3)(6x+2)^{3/2} + C = (1/9)(6x+2)^{3/2} + C$

19. $\int 8x(x^2+3)^3dx = 4\int(x^2+3)^3\cdot 2x\,dx = (x^2+3)^4 + C$

21. $\int(5x^2-1)^{1/3}\cdot x\,dx = (1/10)\int(5x^2-1)^{1/3}\cdot 10x\,dx$
$= (3/40)(5x^2-1)^{4/3} + C$

23. $\int(x^2-1)^4 x\,dx = (1/2)\int(x^2-1)^4\cdot 2x\,dx$
$= (1/10)(x^2-1)^5 + C$

25. $\int(x^2+1)^{-1/2}\cdot 2x\,dx = 2(x^2+1)^{1/2} + C$

27. $\int(3x^2+2)(x^3+2x)^3dx = \int u^3du = \dfrac{u^4}{4} + C$ let $u = x^3 + 2x$
$= (1/4)(x^3+2x)^4 + C$ $du = (3x^2+2)dx$

29. $\int(x^3-4)^{-2}x^2dx = (1/3)\int(x^3-4)^{-2}3x^2dx = \dfrac{-1}{3(x^3-4)} + C$

31. $\int(5x^2-x)^{1/2}(10x-1)dx = (2/3)(5x^2-x)^{3/2} + C$

33. $\int(x^2+x)^{-1/2}(2x+1)dx = 2(x^2+x)^{1/2} + C$

35. $\int(2x+3)^2dx = (1/2)\int(2x+3)^2\cdot 2\,dx = (1/6)(2x+3)^3 + C$

37. $(1/2)\int(2x-1)^4\cdot 2\,dx = (1/10)(2x-1)^5 + C$

39. $\int (x^6 + 3x^4 + 3x^2 + 1)\,dx = (1/7)x^7 + (3/5)x^5 + x^3 + x + C$

41. $2\int (x^2 + 1)^3 \cdot 2x\,dx = (1/2)(x^2 + 1)^4 + C$

43. $2\int (5x^3 + 1)^4 15x^2\,dx = (2/5)(5x^3 + 1)^5 + C$

45. $2\int (x^3 + 1)^{-1/2} 3x^2\,dx = 4(x^3 + 1)^{1/2} + C$

47. $2\int (x^3 + 3x)^{-1/3}(3x^2 + 3)\,dx = 3(x^3 + 3x)^{2/3} + C$

49. $\int x^{-3}(x - 1)\,dx = \int (x^{-2} - x^{-3})dx = -1/x + 1/(2x^2) + C$

Exercises 24.2

1. $y = \int 3x\,dx = (3/2)x^2 + C$
$1 = (3/2)(0) + C$
$C = 1$ and $y = (3/2)x^2 + 1$

3. $y = \int (3x^2 + 3)\,dx = x^3 + 3x + C$
$2 = (-1)^3 + 3(-1) + C$
$C = 6$ and $y = x^3 + 3x + 6$

5. $y = (1/2)\int (x^2 - 3)^2\, 2x\,dx$
$= (1/6)(x^2 - 3)^3 + C$
$7/6 = (1/6)(2^2 - 3)^3 + C$
$C = 1$
$y = (1/6)(x^2 - 3)^3 + 1$

7. $v = \int a\,dt = \int 3t\,dt = (3/2)t^2 + C_1$
$40 = (3/2)(4)^2 + C_1;\ C_1 = 16$
$v = (3/2)t^2 + 16$
$s = \int v\,dt = \int [(3/2)t^2 + 16]\,dt$
$s = (1/2)t^3 + 16t + C_2$
$86 = (1/2)(2)^3 + 16(2) + C_2$
$C_2 = 50$
$s = (1/2)t^3 + 16t + 50$

9. $v = \int -32\,dt = -32t + C_1$
Since $v = 0$ when $t = 0$, $C_1 = 0$.
$v = -32t$
$s = \int -32t\,dt = -16t^2 + C_2$
Since $s = 100$ when $t = 0$,
$C_2 = 100$; $s = -16t^2 + 100$
At $t = 2$, $s = -16(2)^2 + 100 =$
36 ft from the ground, thus ball has fallen 64 ft. Object hits ground at $s = 0$; $0 = -16t^2 + 100$
$t = 5/2$ s so $v = -32(5/2) = -80$ ft/s (down)

11. $\Delta s = (1/2)\,a(\Delta t)^2$
$3000 = (1/2)\,a(30)^2$
$a = 8$ ft/s
$v = \int a\,dt$
$= \int 8\,dt = 8t + C$
At $t = 0$, $v = 0$
Thus, $C = 0$
$v = 8t\Big|_{t=30} = 240$ ft/s

13. $v = \int -9.80\,dt = -9.80t + C_1$

At $t = 0$, $v = 25$, so $C_1 = 25$

$v = -9.80t + 25$

$s = \int v\,dt = \int(-9.80t + 25)\,dt$; $s = -4.90t^2 + 25t + C_2$

At $t = 0$, $s = 0$ Thus $C_2 = 0$; Thus $s = -4.90t^2 + 25t$

a) Max height occurs, $v = 0$; Thus $t = 2.55$ *s* or $s = -4.90(2.55)^2 + 25(2.55) = 31.9$ m

b) Ball hits ground when $s = 0$, so $0 = -4.90t^2 + 25t = t(-4.90t + 25)$
   $t = 0$ *s* or $t = 5.10$ *s*

c) Ball hits the ground with speed v

$$= -9.80t + 25\Big|_{t=5.1} = -25 \text{ m/s}$$

15. $v = \int a\,dt = \int -32\,dt = -32t + C_1$, At $t = 0$, $v = 30$, so $C_1 = 30$

and $v = -32t + 30$, $s = \int v\,dt = \int(-32t + 30)dt$

$s = -16t^2 + 30t + C_2$

a) At $t = 0$, $s = 200$ ft, so $C_2 = 200$ and $s = -16t^2 + 30t + 200$

b) Hits ground $s = 0$, so $0 = -16t^2 + 30t + 200$ or $t = 4.59$ *s*

17. $\theta = \int \omega\,dt = \int(80 - 12t + 3t^2)dt = t^3 - 6t^2 + 80t + C_1$

At $t = 0$, $\theta = 0$, so $C_1 = 0$. $\theta = t^3 - 6t^2 + 80t\Big|_{t=3s} = 213$ revolutions

19. $V_C = (1/10^{-4})\int\left(\frac{1}{2}t^{1/2} + 0.2\right)dt$

$$= (1/10^{-4})\left[\frac{1}{3}t^{3/2} + (1/5)t\right] + C$$

At $t = 0$, $V = 100$, so $C = 100$

$$V_C = \frac{1}{10^{-4}}\left[(1/3)t^{3/2} + t/5\right] + 100\Big|_{t=0.16}$$

$= 633$ *V*

21. $q = \int i\,dt = \int t(t^2 + 1)^{1/2}\,dt = (1/3)(t^2 + 1)^{3/2} + C$

At $t = 0$, $q = 0$, so $C = -1/3$ and

$$q = (1/3)(t^2 + 1)^{3/2} - 1/3\Big|_{t=1\,s} = (1/3)(2^{3/2} - 1) = \frac{2\sqrt{2} - 1}{3}$$

23. $v = \int a\,dt = \int -32dt = -32t + C_1$; At $t = 0$, $C_1 = v_o$

$s = \int v\,dt = \int(-32t + v_o)dt$

$s = -16t^2 + v_ot + C_2$; At $t = 0$, $C_2 = s_o$.

Thus, $s = -16t^2 + v_ot + s_o$.

Similarly for the other form of this equation.

Exercises 24.3

1. $\int x\,dx = x^2/2 + C$; $A = F(2) - F(0) = 2 - 0 = 2$

3. $\int 2x^2\,dx = (2/3)x^3 + C$; $A = F(3) - F(1) = 18 - 2/3 = 17\frac{1}{3}$

5. $\int(3x^2 - 2x)\,dx = x^3 - x^2 + C$; $A = F(2) - F(1) = 4 - 0 = 4$

7. $\int 3x^{-2}\,dx = -3/x + C$; $A = F(2) - F(1) = -3/2 - (-3) = 3/2$

9. $\int(3x-2)^{1/2}\,dx = (1/3)(2/3)(3x-2)^{3/2} + C$; $A = F(2) - F(1) = 14/9$

11. $\int(4x - x^3)\,dx = 2x^2 - (1/4)x^4 + C$; $A = F(2) - F(0) = 4 - 0 = 4$

13. $\int(1 - x^4)\,dx = x - x^5/5 + C$; $A = F(1) - F(-1) = 4/5 + 4/5 = 8/5$

15. $\int(2x+1)^{1/2}\,dx = (1/2)(2/3)(2x+1)^{3/2} + C$; $A = F(12) - F(4) = 125/3 - 9 = 98/3$

17. $\int(x-1)^{1/3}\,dx = (3/4)(x-1)^{4/3} + C$; $A = F(1) - F(0) = 0 - (-3/4) = 3/4$

19. $\int(1/x^2)\,dx = -1/x + C$; $A = F(5) - F(1) = -1/5 - (-1) = 4/5$

21. $\int(2x-1)^{-2}\,dx = -1/[2(2x-1)] + C$; $A = F(0) - F(-3) = 1/2 - 1/14 = 3/7$

23. $\int(9 - x^2)\,dx = 9x - x^3/3 + C$; $A = F(3) - F(-3) = 18 - (-18) = 36$

25. $\int(2x - x^2)\,dx = x^2 - x^3/3 + C$; $A = F(2) - F(0) = 4/3 - 0 = 4/3$

27. $\int(x^2 - x^3)\,dx = x^3/3 - x^4/4 + C$; $A = F(1) - F(0) = 1/12 - 0 = 1/12$

29. $\int(x^2 - x^4)\,dx = x^3/3 - x^5/5 + C$; $2A = 2[F(1) - F(0)] = 2[2/15 - 0] = 4/15$

Exercises 24.4

1. $(5/2)x^2\Big|_0^1 = 5/2 - 0 = 5/2$

3. $(x^3/3 + 3x)\Big|_1^2 = (8/3 + 6) - (1/3 + 3) = 16/3$

5. $x^4/4 + x\Big|_2^0 = 0 - (4 + 2) = -6$

7. $x^3/3 + x^2/2 + 2x\Big|_{-1}^1 = (1/3 + 1/2 + 2) - (-1/3 + 1/2 - 2) = 14/3$

9. $2x^{3/2} + 2x^{1/2}\Big|_4^9 = (2 \cdot 9^{3/2} + 2 \cdot 9^{1/2}) - (2 \cdot 4^{3/2} + 2 \cdot 4^{1/2}) = 40$

11. $\int_1^9 (x^{1/2} + 3x^{-1/2})dx = (2/3)x^{3/2} + 6x^{1/2}\Big|_1^9 = [(2/3)9^{3/2} + 6 \cdot 9^{1/2}] - [(2/3) \cdot 1^{3/2} + 6 \cdot 1^{1/2}] = 88/3$

13. $(1/15)(3x + 4)^5\Big|_1^2 = (1/15)[10^5 - 7^5] = 5546.2$

15. $(1/3)(2x + 4)^{3/2}\Big|_0^{16} = (1/3)[36^{3/2} - 4^{3/2}] = 208/3$

17. $(1/2)(x^2 - 3)^4\Big|_1^2 = (1/2)[(2^2 - 3)^4 - (1^2 - 3)^4] = -15/2$

19. $(-3/10)(1 - x^2)^{5/2}\Big|_{-1}^0 = (-3/10)[(1 - 0^2)^{5/3} - (1 - (-1)^2)^{5/3}] = -3/10$

21. $(1/3)(x^2 + 1)^{3/2}\Big|_0^1 = (1/3)[2^{3/2} - 1^{3/2}] = (2\sqrt{2} - 1)/3$

23. $6(x^2 + 9)^{1/2}\Big|_0^4 = 6[(4^2 + 9)^{1/2} - (0^2 + 9)^{1/2}] = 12$

25. $2(x^3 + x)^{1/2}\Big|_1^2 = 2\sqrt{10} - 2\sqrt{2}$

Chapter 24 Review

1. $(5/3)x^3 - (1/2)x^2 + C$

2. $(3/8)x^8 + x^2 + 4x + C$

3. $(4/3)x^{9/2} + C$

4. $(12/5)x^{5/3} + C$

5. $\int 3x^{-5} dx = -3/(4x^4) + C$

6. $\int x^{-3/2} dx = -2/\sqrt{x} + C$

7. $\frac{1}{2}\int(3x^4+2x-1)^3\cdot 2(6x^3+1)\,dx = (1/8)(3x^4+2x-1)^4+C$

8. $\frac{1}{2}\int(7x^2+8x+2)^{3/5}\cdot 2(7x+4)\,dx = (5/16)(7x^2+8x+2)^{8/5}+C$

9. $\int(x^2+5x)^{-1/2}(2x+5)\,dx = 2(x^2+5x)^{1/2}+C$

10. $\int(5x^3+4x)^{-2/3}(15x^2+4)\,dx = 3(5x^3+4x)^{1/3}+C$

11. $y = \int 3x^2 dx = x^3 + C$

$-3 = 1 + C$ so $C = -4$ and $y = x^3 - 4$

12. $v = \int -32\,dt = -32t + C_1$ At $t = 0$, $v = 25$, so $C_1 = 25$

Thus $v = -32t + 25$

$s = \int v\,dt = \int(-32t+25)dt$ Thus $s = -16t^2 + 25t + C_2$

At $t = 0$, $s = 100$, so $C_2 = 100$ and thus $s = -16t^2 + 25t + 100$

13. $R = \int(0.009T^2+0.02T-0.7)dt = 0.003T^3 + 0.01T^2 - 0.7T + C_1$

At $T = 0$, $R = 0.2$, then $C_1 = 0.2$

$R = 0.003T^3 + 0.01T^2 - 0.7T + 0.2$; find R when $T = 30$

$R = 0.003(30)^3 + 0.01(30)^2 - 0.7(30) + 0.2 = 69.2\ \Omega$

14. $q = \int i\,dt = \int\frac{3t^2+1}{\sqrt{t^3+t+2}}dt = \int(t^3+t+2)^{-1/2}(3t^2+1)dt$

$= 2(t^3+t+2)^{1/2}+C$

At $t = 0$, $q = 2\sqrt{2}$ so $C = 0$ Thus $q = 2(t^3+t+2)^{1/2}\Big|_{t=0.2}$

$= 2\sqrt{2.208} = 2.97\ C$

15. $\int(x^2+1)\,dx = x^3/3 + x + C$; $A = F(2) - F(0) = (8/3 + 2) - 0 = 14/3$

16. $\int(8+6x^2)\,dx = 8x - 2x^3 + C$; $A = F(1) - F(0) = (8 - 2) - 0 = 6$

17. $\int x^{-5}\,dx = -1/(4x^4) + C$; $A = F(2) - F(1) = (-1/64) - (-1/4) = 15/64$

18. $2\int(x^2+1)^{-2}\cdot 2x\,dx = -2/(x^2+1) + C$; $A = F(1) - F(0) = -1 - (-2) = 1$

19. $(1/5)\int(5x+6)^{1/2}\cdot 5\,dx = (2/15)(5x+6)^{3/2} + C;\ A = F(6) - F(0)$

$= (2/15)[(5\cdot 6+6)^{3/2} - (50+6)^{3/2}] = (144 - 4\sqrt{6})/5$

20. $(1/2)\int(x^2+4)^{1/2}\cdot 2x\,dx = (1/3)(x^2+4)^{3/2} + C;$

$A = F(2) - F(0) = (1/3)[(2^2+4)^{3/2} - (0^2+4)^{3/2}] = (16\sqrt{2}-8)/3$

21. $x^4/4 + (2/3)x^3 + (1/2)x^2\Big|_0^1 = (1/4 + 2/3 + 1/2) - 0 = 17/12$

22. $(3/5)x^5 - (1/2)x^4 + (7/2)x^2\Big|_1^2 = [(3/5)2^5 - (1/2)2^4 + (7/2)2^2] - (3/5 - 1/2 + 7/2) = 108/5$

23. $(1/2)\int_0^2 2x(x^2+1)^2\,dx = (x^2+1)^3/6\Big|_0^2 = 5^3/6 - 1/6 = 62/3$

24. $(5/3)\int_1^2 (x^3+2)^{-2}\cdot 3x^2 dx = -5/[3(x^3+2)]\Big|_1^2 = -5/30 - (-5/9) = 7/18$

25. $(2/3)(x^2+x)^{3/2}\Big|_1^2 = \dfrac{12\sqrt{6}}{3} - \dfrac{4\sqrt{2}}{3} = \dfrac{12\sqrt{6}-4\sqrt{2}}{3}$

26. $(1/3)\int_0^3 (x^3+1)^{-1/2}\cdot 3x^2 dx = (2/3)(x^3+1)^{1/2}\Big|_0^3 = \dfrac{4\sqrt{7}-2}{3}$

27. $\int_2^1 (3x^2 - x + x^{-2})dx = x^3 - (1/2)x^2 - 1/x\Big|_2^1 = (1 - 1/2 - 1) - (8 - 2 - 1/2) = -6$

28. $(3/4)\int_0^{1/2} (2x^2+1/2)^{-1/2}\cdot 4x\,dx = (3/2)(2x^2+1/2)^{1/2}\Big|_0^{1/2} = \dfrac{6-3\sqrt{2}}{4}$

Chapter 25, Exercises 25.1

1.

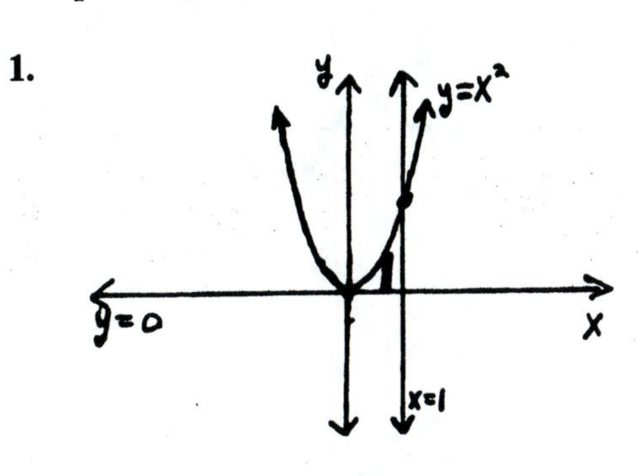

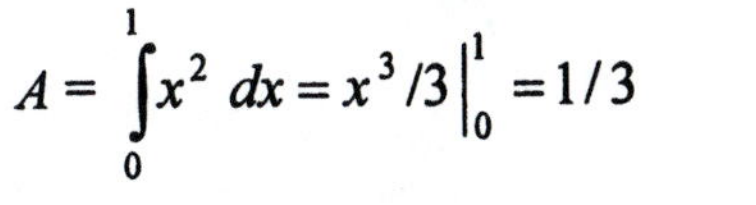

$$A=\int_0^1 x^2\,dx = x^3/3\Big|_0^1 = 1/3$$

3.

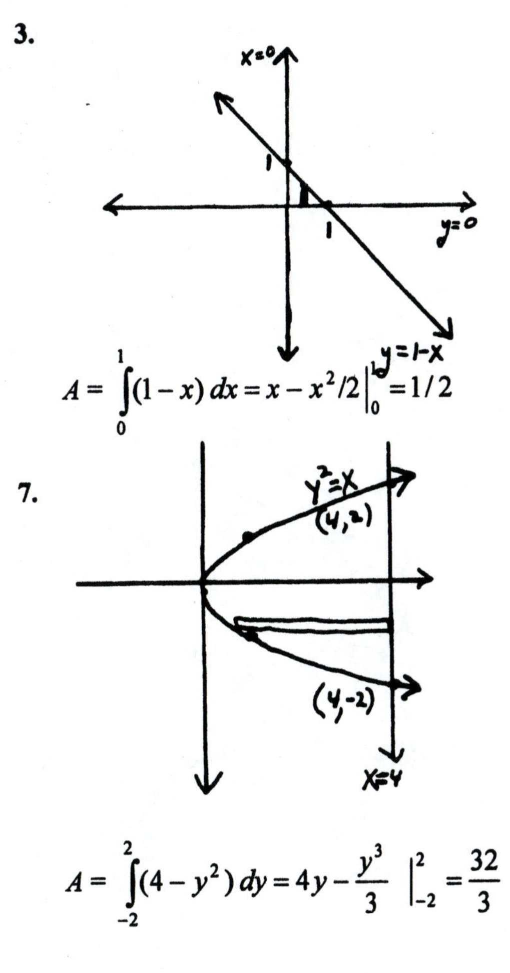

$$A=\int_0^1 (1-x)\,dx = x - x^2/2\Big|_0^1 = 1/2$$

5.

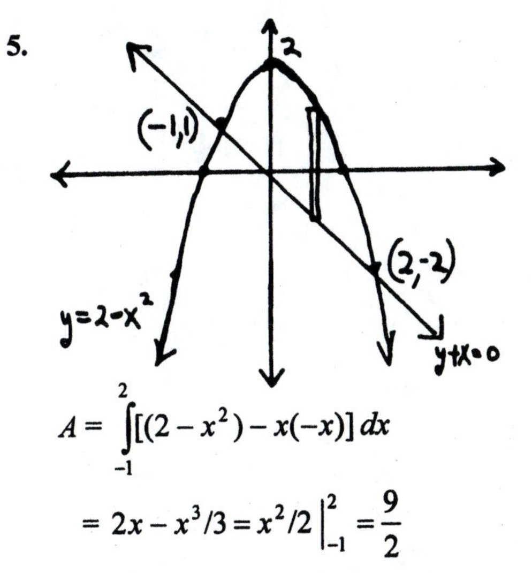

$$A=\int_{-1}^{2}[(2-x^2)-x(-x)]\,dx$$

$$= 2x - x^3/3 = x^2/2\Big|_{-1}^{2} = \frac{9}{2}$$

7.

$$A=\int_{-2}^{2}(4-y^2)\,dy = 4y-\frac{y^3}{3}\Big|_{-2}^{2}=\frac{32}{3}$$

9.

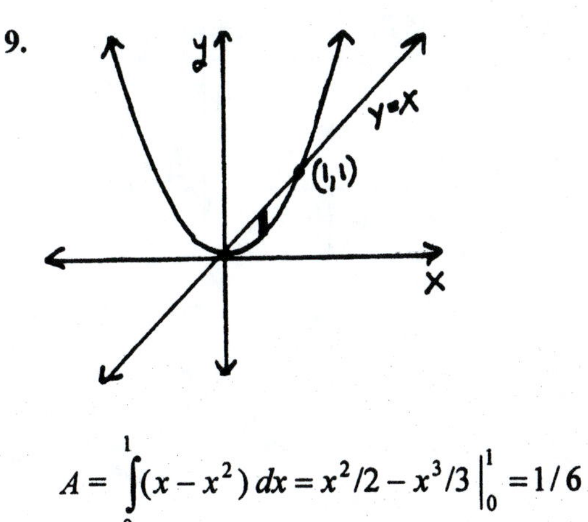

$$A=\int_0^1 (x-x^2)\,dx = x^2/2 - x^3/3\Big|_0^1 = 1/6$$

11.

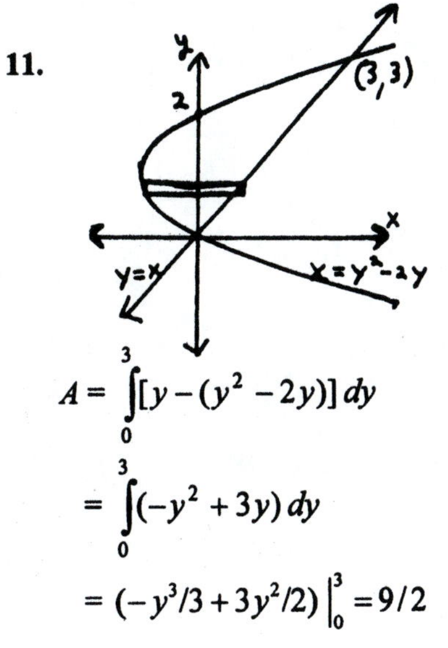

$$A=\int_0^3 [y-(y^2-2y)]\,dy$$

$$= \int_0^3 (-y^2+3y)\,dy$$

$$= (-y^3/3+3y^2/2)\Big|_0^3 = 9/2$$

13. $A = \int_{-1}^{0}(x^3 - x)\,dx + \int_{0}^{1}[0 - (x^3 - x)]\,dx$

$= x^4/4 - x^2/2\,\Big|_{-1}^{0} + (-x^4/4 + x^2/2)\,\Big|_{0}^{1} = 1/2$

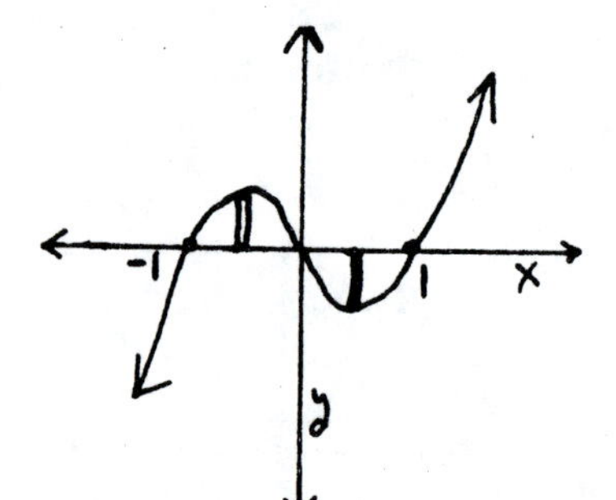

15.

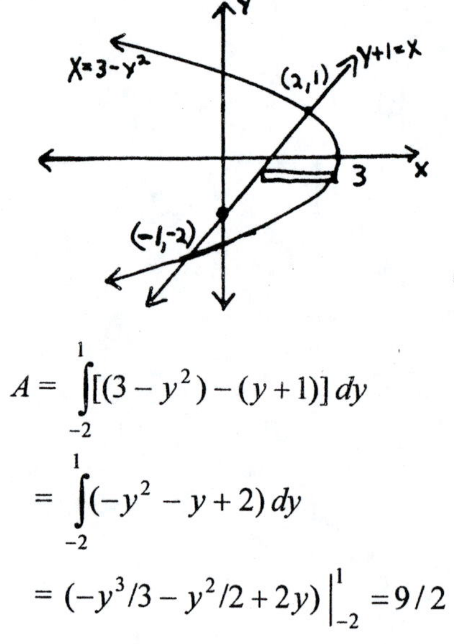

$A = \int_{-2}^{1}[(3 - y^2) - (y + 1)]\,dy$

$= \int_{-2}^{1}(-y^2 - y + 2)\,dy$

$= (-y^3/3 - y^2/2 + 2y)\,\Big|_{-2}^{1} = 9/2$

17.

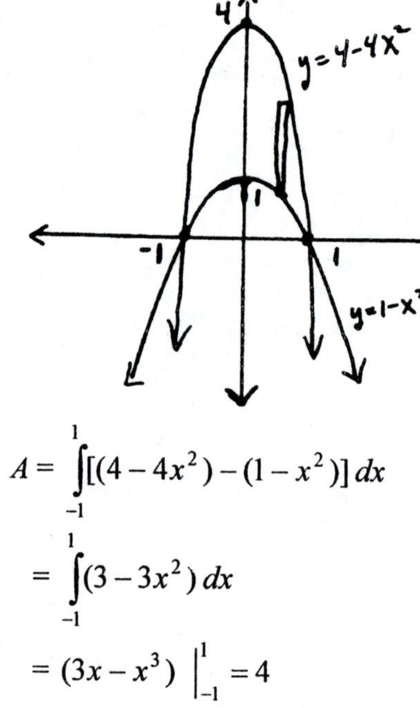

$A = \int_{-1}^{1}[(4 - 4x^2) - (1 - x^2)]\,dx$

$= \int_{-1}^{1}(3 - 3x^2)\,dx$

$= (3x - x^3)\,\Big|_{-1}^{1} = 4$

19.

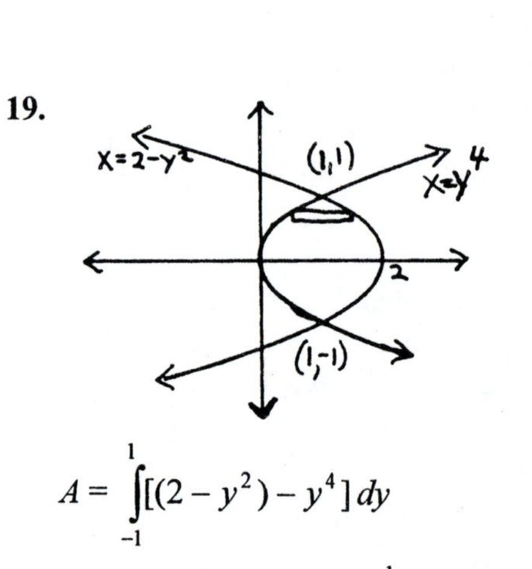

$A = \int_{-1}^{1}[(2 - y^2) - y^4]\,dy$

$= (2y - y^3/3 - y^5/5)\,\Big|_{-1}^{1} = 44/15$

21.

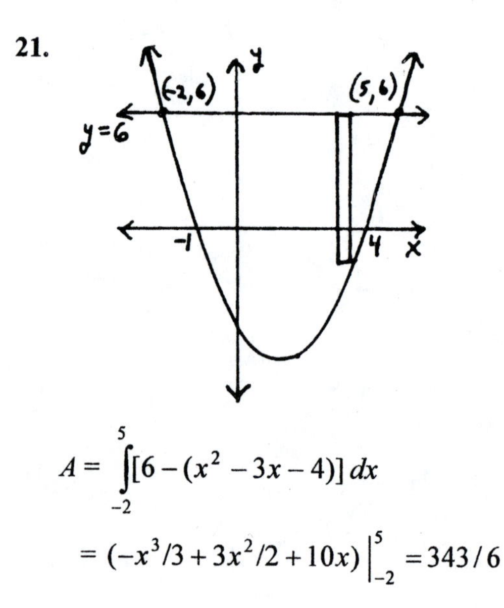

$A = \int_{-2}^{5}[6 - (x^2 - 3x - 4)]\,dx$

$= (-x^3/3 + 3x^2/2 + 10x)\,\Big|_{-2}^{5} = 343/6$

23.

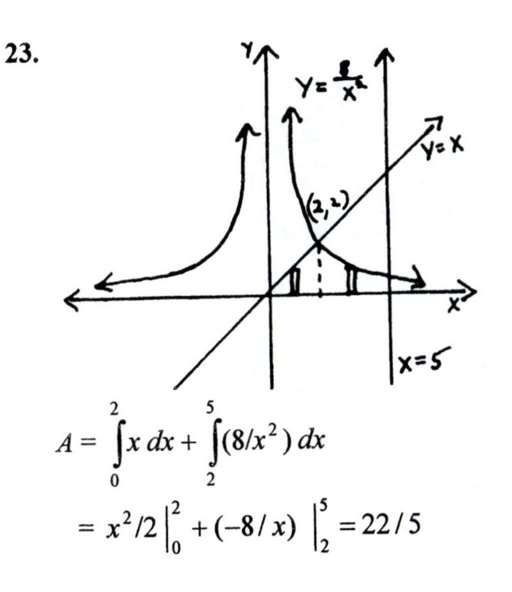

$$A = \int_0^2 x\,dx + \int_2^5 (8/x^2)\,dx$$

$$= x^2/2\Big|_0^2 + (-8/x)\Big|_2^5 = 22/5$$

25.

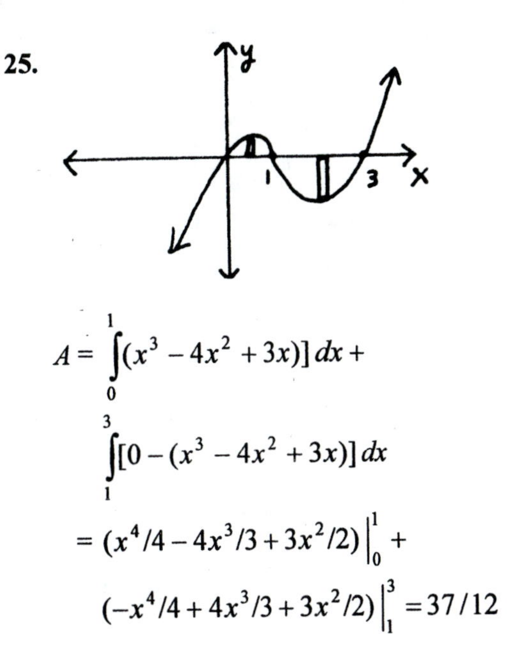

$$A = \int_0^1 (x^3 - 4x^2 + 3x)]\,dx +$$

$$\int_1^3 [0 - (x^3 - 4x^2 + 3x)]\,dx$$

$$= (x^4/4 - 4x^3/3 + 3x^2/2)\Big|_0^1 +$$

$$(-x^4/4 + 4x^3/3 + 3x^2/2)\Big|_1^3 = 37/12$$

Exercises 25.2

1.

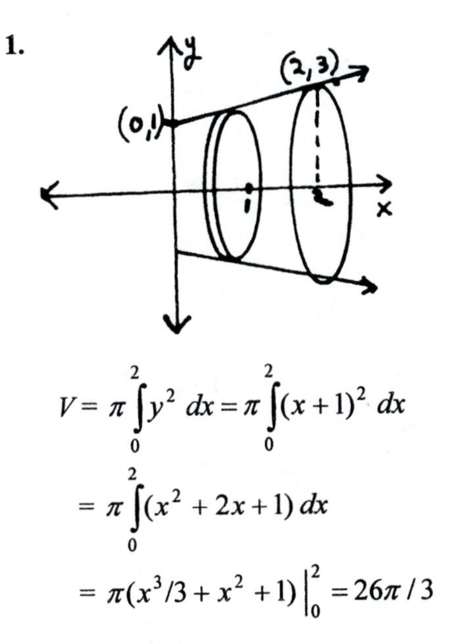

$$V = \pi\int_0^2 y^2\,dx = \pi\int_0^2 (x+1)^2\,dx$$

$$= \pi\int_0^2 (x^2 + 2x + 1)\,dx$$

$$= \pi(x^3/3 + x^2 + 1)\Big|_0^2 = 26\pi/3$$

3.

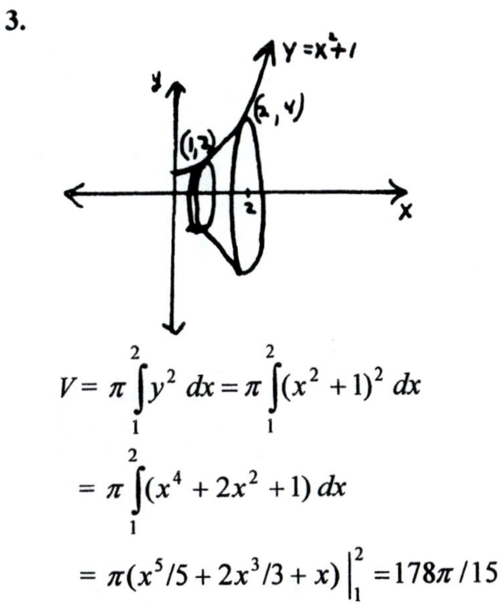

$$V = \pi\int_1^2 y^2\,dx = \pi\int_1^2 (x^2+1)^2\,dx$$

$$= \pi\int_1^2 (x^4 + 2x^2 + 1)\,dx$$

$$= \pi(x^5/5 + 2x^3/3 + x)\Big|_1^2 = 178\pi/15$$

5.

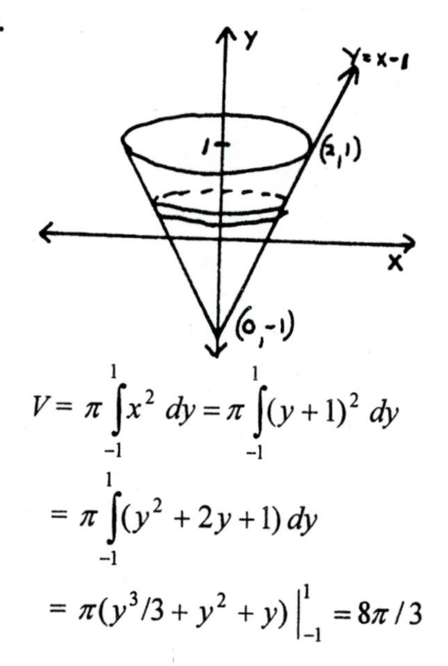

$$V = \pi \int_{-1}^{1} x^2 \, dy = \pi \int_{-1}^{1} (y+1)^2 \, dy$$

$$= \pi \int_{-1}^{1} (y^2 + 2y + 1) \, dy$$

$$= \pi (y^3/3 + y^2 + y) \Big|_{-1}^{1} = 8\pi/3$$

7.

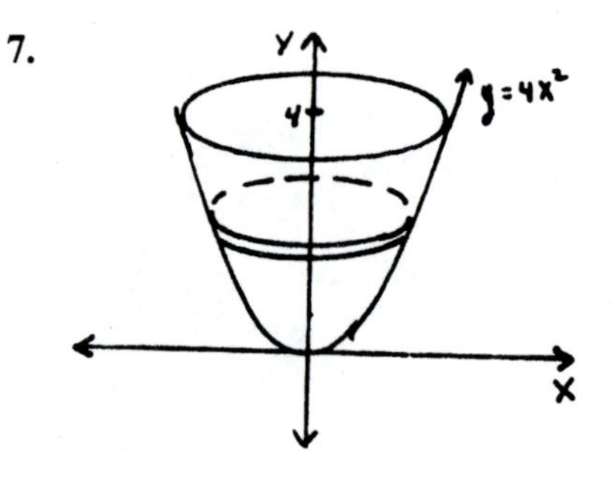

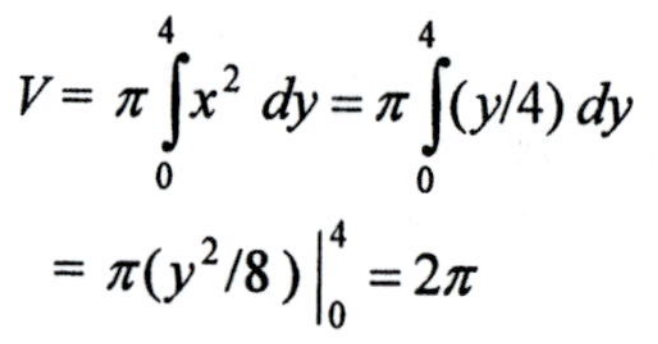

$$V = \pi \int_{0}^{4} x^2 \, dy = \pi \int_{0}^{4} (y/4) \, dy$$

$$= \pi (y^2/8) \Big|_{0}^{4} = 2\pi$$

9.

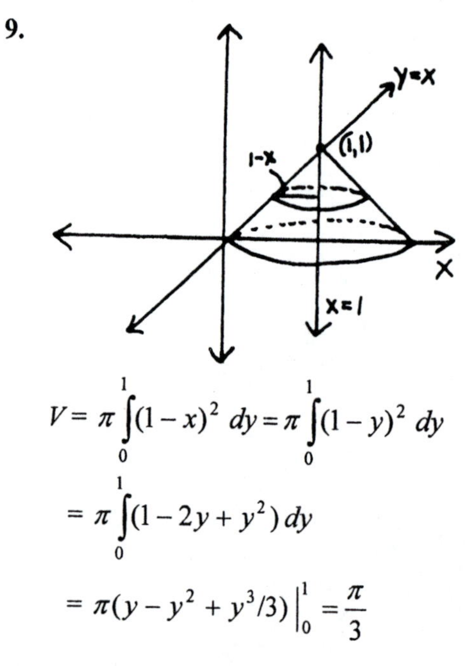

$$V = \pi \int_{0}^{1} (1-x)^2 \, dy = \pi \int_{0}^{1} (1-y)^2 \, dy$$

$$= \pi \int_{0}^{1} (1 - 2y + y^2) \, dy$$

$$= \pi (y - y^2 + y^3/3) \Big|_{0}^{1} = \frac{\pi}{3}$$

11.

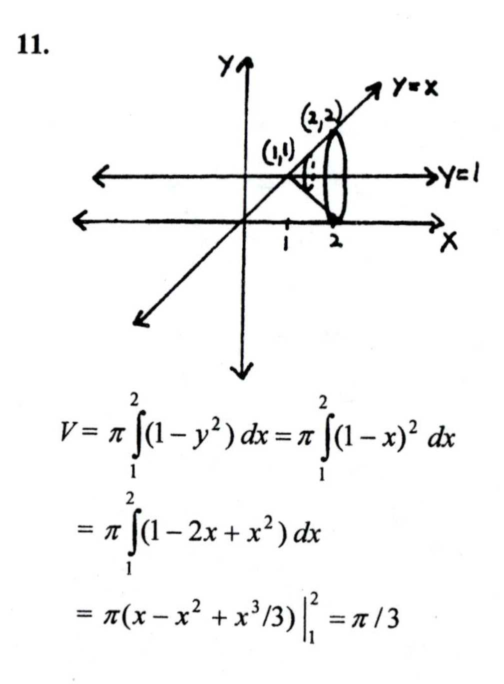

$$V = \pi \int_{1}^{2} (1 - y^2) \, dx = \pi \int_{1}^{2} (1-x)^2 \, dx$$

$$= \pi \int_{1}^{2} (1 - 2x + x^2) \, dx$$

$$= \pi (x - x^2 + x^3/3) \Big|_{1}^{2} = \pi/3$$

13.

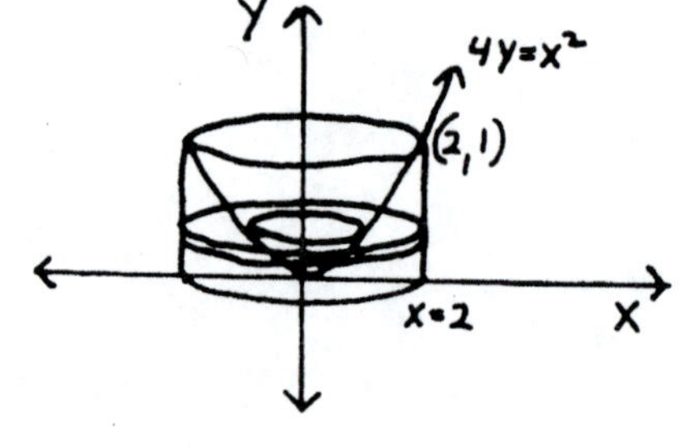

$$V = \pi \int_{0}^{1} (2^2 - x^2) \, dy = \pi \int_{0}^{1} (4 - 4y) \, dy$$

$$= 4\pi (y - y^2/2) \Big|_{0}^{1} = 2\pi$$

15.

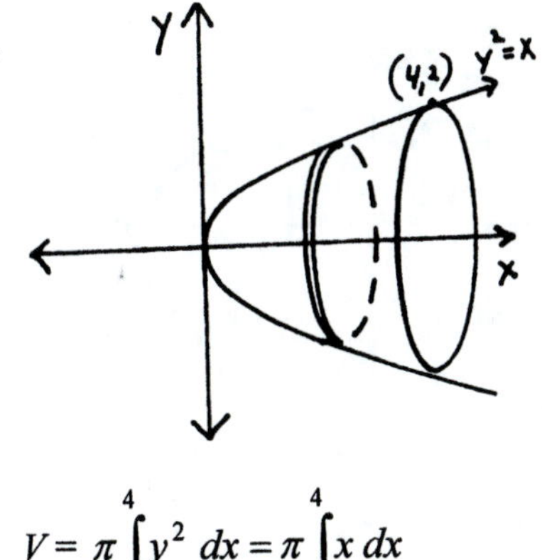

$$V = \pi \int_0^4 y^2\,dx = \pi \int_0^4 x\,dx$$

$$= \pi(x^2/2)\Big|_0^4 = 8\pi$$

17.

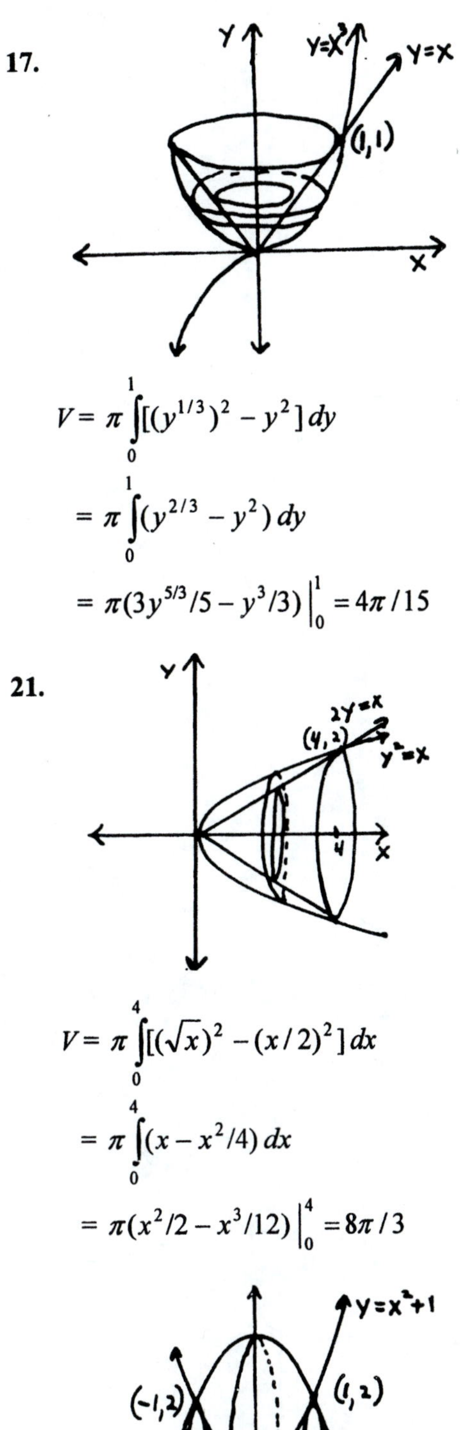

$$V = \pi \int_0^1 [(y^{1/3})^2 - y^2]\,dy$$

$$= \pi \int_0^1 (y^{2/3} - y^2)\,dy$$

$$= \pi(3y^{5/3}/5 - y^3/3)\Big|_0^1 = 4\pi/15$$

19.

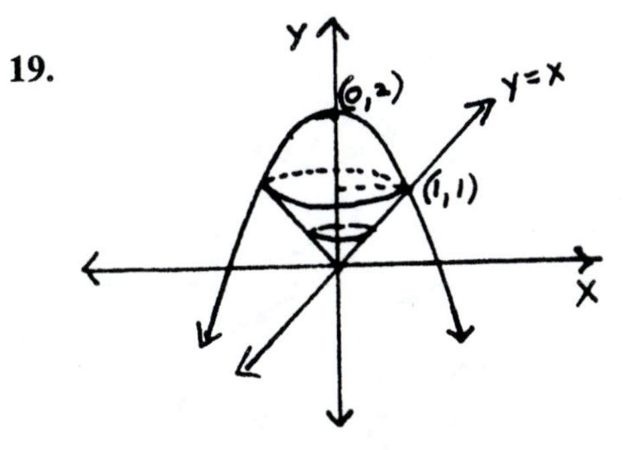

$$V = \pi \int_0^1 y^2\,dy + \pi \int_1^2 (\sqrt{2-y})^2\,dy$$

$$= \pi(y^3/3)\Big|_0^1 + \pi(2y - y^2/2)\Big|_1^2$$

$$= \pi/3 + \pi/2 = 5\pi/6$$

21.

$$V = \pi \int_0^4 [(\sqrt{x})^2 - (x/2)^2]\,dx$$

$$= \pi \int_0^4 (x - x^2/4)\,dx$$

$$= \pi(x^2/2 - x^3/12)\Big|_0^4 = 8\pi/3$$

23. $$V = 2\pi \int_0^1 [(3-x^2)^2 - (x^2+1)^2]\,dx$$

$$= 2\pi \int_0^1 [(9 - 6x^2 + x^4) - (x^4 + 2x^2 + 1)]\,dx$$

$$= 2\pi \int_0^1 (8 - 8x^2)\,dx = 2\pi(8x - 8x^3/3)\Big|_0^1 = 32\pi/3$$

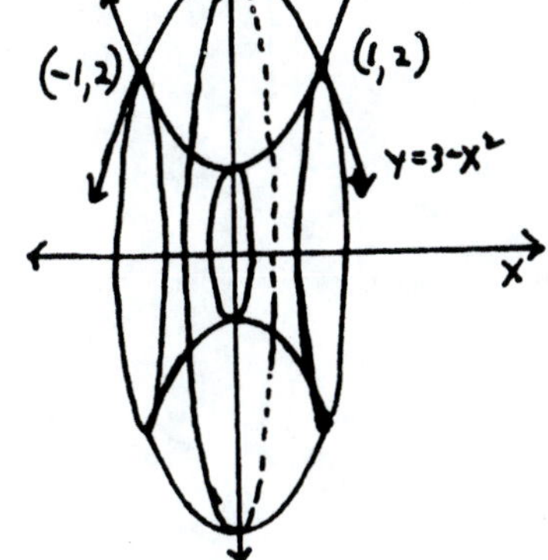

25. $V = 2\pi \int_0^5 y^2\, dx = 2\pi \int_0^5 \frac{225 - 9x^2}{25}\, dx$

$= 2\pi \int_0^5 (9 - 9x^2/25)\, dx$

$= 2\pi (9x - 3x^3/25) \Big|_0^5 = 60\pi$

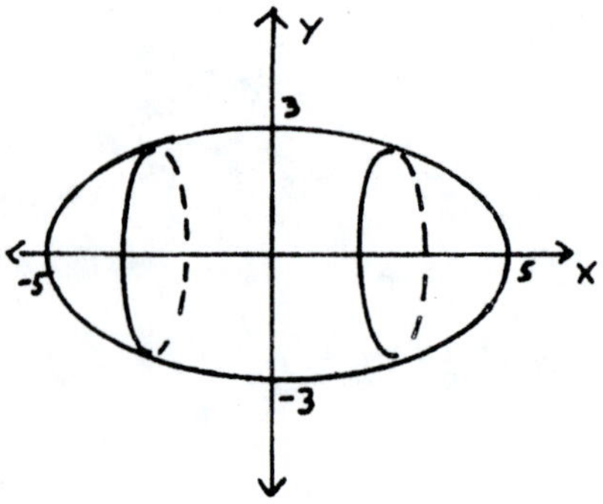

27. intersection of $y = 2$ and

$9x^2 + 25y^2 = 225$

$9x^2 + 25(2)^2 = 225$

$9x^2 = 125$

$x = \pm 5\sqrt{5}/3$

$V = 2\pi \int_0^{5\sqrt{5}/3} (y^2 - 2^2)\, dx$

$= 2\pi \int_0^{5\sqrt{5}/3} \left\{ \frac{225 - 9x^2}{25} - 2^2 \right\} \cdot dx = 2\pi \int_0^{5\sqrt{5}/3} (9 - 9x^2/25 - 4)\, dx$

$= 2\pi \int_0^{5\sqrt{5}/3} (5 - 9x^2/25)\, dx = 2\pi (5x - 3x^3/25) \Big|_0^{5\sqrt{5}/3} = \frac{100\pi\sqrt{5}}{9}$ in^3

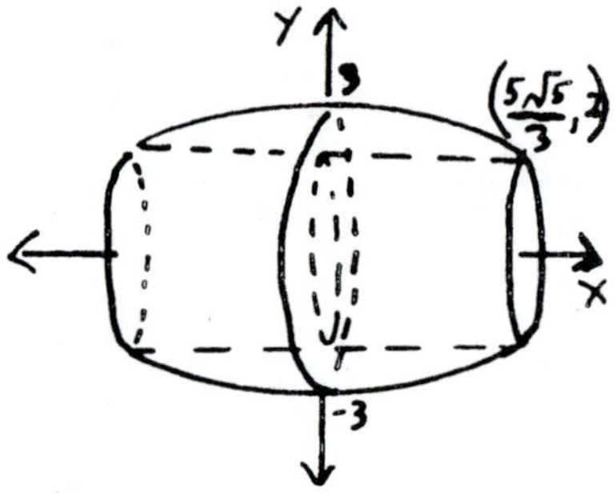

29. $V = 2\pi \int_0^r x^2\, dy = 2\pi \int_0^r (r^2 - y^2)\, dy$

$= 2\pi (r^2 y - y^3/3) \Big|_0^r = 2\pi (r^3 - r^3/3) = \frac{4}{3}\pi r^3$

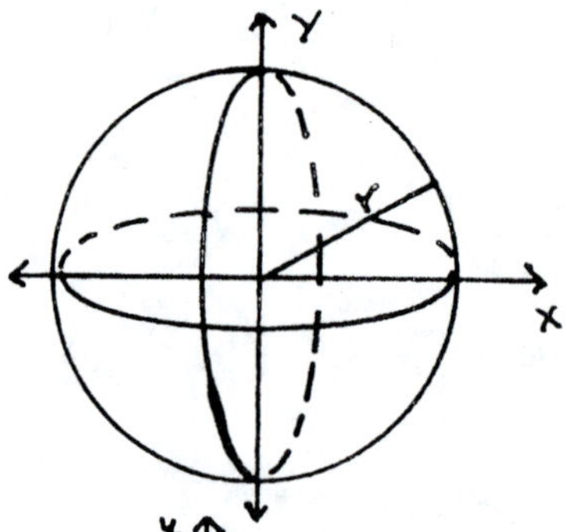

Exercises 25.3

1.

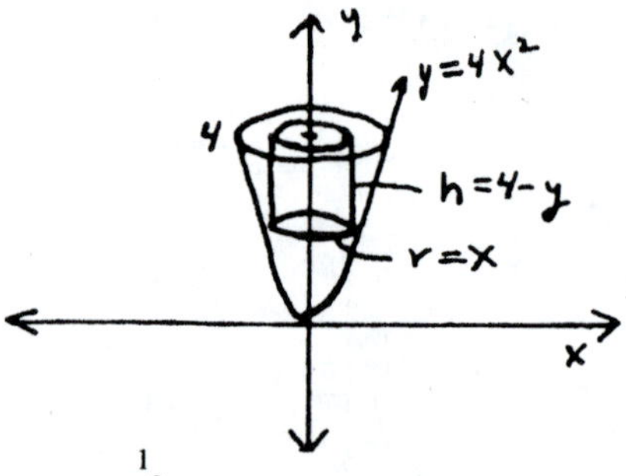

$V = 2\pi \int_0^1 x(4 - 4x^2)\, dx$

$= 2\pi \int_0^4 (4x - 4x^3)\, dx$

$= 2\pi (2x^2 - x^4) \Big|_0^1 = 2\pi$

3.

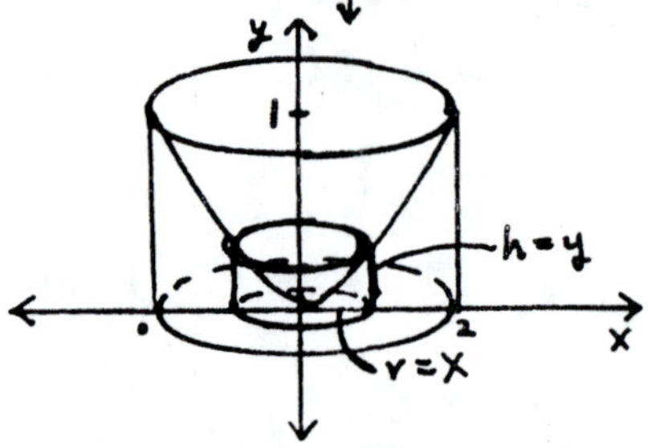

$V = 2\pi \int_0^2 xy\, dx = 2\pi \int_0^2 x(x^2/4)\, dx$

$= 2\pi \int_0^2 (x^3/4)\, dx = 2\pi (x^4/16) \Big|_0^2 = 2\pi$

5.

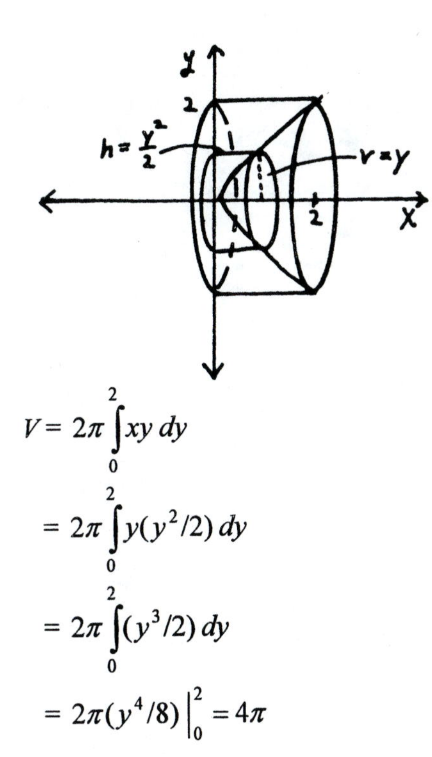

$$V = 2\pi \int_0^2 xy\,dy$$

$$= 2\pi \int_0^2 y(y^2/2)\,dy$$

$$= 2\pi \int_0^2 (y^3/2)\,dy$$

$$= 2\pi(y^4/8)\Big|_0^2 = 4\pi$$

7.

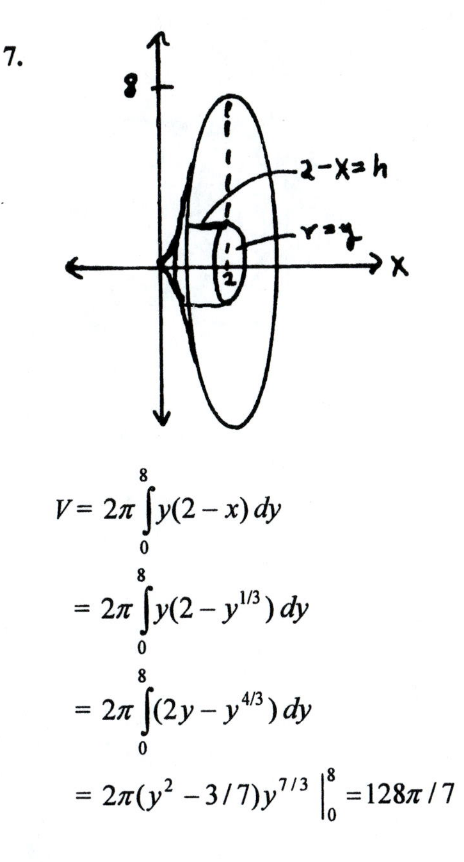

$$V = 2\pi \int_0^8 y(2-x)\,dy$$

$$= 2\pi \int_0^8 y(2-y^{1/3})\,dy$$

$$= 2\pi \int_0^8 (2y-y^{4/3})\,dy$$

$$= 2\pi(y^2 - 3/7)y^{7/3}\Big|_0^8 = 128\pi/7$$

9.

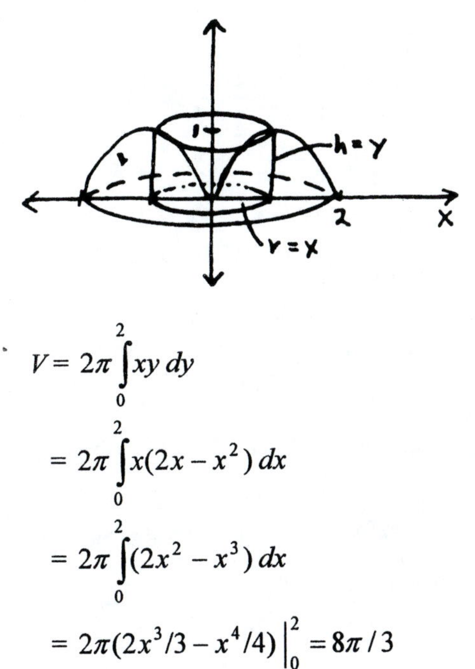

$$V = 2\pi \int_0^2 xy\,dy$$

$$= 2\pi \int_0^2 x(2x-x^2)\,dx$$

$$= 2\pi \int_0^2 (2x^2-x^3)\,dx$$

$$= 2\pi(2x^3/3 - x^4/4)\Big|_0^2 = 8\pi/3$$

11.

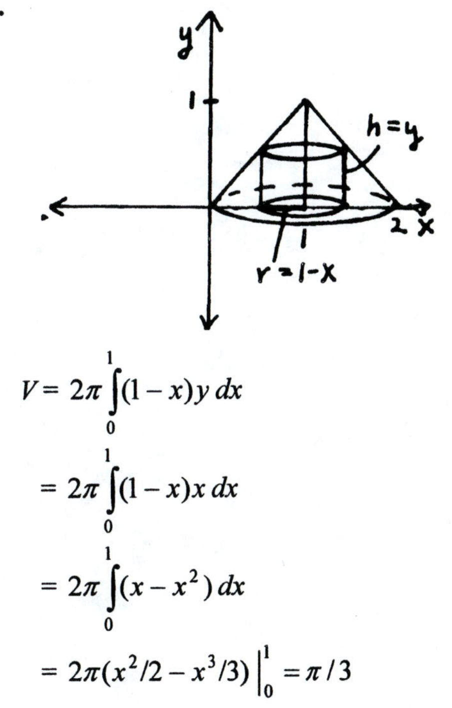

$$V = 2\pi \int_0^1 (1-x)y\,dx$$

$$= 2\pi \int_0^1 (1-x)x\,dx$$

$$= 2\pi \int_0^1 (x-x^2)\,dx$$

$$= 2\pi(x^2/2 - x^3/3)\Big|_0^1 = \pi/3$$

13.

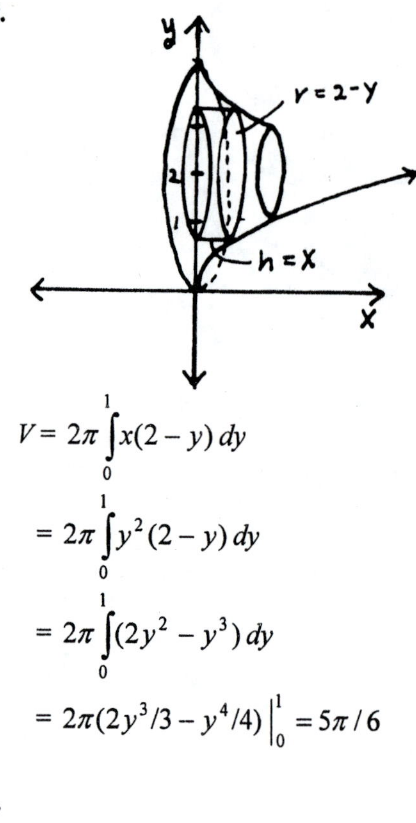

$$V = 2\pi \int_0^1 x(2-y)\,dy$$

$$= 2\pi \int_0^1 y^2(2-y)\,dy$$

$$= 2\pi \int_0^1 (2y^2 - y^3)\,dy$$

$$= 2\pi(2y^3/3 - y^4/4)\Big|_0^1 = 5\pi/6$$

15.

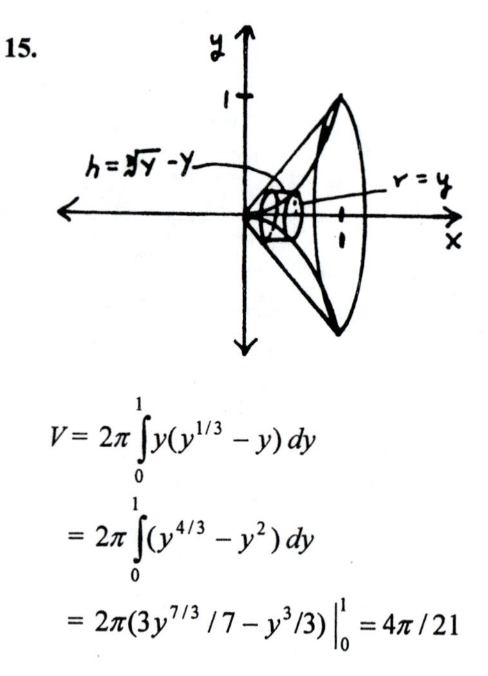

$$V = 2\pi \int_0^1 y(y^{1/3} - y)\,dy$$

$$= 2\pi \int_0^1 (y^{4/3} - y^2)\,dy$$

$$= 2\pi(3y^{7/3}/7 - y^3/3)\Big|_0^1 = 4\pi/21$$

17.

$$V = 2\pi \int_1^2 x(-y)\,dx$$

$$= -2\pi \int_1^2 x(x^2 - 3x + 2)\,dx$$

$$= -2\pi \int_1^2 (x^3 - 3x^2 + 2x)\,dx$$

$$= -2\pi(x^4/4 - x^3 + x^2)\Big|_1^2 = \pi/2$$

19.

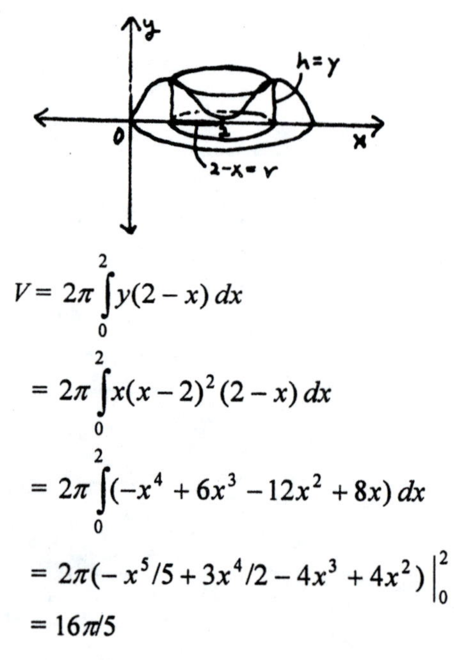

$$V = 2\pi \int_0^2 y(2-x)\,dx$$

$$= 2\pi \int_0^2 x(x-2)^2(2-x)\,dx$$

$$= 2\pi \int_0^2 (-x^4 + 6x^3 - 12x^2 + 8x)\,dx$$

$$= 2\pi(-x^5/5 + 3x^4/2 - 4x^3 + 4x^2)\Big|_0^2$$

$$= 16\pi/5$$

Exercises 25.4

1. $M_o = (3)(-5) + (7)(3) + (4)(6) = 30$; $m = 3 + 7 + 4 = 14$
   $\bar{x} = 30/14 = 15/7$

3. $M_o = (24)(-15) + (15)(-9) + (12)(3) + (9)(6) = -405$
   $m = 24 + 15 + 12 + 9 = 60$; $\bar{x} = -405/60 = -6.75$

5. Given $\bar{x} = 0$, find $(x, 0)$. $M_y = (6)(9) + (18)(-2) + (3)(x)$; $m = 3$
   $\bar{x} = \dfrac{M_y}{m}$ so $0 = \dfrac{3x + 18}{3}$ Thus $x = -6$.

7. Given $\bar{x} = 3$, find $(x, 0)$. $M_y = (24)(-8) + (36)(12) + (9)(x) = 240 + 9x$
   $m = 24 + 36 + 9 = 69$; $\bar{x} = M_y/m$; $3 = \dfrac{240 + 9x}{69}$; $x = -11/3$

9. Given $\bar{x} = 0$, find m. $M_y = (6)(-3) + (9)(12) + m(-6) = 90 - 6m$
   $\bar{x} = M_y/m$; $0 = \dfrac{90 - 6m}{m}$; $m = 15$

11. Given $\bar{x} = 3$, find m. $M_y = (25)(-6) + (45)(8) + (40)(10) + m(-4) = 610 - 4m$
    $m = 25 + 45 + 40 + m = 110 + m$; $\bar{x} = M_y/m$; $3 = \dfrac{610 - 4m}{110 + m}$; $m = 40$

13. $M_y = 75000(0) + 50000(18) + 25000(30) = 1{,}650{,}000$
    $m = 75000 + 50000 + 25000 = 150{,}000$; $\bar{x} = \dfrac{1{,}650{,}000}{150{,}000} = 11$
    Thus locate airport 11 miles north of Flatville.

15. $m = 6 + 3 + 12 = 21$; $M_x = 6(4) + 3(2) + 12(3) = 66$;
    $M_y = 6(1) + 3(6) + 12(3) = 60$; $\bar{x} = M_y/m = 60/21 = 20/7$
    $\bar{y} = M_x/m = 66/21 = 22/7$ Thus center of mass is $(20/7, 22/7)$.

17. $m = 8 + 16 + 20 + 36 = 80$; $M_x = 8(12) + 16(8) + 20(-4) + 36(-20) = -576$
    $M_y = 8(8) + 16(-12) + 20(-16) + 36(4) = -304$
    $\bar{x} = M_y/m -304/80 = -3.8$; $\bar{y} = M_x/m = -576/80 = -7.2$
    Thus center of mass is $(-3.8, -7.2)$.

19. Given $\bar{x} = 0$ and $\bar{y} = 0$, find (x, y); $m = 6 + 9 + 10 = 25$
    $M_x = 6(2) + 9(8) + 10y = 84 + 10y$; $M_y = 6(4) + 9(-5) + 10x = -21 + 10x$
    $\bar{x} = M_y/m$; $0 = \dfrac{-21 + 10x}{25}$; $x = 2.1$; $\bar{y} = M_x/m = \dfrac{84 + 10y}{25}$; $y = -8.4$
    Thus the point is $(2.1, -8.4)$.

21. Given $\bar{x} = -1$ and $\bar{y} = -2$, find m'. $m = 15 + 25 + 40 + m' = 80 + m'$

$M_y = 15(10) + 25(-6) + 40(8) + m'(-5) = 320 - 5m'$

$M_x = 15(3) + 25(-1) + 40(-2) + m'(-3) = -60 - 3m'$

$\bar{x} = M_y/m;\ -1 = \dfrac{320 - 5m'}{80 + m'};\ m' = 100$

23. Place $A(1250)$ at (0, 0), $B(820)$ at (6, −3), and C (520) at (−2, −8); find $(\bar{x}, \bar{y})$, $m = 820 + 520 + 1250 = 2590$;

$M_x = 820(-3) + 520(-8) + 1250(0) = -6620$

$M_y = 820(6) + 520(-2) + 1250(0) = 3880$

$\bar{x} = M_y/m = 3880/2590 = 1.50;\ \bar{y} = M_x/m = -6620/2590 = -2.56$

Thus best location: 1.50 mi east and 2.56 mi. south of A.

Exercises 25.5

1. $\dfrac{\int_0^{20} x\,dx}{\int_0^{20} dx} = \dfrac{\left.\frac{x^2}{2}\right|_0^{20}}{\left.x\right|_0^{20}} = \dfrac{200}{20} = 10$ Thus center is 10 cm from either end.

3. $\bar{x} = \dfrac{M_o}{m} = \dfrac{\int_0^{10}(0.1x)x\,dx}{\int_0^{10} 0.1x\,dx} = \dfrac{\int_0^{10} x^2\,dx}{\int_0^{10} x\,dx} = \dfrac{\left.\frac{x^3}{3}\right|_0^{10}}{\left.\frac{x^2}{2}\right|_0^{10}} = \dfrac{\frac{1000}{3}}{50} = \dfrac{20}{3}$ cm from lighter end

5. $\bar{x} = \dfrac{M_o}{m} = \dfrac{\int_0^{12}(4 + x^2)\,dx}{\int_0^{12}(4 + x^2)\,dx} = \dfrac{\left.\frac{1}{4}(x^2 + 4)^2\right|_0^{12}}{\left.(4x + x^3/3)\right|_0^{12}} = \dfrac{5472}{624} = 8.77$ cm from lighter end

7. $\bar{x} = \dfrac{M_o}{m} = \dfrac{\int_0^{6}(kx)x\,dx}{\int_0^{6} kx\,dx} = \dfrac{\int_0^{6} x^2\,dx}{\int_0^{6} x\,dx} = \dfrac{\left.\frac{x^3}{3}\right|_0^{6}}{\left.\frac{x^2}{2}\right|_0^{6}} = \dfrac{72}{18} = 4$ cm from given end

9. center of rectangle is (4, 6); $A = 32$

center of square is (6, 2); $A = 16$

$m = 32 + 16 = 48$

$M_y = 32(4) + 16(6) = 224$

$M_x = 32(6) + 16(2) = 224$

$\bar{x} = \dfrac{M_y}{m} = \dfrac{224}{48} = 4\dfrac{2}{3};\ \bar{y} = \dfrac{M_x}{m} = \dfrac{224}{48} = 4\dfrac{2}{3}$

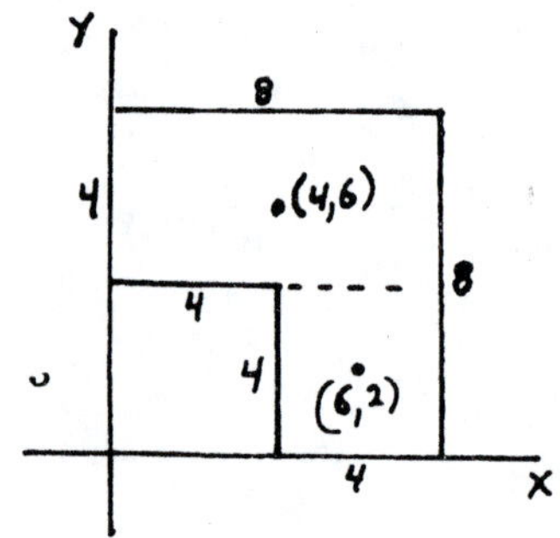

11. center of rectangle 1 is (2, 8); $A = 32$
center of rectangle 2 is (8, 2); $A = 56$
center of rectangle 3 is (14, –4); $A = 32$
$m = 32 + 56 + 32 = 120$
$M_y = 32(2) + 56(8) + 32(14) = 960$
$M_x = 32(8) + 56(2) + 32(-4) = 240$

$$\bar{x} = \frac{M_y}{m} = \frac{960}{120} = 8;\ \bar{y} = \frac{M_x}{120} = 2$$

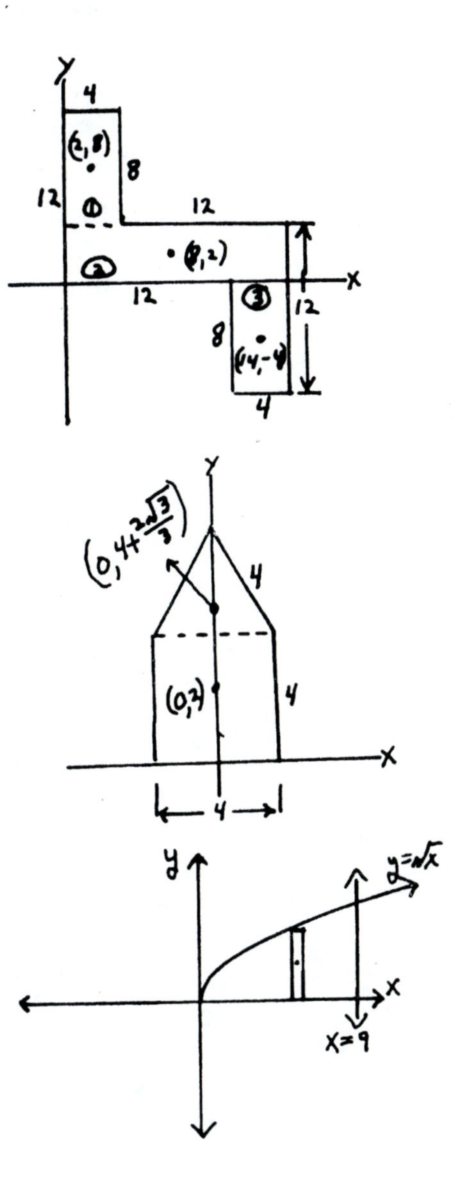

13. center of triangle is $(0, 4 + 2\sqrt{3}/3)$; $A = 4\sqrt{3}$
center of square is (0, 2); $A = 16$
$m = 4\sqrt{3} + 16 = 22.93$
$M_y = 4\sqrt{3}(0) + 16(0) = 0$
$M_x = 4\sqrt{3}(4 + 2\sqrt{3}/3) + 16(2) = 67.71$

$$\bar{x} = \frac{M_y}{m} = 0;\ \bar{y} = \frac{M_x}{m} = \frac{67.71}{22.93} = 2.95$$

15. $A = \int_0^9 \sqrt{x}\, dx = \frac{2}{3} x^{3/2} \Big|_0^9 = 18$

$$\bar{x} = \frac{\int_0^9 x\sqrt{x}\, dx}{18} = \frac{\frac{2}{5} \cdot x^{5/2} \Big|_0^9}{18} = \frac{97.2}{18} = 5.4$$

$$\bar{y} = \frac{\frac{1}{2}\int_0^9 (\sqrt{x})^2\, dx}{18} = \frac{\frac{1}{2} \cdot \frac{x^2}{2} \Big|_0^9}{18} = \frac{20.25}{18} = \frac{9}{8}$$

17. $A = \int_0^2 -(x^2 - 2x)\, dx = -(x^3/3 - x^2) \Big|_0^2 = \frac{4}{3}$

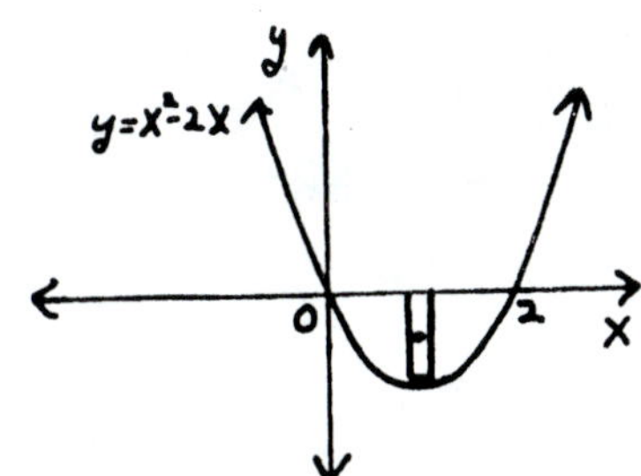

$$\bar{x} = \frac{\int_0^2 x[0 - (x^2 - 2x)]\, dx}{4/3} = \frac{\int_0^2 (-x^3 + 2x^2)\, dx}{4/3}$$

$$= (3/4)(-x^4 + 2x^3/3) \Big|_0^2 = 1$$

$$\bar{y} = \frac{\frac{1}{2}\int_0^2 [0^2 - (x^2 - 2x)^2]dx}{4/3} = 3/8 \int_0^2 -(x^4 - 4x^3 + 4x^2)dx$$

$$= (-3/8)(x^5/5 - x^4 + 4x^3/3 \Big|_0^2 = -2/5$$

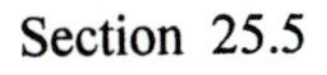

19. $A = \int_{-2}^{2}(4-x^2)\,dx = (4x - x^3/3)\Big|_{-2}^{2} = 32/3$

$\bar{x} = 0$ from sketch

$$\bar{y} = \frac{\frac{1}{2}\int_{-2}^{2}(4-x^2)^2\,dx}{32/3} = \frac{3}{64}\int_{-2}^{2}(16 - 8x^2 + x^4)\,dx$$

$$= (3/64)(16x - 8x^3/3 + x^5/5)\Big|_{-2}^{2} = 8/5$$

Thus (0, 8/5).

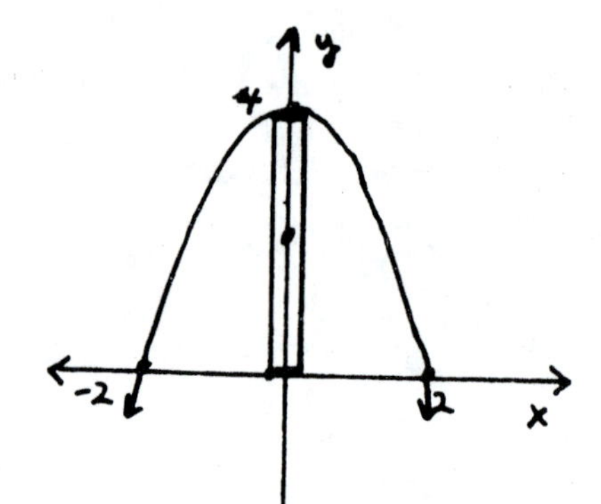

21. $A = \int_{0}^{1}(x - x^3)\,dx = (x^2/2 - x^4/4)\Big|_{0}^{1} = 1/4$

$$\bar{x} = \frac{\int_{0}^{1}x(x - x^3)\,dx}{1/4} = 4(x^3/3 - x^5/5)\Big|_{0}^{1} = 8/15$$

$$\bar{y} = \frac{\frac{1}{2}\int_{0}^{1}(x^2 - x^6)\,dx}{1/4} = 2(x^3/3 - x^7/7)\Big|_{0}^{1} = 8/21$$

Thus (8/15, 8/21).

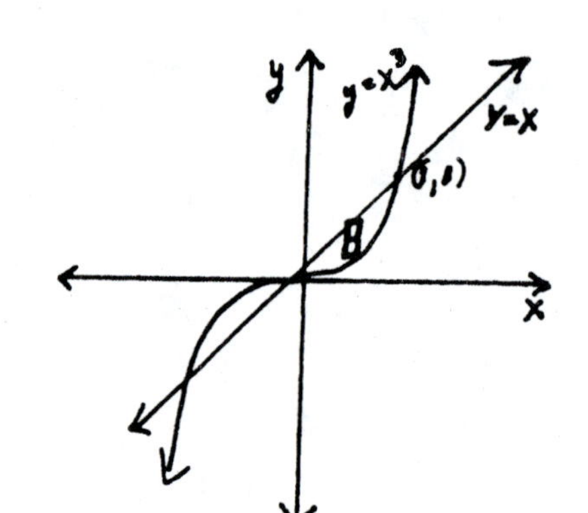

23. $\bar{x} = 0$; $A = (1/2)\pi r^2 = \pi/2$

$$\bar{y} = \frac{\frac{1}{2}\int_{-1}^{1}(\sqrt{1-x^2})^2\,dx}{\pi/2} = \frac{1}{\pi}\int_{-1}^{1}(1-x^2)\,dx$$

$$= (1/\pi)(x - x^3/3)\Big|_{-1}^{1} = \frac{4}{3\pi}$$

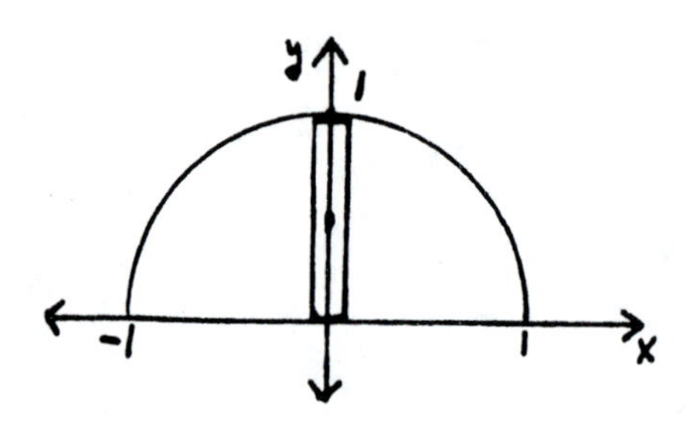

25. $$\bar{x} = \frac{\int_{0}^{1}x(x^3)^2\,dx}{\int_{0}^{1}(x^3)^2\,dx} = \frac{x^8/8\Big|_{0}^{1}}{x^7/7\Big|_{0}^{1}} = \frac{1/8}{1/7} = \frac{7}{8}$$

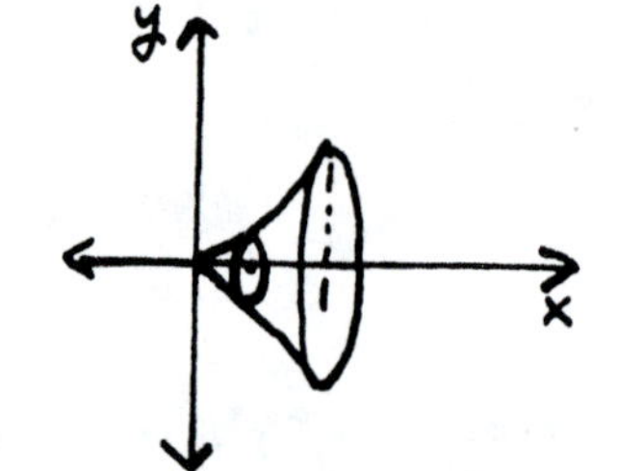

27. $$\bar{x} = \frac{\int_{0}^{3}x(3-x)^2\,dx}{\int_{0}^{3}(3-x)^2\,dx} = \frac{(9x^2/2 - 2x^3 + x^4/4)\Big|_{0}^{3}}{(9x - 3x^2 + x^3/3)\Big|_{0}^{3}} = \frac{27/4}{9} = \frac{3}{4};\ \bar{y} = 0$$

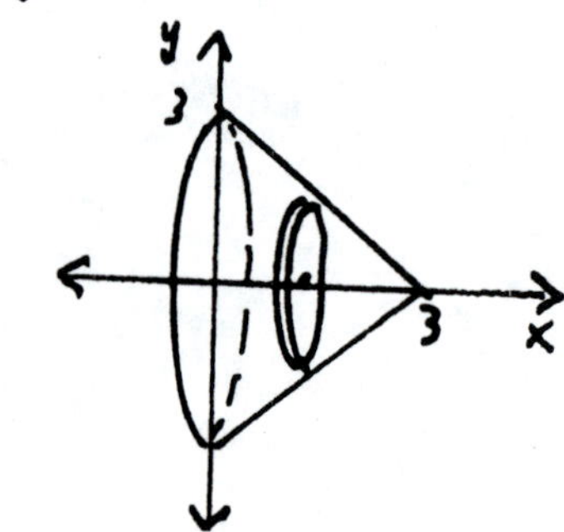

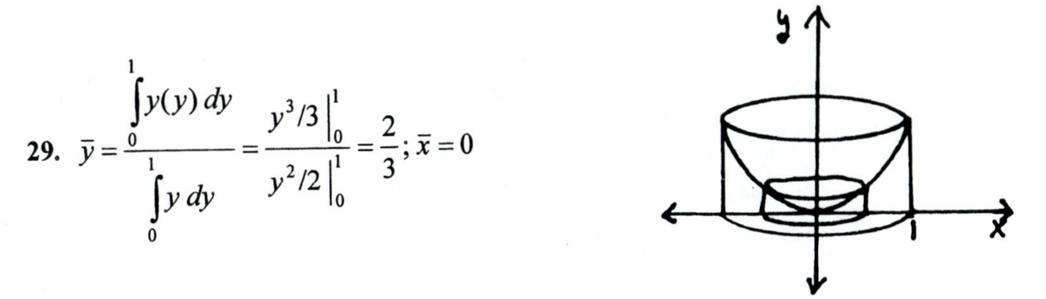

29. $\bar{y} = \dfrac{\int_0^1 y(y)\,dy}{\int_0^1 y\,dy} = \dfrac{y^3/3\,\Big|_0^1}{y^2/2\,\Big|_0^1} = \dfrac{2}{3};\ \bar{x} = 0$

Exercises 25.6

1. $I_y = 9(3)^2 + 12(5)^2 + 15(3)^2 = 516;\ m = 36;\ R = \sqrt{516/36} = 3.79$

3. $I_y = 15(3)^2 + 10(6)^2 + 18(9)^2 + 12(1)^2 = 1965;\ m = 55;\ R = \sqrt{1965/55} = 5.98$

5. $I_x = 9(-2)^2 + 12(4)^2 + 15(7)^2 = 963;\ m = 36;\ R = \sqrt{963/36} = 5.17$

7. $I_x = 9(-9)^2 + 5(-2)^2 + 8(0)^2 + 10(1)^2 = 759;\ m = 32;\ R = \sqrt{759/32} = 4.87$

9. $I_y = 5\int_0^2 x^2(4 - x^2)\,dx = 5\int_0^2 (4x^2 - x^4)\,dx = 5(4x^3/3 - x^5/5)\,\Big|_0^2 = 64/3$

$m = 5\int_0^2 (4 - x^2)\,dx = 5(4x - x^3/3)\,\Big|_0^2 = 80/3;\ R = \sqrt{(64/3)/(80/3)} = 0.894$

11. $I_x = 4\int_0^1 y^2(\sqrt{y} - y)\,dy = 4\int_0^1 (y^{5/2} - y^3)\,dy = 4(2y^{7/2}/7 - y^4/4)\,\Big|_0^1 = 1/7$

$m = 4\int_0^1 (\sqrt{y} - y)\,dy = 4(2y^{3/2}/3 - y^2/2)\,\Big|_0^1 = \dfrac{2}{3};\ R = \sqrt{(1/7)/(2/3)} = 0.463$

13. $I_y = 3\int_0^2 x^2[(5 - x^2) - 1]\,dx = 3\int_0^2 (4x^2 - x^4)\,dx = 3(4x^3/3 - x^5/5)\,\Big|_0^2 = 64/5$

$m = 3\int_0^2 [(5 - x^2) - 1]\,dx = 3(4x - x^3/3)\,\Big|_0^2 = 16\quad R = \sqrt{(64/5)/16} = 0.894$

15. $I_y = 2\int_1^2 x^2(1/x^2)\,dx = 2\int_1^2 dx = 2x\,\Big|_1^2 = 2;$

$m = 2\int_1^2 x^{-2}\,dx = 2(-1/x)\,\Big|_1^2 = 1;\ R = \sqrt{2/1} = 1.41$

17. $m = 2\pi(15)\int_0^2 x(3x)\,dx = 30\pi\int_0^2 3x^2\,dx = 30\pi x^3\Big|_0^2 = 240\pi$

$I_y = 2\pi(15)\int_0^2 x^3(3x)\,dx = 30\pi\int_0^2 3x^4\,dx = 18\pi x^5\,\Big|_0^2 = 576\pi;\ R = \sqrt{576\pi/240\pi} = 1.55$

Sections 6.5 – 6.6

19. $I_x = 2\pi\int_0^{16} y^3(2 - y^{1/2}/2)\,dy = 2\pi\int_0^{16}(2y^3 - y^{7/2})\,dy = 2\pi(y^4/2 - y^{9/2}/9)\Big|_0^{16} = 7282\pi = 2.29\times 10^4$

$m = 2\pi\int_0^{16} y(2 - y^{1/2}/2)\,dy = 2\pi\int_0^{16}(2y - y^{3/2}/2)\,dy = 2\pi(y^2 - y^{5/2}/5)\Big|_0^{16} = 102.4\pi = 322$;

$R = \sqrt{7282\pi/102.4\pi} = 8.43$

21. $I_y = 2\pi(12)\int_0^{3} x^3(9 - x^2)\,dx = 24\pi\int_0^{3}(9x^3 - x^5)\,dx = 24\pi(9x^4/4 - x^6/6)\Big|_0^{3} = 1458\pi$;

$m = 2\pi(12)\int_0^{3} x(9 - x^2)\,dx = 24\pi\int_0^{3}(9x - x^3)\,dx = 24\pi(9x^2/2 - x^4/4)\Big|_0^{3} = 486\pi$;

$R = \sqrt{1458\pi/486\pi} = 1.73$

23. $I_y = 2\pi(15)\int_0^{4} x^3(4x - x^2)\,dx = 30\pi\int_0^{4}(4x^4 - x^5)\,dx = 30\pi(4x^5/5 - x^6/6)\Big|_0^{4} = 4096\pi$;

$m = 2\pi(15)\int_0^{4} x(4x - x^2)\,dx = 30\pi\int_0^{4}(4x^2 - x^3)\,dx = 30\pi(4x^3/3 - x^4/4)\Big|_0^{4} = 640\pi$;

$R = \sqrt{4096\pi/640\pi} = 2.53$

Exercises 25.7

1. $W = \int_0^{3}(x^3 - x)\,dx = (x^4/4 - x^2/2)\Big|_0^{3} = 63/4$

3. $20 = k(10);\ k = 2;\ W = \int_0^{5} 2x\,dx = x^2\Big|_0^{5} = 25$ in.-lb

5. $150 = 4k;\ k = 75/2;\ W = \int_0^{6}(75x/2)\,dx = (75x^2/4)\Big|_0^{6} = 657$ N · cm or 6.75 J

7. $W = \int_{0.01}^{0.05} 3.62\times 10^{-16}x^{-2}\,dx = 3.62\times 10^{-16}(-1/x)\Big|_{0.01}^{0.05} = 2.896\times 10^{-14}$ J

9. **a)** $W = \int_{40}^{50} 2x\,dx = x^2\Big|_{40}^{50} = 900$ ft-lb

b) $W = \int_{25}^{50} 2x\,dx = x^2\Big|_{40}^{50} = 1875$ ft-lb

c) $W = \int_0^{50} 2x\,dx = x^2\Big|_0^{50} = 2500$ ft-lb

11. $W = \int_0^{12} 62.4\pi(4)^2 x\,dx = 998.4\pi(x^2/2)\Big|_0^{12} = 225{,}800$ ft-lb

13. $W = \int_{10}^{22} 62.4\pi(4)^2 x\,dx = 998.4\pi(x^2/2)\Big|_{10}^{22} = 602{,}200$ ft-lb

15. $\dfrac{5}{12} = \dfrac{r}{12-x}$ thus $r = \dfrac{5(12-x)}{12}$

$F = 62.4\pi\dfrac{25(12-x)^2}{144}\Delta x$

$W = \int_0^{12} 10.83\pi x(12-x)^2\,dx$

$= 10.83\pi\int_0^{12}(x^3 - 24x^2 + 144x)\,dx = 10.83\pi(x^4/4 - 8x^3 + 72x^2)\Big|_0^{12}$

$= 58{,}810$ ft-lb

17. $F = 62.4\int_0^{10}(10-y)(8)\,dy$

$= (62.4)(8)(10y - y^2/2)\Big|_0^{10}$

$= 25{,}000$ lb

19. $F = 9800\int_0^{3}(3-y)(8)\,dy$

$= 9800(8)(3y - y^2/2)\Big|_0^{3}$

$= 352{,}800$ N

Note: $\rho g = 9800$ kg/(m^2 s^2)
3/4 full; $A = 24$ m^2;
depth = 3 m

21. $F = 870(9.80)\int_{-5}^{0}(-y)(2)\sqrt{25-y^2}\,dy$

$= 8526\int_{-5}^{0}(25-y^2)^{1/2}(-2y)\,dy$

$= 8526(2/3)(25-y^2)^{3/2}\Big|_{-5}^{0} = 710{,}500$ N

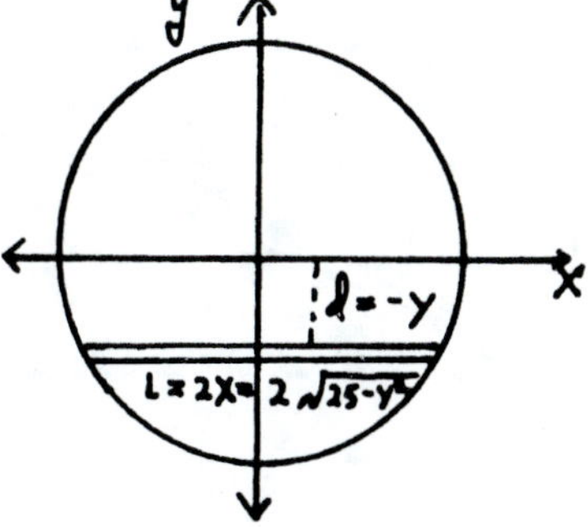

23. $F = 62.4\int_0^2 (2-y)(2y)\,dy$

$= 62.4\int_0^2 (4y - 2y^2)\,dy$

$= 62.4(2y^2 - 2y^3/3)\Big|_0^2 = 166.4$ lb

$\dfrac{4}{2} = \dfrac{L}{Y}$ Thus $L = 2y$.

25. $F = 62.4\int_0^6 (14-y)(16-y)\,dy$

$= 62.4\int_0^6 (224 - 30y + y^2)\,dy$

$= 62.4(224y - 15y^2 - y^3/3)\Big|_0^6 = 54{,}660$ lb

Note: equation of line through (5, 6) and (8, 0) is $y = -2(x-8)$

Thus $L = 2x = 16 - y$.

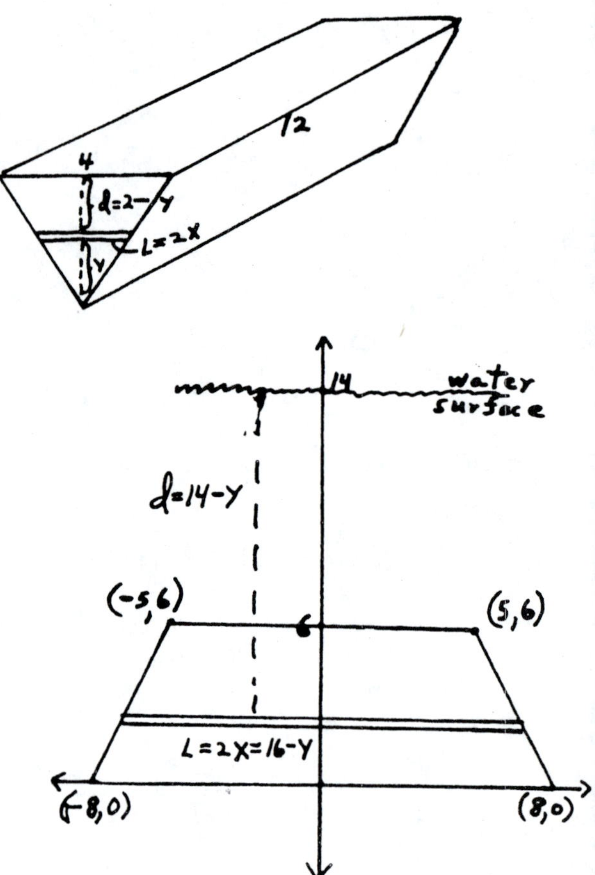

27. $F = 62.4\int_6^9 (9-y)(18)\,dy + 62.4\int_0^6 (9-y)(3y)\,dy$

$= 1123.2\int_6^9 (9-y)\,dy + 187.2\int_0^6 (9y - y^2)\,dy$

$= 1123.2(9y - y^2/2)\Big|_6^9 + 187.2(9y^2/2 - y^3/3)\Big|_0^6 = 5054 + 16{,}848 = 21{,}902$ lb

Note: equation of line through (0, 0) and (18, 6) is $y = x/3$.

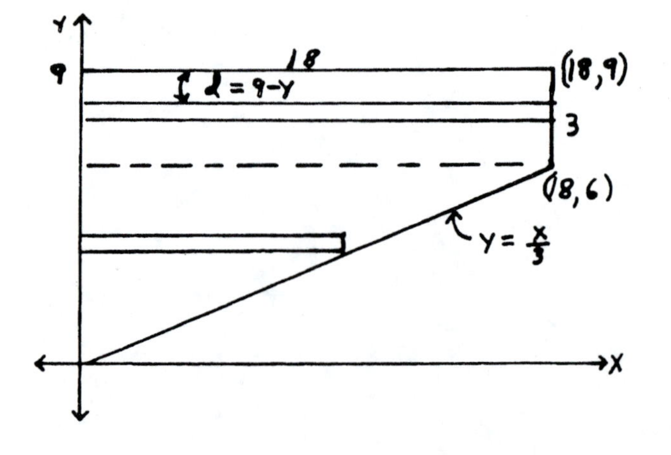

29. $y_{av} = \frac{1}{3-1}\int_1^3 x^2\,dx = (1/2)(x^3/3)\Big|_1^3 = 13/3$

31. $y_{av} = \frac{1}{10-5}\int_5^{10}\frac{dx}{\sqrt{x-1}} = (1/5)\int_5^{10}(x-1)^{-1/2}\,dx = (2/5)(x-1)^{1/2}\Big|_5^{10} = 2/5$

33. $I_{av} = \frac{1}{0.5-0.1}\int_{0.1}^{0.5}(6t-t^2)\,dt = (2.5)(3t^2 - t^3/3)\Big|_{0.1}^{0.5} = 1.70\,A$

Chapter 25 Review

1.

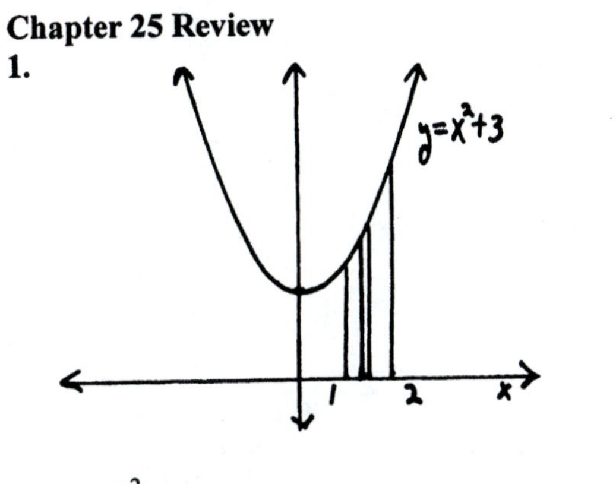

$$A = \int_1^2 (x^2+3)\,dx = (x^3/3+3x)\Big|_1^2 = 16/3$$

2.

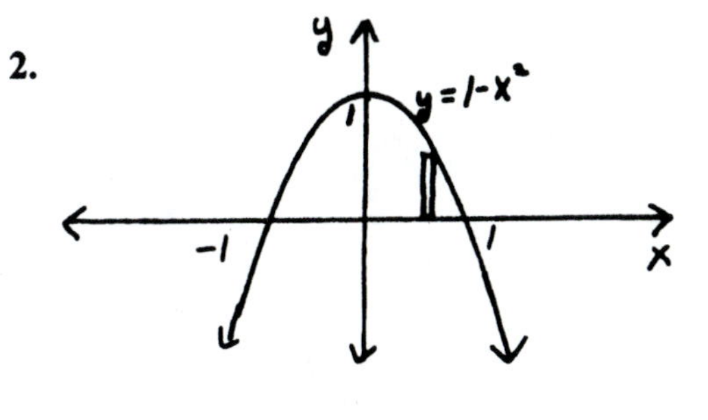

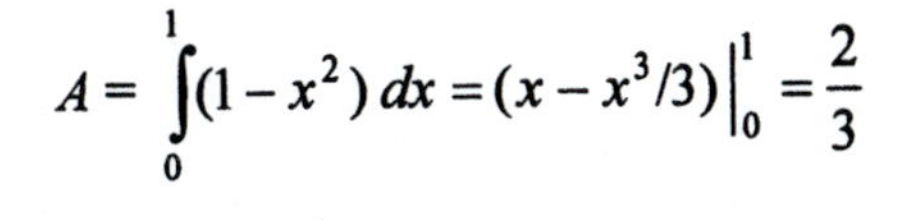

$$A = \int_0^1 (1-x^2)\,dx = (x - x^3/3)\Big|_0^1 = \frac{2}{3}$$

3.

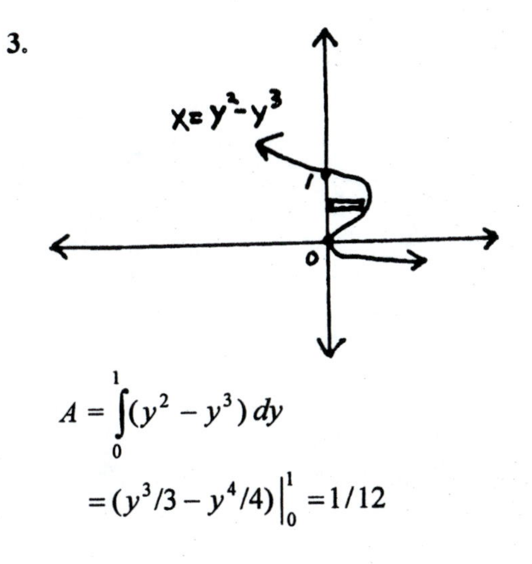

$$A = \int_0^1 (y^2 - y^3)\,dy$$

$$= (y^3/3 - y^4/4)\Big|_0^1 = 1/12$$

4.

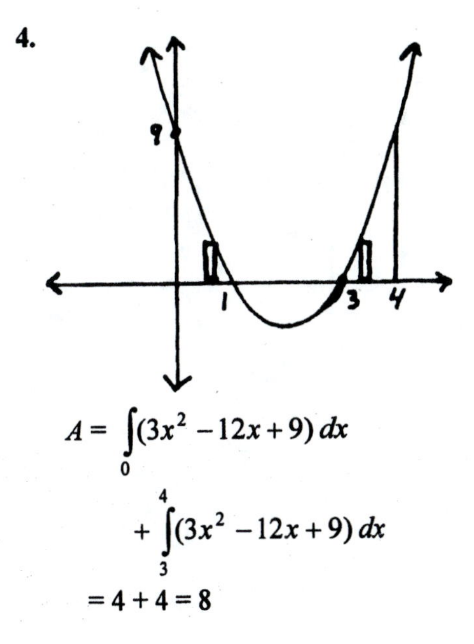

$$A = \int_0 (3x^2 - 12x + 9)\,dx$$

$$+ \int_3^4 (3x^2 - 12x + 9)\,dx$$

$$= 4 + 4 = 8$$

5.

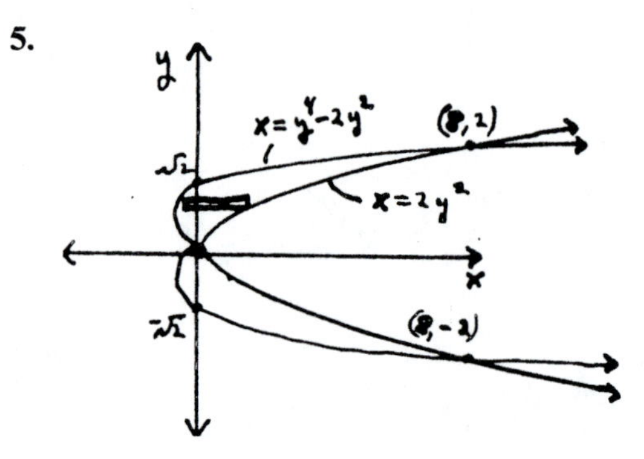

$$A = 2\int_0^2 [2y^2 - (y^4 - 2y^2)]\, dy$$

Note: S: x-axis

$$= 2\int_0^2 (4y^2 - y^4)\, dy$$

$$= 2(4y^3/3 - y^5/5)\Big|_0^2 = 128/15$$

6.

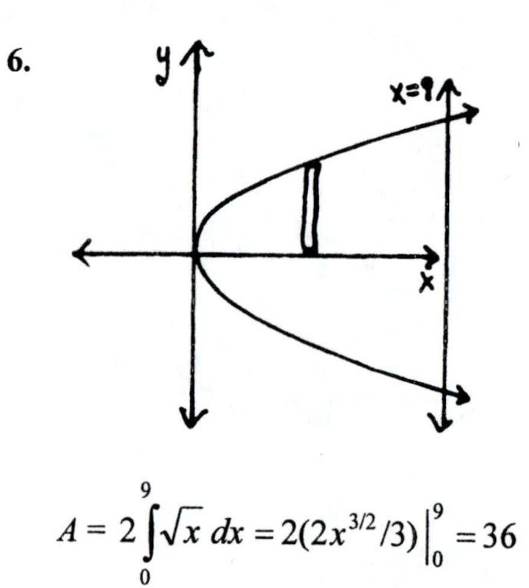

$$A = 2\int_0^9 \sqrt{x}\, dx = 2(2x^{3/2}/3)\Big|_0^9 = 36$$

7. $V = 2\pi \int_0^2 y(4 - y^2)\, dy = 2\pi \int_0^2 (4y - y^3)\, dy$

$$= 2\pi(2y^2 - y^4/4)\Big|_0^2 = 8\pi$$

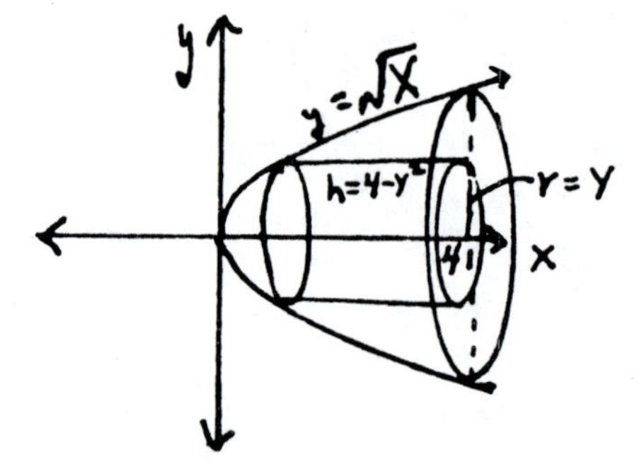

8. (See # 7 diagram)

$$V = \pi \int_0^4 (\sqrt{x})^2\, dx = \pi(x^2/2)\Big|_0^4 = 8\pi$$

9.

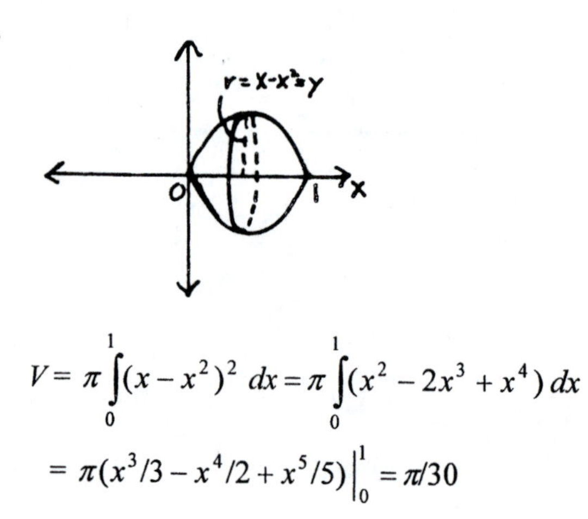

$$V = \pi \int_0^1 (x - x^2)^2\, dx = \pi \int_0^1 (x^2 - 2x^3 + x^4)\, dx$$

$$= \pi(x^3/3 - x^4/2 + x^5/5)\Big|_0^1 = \pi/30$$

10. $V = 2\pi \int_0^2 x[(3x - x^2) - x]\,dx = 2\pi \int_0^2 (2x^2 - x^3)\,dx$

$= 2\pi(2x^3/3 - x^4/4)\Big|_0^2 = \dfrac{8\pi}{3}$

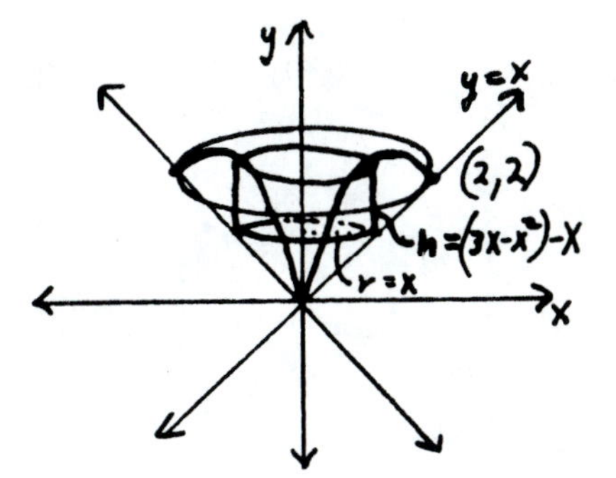

11.

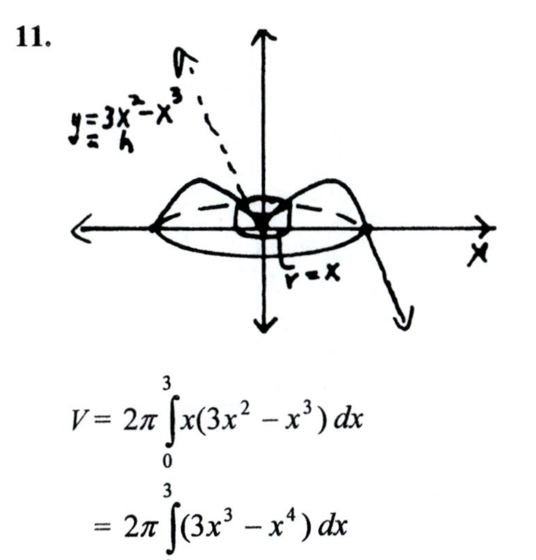

$V = 2\pi \int_0^3 x(3x^2 - x^3)\,dx$

$= 2\pi \int_0^3 (3x^3 - x^4)\,dx$

$= 2\pi(3x^4/4 - x^5/5)\Big|_0^3 = 243\pi/10$

12.

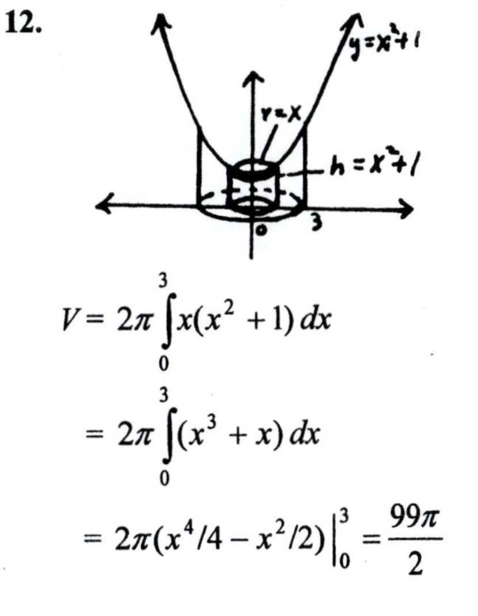

$V = 2\pi \int_0^3 x(x^2 + 1)\,dx$

$= 2\pi \int_0^3 (x^3 + x)\,dx$

$= 2\pi(x^4/4 - x^2/2)\Big|_0^3 = \dfrac{99\pi}{2}$

13.

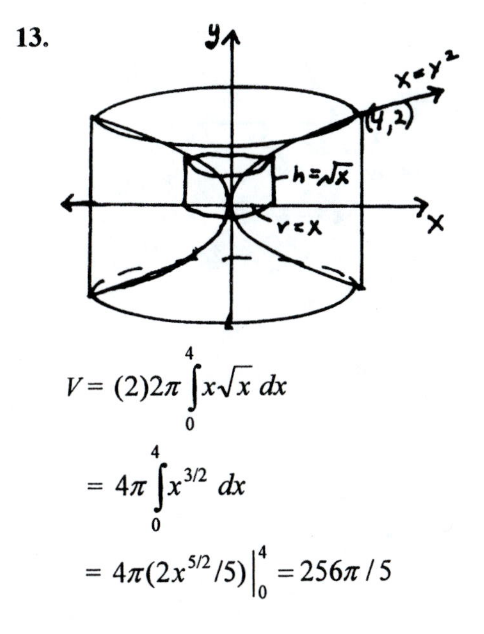

$V = (2)2\pi \int_0^4 x\sqrt{x}\,dx$

$= 4\pi \int_0^4 x^{3/2}\,dx$

$= 4\pi(2x^{5/2}/5)\Big|_0^4 = 256\pi/5$

14.

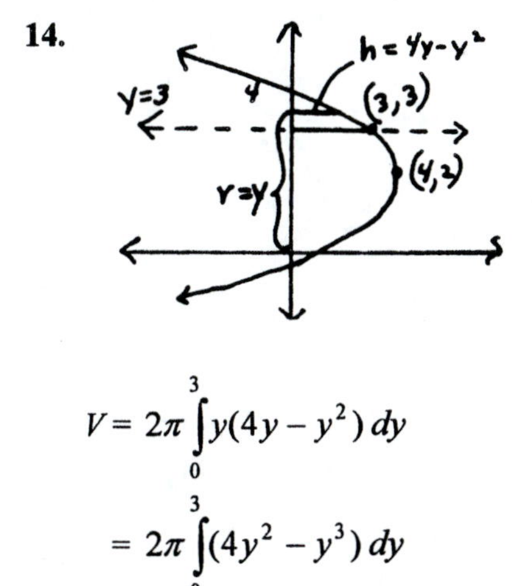

$V = 2\pi \int_0^3 y(4y - y^2)\,dy$

$= 2\pi \int_0^3 (4y^2 - y^3)\,dy$

$= 2\pi(4y^3/3 - y^4/4)\Big|_0^3 = \dfrac{63\pi}{2}$

15. $M_o = 12(-4) + 20(9) + 24(12) = 420$; $m = 12 + 20 + 24 = 56$
$\bar{x} = M_o/m = 420/56 = 7.5$

16. $m = 24 + 36 + 30 = 90$; $M_x = 24(-3) + 36(-15) + 30(0) = -612$
$M_y = 24(11) + 36(-4) + 30(-7) = -90$; $\bar{x} = -90/90 = -1$; $\bar{y} = -612/90 = -6.8$

17. center of top rectangle is (8, 10.5); $A = 40$
center of bottom rectangle is (10, 4); $A = 160$
$M_y = 40(8) + 160(10) = 1920$
$M_x = 40(10.5) + 160(4) = 1060$
$m = 40 + 160 = 200$; $\bar{x} = 1920/200 = 9.6$; $\bar{y} = 1060/200 = 5.3$

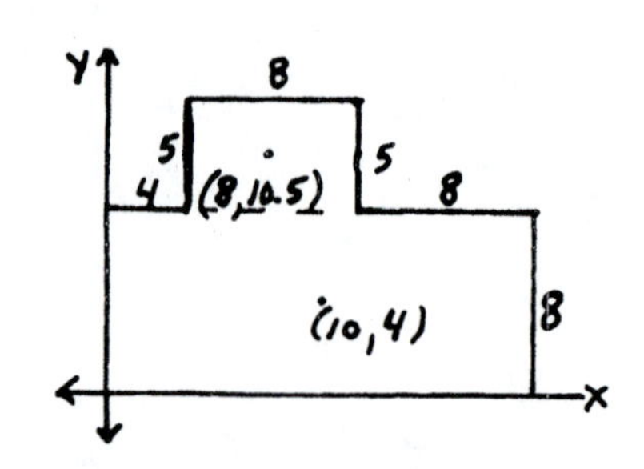

18. $$\bar{x} = \frac{\int_0^4 x(5x)\,dx}{40} = \frac{(5x^3/3)\Big|_0^4}{40} = 2\frac{2}{3}$$

$$\bar{y} = \frac{\frac{1}{2}\int_0^4 (5x)^2\,dx}{40} = \frac{(25x^3/3)\Big|_0^4}{80} = 6\frac{2}{3}$$

$$A = \frac{1}{2}(4)(20) = 40$$

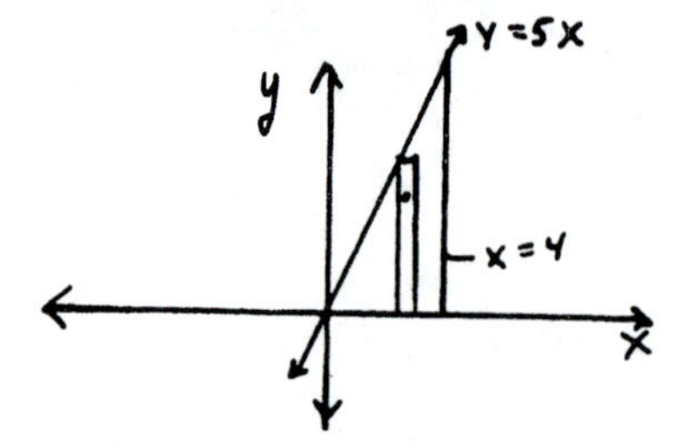

19. $$A = \int_0^3 [(6x - x^2) - 3x]\,dx = \int_0^3 (3x - x^2)\,dx$$

$$= (3x^2/2 - x^3/3)\Big|_0^3 = 9/2$$

$$\bar{x} = \frac{\int_0^3 x[(6x - x^2) - 3x]\,dx}{9/2}$$

$$= \frac{2}{9}\int_0^3 (3x^2 - x^3)\,dx$$

$$= (2/9)(x^3 - x^4/4)\Big|_0^3 = 3/2$$

$$\bar{y} = \frac{\frac{1}{2}\int_0^3 [(6x - x^2)^2 - (3x)^2]\,dx}{9/2} = \frac{1}{9}\int_0^3 (27x^2 - 12x^3 + x^4)\,dx$$

$$= (1/9)(9x^3 - 3x^4 + x^5/3)\Big|_0^3 = 27/5$$

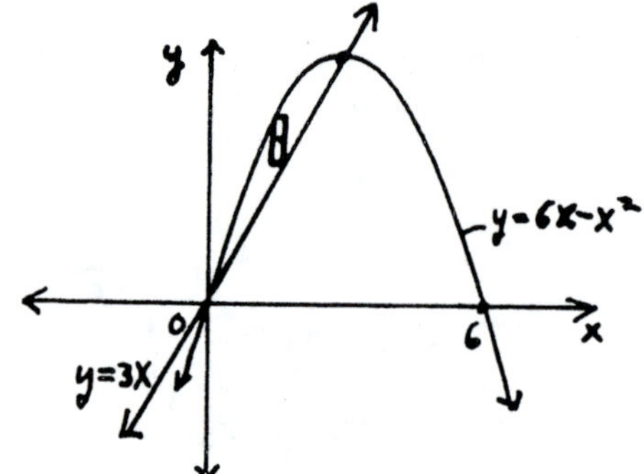

20. $A = \int_0^6 -(x^2 - 6x)\,dx$

$= -(x^3/3 - 3x^2)\Big|_0^6 = 36$

Note: $\bar{x} = 3$

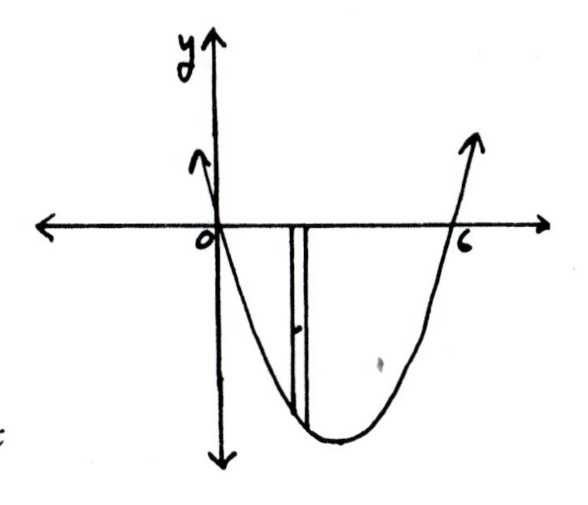

$$\bar{y} = \frac{\frac{1}{2}\int_0^6 [0^2 - (x^2 - 6x)^2]\,dx}{36} = \frac{1}{72}\int_0^6 (-x^4 + 12x^3 - 36x^2)\,dx$$

$= (1/72)(-x^5/5 + 3x^4 - 12x^3)\Big|_0^6 = -3.6$ Thus (3, −3.6).

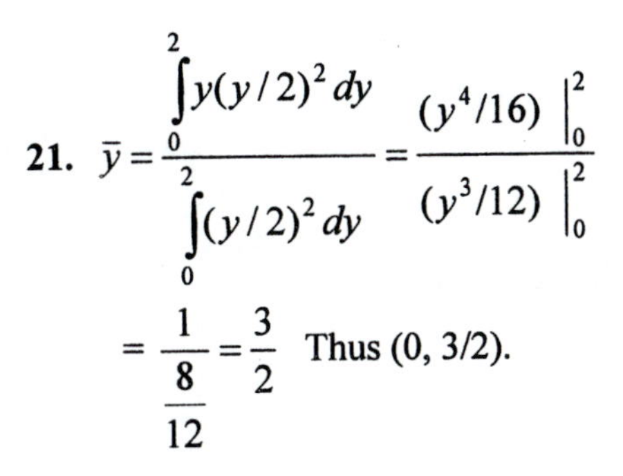

21. $$\bar{y} = \frac{\int_0^2 y(y/2)^2\,dy}{\int_0^2 (y/2)^2\,dy} = \frac{(y^4/16)\Big|_0^2}{(y^3/12)\Big|_0^2}$$

$$= \frac{1}{\frac{8}{12}} = \frac{3}{2}$$ Thus (0, 3/2).

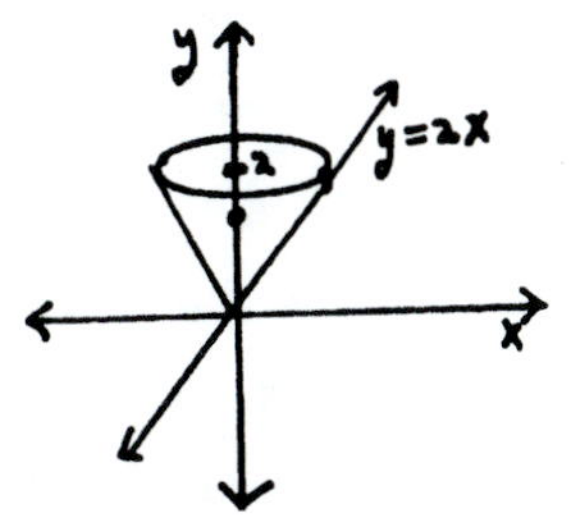

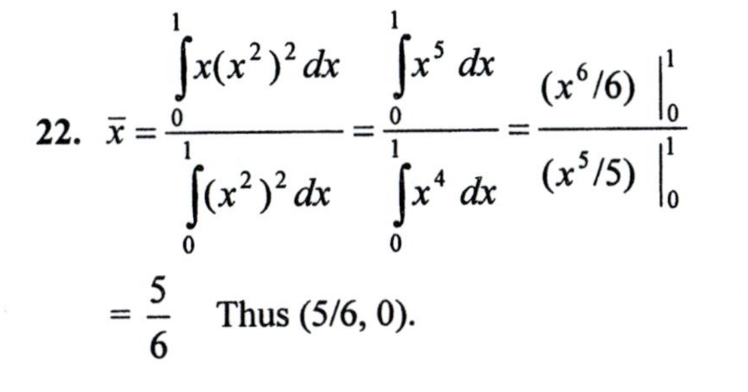

22. $$\bar{x} = \frac{\int_0^1 x(x^2)^2\,dx}{\int_0^1 (x^2)^2\,dx} = \frac{\int_0^1 x^5\,dx}{\int_0^1 x^4\,dx} = \frac{(x^6/6)\Big|_0^1}{(x^5/5)\Big|_0^1}$$

$$= \frac{5}{6}$$ Thus (5/6, 0).

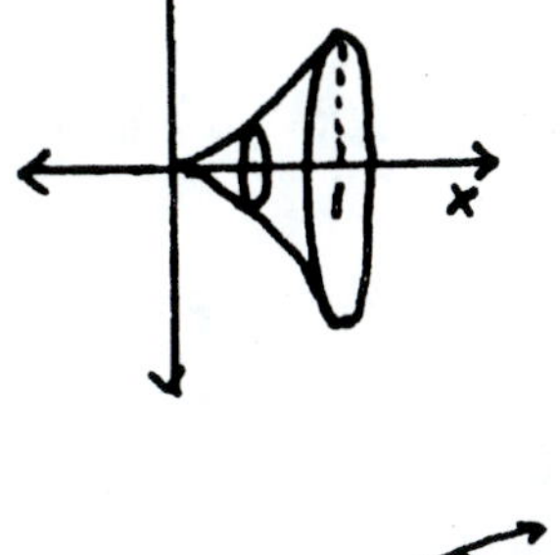

23. $$\bar{y} = \frac{\int_0^4 y(y^2 - 4y)^2\,dy}{\int_0^4 (y^2 - 4y)^2\,dy} = \frac{\int_0^4 (y^5 - 8y^4 + 16y^3)\,dy}{\int_0^4 (y^4 - 8y^3 + 16y^2)\,dy}$$

$$= \frac{(y^6/6 - 8y^5/5 + 4y^4)\Big|_0^4}{(y^5/5 - 2y^4 + 16y^3/3)\Big|_0^4} = \frac{68.2\overline{6}}{34.1\overline{3}} = 2;\ \bar{x} = 0$$

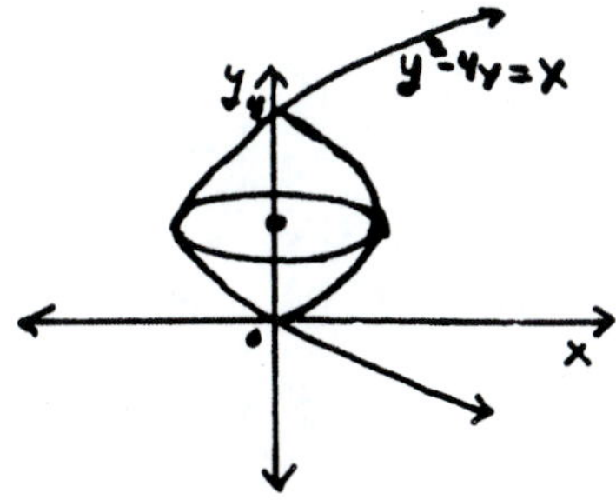

24. $I_x = 10(2)^2 + 6(7)^2 + 8(-4)^2 = 462;\ m = 24;\ R = \sqrt{462/24} = 4.39$

25. $I_y = 1\int_0^4 x^2(3x)\,dx = (3/4)x^4\Big|_0^4 = 192;\ m = 1\int_0^4 3x\,dx = (3x^2/2)\Big|_0^4 = 24$

$R = \sqrt{192/24} = 2.83$

26. $I_x = 1\int_0^{12} y^2(4 - y/3)\,dy = (4y^3/3 - y^4/12)\Big|_0^{12} = 576;\ m = 24$ from #25

$R = \sqrt{576/24} = 4.90$

27. $I_x = 4\int_0^1 y^2(1-y^2)\,dy = 4(y^3/3 - y^5/5)\Big|_0^1 = 8/15$

$m = 4\int_0^1 (1-y^2)\,dy = 4(y - y^3/3)\Big|_0^1 = 8/3;$

$\sqrt{(8/15)/(8/3)};\ R = 0.447$

28. $I_x = 2\pi(4)\int_0^1 y^3(1-y^{1/3})\,dy = 8\pi\int_0^1 (y^3 - y^{10/3})\,dy = 8\pi(y^4/4 - 3y^{13/3}/13)\Big|_0^1 = 2\pi/13$

$m = 2\pi(4)\int_0^1 y(1-y^{1/3})\,dy = 8\pi\int_0^1 (y - y^{4/3})\,dy = 8\pi(y^2/2 - 3y^{7/3}/7)\Big|_0^1 = 4\pi/7$

$R = \sqrt{2\pi/13/4\pi/7} = 0.519$

29. $I_y = 2\pi(4)\int_0^1 x^3(x^3)\,dx = 8\pi\int_0^1 x^6 dx = 8\pi(x^7/7)\Big|_0^1 = 8\pi/7$

$m = 8\pi\int_0^1 x^3 \cdot x\,dx = 8\pi\int_0^1 x^4 dx = 8\pi(x^5/5)\Big|_0^1 = 8\pi/5$

$R = \sqrt{(8\pi/7)/(8\pi/5)} = 0.845$

30. $I_y = 2\pi(3)\int_1^4 x^3(1/x)\,dx = 6\pi(x^3/3)\Big|_1^4 = 126\pi;$

$m = 6\pi\int_1^4 x(1/x)\,dx = 6\pi x\Big|_1^4 = 18\pi;\ \ R = \sqrt{126\pi/18\pi} = 2.65$

31. $16 = 4k;\ k = 4;\ W = \int_0^{10} 4x\,dx = 2x^2\Big|_0^{10} = 200$ in.-lb

32. $W = \int_{0.01}^{0.02} 5.24\times10^{-18}x^{-2}\,dx = 5.24\times10^{-18}(x^{-1}/-1)\Big|_{0.01}^{0.02} = 2.62\times10^{-16}$ J

33. $W = \int_0^{200} 4x\,dx + 250(200) = 2x^2\Big|_0^{200} + 50{,}000 = 130{,}000$ ft-lb

34. $F = 62.4\int_0^{8}(14 - y)(10)\,dy$

$= 624(14y - y^2/2)\Big|_0^{8} = 49{,}920$ lb

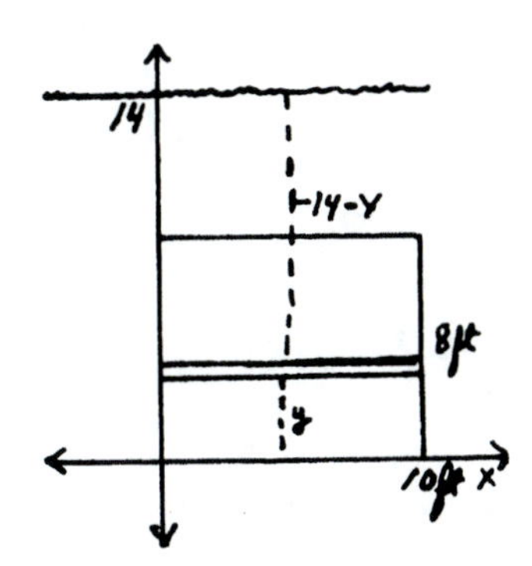

35. $F = 9800\int_{-5}^{0}(-y)(2\sqrt{25 - y^2})\,dy$

$= 9800(2/3)(25 - y^2)^{3/2}\Big|_{-5}^{0}$

816,700 N

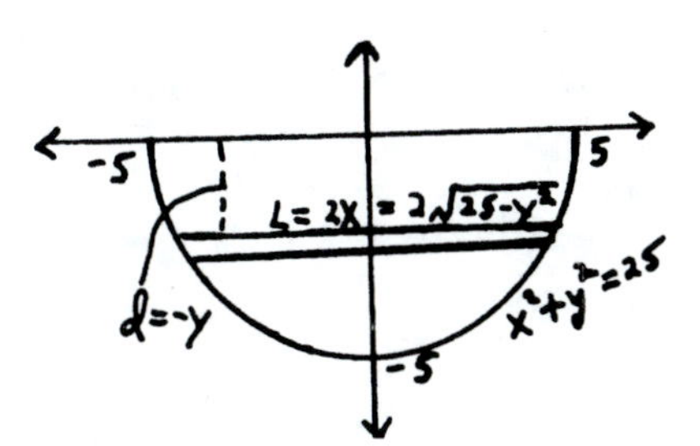

36. $V_{av} = \dfrac{1}{3-0}\int_0^{3}(t^2 + 3t + 2)\,dt = (1/3)(t^3/3 + 3t^2/2 + 2t)\Big|_0^{3} = 9.5$ V

37. $I_{av} = \dfrac{1}{9-4}\int_4^{9}4t^{3/2}\,dt = (1/5)(8t^{5/2}/5)\Big|_4^{9} = 67.52$ A

38. $P_{av} = \dfrac{1}{3-1}\int_1^{3}2t^3\,dt = (1/2)(t^4/2)\Big|_1^{3} = 20$ W

Chapter 26, Exercises 26.1

1. $u = 3x + 2;\ du = 3\,dx;\ (1/3)\int(3x+2)^{1/2}\cdot 3\,dx = (2/9)(3x+2)^{3/2} + C$

3. $u = x + 4;\ du = dx;\ \int(4+x)^{-1/2}\,dx = 2(4+x)^{1/2} + C$

5. $u = x^2 + 4x;\ du = (2x+4)\,dx;\ (1/2)\int(x^2+4x)^{3/4}(x+2)(2)\,dx = (2/7)(x^2+4x)^{7/4} + C$

7. $u = \cos x;\ du = -\sin x\,dx;\ -\int\cos^3 x\ \sin x(-dx) = (-1/4)\cos^4 x + C$

9. $u = \tan 4x;\ du = 4\sec^2 4x\,dx;\ (1/4)\int\tan^3 4x\ \sec^2 4x\cdot 4\,dx = (1/16)\tan^4 4x + c$

11. $u \cos 4x + 1;\ du = -4\sin 4x\,dx;\ (-1/4)\int\sin 4x\,(\cos 4x + 1)(-4)dx = (-1/8)(\cos 4x + 1)^2 + C$

13. $u = 9 + \sec x;\ du = \sec x\tan x\,dx;\ \int(9+\sec x)^{1/2}\sec x\tan x\,dx = (2/3)(9+\sec x)^{3/2} + C$

15. $u = 1 + e^{2x};\ du = 2e^{2x}dx;\ (1/2)\int(1+e^{2x})^{1/2}2e^{2x}dx = (1/3)(1+e^{2x})^{3/2} + C$

17. $u = 1 + e^{x^2};\ du = 2xe^{x^2}\,dx;\ (1/2)\int(1+e^{x^2})^{-1/2}2xe^{x^2}dx = (1+e^{x^2})^{1/2} + C$

19. $u = \ln(3x-5);\ du = \dfrac{3\,dx}{3x-5};\ \dfrac{1}{3}\int\dfrac{\ln(3x-5)}{3x-5}3\,dx = (1/6)\ln^2|3x-5| + C$

21. $u = \ln x;\ du = (1/x)\,dx;\ \int\dfrac{dx}{x\ln^2 x} = \dfrac{\ln^{-2} x\,dx}{x} = -\dfrac{1}{\ln x} + C$

23. $u = \text{Arcsin } 3x;\ du = \dfrac{3\,dx}{\sqrt{1-9x^2}};\ \dfrac{1}{3}\int\dfrac{3\text{ Arcsin } 3x}{\sqrt{1-9x^2}}dx = \dfrac{1}{6}\text{Arcsin}^2 3x + C$

25. $u = \sin x;\ du = \cos x\,dx;\ \int\csc^{-4} x\cot x\,dx = \int\sin^4 x\cdot\dfrac{\cos x}{\sin x}dx = \int\sin^3 x\cos x\,dx = (1/4)\sin^4 x + C$

27. $u = \text{Arctan } x;\ du = \dfrac{dx}{1+x^2};\ \int\dfrac{\text{Arctan}^2 x\,dx}{1+x^2} = (1/3)\text{Arctan}^3 x + C$

29. $u = x^2 - 9;\ du = 2x\,dx;\ \dfrac{1}{2}\int_3^5 2x\sqrt{x^2-9}\,dx = \dfrac{1}{3}(x^2-9)^{3/2}\Big|_3^5 = \dfrac{64}{3}$

31. $u = 1 + e^{3x}$; $du = 3e^{3x}\,dx$; $\dfrac{1}{3}\int_0^1 (1+e^{3x})^{-1/2}3e^{3x}\,dx = \dfrac{2}{3}(1+e^{3x})^{1/2}\Big|_0^1 = (2/3)(\sqrt{1+e^3}) - \sqrt{2})$

33. $u = \ln(2x-1)$; $du = \dfrac{1}{2x-1}\cdot 2\,dx$; $\int_1^2 \dfrac{\ln(2x-1)}{2x-1}\,dx = \dfrac{1}{2}\int_1^2 \dfrac{2\ln(2x-1)dx}{2x-1} = (1/4)\ln^2(2x-1)\Big|_1^2 = \dfrac{\ln^2 3}{4}$

35. $A = \int_0^{\pi/2} \sin^2 x\cos x\,dx = (1/3)\sin^3 x\,\Big|_0^{\pi/2} = 1/3$

Exercises 26.2

1. $u = 3x + 2$; $du = 3\,dx$; $\dfrac{1}{3}\int\dfrac{3\,dx}{3x+2} = (1/3)\ln|3x+2| + C$

3. $u = 1 - 4x$; $du = -4\,dx$; $-\dfrac{1}{4}\int\dfrac{-4\,dx}{1-4x} = (-1/4)\ln|1-4x| + C$

5. $u = 1 - x^2$; $du = -2x\,dx$; $-2\int\dfrac{2x\,dx}{1-x^2} = -2\ln|1-x^2| + C$

7. $u = x^4 - 1$; $du = 4x^3\,dx$; $\dfrac{1}{4}\int\dfrac{4x^3\,dx}{x^4-1} = (1/4)\ln|x^4-1| + C$

9. $u = \cot x$; $du = -\csc^2 x\,\text{du}$; $-\int\dfrac{-\csc^2 x\,dx}{\cot x} = -\ln|\cot x| + C$

11. $u = 1 + \tan 3x$; $du = 3\sec^2 3x\,dx$; $\dfrac{1}{3}\int\dfrac{3\sec^2 3x\,dx}{1+\tan 3x} = \dfrac{1}{3}\ln|1+\tan 3x| + C$

13. $u = 1 + \csc; x$ $du = -\csc x\cot x\,dx$; $-\int\dfrac{-\csc x\cot x\,dx}{1+\csc x} = -\ln|1+\csc x| + C$

15. $u = 1 + \sin x$; $du = \cos x\,dx$; $\int\dfrac{\cos x\,dx}{1+\sin x} = \ln|1+\sin x| + C$

17. $u = \ln x$; $du = (1/x)\,dx$; $\int\dfrac{dx}{x\ln x} = \ln|\ln x| + C$

19. $(1/2)\int e^{2x}\cdot 2\,dx = (1/2)e^{2x} + C$

21. $(-1/4)\int e^{-4x}(-4)\,dx = -\dfrac{1}{4e^{4x}} + C$

23. $(1/2)\int 2xe^{x^2}\,dx = (1/2)e^{x^2} + C$

25. $(-1/2)\int e^{-(x^2+9)}(-2x)\,dx = \dfrac{-1}{2e^{x^2+9}} + C$

27. $-\int -\sin x\, e^{\cos x}\,dx = -e^{\cos x} + C$

29. $\dfrac{1}{2}\int_0^2 2xe^{x^2+2}\,dx = (1/2)e^{x^2+2}\Big|_0^2 = (e^2/2)(e^4-1)$

31. $\dfrac{1}{\ln 4}\int 4^x \ln 4\,dx = \dfrac{4^x}{\ln 4} + C$

33. $u = e^x + 4$; $du = e^x\,dx$; $\int \dfrac{2e^x\,dx}{e^x+4} = 2\ln|e^x+4| + C$

35. $\dfrac{1}{2}\int \dfrac{2x\,dx}{x^2+1} = \dfrac{1}{2}\ln\left|x^2+1\right|\Big|_0^1 = (1/2)\ln 2$ or 0.347

37. $2\ln|2x-1|\Big|_1^5 = 2\ln 9$ or $\ln 81$ or 4.39

39. $2e^{x/2}\Big|_0^2 = 2(e-1)$ or 3.44

41. $\dfrac{1}{3}\int_0^1 3x^2e^{x^3}\,dx = \dfrac{1}{3}e^{x^3}\Big|_0^1 = \dfrac{1}{3}(e-1)$ or 0.573

43. $-\int_0^{\pi/6} \dfrac{-\cos x\,dx}{1-\sin x} = -\ln|1-\sin x|\Big|_0^{\pi/6} = -\ln(1/2) = \ln 2$ or 0.693

45. $A = \int_0^1 \dfrac{dx}{1+2x} = (1/2)\ln|1+2x|\Big|_0^1 = (1/2)\ln 3$ or 0.549

47. $A = \int_0^4 e^{2x}\,dx = (1/2)e^{2x}\Big|_0^4 = (1/2)(e^8-1)$ or 1490

Exercises 26.3

1. $(1/5)\int \sin 5x \cdot 5\,dx = (-1/5)\cos 5x + C$

3. $(1/3)\int \cos(3x-1)\cdot 3\,dx = (1/3)\sin(3x-1) + C$

5. $(1/2)\int 2x\sin(x^2+5)\,dx = (-1/2)\cos(x^2+5) + C$

7. $\sin(x^3 - x^2) + C$

9. $(1/5)\int \csc^2 5x \cdot 5\,dx = (-1/5)\cot 5x + C$

11. $(1/3)\int \sec 3x \tan 3x \cdot 3\,dx = (1/3)\sec 3x + C$

13. $(1/4)\int \sec^2(4x+3) \cdot 4\,dx = (1/4)\tan(4x+3) + C$

15. $(1/2)\int \csc(2x-3)\cot(2x-3) \cdot 2\,dx = (-1/2)\csc(2x-3) + C$

17. $(1/2)\int 2x\sec^2(x^2+3)\,dx = (1/2)\tan(x^2+3) + C$

19. $(1/3)\int 3x^2\csc(x^3-1)\cot(x^3-1)\,dx = (-1/3)\csc(x^3-1) + C$

21. $(1/4)\int 4\tan 4x\,dx = (-1/4)\ln|\cos 4x| + C$

23. $(1/5)\int \sec 5x \cdot 5\,dx = (1/5)\ln|\sec 5x + \tan 5x| + C$

25. $\ln|\sin e^x| + C$

27. $\int (1 + 2\sec x + \sec^2 x)\,dx = x + 2\ln|\sec x + \tan x| + \tan x + C$

29. $\int \frac{5+\sin x}{\cos x}dx = \int (5\sec x + \tan x)\,dx$

$= 5\ln|\sec x + \tan x| + \ln|\sec x| + C$ or $5\ln|\sec x + \tan x| - \ln|\cos x| + C$

31. $\frac{1}{2}\int_0^{\pi/4} \sin 2x \cdot 2\,dx = (-1/2)\cos 2x\Big|_0^{\pi/4} = 1/2$

33. $3\sin(x-\pi/2)\Big|_0^{\pi/2} = 3$

35. $\tan x\Big|_0^{\pi/4} = 1$

37. $(1/2)\sec 2x\Big|_0^{\pi/8} = \frac{\sqrt{2}}{2} - \frac{1}{2} = \frac{\sqrt{2}-1}{2}$

39. $A = \int_0^{\pi} \sin x\,dx = -\cos x\Big|_0^{\pi} = 2$

41. $A = \int_0^{\pi/4} \sec^2 x\,dx = \tan x\Big|_0^{\pi/4} = 1$

43. $A = \int_0^{\pi/4} \tan x\,dx = -\ln|\cos x|\Big|_0^{\pi/4} = \frac{1}{2}\ln 2$

45. $V = \int_0^{\pi/4} \pi\sec^2 x\,dx = \pi\tan x\Big|_0^{\pi/4} = \pi$

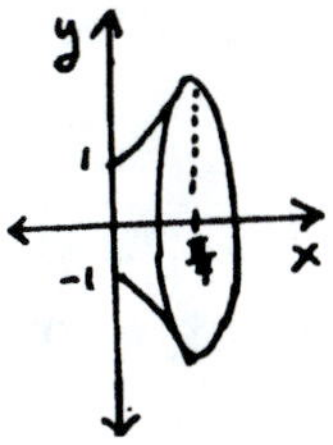

Exercises 26.4

1. $\int \sin^3 x\,dx = \int \sin^2 x \sin x\,dx = \int (1-\cos^2 x)\sin x\,dx = -\cos x + (1/3)\cos^3 x + C$

3. $\int \cos^5 x\,dx = \int \cos^4 x \cos x\,dx = \int (1-\sin^2 x)^2 \cos x\,dx = \int (1-2\sin^2 x + \sin^4 x)\ \cos x\,dx$
$= \sin x - (2/3)\sin^3 x + (1/5)\sin^5 x + C$

5. $\int \sin^2 x \cos dx = (1/3)\sin^3 x + C$

7. $\int \cos^{-3} x \sin x\,dx = \dfrac{-1}{2\cos^2 x} + C$

9. $\int \sin^3 x \cos^2 x\,dx = \int \sin x\,(1-\cos^2 x)\cos^2 x\,dx = \int \sin(\cos^2 x - \cos^4 x)\,dx$
$= (-1/3)\cos^3 x + (1/5)\cos^5 x + C$

11. $\int \sin^2 x\,dx \int (1/2)\,(1-\cos 2x)\,dx = x/2 - (1/4)\sin 2x + C$

13. $\int \cos^4 3x\,dx = \int \left\{\dfrac{1+\cos 6x}{2}\right\}^2 dx = (1/4)\int (1 + 2\cos 6x + \cos^2 6x)\,dx$
$= x/4 + (1/12)\sin 6x + (1/8)\int (1+\cos 12x)\,dx$
$= x/4 + (1/12)\sin 6x + x/8 + (1/96)\sin 12x + C$
$= 3x/8 + (1/12)\sin 6x + (1/96)\sin 12x + C$

15. $\int \sin^2 x \cos^2 x\,dx = (1/4)\int \sin^2 2x\,dx = (1/8)\int (1-\cos 4x)\,dx$
$= x/8 - (1/32)\sin 4x + C$

17. $\int \sin^2 x \cos^4 x\,dx = \int (\sin^2 x \cos^2 x)\cos^2 x\,dx$
$= (1/8)\int \sin^2 2x(1+\cos 2x)\,dx$
$= (1/8)\int \sin^2 2x\,dx + (1/8)\int \sin^2 2x \cos 2x\,dx$
$= (1/16)\int (1-\cos 4x)\,dx + (1/8)\int \sin^2 2x \cos 2x\,dx$
$= x/16 - (1/64)\sin 4x + (1/48)\sin^3 2x + C$

19. $\int \tan^3 x\,dx = \int (\sec^2 x - 1)\tan x\,dx = \int \sec^2 x \tan x\,dx - \int \tan x\,dx = (1/2)\tan^2 x + \ln|\cos x| + C$

21. $\int \cot^4 2x\,dx = \int \cot^2 2x(\csc^2 2x - 1)\,dx = \int \cot^2 2x \csc^2 2x\,dx - \int \cot^2 2x\,dx$
$= \int \cot^2 2x \csc^2 2x\,dx - \int \csc^2 2x\,dx + \int dx = (-1/6)\cot^3 2x + (1/2)\cot 2x + x + C$

23. $\int \sec^6 x\,dx = \int (1+\tan^2 x)^2 \sec^2 x\,dx = \int \sec^2 x\,dx + \int 2\tan^2 x \sec^2 x\,dx + \int \tan^4 x \sec^2 x\,dx$
$= \tan x + (2/3)\tan^3 x + (1/5)\tan^5 x + C$

25. $\int \tan^4 2x\,dx = \int (\sec^2 2x - 1)\tan^2 2x\,dx = (1/2)\int \sec^2 2x \tan^2 2x\,dx - \int \tan^2 2x\,dx$
$= (1/6)\tan^3 2x - (1/2)\tan 2x + x + C$

27. $A = \int_0^{\pi} \sin^2 x\,dx = \frac{1}{2}\int_0^{\pi} (1-\cos 2x)\,dx = (1/2)[x - (1/2)\sin 2x]\Big|_0^{\pi} = \frac{\pi}{2}$

Exercises 26.5

1. $u = 3x$;
$du = 3\,dx$; $\int \frac{dx}{\sqrt{1-9x^2}} = \frac{1}{3}\int \frac{3\,dx}{\sqrt{1-9x^2}} = \frac{1}{3}$ Arcsin $3x + C$
$a = 1$

3. $u = x$;
$du = dx$; $\int \frac{dx}{\sqrt{9-x^2}} =$ Arcsin $\frac{x}{3} + C$
$a = 3$

5. $u = x$;
$du = dx$; $\int \frac{dx}{\sqrt{x^2+25}} = \frac{1}{5}$ Arctan $\frac{x}{5} + C$
$a = 5$

7. $u = 3x$;
$du = 3\,dx$; $\int \frac{dx}{9x^2+4} = \frac{1}{3}\int \frac{3\,dx}{9x^2+4} = \frac{1}{6}$ Arctan $\frac{3x}{2} + C$
$a = 2$

9. $u = 5x$;
$du = 5\,dx$; $\int \frac{dx}{\sqrt{36-25x^2}} = \frac{1}{5}\int \frac{5\,dx}{\sqrt{36-25x^2}} = \frac{1}{5}$ Arcsin $\frac{5x}{6} + C$
$a = 6$

11. $u = \sqrt{12}x$;
$du = \sqrt{12}dx$; $\int \frac{dx}{\sqrt{3-12x^2}} = \frac{1}{\sqrt{12}}\int \frac{\sqrt{12}\,dx}{\sqrt{3-12x^2}} = \frac{1}{\sqrt{12}}$Arcsin$\frac{\sqrt{12}x}{\sqrt{3}} + C = \frac{1}{2\sqrt{3}}$ Arcsin $2x + C$
$a = \sqrt{3}$

13. $u = x - 1$;

$du = dx$; $\int \frac{dx}{4 + (x-1)^2} = \frac{1}{2}\text{Arctan}\frac{x-1}{2} + C$

$a = 2$

15. $u = x + 3$

$du = dx$; $\int \frac{dx}{(x^2 + 6x + 9) + 16} = \int \frac{dx}{(x+3)^2 + 4^2} = \frac{1}{4}\text{Arctan}\frac{x+3}{4} + C$

$a = 4$

17. $u = e^x$;

$du = e^x\,dx$; $\int \frac{e^x\,dx}{\sqrt{1-(e^x)^2}}$ $\text{Arcsin}\, e^x = C$

$a = 1$

19. $u = \cos x$;

$du = -\sin x\,dx$; $\int \frac{\sin x\,dx}{1 + \cos^2 x} = -\int \frac{-\sin x\,dx}{1 + \cos^2 x} = -\text{Arctan}(\cos x) + C$

$a = 1$

21. $u = x$;

$du = dx$; $\int_0^1 \frac{dx}{1 + x^2} = \text{Arctan}\,x \Big|_0^1 = \pi/4$

$a = 1$

23. $u = 3x$;

$du = 3\,dx$; $\int_0^1 \frac{dx}{\sqrt{25 - 9x^2}} = \frac{1}{3}\int \frac{3\,dx}{\sqrt{25 - 9x^2}} = \frac{1}{3}\text{Arcsin}\frac{3x}{5}\Big|_0^1 = \frac{1}{3}\text{Arcsin}\frac{3}{5} = 0.215$

$a = 5$

25. $u = 2x$;

$du = 2\,dx$; $W = \int_1^2 \frac{100\,dx}{1 + 4x^2} = 50\int_1^2 \frac{2\,dx}{1 + 4x^2} = 50\,\text{Arctan}\,2x\Big|_1^2 = 10.9\text{ N}$

$a = 1$

Exercises 26.6

1. $\dfrac{1}{1-x^2}=\dfrac{A}{1-x}+\dfrac{B}{1+x}$; $1=(A-B)x+(A+B)$; $A-B=0; A+B=1; A=1/2, B=1/2$

$$\int\frac{dx}{1-x^2}=\frac{1}{2}\int\frac{dx}{1-x}+\frac{1}{2}\int\frac{dx}{1+x}=(-1/2)\ln|x-1|+(1/2)\ln|x+1|+C$$

$$=\frac{1}{2}\ln\left|\frac{x+1}{x-1}\right|+C$$

3. $\dfrac{1}{(x+4)(x-2)}=\dfrac{A}{x+4}+\dfrac{B}{x-2}$; $1=(A+B)x+(-2A+4B)$; $A+B=0; -2A+4B=1;$

$A=-1/6, B=1/6$

$$\int\frac{dx}{(x+4)(x-2)}=-\frac{1}{6}\int\frac{dx}{x+4}+\frac{1}{6}\int\frac{dx}{x-2}=-\frac{1}{6}\ln|x+4|+\frac{1}{6}\ln|x-2|+C=\frac{1}{6}\ln\left|\frac{x-2}{x+4}\right|+C$$

5. $\dfrac{1}{(x-2)(x-1)}=\dfrac{A}{x-2}+\dfrac{B}{x-1}$; $x=A(x-1)+B(x-2)$; $x=(A+B)x+(-A-2B)$;

$A+B=1; -A-2B=0; A=2, B=-1$

$$\int\frac{xdx}{(x-2)(x-1)}=\int\frac{2\,dx}{x-2}-\int\frac{dx}{x-1}=2\ln|x-2|-\ln|x-1|+C=\ln\left|\frac{(x-2)^2}{x-1}\right|+C$$

7. $\dfrac{x+1}{(x+5)(x-1)}=\dfrac{A}{x+5}+\dfrac{B}{x-1}$; $x+1=A(x-1)+B(x+5)$; $x+1=(A+B)x+(-A+5B)$;

$A+B=1; -A+5B=1; A=2/3, B=1/3$

$$\int\frac{(x+1)dx}{x^2+4x-5}=\frac{2}{3}\int\frac{dx}{x+5}+\frac{1}{3}\int\frac{dx}{x-1}=\frac{2}{3}\ln|x+5|+\frac{1}{3}\ln|x-1|+C$$

9. $\dfrac{1}{x(x+1)^2}=\dfrac{A}{x}=\dfrac{B}{B+1}+\dfrac{C}{(x+1)^2}$; $1=A(x+1)^2+Bx(x+1)+Cx$

$1=Ax^2+2Ax+A+Bx^2+Bx+Cx$; $1=(A+B)x^2+(2A+B+C)x+A$;

$A+B=0; 2A+B+C=0; A=1$; Thus $A=1, B=-1, C=-1$

$$\int\frac{dx}{x(x+1)^2}=\int\frac{dx}{x}-\int\frac{dx}{x+1}-\int\frac{dx}{(x+1)^2}$$

$$=\ln|x|-\ln|x+1|+\frac{1}{x+1}+C=\ln\left|\frac{x}{x+1}\right|+\frac{1}{x+1}+C$$

11. $\dfrac{2x^2+x+3}{x^2(x+3)}=\dfrac{A}{x}+\dfrac{B}{x^2}+\dfrac{C}{x+3}$; $2x^2+x+3=Ax(x+3)+B(x+3)+Cx^2$

$2x^2+x+3=Ax^2+3Ax+Bx+3B+Cx^2$;

$2x^2+x+3=(A+C)x^2+(3A+B)x+3B$; $A+C=2$;

$3A+B=1$; $3B=3$; Thus, $A=0, B=1, C=2$

$$\int\frac{2x^2+x+3}{x^2(x+3)}dx=\int\frac{dx}{x^2}+\int\frac{2\,dx}{x+3}=-\frac{1}{x}+2\ln|x+3|+C$$

13. $\dfrac{x^3}{x^2+3x+2} = x - 3 + \dfrac{7x+6}{x^2+3x+2}; \dfrac{7x+6}{x^2+3x+6} = \dfrac{A}{x+1} + \dfrac{B}{x+2};$

$7x + 6 = A(x+2) + B(x+1);\ 7x + 6 = (A+B)x + (2A+B);$

$A + B = 7,\ 2A + B = 6,\ A = -1,\ B = 8$

$$\int \frac{x^3\,dx}{x^2+3x+2} = \int\left(x - 3 - \frac{1}{x+1} + \frac{8}{x+2}\right)dx = \frac{x^2}{2} - 3x - \ln|x+1| + 8\ln|x+2| + C$$

$$= \frac{x^2}{2} - 3x + \ln\left|\frac{(x+2)^8}{x+1}\right| + C$$

15. $\dfrac{x^2-2}{(x^2+1)x} = \dfrac{Ax+B}{x^2+1} + \dfrac{C}{x};\ x^2 - 2 = (Ax+B)x + C(x^2+1);$

$x^2 - 2 = (A+C)x^2 + Bx + C;\ A + C = 1;\ B = 0;\ C = -2;$ Thus $A = 3$

$$\int \frac{(x^2-2)dx}{(x^2+1)x} = \int\left\{\frac{3x}{x^2+1} - \frac{2}{x}\right\}dx = \frac{3}{2}\ln|x^2+1| - 2\ln|x| + C$$

17. $\dfrac{x^3+2x^2-9}{x^2(x^2+9)} = \dfrac{A}{x} + \dfrac{B}{x^2} + \dfrac{Cx+D}{x^2+9};\ x^3 + 2x^2 - 9 = Ax(x^2+9) + B(x^2+9) + (Cx+D)x^2;$

$x^3 + 2x^2 - 9 = (A+C)x^3 + (B+D)x^2 + 9Ax + 9B;\ A + C = 1;\ B + D = 2;\ 9A = 0;\ 9B = -9;$

Thus $A = 0,\ B = -1,\ C = 1,\ D = 3$

$$\int \frac{x^3+2x^2-9}{x^2(x^2+9)}dx = \int\left\{\frac{-1}{x^2} + \frac{x+3}{x^2+9}\right\}dx$$

$$= -\int\frac{dx}{x^2} + \int\frac{x\,dx}{x^2+9}\int\frac{3\,dx}{x^2+9} = \frac{1}{x} + \frac{1}{2}\ln|x^2+9| + \text{Arctan}\,\frac{x}{3} + C$$

19. $\dfrac{x^3}{(x^2+1)^2} = \dfrac{Ax+B}{x^2+1} + \dfrac{Cx+D}{(x^2+1)^2};\ x^3 = Ax^3 + Bx^2 + (A+C)x + (B+D)$

$A = 1;\ B = 0;\ A + C = 0;\ B + D = 0;\ C = -1,\ D = 0$

$$\int\frac{x^3dx}{(x^2+1)^2} = \int\left\{\frac{x}{x^2+1} - \frac{x}{(x^2+1)^2}\right\}dx = \frac{1}{2}\ln|x^2+1| + \frac{1}{2(x^2+1)} + C$$

21. $\dfrac{3}{(1-x)(1+x)} = \dfrac{A}{1-x} + \dfrac{B}{1+x};\ 3 = A(1+x) + B(1-x);$

$3 = (A-B)x + (A+B);\ A - B = 0;\ A + B = 3;\ A = 3/2,\ B = 3/2$

$$\int_2^3\frac{3dx}{1-x^2} = \int_2^3\left\{\frac{3/2}{1-x} + \frac{3/2}{1+x}\right\}dx = -\frac{3}{2}\ln|1-x| + \frac{3}{2}\ln|1+x|\Big|_2^3$$

$$= -\frac{3}{2}\ln 2 + \frac{3}{2}\ln 4 - \left(-\frac{3}{2}\ln 1 + \frac{3}{2}\ln 3\right) = -\frac{3}{2}\ln\frac{3}{2} \ or\ \frac{3}{2}\ln\frac{2}{3}$$

23. $\dfrac{x}{x^2+4x-5}=\dfrac{A}{x+5}+\dfrac{B}{x-1}$; $x=A(x-1)+B(x+5)$; $A+B=1$; $-A+5B=0$

Thus $A=5/6$, $B=1/6$

$$\int_2^4\frac{xdx}{x^2+4x-5}=\int_2^4\left\{\frac{5/6}{x+5}+\frac{1/6}{x-1}\right\}dx=\frac{5}{6}\ \ln|x+5|+\frac{1}{6}\ln|x-1|\ \Big|_2^4$$

$$=\frac{5}{6}\ln 9+\frac{1}{6}\ln 3-\left(\frac{5}{6}\ln 7+\frac{1}{6}\ln 1\right)=\frac{11}{6}\ln 3-\frac{5}{6}\ \ln 7 \text{ or } 0.393$$

25. $\dfrac{4x}{(x+3)(x-1)}=\dfrac{A}{x+3}+\dfrac{B}{x-1}$; $4x=A(x-1)+B(x+3)$; $A+B=4$; $-A+3B=0$; Thus $A=3$, $B=1$

$$\text{Area}=\int_2^4\frac{4x\,dx}{x^2+2x-3}=\int_2^4\left\{\frac{3}{x+3}+\frac{1}{x-1}\right\}dx=3\ln|x+3|+\ln|x-1|\ \Big|_2^4$$

$$=3\ln 7+\ln 3-(3\ln 5+\ln 1)=3\ln\frac{7}{5}+\ln 3=2.108$$

Exercises 26.7

1. $u=\ln x; dv=dx$

$$\int \ln x\,dx=x\ln x-\int x(1/x)dx=x\ln|x|-x+C$$

$du=(1/x)\,dx$; $v=x$

3. $u=x$; $dv=e^x\,dx$

$$\int xe^x\,dx=xe^x-\int e^x dx=xe^x-e^x+C$$

$du=dx$; $v=e^x$

5. $u=\ln x; dv=x^{1/2}\,dx$ $\quad \int\sqrt{x}\ln x\,dx=(2/3)x^{3/2}\ln|x|$

$du=(1/x)\,dx$; $v=(2/3)x^{3/2}-\int(2/3)x^{3/2}(1/x)\,dx=(2/3)x^{3/2}\ln|x|-(2/3)\int x^{1/2}\,dx$

$=(2/3)x^{3/2}\ln|x|-(4/9)x^{3/2}+C$

7. $\int\ln x^2\,dx=2\int\ln x\,dx=2[x\ln|x|-x]+C$ (from Exercise 1)

$=2x\ln|x|-2x+C$

9. $u=\text{Arccos}\,x$; $dv=dx$ $\quad \int\text{Arccos}\,x\,dx=x\,\text{Arccos}\,x-\int x\dfrac{-1}{\sqrt{1-x^2}}dx$

$du=-\dfrac{dx}{\sqrt{1-x^2}}$; $v=x$

$$=x\,\text{Arccos}\,x+\int\frac{x\,dx}{\sqrt{1-x^2}}=x\,\text{Arccos}\,x-\frac{1}{2}\int(1-x^2)^{-1/2}(-2x)\,dx$$

$$=x\,\text{Arccos}\,x-\sqrt{1-x^2}+C$$

$u=1-x^2$

$du=-2x\,dx$

11. $u = e^x$; $dv = \cos x\, dx$ $u = e^x$; $dv = \sin x\, dx$

$du = e^x\, dx$; $v = \sin x$; $du = e^x\, dx$ $v = -\cos x$

$$\int e^x \cos dx = e^x \sin x - \int \sin x \cdot e^x dx = e^x \sin x - [e^x(-\cos x) - \int -\cos x\, e^x dx]$$

Thus $\int e^x \cos x\, dx = e^x \sin x + e^x \cos x - \int e^x \cos x\, dx$

$$2\int e^x \cos x\, dx = e^x(\sin x + \cos x)$$

$$\int e^x \cos x\, dx = (e^x/2)(\sin x + \cos x) + C$$

13. $u = x^2$; $dv = \cos x\, dv$ $u = 2x$; $dv = \sin x$

$du = 2x\, dx$; $v = \sin x$ $du = 2\, dx$; $v = -\cos x$

$$\int x^2 \cos dx = x^2 \sin x - \int 2x \sin x\, dx = x^2 \sin x - 2\int x \sin x\, dx$$

$$= x^2 \sin x - [-2x\cos x - \int -\cos x \cdot 2x\, dx] = x^2 \sin x + 2x \cos x - 2 \sin x + C$$

15. $u = x$; $dv = \sec^2 x\, dx$

$du = dx$; $v = \tan x$

$$\int x \sec^2 x dx = x \tan x - \int \tan x\, dx = x \tan x + \ln|\cos x| + C$$

17. $u = \ln^2 x$; $dv = dx$

$du = (2/x) \ln x\, dx$; $v = x$

$$\int \ln^2 x\, dx = x \ln^2|x| - \int x(2/x) \ln x\, dx = x \ln^2|x| - 2[x \ln|x| - x] + C$$

$$= x \ln^2|x| - 2x \ln|x| + 2x + C$$

19. $u = x$; $dv = \sec x \tan x\, dx$

$du = dx$; $v = \sec x$

$$\int x \sec x \tan x\, dx = x \sec x - \int \sec x\, dx = x \sec x - \ln|\sec x + \tan x| + C$$

21. $u = x$; $dv = e^{3x}\, dx$

$du = dx$; $v = (1/3)e^{3x}$

$$\int x e^{3x} dx = (1/3)xe^{3x} - \int (1/3)e^{3x} dx$$

$$= [(1/3)xe^{3x} - (1/9)e^{3x}]\Big|_0^1$$

$$= \frac{1}{9}(2e^3 + 1)$$

23. $u = x$; $dv = (x-1)^{1/2}\, dx$

$du = dx$; $v = (2/3)(x-1)^{3/2}$

$$\int x\sqrt{x-1}\, dx = (2/3)x(\sqrt{x-1})^3 - (2/3)\int (x-1)^{3/2} dx = (2x/3)(x-1)^{3/2} - (4/15)(x-1)^{5/2} + C$$

$$\int_1^2 x\sqrt{x-1}\, dx = [(2x/3)(x-1)^{3/2}) - (4/15)(x-1)^{5/2}]\Big|_1^2 = 16/15$$

25. $u = \ln(x+1)$; $\quad dv = dx \qquad \int \ln(x+1)\,dx = x\ln(x+1) - \int x\dfrac{dx}{x+1} = x\ln(x+1) - x + \ln|x+1| + C$

$du = \dfrac{dx}{x+1}$; $\qquad v = x$

$$\int_1^2 \ln(x+1)\,dx = [x\ln(x+1) - x + \ln(x+1)]\Big|_1^2 = \ln\frac{27}{4} - 1$$

27. $u = \ln 2x$; $\qquad dv = dx$
$du = (1/x)\,dx \qquad v = x$

$$A = \int_{1/2}^{1} \ln 2x\,dx = x\ln x - \int x(1/x)dx = x\ln 2x - x\Big|_{1/2}^{1} = \ln 2 - 1/2$$

29. $u = x$; $\qquad du = e^x\,dx$
$du = dx$; $\; v = e^x$

$$V = 2\pi\int_0^1 xe^x\,dx = 2\pi[xe^x - \int e^x dx] = 2\pi[xe^x - e^x]\Big|_0^1 = 2\pi$$

Exercises 26.8

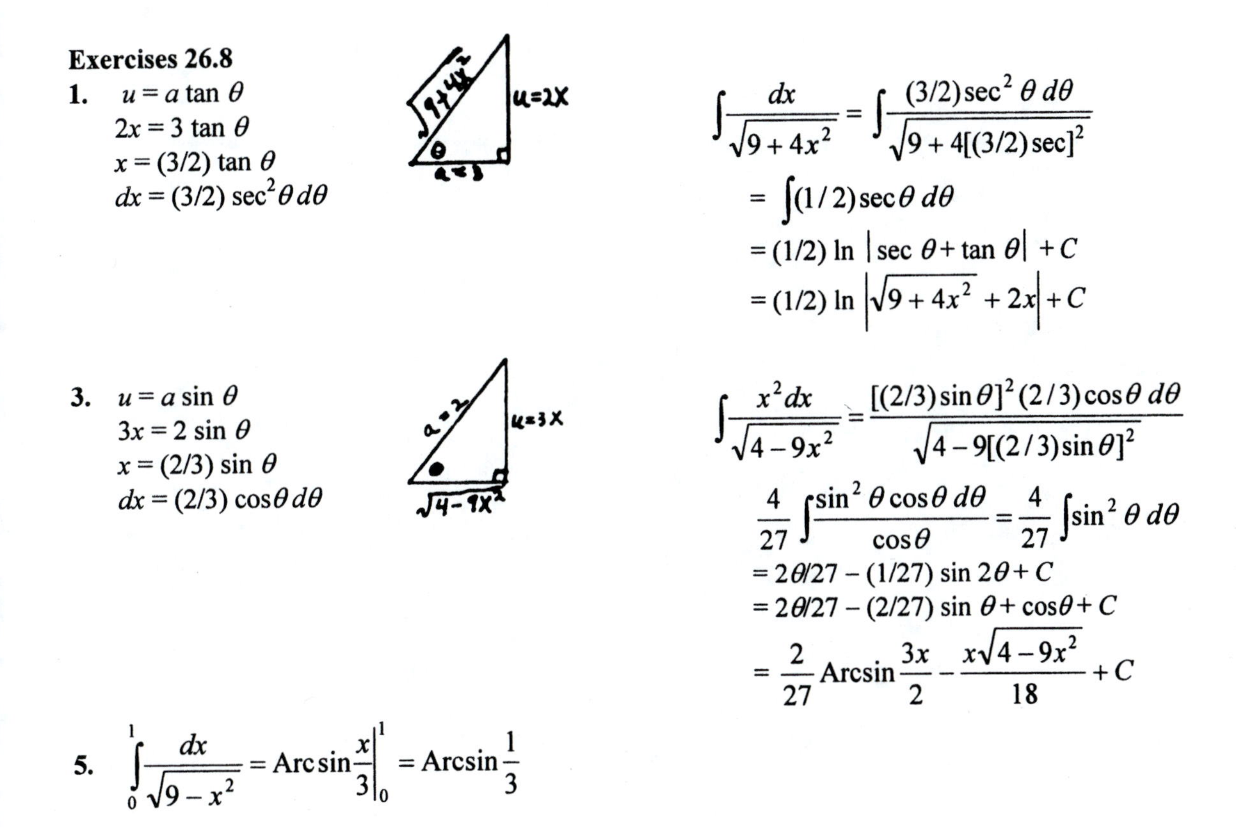

1. $u = a\tan\theta$
$2x = 3\tan\theta$
$x = (3/2)\tan\theta$
$dx = (3/2)\sec^2\theta\,d\theta$

$$\int\frac{dx}{\sqrt{9+4x^2}} = \int\frac{(3/2)\sec^2\theta\,d\theta}{\sqrt{9+4[(3/2)\sec]^2}}$$
$$= \int(1/2)\sec\theta\,d\theta$$
$$= (1/2)\ln|\sec\theta + \tan\theta| + C$$
$$= (1/2)\ln\left|\sqrt{9+4x^2} + 2x\right| + C$$

3. $u = a\sin\theta$
$3x = 2\sin\theta$
$x = (2/3)\sin\theta$
$dx = (2/3)\cos\theta\,d\theta$

$$\int\frac{x^2dx}{\sqrt{4-9x^2}} = \frac{[(2/3)\sin\theta]^2(2/3)\cos\theta\,d\theta}{\sqrt{4-9[(2/3)\sin\theta]^2}}$$
$$\frac{4}{27}\int\frac{\sin^2\theta\cos\theta\,d\theta}{\cos\theta} = \frac{4}{27}\int\sin^2\theta\,d\theta$$
$$= 2\theta/27 - (1/27)\sin 2\theta + C$$
$$= 2\theta/27 - (2/27)\sin\theta + \cos\theta + C$$
$$= \frac{2}{27}\text{Arcsin}\frac{3x}{2} - \frac{x\sqrt{4-9x^2}}{18} + C$$

5. $$\int_0^1\frac{dx}{\sqrt{9-x^2}} = \text{Arcsin}\frac{x}{3}\Big|_0^1 = \text{Arcsin}\frac{1}{3}$$

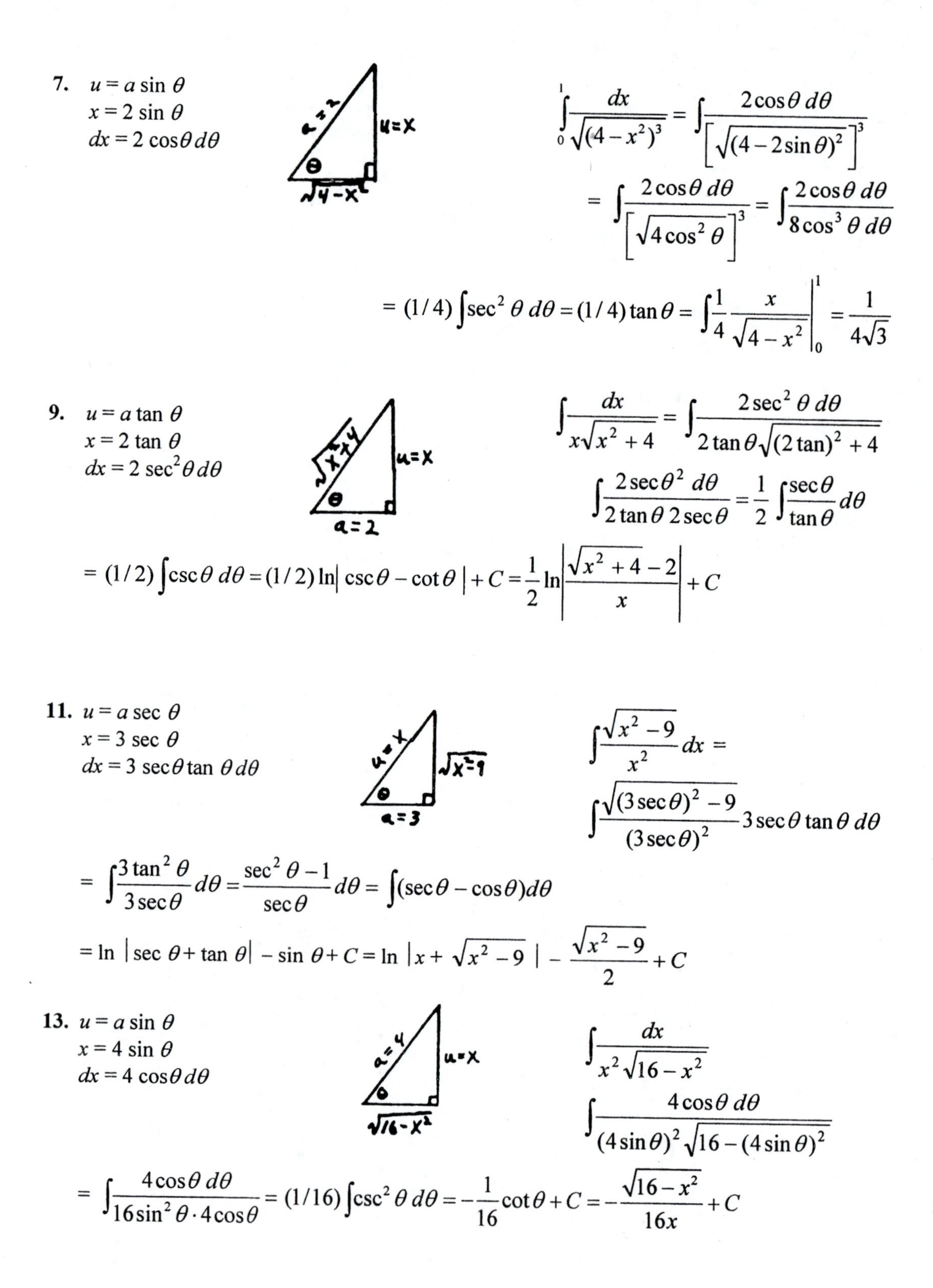

7. $u = a\sin\theta$
$x = 2\sin\theta$
$dx = 2\cos\theta\, d\theta$

$$\int_0^1 \frac{dx}{\sqrt{(4-x^2)^3}} = \int \frac{2\cos\theta\, d\theta}{\left[\sqrt{(4-2\sin\theta)^2}\right]^3}$$

$$= \int \frac{2\cos\theta\, d\theta}{\left[\sqrt{4\cos^2\theta}\right]^3} = \int \frac{2\cos\theta\, d\theta}{8\cos^3\theta\, d\theta}$$

$$= (1/4)\int \sec^2\theta\, d\theta = (1/4)\tan\theta = \int \frac{1}{4}\frac{x}{\sqrt{4-x^2}}\Bigg|_0^1 = \frac{1}{4\sqrt{3}}$$

9. $u = a\tan\theta$
$x = 2\tan\theta$
$dx = 2\sec^2\theta\, d\theta$

$$\int \frac{dx}{x\sqrt{x^2+4}} = \int \frac{2\sec^2\theta\, d\theta}{2\tan\theta\sqrt{(2\tan)^2+4}}$$

$$\int \frac{2\sec\theta^2\, d\theta}{2\tan\theta\, 2\sec\theta} = \frac{1}{2}\int \frac{\sec\theta}{\tan\theta} d\theta$$

$$= (1/2)\int \csc\theta\, d\theta = (1/2)\ln|\csc\theta - \cot\theta| + C = \frac{1}{2}\ln\left|\frac{\sqrt{x^2+4}-2}{x}\right| + C$$

11. $u = a\sec\theta$
$x = 3\sec\theta$
$dx = 3\sec\theta\tan\theta\, d\theta$

$$\int \frac{\sqrt{x^2-9}}{x^2} dx =$$

$$\int \frac{\sqrt{(3\sec\theta)^2-9}}{(3\sec\theta)^2} 3\sec\theta\tan\theta\, d\theta$$

$$= \int \frac{3\tan^2\theta}{3\sec\theta} d\theta = \frac{\sec^2\theta - 1}{\sec\theta} d\theta = \int (\sec\theta - \cos\theta) d\theta$$

$$= \ln|\sec\theta + \tan\theta| - \sin\theta + C = \ln|x + \sqrt{x^2-9}| - \frac{\sqrt{x^2-9}}{2} + C$$

13. $u = a\sin\theta$
$x = 4\sin\theta$
$dx = 4\cos\theta\, d\theta$

$$\int \frac{dx}{x^2\sqrt{16-x^2}}$$

$$\int \frac{4\cos\theta\, d\theta}{(4\sin\theta)^2\sqrt{16-(4\sin\theta)^2}}$$

$$= \int \frac{4\cos\theta\, d\theta}{16\sin^2\theta \cdot 4\cos\theta} = (1/16)\int \csc^2\theta\, d\theta = -\frac{1}{16}\cot\theta + C = -\frac{\sqrt{16-x^2}}{16x} + C$$

15. $u = \tan\theta$
$x = 3\tan\theta$
$dx = 3\sec^2\theta\, d\theta$

u=x, a=3, $\sqrt{9+x^2}$

$$\int \frac{\sqrt{9+x^2}}{x}dx = \int \frac{\sqrt{9+(3\tan\theta)^2}\,3\sec^2\theta\, d\theta}{3\tan\theta} =$$

$$\int \frac{3\sec\theta\, 3\sec^2 d\theta}{3\tan\theta} = 3\int \sec^2\theta\csc\theta\, d\theta \text{ by parts:}$$

$u = \csc\theta$; $dv = \sec^2\theta\, d\theta$;
$du = -\csc\theta\cot\theta\, d\theta$; $v = \tan\theta$

$$= 3[\csc\theta\tan\theta - \int \tan\theta(-\csc\theta\cot\theta)d\theta] = 3\csc\theta\tan\theta + 3\int \csc\theta\, d\theta$$

$$= 3\sec\theta + 3\ln|\csc\theta - \cot\theta| + C = 3\frac{\sqrt{9+x^2}}{3} + \ln\left|\frac{\sqrt{9+x^2}}{3} - \frac{3}{x}\right| + C$$

or $\sqrt{9+x^2} - 3\ln\left|\frac{\sqrt{9+x^2}+3}{x}\right| + C$ if use $\int \csc\theta\, d\theta = -\ln|\csc\theta + \cot\theta|$.

17. $u = a\sin\theta$
$x = 5\sin\theta$
$dx = 5\cos\theta\, d\theta$

a=5, u=x, $\sqrt{25-x^2}$

$$\int \frac{dx}{(25-x^2)^{3/2}} = \frac{5\cos\theta\, d\theta}{(25-25\sin^2\theta)^{3/2}}$$

$$\int \frac{5\cos\theta\, d\theta}{(5\cos\theta)^3} = (1/25)\int \sec^2\theta\, d\theta$$

$$= (1/25)\tan\theta + C = \frac{1}{25}\frac{x}{\sqrt{25-x^2}} + C$$

19. $u = a\sec\theta$
$x = 3\sec\theta$
$dx = 3\sec\theta\tan\theta\, d\theta$

u=x, a=3, $\sqrt{x^2-9}$

$$\int \frac{dx}{\sqrt{x^2-9}} = \int \frac{3\sec\theta\tan\theta\, d\theta}{\sqrt{(3\sec\theta)^2-9}}$$

$$= \int \frac{3\sec\theta\tan\theta\, d\theta}{3\tan\theta} = \int \sec\theta\, d\theta$$

$$\ln|\sec\theta + \tan\theta| + C = \ln\left|\frac{x}{3} + \frac{\sqrt{x^2-9}}{3}\right| + C = \ln\left|x + \sqrt{x^2-9}\right| + C$$

21. $u = a\tan\theta$
$3x = 2\tan\theta$
$dx = (2/3)\sec^2\theta\, d\theta$

u=3x, a=2, $\sqrt{9x^2+4}$

$$\int \frac{x^3 dx}{\sqrt{9x^2+4}}$$

$$= \int \frac{[(2/3)\tan\theta]^3}{\sqrt{9[(2/3)\tan\theta]^2+4}}(2/3)\sec^2\theta\, d\theta$$

$$= \frac{8}{81}\int \frac{\tan^3\theta\sec^2\theta\, d\theta}{\sec\theta} = (8/81)\int \tan^3\theta\sec\theta\, d\theta$$

$$= (8/81)\int (\sec^2\theta - 1)\sec\theta\tan\theta\, d\theta = \frac{8}{81}\left(\frac{\sec^3\theta}{3} - \sec\theta\right) + C$$

$$= \frac{8}{81}\left\{\frac{(\sqrt{9x^2+4})^3}{3\cdot 8} - \frac{\sqrt{9x^2+4}}{2}\right\} + C = \frac{8}{81}\frac{\sqrt{9x^2+4}}{2}\left\{\frac{9x^2+4}{12} - 1\right\}. = \frac{9x^2-8}{243}\sqrt{9x^2+4} + C$$

23. $u = a\sec\theta$

$x - 3 = \sec\theta$

$dx = \sec\theta\,\tan\theta\,d\theta$

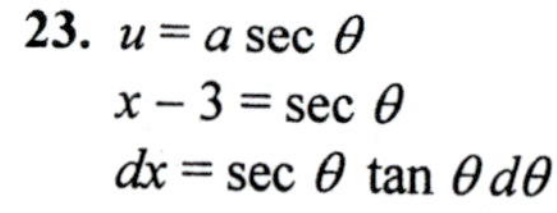

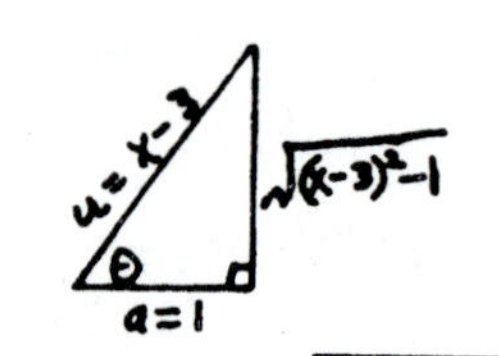

$$\int\frac{dx}{\sqrt{(x-3)^2-1}} = \int\frac{\sec\theta\tan\theta\,d\theta}{\sqrt{\sec^2\theta-1}}$$

$$= \int\frac{\sec\theta\tan\theta\,d\theta}{\tan\theta} = \int\sec\theta\,d\theta$$

$$= \ln\left|\sec\theta + \tan\theta\right| + C = \ln\left|x - 3 + \sqrt{x^2-6x+8}\right| + C$$

25. $u = a\sec\theta$

$x + 4 = \sec\theta$

$dx = \sec\theta\tan\theta\,d\theta$

$$\int\frac{dx}{(x^2+8x+15)^{3/2}} = \int\frac{dx}{[(x+4)^2-1]^{3/2}}$$

$$= \int\frac{\sec\theta\tan\theta\,d\theta}{[\sec^2\theta-1]^{3/2}} = \int\frac{\sec\theta\tan\theta\,d\theta}{\tan^3\theta}$$

$$= \int\sin^{-2}\theta\cos\theta\,d\theta = -\frac{1}{\sin\theta} + C = -\frac{x+4}{\sqrt{x^2+8x+15}} + C$$

27. $u = a\tan\theta$

$x = 2\tan\theta$

$dx = 2\sec^2\theta\,d\theta$

$$A = \int_0^2\frac{dx}{\sqrt{x^2+4}} = \int\frac{2\sec^2\theta\,d\theta}{\sqrt{(2\tan\theta)^2+4}}$$

$$= \int\frac{2\sec^2\theta\,d\theta}{2\sec\theta} = \int\sec\theta\,d\theta$$

$$= \ln\left|\sec\theta + \tan\theta\right| = \ln\left|\frac{\sqrt{x^2+4}}{2} + \frac{x}{2}\right|\Bigg|_0^2 = \ln(\sqrt{2}+1)$$

Exercises 26.9

1. $u = x, a = 5, b = 1$; Formula 15; $\dfrac{1}{\sqrt{5}}\ln\left|\dfrac{\sqrt{x+5}-\sqrt{5}}{\sqrt{x+5}+\sqrt{5}}\right| + C$

3. $u = x, a = 2$, Formula 35; $\ln\left|x + \sqrt{x^2-4}\right| + C$

5. $u = x, a = 3, b = 2$, Formula 13; $\dfrac{-(3-x)\sqrt{3+2x}}{2} + C$

7. $m = 7, n = 3$, Formula 75; $-\dfrac{\sin 10x}{2\cdot 10} + \dfrac{\sin 4x}{2\cdot 4} + C = \dfrac{1}{4}\left\{\dfrac{\sin 4x}{2} - \dfrac{\sin 10x}{5}\right\}$

9. $u = x, a = 3$, Formula 24; $-\dfrac{x}{2}\sqrt{9-x^2} + \dfrac{9}{2}\text{Arcsin}\dfrac{x}{3} + C$

11. $a = 1, b = 9, u = x$, Formula 9; $\dfrac{1}{1+9x} + \ln\left|\dfrac{x}{1+9x}\right| + C$

13. $u = x$, $a = 5$, Formula 20; $\frac{1}{10}\ln\left|\frac{x-5}{x+5}\right| + C$

15. $u = x$, $a = 2$, Formula 30;

$$(1/2)\left[x\sqrt{x^2+4} + 4\ln\left| x + \sqrt{x^2+4}\right|\right] + C = (x/2)\sqrt{x^2+4} + 2\ln\left| x + \sqrt{x^2+4}\right| + C$$

17. $u = x$, $a = 3$, $b = 4$, Formula 7; $\frac{1}{16}\left\{\ln|3+4x| + \frac{3}{3+4x}\right\} + C$

19. $u = 3x$; $a = 4$; $du = 3\ dx$; Formula 36 $\int\frac{dx}{x\sqrt{9x^2-16}} = \int\frac{3dx}{3x\sqrt{9x^2-16}} = \frac{1}{4}\text{Arccos}\frac{4}{3x} + C$

21. $u = x$, $a = 3$, $n = 4$, Formula 59; $\frac{e^{3x}(3\sin 4x - 4\cos 4x)}{25} + C$

23. $u = 2x - 3$; $du = 2\ dx$; Formula 80; $\int(2x-3)\sin(2x-3)dx$

$= (1/2)\int(2x-3)\ \sin(2x-3)\cdot 2\ dx = (1/2)\sin(2x-3) - (1/2)(2x-3)\cos(2x-3) + C$

25. $u = x$, $n = 4$, Formula 83: $\int\sin^4 x\ dx = (-1/4)\sin^3 x\cos x + (3/4)\int\sin^2 x\ dx =$

$= (-1/4)\sin^3 x\cos x + (3/8)\int(1-\cos 2x)\ dx$

$= (-1/4)\sin^3 x\cos x + 3x/8 - (3/16)\sin 2x + C$

$= (-1/4)\sin^3 x\cos x + 3x/8 - (3/8)\sin x\cos x + C$

27. $u = 3x$, $du = 3\ dx$, $a = 4$, Formula 33;

$$\int\frac{\sqrt{9x^2-16}}{x}dx = \int\frac{\sqrt{9x^2-16}}{3x}3dx = \sqrt{9x^2-16} - 4\ \text{Arccos}\frac{4}{3x} + C$$

Exercises 26.10

1. $\frac{2/4}{2}[f(1) + 2f(3/2) + 2f(2) + 2f(5/2) + f(3)] = (1/4)[1 + 4/3 + 1 + 4/5 + 1/3] = 1.117$

3. $\frac{1/4}{2}[f(0) + 2f(1/4) + 2f(1/2) + 2f(3/4) + f(1)] = (1/8)[1 + 32/17 + 8/5 + 32/25 + 1/2] = 0.783$

5. $\frac{1/10}{2}[f(0) + 2f(0.1) + 2f(0.2) + 2f(0.3) + 2f(0.4) + 2f(0.5) + 2f(0.6)$

$+ 2f(0.7) + 2f(0.8) + 2f(0.9) + f(1)]$

$= (1/20)[2 + 3.995 + 3.980 + 3.955 + 3.919 + 3.873 + 3.816$

$+ 3.747 + 3.666 + 3.572 + 1.732] = 1.913$

7. $\frac{1/2}{2}\,[f(0) + 2f(1/2) + 2f(1) + 2f(3/2) + 2f(2) + 2f(5/2) + 2f(3)\,]$

$= (1/4)[1 + 2.121 + 2.828 + 4.183 + 6 + 8.155 + 5.292] = 7.395$

9. $\frac{\pi/24}{2}\,[f(0) + 2f(\pi/24) + 2f(\pi/12) + 2f(\pi/8) + 2f(\pi/6) + 2f(5\pi/24) + 2f(\pi/4) + 2f(7\pi/24) + f(\pi/3)]$

$= (\pi/48)[1 + 2 + 1.995 + 1.976 + 1.925 + 1.819 + 1.631 + 1.336 + 0.457] = 0.925$

11. $(2/2)[24 + 2(21) + 2(18) + 2(17) + 2(15) + 2(12) + 2(10) + 9] = 219$ ft-lb

13. $(3/2)[12 + 2(11) + 2(18) + 2(25) + 2(19) + 2(6) + 10] = 270$

15. $\frac{1/2}{3}\,[f(0) + 4f(1/2) + 2f(1) + 4f(3/2) + f(2)\,]$

$= (1/6)[1 + 3.578 + 1.414 + 2.219 + 0.447] = 1.443$

17. $\frac{1/2}{3}\,[f(2) + 4f(5/2) + 2f(3) + 4f(7/2) + 2f(4) + 4f(9/2) + 2f(5) + 4f(11/2) + f(6)]$

$= (1/6)[0.111 + 0.241 + 0.071 + 0.091 + 0.031 + 0.043 + 0.016 + 0.024 + 0.005] = 0.105$

19. $\frac{1/4}{3}\,[f(0) + 4f(1/4) + 2f(1/2) + 4f(3/4) + f(1)]$

$= (1/12)[1 + 4.258 + 2.568 + 7.020 + 2.718] = 1.464$

21. $\frac{\pi/16}{3}\,[f(0) + 4f(\pi/16) + 2f(\pi/8) + 4f(3\pi/16) + f(\pi/4)]$

$= (\pi/48)[0 + 0.156 + 0.325 + 1.574 + 0.785] = 0.186$

23. $\frac{\pi/12}{3}\,[f(0) + 4f(\pi/12) + 2f(\pi/6) + 4f(\pi/4) + 2f(\pi/3) + 4f(5\pi/12) + f(\pi/2)]$

$= (\pi/36)[1.414 + 5.561 + 2.646 + 4.899 + 2.236 + 4.132 + 1] = 1.910$

25. a) $\frac{1/2}{2}\,[f(0) + 2f(1/2) + 2f(1) + 2f(3/2) + 2f(2) + 2f(5/2) + f(3)]$

$= (1/4)[3 + 5.916 + 5.657 + 5.196 + 4.742 + 3.317 + 0] = 6.889$

b) $\frac{1/2}{3}\,[f(0) + 4f(1/2) + 2f(1) + 4f(3/2) + 2f(2) + 4f(5/2) + f(3)]$

$= (1/6)[3 + 11.832 + 5.657 + 10.392 + 4.472 + 6.633 + 0] = 6.998$

27. a) $\frac{1/4}{2}\,[f(0) + 2f(1/4) + 2f(1/2) + 2f(3/4) + f(1)]$

$= (1/8)[0 + 0.642 + 1.649 + 3.176 + 2.718] = 1.023$

b) $\frac{1/4}{3}\,[f(0) + 4f(1/4) + 2f(1/2) + 4f(3/4) + f(1)]$

$= (1/12)[0 + 1.284 + 1.649 + 6.351 + 2.718] = 1.000$

29. a) $\dfrac{\pi/12}{2}\,[f(0)+2f(\pi/12)+2f(\pi/6)+2f(\pi/4)+2f(\pi/3)+2f(5\pi/12)+f(\pi/2)]$

$= (\pi/24)[1 + 1.744 + 1.499 + 1.265 + 1.041 + 0.828 + 0.312] = 1.006$

$\dfrac{\pi/12}{3}\,[f(0)+4f(\pi/12)+2f(\pi/6)+4f(\pi/4)+2f(\pi/3)+4f(5\pi/12)+f(\pi/2)]$

$= (\pi/36)[1 + 3.488 + 1.499 + 2.529 + 1.041 + 1.655 + 0.312] = 1.006$

31. a) $A = \dfrac{150}{2}[475 + 2(417) + 2(275) + 2(353) + 2(405) + 2(521) + 565] = 373{,}650 \text{ ft}^2$

$\approx 374{,}000 \text{ ft}^2$

b) $A = \dfrac{150}{3}[475 + 4(417) + 2(275) + 4(353) + 2(405) + 4(521) + 565] = 378{,}200 \text{ ft}^2$

$\approx 378{,}000 \text{ ft}^2$

Exercises 26.11

1. $A = \dfrac{1}{2}\displaystyle\int_0^{2\pi/3} r^2 d\theta = \dfrac{1}{2}\int_0^{2\pi/3} 4\,d\theta = 2\theta \,\Big|_0^{2\pi/3} = \dfrac{4\pi}{3}$

3. $A = \dfrac{1}{2}\displaystyle\int_0^{\pi} 9\cos^2\theta\,d\theta = \dfrac{9}{4}\int(1+\cos 2\theta)d\theta = \dfrac{9}{4}(\theta + \dfrac{1}{2}\sin 2\theta)\,\Big|_0^{\pi} = \dfrac{9\pi}{4}$

5. $A = 2\left\{\dfrac{1}{2}\displaystyle\int_0^{\pi/2} 9\sin 2\theta\,d\theta\right\} = -\dfrac{9}{2}\cos 2\theta\,\Big|_0^{\pi/2} = 9$

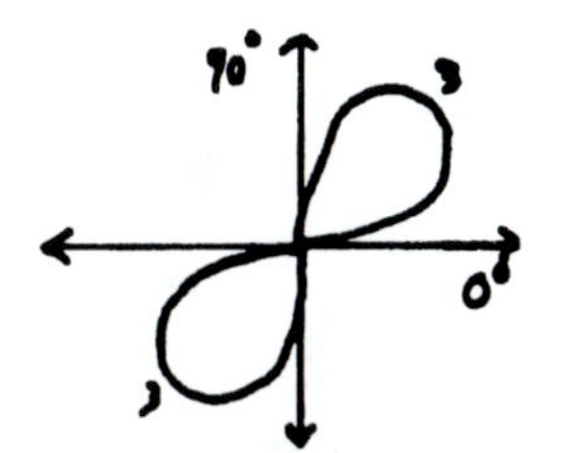

7. $A = 2\left\{\dfrac{1}{2}\displaystyle\int_0^{\pi}(1+\cos\theta)^2\,d\theta\right\} = \int_0^{\pi}(1+2\cos\theta+\cos^2\theta)d\theta$

$= \displaystyle\int_0^{\pi}(3/2 + 2\cos\theta - (1/2)\cos 2\theta)d\theta$

$= \dfrac{3\theta}{2} + 2\sin\theta - \dfrac{1}{4}\sin 2\theta\,\Big|_0^{\pi} = \dfrac{3\pi}{2}$

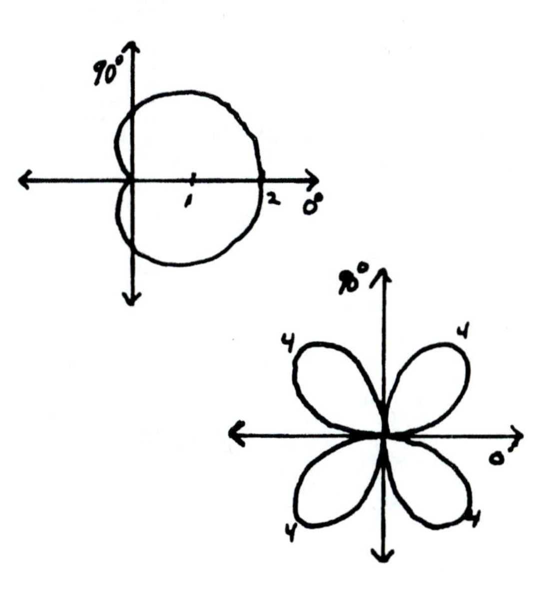

9. $A = 4\left\{\dfrac{1}{2}\displaystyle\int_0^{\pi/2}(4\sin 2\theta)^2\,d\theta\right\} = 16\int_0^{\pi/2}(1-\cos 4\theta)\,d\theta$

$= 16(\theta - (1/4)\sin 4\theta)\,\Big|_0^{\pi/2} = 8\pi$

11. $A = 4\left\{\frac{1}{2}\int_0^{\pi/2} 16\cos^2\theta\, d\theta\right\} = 32\int_0^{\pi/2}\cos^2\theta\, d\theta$

$= 16\int_0^{\pi/2}(1+\cos 4\theta)\, d\theta = 16(\theta + (1/4)\sin 4\theta \Big|_0^{\pi/2} = 8\pi$

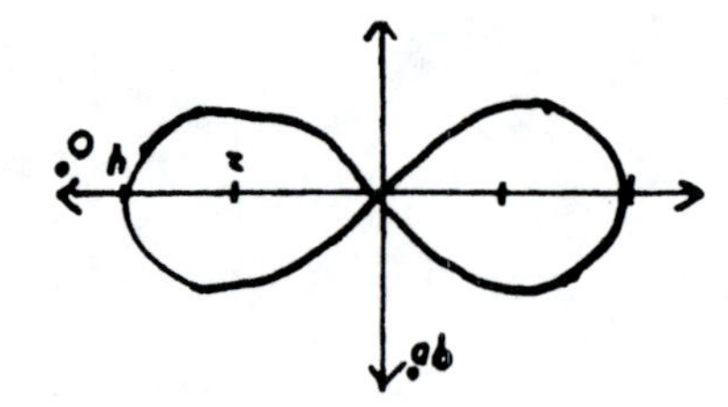

13. $A = 2\left\{\frac{1}{2}\int_0^{\pi}(4+3\cos\theta)^2\, d\theta\right\} = \int_0^{\pi}(16+24\cos\theta+9\cos^2\theta)\, d\theta$

$= \int_0^{\pi}[41/2 + 24\cos\theta + (9/2)\cos 2\theta]\, d\theta$

$= \frac{41\theta}{2} + 24\sin\theta + \frac{9}{4}\sin 2\theta \Big|_0^{\pi} = \frac{41\pi}{2}$

15. $A = \frac{1}{2}\int_0^{\pi} e^{2\theta}\, d\theta = \frac{1}{4}e^{2\theta}\Big|_0^{\pi} = \frac{1}{4}(e^{2\pi} - 1)$

17. $A = 2\left\{\frac{1}{2}\int_0^{\pi/3}(1-2\cos\theta)^2\, d\theta\right\}$

$= \int_0^{\pi/3}(1 - 4\cos\theta + 4\cos^2\theta)\, d\theta$

$= \int_0^{\pi/3}(3 - 4\cos\theta + 2\cos 2\theta\, d\theta = 3\theta - 4\sin\theta + \sin 2\theta \Big|_0^{\pi/3}$

$= \pi - \frac{3\sqrt{3}}{2}$

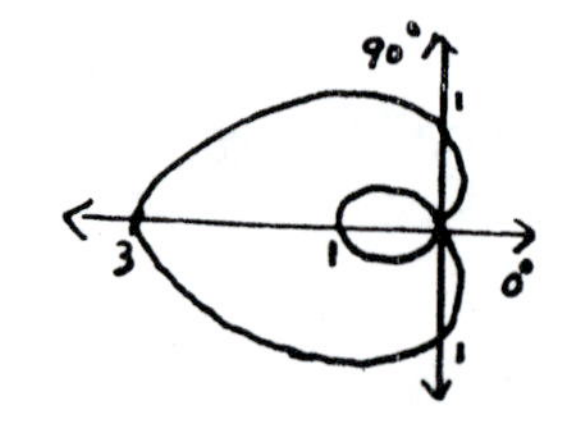

19. $2\sin\theta + 2\cos\theta = 0;$
$\tan\theta = -1;\ \theta = \pi/4, -\pi/4$

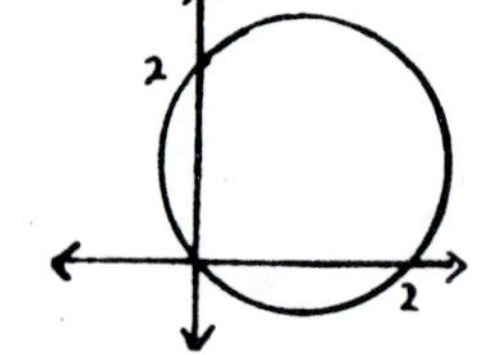

$A = 2\left\{\frac{1}{2}\int_{-\pi/4}^{\pi/4}(2\sin\theta + 2\cos\theta)^2\, d\theta\right.$

$= 4\int_{-\pi/4}^{\pi/4}(\sin^2\theta + \cos^2\theta + 2\sin\theta\cos\theta)\, d\theta = 4\int_{-\pi/4}^{\pi/4}(1+\sin 2\theta)\, d\theta$

$= 4[\theta - (1/2)\cos 2\theta]\Big|_{-\pi/4}^{\pi/4} = 2\pi$

21. $\sin\theta = \sin 2\theta$; $\sin\theta = 2\sin\theta\cos\theta$

$\sin\theta(2\cos\theta - 1) = 0$

$\sin\theta = 0$; $\cos\theta = 1/2$

$\theta = 0, \pi, \pi/3, 5\pi/3$

$$A = 2\left\{\frac{1}{2}\int_0^{\pi/3}\sin^2\theta\, d\theta + \frac{1}{2}\int_{\pi/3}^{\pi/2}\sin^2 2\theta\, d\theta\right\}$$

$$= \frac{1}{2}\left\{\int_0^{\pi/3}(1-\cos 2\theta)\, d\theta + \int_{\pi/3}^{\pi/2}(1-\cos 4\theta)\, d\theta\right\}$$

$$= \frac{1}{2}\left\{[\theta - (1/2)\sin 2\theta]\Big|_0^{\pi/3} + [\theta - (1/4)\sin 4\theta]\Big|_{\pi/3}^{\pi/2} = \frac{\pi}{4} - \frac{3\sqrt{3}}{16}\right.$$

23. $2\cos 2\theta = 1$; $\cos 2\theta = 1/2$;

$2\theta = \pi/3, 5\pi/3, 7\pi/3, 11\pi/3$;

$\theta = \pi/6, 5\pi/6, 7\pi/6, 11\pi/6$

$$A = 8\left\{\frac{1}{2}\int_0^{\pi/6}[(2\cos 2\theta)^2 - 1^2]\, d\theta = 4\int_0^{\pi/6}(4\cos^2 2\theta - 1)\, d\theta\right.$$

$$= 4\int_0^{\pi/6}(1 + 2\cos 4\theta)\, d\theta = 4[\theta + (1/2)\sin 4\theta]\Big|_0^{\pi/6} = \frac{2\pi}{3} + \sqrt{3}$$

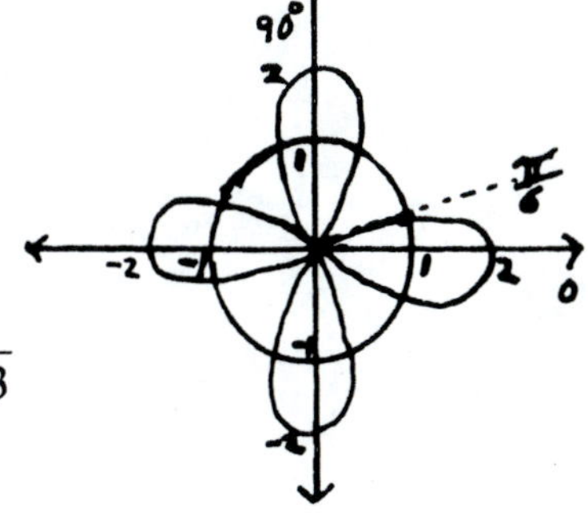

25. $5\sin\theta = 2 + \sin\theta$;

$4\sin\theta = 2$; $\sin\theta = 1/2$;

$\theta = \pi/6, 5\pi/6$

$$A = 2\left\{\frac{1}{2}\int_{-\pi/2}^{0}(2+\sin\theta)^2\, d\theta + \frac{1}{2}\int_0^{\pi/6}[(2+\sin\theta)^2 - (5\sin\theta)^2]d\theta\right\}$$

$$= \int_{-\pi/2}^{0}(4 + 4\sin\theta + \sin^2\theta)^2\, d\theta + \int_0^{\pi/6}(4 + 4\sin\theta - 24\sin^2\theta)\, d\theta$$

$$= \int_{-\pi/2}^{0}[9/2 + 4\sin\theta - (1/2)\cos 2\theta]\, d\theta \int_0^{\pi/6}(-8 + 4\sin\theta + 12\cos 2\theta)\, d\theta$$

$$= \frac{9\theta}{2} - 4\cos\theta - \frac{1}{4}\sin 2\theta\Big|_{-\pi/2}^{0} + (-80 - 4\cos\theta + 6\sin 2\theta)\Big|_0^{\pi/6} = \frac{11\pi}{12} + \sqrt{3}$$

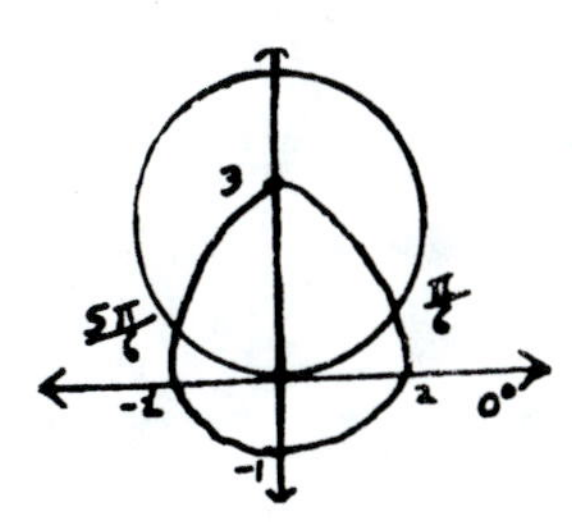

27. $3 + 3\cos\theta = 3 + 3\sin\theta$

$\tan\theta = 1$; $\theta = \pi/4, 5\pi/4$

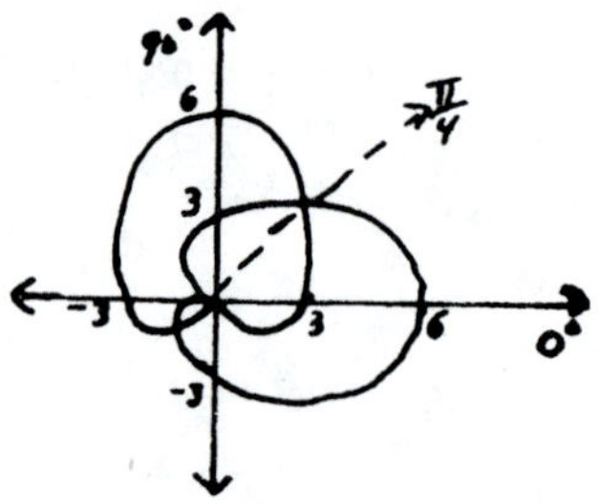

$$A = \frac{1}{2}\int_{\pi}^{3\pi/2}(3+3\cos\theta)^2\,d\theta$$

$$+\frac{1}{2}\int_{3\pi/2}^{\pi/4}[(3+3\cos\theta)^2-(3+3\sin\theta)^2]\,d\theta$$

$$=\frac{9}{2}\int_{\pi}^{3\pi/2}(1+2\cos\theta+\cos^2\theta)\,d\theta+\frac{9}{2}\int_{3\pi/2}^{\pi/4}(2\cos\theta-2\sin\theta+\cos 2\theta)\,d\theta$$

$$=\frac{9}{2}\left\{[3\theta/2+2\sin\theta+(1/4)\sin 2\theta]\,\Big|_{\pi}^{3\pi/2}+[2\sin\theta+2\cos\theta+(1/2)\sin 2\theta]\,\Big|_{3\pi/2}^{\pi/4}=\frac{27\pi}{8}+\frac{9}{4}+9\sqrt{2}\right.$$

Exercises 26.12

1. $\int_1^{\infty}\frac{2}{x^3}dx$

$= \lim_{t\to\infty}\int_1^t 2x^{-3}dx$

$= \lim_{t\to\infty}\frac{2x^{-2}}{-2}\Big|_1^t$

$= \lim_{t\to\infty}\left(-x^2\right)\Big|_1^t$

$= \lim_{t\to\infty}\left(-t^{-2}-\left[-1^{-2}\right]\right)$

$= \lim_{t\to\infty}\left(-\frac{-1}{t^2}+1\right)=1$

3. $\int_0^{\infty}e^{-3x}dx$

$= \lim_{t\to\infty}\int_0^t\left(-\frac{1}{3}\right)e^{-3x}(-3)dx$

$= \lim_{t\to\infty}\frac{-1}{3}e^{-3x}\Big|_0^t$

$= \lim_{t\to\infty}\left(\frac{-e^{-3t}}{3}-\left[-\frac{1}{3}e^0\right]\right)$

$= \lim_{t\to\infty}\left(\frac{-1}{3e^{3t}}+\frac{1}{3}\right)=\frac{1}{3}$

5. $\int_1^{\infty}\frac{1}{\sqrt{x}}dx$

$= \lim_{t\to\infty}\int_1^t x^{\frac{-1}{2}}dx$

$= \lim_{t\to\infty}\frac{-2x^{\frac{1}{2}}}{\frac{1}{2}}\Bigg|_1^t$

$= \lim_{t\to\infty}-4t^{\frac{1}{2}}-\left(-4\left(1^{\frac{1}{2}}\right)\right)$

$= \lim_{t\to\infty}-4\sqrt{t}+4\sqrt{1}=\infty$

diverges

7. $\int_{-\infty}^{0}e^{5x}dx$

$= \lim_{t\to-\infty}\int_t^0 e^{5x}dx$

$= \lim_{t\to-\infty}\frac{1}{5}e^{5x}\Big|_t^0$

$= \lim_{t\to-\infty}\frac{1}{5}e^0-\frac{1}{5}e^{5t}=\frac{1}{5}$

9. $\int_0^1 x^{-2/3}dx$

$= \lim_{t\to0^+}\int_t^1 x^{-2/3}dx$

$= \lim_{t\to0^+} 3x^{1/3}\Big|_t^1$

$= \lim_{t\to0^+}\left(3\left(1^{1/3}\right)-3\left(t^{1/3}\right)\right)$

$= \lim_{t\to0^+}\left(3-\sqrt[3]{t}\right)=3$

11. $\int_0^5 \frac{1}{(x-5)^2}dx$

$= \lim_{t\to5^-}\int_0^1 (x-5)^{-2}dx$

$= \lim_{t\to5^-}\left[-1(x-5)^{-1}\right]\Big|_0^t$

$= \lim_{t\to5^-}\left[\frac{-1}{t-5}+\frac{1}{0-5}\right]=\infty$

diverges

13. $\int_0^\infty xe^{-x^2}dx$

$= \lim_{t\to\infty}\int_0^t xe^{-x^2}dx$

$= \lim_{t\to\infty}\left(-\frac{1}{2}\right)e^{-x^2}\Big|_0^1$

$= \lim_{t\to\infty}\left(\frac{-1}{2}e^{-t^2}+\frac{1}{2}e^0\right)=\frac{1}{2}$

15. $\int_{-\infty}^0 -2xe^x dx$

$= \lim_{t\to-\infty}\int_t^0 -2xe^x dx$

let $u = x$ $\qquad dv = 2e^x dx$

then $du = dx$ $\qquad v = -2e^x$

$= \lim_{t\to-\infty}\left(-2xe^x\Big|_t^0 - \int_t^0 -2e^x dx\right)$

$= \lim_{t\to-\infty}\left(-0+2te^t+2e^x\Big|_t^0\right)$

$= \lim_{t\to-\infty}\left(2te^t+2e^0-2e^t\right)$

$= \lim_{t\to-\infty}\left(2te^t\right)+2-0$

$= \lim_{t\to-\infty}\frac{2t}{e^{-t}}+2$

$= \lim_{t\to-\infty}\frac{2}{-e^{-t}}+2$ using 1′ Hopital's Rule

$= 2$

17. $\int_0^1 -x\ln x\,dx$

$$= \lim_{t\to 0^+}\int_t^1 -x\ln x\,dx$$

$$= \lim_{t\to 0^+}\left(\frac{-1}{2}x^2\ln x\Big|_t^1 - \int_t^1 -\frac{1}{2}x\,dx\right)$$

$$= \lim_{t\to 0^+}\left(\frac{-1}{2}(1^2)\ln 1+\frac{1}{2}t^2\ln t+\frac{1}{2}x^2\left(\frac{1}{2}\right)\Big|_t^1\right)$$

$$= \lim_{t\to 0^+}\left(\frac{-1}{2}t^2\ln t+\frac{1}{4}(1^2)-\frac{1}{4}t^2\right)$$

$$= \lim_{t\to 0^+}\left(\frac{1}{2}t^2\ln t\right)+\frac{1}{4}-0$$

$$= \lim_{t\to 0^+}\left(\frac{\ln t}{2t^{-2}}\right)+\frac{1}{4}$$

$$= \lim_{t\to 0^+}\frac{\frac{1}{t}}{-4t^{-3}}+\frac{1}{4}$$ using 1′ Hopital's Rule

$$= \lim_{t\to 0^+}\frac{t^2}{-4}+\frac{1}{4}$$

$$= \frac{1}{4}$$

19. $\int_1^\infty \frac{4}{x^2+1}dx$

$$= \lim_{t\to\infty}\int_1^t \frac{4}{x^2+1}dx$$

$$= \lim_{t\to\infty} 4\arctan(x)\Big|_1^t$$

$$= \lim_{t\to\infty}\left(4\arctan(t)-4\arctan 1\right)$$

$$= 4\left(\frac{\pi}{2}\right)-4\left(\frac{\pi}{4}\right)=2\pi-\pi=\pi$$

Chapter 26 Review

1. $u = 2+\sin 3x$;

$du = 3\cos x\,dx$;

$$\frac{1}{3}\int(2+\sin 3x)^{-1/2}3\cos 3x\,dx$$

$$= \frac{2}{3}\sqrt{2+\sin 3x}+C$$

2. $u = 5+\tan 2x$;

$du = 2\sec^2 2x\,dx$

$$\frac{1}{2}\int(5+\tan 2x)^3\,2\sec^2 2x\,dx = \frac{1}{8}(5+\tan 2x)^4+C$$

3. $(1/3)\int\cos 3x\cdot 3\,dx = (1/3)\sin 3x + C$

4. $u = x^2-5$;

$du = 2x\,dx$;

$$\frac{1}{2}\int\frac{2x\,dx}{x^2-5}=\frac{1}{2}\ln\,|x^2-5|+C$$

5. $u = 3x^2$; $du = 6x\,dx$; $(1/6)\int 6xe^{3x^2}\,dx = (1/6)e^{3x^2} + C$

6. $u = 2x$; $du = 2\,dx$; $a = 3$; $(1/3)$ Arctan $2x/3 + C$

7. $u = x$; $du = dx$; $a = 4$; Arcsin $x/4 + C$

8. $u = 7x + 2$; $du = 7\,dx$; $(1/7)\int \sec^2(7x+2)\cdot 7\,dx = (1/7)\tan(7x+2) + C$

9. $u = 3 + 5\tan x$;

$$\frac{1}{5}\int\frac{5\sec^2 x\,dx}{3+5\tan x} = (1/5)\ln|3 + 5\tan x| + C$$

$du = 5\sec^2 x\,dx$;

10. $u = x^3 + 4$; $du = 3x^2\,dx$; $(1/3)\int 3x^2\sin(x^3+4)\,dx = (-1/3)\cos(x^3+4) + C$

11. $u = 3x$; $du = 3\,dx$; $a = 4$; $\frac{1}{4}\text{Arctan}\frac{3x}{4}\Big|_0^1 = \frac{1}{4}\text{Arctan}\frac{3}{4}$

12. $\frac{1}{\pi}\int_0^{1/2}\sin\pi x\cdot\pi\,dx = -\frac{1}{\pi}\cos\pi x\Big|_0^{1/2} = \frac{1}{\pi}$

13. $\int_0^{\pi/4}\tan x\,dx = -\ln|\cos x|\Big|_0^{\pi/4} = -\ln(1/\sqrt{2})$ or $\ln\sqrt{2}$

14. $u = 2x$;

$du = 2\,dx$; $\quad \frac{1}{2}\int_0^1\frac{2\,dx}{\sqrt{9-4x^2}} = \frac{1}{2}\text{Arcsin}\frac{2x}{3}\Big|_0^1 = \frac{1}{2}\text{Arcsin}\frac{2}{3}$

$a = 3$

15. $u = 4x$

$du = 4\,dx$; $\quad \int\frac{4\,dx}{4x\sqrt{16x^2-9}} = \frac{1}{3}\text{Arcsec}\frac{4x}{3} + C$ or $\frac{1}{3}\text{Arccos}\frac{3}{4x} + C$

$a = 3$

16. $u = \cos 3x$;

$$\int\frac{\tan 3x\,dx}{\sec^4 3x} = -\frac{1}{3}\int\cos^3 3x\ \sin 3x\,(-3)\,dx = \frac{-\cos^4 3x}{12} + C$$

$d = -3\sin 3x\,dx$

17. $u = $ Arctan $3x$;

$$\frac{1}{3}\int\frac{\text{Arctan}\,3x(3)\,dx}{1+9x^2} = \frac{1}{6}\text{Arctan}^2 3x + C$$

$du = \frac{3}{1+9x^2}dx$

18. $u = 5x$;

$du = 5\ dx$;

$(1/5)\ln|\sec 5x + \tan 5x| + C$

19. $(1/2)\int 2x\tan x^2\,dx = (-1/2)\ln|\cos x^2| + C$

20. $\int \sin^5 2x\cos^2 2x\,dx = \int \sin 2x(1-\cos^2 2x)^2\cos^2 2x\,dx$

$= \int(\cos^6 2x\sin 2x - 2\cos^4 2x\sin 2x + \cos^2 2x\sin 2x)\,dx$

$= (-1/2)\int \cos^6 2x(-2\sin 2x) - 2\cos^4 2x(-2\sin 2x) + \cos^2 2x(-2\sin 2x)dx$

$= (-1/14)\cos^7 2x + (1/5)\cos^5 2x - (1/6)\cos^3 2x + C$

21. $\int \cos^4 3x\sin^2 3x\,dx = \int[(1/2)(1+\cos 6x)]^2[(1/2)(1-\cos 6x)]\,dx$

$= (1/8)\int(1 + 2\cos 6x + \cos^2 6x)(1-\cos 6x)\,dx$

$= (1/8)\int(1 + \cos 6x - \cos^2 6x - \cos^3 6x)\,dx$

$= (1/8)\int(1/2 - (1/2)\cos 12x + \sin^2 6x\cos 6x)\,dx$

$= x/16 - (1/192)\sin 12x + (1/144)\sin^3 6x + C$

$1 = A(x-1) + B(x+3)$

22. $u = \cos 4x$; $dv = e^{3x}\,dx$ Then: $u = \sin 4x$; $dv = e^{3x}$

$du = -4\sin 4x\,dx$; $v = (1/3)e^{3x}$ $du = 4\cos 4x\,dx$; $v = (1/3)e^{3x}$

$\int e^{3x}\cos 4x\,dx = (1/3)e^{3x}\cos 4x + (4/3)\int e^{3x}\sin 4x\,dx$

$= (1/3)e^{3x}\cos 4x + (4/3)[(1/3)e^{3x}\sin 4x - \int(1/3)e^{3x}4\cos 4x\,dx]$

$= (1/3)e^{3x}\cos 4x + (4/9)e^{3x}\sin 4x - (16/9)\int e^{3x}4\cos 4x\,dx$

Thus $(25/9)\int e^{3x}\cos 4x\,dx = (1/3)e^{3x}\cos 4x + (4/9)e^{3x}\sin 4x$

And, $\int e^{3x}\cos 4x\,dx = \dfrac{3e^{3x}\cos 4x + 4e^{3x}\sin x}{25} + C = (e^{3x}/25)(3\cos 4x + 4\sin 4x) + C$

23. $u = \cos x$; $-\int -\sin x\ln(\cos x)\,dx = -\cos x\ln|\cos x| + \cos x + C = \cos x(1 - \ln|\cos x|) + C$

24. $\int \tan^4 x\,dx = \int(\sec^2 x - 1)\tan^2 x\,dx = \int(\sec^2 x\tan^2 x - \tan^2 x)dx$

$= \int(\sec^2 x\tan^2 x - \sec^2 x + 1)\,dx = (1/3)\tan^3 x - \tan x + x + C$

25. $\int \cos^2 5x\,dx = \int[(1/2) + (1/2)\cos 10x]dx = x/2 + (1/20)\sin 10x + C$

26. $u = \ln x;\ dv = x^{1/2}\,dx$

$du = (1/x)\,dx;$

$v = (2/3)x^{3/2}$

$$\int \sqrt{x} \ln x\, dx = (2/3)x^{3/2} \ln x - \int (2/3)x^{3/2}(1/x)dx$$

$$= (2/3)x^{3/2} \ln |x| - (4/9)x^{3/2} + C = (2/3)x^{3/2} (\ln |x| - 2/3) + C$$

27. $u = \cos e^{-x};\quad \int e^{-x} (\cos^2 e^{-x})(\sin e^{-x})\, dx = (1/3) \cos^3 e^{-x} + C$

$du = e^{-x} \sin e^{-x}\, dx$

28. $u = x^2;\quad dv = \sin x\, dx;\quad u = 2x;\quad dv = \cos x\, dx;$

$du = 2x\, dx;\quad v = -\cos x;\quad du = 2\, dx;\quad v = \sin x$

$$\int x^2 \sin dx = -x^2 \cos x + \int \cos x \cdot 2x\, dx = -x^2 \cos x + 2x \sin x - \int 2 \sin x\, dx$$

$$= -x^2 \cos x + 2x \sin x + 2 \cos x + C$$

29. $u = \text{Arcsin } 5x;$

$$\frac{1}{5}\int \frac{5 \text{ Arcsin } 5x\, dx}{\sqrt{1-25x^2}} = \frac{1}{10} \text{ Arcsin}^2\, 5x + C$$

$$du = \frac{5\, dx}{\sqrt{1-25x^2}}$$

30. $u = x;$

$$\int_2^3 \frac{dx}{x\sqrt{x-1^2}} = \text{Arcsec } x \Big|_2^3 = \text{Arcsec } 3 - \text{Arcsec } 2$$

$du = dx;$

31. $u = x;\quad dv = e^{4x}\, dx;$

$$\int_0^1 xe^{4x} dx = \frac{1}{4} xe^{4x} - \int \frac{1}{4} e^{4x} dx$$

$du = dx;$

$v = (1/4)e^{4x};$

$$= \frac{1}{4} xe^{4x} - \frac{1}{16} e^{4x} \Big|_0^1 = \frac{3e^4 + 1}{16}$$

32. $$\int_0^{\pi/8} 4 \tan 2x\, dx = -2 \ln|\cos 2x| \Big|_0^{\pi/8} = -2 \ln(1/\sqrt{2}) = \ln 2$$

33. $u = a\sin\theta$;

$3x = 4\sin\theta$; $\int \frac{dx}{x\sqrt{16-9x^2}} = \int \frac{(4/3)\cos\theta\, d\theta}{[(4/3)\sin\theta)\sqrt{16(1-\sin^2\theta)}]}$

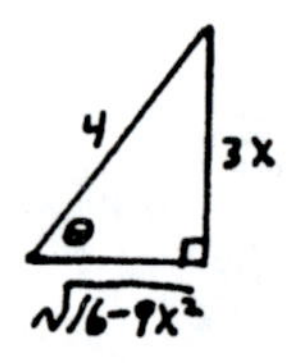

$dx = (4/3)\cos\theta\; d\theta$

$$= \int \frac{\cos\theta\, d\theta}{\sin\theta 4\cos\theta} = \frac{1}{4}\int \csc\theta\, d\theta = \frac{1}{4}\ln|\csc\theta - \cot\theta| + C$$

$$= \frac{1}{4}\ln\left|\frac{4}{3x} - \frac{\sqrt{16-9x^2}}{3x}\right| + C = \frac{1}{4}\ln\left|\frac{4-\sqrt{16-9x^2}}{3x}\right| + C$$

34. $A = \int_2^4 \frac{1}{x-1}dx = \ln|x-1|\Big|_2^4 = \ln 3$

35. $A = \int_1^3 e^{x+2}dx = e^{x+2}\Big|_1^3 = e^5 - e^3 = e^3(e^2-1)$

36. $A = \int_0^1 e^{2x}dx = \frac{1}{2}\int 2e^{2x}dx = \frac{1}{2}e^{2x}\Big|_0^1 = \frac{1}{2}e^2 - \frac{1}{2} = \frac{1}{2}(e^2-1)$

37. $A = \int_0^1 \frac{1}{1+x^2}dx = \text{Arctan } x\Big|_0^1 = \text{Arctan } 1 = \frac{\pi}{4}$

38. $A = \int_0^1 \frac{dx}{\sqrt{4-x^2}} = \text{Arcsin}\frac{x}{2}\Big|_0^1 = \frac{\pi}{6}$

39. $A = \int_0^1 x\,dx + \int_1^2 \frac{1}{x}dx = \frac{x^2}{2}\Big|_0^1 + \ln|x|\Big|_1^2 = \frac{1}{2} + \ln 2$

40. $A = \int_0^{\pi/4} \sec^2 x\,dx = \tan x\Big|_0^{\pi/4} = 1$

41. $A = \int_0^2 \frac{1}{4+x^2}dx = \frac{1}{2}\text{Arctan}\frac{x}{2}\Big|_0^2 = \frac{\pi}{8}$

42. $A = \int_0^{\pi/4} \tan x\,dx = -\ln|\cos x|\Big|_0^{\pi/4} = -\ln(1/\sqrt{2}) = \ln\sqrt{2} = \frac{1}{2}\ln 2$

43. $u = \ln^2 x$; $dv = dx$; $u = \ln x$; $dv = dx$

$du = 2\ln x \cdot (1/x)\,dx$; $v = x$; $du = (1/x)\,dx$; $v = x$;

$$V = \pi \int_1^2 \ln^2 x\,dx = \pi\left\{x\ln^2 x - \int x(2/x)\ln x\,dx\right\} = \pi(x\ln^2 x - 2\int \ln x\,dx)$$

$$= \pi[x\ln^2 x - 2(x\ln x - \int x(1/x)\,dx)]$$

$$= \pi(x\ln^2 x - 2x\ln x + 2x)\Big|_1^2 = \pi(2\ln^2 2 - 4\ln 2 + 2)$$

44. $V = \pi \int_0^1 e^{2x}\,dx = \frac{\pi}{2}e^{2x}\Big|_0^1 = \frac{\pi}{2}(e^2 - 1)$

45. $u = t$; $dv = \sin 3t\,dt$

$du = dt$; $v = (-1/3)\cos 3t$

$q = \int 4t\sin 3t\,dt = 4[(-1/3)t\cos 3t + \int (1/3)\cot 3t\,dt = (-4/3)t\cos 3t + (4/9)\sin 3t + C$

46. $u = x^2$

$du = 2x\,dx$; $F = \frac{1}{2}\int_1^2 2xe^{x^2}\,dx = \frac{1}{2}e^{x^2}\Big|_1^2 = \frac{1}{2}(e^4 - e) = \frac{e}{2}(e^3 - 1)$

47. Entry 5. $u = x, a = 5, b = 3$

$$\int \frac{dx}{x(5+3x)} = \frac{1}{5}\ln\left|\frac{x}{5+3x}\right| + C$$

48. Entry 29. $u = x, a = 4$.

$$\int \frac{\sqrt{16-x^2}}{x^2}\,dx = \frac{-\sqrt{16-x^2}}{x} - \arcsin\frac{x}{4} + C$$

49. Entry 47. $u = x, a = 3, b = 6, c = 1$

$$\int \frac{dx}{\sqrt{3+6x+x^2}} = \ln|2x + 6 + 2\sqrt{3+6x+x^2} + C$$

50. Entry 78. $u = x, a = 1, n = 2$

$$\int e^x \sin 2x\,dx = \frac{e^x}{5}(\sin 2x - 2\cos 2x) + C$$

51. Entry 33. $u = 2x, du = 2dx, a = 3$

$$\int \frac{\sqrt{4x^2-9}}{x}\,dx = \int \frac{\sqrt{(2x)^2-9}}{2x}\cdot 2dx$$

$$= \sqrt{4x^2-9} - 3\arccos\frac{3}{2x} + C$$

$$= \sqrt{4x^2-9} - 3\,\text{arcsec}\frac{2x}{3} + C$$

52. Entry 73. $u = 3x \qquad du = 3\,dx, \qquad n = 4$

$$\int \cos^4 3x \sin 3x\, dx = \frac{1}{3}\int \cos^4 3x \sin 3x\; 3dx$$

$$= \frac{1}{3}\left(\frac{-\cos^5 3x}{5}\right) + C = -\frac{1}{15}\cos^5 3x + C$$

53. Entry 17. $u = x, a = 9, b = 4$

$$\int \frac{\sqrt{9+4x}}{x}\,dx = 2\sqrt{9+4x} + 9\int \frac{dx}{x\sqrt{9+4x}} + C$$

Now use entry 15. $u = x, a = 9, b = 4$

$$= 2\sqrt{9+4x} + 9\left\{\frac{1}{3}\ln\left|\frac{\sqrt{9+4x}-3}{\sqrt{9+4x}+3}\right|\right\} + C$$

$$= 2\sqrt{9+4x} + 3\ln\left|\frac{\sqrt{9+4x}-3}{\sqrt{9+4x}+3}\right| + C$$

54. Entry 85 (three times). $u = x, n = 6$ for first time.

$$\int \tan^6 x\, dx = \frac{\tan^5 x}{5} - \int \tan^4 x\, dx + C$$

$$= \frac{1}{5}\tan^5 x - \left[\frac{1}{3}\tan^3 x\, dx - \int \tan^2 x\, dx\right] + C$$

$$= \frac{1}{5}\tan^5 x - \frac{1}{3}\tan^3 x + \tan x - \int dx + C$$

$$= \frac{1}{5}\tan^5 x - \frac{1}{3}\tan^3 x + \tan x - x + C$$

55. $\frac{1/2}{2}[f(1) + 2f(3/2) + 2f(2) + 2f(5/2) + 2f(3) + 2f(7/2) + f(4)]$

$= (1/4)[1 + 1 + 2/3 + 1/2 + 2/5 + 1/3 + 1/7] = 1.011$

56. $\frac{1/2}{2}[f(1) + 2f(3/2) + 2f(2) + 2f(5/2) + f(3)]$

$= (1/4)[0.1 + 0.178 + 0.154 + 0.131 + 0.056] = 0.155$

57. $\frac{1/2}{2}[f(0) + 2f(1/2) + 2f(1) + 2f(3/2) + 2f(2) + 2f(5/2) + 2f(3) + 2f(7/2) + f(4)]$

$= (1/4)[4 + 7.937 + 7.746 + 7.416 + 6.928 + 6.245 + 5.292 + 3.873 + 0] = 12.395$

58. $\frac{1/2}{2}[f(1) + 2f(3/2) + 2f(2) + 2f(5/2) + 2f(3) + 2f(7/2) + f(4)]$

$= (1/4)[2.571 + 5.013 + 4.820 + 4.547 + 4.160 + 3.583 + 1.260] = 6.489$

59. $(2/2)[2.3 + 2(2.8) + 2(3.4) + 2.1] = 22.2$

60. $(0.2/2)[0.9 + 2(0.7) + 2(0.8) + 2(1.1) + 2(1.3) + 0.9] = 0.96$

61. $\frac{1/2}{2}[f(0)+4f(1/2)+2f(1)+4f(3/2)+2f(2)+4f(5/2)+2f(3)+4f(7/2)+f(4)]$

$= (1/6)[4 + 16.125 + 8.246 + 17.088 + 8.944 + 18.868 + 10 + 21.260 + 5.657] = 18.365$

62. $(2/3)[f(0)+4f(2)+2f(4)+4f(6)+2f(8)+4f(10)+f(12)]$

$= (2/3)[4 + 3.2 + 0.471 + 0.432 + 0.123 + 0.158 + 0.028] = 5.608$

63. $\frac{\pi/8}{3}[f(0)+4f(\pi/8)+2f(\pi/4)+4f(3\pi/8)+f(\pi/2)]$

$= (\pi/24)(1/2 + 1.679 + 0.739 + 1.368 + 1.3) = 0.605$

64. $(1/3)[f(0)+4f(1)+2f(2)+4f(3)+2f(4)+4f(5)+2f(6)+4f(7)+f(8)]$

$= (1/3)[0 + 2.828 + 5.333 + 20.410 + 15.876 + 45.544 + 29.326 + 73.973 + 22.605] = 71.632$

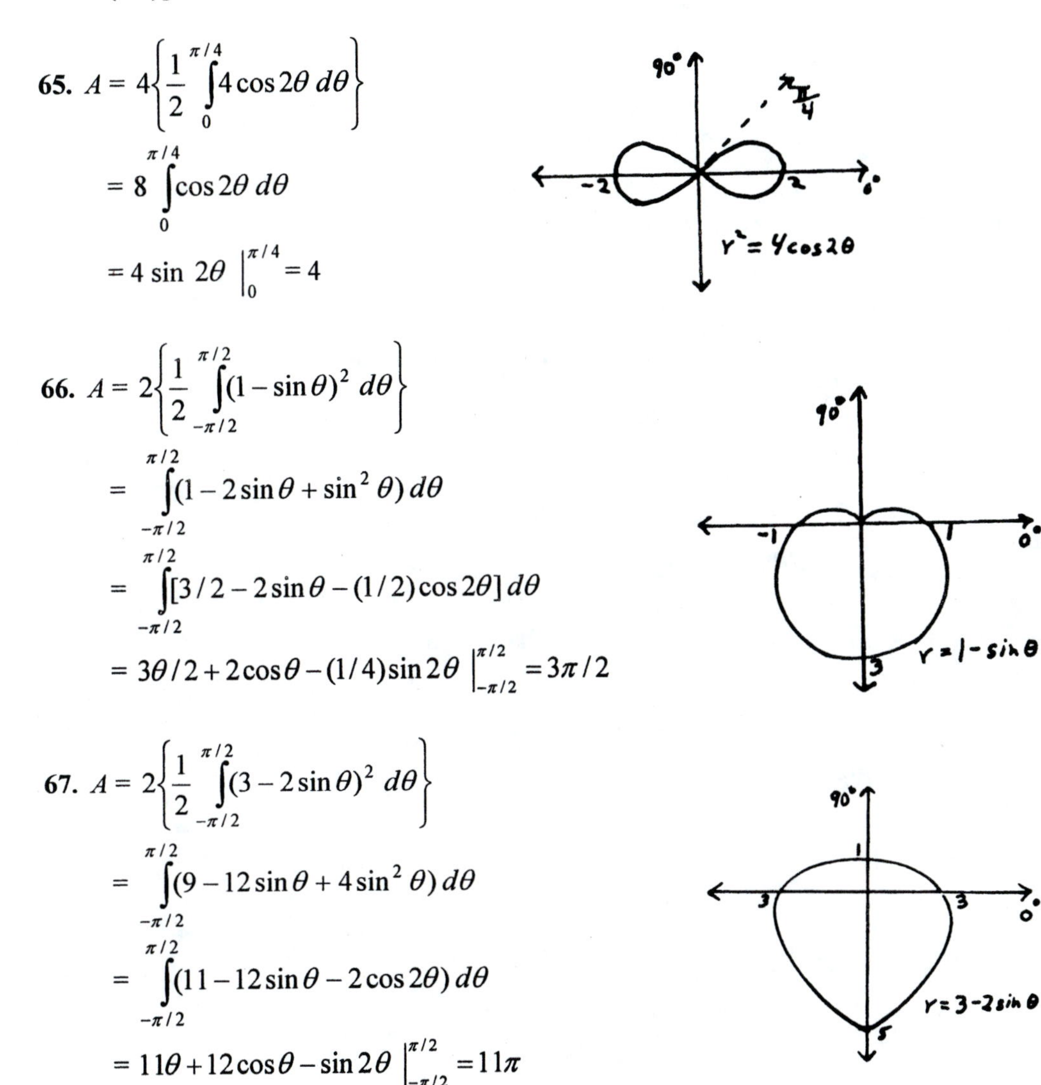

65. $A = 4\left\{\frac{1}{2}\int_0^{\pi/4} 4\cos 2\theta \, d\theta\right\}$

$= 8\int_0^{\pi/4}\cos 2\theta \, d\theta$

$= 4\sin 2\theta \Big|_0^{\pi/4} = 4$

66. $A = 2\left\{\frac{1}{2}\int_{-\pi/2}^{\pi/2}(1-\sin\theta)^2 \, d\theta\right\}$

$= \int_{-\pi/2}^{\pi/2}(1-2\sin\theta+\sin^2\theta)\, d\theta$

$= \int_{-\pi/2}^{\pi/2}[3/2-2\sin\theta-(1/2)\cos 2\theta]\, d\theta$

$= 3\theta/2+2\cos\theta-(1/4)\sin 2\theta \Big|_{-\pi/2}^{\pi/2} = 3\pi/2$

67. $A = 2\left\{\frac{1}{2}\int_{-\pi/2}^{\pi/2}(3-2\sin\theta)^2 \, d\theta\right\}$

$= \int_{-\pi/2}^{\pi/2}(9-12\sin\theta+4\sin^2\theta)\, d\theta$

$= \int_{-\pi/2}^{\pi/2}(11-12\sin\theta-2\cos 2\theta)\, d\theta$

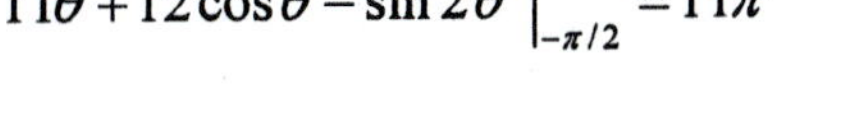

$= 11\theta+12\cos\theta-\sin 2\theta \Big|_{-\pi/2}^{\pi/2} = 11\pi$

Chapter 26 Review

68. $A = 6\left\{\frac{1}{2}\int_0^{\pi/6}[(1/2)\cos 3\theta]^2\,d\theta\right\}$

$= \frac{3}{4}\int_0^{\pi/6}\cos^2 3\theta\,d\theta$

$= \frac{3}{8}\int_0^{\pi/6}(1+\cos 6\theta)\,d\theta = \frac{3}{8}[\theta + (1/6)\sin 6\theta\ \Big|_0^{\pi/6} = \frac{\pi}{16}$

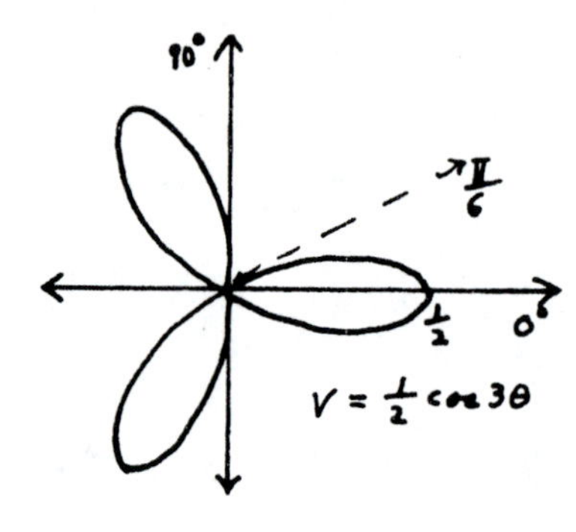

69. $A = \frac{1}{2}\int_0^{\pi}\{[(1/2)(\theta+\pi)]^2 - \theta^2\}\,d\theta = \frac{1}{2}\int_0^{\pi}\left\{-\frac{3}{4}\theta^2 + \frac{1}{2}\pi\theta + \frac{\pi^2}{4}\right\}d\theta$

$= \frac{1}{2}\left\{-\frac{1}{4}\theta^3 + \frac{1}{4}\pi\theta^2 + \frac{\pi^2}{4}\theta\right\}\Bigg|_0^{\pi} = \frac{\pi^3}{8}$

70. $1 = 2\cos\theta$; $\cos\theta = 1/2$;
$\theta = \pi/3, 5\pi/3$

$A = 2\left\{\frac{1}{2}\int_0^{\pi/3}[(2\cos\theta)^2 - 1^2]\,d\theta\right\} = \int_0^{\pi/3}(4\cos^2\theta - 1)\,d\theta$

$= \int_0^{\pi/3}(1 + 2\cos 2\theta)\,d\theta = (\theta + \sin 2\theta)\ \Big|_0^{\pi/3} = \frac{\pi}{3} = \frac{\sqrt{3}}{2}$

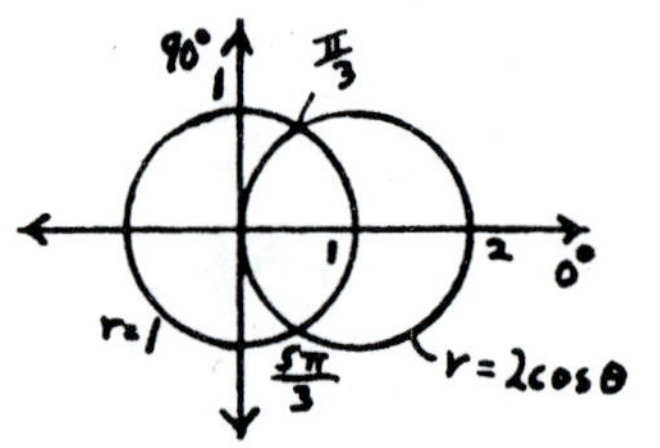

71. $A = 2\left\{\frac{1}{2}\int_0^{\pi/3}(2-4\cos\theta)^2\,d\theta\right\} = 4\int_0^{\pi/3}(1 - 4\cos\theta + 4\cos^2\theta)\,d\theta$

$= 4\int_0^{\pi/3}(3 - 4\cos\theta + 2\cos 2\theta)\,d\theta = 4(3\theta - 4\sin\theta + \sin 2\theta)\ \Big|_0^{\pi/3}$

$= 4\pi - 6\sqrt{3}$

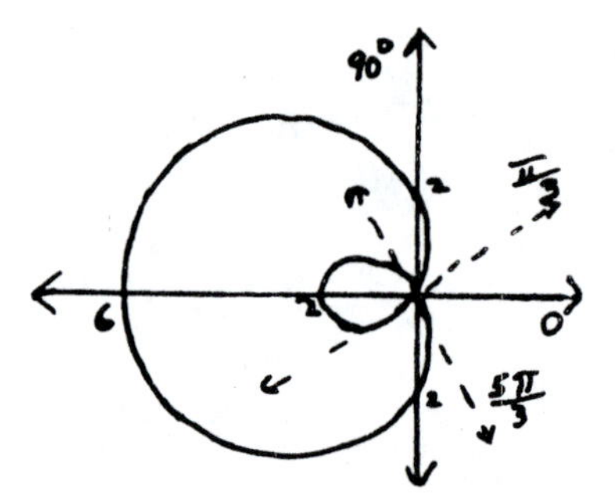

72. $A = 2\left\{\frac{1}{2}\int_{-\pi/2}^{0}[1^2 - (1+\sin\theta)^2]\,d\theta\right\} = \int_{-\pi/2}^{0}(-2\sin\theta - \sin^2\theta)\,d\theta$

$= \int_{-\pi/2}^{0}[-2\sin\theta - (1/2) + (1/2)\cos 2\theta]\,d\theta$

$= [2\cos\theta - \theta/2 + (1/4)\sin 2\theta]\ \Big|_{-\pi/2}^{0} = 2 - \frac{\pi}{4} = \frac{8-\pi}{4}$

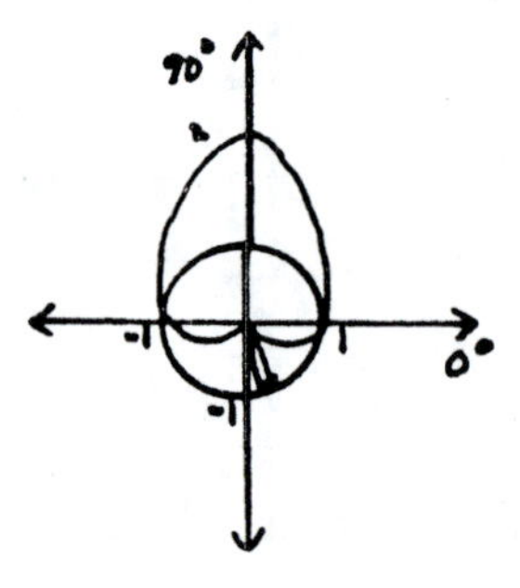

73. $r = \sqrt{8\sin 2\theta} = 2;$

$\sin 2\theta = 4/8 = 1/2$

$2\theta = \pi/6, 5\pi/6, 13\pi/6, 17\pi/6$

$\theta = \pi/12, 5\pi/12, 13\pi/12, 17\pi/12$

$$A = 4\left\{\frac{1}{2}\int_{\pi/4}^{5\pi/12}(8\sin 2\theta - 2^2)\,d\theta\right\} = 8\int_{\pi/4}^{5\pi/12}(2\sin 2\theta - 1)\,d\theta$$

$$= 8(-\cos 2\theta - \theta)\Big|_{\pi/4}^{5\pi/12} = 4\sqrt{3} - 4\pi/3$$

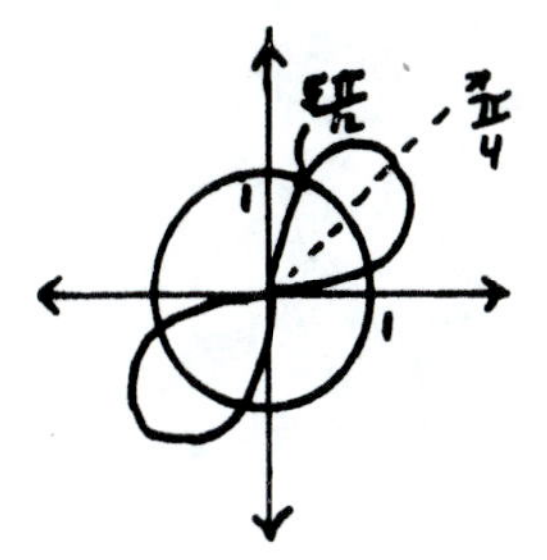

74. $\cos 2\theta = \sin 2\theta;\ \tan 2\theta = 1$

$2\theta = \pi/4, 5\pi/4, \ldots$

$\theta = \pi/8, 5\pi/8, \ldots$

$$A = 2\left\{\frac{1}{2}\int_{0}^{\pi/8}\sin 2\theta\,d\theta + \frac{1}{2}\int_{\pi/8}^{\pi/4}\cos 2\theta\,d\theta\right\} = -\frac{1}{2}\cos 2\theta\,\Big|_{0}^{\pi/8} + \frac{1}{2}\sin 2\theta\,\Big|_{\pi/8}^{\pi/4} = 1 - \frac{\sqrt{2}}{2}$$

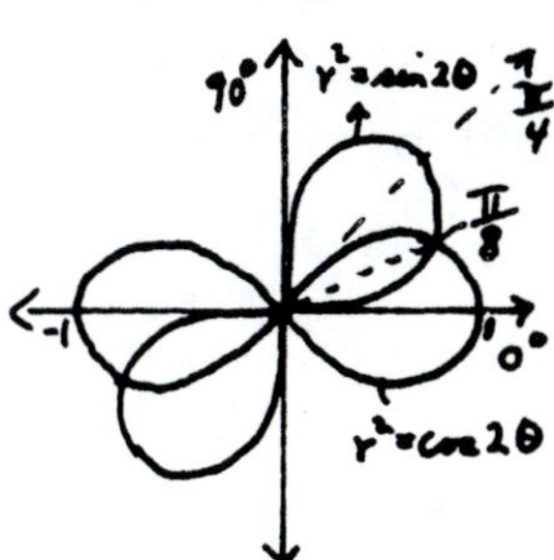

75. $\int_1^{\infty}\frac{1}{x^{4/3}}dx =$

$\lim\limits_{t\to\infty}\int_1^t x^{-4/3}dx =$

$\lim\limits_{t\to\infty}\left(-3x^{-1/3}\right)\Big|_1^t =$

$\lim\limits_{t\to\infty}\left(-3t^{-1/3} + 3(1)^{-1/3}\right) =$

$\lim\limits_{t\to\infty}\left(\frac{-3}{t^{1/3}} + 3\right) = 3$

76. $\int_1^{\infty}\frac{1}{x^{2/3}}dx =$

$\lim\limits_{t\to\infty}\int_1^t x^{-2/3}dx =$

$\lim\limits_{t\to\infty}3x^{1/3}\Big|_1^t =$

$\lim\limits_{t\to\infty}\left(3t^{1/3} - 3\right) = \infty$

diverges

77. $\int_{-\infty}^{0} e^{x/2} dx =$

$\lim_{t\to-\infty} \int_t^0 2e^{x/2}\left(\frac{1}{2}\right)dx =$

$\lim_{t\to-\infty} \left(2e^{x/2}\right)\Big|_t^0 =$

$\lim_{t\to-\infty} \left(2-2e^{t/2}\right) = 2$

78. $\int_{-\infty}^{0} xe^{x} dx$

$= \lim_{t\to-\infty} \int_t^0 xe^x dx$

let $u = x \qquad dv = e^x\, dx$

then $du = dx \qquad v = e^x$

$= \lim_{t\to-\infty} \left(xe^x\Big|_t^0 - \int_t^0 e^x dx\right)$

$= \lim_{t\to-\infty} \left(0 - te^t - (e^0 - e^t)\right)$

$= \lim_{t\to-\infty} \left(-te^t - 1 + e^t\right)$

$= \lim_{t\to-\infty} \left(-te^t\right) - 1$

$= \lim_{t\to-\infty} \left(\frac{-t}{e^{-t}}\right) - 1$ using 1′ Hopital's Rule

$= \lim_{t\to-\infty} \frac{-1}{-e^{-t}} - 1$

$= -1$

79. $\int_3^4 \frac{1}{\sqrt{x-3}} dx =$

$\lim_{t\to3^+} \int_1^4 (x-3)^{-1/2} dx =$

$\lim_{t\to3^+} \left(2(x-3)^{1/2}\right)\Big|_t^4 =$

$\lim_{t\to3^+} \left(2(1) - 2(t-3)^{1/2}\right) = 2$

80. $\int_1^2 \frac{1}{x\ln x} dx$

$\lim_{t\to1^+} \int_1^2 (\ln x)^{-1}\left(\frac{1}{x}\right)dx$

$\lim_{t\to1^+} \left(\ln(\ln x)\right)\Big|_t^2$

$\lim_{t\to1^+} \left(\ln\ln 2 - \ln(\ln t)\right)$

$\ln\ln 2 - (-\infty) = \infty$

diverges

Chapter 27, Exercises 27.1

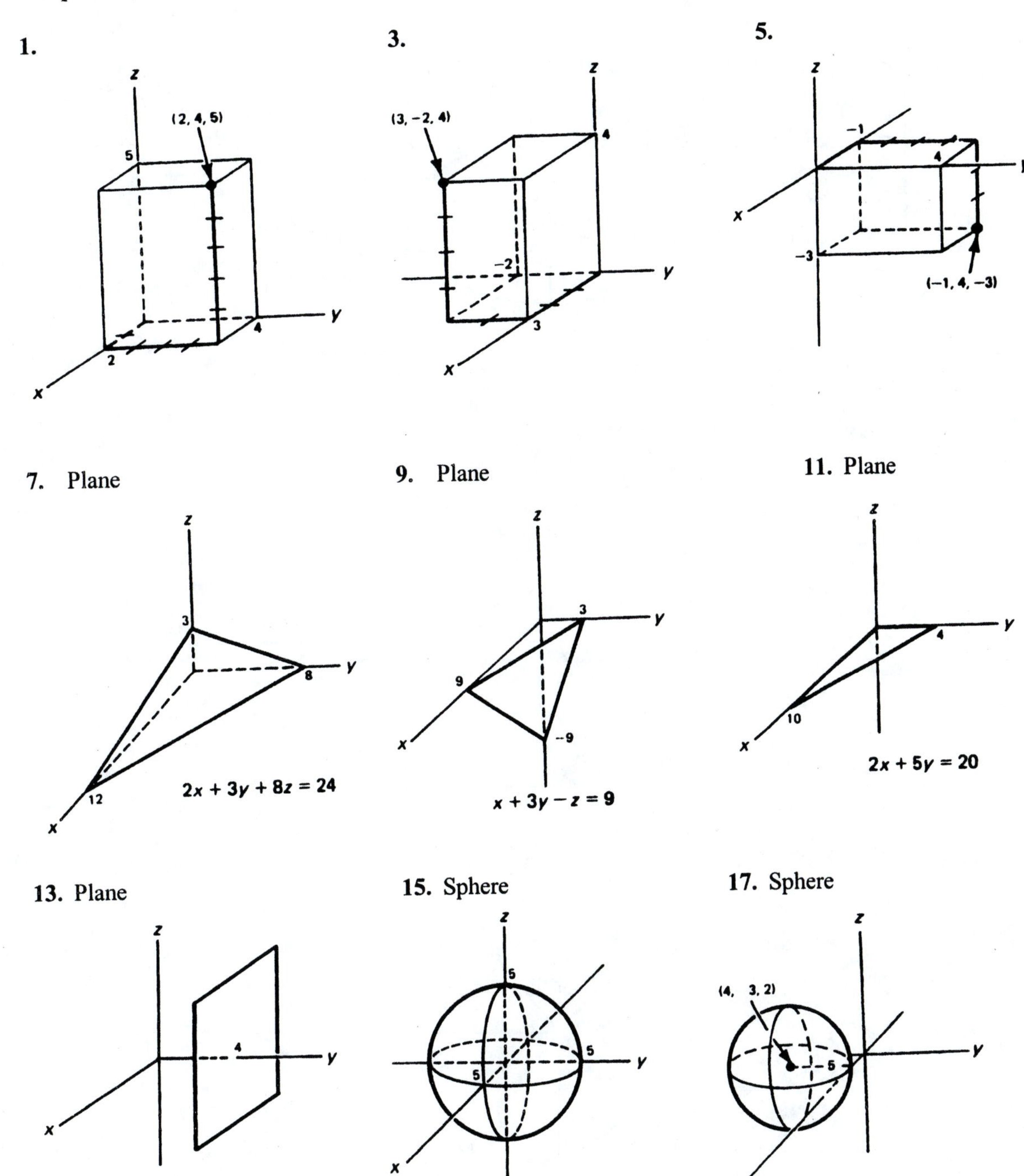

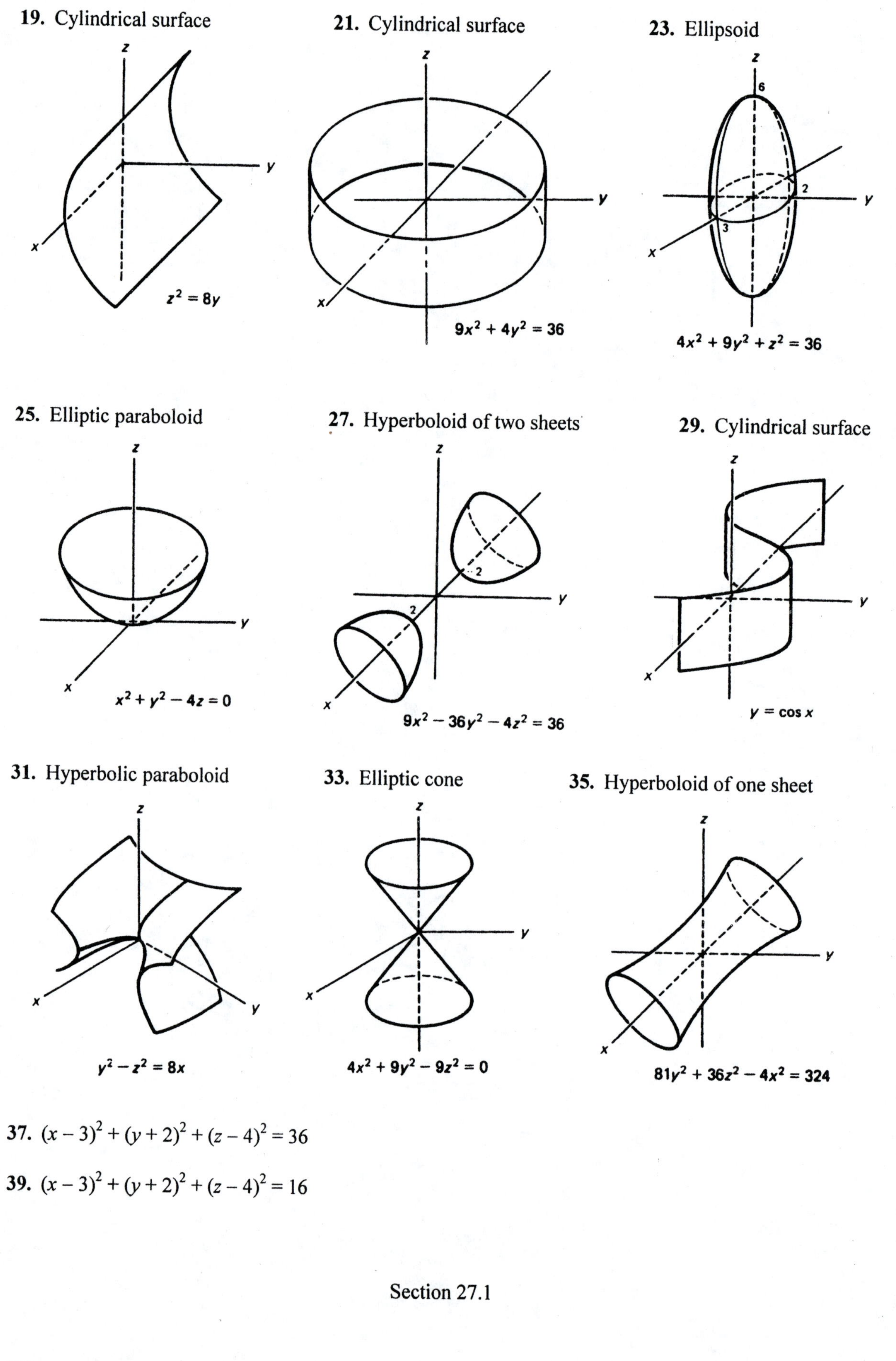

37. $(x-3)^2 + (y+2)^2 + (z-4)^2 = 36$

39. $(x-3)^2 + (y+2)^2 + (z-4)^2 = 16$

Section 27.1

41. Consider two points $P_1(x_1, y_1, z_1)$ and $P_2(x_2, y_2, z_2)$ as shown in the figure at the right. These points determine a rectangular box with P_1 and P_2 as opposite vertices and with edges parallel to the coordinate axes. Triangles P_1RQ and P_1QP_2 are right triangles. By the Pythagorean theorem:

$d^2 = |P_1Q|^2 + |QP_2|^2$ and $|P_1Q|^2 = |P_1R|^2 + |RQ|^2$

Thus $d^2 = |P_1R|^2 + |RQ|^2 + |QP_2|^2$

$= (x_2 - x_1)^2 + (y_2 - y_1)^2 + (z_2 - z_1)^2$

And, $d = \sqrt{(x_2 - x_1)^2 + (y_2 - y_1)^2 + (z_2 - z_1)^2}$

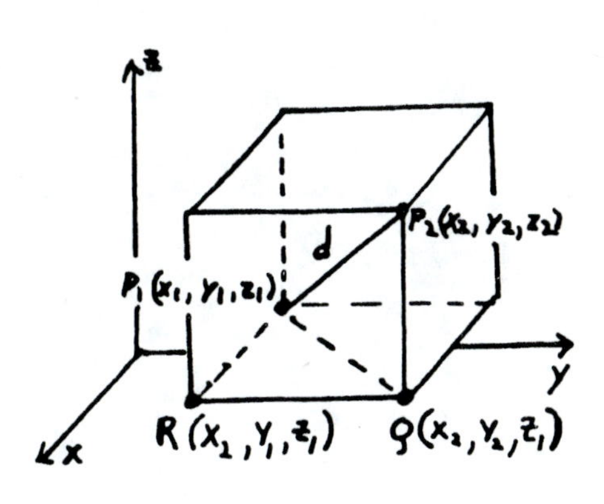

43. $d = \sqrt{(3-5)^2 + (1-(-3))^2 + (-2-2)^2} = \sqrt{36} = 6$

45. $d_1 = \sqrt{(5-2)^2 + (6-8)^2 + (3-4)^2} = \sqrt{14}$

$d_2 = \sqrt{(5-3)^2 + (6-5)^2 + (3-6)^2} = \sqrt{14}$

$d_3 = \sqrt{(2-3)^2 + (8-5)^2 + (4-6)^2} = \sqrt{14}$

Thus the triangle is equilateral.

Exercises 27.2

1. **a)** $\partial z/\partial x = 12x^2y^2$ **b)** $\partial z/\partial y = 8x^3y$

3. **a)** $\partial z/\partial x = 12xy^4 + 2y^2$ **b)** $\partial z/\partial y = 24x^2y^3 = 4xy$

5. **a)** $\partial z/\partial x = (1/2)(x^2 + y^2)^{-1/2}(2x) = \dfrac{x}{\sqrt{x^2 + y^2}}$ **b)** $\dfrac{\partial z}{\partial y} = \dfrac{y}{\sqrt{x^2 + y^2}}$

7. **a)** $\dfrac{\partial z}{\partial x} = \dfrac{(2xy)(2x) - (x^2 - y^2)(2y)}{(2xy)^2} = \dfrac{4x^2y - 2x^2y + 2y^3}{4x^2y^2} = \dfrac{2y(x^2 + y^2)}{4x^2y^2} = \dfrac{x^2 + y^2}{2x^2y}$

b) $\dfrac{\partial z}{\partial y} = \dfrac{(2xy)(-2y) - (x^2 - y^2)(2x)}{(2xy)^2} = \dfrac{-4xy^2 - 2x^3 + 2xy^2}{4x^2y^2} = \dfrac{2x(-y^2 - x^2)}{4x^2y^2} = \dfrac{-y^2 - x^2}{2xy^2}$

9. **a)** $\partial z/\partial x = ay$ **b)** $\partial z/\partial x = ax$

11. **a)** $\dfrac{\partial z}{\partial x} = \dfrac{y}{x} \cdot \dfrac{1}{y} = \dfrac{1}{x}$ **b)** $\dfrac{\partial z}{\partial y} = \dfrac{y}{x}(-xy^{-2}) = -\dfrac{1}{y}$

13. **a)** $\partial z/\partial x = \sec^2(x - y)$ **b)** $\partial z/\partial y = -\sec^2(x - y)$

15. **a)** $\partial z/\partial x = e^{3x}y \cos xy + 3e^{3x} \sin xy = e^{3x}(y \cos xy + 3 \sin xy)$

b) $\partial z/\partial y = e^{3x}x \cos xy$

17. a) $\dfrac{\partial z}{\partial x}=\dfrac{xy\,e^{xy}\sin y-e^{xy}\sin y}{(x\sin y)^2}=\dfrac{e^{xy}\sin y(xy-1)}{x^2\sin^2 y}=\dfrac{e^{xy}(xy-1)}{x^2\sin y}$

b) $\dfrac{\partial z}{\partial y}=\dfrac{x^2\,e^{xy}\sin y-e^{xy}x\cos y}{(x\sin y)^2}=\dfrac{x\,e^{xy}(x\sin y-\cos y)}{x^2\sin^2 y}=\dfrac{e^{xy}(x\sin y-\cos y)}{x\sin^2 y}$

19. a) $\partial z/\partial x=-2\sin x\sin y$ **b)** $\partial z/\partial y=2\cos x\cos y$

21. a) $\partial z/\partial x=xy^2\sec^2 xy+y\tan xy$ **b)** $\partial z/\partial y=x^2y\sec^2 xy+x\tan xy$

23. $\partial P/\partial I=2IR$ 25. $\dfrac{\partial I}{\partial R}=\dfrac{-E}{(R+r)^2}$ 27. $\dfrac{\partial z}{\partial R}=\dfrac{R}{\sqrt{R^2+X_L^2}}$

29. $\partial e/\partial t=2\pi fE\cos 2\pi ft$ 31. $\partial E/\partial I_2=R_2+R_3$

33. $\dfrac{\partial I}{\partial t}=\dfrac{-E}{R^2C}e^{-t/RC}$

35. $\dfrac{\partial q}{\partial C}=\dfrac{Et}{RC}e^{-t/RC}+E\,e^{-t/RC}=E\,e^{-t/RC}\left[\dfrac{t}{RC}+1\right]$

37. $\dfrac{\partial \phi}{\partial R}=\dfrac{1}{(X_L/R^2)+1}\cdot\dfrac{-X_L}{R^2}=\dfrac{R^2}{X_L^2+R^2}\cdot\dfrac{-X_L}{R^2}=\dfrac{-X_L}{X_L^2+R^2}$

39. a) $\partial z/\partial x=18x\Big|_{x=1}=18$ **b)** $\partial z/\partial y=8y\Big|_{y=-2}=-16$

41. a) $\dfrac{\partial z}{\partial x}=\dfrac{25x}{\sqrt{25x^2+36y^2+164}}\Bigg|_{(1,-1,15)}=\dfrac{25}{15}=\dfrac{5}{3}$

b) $\dfrac{\partial z}{\partial y}=\dfrac{36y}{\sqrt{25x^2+36y^2+164}}\Bigg|_{(1,-1,15)}=-\dfrac{36}{15}=-\dfrac{12}{5}$

43. $\dfrac{\partial P}{\partial V}\cdot\dfrac{\partial V}{\partial T}\cdot\dfrac{\partial T}{\partial P}=\dfrac{-nRT}{V^2}\cdot\dfrac{nR}{P}\cdot\dfrac{V}{nR}=\dfrac{-nRT}{VP}=-\dfrac{V}{V}=-1.$

Note: $V=\dfrac{nRT}{P}$

45. $\partial V/\partial r=16\pi r\Big|_{r=6}=96\pi\text{ cm}^3$

47. $s^2\,\partial w/\partial s + t\,\partial w/\partial t = s^2[(-t\,e^{1/s}/s^2)\sec^2 te^{1/s}] + t(2t + e^{1/s}\sec^2 te^{1/s})$
$= -t\,e^{1/s}\sec^2 te^{1/s} + 2t^2 + t\,e^{1/s}\sec^2 te^{1/s} = 2t^2$

Exercises 27.3

1. $dz = (6x + 4y)\,dx + (4x + 3y^2)\,dy$

3. $dz = 2x\cos y\,dx - x^2\sin y\,dy$

5. $dz = \dfrac{(xy)(1) - (x-y)y}{(xy)^2}dx + \dfrac{(xy)(-1) - (x-y)x}{(xy)^2}dy = \dfrac{1}{x^2}dx - \dfrac{1}{y^2}dy$

7. $dz = \dfrac{1}{\sqrt{1+xy}}\cdot\dfrac{1}{2}(1+xy)^{-1/2}(y)dx + \dfrac{1}{\sqrt{1+xy}}\cdot\dfrac{1}{2}(1+xy)^{-1/2}(x)dy = \dfrac{y}{2(1+xy)}dx + \dfrac{x}{2(1+xy)}dy$

9. $dV = lw\,dh + lh\,dw + wh\,dl = (26)(26)(0.15) + (26)(12)(0.15) + (26)(12)(0.15) = 195\text{ cm}^3$

11. $dR = \dfrac{R_2^2}{(R_1+R_2)^2}dR_1 + \dfrac{R_1^2}{(R_1+R_2)^2}dR_2 = \dfrac{(600)^2}{(400+600)^2}(25) + \dfrac{(400)^2(50)}{(400+600)^2} = 17\Omega$

13. $dV = (2/3)\pi rh\,dr + (1/3)\pi r^2\,dh = (2/3)\pi(12.00)(21.00)(-0.15) + (1/3)\pi(12.00)^2(0.10) = -64.1\text{ cm}^3$

15. $f(x, y) = z = 9 + 6x - 8y - 3x^2 - 2y^2$

$f_x(x, y) = 6 - 6x$; $f_y(x, y) = -8 - 4y$

Set each expression equal to zero and solve the resulting system of equations.

$6 - 6x = 0$

$-8 - 4y = 0$

Thus $x = 1$ and $y = -2$. Substitute these values in $z = f(x, y)$ to find

$z = 9 + 6(1) - 8(-2) - 3(1)^2 - 2(-2)^2 = 20.$

Next determine if $(1, -2, 20)$ is a relative maximum or a relative minimum.

$f_{xx}(x, y) = -6$; $f_{yy}(x, y) = -4$; $f_{xy}(x, y) = f_{yx}(x, y) = 0.$

$D(x, y) = f_{xx}(x, y)\,f_{yy}(x, y) - [f_{xy}(x, y)]^2$

$D(1, -2) = -6(-4) - 0 = 24$

Since $D(1, -2) = 24 > 0$ and $f_{xx}(1, 2) = -6 < 0$, then $(1, -2, 20)$ is a relative maximum point.

17. $f(x, y) = z = \dfrac{1}{x} + \dfrac{1}{y} + xy$

$f_x(x, y) = -x^{-2} + y$; $f_y(x, y) = -y^{-2} + x$

Set each expression equal to zero and solve the resulting system of equations.

$\dfrac{-1}{x^2} + y = 0$

$-\dfrac{1}{y^2} + x = 0$

$x = 0$ or 1, but $f(x, y)$ is not defined as $x = 0$.

So $x = 1$ and $y = 1$. Thus $z = \frac{1}{1} + \frac{1}{1} + (1) = 3$.

Next determine if (1, 1, 3) is a relative maximum or a relative minimum.

$f_{xx}(x, y) = 2x^{-3}$; $f_{yy}(x, y) = 2y^{-3}$

$f_{xy}(x, y) = f_{yx}(x, y) = 1$

$D(x, y) = f_{xx}(x, y)f_{yy}(x, y) - [f_{xy}(x, y)]^2$

$D(1, 1) = 2(2) - 1(1) = 3$

Since $D(1, 1) = 3 > 0$ and $f_{xx}(1, 1) = 2 > 0$,

(1, 1, 3) is a relative minimum point.

19. $f(x, y) = z = x^2 - y^2 - 2x - 4y - 4$

$f_x(x, y) = 2x - 2$; $f_y(x, y) = -2y - 4$

Set each expression equal to zero and solve the resulting system of equations.

$$2x - 2 = 0$$
$$-2y - 4 = 0$$

Thus $x = 1$ and $y = -2$. $z = 1^2 - (-2)^2 - 2(1) - 4(-2) - 4 = -1$.

Next determine if (1, −2, −1) is a relative maximum or a relative minimum.

$f_{xx}(x, y) = 2$; $f_{yy}(x, y) = -2$; $f_{xy}(x, y) = 0$

$D(x, y) = f_{xx}(x, y)f_{yy}(x, y) - [f_{xy}(x, y)]^2$

$D(1, -2) = 2(-2) - 0^2 = -4$.

Since $D(1, -2) = -4 < 0$, (1, −2,−1) is a saddle point.

21. $f(x, y) = z = x^2 - y^2 - 6x + 4y$

$f_x(x, y) = 2x - 6$

$f_y(x, y) = -2y + 4$

Set each expression equal to zero and solve the resulting system of equations.

$$2x - 6 = 0$$
$$-2y + 4 = 0$$

Thus, $x = 3$ and $y = 2$. Substitute these values in $z = f(x, y)$ to find

$z = 3^2 - 2^2 - 6(3) + 4(2) = -5$.

Next determine if (3, 2, −5) is a relative maximum or minimum.

$f_{xx}(x, y) = 2$; $f_{yy}(x, y) = -2$, $f_{xy}(x, y) = f_{xy}(x, y) = 0$

$D(x, y) = f_{xx}(x, y)\, f_{yy}(x, y) - [f_{xy}(x, y)]^2$

$= 2(-2) - 0 = -4$

Since $D(3, 2) = -4 < 0$, (3, 2, −5) is a saddle point.

23. $f(x, y) = z = 4x^3 + y^2 - 12x^2 - 36x - 2y$

$f_x(x, y) = 12x^2 - 24x - 36$; $f_y(x, y) = 2y - 2$

Set each expression equal to zero and solve the resulting system of equations.

$$12x^2 - 24x - 36 = 0$$
$$2y - 2 = 0$$

Thus $x = 3$ or -1 and $y = 1$

For $x = 3$ and $y = 1$

$z = 4(3)^3 + 1^2 - 12(3)^2 - 36(3) - 2(1) = -109$.

For $x = -1$ and $y = 1$,

$z = 4(-1) + 1^2 - 12(1) - 36(-1) - 2 = 19.$

Thus (3, 1, −109) and (−1, 1, 19) are critical points.

$f_{xx}(x, y) = 24x - 24; f_{yy}(x, y) = 2, f_{xy}(x, y) = 0$

$D(x, y) = f_{xx}(x, y) f_{yy}(x, y) - [f_{xy}(x, y)]^2$

$D(x, y) = 2(24x - 24) = 48x - 48$

$D(3, 1) = 48(3) - 48 = 96 > 0$

$f_{xx}(3, 1) = 24(3) - 24 = 48 > 0$, thus (3, 1, −109) is a relative minimum point.

$D(-1, 1) = 48(-1) - 48 = -96 < 0$, thus (−1, 1, 19) is a saddle point.

25. $500 = \ell wh$ and $\ell w + 2\ell h + 2wh = M$; $h = 500/\ell w$;

$M = \ell w + 2\ell\,(500/\ell w) + 2\ell(500/\ell w) = \ell w + 1000/\ell + 1000/\ell$

$\partial M/\partial \ell = w - 1000/\ell^2 = 0$; $\partial M/\partial w = \ell - 1000/w^2 = 0$;

$w\ell^2 = 1000$; $\ell w^2 = 1000$; $\ell = 10$ cm, $w = 10$ cm, $h = 5$ cm

27. $x + y + z = 30$; $M = xyz = xy(30 - x - y) = 30xy - x^2y - xy^2$

$\partial M/\partial x = 30y - 2xy - y^2 = y(30 - 2x - y) = 0$

$\partial M/\partial y = 30x - x^2 - 2xy = x(30 - x - 2y) = 0$

$x = 0, y = 0$ gives a min.

$2x + y = 30$; $x + 2y = 30$; $x = 10, y = 10, z = 10$

Exercises 27.4

1. $\int_0^2\int_0^{3x}(x + y)\,dy\,dx = \int_0^2 (xy + y^2/2)\Big|_0^{3x} dx = \int_0^2 (3x^2 + 9x^2/2)\,dx = \int_0^2 15x^2/2\,dx = 5x^3/2\Big|_0^2 = 20$

3. $\int_0^1\int_y^{2y}(3x^2 + xy)\,dx\,dy = \int_0^1 (x^3 + (1/2)x^2 y\Big|_y^{2y} dy = \int_0^1 17y^3/2 dy = 17y^4/8\Big|_0^1 = 17/8$

5. $\int_{-1}^1\int_x^{x^2}(4xy + 9x^2 + 6y)\,dy\,dx = \int_{-1}^1 (2xy^2 + 9x^2y + 3y^2)\Big|_x^{x^2} dx$

$= \int_{-1}^1 (2x^5 + 12x^4 - 11x^3 - 3x^2)\,dx = (x^6/3 + 12x^5/5 - 11x^4/4 - x^3)\Big|_{-1}^1 = \frac{14}{5}$

7. $\int_0^1\int_0^{\sqrt{1-y^2}}(x + y)\,dx\,dy = \int_0^1 (x^2/2 + xy\Big|_0^{\sqrt{1-y^2}} dy = \int_0^1\left[\left(\frac{1 - y^2}{2} + y\sqrt{1 - y^2}\right)\right]dy$

$= [y/2 - y^3/6 - (1/3)(1 - y^2)^{3/2}]\Big|_0^1 = \frac{2}{3}$

9. $\int_0^1\int_0^x e^{x+y}\,dydx = \int_0^1 e^{x+y}\Big|_0^x dx = \int_0^1 (e^{2x} - e^x)dx = (1/2)e^{2x} - e^x\Big|_0^1 = e^2/2 - e + 1/2$

11. $\int_0^3\int_0^{4y}\sqrt{y^2+16}\,dx\,dy = \int_0^3 x\sqrt{y^2+16}\Big|_0^{4y} dy = \int_0^3 4y\sqrt{y^2+16}\,dy = 2[(2/3)(y^2+16)^{3/2}]\Big|_0^3$

$= (4/3)(25)^{3/2} - (4/3)(16)^{3/2} = 244/3$

13. $\int_0^{\pi/2}\int_0^{x}\cos x \sin y\,dy\,dx \quad \int_0^{\pi/2}(-\cos x\cos y)\Big|_0^x dx = \int_0^{\pi/2}(-\cos^2 x + \cos x)\,dx$

$= \int_0^{\pi/2}(-1/2 - (1/2)\cos 2x + \cos x)\,dx = (-x/2 - (1/4)\sin 2x + \sin x)\Big|_0^{\pi/2} = 1 - \pi/4$

15. Region bounded in xy-plane by $x = 0$, $y = 0$, $y = 12 - 3x$.

$V = \int_0^4\int_0^{12-3x}(2 - x/2 - y/6)\,dy\,dx = \int_0^4(2y - xy/2 - y^2/12)\Big|_0^{12-3x} dx$

$= \int_0^4(12 - 6x + 3x^2/4)dx = (12x - 3x^2 + x^3/4)\Big|_0^4 = 16$

17. Region bounded in xy-plane by $x = 2$, $y = x$, $y = 0$.

$V = \int_0^2\int_0^x xy\,dy\,dx = \int_0^2(1/2)xy^2\Big|_0^x dx = \int_0^2 x^3/2\,dx = x^4/8\Big|_0^2 = 2$

19. $V = 2\int_0^2\int_0^{\sqrt{4-x^2}}(2x+3y)\,dy\,dx = \int_0^2(4xy+3y)^2\Big|_0^{\sqrt{4-x^2}} dx$

$= \int_0^2[4x\sqrt{4-x^2} + 3(4-x^2)]dx = [(-4/3)(4-x^2)^{3/2} + 12x - x^3]\Big|_0^2 = 16$

Chapter 27 Review

1. Plane **2.** Elliptic paraboloid **3.** Hyperboloid of one sheet

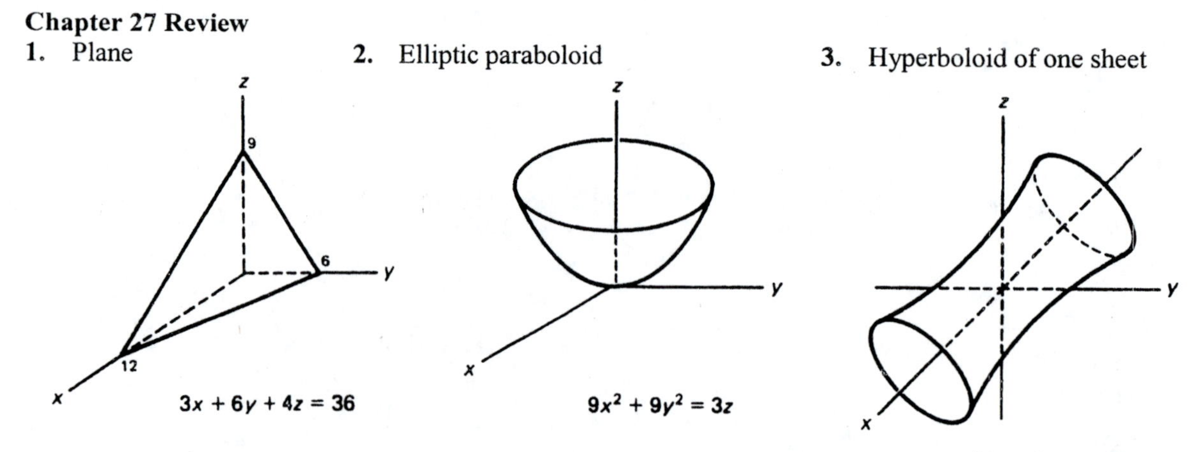

4. Ellipsoid

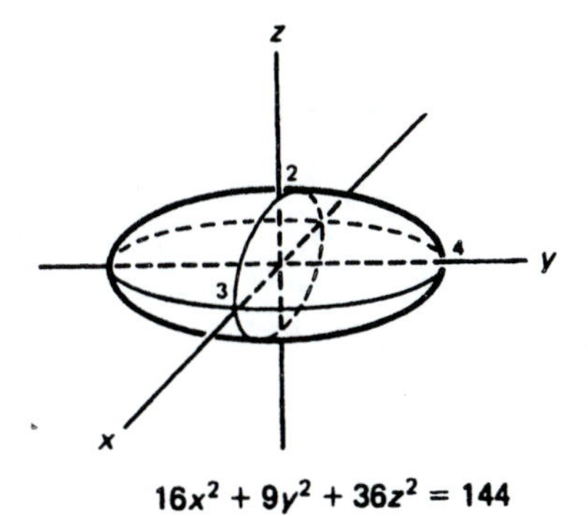

5. Cylindrical surface

$x^2 = -12y$

6. Hyperbolic paraboloid

$9x^2 - 9y^2 = 3z$

7. Hyperboloid of two sheets

$9y^2 - 36x^2 - 16z^2 = 144$

8. Elliptic cone

$9x^2 + 36y^2 - 16z^2 = 0$

9. $(x+3)^2 + (y-2)^2 + (z-1)^2 = 9$

10. $d = \sqrt{(1-4)^2 + (2+2)^2 + (9-4)^2} = 5\sqrt{2}$

11. a) $\partial z/\partial x = 3x^2 + 6xy$ **b)** $\partial z/\partial y = 3x^2 + 4y$

12. a) $\partial z/\partial x = 6x\, e^{2y}$ **b)** $\partial z/\partial y = 6x^2\, e^{2y}$

13. a) $\dfrac{\partial z}{\partial x} = \dfrac{1}{3x^2 y}(6xy) = \dfrac{2}{x}$ **b)** $\dfrac{\partial z}{\partial y} = \dfrac{1}{3x^2 y}(3x^2) = \dfrac{1}{y}$

14. a) $\partial z/\partial x = 3 \sin 3y \cos 3x$ **b)** $\partial z/\partial y = 3 \sin 3x \cos 3y$

15. a) $$\frac{\partial z}{\partial x} = \frac{(x\ln y)2xy\, e^{x^2} - y\, e^{x^2} \ln y}{(x\ln y)^2} = \frac{y\, e^{x^2} \ln y(2x^2-1)}{x^2 \ln^2 y} = \frac{y\, e^{x^2}(2x^2-1)}{x^2 \ln y}$$

b) $$\frac{\partial z}{\partial y} = \frac{(x\ln y)\, e^{x^2} - y\, e^{x^2}(x/y)}{(x\ln y)^2} = \frac{x\, e^{x^2}(\ln y - 1)}{x^2 \ln^2 y} = \frac{e^{x^2}(\ln y - 1)}{x \ln^2 y}$$

16. a) $$\frac{\partial z}{\partial x} = \frac{y\sin x(e^x \sin y) - e^x \sin y(y\cos x)}{y^2 \sin^2 x} = \frac{e^x \sin y(\sin x - \cos x)}{y\sin^2 x}$$

b) $$\frac{\partial z}{\partial y} = \frac{y\sin x(e^x \cos y) - e^x \sin y(\sin x)}{y^2 \sin^2 x} = \frac{e^x(y\cos y - \sin y)}{y^2 \sin x}$$

17. $\partial v/\partial w = (1/2)(p/w)^{-1/2}(-p/w^2) = -p^{1/2}/(2w^{3/2})$

18. $\partial U/\partial I = 2[(1/2)LI] = LI$

19. $$\frac{\partial Z}{\partial X_C} = X_C / \sqrt{R^2 + X_C^2}$$

20. $$\frac{\partial V}{\partial R} = \frac{(R+r)E - RE(1)}{(R+r)^2} = \frac{rE}{(R+r)^2}$$

21. a) $\partial z/\partial x = y + 6x - 4 \Big|_{(2,\,0,\,4)} = 8$

b) $\partial z/\partial y = 2y + x \Big|_{(2,\,0,\,4)} = 2$

22. $$dz = \frac{1}{\sqrt{x^2+xy}} \cdot \frac{1}{2}(x^2+xy)^{-1/2}(2x+y)\,dx + \frac{1}{\sqrt{x^2+xy}} \cdot \frac{1}{2}(x^2+xy)^{-1/2}\,x\,dy$$
$$= \frac{2x+y}{2(x^2+xy)}dx + \frac{x}{2(x^2+xy)}dy$$

23. $$dz = \frac{(x+y)(1)-(x-y)(1)}{(x+y)^2}dx + \frac{(x+y)(-1)-(x-y)(1)}{(x+y)^2}dy = \frac{2y}{(x+y)^2}dx - \frac{2x}{(x+y)^2}dy$$

24. $dV = (\partial V/\partial r)\,dr + (\partial V/\partial h)\,dh = 2\pi rh\,dr + \pi r^2\,dh\,(2)$ [(2) for top and bottom]
$= 2\pi(2)(10)(0.003) + \pi(2)^2(0.003)(2) = 0.452\text{ m}^3 = 452\text{ L}$
Note: $1\text{ m}^3 = 1000\text{ L}$

25. $$di = -50\,e^{-t/50R} \cdot \frac{tR^{-2}}{50}dr + (-50)e^{-t/50R} \cdot \frac{-1}{50R}dt$$
$$= (e^{-t/50R}/R)[dt - (t/R)\,dr] = \frac{e^{(-6)/[50\cdot 150]}}{150}[0.5 - (6/150)(10)] = 0.000666\,A$$

26. $f(x, y) = z = x^2 + 2xy - y^2 - 14x - 6y + 8$
$f_x(x, y) = 2x + 2y - 14$; $f_y(x, y) = 2x - 2y - 6$
Set each expression equal to zero and solve the resulting system of equations.
$2x + 2y - 14 = 0$
$2x - 2y - 6 = 0$
Thus $x = 5$ and $y = 2$
$z = 5^2 + 2(5)(2) - 2^2 - 14(5) - 6(2) + 8 = -33.$

So $(5, 2, -33)$ is a critical point.
$f_{xx}(x, y) = 2$; $f_{yy}(x, y) = -2$; $f_{xy}(x, y) = 2$
$D(x, y) = f_{xx}(x, y)\ f_{yy}(x, y) - [f_{xy}(x, y)]^2$
$D(5, 2) = 2(-2) - (2)^2 = -8 < 0$, thus
$(5, 2, -33)$ is a saddle point.

27. $f(x, y) = z = y^2 - xy - x^2 + 4x - 3y - 6$
$f_x(x, y) = -y - 2x + 4$; $f_y(x, y) = 2y - x - 3$
Set each expression equal to zero and solve the resulting system of equations.
$-y - 2x + 4 = 0$
$2y - x - 3 = 0$
Thus $x = 1$ and $y = 2$.
$z = 2^2 - 1(2) - (1)^2 + 4(1) - 3(2) - 6 = -7$
$(1, 2, -7)$ is a critical point.
$f_{xx}(x, y) = -2$; $f_{yy}(x, y) = 2$; $f_{xy}(x, y) = -1$
$D(x, y) = f_{xx}(x, y)\ f_{yy}(x, y) - [f_{xy}(x, y)]^2$
$D(1, 2) = -2(2) - (-1)^2 = -5 < 0$, thus
$(1, 2, -7)$ is a saddle point.

28. $f(x, y) = z = -x^2 + xy - y^2 + 4x - 8y + 9$
$f_x(x, y) = -2x + y + 4$; $f_y(x, y) = x - 2y - 8$
Set each expression equal to zero and solve the resulting system of equations.
$-2x + y + 4 = 0$
$x - 2y - 8 = 0$
Thus $x = 0$ and $y = -4$. Then
$z = 0 + 0 - (-4)^2 + 0 - 8(-4) + 9 = 25$.
$(0, -4, 25)$ is a critical point.
$f_{xx}(x, y) = -2$; $f_{yy}(x, y) = -2$; $f_{xy}(x, y) = 1$
$D(x, y) = f_{xx}(x, y)\ f_{yy}(x, y) - [f_{xy}(x, y)]^2$
$D(0, -4) = -2(-2) - (1)^2 = 3 > 0$
$f_{xx}(0, -4) = -2 < 0$, thus $(0, -4, 25)$ is a relative maximum point.

29. $V = 8 = \ell wh$ and $C = 2(\ell w) + 2(\ell h) + 1(2wh)$
$h = 8/\ell w$ $\qquad C = 2\ell w + 16/w + 16/\ell$

$\partial C/\partial \ell = 2w - 16/\ell^2 = 0$; $w\ell^2 = 8$
$\partial C/\partial w = 2\ell - 16/w^2 = 0$; $\ell w^2 = 8$ Thus $\ell = 2, w = 2, h = 2$

30. $\int_0^1\int_0^{x^2}(x-2y)\,dy\,dx = \int_0^1 (xy-y^2)\Big|_0^{x^2} dx = \int_0^1 (x^3-x^4)\,dx = \frac{x^4}{4}-\frac{x^5}{5}\Big|_0^1 = \frac{1}{20}$

31. $\int_0^2\int_y^{4y}(4xy+6x^2-9y^2)\,dx\,dy = \int_0^2 (2x^2y+2x^3-9xy^2)\Big|_y^{4y} = \int_0^2 129y^3\,dy = (129/4)y^4\Big|_0^2 = 516$

32. $\int_1^3\int_0^{\ln y} y\,e^x\,dx\,dy = \int_1^3 y\,e^x\Big|_0^{\ln y} dy = \int_1^3 (y\,e^{\ln y}-y)\,dy$

$= \int_1^3 (y^2-y)dy = y^3/3 - y^2/2\Big|_1^3 = 14/3$

33. $\int_0^{\pi/4}\int_0^{x}\sec^2 y\,dy\,dx = \int_0^{\pi/4} \tan y\Big|_0^x dx = \int_0^{\pi/4} \tan x\,dx = \ln|\sec x|\Big|_0^{\pi/4} = \ln\sqrt{2} = (1/2)\ln 2$

34. The region is bounded in the xy-plane by $x = 0$, $y = 0$, $y = b(1-x/a)$.

$V = \int_0^a \int_0^{b(1-x/a)} c(1-x/a-y/b)dy\,dx$

First, find the first integral:

$\int_0^{b(1-x/a)} c(1-x/a-y/b)dy = c[(1-x/a)y - y^2/2b]\Big|_0^{b(1-x/a)} = \frac{cy}{2b}[2b(1-x/a)-y]\Big|_0^{b(1-x/a)}$

$= \frac{c}{2b}\cdot b(1-x/a)\,[2b(1-x/a)-b(1-x/a)] = (c/2)(1-x/a)\cdot b(1-x/a) = (bc/2)(1-x/a)^2.$

Now integrate *wrt* x:

$V = \int_0^a (bc/2)(1-x/a)^2\,dx = (bc/2)(-a)(1/3)(1-x/a)^3\Big|_0^a = (-abc/6)(1-x/a)^3\Big|_0^a = (1/6)abc$

35. $V = \int_0^1\int_0^{\sqrt{1-x^2}} 2x\,dy\,dx = \int_0^1 2xy\Big|_0^{\sqrt{1-x^2}} dx = \int_0^1 2x\sqrt{1-x^2}\,dx = (-2/3)(1-x^2)^{3/2}\Big|_0^1 = 2/3$

36. $V = \int_0^1\int_0^{1-x^2}(1-y-x^2)dy\,dx = \int_0^1 (y-y^2/2-x^2y)\Big|_0^{1-x^2} dx$

$= \int_0^1 (1/2-x^2+x^4/2)dx = (x/2-x^3/3+x^5/10)\Big|_0^1 = 4/15$

Chapter 28, Exercises 28.1

1. $5+9+13+17+21+25$

3. $10+17+26+37+50+65$

5. $\frac{1}{2}+\frac{4}{3}+\frac{9}{4}+\frac{16}{5}+\cdots+\frac{n^2}{n+1}$

7. $-1+\frac{1}{4}-\frac{1}{9}+\frac{1}{16}-\frac{1}{25}+\cdots$

9. $\sum_{n=1}^{12} n$

11. $\sum_{n=1}^{50} 2n$

13. $\sum_{n=1}^{n}(2n-1)$

15. $\sum_{n=3}^{n}(n^2+1)$

17. Diverges because $\lim_{n\to\infty} S_n = \infty$

19. Diverges because $\lim_{n\to\infty}\frac{2n}{n-1}=\lim_{n\to\infty}\frac{2}{1-1/n}=2$

21. Diverges (p-series with $p = 1/4$)

23. Converges (p-series with $p = 2$)

25. Compare $\sum 1/n^2$ and $\sum 1/(n+1)^2$. Since $\sum 1/n^2$ converges and $\frac{1}{(n+1)^2}<\frac{1}{n^2}$ for $n \ge 1$, by the comparison test $\sum 1/(n+1)^2$ also converges.

27. Compare $\sum 1/n^2$ and $\sum\frac{1}{n(n+1)}$. Since $\sum 1/n^2$ converges and

$\lim_{n\to\infty}\frac{1/n^2}{1/(n^2+n)}=\lim_{n\to\infty}\frac{n^2+n}{n^2}=\lim_{n\to\infty}(1+1/n)=1$, both series have the same order of magnitude and by the limit comparison test both series converge.

29. Compare $\sum 1/n$ and $\sum 1/(2n)$. Since $\sum 1/n$ diverges and

$\lim_{n\to\infty}\frac{1/n}{1/(2n)}=\lim_{n\to\infty}\frac{2n}{n}=2$, both series have the same order of magnitude and by the limit comparison test both series diverge.

31. Compare $\sum 1/(2n-1)^2$ and $\sum 1/n^2$. Since $\sum 1/n^2$ converges and $\frac{1}{(2n-1)^2}\le\frac{1}{n^2}$ for $n \ge 1$, then by the comparison test $\sum\frac{1}{(2n-1)^2}$ converges.

33. Compare $\sum 1/\sqrt{n^2+1}$ and $\sum 1/n = \sum 1/\sqrt{n^2}$. Since $\sum 1/n$ diverges and
$\lim_{n\to\infty} \frac{1/\sqrt{n^2+1}}{1/\sqrt{n^2}} = \lim_{n\to\infty} \sqrt{\frac{n^2}{n^2+1}} = \lim_{n\to\infty} \sqrt{\frac{1}{1+1/n^2}}$ = 1, both series have the same order of magnitude and by the limit comparison test both diverge.

35. Compare $\sum \frac{1}{\sqrt{n}(n+1)}$ and $\sum \frac{1}{n^{3/2}}$. Since $\sum \frac{1}{n^{3/2}}$ is a convergent *p*-series $\frac{1}{\sqrt{n}(n+1)} < \frac{1}{n^{3/2}}$ for $n \geq 1$, by the comparison test the given series also converges.

37. Compare $\sum \frac{1}{2^n + 2n}$ and $\sum \left(\frac{1}{2}\right)^n$. Since $\sum \left(\frac{1}{2}\right)^n$ is a geometric convergent series and
$\frac{1}{2^n + 2n} < \frac{1}{2^n}$ for $n > 1$, the given series converges by the comparison test.

39. Compare $\sum 1/\ln n$ and $\sum 1/n$. Since $0 < \ln n < n$ for $n > 1$, $1/\ln n > 1/n$ for $n \geq 2$. Since $\sum 1/n$ diverges, by the comparison test the given series also diverges.

41. Compare $\sum \frac{1+\sin n\pi}{n^2}$ and $\sum 2/n^2$. Since $\sum 2/n^2$ is a convergent *p*-series and
$\frac{1+\sin n\pi}{n^2} < \frac{2}{n^2}$, by the comparison test the given series also converges.

43. Compare $\sum \frac{1}{\sqrt{n(n+1)}}$ and $\sum 1/n$. Since $\sum 1/n$ is a divergent *p*-series and
$\lim_{n\to\infty} \frac{1/\sqrt{n^2+n}}{1/n} = \lim_{n\to\infty} \sqrt{\frac{n^2}{n^2+n}}$ = 1, both series have the same order of magnitude and by the limit comparison test both series diverge.

Exercises 28.2

1. $r = \lim_{n\to\infty} \frac{\frac{n+2}{(n+1)3^{n+1}}}{\frac{n+1}{n\cdot 3^n}} = \lim_{n\to\infty} \frac{n\,3^n(n+2)}{(n+1)^2 3^{n+1}} = \lim_{n\to\infty} \frac{n(n+2)}{3(n+1)^2} = \lim_{n\to\infty} \frac{1(1+2/n)}{3(1+1/n)^2} = \frac{1}{3} < 1$. The given series converges by the ratio test.

3. $r = \lim_{n\to\infty} \frac{1/(n+1)!}{1/n!} = \lim_{n\to\infty} n!/(n+1)! = \lim_{n\to\infty} \frac{1}{n+1} = 0 < 1$.
The given series converges by the ratio test.

5. $r = \lim_{n\to\infty} \dfrac{\dfrac{(n+1)^2}{(n+1)!}}{\dfrac{n^2}{n!}} = \lim_{n\to\infty} \dfrac{(n+1)^2 n!}{(n+1)!n^2} = \lim_{n\to\infty} \dfrac{(n+1)}{n^2} = \lim_{n\to\infty}(1/n + 1/n^2) = 0.$ The given series converges by the ratio test.

7. $r = \lim_{n\to\infty} \dfrac{\dfrac{3^{n+1}}{(n+1)2^{n+1}}}{\dfrac{3^n}{n\,2^n}} = \lim_{n\to\infty} \dfrac{n\,2^n\,3^{n+1}}{(n+1)2^{n+1}3^n} = \lim_{n\to\infty} \dfrac{3n}{2(n+1)} = \dfrac{3}{2} > 1.$

The given series diverges by the ratio test.

9. $r = \lim_{n\to\infty} \dfrac{\dfrac{2n+5}{2^{n+1}}}{\dfrac{2n+3}{2^n}} = \lim_{n\to\infty} \dfrac{(2n+5)2^n}{(2n+3)2^{n+1}} = \lim_{n\to\infty} \dfrac{2n+5}{2(2n+3)} = \lim_{n\to\infty} \dfrac{2+5/n}{4+6/n} = \dfrac{1}{2} < 1.$ The given series converges by the ratio test.

11. $\displaystyle\int_1^\infty \frac{dx}{2x+1} = \lim_{b\to\infty} \int_1^b \frac{dx}{2x+1} = \lim_{b\to\infty} \frac{1}{2} \ln|2x+1| \Big|_1^b = \lim_{b\to\infty} \frac{1}{2} [\ln |2b+1| - \ln 3] = \infty.$ The integral does not exist and the given series diverges by the integral test.

13. $\displaystyle\int_2^\infty \frac{dx}{x \ln x^{1/2}} = \lim_{b\to\infty} \int_2^b \frac{dx}{x \ln x^{1/2}} = \lim_{b\to\infty} 2\sqrt{\ln x} \Big|_2^b = \lim_{b\to\infty} 2(\sqrt{\ln b} - \sqrt{\ln 2}) = \infty.$ The integral does not exist and the series diverges by the integral test.

15. $\displaystyle\int_1^\infty \frac{dx}{2x-1} = \lim_{b\to\infty} \int_1^b \frac{dx}{2x-1} = \lim_{b\to\infty} \frac{1}{2} \ln|2x-1| \Big|_1^b = \lim_{b\to\infty} \frac{1}{2} [\ln |2b-1|] = \infty.$ The integral does not exist and the series diverges by the integral test.

17. $\displaystyle\int_1^\infty \frac{x\,dx}{x^2+1} = \lim_{b\to\infty} \int_1^b \frac{x\,dx}{x^2+1} = \lim_{b\to\infty} \frac{1}{2} \ln|x^2+1| \Big|_1^b = \lim_{b\to\infty} \frac{1}{2} [\ln |b^2+1| - \ln 2] = \infty.$ The integral does not exist and the series diverges by the integral test.

19. $\displaystyle\int_1^\infty (x^2/e^x)\,dx = \lim_{b\to\infty} \int_1^b x^2 e^{-x}\,dx = \lim_{b\to\infty} - e^{-x}(x^2+2x+2) \Big|_1^b = \lim_{b\to\infty} - \frac{b^2+2b+2}{e^b} + \frac{5}{e} = \frac{5}{e}.$ The integral exists and the series converges by the integral test.

Exercises 28.3

1. Converges conditionally: given alternating series converges because

$$a_{n+1} = \frac{1}{2n+3} < \frac{1}{2n+1} = a_n \text{ and } \lim_{n\to\infty} \frac{1}{2n+1} = 0\text{; but } \sum \frac{1}{2n+1} \text{ diverges.}$$

3. Converges absolutely: $\sum \frac{1}{(2n)^2}$ converges by limit comparison test with $\sum \frac{1}{n^2}$, which converges.

5. Diverges: $\lim\limits_{n\to\infty} \frac{2n}{2n-1} = \lim\limits_{n\to\infty} \frac{2}{2-1/n} = 1$

7. Converges conditionally: given alternating series converges because

$$a_{n+1} = \frac{1}{\ln(n+1)} < \frac{1}{\ln n} = a_n \text{ and } \lim_{n\to\infty} \frac{1}{\ln n} = 0\text{; but } \sum \frac{1}{\ln n} \text{ diverges because } \frac{1}{\ln n} > \frac{1}{n} \text{ and}$$

$\sum \frac{1}{n}$ diverges.

9. Converges absolutely: $\sum (n^2/2^n)$ converges (See Exercise 6, Section 10.2).

11. Diverges: $\lim\limits_{n\to\infty} \frac{n^2}{n^2+1} = \lim\limits_{n\to\infty} \frac{1}{(1+1/n^2)} = 1$

13. Diverges: $\lim\limits_{n\to\infty} \frac{n!}{3^n} = \lim\limits_{n\to\infty} \frac{n}{3} \cdot \frac{n-1}{3} \cdots \frac{2}{3} \cdot \frac{1}{3} > \lim\limits_{n\to\infty} \frac{n}{3} = \infty$

15. Converges conditionally: given alternating series converges because

$$\lim_{n\to\infty} \frac{2n+1}{n^2} = \lim_{n\to\infty}(2/n + 1/n^2) = 0 \text{ and}$$

$$a_{n+1} = \frac{2n+3}{(n+1)^2} < \frac{2n+1}{n^2} = a_n$$

Since $\frac{2n+1}{n^2} > \frac{2n}{n^2} = \frac{2}{n}$ and $\sum 2/n$ diverges, $\sum \frac{2n+1}{n^2}$ diverges.

17. Diverges: $\lim\limits_{n\to\infty} \frac{n}{\ln n} = \lim\limits_{n\to\infty} \frac{1}{1/n}$ by l'Hopital $= \lim\limits_{n\to\infty} n = \infty$

19. Converges absolutely: $|\cos n| \le 1$ so the given series converges by the comparison test with the convergent p-series $\sum 1/n^2$.

Exercises 28.4

1. This geometric series converges for $\left|\frac{x}{2}\right| < 1$ or $-2 < x < 2$.

3. $\lim\limits_{n\to\infty}\left|\frac{a_{n+1}}{a_n}\right| = \lim\limits_{n\to\infty}\left|\frac{(4n+4)!(x/2)^{n+1}}{(4n)!(x/2)^n}\right| = \lim\limits_{n\to\infty}\left|(4n+4)(4n+3)(4n+2)(4n+1)\frac{x}{2}\right| = \infty$

Thus the series converges only for $x = 0$.

5. $\lim\limits_{n\to\infty}\left|\frac{(4x)^{n+1}(2n)!}{(2n+2)!(4x)^n}\right| = \lim\limits_{n\to\infty}\frac{4x}{(2n+2)(2n+1)} = 0$

Thus the interval of convergence is $-\infty < x < \infty$.

7. $\lim\limits_{n\to\infty}\left|\frac{\dfrac{x^{n+1}}{(n+2)(n+3)}}{\dfrac{x^n}{(n+1)(n+2)}}\right| = |x|\lim\limits_{n\to\infty}\frac{n+1}{n+3} = |x| < 1$ or $-1 < x < 1$

For $x = 1$, the series $\sum\frac{-1}{(n+1)(n+2)}$ converges. For $x = -1$, the series $\sum\frac{1}{(n+1)(n+2)}$ converges. Thus the interval of convergence is $-1 \le x \le 1$.

9. $\lim\limits_{n\to\infty}\left|\frac{(n+1)x^{n+1}}{(n+2)^2}\cdot\frac{(n+1)^2}{n\,x^n}\right| = |x|\lim\limits_{n\to\infty}\frac{(n+1)^3}{n(n+2)^2} = |x| < 1$ or $-1 < x < 1$. For $x = 1$, the series $\sum\frac{n}{(n+1)^2}$ diverges with comparison with $\sum 1/n$. For $x = -1$, the series $\sum\frac{n(-1)^n}{(n+1)^2}$ converges by the alternating series test. Thus the interval of convergence is $-1 \le x < 1$.

11. $\lim\limits_{n\to\infty}\left|\frac{2^{n+1}x^{n+1}}{3^{n+1}}\cdot\frac{3^n}{2^n\,x^n}\right| = |x|\lim\limits_{n\to\infty}\frac{2}{3} = \frac{2}{3}\;|x| < 1$ or $-1 < \frac{2}{3}x < 1$ or $-3/2 < x < 3/2$.

For $x = \pm 3/2$, each series clearly diverges.

13. $\lim\limits_{n\to\infty}\left|\frac{x^{n+1}}{(n+1)2^{n+1}}\cdot\frac{n\,2^n}{x^n}\right| = \left|\frac{x}{2}\right|\lim\limits_{n\to\infty}\frac{n}{n+1} = \left|\frac{x}{2}\right| < 1$ or $-2 < x < 2$

For $x = 2$, the alternating series $\sum\frac{(-1)^{n+1}}{n}$ converges. For $x = -2$, the series $\sum\frac{(-1)^{2n+1}}{n} = \sum\frac{-1}{n}$ diverges. Thus the interval of convergence is $-2 < x \le 2$.

15. $\lim\limits_{n\to\infty}\left|\frac{(x-2)^{n+1}}{\sqrt{n+1}}\cdot\frac{\sqrt{n}}{(x-2)^n}\right| = |x-2|\lim\limits_{n\to\infty}\frac{\sqrt{n}}{\sqrt{n+1}} = |x-2| < 1$ or $1 < x < 3$. For $x = 3$, the alternating series converges. For $x = 1$, the series $\sum 1/\sqrt{n}$ is a divergent p-series. Thus the interval of convergence is $1 < x \le 3$.

17. $\lim_{n\to\infty}\left|\frac{x^{n+1}}{(n+1)^2}\cdot\frac{n^2}{x^n}\right| = |x|\lim_{n\to\infty}\frac{n^2}{(n+1)^2} = |x| < 1$ or $-1 < x < 1$. For $x = \pm 1$, we have a convergent p-series. Thus the interval of convergence is $-1 \le x \le 1$.

19. $\lim_{n\to\infty}\left|\frac{x^{2n+2}}{(n+1)!}\cdot\frac{n!}{x^{2n}}\right| = x^2\lim_{n\to\infty}\frac{1}{n+1} = 0$. Thus the interval of convergence is $-\infty < x < \infty$.

21. $\lim_{n\to\infty}\left|\frac{2^{n+1}x^{n+2}}{(n+1)3^{n+2}}\cdot\frac{n\,3^{n+1}}{2^n x^{n+1}}\right| = \left|\frac{2x}{3}\right|\lim_{n\to\infty}\frac{n}{n+1} = \left|\frac{2}{3}x\right| < 1$ or $-3/2 < x < 3/2$. For $x = 3/2$, the series $\sum 1/(2n)$ diverges. For $x = -3/2$, the series $\sum\frac{(-1)^{n+1}}{2n}$ converges. Thus the interval of convergence is $-3/2 \le x < 3/2$.

23. $\lim_{n\to\infty}\left|\frac{(2x-5)^{n+1}}{(n+1)^2}\cdot\frac{n^2}{(2x-5)^n}\right| = |2x-5|\lim_{n\to\infty}\frac{n^2}{(n+1)^2} = |2x-5| < 1$ or $2 < x < 3$. For $x = 2$, the series $\sum\frac{(-1)^n}{n^2}$ converges. For $x = 3$, the series $\sum 1/n^2$ converges. Thus the interval of convergence is $2 \le x \le 3$.

Exercises 28.5

1. $f(x) = \sin x; f(0) = 0$
$f'(x) = \cos x; f'(0) = 1$
$f''(x) = -\sin x; f''(0) = 0$
$f'''(x) = -\cos x; f'''(0) = -1$
$f^{(4)}(x) = \sin x; f^{(4)}(0) = 0$

$$\sin x = x - \frac{x^3}{3!} + \frac{x^5}{5!} - \cdots$$

3. $f(x) = e^{-x}; f(0) = 1$
$f'(x) = -e^{-x}; f'(0) = -1$
$f''(x) = e^{-x}; f''(0) = 1$
$f'''(x) = -e^{-x}; f'''(0) = -1$
$f^{(4)}(x) = e^{-x}; f^{(4)}(0) = 1$

$$e^{-x} = 1 - x + \frac{x^2}{2!} - \frac{x^3}{3!} + \frac{x^4}{4!} - \cdots$$

5. $f(x) = \ln(1+x); f(0) = 0$

$f'(x) = \frac{1}{1+x}; f'(0) = 1$

$f''(x) = \frac{-1}{(1+x)^2}; f''(0) = -1$

$f'''(x) = \frac{2!}{(1+x)^3}; f'''(0) = 2$

$f^{(4)}(x) = \frac{-3!}{(1+x)^4}; f^{(4)}(0) = -3!$

$$\ln(1+x) = x - \frac{x^2}{2!} + 2!\frac{x^3}{3!} - 3!\frac{x^4}{4!} + \cdots$$
$$= x - \frac{1}{2}x^2 + \frac{1}{3}x^3 - \frac{1}{4}x^4 + \cdots$$

7. $f(x) = \cos 2x; f(0) = 1$
$f'(x) = -2\sin 2x; f'(0) = 0$
$f''(x) = -4\cos 2x; f''(0) = -4$
$f'''(x) = 8\sin 2x; f'''(0) = 0$
$f^{(4)}(x) = 16\cos 2x; f^{(4)}(0) = 16$

$$\cos 2x = 1 - 2x^2 + \frac{16}{4!}x^4 - \cdots$$

9. $f(x) = xe^x; f(0) = 0$
$f'(x) = e^x(x + 1); f'(0) = 1$
$f''(x) = e^x(x + 2); f''(0) = 2$
$f'''(x) = e^x(x + 3); f'''(0) = 3$
$f^{(4)}(x) = e^x(x + 4); f^{(4)}(0) = 4$

$$x\,e^x = x + x^2 + \frac{x^3}{2!} + \frac{x^4}{3!} + \cdots$$

11. $f(x) = (4 - x)^{1/2}; f(0) = 2$
$f'(x) = (-1/2)(4 - x)^{-1/2}; f'(0) = -1/4$
$f''(x) = (-1/4)(4 - x)^{-3/2}; f''(0) = -1/32$
$f'''(x) = (-3/8)(4 - x)^{-5/2}; f'''(0) = \dfrac{-3}{256}$

$$\sqrt{4 - x} = 2 - \frac{x}{4} - \frac{x^2}{32(2!)} - \frac{3x^3}{256(3!)} - \cdots$$

13. $f(x) = \sin(x - \pi/2) = \sin[-(\pi/2 - x)] = -\sin(\pi/2 - x) = -\cos x;\ f(0) = -1$
$f'(x) = \sin x; f'(0) = 0$
$f''(x) = \cos x; f''(0) = 1$
$f'''(x) = -\sin x; f'''(0) = 0$
$f^{(4)}(x) = -\cos x; f^{(4)}(0) = -1$ Thus $\sin(x - \pi/2) = -1 + x^2/2! - x^4/4! + \cdots$

15. $f(x) = 1(1 - x)^2;\ f(0) = 1$
$f'(x) = 2(1 - x)^{-3}; f'(0) = 2$
$f''(x) = 6(1 - x)^{-4}; f''(0) = 6$
$f'''(x) = 24(1 - x)^{-5}; f'''(0) = 24$
$f^{(4)}(x) = 120(1 - x)^{-6}; f^{(4)}(0) = 120$
$1/(1 - x)^2 = 1 + 2x + 6x^2/2! + 24x^3/3! + 120x^4/4! + \cdots$

17. $f(x) = (1 + x)^5;\ f(0) = 1$
$f'(x) = 5(1 + x)^4; f'(0) = 5$
$f''(x) = 20(1 + x)^3; f''(0) = 20$
$f'''(x) = 60(1 + x)^2; f'''(0) = 60$
$f^{(4)}(x) = 120(1 + x); f^{(4)}(0) = 120$
$f^{(5)}(x) = 120; f^{(5)}(0) = 120$
$(1 + x)^5 = 1 + 5x + 10x^2 + 10x^3 + 5x^4 + x^5$ (Sum is finite.)

19. $f(x) = e^{-x}\sin x; f(0) = 0$
$f'(x) = e^{-x}(\cos x - \sin x); f'(0) = 1$
$f''(x) = -2e^{-x}\cos x; f''(0) = -2$
$f'''(x) = 2e^{-x}(\sin x + \cos x); f'''(0) = 2$
$f^{(4)}(x) = -4e^{-x}\sin x; f^{(4)}(0) = 0$
$f^{(5)}(x) = 4e^{-x}(\sin x - \cos x);\ f^{(5)}(0) = -4$
Thus $e^{-x}\sin x = x - x^2 + x^3/3 - x^5/30 + \cdots$

Exercises 28.6

1. Substitute $-x$ for x in equation A.

$$f(x) = e^{-x} = 1 - x + \frac{x^2}{2!} + \frac{x^3}{3!} + \frac{x^4}{4!} - \cdots$$

3. Substitute x^2 for x in equation A.

$$f(x) = e^{x^2} = 1 + x^2 + \frac{x^4}{2!} + \frac{x^6}{3!} + \cdots$$

5. Substitute $-x$ for x in equation D.

$$f(x) = \ln(1-x) = -x - \frac{x^2}{2} - \frac{x^3}{3} - \frac{x^4}{4} - \cdots$$

7. Substitute $5x^2$ for x in equation C.

$$f(x) = \cos 5x^2 = 1 - \frac{25x^4}{2!} + \frac{625x^8}{3!} - \cdots$$

9. Substitute x^3 for x in equation D.

$$f(x) = \sin x^3 = x^3 - \frac{x^9}{3!} + \frac{x^{15}}{5!} - \cdots$$

11. $f(x) = x\,e^x = x\left(1 + x + \dfrac{x^2}{2!} + \dfrac{x^3}{3!} + \cdots\right) = x + x^2 + \dfrac{x^3}{2!} + \dfrac{x^4}{3!} + \cdots$

13. $\dfrac{\cos x - 1}{x} = \dfrac{[1 - x^2/2! + x^4/4! - x^6/6! + \cdots] - 1}{x} = \dfrac{-x}{2!} + \dfrac{x^3}{4!} - \dfrac{x^5}{6!} + \cdots$

15. $\displaystyle\int_0^1 e^{-x^2}\,dx \int_0^1 (1 - x^2 + x^4/2! - x^6/3! + \cdots)\,dx$

$$= (x - x^3/3 + x^5/10 - x^7/42)\Big|_0^1 = 1 - 1/3 + 1/10 - 1/42 - (0) = 0.743$$

17. Let $u = x - 1$; $\dfrac{e^{x-1}}{x-1} = \dfrac{e^u}{u} = \dfrac{1 + u + u^2/2! + u^3/3! + \cdots}{u} = \dfrac{1}{u} + 1 + \dfrac{u}{2!} + \dfrac{u^2}{3!} + \cdots$

$$\int_2^3 \frac{e^{x-1}}{x-1} = \int_2^3 \left[\frac{1}{x-1} + 1 + \frac{x-1}{2!}\right] dx = \ln(x-1) + x + \frac{(x-1)^2}{4}\Bigg|_2^3 = \ln 2 + \frac{7}{4}$$

19. $\displaystyle\int_0^1 \sin\sqrt{x}\,dx \int_0^1 \left(x^{1/2} - \frac{x^{3/2}}{3!} + \frac{x^{5/2}}{5!}\right) dx = (2/3)x^{3/2} - (1/15)x^{5/2} + (1/420)x^{7/2}\Big|_0^1 = 0.602$

21. $\sinh x = (1/2)(e^x - e^{-x}) = (1/2)\left(2x + \dfrac{2x^3}{3!} + \dfrac{2x^5}{5!} + \cdots\right) = x + \dfrac{x^3}{3!} + \dfrac{x^5}{5!} + \cdots$

$$e^x = 1 + x + \frac{x^2}{2!} + \frac{x^3}{3!} + \frac{x^4}{4!} + \cdots$$

$$e^{-x} = 1 - x + \frac{x^2}{2!} + \frac{x^3}{3!} + \frac{x^4}{4!} - \cdots$$

$$e^x - e^{-x} = 2x + \frac{2x^3}{3!} + \frac{2x^5}{5!} + \cdots$$

Section 28.6

23. $q = \int_0^{0.5} \sin t^2 \, dt \int_0^{0.5} \left(t^2 - \frac{t^6}{3!} + \frac{t^{10}}{5!} - \frac{t^{14}}{7!} \right) dt = \frac{t^3}{3} - \frac{t^7}{3!(7)} + \frac{t^{11}}{5!(11)} - \frac{t^{15}}{7!(15)} \Bigg|_0^{0.5} = 0.041481\,C$

25.

$$e^{jx} = 1 + jx + \frac{(jx)^2}{2!} + \frac{(jx)^3}{3!} + \cdots$$

$$e^{-jx} = 1 - jx + \frac{(-jx)^2}{2!} + \frac{(-jx)^3}{3!} + \cdots$$

$$e^{jx} = e^{-jx} = 2jx \qquad - \frac{2jx}{3!} + \cdots$$

$$\frac{e^{jx} - e^{-jx}}{2j} = x - \frac{x}{3!} + \frac{x^5}{5!} - \cdots = \sin x$$

Exercises 28.7

1. $f(x) = \cos x; = f(\pi/2) = 0$
$f'(x) = -\sin x; f'(\pi/2) = -1$
$f''(x) = -\cos x; f''(\pi/2) = 0$
$f'''(x) = \sin x; f'''(\pi/2) = 1$
$f^{(4)}(x) = \cos x; f^{(4)}(\pi/2) = 0$
$f^{(5)}(x) = -\sin x; f^{(5)}(\pi/2) = -1$

Thus $f(x) = -(x - \pi/2) + \frac{(x - \pi/2)^3}{3!} - \frac{(x - \pi/2)^5}{5!} + \cdots$

3. $f(x) = e^x = f'(x) = f''(x) = f'''(x) = \cdots$
$f(2) = e^2 = f'(2) = f''(2) = f'''(2) = \cdots$

Thus $f(x) = e^2 + e^2(x - 2) + \frac{e^2}{2!}(x - 2)^2 + \frac{e^2}{3!}(x - 2)^3 + \cdots$

$$= e^2 \left[1 + (x - 2) + \frac{(x - 2)^2}{2!} + \frac{(x - 2)^3}{3!} + \cdots \right]$$

5. $f(x) = x^{1/2}; f(9) = 3$
$f'(x) = (1/2)x^{-1/2}; f'(9) = 1/6$
$f''(x) = (-1/4)x^{-3/2}; f''(9) = -1/108$
$f'''(x) = (3/8)x^{-5/2}; f'''(9) = 1/648$

Thus $f(x) = 3 + \frac{x - 9}{6} - \frac{(x - 9)^2}{2!(108)} + \frac{(x - 9)^3}{3!(648)} + \cdots$

7. $f(x) = x^{-1}; f(2) = 1/2$
$f'(x) = -x^{-2}; f'(2) = -1/4$
$f''(x) = 2x^{-3}; f''(2) = 1/4$
$f'''(x) = -6x^{-4}; f'''(2) = -3/8$
$$f(x) = \frac{1}{2} - \frac{1}{4}(x-2) + \frac{(x-2)^2}{4(2!)} - \frac{3(x-2)^3}{8(3!)} + \cdots$$

9. $f(x) = \ln x; f(1) = 0$
$f'(x) = 1/x; f'(1) = 1$
$f''(x) = -x^{-2}; f''(1) = -1$
$f'''(x) = 2x^{-3}; f'''(1) = 2$
$f^{(4)}(x) = -6x^{-4};\ f^{(4)}(1) = -6$
$$\text{Thus } f(x) = (x-1) - \frac{(x-1)^2}{2!} + \frac{2(x-1)^3}{3!} - \frac{6(x-1)^4}{4!} + \cdots$$

11. $f(x) = x^{-1/2}; f(1) = 1$
$f'(x) = (-1/2)x^{-3/2}; f'(1) = -1/2$
$f''(x) = (3/4)x^{-5/2}; f''(1) = 3/4$
$f'''(x) = (-15/8)x^{-7/2}; f'''(1) = -15/8$
$f^{(4)}(x) = (105/16)x^{-9/2};\ f^{(4)}(1) = 105/16$
Thus $f(x) = 1 - (1/2)(x-1) + (3/8)(x-1)^2 - (5/16)(x-1)^3 + \cdots$

13. $f(x) = x^{-2}; f(1) = 1$
$f'(x) = -2x^{-3}; f'(1) = -2$
$f''(x) = 6x^{-4}; f''(1) = 6$
$f'''(x) = -24x^{-5}; f'''(1) = -24$
$$\text{Thus } f(x) = 1 - 2(x-1) + \frac{6(x-1)^2}{2!} - \frac{24(x-1)^3}{3!} + \cdots$$
$$= 1 - 2(x-1) + 3(x-1)^2 - 4(x-1)^3 + \cdots$$

15. $f(x) = \cos x; f(\pi) = -1$
$f'(x) = -\sin x; f'(\pi) = 0$
$f''(x) = -\cos x; f''(\pi) = 1$
$f'''(x) = \sin x; f'''(\pi) = 0$
$f^{(4)}(x) = \cos x;\ f^{(4)}(\pi) = -1$
$$\text{Thus } f(x) = -1 + \frac{(x-\pi)^2}{2!} - \frac{(x-\pi)^4}{4!} + \cdots$$

Note: Due to space restrictions in this manual, answers only are provided in Sections 28.8 and 28.9.

Exercises 28.8

1. 1.10517 **3.** 0.99985 **5.** −0.68229 **7.** 1.0488 **9.** 3.66832

11. 0.48481 **13.** 0.029996

Exercises 28.9

1. $f(x) = -\pi + 2\sin x + \sin 2x + (2/3)\sin 2x/3 + \cdots$

$f(x) = -x,\ 0 \le x < 2\pi$

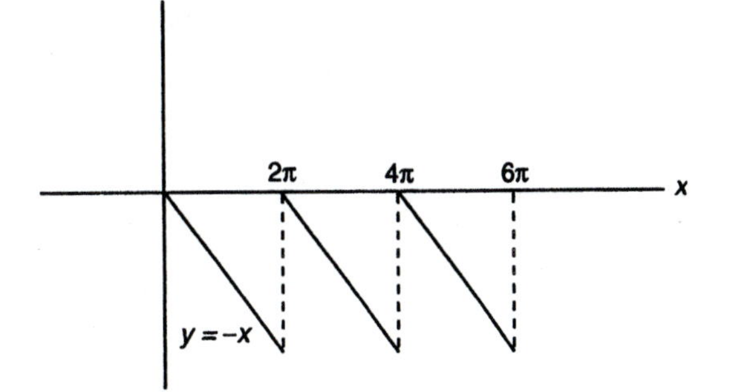

3. $f(x) = \pi/3 - (2/3)\sin x - (1/3)\sin 2x - (2/9)\sin 3x - \cdots$

$f(x) = \dfrac{1}{3}x,\ 0 \le x < 2\pi$

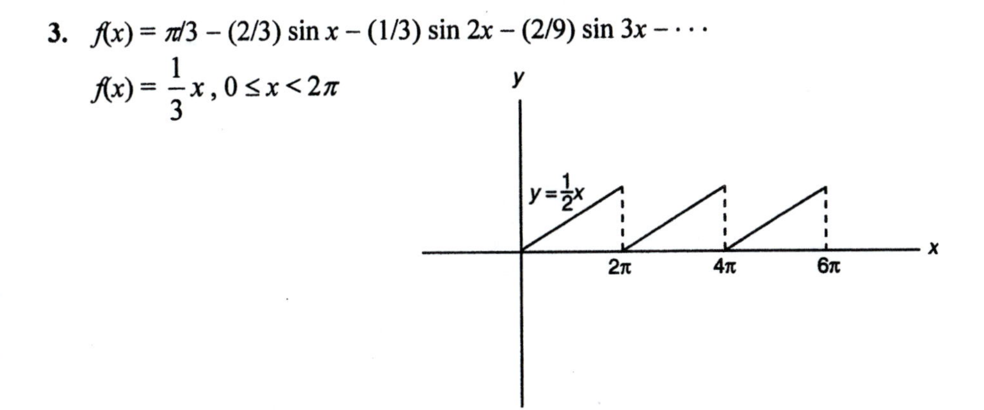

5. $f(x) = 1/2 - (2/\pi)\sin x - (2/3\pi)\sin 3x - (2/5\pi)\sin 5x - \cdots$

$$f(x) = \begin{cases} 0 \le x < \pi \\ 1\pi \le x < 2\pi \end{cases}$$

7. $f(x) = (4/\pi)\sin x + (4/3\pi)\sin 3x + (4/5\pi)\sin 5x + \cdots$

$$f(x) = \begin{cases} 1 & 0 \le x < \pi \\ -1 & \pi \le x < 2\pi \end{cases}$$

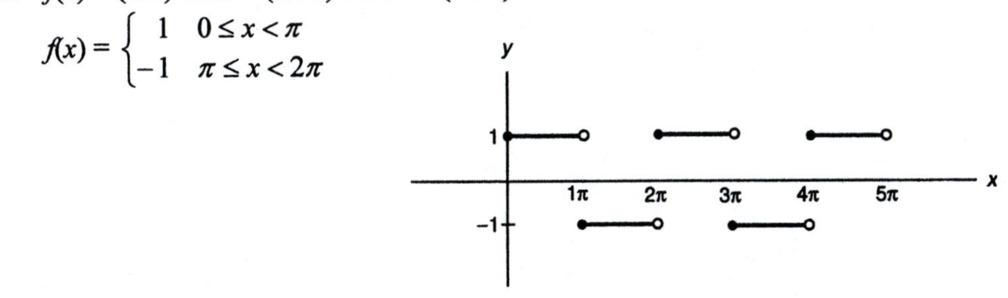

9. $f(x) = 3 + \frac{12}{\pi}\sin\frac{\pi x}{5} + \frac{4}{\pi}\sin\frac{3\pi x}{5} + \frac{12}{5\pi}\sin \pi x + \frac{12}{7\pi}\sin\frac{7\pi x}{5} + \cdots$

$f(x) = \begin{cases} 0 & -5 \le x < 0 \\ 6 & 0 \le x < 5 \end{cases}$

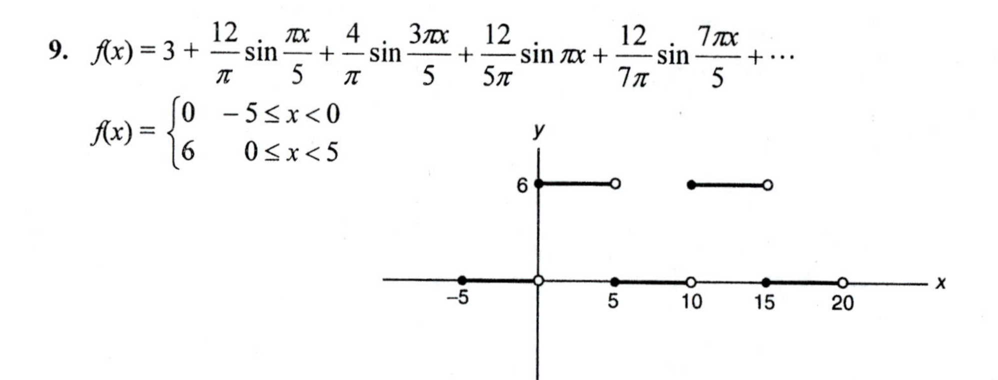

11. $f(x) = \pi/2 - (4/\pi)\cos x - (4/9\pi)\cos 3x - (4/25\pi)\cos 5x - \cdots$

$f(x) = \begin{cases} x & 0 \le x < \pi \\ 2\pi - x, & \pi \le x < 2\pi \end{cases}$

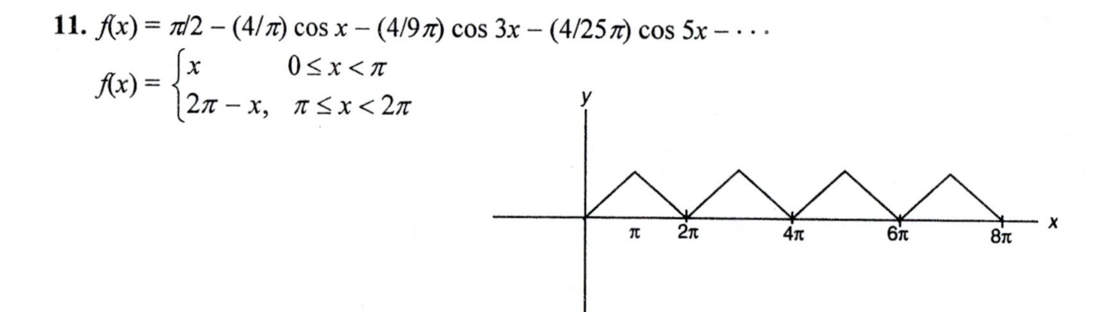

13. $f(x) = \frac{e^{2\pi} - 1}{2\pi} + \frac{e^{2\pi} - 1}{2\pi}\cos x + \frac{e^{2\pi} - 1}{5\pi}\cos 2x + \frac{e^{2\pi} - 1}{10\pi}$

$\cos 3x + \cdots + \frac{1}{\pi}\cdot\frac{-e^{2\pi} + 1}{2}\sin x + \frac{1}{\pi}\cdot\frac{-e^{2\pi} + 2}{5}\sin 2x + \cdots$

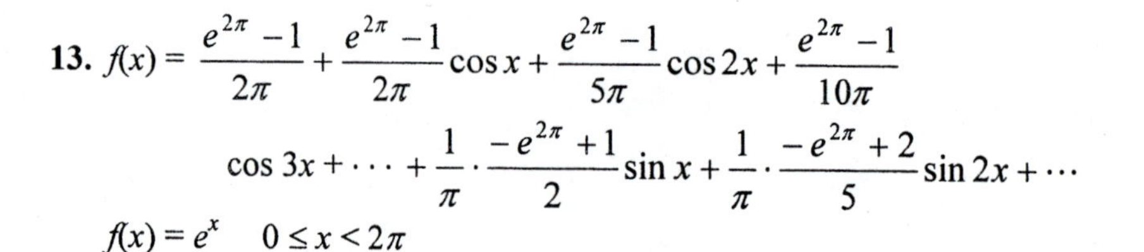

$f(x) = e^x \quad 0 \le x < 2\pi$

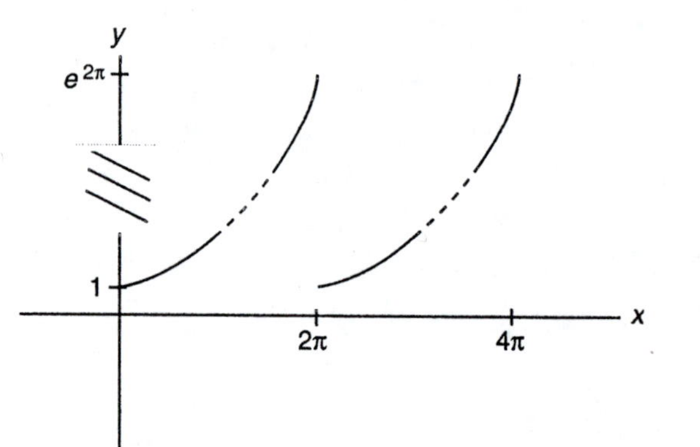

Section 28.9

15. $\frac{1}{\pi}+\frac{1}{2}\sin x-\frac{2}{\pi}\left\{\frac{1}{3}\cos 2x+\frac{1}{15}\cos 4x+\cdots\right\}$

$$f(x)=\begin{cases}\sin x & 0\le x<\pi\\ 0 & \pi\le x<2\pi\end{cases}$$

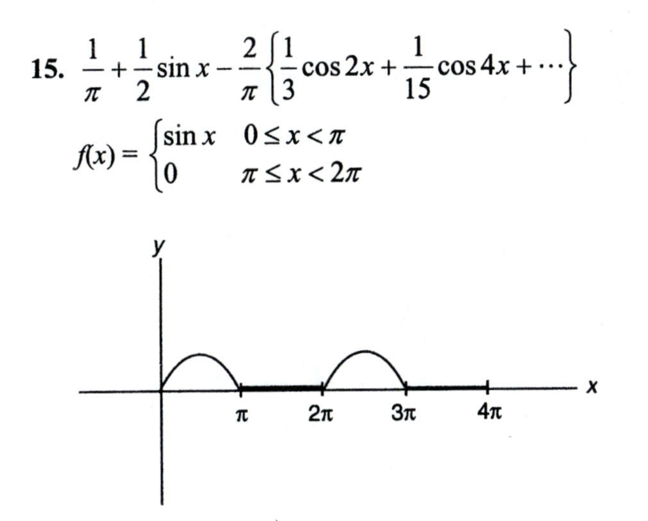

Chapter 28 Review

1. $-2-5-8-11-14-17$

2. $2+3/2+4/3+5/4+\cdots+(n+1)/n$

3. $\sum_{n=1}^{7}1/3^{n}$

4. $\sum_{n=1}^{10}\frac{n}{n+3}$

5. Converges; p-series with $p=3$

6. Diverges; p-series with $p=1/4$

7. Converges; $\lim_{n\to\infty}\frac{1/(6n^2+2)}{1/n^2}=\lim_{n\to\infty}\frac{n^2}{6n^2+2}=\frac{1}{6}$ Since the given series and $\sum 1/n^2$ have the same order of magnitude and $\sum 1/n^2$ converges, the given series also converges by the limit comparison test.

8. Converges; Compare with $\sum\sqrt{n}/n^2=\sum 1/n^{3/2}$, which is a convergent p-series.

$\lim_{n\to\infty}\frac{\sqrt{n}/(n^2-1)}{\sqrt{n}/n^2}=\lim_{n\to\infty}\frac{n^2}{n^2-1}=1$ Since both series have the same order of magnitude, both series converge by the limit comparison test.

9. Diverges; using the integral test, $\lim_{b\to\infty}\int_2^b\frac{\ln x\,dx}{x}=\lim_{b\to\infty}\frac{\ln^2 x}{2}\bigg|_2^b=\lim_{b\to\infty}\left\{\frac{\ln^2 b}{2}-\frac{\ln^2 2}{2}\right\}=\infty$

10. Converges; $r=\lim_{n\to\infty}\frac{5n+7}{(3n+4)4^{n+1}}\cdot\frac{(3n+1)4^n}{5n+2}=\lim_{n\to\infty}\frac{(5n+7)(3n+1)}{4(3n+4)(5n+2)}$

$$=\lim_{n\to\infty}\frac{(5+7/n)(3+1/n)}{4(3+4/n)(5+2/n)}=\frac{1}{4}<1$$

The series converges by the ratio test.

11. Converges; $r = \lim_{n\to\infty} \frac{(n+1)^3}{2^{n+1}} \cdot \frac{2}{n^3} = \lim_{n\to\infty} \frac{(n+1)^3}{2n^3} = \frac{1}{2} < 1$
This series converges by the ratio test.

12. Diverges; $\lim_{n\to\infty} \frac{3+1/n}{4-5/n} = \frac{3}{4}$

13. Diverges; compare with $\sum 1/n$, which diverges. $\lim_{n\to\infty} \frac{(n+1)/(n^2+4n)}{1/n} = \lim_{n\to\infty} \frac{n+1}{n+4} = 1$
Since both series have the same order of magnitude, both series diverge by the limit comparison test.

14. Converges; compare with $\sum 1/n^2$. $\frac{|\sin n|}{n^2} \le \frac{1}{n^2}$ Thus the given series converges by the comparison test.

15. Converges absolutely; $r = \lim_{n\to\infty} \frac{3^{n+1}}{(n+1)!} \cdot \frac{n!}{3^n} = \lim_{n\to\infty} \frac{3}{n+1} = 0 < 1$

Thus the series $\sum 3^n/n!$ converges by the ratio test.

16. Converges absolutely; $r = \lim_{n\to\infty} \frac{2^{n+1}}{5^{n+1}(n+2)} \cdot \frac{5^n(n+1)}{2^n} = \lim_{n\to\infty} \frac{2(n+1)}{5(n+2)} = \frac{2}{5} < 1$ Thus the given series

$\frac{2^n}{5^n(n+1)}$ converges by the ratio test.

17. Diverges; $\lim_{n\to\infty} \frac{n+1}{n-1} = 1$

18. Conditionally convergent; the alternating series converges by the alternating series test. The corresponding series of positive terms diverges by the limit comparison test with $\sum 1/n$.

19. $\lim_{n\to\infty} \left| \frac{[(n+1)/2](x-2)^{n+1}}{(n/2)(x-2)^n} \right| = |x-2| \lim_{n\to\infty} \frac{n+1}{n} = |x-2| < 1$ or $1 < x < 3$. For $x = 3$, $\sum n/2$ diverges. For $x = 1$, $\sum (-1)^n (n/2)$ diverges. Thus the interval of convergence is $1 < x < 3$.

20. $\lim_{n\to\infty} \left| \frac{(n+1)^2(x-3)^{n+1}}{n^2(x-3)^n} \right| = |x-3| \lim_{n\to\infty} \frac{(n+1)^2}{n^2} = |x-3| < 1$ or $2 < x < 4$. For $x = 4$, the series $\sum n^2$ diverges. For $x = 2$, the series $\sum (-1)^n n^2$ diverges. The interval of convergence is $2 < x < 4$.

21. $\lim_{n\to\infty}\left|\frac{(x-1)^{n+1}}{(n+1)!}\cdot\frac{n!}{(x-1)^n}\right| = |x-1|\lim_{n\to\infty}\frac{1}{n+1} = 0$

Thus the interval of convergence is $-\infty < x < \infty$.

22. $\lim_{n\to\infty}\left|\frac{3^{n+1}(x-4)^{n+1}}{(n+1)^2}\cdot\frac{n^2}{3^n(x-4)^n}\right| = 3|x-4|\lim_{n\to\infty}\frac{n^2}{(n+1)^2} = 3|x-4| < 1$ or $11/3 < x < 13/3$.

For $x = 13/3$, the series $\sum 1/n^2$ converges. For $x = 11/3$, the series $\sum(-1)^n/n^2$ also converges.

Thus the interval of convergence is $11/3 \le x \le 13/3$.

23. $f(x) = 1/(1-x); f(0) = 1$
$f'(x) = 1/(1-x)^2; f'(0) = 1$
$f''(x) = 2/(1-x)^3; f''(0) = 2$
$f'''(x) = 6/(1-x)^4; f'''(0) = 6$
Thus $\frac{1}{1-x} = 1 + x + x^2 + x^3 + \cdots$

24. $f(x) = (x+1)^{1/2}; f(0) = 1$
$f'(x) = (1/2)(x+1)^{-1/2}; f'(0) = 1/2$
$f''(x) = (-1/4)(x+1)^{-3/2}; f''(0) = -1/4$
$f'''(x) = (3/8)(x+1)^{-5/2}; f'''(0) = 3/8$
$f^{(4)}(x) = (-15/16)(x+1)^{-7/2}; f^{(4)}(0) = -15/16$
Thus $\sqrt{x+1} = 1 + x/2 - x^2/8 + x^3/16 - 5x^4/128 + \cdots$

25. $f(x) = \sin x + \cos x; f(0) = 1$
$f'(x) = \cos x - \sin x; f'(0) = 1$
$f''(x) = -\sin x - \cos x; f''(0) = -1$
$f'''(x) = -\cos x + \sin x; f'''(0) = -1$
$f^{(4)}(x) = \sin x + \cos x; f^{(4)}(0) = 1$
Thus $\sin x + \cos x = 1 + x - x^2/2! - x^3/3! + x^4/4! + \cdots$

26. $f(x) = e^x \sin x; f(0) = 0$
$f'(x) = e^x(\cos x + \sin x); f'(0) = 1$
$f''(x) = 2e^x \cos x; f''(0) = 2$
$f'''(x) = 2e^x(\cos x - \sin x); f'''(0) = 2$
Thus $e^x \sin x = 0 + x + 2x^2/2! + 2x^3/3! + \cdots = x + x^2 + x^3/3 + \cdots$

27. $\frac{1-e^x}{x} = \frac{1}{x} - \frac{e^x}{x} = \frac{1}{x} - \left[\frac{1+x+x^2/2!+x^3/3!+\cdots}{x}\right] = \frac{1}{x} - \left[\frac{1}{x}+1+\frac{x}{2!}+\frac{x^2}{3!}+\cdots\right] = -1-\frac{x}{2!}-\frac{x^2}{3!}-\cdots$

28. Substitute x^2 for x in equation (3)in Section 28.6.
$\cos x^2 = 1 - x^4/2! + x^8/4! - x^{12}/6! + \cdots$

29. Substitute $3x$ for x in equation (2) in Section 28.6.
$\sin 3x = 3x - 9x^3/2 + 81x^5/40 - \cdots$

30. Substitute sin x for x in equation (1) in Section 28.6.

$$e^{\sin x} = 1 + \sin x + \frac{\sin^2 x}{2!} + \frac{\sin^3 x}{3!} + \cdots$$

31. Use equation (4) in Section 28.6.

$$\int_0^{0.1} \frac{\ln(x+1)}{x} dx = \int_0^{0.1} (1 - x/2 + x^2/3 - x^3/4) dx = (x - x^2/4 + x^3/9 - x^4/16) \Big|_0^{0.1} = 0.09772$$

32. Use equation (2) in Section 10.6.

$$\int_0^{0.1} \frac{\sin t}{t} dt = \int_0^{0.1} (1 - t^2/3! + t^4/5!) dt = \left[t - \frac{t^3}{3!(3)} + \frac{t^5}{5!(5)} \right] \Bigg|_0^{0.1} = 0.09994$$

33. $f(x) = \cos 2x; f(\pi/6) = 1/2$

$f'(x) = -2 \sin 2x; f'(\pi/6) = -\sqrt{3}$

$f''(x) = -4 \cos 2x; f''(\pi/6) = -2$

$f'''(x) = 8 \sin 2x; f'''(\pi/6) = 4\sqrt{3}$

$f^{(4)}(x) = 16 \cos 2x; f^{(4)}(\pi/6) = 8$

$$\text{Thus } f(x) = \frac{1}{2} - \sqrt{3}\left(x - \frac{\pi}{6}\right) - 2\left(x - \frac{\pi}{6}\right)^2 + 4\sqrt{3}\left(x - \frac{\pi}{6}\right)^3 + 8\left(x - \frac{\pi}{6}\right)^4 + \cdots$$

34. $f(x) = \ln x; f(4) = \ln 4$

$f'(x) = 1/x; f'(4) = 1/4$

$f''(x) = -1/x^2; f''(4) = -1/16$

$f'''(x) = 2/x^3; f'''(4) = 1/32$

$$\text{Thus } f(x) = \ln 4 + \frac{x-4}{4} - \frac{(x-4)^2}{32} + \frac{(x-4)^3}{192} - \cdots$$

35. $f(x) = e^{x^2}; f(1) = e$

$f'(x) = 2x\, e^{x^2}; f'(1) = 2e$

$f''(x) = e^{x^2}(4x^2 + 2); f''(1) = 6e$

$f'''(x) = e^{x^2}(8x^3 + 12x); f'''(1) = 20e$

$$\text{Thus } f(x) = e\left[1 + 2(x-1) + \frac{6(x-1)^2}{2!} + \frac{20(x-1)^3}{3!} + \cdots\right]$$

36. $f(x) = \sin x; f(3\pi/2) = -1$

$f'(x) = \cos x; f'(3\pi/2) = 0$

$f''(x) = -\sin x; f''(3\pi/2) = 1$

$f'''(x) = -\cos x; f'''(3\pi/2) = 0$

$f^{(4)}(x) = \sin x; f^{(4)}(3\pi/2) = -1$

$$\text{Thus } f(x) = -1 + \frac{(x - 3\pi/2)^2}{2!} - \frac{(x - 3\pi/2)^3}{4!} + \cdots$$

37. $(31° = 31\pi/180)$ Use equation (2) in Section 10.6.

$$\sin 31° = 31\ \pi/180 - \frac{(31\pi/180)^3}{3!} + \frac{(31\pi/180)^5}{5!} = 0.5150$$

38. $e^{x-1} = e\left[1 + (x-1) + \frac{(x-1)^2}{2!} + \frac{(x-1)^3}{3!}\right]$ (used Taylor)

$= e[1 + 0.2 + (0.2)^2/2! + (0.2)^3/3!] = 3.3199$

39. Use equation (4) in Section 10.6 and let $x = 0.2$.
$\ln 1.2 = \ln(1 + 0.2) = 0.2 - (0.2)^2/2 + (0.2)^3/3 - (0.2)^4/4 = 0.18227$

40. Use Exercise 4 in Section 10.7 with $x - a = 4.1 - 4 = 0.1$

$$\sqrt{4.1} = 2 + \frac{0.1}{4} - \frac{(0.1)^2}{2!(32)} + \frac{3(0.1)^3}{3!(256)} = 2.024846$$

41. $f(x) = -1/2 + (2/\pi)\sin x + (2/3\pi)\sin 3x + (2/5\pi)\sin 5x + \cdots$

$$f(x) = \begin{cases} 0 & 0 \le x < \pi \\ -1 & \pi \le x < 2\pi \end{cases}$$

42. $f(x) = \pi^2/6 - 2\cos x + (1/2)\cos 2x - (2/9)\cos 3x + (1/8)\cos 4x -$

$$\cdots + \frac{\pi^2 - 4}{\pi}\sin x - \frac{\pi}{2}\sin 2x + \frac{9\pi^2 - 4}{27\pi}\sin 3x - \cdots$$

$$f(x) = \begin{cases} x^2 & 0 \le x < \pi \\ 0 & \pi \le x < 2\pi \end{cases}$$

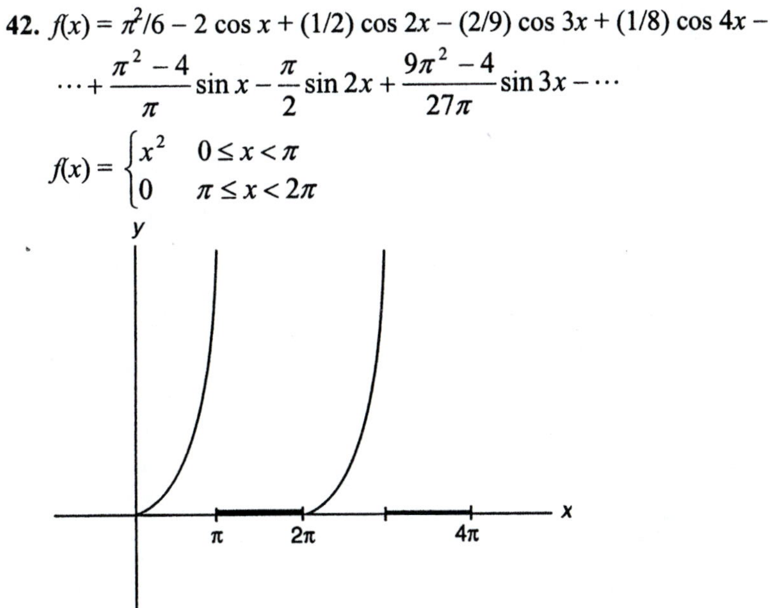

Chapter 29, Exercises 29.1

1. order 1, degree 1

3. order 2, degree 1

5. order 3, degree 1

7. order 2, degree 3

9. $y' = 3 = dy/dx$

11. $y' = 2x - 4;\ xy' - 2y = 4x;\ x(2x - 4) - 2(x^2 - 4) = 4x;\ 4x = 4x$

13. $y' = -(x + 2)e^{-x} + e^{-x}; y' + y = e^{-x};$
$[-(x + 2)e^{-x} + e^{-x}] + (x + 2)e^{-x} = e^{-x}$
$e^{-x} = e^{-x}$

15. $y' = 4C_1 \cos 4x - 4C_2 \sin 4x;\ y'' = -16C_1 \sin 4x - 16C_2 \cos 4x;$
$y'' + 16y = 0; (-16C_1 \sin 4x - 16C_2 \cos 4x) + 16(C_1 \sin 4x + C_2 \cos 4x) = 0$
$0 = 0$

17. $y' = \cos x - \sin x + e^{-x};\ y' + y - 2 \cos x = 0;$
$(\cos x - \sin x + e^{-x}) + (\sin x + \cos x - e^{-x}) - 2 \cos x = 0;$
$0 = 0$

19. $y' = \cos^2 x - \sin^2 x;\ y'' = -4 \sin x \cos x;\ (y'')^2 + 4(y')^2 = 4;$
$(-4 \sin x \cos x)^2 + 4(\cos^2 x - \sin^2 x)^2 = 4;$
$16 \sin^2 x \cos^2 x + 4 \cos^4 x - 8 \cos^2 x \sin^2 x + 4 \sin^4 x = 4;$
$4 \cos^4 x + 8 \sin^2 x \cos^2 x + 4 \sin^4 x = 4;$
$4(\cos^4 x + 2 \sin^2 x \cos^2 x + 4 \sin^4 x) = 4;$
$4(\cos^2 x + \sin^2 x)^2 = 4;\ 4 = 4$

21. $y' = 4e^{4x};\ y'' = 16e^{4x};\ y'' - 5y' + 4y = 0;$
$16e^{4x} - 5(4e^{4x}) + 4e^{4x} = 0;\ 0 = 0$

23. $y' = e^{-x}(1 - x);\ y'' = -e^{-x}(2 - x);\ y'' + 2y' + y = 0;$
$[-e^{-x}(2 - x)] + 2[e^{-x}(1 - x)] + xe^{-x} = 0;$
$-2e^{-x} + xe^{-x} + 2e^{-x} - 2xe^{-x} + xe^{-x} = 0;\ 0 = 0$

Exercises 29.2

1. $x\,dy - y^2\,dx = 0;\ \int dy/y^2 = \int dx/x; -1/y = \ln x - C;$
$y \ln x + 1 = Cy$

3. $x\,dy + y\,dx = 0;\ \int dy/y + \int dx/x = 0;\ \ln y + \ln x = \ln C;$
$\ln yx = \ln C; xy = C$

5. $dy/dx = y^{-3/2}; = \int y^{-3/2} dy = \int dx;\ -2y^{-1/2} = x + C$

7. $dy/dx = x^2(1 + y^2)$; $\int dy/(1+y^2) = \int x^2 dx$; Arctan $y = x^3/3 + C$

9. $x\,dy/dx + y = 3$; $x\,dy = (3 - y)dx$; $\int dy/(3-y) = \int dx/x$;
$-\ln(3 - y) = \ln x - \ln C$; $\ln C = \ln x(3 - y)$; $x(3 - y) = C$

11. $dy/dx = x^2/y$; $\int y\,dy = \int x^2 dx$; $y^2/2 = x^3/3 + C$; $3y^2 = 2x^3 + C$

13. $dy/dx + y^3 \cos x = 0$; $\int dy/y^3 + \int \cos x\,dx = 0$;
$-1/(2y^2) + \sin x + C = 0$; $1 = 2y^2(\sin x + C)$

15. $e^{3x}\,dy/dx - e^x = 0$; $\int dy = \int e^{-2x} dx$; $y = (-1/2)\,e^{-2x} + C$; $2y = -e^{-2x} + C$

17. $(1 + x^2)\,dy - dx = 0$; $\int dy = \int dx/(1+x^2)$; $y =$ Arctan $x + C$

19. $dy/dx = 1 + x^2 + y^2(1 + x^2)$; $dy/dx = (1 + x^2)(1 + y^2)$
$\int dy/(1+y^2) = \int (1+x^2)dx$; Arctan $y = x + x^3/3 + C$

21. $dy/dx = e^x \cdot e^{-y}$; $\int e^y dy = \int e^x dx$; $e^y = e^x + C$; $e^y - e^x = C$

23. $(x + 1)\,dy/dx = y^2 + 4$; $\int dy/(y^2+4) = \int dy/(x+1)$;
$(1/2)$ Arctan $(y/2) = \ln(x + 1) + C$; Arctan $(y/2) = 2\ln(x + 1) + C$

25. $(4xy + 12x)\,dx = (5x^2 + 5)\,dy$; $4x(y + 3)\,dx = 5(x^2 + 1)\,dy$;
$\int \frac{4x\,dx}{x^2+1} = \int \frac{5\,dy}{y+3}$; $2\ln(x^2 + 1) = 5\ln(y + 3) + \ln C$;
$\ln(x^2 + 1)^2 = \ln(y + 3)^5 + \ln C$; $\ln \frac{(x^2+1)^2}{(y+3)^5} = \ln C$; $(x^2 + 1)^2 = C(y + 3)^5$

27. $dy/dx = x^2y^4$; $\int dy/y^4 = \int x^2 dx$; $(-1/3)y^{-3} = (1/3)x^3 + C$;
for $x = 1, y = 1, C = -2/3$; Thus $0 = x^3y^3 + 1 - 2y^3$; $y^3(2 - x^3) = 1$

29. $\frac{dy}{dx} = \frac{2x}{y + x^2y}$; $\int y\,dy = \int \frac{2x\,dx}{x^2+1}$; $y^2/2 = \ln(x^2 + 1) + C$;
for $x = 0, y = 4, C = 8$; Thus $y^2 = 2\ln(x^2 + 1) + 16$

31. $y\,dy/dx = e^x$; $\int y\,dy = \int e^x dx$; $y^2/2 = e^x + C$; for $x = 0, y = 6, C = 17$ Thus $y^2 = 2e^x + 34$

33. $\sqrt{x} + \sqrt{y}\, dy/dx = 0$; $\int x^{1/2}dx + \int y^{1/2}dy = 0$;
$(2/3)x^{3/2} + (2/3)y^{3/2} = C$; for $x = 1, y = 4, C = 6$
Thus $x^{3/2} + y^{3/2} = 9$

35. $xy\, dy/dx = \ln x$; $\int y\, dy = \int \frac{\ln x}{x}dx$; $y^2/2 = (1/2)\ln^2 x + C$;
$y^2 = \ln^2 x + C$; for $x = 1, y = 0, C = 0$; Thus $y^2 = \ln^2 x$

Exercises 29.3

1. $x\, dy + y\, dx = y^2\, dy$; $\int d(xy) = \int y^2 dy$; $xy = y^3/3 + C$
$3xy = y^3 + C$

3. $x\, dy - y\, dx = 5x^2\, dy$; $\frac{x\, dy - y\, dx}{x^2} = 5\, dy$; $\int d\,(y/x) = 5\int dy$;
$y/x = 5y + C$; $y = 5xy + Cx$

5. $y\, dx - x\, dy + y^2\, dx = 3\, dy$; $\frac{y\, dx - x\, dy}{y^2} + dx = \frac{3}{y^2}dy$;
$\int d\,(x/y) + \int dx = 3\int y^{-2}dy$; $x/y + x = -3/y - C$;
$x + xy = -3 - Cy$; $x + xy + Cy + 3 = 0$

7. $x\sqrt{x^2 + y^2}\, dx - 2x\, dx = 2y\, dy$; $x\sqrt{x^2 + y^2}\, dx = 2(x\, dx + y\, dy)$;
$x\, dx = 2\frac{x\, dx + y\, dy}{\sqrt{x^2 + y^2}}$; $\int x\, dx = \int \frac{d(x^2 + y^2)}{\sqrt{x^2 + y^2}}$;
$x^2/2 = 2(x^2 + y^2)^{1/2} + C$; $x^2 = 4\sqrt{x^2 + y^2} + C$

9. $x\, dx + y\, dy = x(x^2 + y^2)\, dy + y(x^2 + y^2)\, dx$; $\frac{x\, dx + y\, dy}{x^2 + y^2} = x\, dy + y\, dx$;
$\int d(\ln\sqrt{x^2 + y^2}) = \int d(xy)$; $\ln\sqrt{x^2 + y^2} = xy + C$

11. $x\, dy + y\, dx = 2(x\, dx + y\, dy)$; $\int d(xy) = \int d(x^2 + y^2)$;
$xy = x^2 + y^2 + C$; for $x = 0, y = 1, C = -1$;
Thus $xy = x^2 + y^2 - 1$

13. $x\, dy - y\, dx = (x^3 + y^2x)\, dy + (x^2y + y^3)\, dx$;
$x\, dy - y\, dx = x(x^2 + y^2)dy + y(x^2 + y^2)\, dx$; $\frac{x\, dy - y\, dx}{x^2 + y^2} = x\, dy + y\, dx$;
$\int d(\text{Arctan } y/x) = \int d(xy)$; $\text{Arctan } y/x = xy + C$; for $x = 2, y = 2$, $C = \frac{\pi - 16}{4}$;
$\text{Arctan } \frac{y}{x} = xy + \frac{\pi - 16}{4}$

Sections 11.2 – 11.3

Exercises 29.4

1. $dy/dx - 5y = e^{3x}$; $dy - 5y\,dx = e^{3x}\,dx$; $P(x) = -5$, $Q(x) = e^{3x}$,
$\int P(x)\,dx = -5x$, $ye^{-5x} = \int e^{3x}e^{-5x}dx$;
$ye^{-5x} = (-1/2)e^{-2x} + C$; $2y + e^{3x} = Ce^{5x}$

3. $dy/dx + 3y/x = x^3 - 2$; $P(x) = 3/x$, $Q(x) = x^3 - 2$,
$\int P(x)\,dx = \int (3/x)dx = 3\ln x = \ln x^3$;
$ye^{\ln x^3} = \int (x^3 - 2)x^3 dx$; $yx^3 = y^7/7 - x^4/2 + C$;
$14yx^3 = 2x^7 - 7x^4 + C$

5. $dy/dx + 2xy = e^{3x}(3 + 2x)$; $P(x) = 2x$, $Q(x) = e^{3x}(3 + 2x)$,
$\int P(x)\,dx = x^2$;
$ye^{x^2} = \int e^{3x}(3 + 2x)e^{x^2}dx = \int e^{x^2+3x}(3 + 2x)\,dx = e^{x^2+3x} + C$; $y = e^{3x} + Ce^{-x^2}$

7. $dy - 4y\,dx = x^2 e^{4x}\,dx$; $P(x) = -4$, $Q(x) = x^2 e^{4x}$,
$\int P(x)\,dx = \int -4x$, $ye^{-4x} = \int x^2e^{4x}e^{-4x}dx = \int x^2dx = x^3/3 + C$; $3y = x^3 e^{4x} + Ce^{4x}$

9. $x\,dy - 5y\,dx = (x^6 + 4x)\,dx$; $dy - (5/x)\,y\,dx = (x^5 + 4)\,dx$;
$P(x) = -5/x$, $Q(x) = x^5 + 4$, $\int P(x)\,dx = \int (-5/x)\,dx = -5\ln/x = \ln x^{-5}$, $e^{P(x)\,dx} = e^{\ln x^{-5}} = x^{-5}$;
$yx^{-5} = \int (x^5 + 4)x^{-5}dx = \int (1 + 4x^{-5})\,dx = x - 1/x^4 + C$; $y = x^6 - x + Cx^5$

11. $(1 + x^2)\,dy + 2xy\,dx = 3x^2\,dx$; $P(x) = \dfrac{2x}{x^2+1}$;
$Q(x) = \dfrac{3x^2}{x^2+1}$; $\int P(x)\,dx = \int \dfrac{2x\,dx}{x^2+1} = \ln(x^2 + 1)$;
$ye^{\ln(x^2+1)} = \int \dfrac{3x^2}{x^2+1}(x^2 + 1)\,dx = x^3 + C$;
$y(x^2 + 1) = x^3 + C$

13. $dy/dx + (2/x)\,y = (x^2 - 7/x^2)$; $P(x) = 2/x$, $Q(x) = x^2 - 7x^{-2}$,
$\int P(x)\,dx = \int (2/x)\,dx = 2\ln x = \ln x^2$; $e^{\int P(x)(dx)} = e^{\ln x^2} = x^2$;
$yx^2 = \int (x^4 - 7)dx = x^5/5 - 7x + C$; $5yx^2 = x^5 - 35x + C$

15. $dy/dx + 2y = e^{-x}$; $P(x) = 2$, $Q(x) = e^{-x}$, $\int P(x)\,dx = 2x$;
$ye^{2x} = \int e^{-x}e^{2x}\,dx = \int e^{x}dx = e^{x} + C$; $y = e^{-x} + Ce^{-2x}$

17. $dy/dx - (1/x)\,y = 3x;\ \ P(x) = -1/x,\ Q(x) = 3x,$

$\int P(x)\,dx = \int(-1/x)\,dx = -\ln x = \ln x^{-1}; e^{\ln x^{-1}} = x^{-1};$

$yx^{-1} = \int 3x\,x^{-1}dx = \int 3\,dx = 3x + C\,;\, y = 3x^2 + Cx$

19. $dy/dx + y\cos x = \cos x;\ P(x) = \cos x;\ Q(x) = \cos x;$

$\int P(x)\,dx = \int \cos x\,dx = \sin x; ye^{\sin x} = \int \cos x\,e^{\sin x}dx = e^{\sin x} + C; (y-1)e^{\sin x} = C$

21. $dy/dx - 3y = e^{2x};\ P(x) = -3,\ Q(x) = e^{2x},\ \int P(x)\,dx = -3x;$

$ye^{-3x} = \int e^{2x}e^{-3x}dx = \int e^{-x}dx = -e^{-x} + C;$

$y = -e^{2x} + Ce^{3x};$ for $x = 0,\ y = 2,\ C = 3;$
$y = -e^{2x} + 3e^{3x} = e^{2x}(3e^{x} - 1)$

23. $dy/dx + y\cot x = \csc x;\ P(x) = \cot x,\ Q(x) = \csc x,\ \int P(x)\,dx = \int \cot x\,dx = \ln(\sin x);$

$ye^{\ln(\sin x)} = \int \csc x\,e^{\ln(\sin x)}dx\,;$

$y\sin x = \int \csc x\sin x\,dx = \int dx = x + C;\ y\sin x = x + C;$

for $x = \pi/2,\ y = 3\pi/2,\ C = \pi;\ y\sin x = x + \pi$

25. $dy/dx + y = e^{x};\ P(x) = 1,\ Q(x) = e^{x},\ \int P(x)\,dx = x;$

$ye^{x} = \int e^{x}e^{x}dx = \int e^{2x}dx = (1/2)e^{2x} + C;$ for $x = 0,\ y = 3/2,\ C = 1;$

$ye^{x} = (1/2)e^{2x} + 1;\ y = e^{x}/2 + e^{-x}$

27. $dy/dx + (1/x)\,y = 3/x;\ \ P(x) = 1/x,\ Q(x) = 3/x,$

$\int P(x)\,dx = \int(1/x)dx = \ln x;\ ye^{\ln x} = \int(3/x)x\,dx = 3x + C;$

$yx = 3x + C;$ for $x = 1,\ y = -2,\ C = -5,\ yx = 3x - 5;$
$y = 3 - 5/x$

29. $dy/dx + (1/x)y = 4x^2;\ \ P(x) = 1/x,\ Q(x) = 4x^2,\ \int P(x)\,dx = \int(1/x)\,dx = \ln x$

$ye^{\ln x} = \int 4x^2x\,dx\,;\ yx = x^4 + C;$ for $x = 2,\ y = 3,\ C = -10;$

$xy = x^4 - 10$

Exercises 29.5

1. $v = \int dv = \int a\,dt = \int 5\,dt = 4t + C;\ \ v = 10$ m/s at $t = 0$ so $C = 10;$

$v = 5t + 10\Big|_{t=3} = 25$ m/s

3. $dy/dx = (x^2 - y)/x;\ x\,dy = x^2\,dx - y\,dx;\ x\,dy + y\,dx = x^2\,dx;$

$\int d(xy) = \int x^2dx;\ \ xy = x^3/3 + C;$ at $(1, 1),\ C = 2/3,\ xy = x^3/3 + 2/3;\ \ 3xy = x^3 + 2$

5. $di/dt + (R/L)i = V/L$; $di/dt + 800i = 1200$ since $L = 0.1H$, $R = 80\ \Omega$, $V = 120\ V$. $P(t) = 800$, $Q(t) = 1200$, $\int P(t)\,dt = 800t$;

$ie^{800t} = \int 1200e^{800t}\,dt = (3/2)\,e^{800t} + C$; at $t = 0$; $i = 2A$

Thus $C = 1/2$; $ie^{800t} = (3/2)\,e^{800t} + 1/2$; $i = (1/2)(3 + e^{-800t})$

7. $\frac{di}{dt} + \left(R/L\right)i = V/L$; $L = 1H, R = 4\Omega, V = 10 \sin 2t$ volts

$\frac{di}{dt} + \frac{4}{1}i = 10\sin 2t$

$ie^{\int 4dt} = \int 10\sin 2te^{\int 4dt}\,dt$

$ie^{4t} = 10\int \sin 2t\left(e^{4t}\right)dt$

First find $\int e^{4t}\ \sin 2t\ dt$

Let $u = \sin 2t \quad dv = e^{4t}\,dt$

$du = 2\cos 2t\,dt \qquad v = \frac{1}{4}e^{4t}$

$\int e^{4t}\ \sin 2t\ dt = \frac{1}{4}e^{4t}\sin 2t - \int \frac{1}{2}e^{4t}\cos 2t\,dt$

Let $\quad u = \cos 2t \quad dv = \frac{1}{2}e^{4t}dt$

$du = -2\sin 2t \cdot v = \frac{1}{8}e^{4t}$

$\int e^{4t}\sin 2t\,dt = \frac{1}{4}e^{4t}\sin 2t - \left[\frac{1}{8}e^{4t}\cos 2t - \int \frac{-1}{4}e^{4t}\sin 2t\,dt\right]$

$\int e^{4t}\sin 2t\,dt = \frac{1}{4}e^{4t}\sin 2t - \frac{1}{8}e^{4t}\cos 2t - \frac{1}{4}\int e^{4t}\sin 2t\,dt$

$\frac{5}{4}\int e^{4t}\sin 2t\,dt = \frac{1}{4}e^{4t}\sin 2t - \frac{1}{8}e^{4t}\cos 2t$

thus: $\int e^{4t}\sin 2t\,dt = \frac{1}{5}e^{4t}\sin 2t - \frac{1}{10}e^{4t}\cos 2t$

So $ie^{4t} = 10\int e^{4t}\sin 2t\,dt$ becomes

$ie^{4t} = 10\left[\frac{1}{5}e^{4t}\sin 2t - \frac{1}{10}e^{4t}\cos 2t\right]$

$ie^{4t} = 2e^{4t}\sin 2t - e^{4t}\cos 2t + C$

$i = 2\sin 2t - \cos 2t \ + Ce^{-4t}$

At $t = 0$, $i_o = 0$, so

$0 = 2\sin 0 - \cos 0 + Ce^0$

$0 = -1 + C$ thus $C = 1$

Solution is: $i = 2\sin 2t - \cos 2t + e^{-4t}$

9. $dQ/dt = kQ$; $Q = Q_o e^{kt}$; $Q_o = 1$ g, $Q = 0.5$ g when $t = 4.5 \times 10^9$ yr

$0.5 = 1\,e^{(4.5\times10^9)k}$; $k = \dfrac{\ln 0.5}{4.5\times10^9} = -1.54 \times 10^{-10}$;

$Q = 1\,e^{-1.54\times10^{-10}t}$

11. $dQ/dt = kQ$; $Q = Q_o e^{kt}$; at $t = 0$, $Q_o = 5$ g, $Q = 5e^{kt}$; at $t = 36$.

$Q = 4.5$ so $4.5 = 5e^{36k}$; $k = \dfrac{\ln 0.9}{36} = -0.00293$;

Half-life: $t = \dfrac{-\ln 2}{k} = \dfrac{-\ln 2}{-0.00293} = 237$ yr

13. Let Q = amt of salt at time t. dQ/dt = rate of gain – rate of loss.
(2 lb/gal)(2 gal/min) – (Q lb/50 gal)(2 gal/min) = 4 – Q/25;

$\dfrac{dQ}{dt} = \dfrac{100 - Q}{25}$; $\displaystyle\int \frac{dQ}{100 - Q} = \int \frac{dt}{25}$; $Q = Ce^{-0.04t} + 100$; $t = 0$,

$Q = 10$, $C = -90$; $Q = -90^{-0.04t} + 100$; find Q at $t = 30$ min;
$Q = -90e^{(-0.04)(30)} + 100 = 72.9$ lb

15. $dT/dt = k(T - T_o) = k(T - 10)$; $\displaystyle\int dT/(T-10) = \int k\,dt$;

$\ln(T - 10) = kt + \ln C$; $\ln\dfrac{T-10}{C} = kt$; $\dfrac{T-10}{C} = e^{kt}$;

$T = Ce^{kt} + 10$; At $t = 0$, $T = 90$ so $90 = Ce^{k\cdot 0} + 10$; $C = 80$;
$T = 80e^{kt} + 10$; find k when $t = 5$ and $T = 70$: $70 = 80e^{5k} + 10$;
$k = -0.0575$; then at $t = 30$: $T = 80e^{(-0.0575)(30)} + 10 = 24.3°$ C

17. $y' = ky$; $\displaystyle\int dy/y = \int k\,dt$; $\ln y = kt + \ln C$; $y = Ce^{kt}$; at $t = 0$, $y = 2{,}000{,}000 = C$
Since the population doubles in 20 yr, $2 = e^{20k}$; $k = 0.0347$;
in 80 yr, $y = 2{,}000{,}000\, e^{(0.0347)(80)} = 3.2 \times 10^7$

19. $dp/dV = kp/V$; $dp/p = k\,dV/V$; $\displaystyle\int dp/p = k\int dV/V$;
$\ln p = k \ln V + \ln C$; $\ln p = \ln CV^k$; Thus $p = CV^k$

Chapter 29 Review

1. order 2, degree 1

2. order 1, degree 2

3. order 2, degree 3

4. order 1, degree 2

5. $y = x^3 - 2x$; $y' = 3x^2 - 2$; $y'' = 6x$; $y'' + 3y = 3x^3$;
$6x + 3(x^3 - 2x) = 3x^3$; $3x^3 = 3x^3$

6. $y = 2e^x - 3xe^x + e^{-x}$; $y' = -e^x - 3xe^x - e^{-x}$;
$y'' = -4e^x - 3xe^x + e^{-x}$;
$y'' - 2y' + y = 4e^{-x}$;
$(-4e^x - 3xe^x + e^{-x}) - 2(-e^x - 3xe^x - e^{-x}) + (2e^x - 3xe^x + e^{-x}) = 4e^{-x}$; $4e^{-x} = 4e^{-x}$

7. $y = \sin x + x^2$; $y' = \cos x + 2x$; $y'' = -\sin x + 2$;
$y'' + y = x^2 + 2$; $(-\sin x + 2) + (\sin x + x^2) = x^2 + 2$;
$x^2 + 2 = x^2 + 2$

8. $y = e^{2x}(x^5 - 1)$; $y' = 2e^{2x}(x^5 - 1) + 5x^4 e^{2x}$; $y' - 2y = 5x^4e^{2x}$;
$2e^{2x}(x^5 - 1) + 5x^4e^{2x} - 2e^{2x}(x^5 - 1) = 5x^4 e^{2x}$;
$5x^4 e^{2x} = 5x^4 e^{2x}$

9. $\int dy/y = \int dx/x^3$; $\ln y = -1/(2x^2) + C$; $2x^2 \ln y = -1 + Cx^2$

10. $\int dy/(3y) = \int e^{-2x} dx$; $(1/3) \ln y = (-1/2) e^{-2x} + C$; $2 \ln y + 3e^{-2x} = C$

11. $\int y^2 dy + \int \frac{x\,dx}{x^2 + 9} = 0$; $y^3/3 + (1/2) \ln (x^2 + 9) = C$;
$2y^3 + 3 \ln(x^2 + 9) = C$

12. $\int e^{-y} dy = \int \sec^2 x\,dx$; $-e^{-y} = \tan x + C$; $\tan x + e^{-y} = C$

13. $\frac{x\,dy - y\,dx}{x^2} = 3x^2 dx$; $\int d(y/x) = 3 \int x^2 dx$; $y/x = x^3 + C$; $y = x^4 + Cx$

14. $dy/dx + 6x^2y = 12x^2$; $P(x) = 6x^2$, $Q(x) = 12x^2$, $\int P(x)\,dx = 2x^3$;
$ye^{2x^3} = \int 12x^2 e^{2x^3} dx = 2e^{2x^3} + C$; $(y - 2)e^{2x^3} = C$

15. $dy/dx - (1/x^2)\,y = 5/x^2$; $P(x) = -1/x^2$, $Q(x) = 5/x^2$,
$\int P(x)\,dx = 1/x$; $ye^{1/x} = \int (5/x^2)\,e^{1/x} dx = -5e^{1/x} + C$; $(y + 5)e^{1/x} = C$

16. $\frac{x\,dx + y\,dy}{x^2 + y^2} = y\,dy$; $\int d(\ln \sqrt{x^2 + y^2}) = \int y\,dy$; $\ln \sqrt{x^2 + y^2} = y^2/2 + C$;
$(1/2) \ln (x^2 + y^2) = (1/2)y^2 + C$; $\ln (x^2 + y^2) = y^2 + C$

17. $\frac{x\,dy - y\,dx}{x^2 + y^2} = x^2 dx$; $\int d\,(\text{Arctan}\ y/x) = \int x^2 dx$; $3\ \text{Arctan}\ y/x = x^3 + C$

18. $x\,dy + y\,dx = 14x^5\,dx$; $\int d(xy) = \int 14x^5 dx$; $xy = (7/3)x^6 + C$; $3xy = 7x^6 + C$

19. $dy/dx + 3y = e^{-2x}$; $P(x) = 3$; $Q(x) = e^{-2x}$, $\int P(x)\,dx = 3x$;
$ye^{3x} = \int e^{-2x} e^{3x} dx = e^x + C$; $y = e^{-2x} + Ce^{-3x}$

Chapter 29 Review

20. $dy/dx - (5/x)y = x^3 + 7$; $P(x) = -5/x$, $Q(x) = x^3 + 7$,

$\int P(x)\,dx \int(-5/x)\,dx = -5\ln x = \ln x^{-5}$;

$ye^{\ln x^{-5}} = \int(x^3 + 7)e^{\ln x^{-5}}dx$; $yx^{-5} = \int(x^{-2} + 7x^{-5})dx =$

$-x^{-1} - (7/4)x^{-4} + C$; $4y = -4x^4 - 7x + Cx^5$

21. $x\,dy - 3y\,dx = (4x^3 - x^2)\,dx$; $dy/dx - (3/x)\,y = 4x^2 - x$;

$P(x) = -3/x$, $Q(x) = 4x^2 - x$, $\int P(x)\,dx \int(-3/x)\,dx = \ln x^{-3}$; $ye^{\ln x^{-3}} = \int(4x^2 - x)e^{\ln x^{-3}}dx$;

$yx^{-3} = \int(4x^2 - x)x^{-3}dx = \int(4/x - 1/x^2)dx = 4\ln x + 1/x + C$;

$y = 4x^3\ln x - x^2 + Cx^3$

22. $dy/dx + y\cot x = \cos x$; $P(x) = \cot x$, $Q(x) = \cos x$, $\int P(x)\,dx = \ln(\sin x)$;

$ye^{\ln(\sin x)} = \int\cos x \cdot e^{\ln(\sin x)}dx$; $y\sin x = \int\cos x\sin x\,dx = \frac{1}{2}\sin^2 x + C$

$2y\sin x = \sin^2 x + C$

23. $\int 3x\,dx = \int dy/y^2$; $3x^2/2 = 1/y + C$; for $x = -2$,

$y = -1$, $C = 5$; $3x^2/2 = -1/y + 5$; $3x^2y - 10y + 2 = 0$

24. $\int y\,dy = -3\int e^{-2x}dx$; $y^2/2 = (3/2)\,e^{-2x} + C$; for $x = 0$,

$y = -2$, $C = 1/2$; $y^2/2 = (3/2)e^{-2x} + 1/2$; $y^2 = 3e^{-2x} + 1$

25. $\dfrac{x\,dy - y\,dx}{x^2} = x^3dx$; $\int d(y/x) = \int x^3dx$;

$y/x + x^4/4 + C$; for $f(1) = 1$, $C = 3/4$; $y/x = x^4/4 + 3/4$;

$4y = x^5 + 3x$

26. $x\,dy + y\,dx = x\ln x\,dx$; $\int d(xy) = \int x\ln x\,dx$;

$xy = (x^2/2)\ln x - \int(x^2/2)(1/x)\,dx = (x^2/2)\ln x - (1/4)x^2 + C$;

For $f(1) = 3/4$, $C = 1$; $xy = (x^2/2)\ln x - x^2/4 + 1$;

$xy = (x^2/4)(2\ln x - 1) + 1$

27. $dy/dx - y = e^{5x}$; $P(x) = -1$, $Q(x) = e^{5x}$, $\int P(x)\,dx = -x$;

$ye^{-x} = \int e^{5x}\,e^{-x}\,dx = (1/4)e^{4x} + C$; $4y = e^{5x} + Ce^x$;

for $f(0) = -3$, $C = -13$, $4y = e^{5x} - 13e^x$

28. $dy/dx - (2/x)\,y = x^3 - 5x;\ P(x) = -2/x,\ Q(x) = x^3 - 5x,$

$\int P(x)\,dx = \int(-2/x)\,dx = \ln x^{-2};\ ye^{\ln x^{-2}} = \int(x^3 - 5x)e^{\ln x^{-2}}\,dx;$

$yx^{-2} = \int(x - 5/x)\,dx = x^2/2 - 5\ln x + C;$

$y = x^4/2 - 5x^2\ln x + Cx^2;$ for $f(1) = 3,\ C = 5/2$

$2y = x^4 - 10x^2\ln x + 5x^2$

29. $di/dt + (R/L)i = V/L;\ L = 0.2\ H,\ R = 60\ \Omega,\ V = 120\ V;$

$di/dt + 300i = 600;\ P(t) = 300,\ Q(t) = 600,\ \int P(t)\,dt = 300\,t;$

$ie^{300t} = \int 600e^{300t}\,dt = 2e^{300t} + C;$ at $t = 0;\ i = 1,\ C = -1;$

$ie^{300t} = 2e^{300t} - 1;\ i = 2 - e^{-300t}$

30. $dQ/dt = kQ;\ Q = Q_oe^{kt};$ at $t = 0,\ Q = Q_o = 1$ and $Q = e^{kt};$

$k = \dfrac{-\ln 2}{8.8\times 10^8} = -7.88\times 10^{-10};\ Q = e^{-7.88}\times 10^{-10}t$

31. dQ/dt = rate of gain – rate of loss

(0 lb/L)(1 L/min) – (Q lb/200 L)(1 L/min) = $-Q/200$;

$dQ/dt = -Q/200;\ 200\int dQ/Q = -\int dt;\ 200\ln Q = -t + \ln C;$

$Q = Ce^{-t/200};$ at $t = 0,\ Q = 5,\ C = 5,\ Q = 5e^{-t/200};$

at $t = 20$ min, $Q = 5e^{(-20/200)} = 4.524$ kg

32. $dT/dt = k(T - 12);\ \int dT/(T - 12) = k\int dt;$

$\ln(T - 12) = kt + \ln C;\ \ln\dfrac{T-12}{C} = kt;\ T = Ce^{kt} + 12;$

at $t = 0,\ T = 30,\ C = 18,\ T = 18e^{kt} + 12;$ find k if $T = 25$

when $t = 20;\ 25 = 18e^{20k} + 12;\ k = \dfrac{\ln(13/18)}{20} = -0.0163$

$T = 18e^{-0.0163t} + 12;$ at $t = 45$ min, $T = 18e^{(-0.0163)(45)} + 12 = 20.6°$ C

33. $m = dy/dx = x^2y;\ \int dy/y = \int x^2\,dx;\ \ln y = x^3/3 + C;$ for $x = 0,$

$y = 2,$ and $C = \ln 2;\ \ln y = x^3/3 + \ln 2;\ \ln(y/2) = x^3/3;$

$y = 2e^{x^3/3}$ or $y^3 = 8e^{x^3}$

34. $a = \int dv = \int a\,dt = \int 4t^3\,dt = t^4 + C;$ at $t = 0,\ v = 0,\ C = 0$

$v = t^4\Big|_{t=3\text{ s}} = 81\text{ m/s}^2$

Chapter 30, Exercises 30.1

1. Homogeneous; 4

3. Homogeneous; 3

5. Homogeneous; 2

7. Nonhomogeneous; 3

9. $m^2 - 5m - 14 = 0$; $(m - 7)(m + 2) = 0$; $m = -2, 7$;
$y = k_1e^{7x} + k_2e^{-2x}$

11. $m^2 - 2m - 8 = 0$; $(m - 4)(m + 2) = 0$; $m = -2, 4$
$y = k_1e^{4x} + k_2e^{-2x}$

13. $m^2 - 1 = 0$; $(m - 1)(m + 1) = 0$; $m = -1, 1$; $y = k_1e^{x} + k_2e^{-x}$

15. $m^2 - 3m = 0$; $m(m - 3) = 0$; $m = 0, 3$; $y = k_1 + k_2e^{3x}$

17. $2m^2 - 13m + 15 = 0$; $(2m - 3)(m - 5) = 0$; $m = 3/2, 5$;
$y = k_1e^{3x/2} + k_2e^{5x}$

19. $3m^2 - 7m + 2 = 0$; $(3m - 1)(m - 2) = 0$; $m = 2, 1/3$;
$y = k_1e^{2x} + k_2e^{x/3}$

21. $m^2 - 4m = 0$; $m(m - 4) = 0$; $m = 0, 4$; $y = k_1 + k_2e^{4x}$;
$y' = 4k_2e^{4x}$ substitute $y = 3, x = 0$, and $y' = 4$ into y and y' equations and solve the resulting system of equations:
$3 = k_1 + k_2e^{4(0)}$; $4 = 4k_2e^{4(0)}$;
$k_1 = 2, k_2 = 1$; Thus $y = 2 + e^{4x}$

23. $m^2 - m - 2 = 0$; $(m - 2)(m + 1) = 0$; $m = -1, 2$;
$y = k_1e^{2x} + k_2e^{-x}$; $y' = 2k_1e^{2x} - k_2e^{-x}$; substitute $y = 2$,
$x = 0, y' = 1$; $2 = k_1 + k_2$; $1 = 2k_1 - k_2$; $k_1 = 1$; $k_2 = 1$
Thus $y = e^{2x} + e^{-x}$

25. $m^2 - 8m + 15 = 0$; $(m - 5)(m - 3) = 0$; $m = 3, 5$;
$y = k_1e^{3x} + k_2e^{5x}$; $y' = 3k_1e^{3x} + 5k_2e^{5x}$; substitute $y = 4, x = 0$,
$y' = 2$; $4 = k_1 + k_2$; $2 = 3k_1 + 5k_2$; $k_1 = 9$; $k_2 = -5$,
Thus $y = 9e^{3x} - 5e^{5x}$

Exercises 30.2

1. $m^2 - 4m + 4 = 0$; $(m - 2)^2 = 0$; $m = 2, 2$; $y = k_1e^{2x} + k_2xe^{2x}$;
$y = e^{2x}(k_1 + k_2x)$

3. $m^2 - 4m + 5 = 0$; $m = 2 \pm j$; $y = e^{2x}(k_1 \sin x + k_2 \cos x)$

5. $4m^2 - 4m + 1 = 0$; $(2m - 1)^2 = 0$; $m = 1/2, 1/2$;
$y = e^{x/2}(k_1 + k_2x)$

7. $m^2 - 4m + 13 = 0$; $m = 2 \pm 3j$; $y = e^{2x}(k_1 \sin 3x + k_2 \cos 3x)$

9. $m^2 - 10m + 25 = 0$; $(m-5)^2 = 0$; $m = 5, 5$; $y = e^{5x}(k_1 + k_2x)$

11. $m^2 + 9 = 0$; $m = \pm 3j$; $y = k_1 \sin 3x + k_2 \cos 3x$

13. $m^2 = 0$; $m = 0, 0$; $y = k_1 + k_2x$

15. $m^2 - 6m + 9 = 0$; $(m-3)^2 = 0$; $m = 3, 3$; $y = k_1e^{3x} + k_2xe^{3x}$;
$y' = 3k_1e^{3x} + k_2e^{3x} + 3k_2xe^{3x}$; substitute $y = 2, x = 0, y' = 4$;
$2 = k_1, 4 = 3k_1 + k_2$; $k_1 = 2, k_2 = -2$; $y = 2e^{3x} - 2xe^{3x} = 2e^{3x}(1-x)$

17. $m^2 + 25 = 0$; $m = \pm 5j$; $y = k_1 \sin 5x + k_2 \cos 5x$;
$y' = 5k_1 \cos 5x - 5k_2 \sin 5x$; substitute $y = 2, x = 0, y' = 0$;
$2 = k_2, 0 = k_1, y = 2 \cos 5x$

19. $m^2 - 12m + 36 = 0$; $(m-6)^2 = 0$; $m = 6, 6$; $y = k_1e^{6x} + k_2xe^{6x}$;
$y' = 6k_1e^{6x} + k_2e^{6x} + 6k_2xe^{6x}$; substitute $y = 1, x = 0, y' = 0$;
$1 = k_1$; $0 = 6k_1 + k_2$; $k_1 = 1$; $k_2 = -6$;
$y = e^{6x} - 6xe^{6x} = e^{6x}(1 - 6x)$

Exercises 30.3

1. yc: $m^2 + m = 0$; $m(m+1) = 0$; $m = 0, -1$; $y_c = k_1 + k_2e^{-x}$;
y_p: $y_p = A \sin x + B \cos x$; $y'_p = A \cos x - B \sin x$;
$y''_p = -A \sin x - B \cos x$; substitute into the given differential equation:
$(-A \sin x - B \cos x) + (A \cos x - B \sin x) = \sin x$
$(-A - B) \sin x + (A - B) \cos x = \sin x$; equate coefficients:
$-A - B = 1, A - B = 0; A = -1/2, B = -1/2$;
$y = k_1 + k_2e^{-x} - (1/2) \sin x - (1/2) \cos x$

3. yc: $m^2 - m - 2 = 0$; $(m-2)(m+1) = 0$; $m = -1, 2$;
$y_c = k_1e^{2x} + k_2e^{-x}$
y_p: $y_p = A + Bx$; $y'_p = B$; $y_p'' = 0$; substitute into the given differential equation:
$0 - B - 2(A + Bx) = 4x$;
$(-2A - B) - 2Bx = 4x$; equate coefficients: $-2A - B = 0$,
$-2B = 4; B = -2; A = 1, y = k_1e^{2x} + k_2e^{-x} - 2x + 1$

5. y_c: $m^2 - 10m + 25 = 0$; $(m-5)^2 = 0$; $m = 5, 5$; $y_c = e^{5x}(k_1 + k_2x)$
y_p: $y_p = Ax + B$; $y'_p = A$; $y_p'' = 0$; substitute into given D.E.
$0 - 10A + 25(Ax + B) = x$; $25Ax + (-10A + 25B) = x$;
equate coefficients: $25A = 1, -10A + 25B = 0$; $A = 1/25, B = 2/125$;
$y = e^{5x}(k_1 + k_2x) + x/25 + 2/125$

7. y_c: $m^2 - 1 = 0$; $m = \pm 1$; $y_c = k_1e^x + k_2e^{-x}$
y_p: $y_p = Ax^2 + Bx + C$; $y'_p = 2Ax + B$; $y''_p = 2A$; substitute into D.E.
$2A - (Ax^2 + Bx + C) = x^2$; $-Ax^2 - Bx + (2A - C) = x^2$;
equate coefficients; $-A = 1, -B = 0, 2A - C = 0; A = -1, B = 0, C = -2$;
$y = k_1e^x + k_2e^{-x} - x^2 - 2$

9. y_c: $m^2 + 4 = 0$; $m = \pm 2j$; $y_c = k_1 \sin 2x + k_2 \cos 2x$
y_p: $y_p = Ae^x + B$; $y'_p = Ae^x$; y''_p; substitute into D.E.
$Ae^x + 4(Ae^x + B) = e^x - 2$; $5Ae^x + 4B = e^x - 2$; $A = 1/5$,
$B = -1/2$; $y = k_1 \sin 2x + k_2 \cos 2x + (1/5)\, e^x - 1/2$

11. y_c: $m^2 - 3m - 4 = 0$; $(m-4)(m+1) = 0$; $m = -1, 4$;
$y_c = k_1 e^{4x} + k_2 e^{-x}$; y_p: $y_p = Ae^x = y'_p = y''_p$; substitute into D.E.
$Ae^x - 3Ae^x - 4Ae^x = 6e^x$; $-6Ae^x = 6e^x$; $A = -1$;
$y = k_1 e^{4x} + k_2 e^{-x} - ex$

13. y_c: $m^2 + 1 = 0$; $m = \pm j$; $y_c = k_1 \sin x + k_2 \cos x$
y_p: $y_p = A + B \cos 3x + C \sin 3x$; $y'_p = -3B \sin 3x + 3C \cos 3x$;
$y''_p = -9B \cos 3x - 9C \sin 3x$; substitute into D.E.
$(-9B \cos 3x - 9C \sin 3x) + (A + B \cos 3x + C \sin 3x) = 5 + \sin 3x$;
$A - 8B \cos 3x - 8c \sin 3x = 5 + \sin 3x$;
$A = 5, B = 0, C = -1/8$
$y = k_1 \sin x + k_2 \cos x + 5 - (1/8) \sin 3x$

15. y_c: $m^2 - 1 = 0$; $m = \pm 1$; $y_c = k_1 e^x + k_2 e^{-x}$; note $g(x) = e^x$ is a solution of the homogeneous equation $y'' - y = 0$; use note 2 from table: y_p: $y_p = Axe^x$; $y'_p = Axe^x + Ae^x$; $y''_p = Axe^x + 2Ae^x$;
substitute into D.E. $(Axe^x + 2Ae^x) - Axe^x = e^x$; $2Ae^x = e^x$;
$2A = 1$; $A = 1/2$; $y_p = (1/2)xe^x$; thus $y = k_1 e^x + k_2 e^{-x} + (1/2)xe^x$

17. y_c: $m^2 + 4 = 0$; $m = \pm 2j$; $y_c = k_1 \sin 2x + k_2 \cos 2x$;
note $g(x) = \cos 2x$ is a solution of the homogeneous equation
$y'' + 4y = 0$; use note 2 from table: y_p: $y_p = Ax \cos 2x + Bx \sin 2x$;
$y'_p = A \cos 2x - 2Ax \sin 2x + B \sin 2x + 2Bx \cos 2x$;
$y''_p = -4A \sin 2x - 4Bx \sin 2x - 4Ax \cos 2x + 4B \cos 2x$; substitute into D.E:
$(-4A \sin 2x - 4Bx \sin 2x - 4Ax \cos 2x + 4B \cos 2x)$
$-4(Ax \cos 2x + Bx \sin 2x) = \cos 2x$;
$-4A \sin 2x + 4B \cos 2x = \cos 2x$; $-4A = 0, 4B = 1$; $A = 0, B = 1/4$;
$y = k_1 \sin 2x + k_2 \cos 2x + (1/4)x \sin 2x$

19. y_c: $m^2 + 1 = 0$; $m = \pm j$; $y_c = k_1 \sin x + k_2 \cos x$
y_p: $y_p = Ae^{2x}$; $y'_p = 2Ae^{2x}$; $y''_p = 4Ae^{2x}$; substitute into D.E.
$4Ae^{2x} + Ae^{2x} = 10e^{2x}$; $A = 2$; $y = k_1 \sin x + k_2 \cos x + 2e^{2x}$;
$y' = k_1 \cos x - k_2 \sin x + 4e^{2x}$; substitute $y = 0, x = 0, y' = 0$;
$0 = k_2 + 2$; $0 = k_1 + 4$; $k_1 = -4, k_2 = -2$
$y = -4 \sin x - 2 \cos x + 2e^{2x}$

21. y_c: $m^2 + 1 = 0$; $m = \pm j$; $y_c = k_1 \sin x + k_2 \cos x$
y_p: $y_p = Ae^x = y'_p = y''_p$; substitute into D.E.: $Ae^x + Ae^x = e^x$;
$A = 1/2$; $y = k_1 \sin x + k_2 \cos x + (1/2)\, e^x$;
$y' = k_1 \cos x - k_2 \sin x + (1/2)\, e^x$; substitute $y = 0, x = 0$
$y' = 3$; $k_2 + 1/2 = 0$; $k_1 + 1/2 = 3$; $k_1 = 5/2, k_2 = -1/2$;
$y = (5/2) \sin x - (1/2) \cos x + (1/2)\, e^x$ or
$y = (1/2)(5 \sin x - \cos x + e^x)$

Exercises 30.4

1. $m = \frac{W}{g} = \frac{4\text{ lb}}{32\text{ ft/s}^2} = \frac{1}{8}$ lb-s^2/ft; $k = \frac{F}{s} = \frac{4\text{ lb}}{(1/4)\text{ ft}} = 16$ lb/ft;

$x = C_1 \sin 8\sqrt{2}\,t + C_2 \cos 8\sqrt{2}\,t$;

$dx/dt = 8\sqrt{2}\,C_1 \cos 8\sqrt{2}\,t - 8\sqrt{2}\,C_2 \sin 8\sqrt{2}\,t$

substitute $x = 2$ in. $= 1/6$ ft, $dx/dt = 0$, $t = 0$;

$1/6 = C_1 \sin 0 + C_2 \cos 0$; $0 = 8\sqrt{2}\,C_1 \cos 0 - \sqrt{8}\,C_2 \sin 0$;

$C_1 = 0$, $C_2 = 1/6$; $x = (1/6) \cos 8\sqrt{2}\,t$

3. $p = \frac{F_{\text{resistant}}}{dx/dt} = \frac{5\text{ lb}}{1/3\text{ ft/s}} = 15$ lb-s^2/ft; $m = \frac{W}{g} = \frac{20\text{ lb}}{32\text{ ft/s}^2} = 0.625$ lb-s^2/ft

$k = \frac{F}{s} = \frac{20\text{ lb}}{1/2\text{ ft}} = 40$ lb/ft;

$d = \left\{\frac{p}{2m}\right\}^2 - \frac{k}{m} = \left\{\frac{15}{2(0.625)}\right\}^2 - \frac{40}{0.625} = 80$; since $d > 0$, we have two real roots of the auxiliary equation

$\overline{m}_1 = (-p/2m) + \sqrt{d} = -12 + \sqrt{80} = -3.06$

$\overline{m}_2 = -12 - \sqrt{80} = -20.9$

$x = C_1 e^{-3.06t} + C_2 e^{-20.9t}$ (Case 1--overdamped)

5. Let x = downward displacement; the change in submerged volume is $\pi(1.5)^2x$; the buoyant force is $62.4\pi(1.5)^2x$; and W = weight of buoy in lbs; $-62.4\pi(1.5)^2x = \frac{W}{g}\frac{d^2x}{dt^2}$ or $\frac{d^2x}{dt^2} + \frac{4521\pi}{W}x = 0$;

$x = C_1 \sin \sqrt{4521\pi/W}\,t$; $C_2 \cos \sqrt{4521\pi/W}\,t$; since $p = 6 = \frac{2\pi}{\sqrt{4521\pi/W}}$

$\pi/3 = \sqrt{4521\pi/W}$; $W = 9(4521)/\pi = 13{,}000$ lb

7. Let x = downward displacement, πr^2x = change in submerged volume, $62.4\pi r^2x$ = the buoyant force, $W = 1000$ lb, find r.

$-62.4\pi r^2x = \frac{1000}{g}\frac{d^2x}{dt^2}$; $\frac{d^2x}{dt^2} - \frac{2009\pi r^2}{1000}x = 0$

$x = C_1 \sin \sqrt{2009\pi r^2/1000}\,t + C_2 \cos \sqrt{2009\pi r^2/1000}\,t$

$p = 2 = \frac{2\pi}{\sqrt{2009\pi r^2/1000}}$; $\pi^2 = 2009\pi r^2/1000$

$r = \sqrt{1000\pi/2009} = 1.25$ ft

9. $L = 0.1\ H, R = 50\ \Omega, C = 2 \times 10^{-4} F;\ d = \frac{R^2}{4L^2} - \frac{1}{CL} = \frac{50^2}{4(0.1)^2} - \frac{1}{(2\times10^{-4})(0.1)} = 12{,}500;$

$-\frac{R}{2L} = -250$; since $d > 0$, $m_1 = -\frac{R}{2L} + \sqrt{d} = -250 + \sqrt{12500} = -138$; $m_2 = -250 - \sqrt{12500}$ $= -362$ Thus $i = k_1 e^{-138t} + k_2 e^{-362t}$

11. $L = 0.4\ H, R = 200\ \Omega, C = 5 \times 10^{-5} F;$

$$d = \frac{R^2}{4L^2} - \frac{1}{CL} = \frac{(200)^2}{4(0.4)^2} - \frac{1}{(5\times10^{-5})(0.4)} = 12500; \quad -\frac{R}{2L} = \frac{-200}{2(0.4)} = -250;$$

since $d > 0$, $m_1 = -250 + \sqrt{12500} = -140$;
$m_2 = -250 - \sqrt{12500} = -360$; Thus $i = k_1 e^{-140t} + k_2 e^{-360t}$;
$di/dt = -140k_1e^{-140t} - 360k_2e^{-360t}$;
Since $i = 0$ at $t = 0$, $0 = k_1 + k_2$;

using $L\, di/dt + Ri + (1/C)\int_0^t i\, dt = V$;

$(0.4)\, di/dt + R(0) + (1/C)\int_0^t 0\, dt = 12$; $0.4\, di/dt = 12$; $di/dt = 30$;

$30 = -140k_1e^{-140t} - 360k_2e^{-360t}$; at $t = 0$, $i = 0$;
$30 = -140k_1 - 360k_2$; solve system with $0 = k_1 + k_2$ from above;
$k_1 = 0.136$, $k_2 = -0.136$; Thus $i = 0.136e^{-140t} - 0.136e^{-360t}$

Exercises 30.5

$\mathcal{L}$ is our notation for the Laplace transform.

1. $\mathcal{L}[f(t)] = \int_0^\infty e^{-st} \cdot t\, dt = \lim_{t\to\infty} \int_0^t e^{-st} \cdot t\, dt$

$$= \lim_{t\to\infty} \frac{[e^{-st}(-st-1)]}{s^2}\Bigg|_0^t = \lim_{t\to\infty}\left\{\left[\frac{e^{-st}}{s^2}(-st-1)\right] - \left[\frac{e^0}{s^2}(-0-1)\right]\right\}$$

$$= \lim_{t\to\infty}\left\{\left[\frac{-1}{e^{st}s^2}(st+1)\right] + \frac{1}{s^2}\right\} = \frac{1}{s^2}$$

3. #8; $\frac{3}{s^2+9}$

5. #13; $\frac{-4}{s(s-4)}$

7. #6; $\mathcal{L}(t)^2 = 2\,\mathcal{L}(t^2/2) = 2/s^3$

9. #20; $\frac{2\cdot 2^3}{(s^2+4)^2} = \frac{16}{(s^2+4)^2}$

11. #5, 7; $\mathcal{L}(t - e^{2t}) = \mathcal{L}(t) - \mathcal{L}(e^{2t}) = \frac{1}{s^2} - \frac{1}{s-2} = \frac{-s^2+s-2}{s^2(s-2)}$

13. #18; $\dfrac{8s}{(s^2+16)^2}$ **15.** #11; $\dfrac{s+3}{(s+3)^2+25} = (s+3)/(s^2+6s+34)$

17. #5, 6; $\mathscr{L}(8t+4t^3) = 8\,\mathscr{L}(t) + 24\,\mathscr{L}(t^3/3!) = 8/s^2 + 24/s^4 = \dfrac{8s^2+24}{s^4}$

19. $\mathscr{L}(y''-3y') = \mathscr{L}(y'') - 3\,\mathscr{L}(y') = [s^2\mathscr{L}(y) - s\,y(0) - y'(0)] - 3[s\,\mathscr{L}(y) - y(0)] = (s^2-3s)\ \mathscr{L}(y)$
Since $y(0) = y'(0) = 0$

21. $\mathscr{L}(y''+y'+y) = [s^2\ \mathscr{L}(y) - s\,y(0) - y'(0)] + [s\ \mathscr{L}(y) - y(0)] + \mathscr{L}(y)\ \ [y(0)=0, y'(0)=1]$
$= s^2\ \mathscr{L}(y) - 1 + s\ \mathscr{L}(y) + \mathscr{L}(y) = (s^2+s+1)\ \mathscr{L}(y) - 1$

23. $\mathscr{L}(y''-3y'+y) = \mathscr{L}(y'') - 3\ \mathscr{L}(y') + \mathscr{L}(y)\ [y(0)=1, y'(0)=0]$
$= [s^2\ \mathscr{L}(y) - s\,y(0) - y'(0)] - 3[s\ \mathscr{L}(y) - y(0)] + \mathscr{L}(y) = (s^2-3s+1)\ \mathscr{L}(y) - s + 3$

25. $\mathscr{L}(y''+8y'+2y) = \mathscr{L}(y'') + 8\ \mathscr{L}(y') + 2\ \mathscr{L}(y)\ [y(0)=4, y'(0)=6]$
$= [s^2\ \mathscr{L}(y) - s\,y(0) - y'(0)] + 8[s\ \mathscr{L}(y) - y(0)] + 2\,\mathscr{L}(y)$
$= s^2\ \mathscr{L}(y) - 4s - 6 + 8s\ \mathscr{L}(y) - 32 + 2\ \mathscr{L}(y)$
$= (s^2+8s+2)\ \mathscr{L}(y) - 4s - 38$

27. $\mathscr{L}(y''-6y') = \mathscr{L}(y'') - 6\ \mathscr{L}(y')\ \ [y(0)=3, y'(0)=7]$
$= [s^2\ \mathscr{L}(y) - s\,y(0) - y'(0)] - 6[s\ \mathscr{L}(y) - y(0)]$
$= s^2\ \mathscr{L}(y) - 3s - 7 - 6s\ \mathscr{L}(y) + 18 = (s^2-6s)\ \mathscr{L}(y) - 3s + 11$

29. $\mathscr{L}(y''+8y'-3y) = \mathscr{L}(y'') + 8\ \mathscr{L}(y') - 3\ \mathscr{L}(y)\ [y(0)=-6, y'(0)=2]$
$= [s^2\ \mathscr{L}(y) - s\,y(0) - y'(0)] + 8[s\ \mathscr{L}(y) - y(0)] - 3\ \mathscr{L}(y)$
$= s^2\ \mathscr{L}(y) + 6s - 2 + 8s\ \mathscr{L}(y) + 48 - 3\ \mathscr{L}(y)$
$= (s^2+8s-3)\ \mathscr{L}(y) + 6s + 46$

31. #4; 1 **33.** #15; te^{5t} **35.** #10; $\cos 8t$ **37.** #12; $e^{6t} - e^{2t}$

39. #9; $\mathscr{L}^{-1}\left\{\dfrac{2}{(s-3)^2+4}\right\} = e^{3t}\sin 2t$ **41.** #20; $\sin t - t\cos t$

43. use partial fractions: $\mathscr{L}^{-1}\left\{\dfrac{1}{s-1} - \dfrac{s+1}{s^2+1}\right\}$ #7, 8, 10
$= \mathscr{L}^{-1}\left\{\dfrac{1}{s-1}\right\} - \mathscr{L}^{-1}\left\{\dfrac{s}{s^2+1}\right\} - \mathscr{L}^{-1}\left\{\dfrac{s}{s^2+1}\right\} = e^t - \cos t - \sin t$

45. use partial fractions: $6\ \mathscr{L}^{-1}\left\{\frac{1/2}{s}+\frac{1/6}{s+6}+\frac{1/3}{s+3}\right\}$ #4, 7

$$= 3\ \mathscr{L}^{-1}\left\{\frac{1}{s}\right\} + \mathscr{L}^{-1}\left\{\frac{1}{s+6}\right\} + 2\ \mathscr{L}^{-1}\left\{\frac{1}{s+3}\right\} = 3 + e^{-6t} + 2e^{-3t}$$

47. #9, 11; $\mathscr{L}^{-1}\left\{\frac{s+11}{(s+5)^2+2^2}\right\} = \mathscr{L}^{-1}\left\{\frac{(s+5)+6}{[s-(-5)]^2+2^2}\right\}$

$$= \mathscr{L}^{-1}\left\{\frac{s+5}{(s+5)^2+2^2}\right\} + 3\ \mathscr{L}^{-1}\left\{\frac{2}{(s+5)^2+2^2}\right\} = e^{-5t}\cos 2t + 3e^{-5t}\sin 2t = e^{-5t}(\cos 2t + 3\sin 2t)$$

49. $\mathscr{L}[f''(t)] = \int_0^\infty e^{-st}\ f''(t)dt$. Integrate by parts.

$$u = e^{-st} \qquad dv = f''(t)dt$$
$$du = -se^{-st}dt \qquad v = f'(t)$$

$$\int_0^\infty e^{-st}\ f''(t)dt = e^{-st}f'(t)\Big|_0^\infty - \int_0^\infty f'(t)(-se^{-st}dt)$$

$$= -f'(0) + s\int_0^\infty e^{-st}f'(t)dt$$

$$= -f'(0) + s\ \mathscr{L}[f'(t)]$$

$$= -f'(0) + s\{s\ \mathscr{L}[f(t)] - f(0)\}$$

$$= s^2\ \mathscr{L}[f(t)] - s f(0) - f'(0)$$

Exercises 30.6

1. $\mathscr{L}(y') - \mathscr{L}(y) = 0$; $s\ \mathscr{L}(y) - y(0) - \mathscr{L}(y) = 0$;

$y(0) = 2$; $s\ \mathscr{L}(y) - 2 - \mathscr{L}(y) = 0$; $(s-1)\ \mathscr{L}(y) - 2 = 0$;

$\mathscr{L}(y) = 2/(s-1)$; #7, $y = 2e^t$

3. $4\ \mathscr{L}(y') + 3\ \mathscr{L}(y) = 0$; $4[s\ \mathscr{L}(y) - y(0)] + 3\ \mathscr{L}(y) = 0$; $y(0) = 1$;

$4s\ \mathscr{L}(y) - 4 + 3\ \mathscr{L}(y) = 0$; $(4s+3)\ \mathscr{L}(y) = 4$;

$\mathscr{L}(y) = \frac{4}{4s+3} = \frac{1}{s+3/4}$ #7; $y = e^{(-3/4)t}$

5. $\mathscr{L}(y') - 7\,\mathscr{L}(y) = \mathscr{L}(e^t);\ s\,\mathscr{L}(y) - y(0) - 7\,\mathscr{L}(y) = \dfrac{1}{s-1}:$

$y(0) = 5;\ s\ \mathscr{L}(y) - 5 - 7\,\mathscr{L}(y) = \dfrac{1}{s-1};\ (s-7)\ \mathscr{L}(y) = \dfrac{1}{s-1} + 5;$

$$\mathscr{L}(y) = \frac{1}{(s-7)(s-1)} + \frac{5}{s-7};$$

$$y = \frac{1}{6}\ \mathscr{L}^{-1}\left\{\frac{6}{(s-7)(s-1)}\right\} + 5\ \mathscr{L}^{-1}\left\{\frac{1}{s-7}\right\}$$

$y = (1/6)[e^{7t} - e^t] + 5e^{7t}$ or $6y = 31e^{7t} - e^t$ (#7, #12)

7. $\mathscr{L}(y'') + \mathscr{L}(y) = 0;\ s^2\,\mathscr{L}(y) - s\,y(0) - y'(0) + \mathscr{L}(y) = 0;$

$y(0) = 1, y'(0) = 0;\ (s^2 + 1)\ \mathscr{L}(y) = s;\ \mathscr{L}(y) = \dfrac{s}{s^2+1};$

$y = \cos t$ (#10)

9. $\mathscr{L}(y'') - 2\,\mathscr{L}(y') = 0;\ [s^2\,\mathscr{L}(y) - s\,y(0) - y'(0)] - 2[s\,\mathscr{L}(y) - y(0)] = 0;$

$y(0) = 1, y'(0) = -1;$

$[s^2\,\mathscr{L}(y) - s + 1] - 2\,[\,s\,\mathscr{L}(y) - 1] = 0;\ \mathscr{L}(y) = \dfrac{s-3}{s^2-2s};$

$$y = \frac{1}{2}\ \mathscr{L}^{-1}\left\{\frac{2s}{(s-0)((s-2)}\right\} + \frac{3}{2}\ \mathscr{L}^{-1}\left\{\frac{(-3)(2/3)}{(s-0)(s-2)}\right\}$$

$= e^{2t} + \dfrac{3}{2} - \dfrac{3}{2}e^{2t}$ or $2y = 3 - e^{2t}$ (#13, #14)

11. $\mathscr{L}(y'') + 2\,\mathscr{L}(y') + \mathscr{L}(y) = 0;\ [s^2\,\mathscr{L}(y) - s\,y(0) - y'(0)] - 2[s\,\mathscr{L}(y) - y(0)] + \mathscr{L}(y) = 0;$

$y(0) = 1, y'(0) = 0;$

$(s^2 + 2s + 1)\,\mathscr{L}(y) - s - 2 = 0;\ \mathscr{L}(y) = \dfrac{s+2}{s^2+2s+1}$

$$y = \mathscr{L}^{-1}\left\{\frac{s+2}{(s+1)^2}\right\} = \mathscr{L}^{-1}\left\{\frac{s}{(s+1)^2}\right\} + 2\ \mathscr{L}^{-1}\left\{\frac{1}{(s+1)^2}\right\}$$

$= e^{-t}(1-t) + 2te^{-t};\ y = e^{-t} + te^{-t} = e^{-t}(1+t)$

13. $\mathscr{L}(y'') - 4\,\mathscr{L}(y') + 4\,\mathscr{L}(y) = \mathscr{L}(te^{2t});\ y(0) = 0, y'(0) = 0$

$(s^2\,\mathscr{L}(y) - s\,y(0) - y'(0)] - 4[s\,\mathscr{L}(y) - y(0)] + 4\,\mathscr{L}(y) = \dfrac{1}{(s-2)^2};$

$(s^2 - 4s + 4)\,\mathscr{L}(y) = \dfrac{1}{(s-2)^2};\ \mathscr{L}(y) = \dfrac{1}{(s-2)^4};$

$$y = \mathscr{L}^{-1}\left\{\frac{1}{(s-2)^4}\right\} = \frac{t^3e^{2t}}{3!} = (1/6)t^3e^{2t} \text{ (\#16)}$$

15. $\mathscr{L}(y'') + 2\,\mathscr{L}(y') + \mathscr{L}(y) = \mathscr{L}(3te^{-t});\ y(0) = 4,\ y'(0) = 2;$

$$[s^2\,\mathscr{L}(y) - s\,y(0) - y'(0)] + 2[s\,\mathscr{L}(y) - y(0)] + \mathscr{L}(y) = \frac{3}{(s+1)^3}$$

$$(s^2 + 2s + 1)\,\mathscr{L}(y) - 4s - 10 = \frac{3}{(s+1)^2};\ \mathscr{L}(y) = \frac{4s+10}{(s+1)^2} + \frac{3}{(s+1)^4};$$

$$y = \mathscr{L}^{-1}\left\{\frac{4s+10}{(s+1)^2}\right\} + \mathscr{L}^{-1}\left\{\frac{3}{(s+1)^4}\right\}$$

$$= \mathscr{L}^{-1}\left\{\frac{4s}{(s+1)^2}\right\} + \mathscr{L}^{-1}\left\{\frac{10}{(s+1)^2}\right\} + \mathscr{L}^{-1}\left\{\frac{3}{(s+1)^4}\right\} \quad (\#17, 15, 16)$$

$$= 4e^{t}(1 - t) + 10te^{-t} + \frac{3t^3e^{-t}}{6} = 4e^{-t} + 6te^{-t} + (1/2)t^3e^{-t} \text{ or } y = e^{-t}(4 + 6t + (1/2)t^3)$$

17. $\mathscr{L}(y'') + 3\,\mathscr{L}(y') - 4\,\mathscr{L}(y) = 0;\ [y(0) = 1,\ y'(0) = -2]$

$$[s^2\,\mathscr{L}(y) - s\,y(0) - y'(0)] + 3[s\,\mathscr{L}(y) - y(0)] - 4\,\mathscr{L}(y) = 0;$$

$$(s^2 + 3s - 4)\,\mathscr{L}(y) - s - 1 = 0;\ \mathscr{L}(y) = \frac{s+1}{s^2+3s-4} = \frac{s+1}{(s+4)(s-1)}$$

$$y = \mathscr{L}^{-1}\left\{\frac{s+1}{(s^2+4)(s-1)}\right\} = \mathscr{L}^{-1}\left\{\frac{3/5}{s+4} + \frac{2/5}{s-1}\right\} \text{ using partial fractions}$$

$y = (3/5)e^{-4t} + (2/5)e^{t}$ (#7) or $5y = 3e^{-4t} + 2e^{t}$

19. $L\dfrac{di}{dt} + Ri = V$

$$0.2\frac{di}{dt} + 20i = 12$$

$$\frac{di}{dt} + 100i = 60$$

$$\mathscr{L}\left(\frac{di}{dt} + 100i\right) = \mathscr{L}(60)$$

$$s\,\mathscr{L}(i) - 3 + 100\,\mathscr{L}(i) = \frac{60}{s} \qquad (i = 3 \text{ at } t = 0)$$

$$(s + 100)\,\mathscr{L}(i) = \frac{60}{s} + 3$$

$$\mathscr{L}(i) = \left(\frac{60+3s}{s}\right)\left(\frac{1}{s+100}\right)$$

$$\mathscr{L}(i) = \frac{60+3s}{s(s+100)}$$

$$\mathscr{L}(i) = \frac{0.6}{s} + \frac{2.4}{s+100}$$

$I = 0.6 + 2.4e^{-100t}$

21. $L\frac{di}{dt}+Ri+\frac{1}{C}q=V$

$*\quad 0.1\frac{di}{dt}+45i+\frac{1}{2(10^{-4})}q=12$

$0.1\frac{di}{dt}+45(0)+\frac{1}{2(10^{-4})}(0)=12$ note: $\frac{di}{dt}=120$ when $t=0$

Differentate both sides of * with respect to t.

$$0.1\frac{d^2i}{d^2t}+45\frac{di}{dt}+\frac{1}{2(10^{-4})}i=0$$

$$\frac{d^2i}{dt^2}+450\frac{di}{dt}+\frac{10}{2(10^{-4})}i=0$$

$$\mathscr{L}\left(\frac{d^2i}{dt^2}+450\frac{di}{dt}+50000i\right)=\mathscr{L}(0)$$

$$s^2\mathscr{L}(i)-s(0)-120+450s\mathscr{L}(i)-0+50000\mathscr{L}(i)=0$$

$$(s^2+450s+50000)\mathscr{L}(i)=120$$

$$\mathscr{L}(i)=\frac{120}{(s+200)(s+250)}=\frac{2.4}{s+200}+\frac{-2.4}{s+250}$$

$$I=2.4e^{-200t}-2.4e^{-250t}$$

Chapter 30 Review

1. homogeneous, 2

2. nonhomogeneous, 2

3. nonhomogeneous, 1

4. homogeneous, 3

5. $m^2+4m-5=0$; $(m+5)(m-1)=0$; $m=-5, 1$;
$y=k_1e^{-5x}+k_2e^x$

6. $m^2-5m+6=0$; $(m-3)(m-2)=0$; $m=2, 3$;
$y=k_1e^{2x}+k_2e^{3x}$

7. $m^2-6m=0$; $m(m-6)=0$; $m=0, 6$; $y=k_1+k_2e^{6x}$

8. $2m^2-m-3=0$; $(2m-3)(m+1)=0$; $m=3/2, -1$;
$y=k_1e^{3x/2}+k_2e^{-x}$

9. $m^2-6m+9=0$; $(m-3)^2=0$; $m=3, 3$; $y=k_1e^{3x}+k_2xe^{3x}$

10. $m^2+10m+25=0$; $(m+5)^2=0$; $m=-5, -5$; $y=k_1e^{-5x}+k_2xe^{-5x}$

11. $m^2-2m+1=0$; $(m-1)^2=0$; $m=1, 1$; $y=k_1e^x+k_2xe^x$

12. $9m^2 - 6m + 1 = 0$; $(3m - 1)^2 = 0$; $m = \frac{1}{3}, \frac{1}{3}$; $y = k_1e^{x/3} = k_2xe^{x/3}$

13. $m^2 + 16 = 0$; $m = \pm 4j$; $y = k_1 \sin 4x + k_2 \cos 4x$

14. $4m^2 + 25 = 0$; $m = \pm(5/2)j$; $y = k_1 \sin 5x/2 + k_2 \cos 5x/2$

15. $m^2 - 2m + 3 = 0$; $m = 1 \pm \sqrt{2}\,j$; $y = e^x (k_1 \sin \sqrt{2}\,x + k_2 \cos \sqrt{2}\,x)$

16. $m^2 - 3m + 8 = 0$; $m = 3/2 \pm (\sqrt{23}/2)j$;
$y = e^{3x/2}[k_1 \sin (\sqrt{23}/2)x + k_2 \cos (\sqrt{23}/2)x]$

17. $m^2 + m - 2 = 0$; $(m + 2)(m - 1) = 0$; $m = -2, 1$;
$y_c = k_1e^x = k_2e^{-2x}$
y_p: $y_p = Ax + B$; $y'_p = A$; $y''_p = 0$; $A - 2(Ax + B) = x$;
$-2Ax + A - 2B = x$; $-2A = 1$; $A - 2B = 0$; $A = -1/2, B = -1/4$;
$y = k_1e^x + k_2e^{-2x} - x/2 - 1/4$

18. $m^2 - 6m + 9 = 0$; $(m - 3)^2 = 0$; $m = 3, 3$; $y_c = k_1e^{3x} + k_2xe^{3x}$;
y_p: $y_p = Ae^x = y'_p = y''_p$; $Ae^x - 6Ae^x + 9Ae^x = e^x$;
Thus $4Ae^x = e^x$; $A = 1/4$; $y = k_1e^{3x} + k_2xe^{3x} + (1/4)e^x$

19. $m^2 + 4 = 0$; $m = \pm 2j$; $y_c = k_1 \sin 2x + k_2 \cos 2x$;
y_p: $y_p = A \cos x + B \sin x$; $y'_p = -A \sin x + B \cos x$;
$y''_p = -A \cos x - B \sin x$
$-A \cos x - B \sin x + 4(A \cos x + B \sin x) = \cos x$;
$3A \cos x + 3B \sin x = \cos x$; $A = 1/3, B = 0$;
$y = k_1 \sin 2x + k_2 \cos 2x + (1/3) \cos x$

20. $m^2 - 2m + 3 = 0$; $m = 1 \pm \sqrt{2}\,j$; $y_c = e^x(k_1 \sin \sqrt{2}x + k_2 \cos \sqrt{2}\,x)$
y_p: $y_p = Ae^{2x}$; $y'_p = 2Ae^{2x}$; $y'' = 4Ae^{2x}$; $4Ae^{2x} - 2(2Ae^{2x}) + 3A^{2x} = 6e^{2x}$;
$3A = 6$; $A = 2$; $y = e^x(k_1 \sin \sqrt{2}\,x + k_2 \cos \sqrt{2}\,x) + 2e^{2x}$

21. $m^2 + 2m - 8 = 0$; $(m + 4)(m - 2) = 0$; $m = -4, 2$;
$y = k_1e^{-4x} + k_2e^{2x}$; $y' = -4k_1e^{-4x} + 2k_2e^{2x}$; substitute $y = 6$,
$x = 0, y' = 0$; $k_1 + k_2 = 6$; $-4k_1 + 2k_2 = 0$; $k_1 = 2, k_2 = 4$;
$y = 2e^{-4x} + 4e^{2x}$

22. $m^2 - 3m = 0$; $m(m - 3) = 0$; $m = 0, 3$; $y = k_1 + k_2e^{3x}$;
$y' = 3k_2e^{3x}$; substitute $y = 3, x = 0, y' = 6$; $k_1 + k_2 = 3$; $3k_2 = 6$; $k_1 = 1, k_2 = 2$; $y = 1 + 2e^{3x}$

23. $m^2 - 4m + 4 = 0$; $(m - 2)^2 = 0$; $m = 2, 2$; $y = k_1e^{2x} + k_2xe^{2x}$;
$y' = 2k_1e^{2x} + k_2e^{2x} + 2k_2xe^{2x}$; substitute $y = 0, x = 0, y' = 3$;
$k_1 = 0$; $2k_1 + k_2 = 3$; $k_1 = 0, k_2 = 3$; $y = 3xe^{2x}$

Chapter 30 Review

24. $m^2 + 6m + 9 = 0$; $(m + 3)^2 = 0$; $m = -3, -3$;
$y = k_1e^{-3x} + k_2xe^{-3x}$; $y' = -3k_1e^{-3x} + 3k_2xe^{-3x} + k_2e^{-3x}$;
substitute $y = 8, x = 0, y' = 0$; $k_1 = 8$; $-3k_1 + k_2 = 0$; $k_1 = 8, k_2 = 24$;
$y = 8e^{-3x} + 24xe^{-3x}$; $y = 8e^{-3x}(1 + 3x)$

25. $m^2 + 4 = 0$; $m = \pm 2j$; $y = k_1 \sin 2x + k_2 \cos 2x$;
$y' = 2k_1 \cos 2x - 2k_2 \sin 2x$; substitute $y = 1, x = 0, y' = -4$;
$k_2 = 1$; $2k_1 = -4$; $k_1 = -2, k_2 = 1$; $y = -2 \sin 2x + \cos 2x$

26. $m^2 - 8m + 25 = 0$; $m = 4 \pm 3j$; $y = e^{4x}(k_1 \sin 3x + k_2 \cos 3x)$
$y' = (3e^{4x}k_1 + 4e^{4x}k_2) \cos 3x + (4e^{4x}k_1 - 3e^{4x}k_2) \sin 3x$;
substitute $y = 2, x = 0, y' = 11$; $k_2 = 2$; $3k_1 + 4k_2 = 11$;
$k_1 = 1, k_2 = 2$; $y = e^{4x}(\sin 3x + 2 \cos 3x)$

27. $m^2 - m = 0$; $m(m - 1) = 0$; $m = 0, 1$; $y_c = k_1 + k_2e^x$;
y_p: $y_p = Ae^{2x}$; $y'_p = 2Ae^{2x}$; $y''_p = 4Ae^{2x}$; $4Ae^{2x} - 2Ae^{2x} = e^{2x}$; $2Ae^{2x} = e^{2x}$;
$A = 1/2$; $y = k_1 + k_2e^x + (1/2)e^{2x}$; $y' = k_2e^x + e^{2x}$;
substitute $y = 1, x = 0, y' = 0$; $1 = k_1 + k_2 + 1/2$;
$0 = k_2 + 1$; $k_2 = 3/2, k_2 = -1$; $y = 3/2 - e^x + (1/2)e^{2x}$ or $2y = e^{2x} - 2e^x + 3$

28. $m^2 + 4 = 0$; $m = \pm 2j$; $y_c = k_1 \sin 2x + k_2 \cos 2x$
y_p: $y_p = A \sin x + B \cos x$; $y'_p = A \cos x - B \sin x$;
$y''_p = -A \sin x - B \cos x$;
$-A \sin x - B \cos x + 4(A \sin x + B \cos x) = \sin x$;
$3A \sin x + 3B \cos x = \sin x$; $3A = 1$; $3B = 0$; $A = 1/3, B = 0$;
$y = k_1 \sin 2x + k_2 \cos 2x + (1/3) \sin x$;
$y' = 2k_1 \cos 2x - 2k_2 \sin 2x + (1/3) \cos x$; substitute $y = 2$,
$x = 0, y' = 7/3$; $2 = k_2$; $7/3 = 2k_1 + 1/3$; $k_1 = 1$; $k_2 = 2$;
$y = \sin 2x + 2 \cos 2x + (1/3) \sin x$

29. $m = \dfrac{W}{g} = \dfrac{16\text{ lb}}{32\text{ ft/s}^2} = \dfrac{1}{2}$ lb-s²/ft; $k = \dfrac{F}{s} = \dfrac{16\text{ lb}}{1/3\text{ ft}} = 48$ lb/ft;
$x = C_1 \sin 4\sqrt{6}\, t + C_2 \cos 4\sqrt{6}\, t$;
$dx/dt = 4\sqrt{6}\, C_1 \cos 4\sqrt{6}\, t - 4\sqrt{6}\, C_2 \sin 4\sqrt{6}\, t$ substitute $x = 6$ in. $= 1/2$ ft, $dx/dt = 0, t = 0$;
$1/2 = C_2$; $0 = 4\sqrt{6}\, C_1$; $C_1 = 0$; $x = (1/2) \cos 4\sqrt{6}\, t$

30. $m = \dfrac{W}{g} = \dfrac{12\text{ lb}}{32\text{ ft/s}^2} = 3/8$ lb-s²/ft; $k = \dfrac{F}{s} = \dfrac{12\text{ lb}}{1/6\text{ ft}} = 72$ lb/ft;

$P = \dfrac{8\text{ lb}}{1/3\text{ ft/s}} = 24$ lb-s/ft; $d = \left\{\dfrac{p}{2m}\right\}^2 - \dfrac{k}{m} = \left\{\dfrac{24}{2(3/8)}\right\}^2 - \dfrac{72}{3/8} = 832 > 0$;
since $d > 0$, we have two real roots of the auxiliary equation
$\overline{m}_1 = -p/2m + \sqrt{d} = -32 + \sqrt{832} = -3.16$; $\overline{m}_2 = -32 - \sqrt{832} = -60.8$;

$y = k_1e^{-3.16t} + k_2e^{-60.8t}$

31. $L = 2$ H, $R = 400\ \Omega$, $C = 10^{-5}$F; from Section 12.4:

$$d = \frac{R^2}{4L^2} - \frac{1}{CL} = \frac{(400)^2}{4(2)^2} - \frac{1}{(10^{-5})(2)} = -4 \times 10^4 < 0;$$

$$-\frac{R}{2L} = -\frac{400}{2(2)} = -100;\ \omega = \sqrt{\left|-4 \times 10^4\right|} = 200;$$

$$i = e^{-100t}(k_1 \sin 200t + k_2 \cos 200t)$$

32. $L = 1\ H$, $R = 2000\ \Omega$, $C = 4 \times 10^{-6}$F; from Section 29.4:

$$d = \frac{R^2}{4L^2} - \frac{1}{CL} = \frac{(2000)^2}{4(1)^2} - \frac{1}{(1)(4 \times 10^{-6})} = 7.5 \times 10^5 > 0;$$

since $d > 0$, $-\frac{R}{2L} = \frac{-2000}{2(1)} = -1000;\ \overline{m}_1 = -1000 + \sqrt{7.5 \times 10^5} = -134;$

$\overline{m}_2 = -1000 + \sqrt{7.5 \times 10^5} = -1866;$

$i = k_1 e^{-134t} + k_2 e^{-1866t}$

33. $\frac{1}{s-6}$; #7

34. $\frac{s+2}{(s+2)^2+9}$; #11

35. $\mathscr{L}(t^3) + \mathscr{L}(\cos t) = \frac{6}{s^4} + \frac{s}{s^2+1}$; #6, 10

36. $3\,\mathscr{L}(t) - \mathscr{L}(e^{5t}) = \frac{3}{s^2} - \frac{1}{s-5}$; #5, 7

37. $\mathscr{L}^{-1}(1/s^2) = t$

38. $-\frac{3}{2}\,\mathscr{L}^{-1}\left\{\frac{-2}{s(s-2)}\right\} = -\frac{3}{2}(1 - e^{2t}) = \frac{3}{2}(e^{2t} - 1)$ #13

39. $-2\,\mathscr{L}^{-1}\left\{\frac{-1}{(s-3)(s-4)}\right\} = -2(e^{3t} - e^{4t}) = 2(e^{4t} - e^{3t})$ #12

40. $\sin 3t$ #8

41. $4\,\mathscr{L}(y') - 5\,\mathscr{L}(y) = 0;\ 4[s\,\mathscr{L}(y) - y(0)] - 5\,\mathscr{L}(y) = 0;\ y(0) = 2;$

$(4s - 5)\,\mathscr{L}(y) - 8 = 0;\ \mathscr{L}(y) = \frac{8}{4s-5};\ y = \mathscr{L}^{-1}\left\{\frac{8}{4s-5}\right\} = \mathscr{L}^{-1}\left\{\frac{2}{s-5/4}\right\} = 2e^{5t/4}$

42. $\mathscr{L}(y'') + 9\,\mathscr{L}(y) = 0;\ y(0) = 3;\ y'(0) = 0;$

$[s^2\,\mathscr{L}(y) - s\,y(0) - y'(0)] + 9\,\mathscr{L}(y) = 0;\ s^2\,\mathscr{L}(y) - 3s + 9\,\mathscr{L}(y) = 0;$

$(s^2 + 9)\,\mathscr{L}(y) = 3s;\ \mathscr{L}(y) = \dfrac{3s}{s^2 + 9};\ y = \mathscr{L}^{-1}\left\{\dfrac{3s}{s^2 + 9}\right\} = 3\cos 3t$

43. $\mathscr{L}(y'') + 5\,\mathscr{L}(y') = 0;\ y(0) = 0;\ y'(0) = 2;$

$[s^2\,\mathscr{L}(y) - s\,y(0) - y'(0)] + 5\,[s\,\mathscr{L}(y) - y(0)] = 0;\ s^2\,\mathscr{L}(y) - 2 + 5s\,\mathscr{L}(y) = 0;$

$(s^2 + 5s)\,\mathscr{L}(y) = 2;\ \mathscr{L}(y) = \dfrac{2}{s(s+5)};\ y = \mathscr{L}^{-1}\left\{\dfrac{2}{s(s+5)}\right\} = \dfrac{2}{5}\,\mathscr{L}^{-1}\left\{\dfrac{5}{s(s+5)}\right\} = \dfrac{2}{5}(1 - e^{-5t})$

44. $\mathscr{L}(y'') + 4\,\mathscr{L}(y') + 4\,\mathscr{L}(y) = \mathscr{L}(e^{-2t});\ y(0) = 0;\ y'(0) = 0;$

$[s^2\,\mathscr{L}(y) - s\,y(0) - y'(0)] + 4[s\,\mathscr{L}(y) - y(0)] + 4\,\mathscr{L}(y) = \dfrac{1}{s+2};$

$(s^2 + 4s + 4)\,\mathscr{L}(y) = \dfrac{1}{s+2};\ \mathscr{L}(y) = \dfrac{1}{(s+2)^3};$

$y = \mathscr{L}^{-1}\left\{\dfrac{1}{(s+2)^3}\right\} = \dfrac{t^2 e^{-2t}}{2}$